Assessing *the* Sustainability and Biological Integrity *of* Water Resources Using Fish Communities

Edited by
Thomas P. Simon

CRC Press
Boca Raton   London   New York   Washington, D.C.

Project Editor: Sylvia Wood
Marketing Manager: Becky McEldowney
Cover design: Dawn Boyd
PrePress: Carlos Esser

Library of Congress Cataloging-in-Publication Data

Assessing the sustainability and biological integrity of water
  resources using fish communities / edited by Thomas P. Simon.
      p. cm.
    Includes bibliographical references and index.
    ISBN 0-8493-4007-1 (alk. paper)
    1. Water quality biological assessment. 2. Freshwater fishes-
-Effect of water quality on—North America. I. Simon, Thomas P.
QH96.8B5A77 1998
597.176—dc21   98-18557
                                                                   CIP

This book contains information obtained from authentic and highly regarded sources. Reprinted material is quoted with permission, and sources are indicated. A wide variety of references are listed. Reasonable efforts have been made to publish reliable data and information, but the author and the publisher cannot assume responsibility for the validity of all materials or for the consequences of their use.

Neither this book nor any part may be reproduced or transmitted in any form or by any means, electronic or mechanical, including photocopying, microfilming, and recording, or by any information storage or retrieval system, without prior permission in writing from the publisher.

All rights reserved. Authorization to photocopy items for internal or personal use, or the personal or internal use of specific clients, may be granted by CRC Press LLC, provided that $.50 per page photocopied is paid directly to Copyright Clearance Center, 27 Congress Street, Salem, MA 01970 USA. The fee code for users of the Transactional Reporting Service is ISBN 0-8493-4007-1/98/$0.00+$.50. The fee is subject to change without notice. For organizations that have been granted a photocopy license by the CCC, a separate system of payment has been arranged.

The consent of CRC Press LLC does not extend to copying for general distribution, for promotion, for creating new works, or for resale. Specific permission must be obtained in writing from CRC Press LLC for such copying.

Direct all inquiries to CRC Press LLC, 2000 Corporate Blvd., N.W., Boca Raton, Florida 33431.

Trademark Notice: Product or corporate names may be trademarks or registered trademarks, and are only used for identification and explanation, without intent to infringe.

© 1999 by CRC Press LLC

No claim to original U.S. Government works
International Standard Book Number 0-8493-4077-1
Library of Congress Card Number QH96.8.B5A77
Printed in the United States of America 1 2 3 4 5 6 7 8 9 0
Printed on acid-free paper

Foreword

This book is the first dedicated to the use of fish assemblages for aquatic biological assessment and criteria development. Long overdue is the concise documentation of the role of fish in the new field of multimetric indices, also referred to as indices of biological integrity. *Assessing the Sustainability and Biological Integrity of Water Resources Using Fish Assemblages* builds on classical basic research ecology and, unlike applied ecology, focuses on what must be done to protect biological resources through appropriate biological monitoring and assessment. Dr. Thomas Simon has been a leader in this discipline for a number of years, and through unwavering dedication — and three children — has edited an impressive collection of works that will be a standard reference for years to come.

Few events can transform the nature of a discipline as has the development and application of the original Index of Biotic Integrity (IBI) by Dr. James R. Karr using fish communities. However, its real success depends on the research, testing, and application by fish biologists and environmental managers over time, throughout this country, and throughout the world. As this book attests, the IBI is a fundamentally sound and critical approach to measuring the health of our waters. Its increased use as a tool for both field biologists and water resource managers in local, state, and federal governments is based on the constant research and development that result in modifications of the original IBI for situations outside of wadeable warmwater streams, such as great rivers, coldwater habitats, lakes, estuaries, coastal areas, and extremely polluted and modified waters. Many of these modifications are well documented by the authors of this book.

Recognition of the importance of protecting our biological resources, and using direct measures of fish and other aquatic life to gauge the effectiveness of achieving our environmental goals, was clearly articulated in the U.S. Clean Water Act of 1972. The Act's only objective is to "restore and maintain the chemical, physical, and biological integrity of the nation's waters." Until recently, the biological integrity objective of the Act was essentially ignored in favor of the easier to measure and understand chemical and toxicological tests. These tests were used, and still are, as surrogates for directly measuring ecological attributes of the biological community, but they are now beginning to be seen as poor and unreliable signals of human influence on biological integrity.

Responding to a more recent U.S. law — called the Government Performance and Results Act of 1993 — the United States Environmental Protection Agency had to reevaluate its measures of environmental success. In 1996, the EPA declared a national goal for clean water stating that "America's rivers, lakes, and coastal waters will support healthy communities of fish, plants, and other aquatic life...," along with the objective to "conserve and enhance aquatic ecosystems." Biological integrity was chosen as the key indicator to reflect this goal and objective. In 1997, the EPA released to the American people its Strategic Plan from the Office of Chief Financial Officer, containing the measures that will be used to document environmental progress. In this plan, the EPA presents three key objectives to meet the goal of clean and safe water, the second of which is to "conserve and enhance the ecological health of the nation's (state, interstate, and tribal) waters and aquatic ecosystems — rivers and streams, lakes, wetlands, estuaries, coastal areas, oceans, and groundwater — so that 75% of waters will support healthy aquatic communities by 2005." This measure of performance can only be accurately determined through multimetric biological monitoring and assessment approaches. This book, and the recent efforts of Dr. Thomas Simon with many authors from this book, (i.e., *Biological Assessment and Criteria: Tools for Water Resource Planning and Decision Making*), are critical resources in this discipline.

To truly understand the restoration and protection requirements for healthy aquatic ecosystems, and to measure the progress in meeting these goals, biological monitoring and assessment must first focus on the ability to discern gradients of human influence — not the natural geographic and temporal variability of aquatic populations. When one recognizes that lesson, as put forth throughout this book, only then can one sit back and let the next generation of aquatic ecologists improve on what is accomplished here.

Wayne S. Davis
Senior Environmental Scientist
Office of Policy, Planning, and Evaluation
United States Environmental Protection Agency
Washington, DC

Preface

As editor of this volume, I have compiled a series of chapters that capture the technical needs of fisheries biologists, ichthyologists, and environmental scientists interested in fish ecology, management, and assessment. Ten papers in this book were presented at the American Fisheries Society National Meeting during 1996 in Dearborn, MI, providing a forum to discuss current knowledge and needs. The science of environmental indicators and biological assessment has experienced rapid growth, compromising the literature necessary to sustain growth in this field and providing training for other professional biologists (Fausch et al., 1990).

The primary purpose of this book is to further the technical knowledge necessary to develop, calibrate, and validate an approach to assess fish assemblages. The use of environmental indicators, such as fish communities, provides a direct measure of environmental quality. Using a variety of indices, cumulatively referred to as the Index of Biotic Integrity, the prediction of a range of disturbances with an impressive degree of precision is possible. The Index of Biotic Integrity was first developed by James R. Karr (1981) to evaluate midwestern stream fish communities. The publication of that landmark paper developed the concepts that have been improved and expanded on in subsequent papers, thus furthering the ideas behind biological integrity (Fausch et al., 1984; Angermeier and Karr, 1986; Angermeier and Schlosser, 1987; Karr et al., 1986) and water resource management (Gibson, 1994; Davis and Simon, 1995).

Current approaches for developing regional indices that accurately reflect the various fish community structure and function characteristics worldwide have magnified a number of information gaps that are not yet met using currently available literature. Simon and Lyons (1995) highlighted a number of these inconsistencies, including the lack of information on framing points such as zoogeographic implications to developing reference conditions, differences between biological integrity of altered environments and natural habitats, inaccurate or misinformation in the guild descriptions and classifications that are the premise behind the various metrics, and application of the index to areas other than small warmwater streams.

The 26 chapters in this book are designed to build on the foundation set by the book entitled, *Biological Assessment and Criteria: Tools for Water Resource Planning and Decision Making* (Davis and Simon, 1995). The book, which Wayne Davis and I edited, developed the foundational concepts and background necessary to fully utilize the ideas presented in this volume. I have divided it into five dominant sections: (1) perspectives, (2) guild and metric determination, (3) regional applications of IBI, (4) application to freshwater resource types other than wadeable warmwater streams, and (5) data validation.

The perspectives section (Chapters 1–5) begins with information on ideas behind ecological health and biological integrity. Following are a series of synthesis chapters that describe the rationale and decision-making processs necessary to implement a program in biological assessment and criteria. The state of Ohio biological assessment program is a national model for groups interested in improving their biological assessment approaches. In addition, a case study is presented that describes the involvement of stakeholders and partnering in the development of a monitoring program in the Red River of the North. In an age of dwindling resources and increasing needs, project completion deadlines have required that a number of partnerships be established between groups that often have differing objectives. For ultimate success, the need to bring diverse resource agencies and the public together must be a win-win situation. The importance of reference conditions for the application of environmental indicators has focused on the importance of historical biogeography and ecology to protect all levels of biological diversity (i.e., genetic to ecosystem). The application of the IBI outside the United States required a review of the worldwide literature on the development and use of the index outside of the United States. Information on the use of the IBI in the United States was recently reviewed by Simon and Lyons (1995).

The guild and metrics determination section (Chapters 6–9) includes literature reviews for reproductive guilds and trophic dynamics of select North American species. These guilds are perhaps the least known, although the improvement of a literature base will strengthen the range of substitution metrics that can be developed. My review of North American species patterns built on the reproductive guilds concept of Balon (1975; 1981) and our trophic framework modified that proposed by Gerking (1994). In addition, two specific metrics are reviewed in detail. The influence of the sucker family is analyzed for its influence on other metrics used in a Great River IBI. Also, representing the lowest extremes of biological integrity is the presence of external deformities, eroded fins, lesions, and tumors (DELTs). The effects on individual health and condition were reviewed, as was the use of the DELT metric in seven case studies in Ohio.

The regional applications of the Index of Biotic Integrity section (Chapters 10–14) provides examples of multimetric and multivariate analyses for Atlantic Coast, Pacific Northwest, New England, Hudson Bay, and Pacific coastal streams. Although the chapters provide information on development of an index for these systems, they also include discussion of several important issues. The effects of landscape disturbance on IBI metrics and stream size was studied in Virgina. Native, nonindigenous, and exotic species influences are discussed in several chapters, since an increase in biological diversity is not always a positive influence on native species and biological integrity. It is difficult to apply a conventional index of biological integrity to systems with naturally depauperate species richness. Rationale and concepts for the advancement of alternatives (i.e., use of biomass metrics, inclusion of alternate organismal assemblages into a single index, and evaluation of multiple applications in various-sized streams) are innovative approaches for jumping these hurdles.

The application of ecological health concepts and biological integrity indices to resource types other than wadeable warmwater streams (Chapters 15–22) explores the development of indices for coldwater streams, Great Lakes nearshore, Great Rivers, reservoirs, and inland lakes. Information included in this section involves aspects of sampling and data validation. In addition, concepts requiring modification for proper calibration and data usage are explored. Of consequence is the presentation of alternative metrics, as well as the validation of the newly constructed indices. The development of ecological health indices for modified systems such as tailwaters and reservoirs is an important consideration.

The last section of the book (Chapters 23–26) deals with aspects of calibration and data validation. Two chapters discuss the role of anthropogenic disturbance and correspondence with other factors. Fausch et al. (1984) first described means of calibrating the Index of Biotic Integrity, which is primarily derived from a line drawn by eye. The line is drawn such that 95% of the data fall below the line. This line is termed the "maximum species richness line." Alternative methods for removing subjective bias from the derivation of the maximum species richness line is generated after subdividing data. Additionally, artificially elevated scores are often observed when too few individuals are collected from a reach. It has been shown that in large systems, the predominance of certain species can mask the subtle patterns (Simon and Emery, 1995). Alternatives for low-end adjusting scores are presented to avoid false negative site assessment.

This book is the beginning of further work using fish as environmental indicators of biological integrity. It is my hope that biologists using this book will benefit from the experiences of the authors who are in the forefront of this field. Although this book puts into perspective the development, application, and facilitation of technical data needs, additional work is yet required, and more research fisheries biologists and ichthyologists are needed to answer the remaining issues. The roles of the fisheries biologists and environmental scientists are merging together. Historically, these fields were easily separated; however, sustaining biological integrity and diversity is an important goal toward which to strive.

Thomas P. Simon

REFERENCES

Angermeier, P.L. and J.R. Karr. 1986. Applying an index of biotic integrity based on streamfish communities: considerations in sampling and interpretations, *North American Journal of Fisheries Management,* 6, 418–429.

Angermeier, P.L. and I.J. Schlosser. 1987. Assessing biological integrity of the fish community in a small Illinois stream, *North American Journal of Fisheries Management,* 7, 331–338.

Balon, E.K. 1975. Reproductive guilds of fishes: a proposal and definition, *Journal of the Fisheries Research Board of Canada,* 32, 821–864.

Balon, E.K. 1981. Addition and amendments to the classification of reproductive styles in fishes, *Environmental Biology of Fishes,* 6, 377–389.

Davis, W.S. and T.P. Simon Eds. 1995. *Biological Assessment and Criteria: Tools for Water Resource Planning and Decision Making.* Lewis, Boca Raton, FL.

Fausch, K.D., J.R. Karr, and P.R. Yant. 1984. Regional application of an index of biotic integrity based on stream fish communities, *Transactions of the American Fisheries Society,* 113, 39–55.

Fausch, K.D., J. Lyons, J.R. Karr, and P.L. Angermeier. 1990. Fish communities as indicators of environmental degradation, in S.M. Adams Ed. *Biological Indicators of Stress in Fish.* American Fisheries Society Symposium 8, Bethesda, MD. 123–144.

Gerking, S.D. 1994. *Feeding Ecology of Fish.* Academic, San Diego.

Gibson, G.R. Ed. 1994. Biological Criteria: Technical Guidance for Streams and Small Rivers. EPA 822-B-94-001. U.S. Environmental Protection Agency, Office of Science and Technology, Washington, DC.

Karr, J.R. 1981. Assessment of biotic integrity using fish communities. *Fisheries,* 6, 21–27.

Karr, J.R., K.D. Fausch, P.L. Angermeier, P.R. Yant, and I.J. Schlosser. 1986. Assessing Biological Integrity in Running Waters: A Method and Its Rationale. Illinois Natural History Survey Special Publication 5. Champaign. IL.

Simon, T.P. and E.B. Emery. 1995. Modification and assessment of an index of biotic integrity to quantify water resource quality in Great Rivers, *Regulated Rivers: Research and Management,* 11, 283–298.

Simon, T.P. and J. Lyons. 1995. Application of the index of biotic integrity to evaluate water resource integrity in freshwater ecosystems, in W.S. Davis and T.P. Simon Eds. *Biological Assessment and Criteria: Tools for Water Resource Planning and Decision Making.* Lewis, Boca Raton, FL. 245–262.

Acknowledgments

This book was made possible by the many contributors to this discipline, both past and present. We gratefully acknowledge the chapter authors for their effort and dedication. We also thank the many listed reviewers who enabled this book to meet scientific peer review standards. Chapters affiliated with the U.S. Environmental Protection Agency, Pacific Ecology Laboratory, Corvallis (where much of the original work was initiated and defined), and various offices of the U.S. Geological Survey, National Water Quality Assessment were part of the official U.S. EPA and USGS peer review process, and those reviewers are acknowledged. I especially want to acknowledge the support of my wife, who provided enormous strength for this project. Without her encouragement and tolerance, I never would have been able to finish it. Special thanks to those involved with the American Fisheries Society Conference in Dearborn, MI, where 10 of these chapters were originally presented. Special thanks to Robert Hughes, Paul Angermeier, Thom Whittier, John Lyons, and Ed Rankin for their assistance, advice, and professionalism. This book was edited by Thomas P. Simon in his private capacity. No official support or endorsement by the Environmental Protection Agency or any other agency of the federal government is intended or should be inferred.

Contributors

Paul Angermeier
Virginia Cooperative Fish and Wildlife Research Unit
Virginia Polytechnic Institute & State University
Blacksburg, VA

Patricia A. Bailey
Minnesota Pollution Control Agency
Water Quality Division
St. Paul, MN

Michael T. Barbour
Tetra-Tech, Inc.
Owings Mills, MD

T. Douglas Beard, Jr.
Bureau of Fisheries Management and Habitat Protection
Wisconsin Department of Natural Resources
Madison, WI

Michael Bozek
U.S. Geological Survey
Wisconsin Cooperative Fishery Research Unit
University of Wisconsin – Stevens Point
Stevens Point, WI

Robert A. Daniels
New York State Museum
Biological Survey
Albany, NY

Michael Ell
North Dakota State Department of Health and Consolidated Laboratories
Bismark, ND

Erich B. Emery
Ohio River Valley Water Sanitation Commission
Cincinnati, OH

Edward E. Emmons
Bureau of Integrated Science Services
Wisconsin Department of Natural Resources
Monona, WI

John Emblom
Minnesota Department of Natural Resources
St. Paul, MN

Don Fago
Bureau of Integrated Science Services
Wisconsin Department of Natural Resources
Monona, WI

Robert M. Goldstein
U.S. Geological Survey
Water Resources Division
Mounds View, MN

David B. Halliwell
Aquatic Resources Conservation Systems
Clinton, ME

Gene R. Hatzenbeler
Wisconsin Cooperative Fishery Research Unit
University of Wisconsin – Stevens Point
Stevens Point, WI

Gary D. Hickman
Aquatic Biology Laboratory
Tennessee Valley Authority
Norris, TN

Robert M. Hughes
Dynamac International, Inc.
Corvallis, OR

Richard Jacobson
Fisheries Division
Connecticut Department of Environmental Protection
Hartford, CT

Martin J. Jennings
Bureau of Integrated Science Services
Wisconsin Department of Natural Resources
Spooner, WI

James P. Kurtenbach
Environmental Science and Assessment
 Division
U.S. Environmental Protection Agency
Edison, NJ

Richard Langdon
Environmental Laboratory
Vermont Department of Environmental
 Conservation
Waterbury, VT

John Lyons
Bureau of Integrated Science Services
Wisconsin Department of Natural Resources
Monona, WI

Thomas A. McDonough
Aquatic Biology Laboratory
Tennessee Valley Authority
Norris, TN

Michael P. Marchetti
Department of Wildlife, Fish, and Conservation
 Biology
University of California — Davis
Davis, CA

Terry R. Maret
Water Resources Division
U.S. Geological Survey
Boise, ID

Robert J. Miltner
Division of Surface Water
Ohio Environmental Protection Agency
Columbus, OH

Carroll L. Missimer
P.H. Glatfelter Company
Spring Grove, PA

Peter B. Moyle
Department of Wildlife, Fish, and Conservation
 Biology
University of California – Davis
Davis, CA

Neal D. Mundahl
Winona State University
Department of Biological Sciences
Winona, MN

Scott Niemela
Minnesota Pollution Control Agency
St. Paul, MN

Thierry Oberdorff
Laboratoire d'Ichthyologie Générale et
 Appliquée
Museum National d'Historie Naturelle
Paris, France

Robert Ovies
Ohio River Valley Water Sanitation
 Commission
Cincinnati, OH

Eric Pearson
North Dakota State Department of Health and
 Consolidated Laboratories
Bismark, ND

Edward T. Rankin
Ohio Environmental Protection Agency
Ecological Assessment Section
Columbus, OH

Robin Reash
Environmental Services Department
American Electric Power
Columbus, OH

Randall E. Sanders
Division of Wildlife, Fish Management and
 Research
Ohio Environmental Protection Agency
Columbus, OH

Lynn Schlueter
North Dakota Department of Game and Fish
Fisheries Division
Devils Lake, ND

Konrad Schmidt
Minnesota Department of Natural Resources
St. Paul, MN

Edward Scott
Aquatic Biology Laboratory
Tennessee Valley Authority
Norris, TN

Thomas P. Simon
U.S. Environmental Protection Agency,
  Region 5
Water Division
Chicago, IL

Timothy D. Simonson
Bureau of Fisheries Management and Habitat
  Protection
Wisconsin Department of Natural Resources
Madison, WI

Marc A. Smith
Ohio Environmental Protection Agency
Ecological Assessment Section
Columbus, OH

Roy Smogor
Illinois Natural History Survey
Illinois Department of Natural Resources
Springfield, IL

Blaine D. Snyder
Tetra-Tech, Inc.
Owings Mills, MD

Rex Meade Strange
Department of Biology
Southeast Missouri State University
Cape Girardeau, MO

James B. Stribling
Tetra-Tech, Inc.
Owings Mills, MD

Roger Thoma
Ohio Environmental Protection Agency
Ecological Assessment Section
Columbus, OH

Thomas R. Whittier
Dynamac International, Inc.
Corvallis, OR

Chris O. Yoder
Ohio Environmental Protection Agency
Ecological Assessment Section
Columbus, OH

Reviewers

Michael T. Barbour
Tetra-Tech, Inc.
Owings Mills, MD

Paul C. Baumann
U.S. Geological Survey
Ohio State University
Columbus, OH

John Brazner
U.S. Environmental Protection Agency
NEEHL, MCD
Duluth, MN

Linda K. Channel
U.S. Geological Survey
Water Resources Division
Boise, ID

Wayne S. Davis
U.S. Environmental Protection Agency
Washington, DC

Melissa Drake
Minnesota Department of Natural
 Resources
St. Paul, MN

Thomas A. Edsall
U.S. Geological Survey
Great Lakes Research Center
Ann Arbor, MI

Shelby D. Gerking
Arizona State University
Department of Zoology
Tempe, AZ

Susan Allen-Gil
National Research Council Associate
U.S. Environmental Protection Agency
Corvallis, OR

Robert M. Goldstein
U.S. Geological Survey
Mounds View, MN

John M. Grizzle
Auburn University
Department of Fisheries and Allied Aquaculture
Auburn University, AL

Martin E. Gurtz
U.S. Geological Survey
Water Resources Division
Raleigh, NC

Robert M. Hughes
Man-Tech Environmental Sciences
Corvallis, OR

David Jude
University of Michigan
Center for Great Lakes and Aquatic Sciences
Ann Arbor, MI

James R. Karr
Institute for Environmental Studies
University of Washington
Engineering Annex — FM 12
Seattle, WA

Arthur Lubin
U.S. Environmental Protection Agency
Office of Strategic Environmental Assessment
Chicago, IL

John Lyons
Wisconsin Department of Natural Resources
Bureau of Research
Monona, WI

Phillip B. Moy
U.S. Army Corps of Engineers
Chicago District
Chicago, IL

William H. Mullins
U.S. Geological Survey
Water Resources Division
Boise, ID

David V. Peck
U.S. Environmental Protection Agency
National Environmental Health and
  Environmental Effects Research Laboratory
Western Ecology Division
Corvallis, OR

Randall A. Sanders
Ohio Environmental Protection Agency
Ecological Assessment Section
Columbus, OH

Jerry Shulte
Ohio River Valley Water Sanitation
  Commission
Cincinnati, OH

Thomas P. Simon
Ogden Dunes, IN

James Smith
Indiana Department of Environmental
  Management
Indianapolis, IN

Paul M. Stewart
U.S. Geological Survey
Lake Michigan Ecology Station
Porter, IN

Robert Wallus
Tennessee Valley Authority
Chattanooga, TN

Thomas Whittier
Dynamac International, Inc.
Corvallis, OR

Chris O. Yoder
Ohio Environmental Protection Agency
Ecological Assessment Section
Columbus, OH

Contents

FOREWORD: Wayne S. Davis

SECTION I: PERSPECTIVES

Chapter 1
Introduction: Biological Integrity and Use of Ecological Health Concepts for Application to Water Resource Characterization 3
Thomas P. Simon

Chapter 2
Using Fish Assemblages in a State Biological Assessment and Criteria Program: Essential Concepts and Considerations 17
Chris O. Yoder and Marc A. Smith

Chapter 3
Collaboration, Compromise, and Conflict: How to Form Partnerships in Environmental Assessment and Monitoring..57
Thomas P. Simon, Robert M. Goldstein, Patricia A. Bailey, Eric Pearson, Michael Ell, Konrad Schmidt, John Emblom, and Lynn Schlueter

Chapter 4
Historical Biogeography, Ecology, and Fish Distributions: Conceptual Issues for Establishing IBI Criteria 65
Rex Meade Strange

Chapter 5
Applications of IBI Concepts and Metrics to Waters Outside the United States and Canada..79
Robert M. Hughes and Thierry Oberdorff

SECTION II: GUILD AND METRICS DETERMINATION

Chapter 6
Assessment of Balon's Reproductive Guilds with Application to Midwestern North American Freshwater Fishes..97
Thomas P. Simon

Chapter 7
Toward a United Definition of Guild Structure for Feeding Ecology of North American Freshwater Fishes..123
Robert M. Goldstein and Thomas P. Simon

Chapter 8
Influence of the Family Catostomidae on the Metrics Developed for a Great Rivers
Index of Biotic Integrity 203
Erich B. Emery, Thomas P. Simon, and Robert Ovies

Chapter 9
The Use of External Deformities, Erosion, Lesions, and Tumors (DELT Anomalies)
in Fish Assemblages for Characterizing Aquatic Resources: A Case Study of
Seven Ohio Streams 225
Randall E. Sanders, Robert J. Miltner, Chris O. Yoder, and Edward T. Rankin

SECTION III: REGIONAL APPLICATIONS OF THE IBI

Chapter 10
Effects of Drainage Basin Size and Anthropogenic Disturbance on Relations Between
Stream Size and IBI Metrics in Virginia 249
Roy A. Smogor and Paul L. Angermeier

Chapter 11
Characteristics of Fish Assemblages and Environmental Conditions in Streams of the
Upper Snake River Basin, in Eastern Idaho and Western Wyoming...273
Terry R. Maret

Chapter 12
Classification of Freshwater Fish Species of the Northeastern United States for Use
in the Development of Indices of Biological Integrity, with Regional Applications 301
*David B. Halliwell, Richard W. Langdon, Robert A. Daniels, James P. Kurtenbach,
and Richard A. Jacobson*

Chapter 13
Development of an Index of Biotic Integrity for the Species Depauperate Lake Agassiz
Plain Eco Region, North Dakota and Minnesota 339
Scott Niemela, Eric Pearson, Thomas P. Simon, Robert M. Goldstein, and Patricia A. Bailey

Chapter 14
Applications of Indices of Biotic Integrity to California Streams and Watersheds 367
Peter B. Moyle and Michael P. Marchetti

**SECTION IV: APPLICATION TO FRESHWATER RESOURCE TYPES OTHER THAN
WADEABLE WARMWATER STREAMS**

Chapter 15
Development and Application of an Index of Biotic Integrity for Coldwater Streams
of the Upper Midwestern United States 383
Neal D. Mundahl and Thomas P. Simon

Chapter 16
Biological Monitoring and An Index of Biotic Integrity for Lake Erie's
Nearshore Waters 417
Roger F. Thoma

Chapter 17
Considerations for Characterizing Midwestern Large River Habitats..463
Robin J. Reash

Chapter 18
Applying an Index of Biotic Integrity Based on Great River Fish Communities:
Considerations in Sampling and Interpretation 475
Thomas P. Simon and Randall E. Sanders

Chapter 19
Tailwater Fish Index (TFI) Development for the Tennessee River Tributary Tailwaters.............507
Edwin M. Scott, Jr.

Chapter 20
Reservoir Fishery Assessment Index Development: A Tool for Assessing Ecological
Health in Tennessee Valley Authority Impoundments 523
Thomas A. McDonough and Gary D. Hickman

Chapter 21
Toward the Development of An Index of Biotic Integrity for Inland Lakes in Wisconsin 541
*Martin J. Jennings, John Lyons, Edward E. Emmons, Gene R. Hatzenbeler, Michael Bozek,
Timothy D. Simonson, T. Douglas Beard, Jr., and Don Fago*

Chapter 22
Development of IBI Metrics for Lakes in Southern New England 563
Thomas R. Whittier

SECTION V: DATA VALIDATION

Chapter 23
Relations Between Fish Metrics and Measures of Anthropogenic Disturbance in Three IBI
Regions in Virginia 585
Roy A. Smogor and Paul L. Angermeier

Chapter 24
Methods for Deriving Maximum Species Richness Lines and Other Threshold Relationships
in Biological Field Data...611
Edward T. Rankin and Chris O. Yoder

Chapter 25
Adjustments to the Index of Biotic Integrity: A Summary of Ohio Experiences and Some
Suggested Modifications 625
Edward T. Rankin and Chris O. Yoder

Chapter 26
Integrating Assessments of Fish and Macroinvertebrate Assemblages and Physical Habitat Conditions in Pennsylvania..639
Blaine D. Snyder, James B. Stribling, Michael T. Barbour, and Carroll L. Missimer

Index...

Section I

Perspectives

1 Introduction: Biological Integrity and Use of Ecological Health Concepts for Application to Water Resource Characterization

Thomas P. Simon

CONTENTS

1.1 Introduction 3
1.2 Use of Fish As Environmental Indicators 4
1.3 Changing Societal Values Have Changed Water Resource Integrity 5
1.4 Biological Monitoring Detects Changes in Biological Condition 6
1.5 Criticisms of the Index of Biotic Integrity 7
    1.5.1 Factors That Affect Biological Integrity 7
    1.5.2 Critical Flow Values and Biological Integrity 8
    1.5.3 Criticisms of Ecological Health and the Index of Biotic Integrity 10
1.6 Conclusion 12
References 13

1.1 INTRODUCTION

Beginning around 1900 and accelerating greatly in the last 20 years, fish community characteristics have been used to measure relative ecosystem health. Recent advances can be attributed to the development of integrative ecological indices that directly relate fish communities to other biotic and abiotic components of the ecosystem (Karr, 1981; Karr et al., 1986; Mundahl and Simon, Chapter 15; Scott, Chapter 19), the delineation of ecoregions that allow explicit consideration of natural differences among fish communities from different geographic areas (Hughes et al., 1986; 1987; Omernik, 1987; Hughes and Larsen, 1988; Strange et al., Chapter 4), and the recognition of the importance of cumulative effects of degradation at the landscape scale (Smogor and Angermeier, Chapters 10 and 23).

    A variety of quantitative indices can be used to define specific biocriteria, including indicator species or guilds; species richness, diversity, and similarity indices; the Index of Well-Being; multivariate ordination and classification; and the Index of Biotic Integrity (reviewed by Fausch et al. 1990). Of these, the most commonly used — and arguably the most effective — has been the Index of Biotic Integrity (IBI).

Karr and Dudley (1981) defined biological integrity as "the capability of supporting and maintaining a balanced, integrated, adaptive community of organisms having a species composition, diversity, and functional organization comparable to that of natural habitats of the region." Aquatic communities have been used as environmental indicators to measure watershed condition (Fausch et al., 1990). The relative health and condition of an aquatic community is a sensitive measure of the site-specific condition (Karr, 1987). Fish communities integrate the direct and indirect effects of stress on the entire aquatic ecosystem. In addition, evaluation of specific characteristics of fish communities can be used to diagnose the ecological significance of the environmental disturbance (Davis and Simon, 1995; Simon and Lyons, 1995).

The use of fish communities as indicators of biological integrity was documented as early as the turn of the century on the Illinois River (Forbes and Richardson, 1913). Numerous studies have observed changes in fish distribution as a result of pollution plumes and sewerage outfalls (Brinley, 1942; Katz and Gaufin, 1953; Karr et al., 1985a); non-point source effects such as siltation (Menzel et al., 1984; Berkman and Rabeni, 1987); and cultural eutrophication (Hartman, 1983; Weaver and Garman, 1994). Guild response and specific characteristics of community structure and function are important attributes of this treatment of biological integrity, since classification ultimately affects biological assessment potential.

1.2 USE OF FISH AS ENVIRONMENTAL INDICATORS

Fish have been and remain a major part of any aquatic study designed to evaluate water quality (Table 1.1). Not only are fish a highly visible component of the aquatic resource, but they are one component that is easily sampled by professional biologists. In this book, the authors are not advocating the exclusive use of fish over any other taxonomic group; however, it is necessary to evaluate fundamental considerations of biological integrity at multiple hierarchical levels of ecosystem integration. Herein, the treatment of organismal groups is limited to fish since parallel efforts have recently been completed for macroinvertebrates (Rosenberg and Reash, 1993).

Response indicators have been proposed and developed for fish. Various indicators include using molecular and biochemical characteristics, physiology and behavior indicators, immunological indicators, histopathological approaches, enzymes, bioenergetic modeling, autopsy-based organismic indices, population, community, and ecosystem levels of response (Adams, 1990). Gammon (1973) developed an Index of Well-Being to evaluate structural components in numbers, biomass, and species richness for assessing water resources. Gammon (1976, 1980, 1983), Gammon and Reidy (1981), Gammon et al. (1981), and Gammon and Riggs (1983) used the index to assess environmental impact in the middle Wabash River and its tributaries in Indiana and Illinois. The Ohio Environmental Protection Agency (1989a) modified this index for Ohio streams by removing tolerant species from the calculation. Hughes and Omernik (1981) and Hughes and Gammon (1987) used the index to evaluate degradation in midwestern streams and the Willamette River in Oregon. Simon (1989) developed a multimetric index for early life stages of fishes, termed the Ichthyoplankton Index.

Although the Index of Biotic Integrity has been the primary focus of fisheries management and water resource biologists, many questions remain concerning the application and implementation of this index. Foremost is the lack of comparative information for resource types other than warm-water streams and rivers (Karr, 1981; Karr et al., 1986; Miller et al., 1988; Simon and Lyons, 1995). Applications of the IBI have been developed for cold-water streams (Oberdorff and Porcher, 1994; Lyons et al., 1996; Halliwell et al., Chapter 12; Mundahl and Simon, Chapter 15), caves (Poulson, 1991; 1992a, b), lentic waters (Dionne and Karr, 1992; Jennings et al., Chapter 21; Whittier, Chapter 22), the Great Lakes (Minns, 1994; Thoma, Chapter 16), the Great Rivers (Simon and Emery, 1995; Emery et al., Chapter 8; Reash, Chapter 17; Simon and Sanders, Chapter 18), and palustrine wetlands (Simon, submitted; Simon and Stewart, submitted). The IBI is considered a family of multimetric indices that change structural characteristics, depending on the geographic area (Simon

TABLE 1.1
Attributes of Fishes That Make Them Desirable Components of Biological Assessment and Monitoring Programs

Goal/Quality	Attribute
Accurate environmental assessment of health	Fish populations and individuals generally remain in the same area during summer seasons.
	Communities are persistent and recover rapidly from natural disturbances. Comparable results can be expected from an unperturbed site at various times.
	Fish have larger ranges and are less affected by natural microhabitat differences than smaller organisms. This makes fish extremely useful for assessing regional and macrohabitat differences.
	Most fish species have long life spans (3–10+ years) and can reflect both long-term and current water resource quality.
	Fish continually inhabit the receiving water and assimilate the chemical, physical, and biological degradation in characteristic response patterns.
Visibility	Fish are a highly visible component of the aquatic community to the public.
	Aquatic life uses and regulatory language are generally characterized in terms of fish (i.e., fishable and swimmable goal of the Clean Water Act).
Ease of use and interpretation	The sampling frequency for trend assessment is less than for short-lived organisms.
	Taxonomy of fishes is well established, allowing professional biologists the ability to reduce laboratory time by identifying many specimens in the field.
	Distribution, life histories, and tolerances to environmental stresses of most North American species are well documented in the literature.

From Simon, T.P. 1991. Development of Index of Biotic Integrity Expectations for the Ecoregions of Indiana. I. Central Corn Belt Plain. EPA-905-9-91/025. U.S. Environmental Protection Agency, Chicago, IL.

and Lyons, 1995; Hughes and Oberdorf, Chapter 5). Numerous modifications of the IBI have been developed for regional use in North America (Steedman, 1988; Goldstein et al., 1994; Lyons et al., 1995) and Europe (Hughes and Oberdorff, 1992; Didier and Kestemont, 1996).

1.3 CHANGING SOCIETAL VALUES HAVE CHANGED WATER RESOURCE INTEGRITY

Fish communities can be used to evaluate societal costs of degradation more directly than other taxa because their economic and aesthetic values are widely recognized. Much of the world's landscape has been converted from a natural ecosystem to one more valued for its cultural and societal values (Ward, 1992). Societal values determine various environmental management perspectives, including the development of philosophies, laws, agencies, and regulations to ensure that the landscape reflects societal goals.

Karr et al. (1985b) found that, despite strong legal mandates and extensive expenditures, signs of continuing degradation in aquatic systems, and especially large rivers, have continued to deteriorate at a much more rapid rate than terrestrial systems. Devastation of water resources is evident with channelization, impoundment by dams, dredging and filling of wetlands, and water withdrawal for irrigation and industrial uses. Loss of biological diversity (Hughes and Noss, 1992; Moyle and Leidy, 1992; William and Neves, 1992), introduction of exotic and non-native species (Courtney and Stauffer, 1984), loss of endangered and threatened species (Carlson and Muth, 1989; Master, 1990; Miller et al., 1989), and declining genetic diversity (Nehlsen et al., 1991) have imperiled aquatic communities and reduced biological integrity.

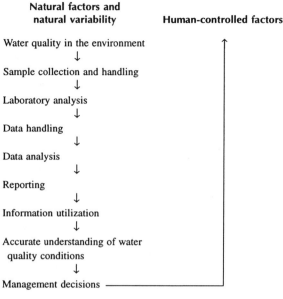

TABLE 1.2
Monitoring of Biological Systems Requires Following a Flow of Information

Note: See Ward, R.C. 1992. *Ecological Indicators.* Vol. I. Elsevier Applied Science. New York. 147–157.

1.4 BIOLOGICAL MONITORING DETECTS CHANGES IN BIOLOGICAL CONDITION

Fundamental to understanding sustainable biological systems is direct measurement of living systems to effectively monitor the "integrity of water" (Ballentine and Guarraia, 1975). Focus on biological evaluations enables integration of water-body change, since living systems assimilate the range of site-specific biotic and abiotic conditions (Table 1.2). Karr et al. (1997) suggest that narrow chemical criteria assume declining water resource conditions have only been caused by chemical contamination. On the contrary, habitat fragmentation, invasion by exotic species, excessive water withdrawal, and overfishing are often more devastating to fish communities than toxins.

Because biological measures are based on living organisms that adapt to conditions around them, biological evaluations can diagnose chemical, physical, and biological impacts as well as cumulative environmental impacts. Karr (1987) and Karr et al. (1997) have shown that, as human-induced stress has become more widespread and complicated, tools have been developed to increase measurement sensitivity to determine the magnitude of environmental impacts. The first step in effective biological management is to realize that the goal is to measure and evaluate the consequences of human actions on biological systems. Appropriate measurement endpoints for biological monitoring become biological condition; detecting change in the endpoint requires comparison of the change to an undisturbed baseline condition. Consideration for accurate assessment of reach-specific conditions requires understanding of zoogeographic constraints (Strange, Chapter 4).

Herricks and Schaefer (1985) evaluated six criteria for determining whether a biological monitoring program meets the objectives of biological integrity. These six criteria are divided into three categories: sensitivity, reproducibility, and variability.

1. *The measures used must be biological.* The Index of Biotic Integrity, the Index of Well-Being, and the Ichthyoplankton Index are based entirely on biological community attributes.
2. *The measures must be interpretable at several trophic levels, or provide a connection to other organisms not directly involved in the monitoring.* Fish community characteristics include functional attributes, such as omnivores, insectivores, and carnivores, providing a measure at multiple trophic levels. Fish communities also provide an indirect measure of the ecosytem dynamics since they reflect lower trophic-level responses.
3. *The measure must be sensitive to the environmental conditions being monitored.* The ability of fish communities to reflect the broad range of site conditions and assimilate these stressors make fish an esteemed environmental indicator. In addition, the Index of Biotic Integrity has redundant measures of community response to ensure that site-specific conditions can be precisely defined and exact conditions placed into one of six narrative classes (Karr, 1981; Karr et al., 1986).
4. *The response range (i.e., sensitivity) of the measure must be suitable for the intended application.* Fish communities and biological indices have shown they have a high degree of sensitivity to small, subtle changes in the environment and a wide variety of environmental disturbance types (Karr et al., 1986). The ability to differentiate among a variety of stream types has been one of the most powerful uses of the Index of Biotic Integrity and the use of fish communities.
5. *The measure must be reproducible and precise within defined and acceptable limits for data collection over space and time.* Fish communities require consistent sampling methods, analytical techniques, and evaluation indices (Angermeier and Karr, 1986; Ohio EPA, 1989). Where standard collection methods are used, the IBI has been shown to produce consistent, reproducible expectations within acceptable limits (Fore et al., 1993; Simon and Sanders, Chapter 18; Karr et al., Chapter 27).
6. *Variability of the measure(s) must be low.* The inherent variabilty associated with the employment of the Index of Biotic Integrity has been quite low (Fore and Karr, 1994). It is only when misapplications due to sampling inconsistencies, seasonal influences, and inappropriate gear types are used that large differences in site classification can result. When the sampling methods are correctly employed, the differences and variation in site scores irrefutably increase with disturbance (Angermeier and Karr, 1986; Simon and Sanders, Chapter 18).

1.5 CRITICISMS OF THE INDEX OF BIOTIC INTEGRITY

1.5.1 Factors That Affect Biological Integrity

Ecosystems that possess or reflect biological integrity can withstand or rapidly recover from most perturbations imposed by natural environmental processes and some of those induced by humans (Cairns, 1975; Karr et al., 1986). The reaction of an aquatic ecosystem to perturbation(s) depends largely on the frequency, magnitude, and duration of the effect and the inherent sensitivity of the system. Thus, biological communities that are degraded and therefore lack integrity have had their capacity to withstand and rapidly recover from perturbation(s) exceeded. Perturbed communities are likely to become even further degraded under incremental increases in stress. In contrast, communities that reflect biological integrity do so because their capacity to withstand stress has not been exceeded, resulting in a temporarily extended degradation of structural and functional organization. A biological system can be considered to have integrity when its inherent potential is realized, its condition is stable, its capacity for self-repair when perturbed is preserved, and minimal external support for management is needed (Karr et al., 1986). This is reflected in aquatic communities that have a high proportion of the extant fauna composed of functionally specialized and environmentally sensitive forms with no one group or guild predominating at the expense of a balanced, integrated, adaptive community. Biological integrity is not necessarily equated with harvestable products of economic or recreational value, although those systems that reflect high integrity often have the best commercial and recreational opportunities.

1.5.2 Critical Flow Values and Biological Integrity

Water quality standards contain rules that define minimum stream flows above which chemical and narrative criteria must be met. This is most commonly the seven-day average flow that has a probability of reoccurring once every 10 years (i.e., $Q_{7,10}$ flow). Other low-flow values can be used (95% duration flow, $Q_{30,10}$ flow) as well, and these can approximate the $Q_{7,10}$ relative to the annual hydrograph for a given stream or river. Because the customary implementation of chemical and narrative criteria is essentially based on a steady-state, dilution-oriented process, a design "critical" flow is necessary. This has been a widely accepted and essentially unquestioned practice in surface water quality regulation for many years. It is an inherently necessary component of the water quality-based approach to limit and control the discharge of toxic substances. However, a direct ecological basis for such flow regulation is lacking and, furthermore, may not be relevant; thus, one flow duration determines ecological health and well-being. It is simplistic and ecologically unrealistic to expect that worst-case biological community performance can only be measured under a $Q_{7,10}$ flow or some facsimile thereof.

There have been efforts to define ecologically critical flow thresholds using a toxicological rationale (U.S. EPA, 1986). This involved making judgments about the number of exceedences of acute and chronic chemical criteria that could occur *without causing harm to the aquatic community*. This effort attempts to establish a minimum flow at which chemical and/or toxic unit limits could be set and not have the aquatic communities in "a perpetual state of recovery" (Stephan et al., 1985). While there has been no direct experimental validation of the maximum exceedence frequency using complex ecological measures in the ambient environment, validation efforts have been directed at using experimental streams (U.S. EPA, 1991). These efforts, while being experimentally valid, still retain many of the basic limitations inherent to surrogate criteria, one of which remains: that a single species serves as a "surrogate" for community health.

Establishing a single critical flow (i.e., $Q_{7,10}$, $Q_{30,10}$, etc.) on an ecological basis, however, is not only improbable under current science, but it is technically inappropriate. There are simply too many additional variables that simultaneously affect the response and resultant conditions of aquatic communities, both spatially and temporally. Some can be estimated (e.g., duration of exposure, chemical fate dynamics, additivity), but many cannot because of the intensive data collection and analysis requirements; other phenomena are simply not adequately understood, yet their influence is integrated into the biological end result.

The ecological ramifications of low-flow conditions (particularly extreme drought) in small streams has probably contributed to much of the attention given to critical low flow. The results of low stream flow alone can be devastating in small watersheds (particularly those that have been modified via wetland destruction) during extended periods of severe drought (Larimore et al., 1957). The principal stressor in these cases is a loss of habitat via desiccation, in which organisms either leave or die during these periods. Ironically enough, the sustaining flow provided by a point-source discharge can mitigate the effects of desiccation provided that chemical conditions are minimally satisfactory for organism function and survival. While this may seem enigmatic in light of current strategies to regionalize wastewater flows, the presence of water with a seemingly marginal chemical quality can successfully mitigate what otherwise would be a total community loss. As was previously mentioned, this is dependent on the frequency, duration, and magnitude of any chemical stresses and local faunal tolerances. Small headwater streams (typically less than 10–20-square-mile drainage areas) commonly experience near-zero flows during extended dry-weather periods, sometimes during several consecutive summers. Given the historical loss of wetlands that functioned to sustain flows during dry periods, strategies such as opting for small wastewater treatment plants in lieu of regionalization need to be considered if the aquatic communities in headwater streams are to be restored and maintained. In this situation, the discharge flow assumes the functional loss of the sustained dry-weather flows formerly produced by wetlands. While this may seem contradictory, the consequences of repeated desiccation are far worse from a biological integrity standpoint.

Numerical chemical applications necessarily have their basis in dilution scenarios. However, these types of simplified analyses are no match for the insights provided into the chemical, physical, and biological dynamics that are "included" in the condition of the resident biota. The resolution of steady-state chemical application techniques suffers when applied to extreme low-flow or high-flow conditions. Site-specific factors that outweigh the importance of flow alone include the availability and quality of permanent pools and other refugia, gradient, organism acclimatization, and riparian characteristics such as canopy cover. Together, these and other factors determine the ability of a biological community to function and resist stress under "worst-case" low-flow conditions and hence retain the essential elements of biological integrity. It would be a serious mistake to draw the conclusion that the only important function of stream flow is to dilute pollutant concentrations when, in fact, the influence on physical habitat, both flow and water volume, is the far more important factor. The misconceptions about the role of stream flow have not only hampered efforts to more accurately manage wastewater flows in small streams, but in some cases have actually led to policies resulting in more devastating ecological impacts than that experienced under the original problems. The most frequently cited concept is that biological data collected during any time other than $Q_{7,10}$ critical flow does not represent the effect of "worst-case" conditions and therefore has limited applications in water quality-based issues. Sampling under worst-case, low-flow conditions is simply not necessary when measuring the condition of communities that have relatively long life spans and carry out all of their life functions in the water body. It is inappropriate to expect biological community condition (which is the integrated result of physical, chemical, and biological factors) to be so dependent on a temporal extreme of a single physical variable.

Since the observed condition of the aquatic biota at any given time is the product of the chemical, physical, and biological processes that have occurred in a water body over time, this should determine how chemical and other criteria are applied. Unlike chemical water quality, the aquatic biota does not respond instantaneously to normal short-term events, unless they are catastrophic in nature. This implies that one variable used in chemical criteria application cannot "make or break" the aquatic biota on its own. It would indeed be a poor survival strategy if aquatic organisms were so affected by such short-term and temporarily extreme events. Harmful, short-term episodes ranging from the onset of rapid lethal conditions to a protracted chronic stress will be manifested in the response of the resident biota. Thus, the biota can reveal the real-world effects of exceedences and consequent harm more precisely than can be predicted or measured on a chemical and/or toxicity basis alone. A finding that biological integrity is being achieved not only reflects a current healthy condition, but also means that the community has withstood and recovered from any short-term stresses as a result of critical low flow (or other temporary events) that may have occurred prior to or during field sampling. Therefore, because biological communities inhabit the receiving waters all of the time and will show the truly harmful effects of past stresses, it is not necessary to conduct sampling coincidental with critical low-flow conditions (or other short-term extremes) to gain a representative picture of community health and well-being. Indeed, more important and significant stresses can and do take place as a result of events that occur at times other than critical low flow. The condition of the aquatic biota is generally representative of environmental conditions, even though maximum stresses might have occurred at times other than the sampling dates. This should have some important ramifications in water resource management — with sound ambient information, it may no longer be necessary to manage from a worst-case basis in every situation.

1.5.3 CRITICISMS OF ECOLOGICAL HEALTH AND THE INDEX OF BIOTIC INTEGRITY

Sutter (1993) critically evaluated ecological health and the IBI, although he states that his "paper does not attack the concept [IBI] but rather the much more limited belief that the best way to use ... biosurvey data is to create an index of heterogenous variables [multimetric approach] and claim that it represents ecosystem health." The following is a list of his criticisms and a response to a potentially limited viewpoint of the IBI.

1. *Ambiguity.* Sutter suggests that using multimetric indices one cannot determine why values are high or low. The IBI utilizes multiple metrics to evaluate the water resource. One of the greatest advantages of IBI is that the site score can be dissected to reveal patterns exhibited at the specific reach compared with the reference community. Overall site quality can be determined from both the composite score and evaluation of each of the individual metrics. This reduces ambiguity compared with single metric indices such as the Shannon-Weiner Diversity Index.

2. *Eclipsing.* The eclipsing of low values of one metric can be dampened by the high values of another. Sutter suggested that the density and disease linkage in epidemiology is an interrelated effect. He suggests that when toxic chemicals are involved, the disease factor may not be reflective of the density or quality of an otherwise unimpacted community. Studies by Ohio EPA (1988) and other authors (e.g., Karr et al., 1985; Karr et al., 1986) have shown that when the IBI is assessed properly, each metric provides relevant information that determines the position along a continuum of water resource quality. Thus, some sites score well in some areas but poorly in other metrics, depending on levels of degradation. Thus, the reference condition is critical in determining the "least impacted" condition for the region.

3. *Arbitrary variance.* Variance demonstrated in indices may be high due to the compounding of individual metric variances. Sutter further suggests that other statistical properties of multimetric variables might be difficult to define. In studies conducted by the Ohio EPA (Rankin and Yoder, 1990; Yoder, 1991), they showed that IBI variability increased at highly degraded and disturbed sites but was low and stable at high-quality sites with increased biological integrity. The amount of variability within any of the component IBI metrics is irrelevant and does not necessarily have to be on the same "scale," assuming that proper metrics are selected and knowledge of how the metrics are applied is assessed by the field biologist. The high degree of resultant variability at sites that exhibit low biological integrity is an important indicator of site structure and function.

4. *Unreality.* Sutter argues that using multimetric approaches results in values with "nonsense units." He suggests that the IBI does not use "real" properties to describe the status of the reach-specific water resource. In contrast, he used an example of dose-response curves or habitat suitability to better predict a real-world property such as the presence of trout in a stream following a defined perturbation. In Sutter's simplistic approach to this complex problem, he fails to recognize the multiple stresses that could potentially limit the possibility of aquatic organism uses of a stream. The water resource manager is not only interested in whether a species is present or absent from a stream reach, but also puts more weight on the species interactions in a web of dynamic interactions. The resultant hyper-niche, defined by not only a single species but multiple species, becomes impossible with the limited amount of chemical-specific information available. Likewise, the modeling of the synergistic and additive effects of multiple stressors suggests that the IBI is *only* sensitive to toxic influences. This has been shown to be a poor assumption, since the IBI can determine poor performance from point source, nonpoint source, and combinations of these effects. The IBI does use "real-world" measures that individually are important attributes of a properly functioning and stable aquatic community. Addi-

tionally, the assessment of the aquatic community is enhanced by the acquisition of appropriate habitat information. It is highly recommended that all assessments include not only biological community information but habitat information (Davis and Simon, 1989).

5. *Post hoc justification.* Sutter suggests that the reduction in IBI values is a tautology since the assessment of poor biological integrity is the result of the reduced score. He further suggests that the IBI will only work if all ecosystems in all cases become unhealthy in the same manner. The IBI metrics are *a priori* assumed to measure a specific attribute of the community. Each metric is not an answer unto itself, and not all measure only attributes of a properly functioning community (e.g., percent disease). The metric must be sensitive to the environmental condition being monitored. The definition of degradation responses *a priori* is justified if clear patterns emerge from specific metrics. Although the probability of all ecosystems becoming unhealthy in the same manner is unrealistic, it is important to note that response signatures are definable (Yoder and Rankin, 1995) based on patterns of specific perturbations.

6. *Unitary response scales.* Sutter suggests that combining multimetric measures into a single index value suggests only a single linear scale of response and therefore only one type of response by ecosystems to disturbance. Sutter fails to recognize that the individual patterns exhibited by the various individual metrics usually reduces to single patterns in the community. For example, whether discussing siltation, reduced dissolved oxygen, or toxic chemical influences, all reduce the sensitive species component of the community and reduce species diversity. Thus, although multiple measures of the individual metrics result in multiple vectors explaining those dynamic patterns, *a priori* predictions of the metric response will result in the biological integrity categories defined by Karr et al. (1986).

7. *No diagnostic results.* Sutter suggests that one of the most important uses of biological survey data is to determine the cause of changes in ecosystem properties. He further suggests that by combining the individual metrics into a single value causes a loss of resolution when attempting to diagnose the responsible entity. This is the same argument raised in the ambiguity discussion above. The greatest use of the IBI is the ability to discern differences in individual metrics and determine cause and effect using additional information such as habitat, chemical water quality, and toxicity information. The inverse, however, is not apparent when attempting to reduce chemical water quality and toxicity test information into simple predictions of biological integrity based on complex interactions.

8. *Disconnected from testing and modeling.* Sutter suggests that the field results determined from the IBI need to be verified in the laboratory using controlled studies such as toxicity tests. This is a narrow viewpoint of the complex nature of the multimetric approach. Seldom does the degradation observed at a site result from a single chemical contaminant. To suggest that a single species or even multiple species (usually run individually) toxicity test can predict an IBI is ridiculous, given that the effects of siltation, habitat modification, and guild and trophic responses cannot be adequately determined in a laboratory beaker. Those aspects of a community that can be tested in the laboratory have validated the individual metric approach, i.e., thermal responses. It is the compilation of the various attributes that gives the IBI a robust measure of the community.

9. *Nonsense results.* Sutter indicates any index based on multiple metrics can produce nonsense results if the index has no interpretable real-world meaning. Sutter suggests that green sunfish (considered a tolerant species in the IBI) may have a greater sensitivity to some chemicals than some "sensitive species," and that the reduction of these contaminants may enable increases in green sunfish populations that result in a reduction in biological integrity. However, Sutter has mistakenly suggested that green sunfish have a

greater position in the community than do sensitive species. Green sunfish and other tolerant species are defined by the species' ability to increase under degraded conditions (Karr et al., 1986; Ohio EPA, 1989). Range extensions and the disruption of evenness in the community often occur at the expense of other sensitive species. This suggests that scoring modifications and other mechanisms for factoring out problems when few individuals are collected are not real-world situations.

10. *Improper analogy to other indexes.* Since environmental health as a concept has been compared to an economic index, several authors have argued that the environmental indexes are not generally comprehensible and require an act of faith to make informed judgments or decisions. The IBI has greatly improved the decision-making process by removing the subjective nature of past biological assessments. By using quantitative criteria (biological criteria) to determine goals of the Clean Water Act (attainable goals and designated uses), the generally comprehensible goals of the IBI enable a linkage between water resource status and biological integrity. This does not require an act of faith; rather, it broadens the tools available to water resource managers for screening waterbody status and trends.

1.6 CONCLUSIONS

The Index of Biotic Integrity has been a widely applied and effective tool for using fish assemblage data to assess the environmental quality of aquatic habitats. The original version of the IBI has been modified in numerous ways for application in many different regions and habitat types, and the IBI is now best thought of as a family of related indices rather than a single index. The commonalities linking all IBI versions are a multimetric approach that rates different aspects of fish community structure and function based on quantitative expectations of what constitutes a fish community with high biotic integrity in a particular region and habitat type. All versions include metrics that address species richness and composition, indicator species, trophic function, reproductive function, and/or overall abundance and individual condition. Different metrics and metric expectations within each of these metric categories are what distinguish different IBI versions. Many approaches have been used to generate metric expectations, and the process of establishing appropriate, sensitive expectations is probably the most difficult step in preparing a new version of the IBI. At present, most existing IBI versions are for wadeable warmwater streams in the central United States. However, versions have also been developed, or are in the process of being developed, for coldwater streams, large unwadeable rivers, lakes, impoundments, and Great Lakes lacustuaries in many different regions of the United States. Despite the large amount of effort that has been directed toward IBI development, much remains to be done, both in terms of generating new versions for different regions and habitat types, and in terms of validating existing versions.

ACKNOWLEDGMENT

This project would not have occurred without the assistance of the following individuals: J.R. Karr, R.M. Hughes, R.M. Goldstein, C. Yoder, E.T. Rankin, R. Thoma, and E.B. Emery. We appreciated critical review comments from anonymous reviewers that greatly improved an earlier draft of this manuscript. The opinions expressed are not necessarily those of the U.S. Environmental Protection Agency, although the manuscript may have been funded wholly or in part by the agency.

REFERENCES

Adams, S.M. (Ed.). 1990. Biological Indicators of Stress in Fish. *American Fisheries Symposium 8*. Bethesda, MD.

Angermeier, P.L. and J.R. Karr. 1986. Applying an index of biotic integrity based on stream fish communities: considerations in sampling and interpretation, *North American Journal of Fisheries Management*, 6, 418–429.

Ballentine, R.K. and L.J. Guarraia (Eds.). 1975. The Integrity of Water: A Symposium. U.S. Environmental Protection Agency, Washington, D.C.

Berkman, H.E. and C.F. Rabeni. 1987. Effect of siltation of stream fish communities, *Environmental Biology of Fishes*, 18, 285–294.

Brinley, F.J. 1942. Biological studies, Ohio River pollution survey: I. Biological zones in a polluted stream, *Sewage Works Journal*, 14, 147–152.

Cairns, J., Jr. 1974. Indicator species vs. the concept of community structure as an index of pollution, *Water Resources Bulletin*, 10, 338–347.

Carlson, C.A. and R.T. Muth. 1989. The Colorado River: lifeline of the American Southwest, *Canadian Special Publication Fisheries and Aquatic Sciences*, 106, 220–239.

Courtney, W.R., Jr. and J.R. Stauffer, Jr. (Eds.). 1984. *Distribution, Biology, and Management of Exotic Fishes*. The Johns Hopkins Press, Baltimore, MD.

Davis, W.S. and T.P. Simon. 1989. Sampling and data evaluation requirements for fish and macroinvertebrate communities, in T.P. Simon, L.L. Holst, and L.J. Shepard (Eds.), *Proceedings of the First National Workshop on Biological Criteria*, Lincolnwood, IL, December 2–4, 1987. EPA 905-9-89-003. U.S. Environmental Protection Agency, Chicago, Illinois, 89–97.

Davis, W.S. and T.P. Simon (Eds.). 1995. *Biological Assessment and Criteria: Tools for Water Resource Planning and Decision Making*. Lewis, Boca Raton, FL.

Didier, J. and P. Kestemont. 1996. Relationships between mesohabitats, ichthyological communities, and IBI metrics adapted to a European river basin (The Meuse, Belgium), *Hydrobiologia*, 341, 133–144.

Dionne, M. and J.R. Karr. 1992. Ecological monitoring of fish assemblages in Tennessee River Reservoirs, in D.H. McKenzie, D.E. Hyatt, and V.J. McDonald (Eds.), *Ecological Indicators*. Vol. 1. Elsevier, New York, 259–281.

Forbes, S.A. and R.E. Richardson. 1913. Studies on the Biology of the Upper Illinois River. *Illinois State Laboratory of Natural History Bulletin*, 9, 1–48.

Fausch, K.D., J. Lyons, J.R. Karr, and P.L. Angermeier. 1990. Fish communities as indicators of environmental degradation. *American Fisheries Society Symposium*, 8, 123–144.

Fore, L.S., J.R. Karr, and L.L. Conquest. 1993. Statistical properties of an Index of Biotic Integrity used to evaluate water resources. *Canadian Journal of Fisheries and Aquatic Sciences*, 51, 1077–1087.

Gammon, J.R. 1976. The Fish Populations of the Middle 340 km of the Wabash River. Purdue University Water Resources Research Center, Technical Report 86. West Lafayette, Indiana.

Gammon, J.R. 1983. Changes in the fish community of the Wabash River following power plant start-up: projected and observed, *American Society for Testing and Materials Special Technical Publication*, 802, 350–366.

Gammon, J.R. and J.M. Reidy. 1981. The role of tributaries during an episode of low dissolved oxygen in the Wabash River, IN., in L.A. Krumholz (Ed.), *The Warmwater Streams Symposium*. American Fisheries Society, Southern Division, Bethesda, MD, 396–407.

Gammon, J.R. and J.R. Riggs. 1983. The fish communities in Big Vermilion River and Sugar Creek, *Proceedings of the Indiana Academy of Science*, 92, 183–190.

Goldstein, R.M., T.P. Simon, P.A. Bailey, M. Ell, K. Schmidt, and J.W. Emblom. 1994. Concepts for the an index of biotic integrity for the streams of the Red River of the North basin, *Proceedings of the North Dakota Academy of Science*, Water Quality Symposium, March 30–31, Fargo, ND.

Hartman, J. 1983. Charr as indicators of pollution, *Fisheries*, 8(6), 10–12.

Herricks, E.E. and D.J. Schaefer. 1985. Can we optimize biomonitoring?, *Environmental Management*, 9, 487–492.

Hughes, R.M. and J.R. Gammon. 1987. Longitudinal changes in fish assemblages and water quality in the Willamette River, Oregon, *Transactions of the American Fisheries Society*, 116, 196–209.

Hughes, R.M. and D.P. Larsen. 1988. Ecoregions: an approach to surface water protection, *Journal Water Pollution Control Federation,* 60, 486–493.

Hughes, R.M. and R.F. Noss. 1992. Biological diversity and biological integrity: current concerns for lakes and streams, *Fisheries,* 17(3), 11–19.

Hughes, R.M. and J.M. Omernik. 1981. A proposed approach to determining regional patterns in aquatic ecosystems, in N.B. Armantrout (Ed.), *Acquisition and Utilization of Aquatic Habitat Inventory Information.* American Fisheries Society, Western Division, Bethesda, MD.

Hughes, R.M., D.P. Larsen, and J. M. Omernik. 1986. Regional reference sites: a method for assessing stream potentials, *Environmental Management,* 10, 629–635.

Karr, J.R. 1981. Assessment of biotic integrity using fish communities, *Fisheries,* 6(6), 21–27.

Karr, J.R. 1987. Biological monitoring and environmental assessment: a conceptual framework, *Environmental Management,* 11, 249–256.

Karr, J.R., L.S. Fore, and E.W. Chu. 1997. Making Biological Monitoring More Effective: Integrating Biological Sampling with Analysis and Interpretation. U.S. Environmental Protection Agency, Washington, DC.

Karr, J.R., R.C. Heidinger, and E.H. Helmer. 1985a. Sensitivity of the index of biotic integrity to changes in chlorine and ammonia levels from wastewater treatment facilities, *Journal of the Water Pollution Control Federation,* 57, 912–915.

Karr, J.R., L.A. Toth, and D.R. Dudley. 1985b. Fish communities of midwestern rivers: a history of degradation, *Bioscience,* 35, 90–95.

Karr, J.R., P.R. Yant, K.D. Fausch, and I.J. Schlosser. 1987. Spatial and temporal variability of the Index of Biotic Integrity in three midwestern streams, *Transactions of the American Fisheries Society,* 116, 1–11.

Karr, J.R., K.D. Fausch, P.L. Angermeier, P.R. Yant, and I.J. Schlosser. 1986. Assessing Biological Integrity in Running Waters: A Method and Its Rationale. Illinois Natural History Survey Special Publication 5.

Katz, M. and A.R. Gaufin. 1953. The effects of sewage pollution on the fish population of a midwestern stream, *Transactions of the American Fisheries Society,* 82, 1, 56–165.

Lyons, J., S. Navarro-Perez, P.A. Cochran, E. Santana, and M. Guzman-Arroyo. 1995. Index of biotic integrity based on fish assemblages for the conservation of streams and rivers in west-central Mexico, *Conservation Biology,* 9, 569–584.

Lyons, J., L. Wang, and T.D. Simonson. 1996. Development and validation of an index of biotic integrity for coldwater streams in Wisconsin, *North American Journal of Fisheries Management,* 16, 241–256.

Master, L. 1990. The imperiled status of North American aquatic animals. *Biodiversity Network News (Nature Conservancy),* 3, 1–2 and 7–8.

Menzel, B.W., J.B. Barnum, and L.M. Antosch. 1984. Ecological alterations of Iowa prairie-agricultural streams, *Iowa State Journal of Research,* 59, 5–30.

Miller, D.L., P.M. Leonard, R.M. Hughes, J.R. Karr, P.B. Moyle, L.H. Schrader, B.A. Thompson, R.A. Daniels, K.D. Kausch, G.A. Fitzhugh, J.R. Gammon, D.B. Halliwell, P.L. Angermeier, and D.J. Orth. 1988. Regional applications of an index of biotic integrity for use in water resource management, *Fisheries,* 13(5), 12–20.

Minns, C.K., V.W. Cairns, R.G. Randall, and J.E. Moore. 1994. An index of biotic integrity (IBI) for fish assemblages in the littoral zone of Great Lakes' areas of concern, *Canadian Journal of Fisheries and Aquatic Sciences,* 51, 1804–1822.

Moyle, P.B. and R.A. Leidy. 1992. Loss of aquatic ecosystems: Evidence from fish faunas, in P.L. Fielder and S.K. Jain (Eds.), *Conservation Biology: The Theory and Practice of Nature Conservation, Preservation and Management.* Chapman and Hall, New York, 127–169.

Nehlsen, W., J.E. Williams, and J.A. Lichatowich. 1991. Pacific salmon at the crossroads: stocks at risk from California, Oregon, Idaho, and Washington, *Fisheries,* 16(2), 4–21.

Oberdorff, T. and R.M. Hughes. 1992. Modification of an index of biotic integrity based on fish assemblages to characterize rivers of the Seine basin, France, *Hydrobiologia,* 228, 117–130.

Oberdorff, T. and J.P. Porcher. 1994. An index of biotic integrity to assess biological impacts of salmonid farm effluents on receiving waters, *Aquaculture,* 119, 219–235.

Ohio Environmental Protection Agency (OEPA). 1989. *Biological Criteria for the Protection of Aquatic Life.* Vol. II. *Users Manual for Biological Field Assessment of Ohio Surface Waters.* Ohio Environmental Protection Agency, Division of Water Quality Planning and Assessment, Columbus, OH.

Omernik, J. M. 1987. Ecoregions of the conterminous United States, *Annuals of the Association of American Geographers,* 77, 118–125.

Poulson, T.L. 1991. Assessing groundwater quality in caves using indices of biological integrity, *Proceedings of the 3rd Conference on Hydrogeology, Ecology, Monitoring and Management of Groundwater in Karst Terranes.* Waterwell Journal Publishing, Dublin, OH.

Poulson, T.L. 1992a. Case studies of groundwater biomonitoring in the Mammoth Cave region, in J.A. Stanfords and J.S. Simon (Eds.), *Proceedings of the 1st International Conference of Groundwater Ecology.* American Water Resources Association, Bethesda, MD.

Poulson, T.L. 1992b. Biological Indicators of Groundwater Quality. National Groundwater Conference, Nashville, TN.

Rankin, E.T. and C.O. Yoder. 1990. The nature of sampling variability in the Index of Biotic Integrity in Ohio streams, in W. S. Davis (Ed.), *Proceedings of the 1990 Midwest Pollution Control Biologists Meeting.* U.S. Environmental Protection Agency, Region 5, Environmental Sciences Division, Chicago, IL. EPA 905/9-90/005, 9–18

Rosenberg, D.M. and V.H. Reash (Eds). 1993. *Freshwater Biomonitoring and Benthic Macroinvertebrates.* Chapman and Hall, New York.

Simon, T.P. 1989. Rationale for a family-level ichthyoplankton index for use in evaluating water quality, in W.S. Davis and T.P. Simon (Eds.), *Proceedings of the 1989 Midwest Pollution Control Biologists Meeting,* Chicago, IL, February 14–17, 1989. EPA 905-9-89-007. U.S. Environmental Protection Agency, Chicago, IL, 41–65.

Simon, T.P. 1991. Development of Index of Biotic Integrity Expectations for the Ecoregions of Indiana. I. Central Corn Belt Plain. EPA-905-9-91/025. U.S. Environmental Protection Agency, Chicago, IL.

Simon, T.P. Modification of an index of biotic integrity and development of reference condition expectations for dunal, palustrine wetland fish communities along the southern shore of Lake Michigan, *Aquatic Ecosystem Health and Restoration,* (in press).

Simon, T.P. and E.B. Emery. 1995. Modification and assessment of an index of biotic integrity to quantify water resource quality in Great Rivers, *Regulated Rivers Research and Management,* 11, 283–298.

Simon, T.P. and J. Lyons. 1995. Application of the index of biotic integrity to evaluate water resource integrity in freshwater ecosystems, in W.S. Davis and T.P. Simon (Eds), *Biological Assessment and Criteria: Tools for Water Resource Planning and Decision Making.* Lewis, Boca Raton, FL, 245–262.

Simon, T.P. and P.M. Stewart. Validation of an index of biotic integrity for dunal, palustrine wetlands: emphasis on assessment of nonpoint source landfill effects on the Grand Calumet Lagoons, *Aquatic Ecosystem Health and Restoration,* (in press).

Steedman, R.J. 1988. Modification and assessment of an index of biotic integrity to quantify stream quality in southern Ontario, *Canadian Journal of Fisheries and Aquatic Sciences,* 45, 492–501.

Stephen, C.E., D.I. Mount, D.J. Hansen, J.H. Gentile, G.A. Chapman, and W.A. Brungs. 1985. Guidelines for Deriving Numerical National Water Quality Criteria for the Protection of Aquatic Organisms and Their Uses. National Technical Information Services #PB85-227049, U.S. Environmental Protection Agency, Office of Research and Development, Washington, DC.

Suter, G.W., II. 1993. A critique of ecosystem health concepts and indexes, *Environmental Toxicology and Chemistry,* 12, 1533–1539.

U.S. Environmental Protection Agency. 1986. Technical Guidance Manual for Performing Wasteload Allocations, in Stream Design and Flow for Steady-State Modeling. U.S. Environmental Protection Agency, Office of Water Regulation and Standards, Office of Water, Washington, DC.

U.S. Environmental Protection Agency. 1991. Technical Support Document for Water Quality-Based Toxics Control. EPA 505-2-90-001. U.S. Environmental Protection Agency, Office of Water, Washington, DC.

Ward, R.C. 1992. Indicator selection: a key element in monitoring system design, in D.H. McKenzie, D.E. Hyatt, and V.J. McDonald (Eds.), *Ecological Indicators.* Vol. I. Elsevier, New York, 147–157.

Weaver, L.A. and G.C. Garman. 1994. Urbanization of a watershed and historical changes in a stream fish assemblage, *Transactions of the American Fisheries Society,* 123, 162–172.

William, J.E. and R.J. Neves. 1992. Biological diversity in aquatic management, *Transactions of the North American Wildlife and Natural Reosurces Conference,* 57, 343–432.

Yoder, C.O. 1991. The integrated biosurvey as a tool for evaluation of aquatic life use attainment and impairment in Ohio surface waters, in *Biological Criteria, Research and Regulation: Proceedings of a Symposium,* 12–13 December 1990, Arlington, Virginia. U.S. Environmental Protection Agency, Office of Water, Washington, DC. EPA 440/5-91/005.

Yoder, C.O. and E.T. Rankin. 1995. Biological response signatures and the area of degradation value: new tools for interpreting mulitmetric data, in W.S. Davis and T.P. Simon (Eds.), *Biological Assessment and Criteria: Tools for Water Resource Planning and Decision Making.* CRC, Boca Raton, FL, 263–286.

2 Using Fish Assemblages in a State Biological Assessment and Criteria Program: Essential Concepts and Considerations

Chris O. Yoder and Marc A. Smith

CONTENTS

2.1 Introduction 18
    2.1.1 The Role of Fish Assemblage Data in an Environmental Indicators Framework..19
    2.1.2 The Value of Fish Assemblage Data to Environmental Monitoring and Assessment 21
2.2 Sampling Methods and Logistics 23
    2.2.1 Initial Decisions and Other Considerations...24
    2.2.2 Field Sampling Methods..24
    2.2.3 Wading Methods..26
    2.2.4 Boat Methods...31
    2.2.5 Fish Handling and Enumeration Procedures...32
    2.2.6 General Cautions Concerning Field Conditions..34
    2.2.7 Data Management and Information Processing 34
    2.2.8 Cost Considerations 34
2.3 Biological Criteria..35
    2.3.1 The Role of Fish Assemblage Data in Biological Criteria 35
    2.3.2 Regionally Referenced Numerical Biological Criteria 36
    2.3.3 New Generation Biological Community Evaluation Mechanisms 36
    2.3.4 The Index of Biotic Integrity (IBI) Modified by the Ohio EPA..........................37
    2.3.5 Modified Index of Well-Being (MIwb) 39
    2.3.6 IBI/MIwb Relationship 40
2.4 Using Fish Assemblage Information in Bioassessments...41
    2.4.1 Evaluating the Effectiveness of Water Quality Management Programs 41
    2.4.2 Scioto River Case Example 41
    2.4.3 Ottawa River Case Example 42
    2.4.4 Habitat Assessment 45
    2.4.5 Statewide Assessments..47
    2.4.6 Long-Term Goal Assessment..49
    2.4.7 Watershed Assessments...49
    2.4.8 Other Uses of Fish Assemblage Information 51
2.5 Conclusion...51
References 52

2.1 INTRODUCTION

State water quality management agencies usually function as the principal custodians for the monitoring and assessment of the quality of surface and ground waters. Biological monitoring, while included in some state programs, historically has received a lower priority than chemical/physical monitoring. Recently, this has changed with a growing number of states now developing and using biological monitoring, criteria, and assessments (U.S. EPA, 1995a), and by recently emphasizing the role of biological assessments in guidance for state 305(b) reports (U.S. EPA, 1997). Macroinvertebrates have historically been the most popular organism group used by state water quality agencies. The use of fish assemblages is recent, by comparison, and this has been spurred on principally by two events: (1) the availability of cost-effective field sampling methods such as pulsed DC electrofishing; and (2) the development of evaluation tools like the Index of Biotic Integrity (IBI) by Karr (1981; also Fausch et al., 1984 and Karr et al., 1986) and the Index of Well-Being (Iwb; Gammon, 1976; Gammon et al., 1981). These developments have served to enhance and make more practical the use of fish assemblages as a cost-effective and relatively rapid bioassessment technique. This chapter is intended as a treatment of the relevant issues and concepts involved in designing and implementing a statewide biological monitoring and biological criteria program using fish assemblages.

Fish have been one of the most studied groups of aquatic organisms since the initial European settlement of the midwestern U.S. in the late 18th and early 19th centuries. Early investigations of the distribution and composition of the fish faunas of the region's rivers and streams were accomplished by Rafinesque (1820), Jordan (1890), Kirsch (1895a,b), Kirtland (1838), and others. These were later followed in the 20th century by the many "fishes of" texts that inventoried the composition and distribution of the fish faunas of many states. One of the first of these statewide texts was published on the fish fauna of Illinois by Forbes and Richardson (1920). This was later followed by similar texts for Ohio (Trautman, 1957; 1981), Indiana (Gerking, 1945), Missouri (Pflieger, 1975), Kentucky (Clay, 1975), the Upper Mississippi River Valley (Eddy and Underhill, 1974), the Great Lakes drainage (Hubbs and Lagler, 1964), Canada (Scott and Crossman, 1973), and more recently in many other states including Illinois (Smith, 1979), Wisconsin (Becker, 1983), New York (Smith, 1985), Arkansas (Robison and Buchanan, 1988), Tennessee (Etnier, 1993), and Virginia (Jenkins and Burkhead, 1993). These not only provided a baseline against which changes through the 19th and 20th centuries were evaluated, but provided the impetus for future developments, including the routine use of fish assemblages as an indicator of the condition of water resources as a whole.

If there were any shortcomings in these pioneering efforts, it was with the principal reliance on presence-absence type of data. Most lacked consistent and reliable relative abundance estimates based on standardized sampling methods. However, in a few instances (e.g., Wisconsin in Lyons, 1992a), the historical fish assemblage database was transferable to the later development of relative abundance-based assessment frameworks that supported the development of biological criteria based on multimetric indices such as the Index of Biotic Integrity (Karr, 1981; Fausch et al., 1984) and the Index of Well-Being (Gammon, 1976; Gammon et al., 1981). It was Gammon's (1973; 1976; 1980) work in the Wabash and Great Miami Rivers that served as the conceptual model for Ohio's EPA use of pulsed DC electrofishing to collect relative abundance information about fish assemblages on a longitudinal, river reach scale. The collection of relative abundance data includes the use of standardized sampling procedures designed to produce a sufficiently representative sample of the fish assemblage at a site with a reasonable sampling effort (i.e., 1 to 3 hours per site). As such, this type of assessment is distinguished from the much more resource-intensive efforts using multiple collection gear and those required to obtain estimates of population, standing crop, or a complete inventory of all species present.

Relative abundance assessments are generally included in the category of "rapid bioassessments," a term that was institutionalized with the release of the U.S. EPA bioassessment guidelines

(Plafkin et al., 1989) in the late 1980s and recently updated in Barbour et al. (1997). The Ohio EPA initiated the use of fish assemblages as part of a statewide assessment of rivers and streams in the late 1970s. While refinements have been made along the way, the original design of the sampling equipment and field procedures has changed little over the past 2 decades. More importantly, extensions to other water bodies have been made and now include procedures for assessing fish assemblages in the Ohio River mainstem (Simon and Emery, 1995; Sanders, 1991; Simon and Sanders, Chapter 18) and the Lake Erie nearshore (Thoma, Chapter 16).

2.1.1 THE ROLE OF FISH ASSEMBLAGE DATA IN AN ENVIRONMENTAL INDICATORS FRAMEWORK

An environmental indicator is defined here as a measure of environmental properties that singly or in combination provide scientifically and managerially useful information about status and trends in environmental quality (ITFM, 1995). True environmental indicators are comprised of chemical, physical, and biological measures and are further stratified according to stressor, exposure, and response roles (U.S. EPA, 1991). Stressor indicators generally include quantitative measures of activities that have the potential to degrade the aquatic environment, such as pollutant discharges (both permitted and unpermitted), land use impacts, and habitat modifications. Exposure indicators are those that measure the potential and realized effects of stressors and can include concentrations of chemicals, measures of whole toxicity (e.g., bioassay tests), chemical residues in tissues, and biomarkers, each of which can provide evidence of an exposure to a stressor or bioaccumulative agent. Response indicators are generally composite measures of the cumulative effects of stress and exposure, and include the more direct measures of biological community and population response that are represented by multimetric biological indices such as the Index of Biotic Integrity (IBI) for fish assemblages and similar mechanisms for assessing macroinvertebrate community information (Plafkin et al., 1989; Kerans and Karr, 1992; DeShon, 1995; Barbour et al., 1996).

Fish assemblage information (along with macroinvertebrate community data) composes one component of the Ohio EPA biological criteria (Yoder and Rankin, 1995a). Other relevant response indicators may include elements of the same database used to develop the multimetric indices for fish assemblages such as target species (i.e., rare, threatened, endangered, special status, and declining species). The different types of indicators described above represent the essential technical elements for watershed-based management approaches. The key is to use the different indicators within the roles that are most appropriate for each. People are beginning to recognize that the inappropriate use of stressor and exposure indicators as substitutes for response indicators has resulted in disparities and inconsistencies between states in terms of the statistics reported to the U.S. EPA for purposes such as the national water quality inventory (Yoder and Rankin, 1998). Generally, the failure to include biological indicators results in overly optimistic assessments of the condition of surface water resources (Ohio EPA, 1997; Yoder and Rankin, 1998).

The U.S. EPA developed a hierarchy of surface water indicators as a framework for developing an environmental indicators-based process for managing and reporting on surface water quality (Figure 2.1; U.S. EPA, 1995b,c). This hierarchy has been used by the Ohio EPA as a means of integrating traditional water quality management programs (e.g., permitting, funding, enforcement, etc.) with ambient monitoring and assessment information (Ohio EPA, 1997). The traditional measures of water quality management program success have been comprised of administrative measures such as number of permits issued, grant dollars awarded, and enforcement actions taken. Now, efforts are underway to link these activity measures with results in the environment. In Ohio, the results of this process are evident in numerous watershed-specific assessments and the biennial Ohio Water Resource Inventory (305[b] report; Ohio EPA, 1997). The linkages made between administrative activity measures and ambient environmental monitoring information provide the basis for setting goals based on environmental quality which is exemplified by the Ohio 2000 goals for surface waters (Ohio EPA, 1997).

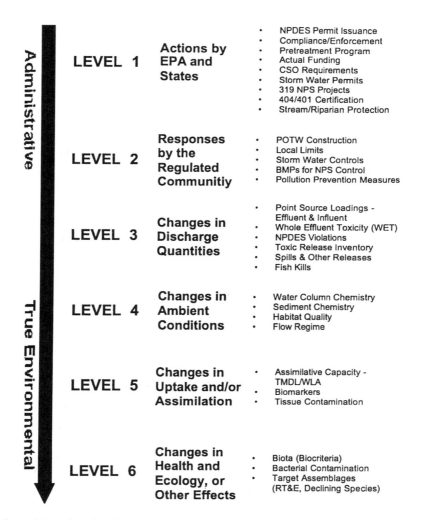

FIGURE 2.1 A hierarchy of water indicators from administrative through true environmental indicators developed by the U.S. EPA (1995a) for evaluating the effectiveness of water quality management program efforts.

Fish assemblage information is a level 6 indicator (Figure 2.1) and, as such, represents a tangible and integrated end result of water quality management efforts. The inherently broad applicability of ambient biological evaluation has been demonstrated in numerous instances (Karr et al., 1986; Ohio EPA, 1990; Yoder and Rankin, 1998). The comparative ability and "power" of the traditional water quality and toxicological assessment tools to measure or reflect key components of the five major factors that determine the integrity of surface water resources (Figure 2.2; after Karr et al., 1986) was compared with ambient biological assessment in the Ohio EPA (1987a). Biological assessment tools have the comparatively unique ability to integrate, and thus reflect, the aggregate condition of a water body. As such, bioassessment represents the broadest single approach for assessing the effect of multiple and varied environmental influences. When used in conjunction with chemical, physical, and toxicological assessment tools, the ability to identify and quantify associated causes and sources is greatly enhanced.

A multiple indicators approach that utilizes each according to its most appropriate role will lead to more effective regulation of pollution sources, improved assessment of diffuse and nonchemical impacts, and improve the ability to implement management strategies for successfully protect-

Using Fish Assemblages in a State Biological Assessment and Criteria Program

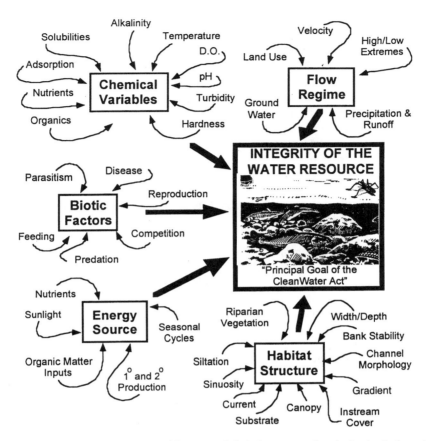

FIGURE 2.2 The five principal factors, with some of their important chemical, physical, and biological components that influence and determine the integrity of surface water resources (modified from Karr et al., 1986).

ing and restoring the ecological integrity of watersheds (Yoder and Rankin, 1998). Some of the attributes that biological indicators should exhibit include: (1) cost-effective collection of data; (2) readily available science; (3) being indicative of and extend to several different trophic levels; (4) integrate multiple effects and exposures; (5) exhibit reasonable response and recovery rates; (6) being precise and reproducible; (7) being responsive to a wide range of perturbations; and (8) being relevant to managerial and programmatic issues. Using fish assemblages as a key biological indicator group can help satisfy these objectives.

2.1.2 THE VALUE OF FISH ASSEMBLAGE DATA TO ENVIRONMENTAL MONITORING AND ASSESSMENT

The decision about which indicator organism groups to monitor is critical to the eventual success of a biological monitoring and assessment effort. While history shows that this can be contentious, experience shows that it is unproductive to advocate a particular organism group as a way to "sell" the need to monitor one group at the expense of all others. This has had the unfortunate consequence of initiating negative comparisons that eventually become counterproductive. The positive and compelling attributes of using a particular organism group should be emphasized in this process. This was exemplified in the indicator selection process of the U.S. EPA Environmental Monitoring and Assessment Program (EMAP) and U.S. Geological Survey National Water Quality Assessment (NAWQA) programs and the steps taken by each to resolve these issues.

Because it is impractical to monitor every organism group present in an aquatic ecosystem, choices need to be made. The ITFM (1992) produced an indicators matrix that is helpful in deciding which indicators to use. Our choice of two organism groups (benthic macroinvertebrates and fish) is consistent with the ITFM (1992) recommendations and was done for a number of reasons. The benefit of having two different groups independently showing the same result is obvious and lends considerable strength to a bioassessment. However, differences in the responses by each group can lead to the definition of problems that might otherwise have gone undetected or underrated in the absence of information from either organism group. For example, representatives of one group may be able to tolerate and metabolize some toxic substances that are highly detrimental to representatives of the other group. The value of such information in a risk management process should be obvious.

It has long been recognized that fish can be one of the most sensitive indicators of aquatic ecosystem quality (Smith, 1971) and reliably demonstrate conditions that are suitable or unsuitable for their existence (Gaufin and Tarzwell, 1953; Doudoroff and Warren, 1957). In a long-running study of the Wabash River, Gammon (1976) observed that "... the populations of fish in a river reflect the overall state of environmental health of the watershed as a whole (because) fish live in water which has previously fallen on the cities, fields, strip mines, grasslands, and forests of the watershed ... has previously been used by man (and) despite the fact that fish represent the end-product of the food chain their use as biological indicators of pollution has received considerably less attention than other riverine groups." The inherently broad applicability of using fish communities to evaluate and characterize the quality of aquatic ecosystems was also demonstrated by early studies of pollution in large rivers such as the Ohio (Krumholz and Minckley, 1964) and Illinois (Mills et al., 1966; Sparks and Starrett, 1975), along the Lake Erie shoreline (White et al., 1975), documenting long-term changes in relation to stream habitat (Larimore and Smith, 1963), classifying streams on a statewide basis (Smith, 1971), characterizing urban impacts (Tramer and Rogers, 1973), and through the use of various diversity indices (Tsai, 1968; 1971; Denoncourt and Stambaugh, 1974; Hill et al., 1975; Teppen and Gammon, 1976).

While these efforts served to highlight the value of fish as indicators of water quality, the evaluations failed to produce the precision and assurance needed to compete with chemical/physical assessments. It was the later development of more quantitative multimetric indices such as the IBI (Karr, 1981; Fausch et al., 1984; Karr et al., 1986) and the development of numerical biological criteria (Ohio EPA, 1987b; Fausch et al., 1990; Lyons, 1992a; Simon and Lyons, 1995; Yoder and Rankin, 1995a,b) that closed the gap in demonstrating the effectiveness of using fish assemblages as reliable environmental indicators. The capability of fish assemblages to accurately reflect environmental quality lies in their integration of the five major factors that determine the integrity of a water resource (Figure 2.2; Karr et al., 1986). When used in concert with the other biological, chemical, and physical assessment tools, the ability to identify and quantify environmental quality and adverse effects is greatly enhanced. Such an approach has led to the more effective regulation of pollution sources (Ohio EPA, 1992), improved the assessment of diffuse and nonchemical impacts (Ohio EPA, 1997), and broadened the ability to implement management strategies for protecting and restoring watersheds (Yoder and Rankin, 1995a; Yoder and Rankin, 1998).

The "popular" belief that fish are too mobile to be used effectively as an indicator group is a conventional-wisdom argument that continues to be raised today. The mobility of fishes compared with the more immotile groups (e.g., Unionidae, other macroinvertebrates) is obvious. However, mobility alone does not disqualify fish assemblages from functioning as a valid indicator. Stream and many riverine fish species are not so excessively mobile as to become unusable as indicators. The majority of species encountered in warmwater rivers and streams can reasonably be expected to be encountered in repeated samples during the summer and fall months. All organism groups have dispersal mechanisms that are critical if the full complement of species is to be sustained. Fish accomplish dispersal through swimming, macroinvertebrates by drifting, winged insects by flying, and the immotile Unionidae via fish (Watters, 1992). Much of the extant literature indicates that warmwater stream fish populations are comparatively sedentary, some species more so than others, and particularly so during the summer–fall period. Gerking (1953; 1959) concluded that most stream fish moved less than 0.5 km in their lifetime in a small Indiana stream, with each species having a small fraction of the population that moved greater distances. Information from other sources (Funk, 1954; White et al., 1975) also suggests that fish are relatively sedentary during the summer and early fall months. Our experience during the past 2 decades seems to corroborate these observations.

Other organism groups such as algae, diatoms, zooplankton, macrophytes, naiad mollusks (Unionidae), and others are not routinely included in our bioassessments. This does not diminish our ability to broadly evaluate and quantify the quality of aquatic ecosystems. Fish and macroinvertebrates are not only among the most sensitive components of the aquatic biota, but their function and overall well-being are dependent on the primary and secondary productivity of the aquatic ecosystem. Thus, problems that first occur in the lower trophic groups will eventually be revealed in the higher organism groups if they are indeed of ecological consequence. It may be necessary on occasion to monitor and use data from other organism groups for specific purposes such as refining our understanding of cause/effect relationships, verifying the presence of specific contaminants, and for considerations involving rare, threatened, and endangered species. An important goal in using bioassessments is to evaluate the attainment and non-attainment of beneficial aquatic life uses as defined in state water quality standards. Focusing on two of the better understood, responsive, and higher organism groups adequately satisfies this important goal.

The common tendency in water quality assessment and management has been to make biological assessments fit the perceptions derived from surrogate indicators, rather than the reverse. This is clearly illogical because the condition of the fish community is the embodiment of the temporal and spatial chemical, physical, and biological dynamics and elements of the aquatic ecosystem (Figure 2.2). Perhaps the historical inability of biologists to agree on a set of empirical measurements of biological integrity, or at least a common framework, has resulted in this situation (Karr et al., 1986). One solution to this deficiency is to employ numerical biological criteria that can quantitatively indicate the degree to which biological integrity is or is not being achieved (Yoder and Rankin, 1995 a,b). Fish community assessment is a critical component of bioassessment programs in a growing number of states (U.S. EPA, 1996a).

2.2 SAMPLING METHODS AND LOGISTICS

The choice of sampling methods is a fundamental decision or "cornerstone" for using fish assemblages as an environmental monitoring and assessment tool. Although a wide variety of possible methods and techniques are available, the choice of which one(s) to use should be dictated by the objectives of the program and the conditions that exist in the particular state or region. Regarding the former, the objective is to have a method that meets the previously described objectives (reasonable cost and effort, relatively rapid, etc.). While the Ohio EPA methods for sampling fish are highlighted here, there exist other equally valid techniques, some of which may work better in some regions than in others. The following descriptions of the Ohio EPA biological methods are intended as an overview. Prospective users need to consult the methods documents (Ohio EPA, 1987a,b; 1989a,b; Rankin, 1989) prior to sampling.

Reconciling the differences in methods currently employed and those proposed for emerging monitoring networks poses some challenges. States that have taken the initiative to invest in biological monitoring in the previous absence of clear federal directives will naturally and somewhat justifiably be reluctant to conform to different methods. Since it is unlikely that bioassessment methods will be completely uniform on a nationwide basis, a more reasonable goal is to have comparable assessments produced by different methods. This is a critical national issue that needs to be addressed by the various interagency groups working on monitoring coordination issues.

In the process of selecting the field and laboratory methods used by the Ohio EPA, there were several considerations. These include the need to produce assessments that: (1) are capable of discriminating the varied impacts that occur in Ohio's surface waters; (2) produce scientifically valid results; and (3) are cost-effective with regard to the two previous objectives. These are inherently competing objectives as elaborate and highly detailed assessments are not likely to be very cost-effective, yet the need for scientific validity prescribes an inherent level of rigor and, hence, complexity of assessment. In contrast, assessments lacking sufficient detail and rigor may cost less, but lack the elements of scientific validity sufficient to discriminate the impacts that actually exist. Given the economic, social, and environmental consequences of the decisions that need to be made by the states, it seems prudent to opt for a sufficiently complex and rigorous assessment.

2.2.1 Initial Decisions and Other Considerations

There are a number of decisions that need to be made prior to adopting sampling methods. This is a critical juncture in the process because the decisions made here will influence the effectiveness of the overall bioassessment effort well into the future. The decisions about which sampling methods and gear to use, seasonal considerations, which organism groups to monitor, which parameters to measure and record, which level of taxonomy to use, etc. all need to be made up-front in the process. If there is any axiom to all of this, it is "... when in doubt choose to take more measurements than seem necessary at the time since information not collected is impossible to retrieve at a later date." This does not apply equally to all variables and factors. For example, seasonality is a well-developed concept; therefore, it is not necessary to sample in multiple seasons for the sake of data redundancy. However, parameters that require little or no extra effort to acquire should be included until enough evidence is amassed to prove its relative worth. One example in Ohio is with external anomalies on fish. It was initially decided to record this information even though it was not apparent what its usefulness would be (this preceded the IBI and other studies on the importance of fish anomalies). This one measure of organism health has proven over time to be one of our most valuable parameters (Sanders et al., Chapter 9). Length data is an example of a parameter that was initially recorded for each and every species collected, but proved useful only on occasion. Now, length data are recorded selectively.

Another important consideration is assuring that qualified and regionally experienced staff are available to conduct the monitoring and assessment activities. Ecological assessment is not unlike many other professions in which skilled and experienced staff are sought to direct, manage, and supervise. However, biological field assessment requires a high level of expertise in the field since many of the critical pieces of information are recorded and, to a degree, interpreted in the field. There is simply no substitute for the intangibles gained by direct experience in the field. This is not a job to be left to technicians alone. In addition, it is only prudent that the same professional staff who collect the field data also interpret and apply the information derived from the data in a "cradle-to-grave" fashion. Thus, the same staff who perform the field work also plan that work, process the data into information, interpret the results, and apply the results via assessment and reporting. Such staff, particularly those with sufficient experience, also contribute to policy and program development.

2.2.2 Field Sampling Methods

The principal method for sampling stream and riverine fish communities in Ohio is pulsed DC electrofishing. Several studies have shown electrofishing to be the single most effective gear for obtaining fish assemblage data in midwestern streams (Funk, 1958; Larimore, 1961; Boccardy and Cooper, 1963; Bayley et al., 1989), rivers (Vincent, 1971; Novotny and Priegel, 1974; Gammon, 1973; 1976; Hendricks et al., 1980; Ohio EPA, 1987b), the Ohio River (Sanders, 1991; Simon and Emery, 1995; Simon and Sanders, Chapter 17), and the Lake Erie shoreline (Thoma, Chapter 16). This gear type has also been applied successfully in other parts of the U.S. (Hughes and Gammon, 1987; Paller, 1994). This is also the method of choice for the other local, state, and federal agencies that collect fish assemblage data in Ohio, midwestern states, and many other parts of the U.S. (U.S. EPA, 1991; Meador et al., 1993).

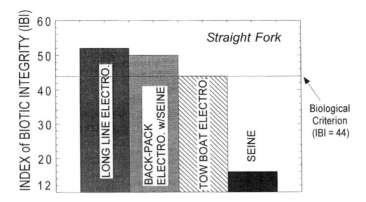

FIGURE 2.3 Comparison of IBI values derived from fish sampling results by four different field crews using seining and various pulsed DC electrofishing gear in a small southeastern Ohio stream.

An initial effort was made to standardize field and data processing procedures among the Ohio agencies, which has enhanced the overall usefulness of the statewide database. The liabilities of failing to use effective sampling methods are illustrated in Figure 2.3. In this case, a small southeastern Ohio stream was sampled by four different entities using different collection methods ranging from seining to a combination of seines and backpack electrofishing to larger generator-powered electrofishing units. The results show that seining substantially underestimated the biological condition of the stream, producing only five species and a poor quality rating based on the IBI; whereas, the electrofishing samples produced more than 25 species and an exceptional quality rating. Seining is made difficult in these streams by the substrate that contains slab boulders and cobbles, which interfere with efficient use of a seine. Pulsed DC electrofishing is able to overcome these difficulties in that fish are drawn out from under the substrate and other submerged objects that otherwise inhibit the effectiveness of seining. The implications of this to the eventual assessment of the condition of the stream are obvious. Thus, the choice of sampling gear is a critical step in the process. These findings are similar to those obtained by Bayley et al. (1989) in small Illinois streams.

Electrofishing may not work equally well in all settings and can require supplemental gear types such as seining, or replacement altogether where conductivity (both high and low) may be limiting. In any case, state programs should strive for a single gear approach that meets the goals of cost effectiveness while generating sufficiently robust information. Ohio EPA separates electrofishing into two broad categories: wading methods and boat-mounted methods (Table 2.1). Criteria for the selection of the appropriate sampler type include stream depth, width, and other factors outlined in the Ohio EPA field sampling protocols (Ohio EPA, 1989a). A flowchart (Figure 2.4) demonstrates the decision tree that is followed in this process, which is also tied to the modified IBI that is used to evaluate the results. Wading methods range from small, backpack electrofishing units powered by a battery or small generator to a generator-powered pulsed DC electrofishing unit as a bank set or towed in a small skiff. Boat-mounted methods employ larger and more powerful generator-powered electrofishing units in custom-designed john boats ranging from 12 to 21 feet in length (Table 2.1). Boat electrofishing is conducted during daylight hours in inland rivers draining less than 6000 sq. mi. and at night on the Ohio River (Sanders, 1991; Simon and Sanders, Chapter 18) and the Lake Erie shoreline (Thoma, Chapter 16).

Although passive gear approaches are also described in the Ohio EPA methods protocols, they are rarely used, if ever, for formal bioassessment purposes. This is consistent with other studies that show that passive gear produce only minimal information in addition to electrofishing (Funk, 1958; Gammon, 1973; Thoma, Chapter 16). An important objective is to use methods and equipment that are powerful enough to secure a sufficiently representative sample, do so with a reasonable

TABLE 2.1
Summary of Pulsed DC Electrofishing Methods Used Routinely by the Ohio EPA to Assess Fish Communities in Rivers and Streams Throughout Ohio and General Factors that Affect the Choice and Effectiveness of Sampling Gear

Category	Wading Methods		Boat Methods		
	Small Streams	Larger Streams	Small Rivers	Large Rivers	Great Rivers[1] Lake Erie[2]
Water body size dimensions[3]	<1.0–10 sq. mi <0.3–0.5 m depth 1–2 m width	10–500 sq. mi. 0.5–1.0 m depth 2–20 m width	150–1000 sq. mi. >1.0 m depth 10–100 m width	1000–6000 sq. mi. >1.0 m depth >50 m width	>6000 sq. mi >1.0 m depth (Ohio River)
Platform	Backpack; bank set	Tow boat; bank set	12–14 ft. boat	14–16 ft. boat	18 ft boat/21 ft. boat
Unit	Battery/generator	Generator	Generator	Generator	Generator
Power source	12 V battery/ 300–1750 W alt.	1750–2500 W alternator	2500–3500 W alternator	3500–5000 W alternator	5000/7500 W alternator
Amperage output Volts DC output	1.5–2A; 2–12 A 100–200	2–12 A 150–300; 300–1000	4–15 A 500–1000	15–20 A 500–1000	15–20 A 500–1000
Anode location	Net ring	Net ring	Boom (droppers)	Boom (droppers)	Boom; spheres[4]
Sampling direction	Upstream	Upstream	Downstream	Downstream	Downstream; downcurrent
Distance sampled	0.15–0.20 km	0.15–0.20 km	0.5 km	0.5 km	0.5–1.0 km
CPUE[5] basis	per 0.3 km	per 0.3 km	per 1.0 km	per 1.0 km	per 1.0 km
Time sampled (typical)[6]	1800–3600 s	1800–3600 s	1600–3500 s	1600–4500 s	2000–3500 s
Time of sampling	Daylight	Daylight	Daylight	Daylight	Twilight/night

[1] Great Rivers generally exceed 6000 squre miles drainage area at the sampling site.
[2] Includes Lake Erie shoreline, inundated river mouths, and harbor areas.
[3] Maximum pool depth in small streams; sampling depth along shoreline in large rivers.
[4] Droppers are used in inland rivers and the Ohio River. Electrosphere design is used on Lake Erie only.
[5] CPUE: catch per unit of effort.
[6] Normal range — sampling time may vary upwards due to factors such as cover and instream obstructions.

effort, and minimize potential bias induced by different operators. Variability is controlled through extensive training and standardized protocols (Ohio EPA, 1989a).

2.2.3 WADING METHODS

Wading methods are used in smaller, wadeable streams that cannot be sampled with boat-mounted methods because the physical limitations of the stream channel. These range in size from the smallest headwater streams (<20 sq. mi. drainage area) to sites of 400 to 500 sq. mi. drainage area that typically range from 5 to 40 m in width and 0.5 to 1.0 m in depth. The wading method employs a

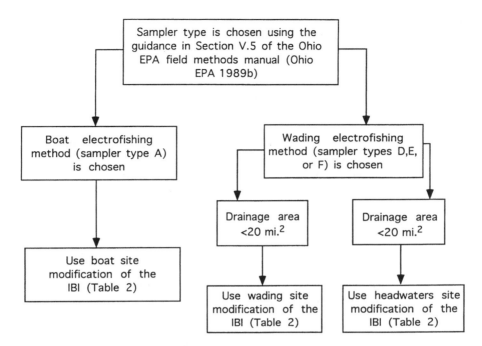

FIGURE 2.4 Schematic of the process by which the appropriate fish sampling method is selected and which modification of the IBI is used for each of three site types.

2.1-m plastic tow boat with the capacity to carry the generator-powered pulsed DC electrofishing unit and 30-gallon plastic live well or a 100-m long electrical cord for the bank-set variation. The electrofishing unit consists of a 1750-watt variable voltage generator that has the capability to supply 150 to 300 V pulsed DC. At sites that have greater pool widths and depths consistently greater than 1 m, the use of a more powerful generator/pulsator (2500 watts, 500 to 1000 V DC) unit rigged for use with the towboat or as a bank set may be required. This arrangement provides the additional power needed to efficiently sample pools that are consistently greater than 1 m deep (average) and wider than 30 to 40 m. The primary dip net ring also serves as the anode (+ electrode). This is an important design feature that deviates from the more traditional anode probes that are held separately from the dip nets. This latter method not only requires additional operators, but is potentially less efficient for capturing stunned fish in flowing water. Having the primary net ring simultaneously serving as the anode provides an efficient and consistent capture technique since most of the fish are attracted directly to the net ring. A positive pressure switch mounted on the net pole must be manually depressed to complete the switch circuit and allow electrical current to the electrodes.

Procedures for sampling require a three-person crew, all wearing chest waders and rubber gloves. The primary netter operates the anode net ring while one crew member guides the towboat or long line and the third crew member assists in capturing fish. The primary netter works the net ring beneath undercut banks, in and around root wads, woody debris, large boulders, and other submerged structures using various techniques to enhance capturing fish. An effective technique for capturing fish under submerged objects is to thrust the anode ring within and/or under the structure with the current off and then quickly withdraw the anode ring in one swift motion while simultaneously activating the current. This has the effect of drawing fish out from under such structures, making their capture easier. In the wider and deeper pools, the anode ring is "cast" ahead of the netter with the current on. This is an effective technique for capturing larger fish and mid-water pool species. In riffle and run areas, the primary netter "rakes" the anode ring from upstream to downstream, allowing it to drift with the current. At the same time, the assist netter blocks off an area downstream of the anode ring with the assist net. This enhances the capture of riffle-dwelling species. These are some examples of the types of techniques and skills that sampling crews must learn to use in order to produce credible and reproducible results. Wading sites are sampled one or two times during the mid-June to mid-October index sampling period. Sampling passes take place at least six weeks apart between individual sampling passes for a two-pass effort.

The use of block nets at either end of the sampling zone is not specified. Sites are arranged to begin and end in riffles or near instream obstructions in order to minimize fish movement out of the sampling zone. It is observed that the majority of the fish, when disturbed by the sampling crew, will seek adjacent cover instead of continuously moving upstream. This approach is consistent with the findings of Simonson and Lyons (1995), who found that single-pass electrofishing samples based on a zone length of 35 times the average stream width were suitable for generating species richness, composition, and relative abundance information for wadeable Wisconsin streams without the use of block nets.

All habitat types are thoroughly sampled in an upstream direction for a distance of at least 150, but no more than 200 m. These distances were established following repeated experimental sampling at a test site. A 300 m long site was subdivided into six consecutive 50-m increments. Each 50-m increment was sampled from downstream to upstream using the standardized wading electrofishing methods (Ohio EPA, 1989a). Sampling occurred weekly during the June 15–October 15 index sampling period. The results from each 50 m increment were recorded separately. Analysis of the results included combining the data in a cumulative manner starting with the first 50 m increment, adding the next 50 m and so on. The cumulative effect of sampling distance on the IBI and number of species was examined (Figure 2.5). In terms of the IBI, an asymptotic relationship was reached between 150 and 200 m. Continuing sampling beyond this distance would produce little additional information relative to the increased sampling effort. For species richness, there was a perceptible, but diminishing, increase up to a cumulative distance of 250 to 300 m. These findings generally agree with Angermeier and Karr (1986), who showed a leveling off of the influence of distance fished on the IBI between 140 and 280 m. This distance also approximates the number of stream widths defined by Lyons (1992b) for Wisconsin streams and that used by the U.S. EPA EMAP program (U.S. EPA, 1991), except that the latter protocol can result in site lengths of up to 500 m.

The EMAP program protocols specify sampling distances of 40 times the average stream width, which the authors had an opportunity to test during the E. Corn Belt Plains ecoregion regional EMAP (REMAP) project. More than 200 wadeable stream sites were sampled using both the EMAP 40 times stream width and 150 to 200-m Ohio EPA protocols. At some sites the 40-times-stream-width method resulted in sampling distances of up to 500 m, well beyond the maximum 200-m length specified by our protocol. While there was not a substantial difference between the median IBI scores (Figure 2.6), the net effect on the IBI was to slightly inflate the score at the sites where more than 200 m was sampled. This was the result of the addition of new species that affected the species richness based metrics. One problem with this comparison is that the IBI was calibrated based on the 150- to 200-m sampling distance, making it vulnerable to the inflating effect of adding species by sampling longer distances. It has not yet been determined whether having such variable distances would induce unintended bias into the IBI calibration process.

The time spent electrofishing, while not the primary arbiter of effort, is important in the sense that a minimum amount of time should be expended to ensure thorough sampling of a site. Plots of time electrofished (seconds) vs. various community parameters (biomass, numbers, species richness) and the IBI from more than 3000 wading electrofishing samples showed that a minimum of 1500 to 2000 s should be spent sampling each site (Ohio EPA, 1987b). The actual time spent is influenced by the relative ease of mobility through a sampling site, the number of fish encountered, amount of debris, depth heterogeneity, habitat variability, and other factors.

One is strongly cautioned about the use of backpack units in that they are under-powered for all except the smallest wadeable streams. It is for this reason that these units are restricted for use in only the smallest headwater streams by our methods protocols (Ohio EPA, 1989b). The more

Using Fish Assemblages in a State Biological Assessment and Criteria Program 29

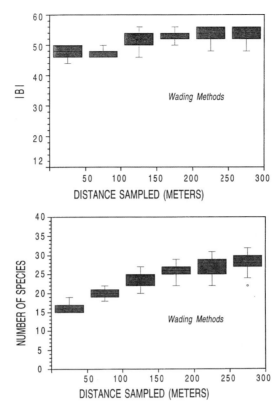

FIGURE 2.5 The effect of the cumulative distance sampled on the Index of Biotic Integrity (IBI; upper) and species richness (lower) for wading methods based on test sampling conducted in Little Darby Creek.

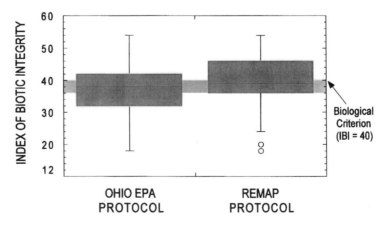

FIGURE 2.6 A comparison of the frequency distribution of IBI scores based on standard sampling distances used by the Ohio EPA and the stream width formula used by the U.S. EPA Regional Environmental Monitoring and Assessment Program (REMAP) at 35 small stream locations in the E. Corn Belt Plains (ECBP) ecoregion of Ohio.

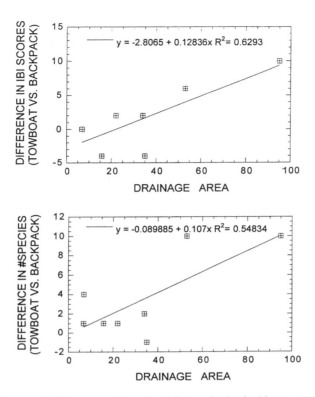

FIGURE 2.7 The difference in IBI scores and species richness obtained with generator-powered electrofishing units and backpack electrofishing units at selected small stream sites in the E. Corn Belt Plains (ECBP) ecoregion of Ohio.

powerful generator-powered electrofishing units are preferred as these generate the power and electric field necessary to effectively sample most wadeable stream fish assemblages. The payback in results more than offsets the heavier weight of these units. A comparison of the results obtained by backpack and generator-powered towboat or long line electrofishing units is illustrated in Figure 2.7 for wadeable Ohio streams ranging in size from less than 10 to 100 sq. mi. There was a significant increase in the differences between IBI scores and species richness with increasing drainage area. This illustrates the relative lack of effectiveness of the less powerful backpack units as wadeable streams increase in size. It also supports our recommendation that generator-powered tow boat or long line units be used in all except the smallest headwater streams and where maximum water depths are less than 0.5 m and widths are less than 2 m (Ohio EPA, 1989a).

In cases where access presents an insurmountable problem, which we rarely encounter in Ohio, backpack units should be supplemented by thorough seining. In our experience, skillful seining in combination with a backpack unit can produce results that are comparable to that produced by a generator-powered unit. However, the skill level needed to consistently produce comparable results is comparatively higher. After experimenting with battery-powered units and supplemental gear such as seines, the Ohio EPA switched entirely to generator-powered units in 1983. The reasoning behind this decision was that the generator-powered units (of at least 1500 to 1750 watts output) yielded a more consistent, representative sample and in less time than was possible with the battery-powered backpack electrofishing/seine combination. Because it is more dependent on the skill and determination of the sampling crew, the latter method was vulnerable to yielding substandard samples that subsequently could not be used in any quantitative analysis. This was most prevalent in streams with pools greater than 0.5 to 1.0 m in depth and/or those containing woody debris, slab boulders, or fractured bedrock substrates. Similar findings were recently obtained by Bayley et al. (1989) using a generator-powered electric seine in Illinois and previously by Funk (1958) in Iowa.

2.2.4 BOAT METHODS

In nonwadeable rivers, a boat-mounted system is used that employs a generator/pulsator combination (Ohio EPA, 1989a). This can include sites on inland rivers with drainage areas as low as 200 sq. mi. up to 6000 sq. mi. to the Ohio River and the Lake Erie shoreline. The boat methods employed by the Ohio EPA include the use of specially constructed john boats ranging in length from 12 to 21 ft. Boats used in inland rivers consist of 12, 14, and 16 ft lengths and sampling is conducted during daylight hours. Larger boats of 19 and 21 ft are used, respectively, in the Ohio River (Sanders, 1991; Simon and Sanders, Chapter 18) and Lake Erie shoreline, harbor areas, and lacustuaries (Thoma, Chapter 16). It was determined that night electrofishing was more effective in the Ohio River (Sanders 1991) and the Lake Erie shoreline (Thoma, Chapter 16) than daytime sampling. The types of electrofishing equipment, design of the electrode array, and sampling procedures generally follow the rationale and procedures outlined in Gammon (1973; 1976) and Novotny and Priegel (1974). Pulsed DC electrofishing units powered by 2500- to 7500-watt variable-voltage (500 to 1000 V DC) gasoline-powered generators are used.

Pulsed DC current is transmitted through the water by an arrangement of anodes and cathodes suspended in the water from the boat. On the 12-, 14-, 16-ft, and Ohio River boats, steel cable anodes are hung from a retractable aluminum boom that extends approximately 2 m in front of the boat. The exact boom dimensions vary according to boat size, being longer and wider on the larger boats. Four anodes are positioned on the front of the boom in a line perpendicular to the length of the boat. Four flexible galvanized steel conduits serve as cathodes and are suspended directly from the bow in a line perpendicular to the length of the boat. Modified electrospheres are used on the Lake Erie boat. Safety equipment includes a positive pressure foot-pedal switch located on the bow deck and an emergency toggle cutoff switch adjacent to the stern seat.

A boat sampling crew consists of a netter and a driver. The netter's primary responsibility is to capture all fish sighted; the driver's responsibility is to maneuver the boat so as to provide the netter with the best opportunity to capture stunned fish (the driver can assist in netting stunned fish that appear near the stern or behind the boat). Both require skill and training, but the boat-maneuvering task requires the most experience to gain adequate proficiency and ensure the overall safe operation of sampling. Each sampling zone is fished in a downstream direction by slowly and steadily maneuvering the electrofishing boat along the shoreline and in and around submerged objects by motoring or, if necessary, rowing or pushing. This usually requires frequent turning, backing, shifting (forward, reverse), changing speed, etc. in areas of moderate to extensive cover. It is absolutely essential to sample carefully, particularly at difficult sites where extensive woody debris or moderately fast to swift current require skillful maneuvering. In zones with extensive woody debris and slow current, it is necessary to maneuver the boat in and out of the "pockets" of habitat formed by the debris. Where water depth approaches 1.5 to 2.0 m, it is frequently necessary to "wait" for fish to appear near the surface. In moderately fast or swift current, it is necessary to conduct fast turns and maneuvers in order to put the netter in the best position to capture stunned fish.

Electrofishing efficiency is enhanced if the boat and electric field can be kept moving downstream at a pace slightly faster than the current velocity. Fish are usually oriented into the current and must either swim into the approaching electrical field or turn sideways to escape downstream. This latter movement presents an increased voltage gradient, making the fish more susceptible to the electric current. It is frequently necessary to pass over the fast water sections of these zones at least two times. Also, portions of zones with continuous fast current can be more effectively sampled by "backing" the boat downstream and occasionally pausing to allow the netter to capture stunned fish. The driver may need to assist with netting when large numbers of fish are stunned. Attempting to sample fast water areas in an upstream direction greatly diminishes sampling efficiency. The electrofishing boat is manuvered by pushing on the transom when the water is too shallow to motor or row. A hand-actuated, positive-pressure cutoff switch located on the inside of the transom is used during this procedure in addition to the bow foot-pedal switch.

A boat electrofishing field crew consists of a minimum of three persons (whenever possible): a boat driver, a netter, and a support vehicle driver. The netter and driver are clad in chest waders and rubber gloves. The netter also wears a jacket-type personal flotation device. Sampling sites are selected along the shoreline offering the most diverse macrohabitat features. This is usually along the gradual outside bends of the larger rivers (Gammon, 1973; 1976), but this is not invariable. In free-flowing habitats a portion of each zone should include at least one riffle/run or fast chute habitat if at all possible.

Boat electrofishing zones generally measure 0.5 kilometers (km) in length, although shorter distances may be necessary in given instances. This is in keeping with the recommendations of Gammon (1976) and confirmed by our own experimental test site sampling. Similar to the wading method test site, a 1.50-km long test reach was subdivided into six consecutive 0.25-km increments. Each 0.25-km increment was sampled in a downstream direction using standardized boat electrofishing methods (Ohio EPA, 1989a). Sampling occurred weekly to biweekly during the June 15 to October 15 index sampling period. The results from each 0.25-km increment were recorded separately. The analysis of the results included combining the data in a cumulative manner, starting with the first 0.25-km increment, adding the next, and so on. The cumulative effect of sampling distance on the IBI and number of species was then examined (Figure 2.8). In terms of the IBI, an asymptotic relationship is reached at 0.50 km, a distance beyond which little additional information is gained relative to the increased sampling effort. For species richness, there was a perceptible, but diminishing increase up to a cumulative distance of 1.25 to 1.50 km. Basing sampling distance on a stream width formula has not been formally proposed for boat-mounted methods, but experimental protocols recently used by the U.S. EPA EMAP program in large western rivers include 100 widths that result in distances of 0.5 to 4.0 km being sampled for at least 1 hour (R. Hughes, personal communication).

Boat sites are sampled two or three times during the mid-June to mid-October index sampling period. Sampling passes take place at least three to four weeks apart for a three-pass effort, and six weeks between individual sampling passes for a two-pass effort. The time spent electrofishing, while not the primary arbiter of effort, is important in the sense that a minimum amount of time should be expended to ensure thorough sampling of a site. This is especially critical at boat sites where the risk of under-sampling is higher. Although sampling is according to zone length, the amount of time spent electrofishing each zone is an equally important consideration. Time fished can vary, depending on current velocity, number of fish collected, and the amounts and types of cover within a zone. An analysis of the effect of time was conducted by comparing various parameters (MIwb, numbers, biomass, species richness), which are all sensitive to the level of effort expended, from more than 1600 boat electrofishing samples. Inspection of the results showed that a minimum of 1600 to 2000 s should be spent sampling a 0.5-km boat electrofishing zone (Ohio EPA, 1987b). Based on the authors' experience, this time should increase to more than 2000 to 2500 s in slower flowing zones that have numerous downed trees, logs, and other submerged structures. In some cases, sampling times may exceed 3000 to 3500 s. Moderately fast to swift flowing zones may also require more than 2000 s to sample since the boat must repeatedly be maneuvered back upstream to cover all areas thoroughly. The key is to sample thoroughly in accordance with the previous qualitative guidelines.

2.2.5 Fish Handling and Enumeration Procedures

Captured fish are immediately placed in a live well for processing. The authors use a system where the live well water is continuously exchanged with ambient water to minimize stress and mortality

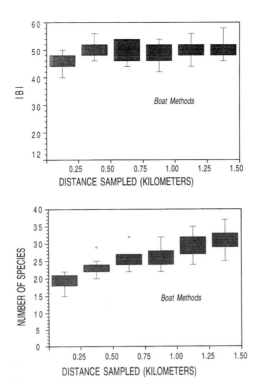

FIGURE 2.8 The effect of the cumulative distance sampled on the Index of Biotic Integrity (IBI; upper) and species richness (lower) for boat methods based on test sampling conducted in Big Darby Creek.

of captured fish. Fish are released immediately after they are identified to species, weighed (except headwater sites), and examined for external anomalies. If necessary, fish are anesthetized to minimize trauma and handling stress. The majority of captured fish are identified to species in the field; however, any uncertainty about the field identification of individual fish requires preservation of voucher specimens for later laboratory identification. Fish are preserved in a borax-buffered 10% formalin solution and labeled by date, river or stream, and river mile. Identification is made to the species level at a minimum and may be done to the subspecific level in certain instances. The collection techniques used are not consistently effective for fish less than 15 to 25 mm in length; thus, inclusion in the sample is not recommended. This follows the recommendations of Angermier and Karr (1986) and Angermier and Schlosser (1988) that fish of this size (young of the year) not be included in IBI calculations as they can unduly bias the resultant bioassessment.

All fish that are weighed, whether done individually, in aggregate, or subsampled (only the fish that are actually weighed), are examined for the presence of gross external anomalies. An external anomaly is defined as the presence of a visible skin, extremity (fin, barbel), or subcutaneous disorder, and is expressed as the weighted percentage of affected fish among all fish weighed. Light and heavy infestations are noted for certain types of anomalies (Sanders et al., Chapter 9). In order to standardize the procedure for counting and identifying anomalies, criteria have been established for their identification and enumeration (Ohio EPA, 1989b; Sanders et al., Chapter 9).

A number of other physical measurements are also taken in the field during fish sampling. Zone distance is measured with a hip chain or a calibrated optical range finder. Temperature and, in some cases, conductivity and dissolved oxygen (D.O.) are recorded. Ohio EPA has recently begun recording secchi disk depth in order to determine how turbidity can affect electrofishing efficiency. A general evaluation of macrohabitat is also made by the fish field crew leader after each location is sampled using the Qualitative Habitat Evaluation Index (QHEI; Rankin, 1989; 1995).

2.2.6 GENERAL CAUTIONS CONCERNING FIELD CONDITIONS

Electrofishing should be conducted only during "normal" water flow and clarity conditions. What constitutes normal can vary considerably from site to site and region to region. Generally, normal water conditions in Ohio occur during well below annual average stream and river flows. Under these conditions, the surface of the water generally will have a placid appearance. Abnormally turbid conditions are to be avoided, as are high water levels and elevated current velocities. All of these conditions can adversely affect sampling efficiency and likely rule out data applicability for bioassessment purposes. Since the ability of the netter to see and capture stunned fish is crucial, sampling should take place only during periods of normal water clarity and flow. Most Ohio surface waters have some background turbidity due to planktonic algae and suspended sediment and very few, if any, are entirely clear. Rainfall and subsequent runoff can cause rapid increases in turbidity due to the increased presence of suspended sediment (clays and silt). In most areas of Ohio, this imparts a light- to medium-thick brown coloration in the water. Floating debris such as twigs, tree limbs, and other trash is usually obvious on the surface during elevated flow events. The depth of visibility under such conditions is seldom more than a few centimeters. Such conditions should be avoided and sampling delayed until the water returns to a "normal" clarity. High flows should also be avoided for obvious safety reasons in addition to a reduction in sampling efficiency. The boat methods are particularly susceptible as it becomes more difficult to maneuver the boat into areas of cover and the fish community is locally displaced by the more severe flood events. It may take several days or even weeks for the community to return to its normal summer–fall distribution patterns. Thus, sampling may need to be delayed by a similar time period if necessary.

2.2.7 DATA MANAGEMENT AND INFORMATION PROCESSING

Once field data is collected, processed, and verified, the next step is to begin the process of analyzing and reducing the data to scientifically and managerially useful information. The principal Ohio EPA data management system for fish, macroinvertebrate, and habitat data is termed Ohio ECOS, which contains data storage, processing, and analysis routines. Once data is tabulated on the field data sheets and verified via chain-of-custody procedures, it is entered directly into the Ohio ECOS databases. Basic information recorded includes the field staff, water body name, date, and time. Site location is indicated by river mile (distance upstream in miles from the mouth) and latitude/longitude as determined from USGS 7.5 minute topographic maps. A basin-river code system is used to electronically identify individual streams, rivers, and lakes. Sampling information includes method or gear type and other information relevant to the use of each. Ohio ECOS provides data summaries and reports for a variety of community measures and individual species analyses.

2.2.8 COST CONSIDERATIONS

Concern has historically been expressed about the practical utility of biological field data and the resources needed to implement such programs (Loftis et al., 1983; U.S. EPA, 1985). Chemical/physical sampling and bioassays have historically been advocated, partly because they are viewed as cheaper and more cost-effective than biological field evaluations (U.S. EPA, 1985). The authors' extensive experience with using a standardized and systematic application of fish and macroinvertebrate monitoring techniques integrated with the traditional chemical/physical and bioassay assessment techniques allows for a detailed comparison of the costs involved with each component, as described elsewhere (Yoder and Rankin, 1995a). Out of nearly 100 WYE (work year equivalents) devoted to surface water monitoring and laboratory activities within the Division of Surface Water in 1987 and 1988, 19.34 WYE or just over 19% of the total was devoted to ambient biological monitoring. When considered on the basis of all Ohio EPA water programs, this percentage is approximately 6%.

An analysis of resources expended during 1987 and 1988 revealed the following about the Ohio EPA's effort to collect fish assemblage data on a statewide basis in Ohio: (1) 8.44 WYE (work year equivalents) were used to collect 1277 samples at 617 sampling sites; (2) an average of 0.014 WYE or 29.1 hr/site were expended to plan, collect, analyze, and interpret data, and produce reports at an average cost of $740 per site and $340 per sample; (3) approximately 1 to 3 hours were required to sample each site with a two- or three-person field crew visiting three to six sites per day working 10 to 14 hours per day (a field crew consists of one full-time biologist and one or two interns); and (4) the post-field season laboratory effort ranged from 2 to 4 weeks for each crew leader. While the dollar amounts are dated, the other statistics remain valid today. These statistics also included the start-up costs for equipment and supplies, which were amortized over 5 or 10 years depending on the item. Start-up equipment and supplies, for many states, could total between $200,000 and $500,000, depending on the number of field crews.

2.3 BIOLOGICAL CRITERIA

A common tendency in water quality assessment has been to make biological measurements fit the perceptions and use of chemical criteria and other surrogate indicators, rather than the reverse. This is clearly illogical because the structure and function of the aquatic community is the embodiment of the temporal and spatial chemical, physical, and biological dynamics of the aquatic environment (see Figure 2.2). Perhaps the past inability of biologists to agree on a set of empirical measurements of biological integrity, or at least a common framework, has resulted in this situation (Karr et al., 1986). A principal reliance by many states and the U.S. EPA on chemical and other surrogate indicators leads to an incomplete assessment and understanding of the status of the nation's waters and the causes and sources associated with impairment of designated aquatic life uses (Ohio EPA, 1995; Yoder and Rankin, 1998). Statistics showing approximately 60 to 70% of the nation's rivers and streams fully attaining designated uses (U.S. EPA, 1996b) are now increasingly regarded as being overly optimistic. One solution to these deficiencies is to base state assessments on regionally referenced numerical biological criteria that can quantitatively indicate the degree to which designated uses are or are not being achieved (Davis and Simon, 1995; Yoder and Rankin, 1995a,b). Traditionally used surrogate indicators such as chemical criteria and bioassay tests will always play an important role in surface water resource management. Their value and accuracy, however, would be greatly enhanced when used in combination with biological criteria.

2.3.1 THE ROLE OF FISH ASSEMBLAGE DATA IN BIOLOGICAL CRITERIA

The lack of adequate and reliable decision criteria for biological data has historically limited the usefulness of biological monitoring and assessment. We developed an initial set of decision criteria for fish and macroinvertebrate assemblages in 1980 that consisted of narrative quality ratings based in part on numerical biological index "guidelines." These were intended to more directly reflect the ecological goals espoused by designated aquatic life uses codified in the Ohio water quality standards (WQS). The early numeric indices were composed of contemporary measures such as taxa richness, the Shannon diversity index, and the Index of Well-Being (Gammon, 1976). Attainable expectations for a set of narrative community attributes were based on Ohio EPA's experience with sampling approximately 150 to 200 sites statewide. This approach was used from 1978 through 1987 and was applied uniformly across the state. No effort was made to account for background variability by using landscape partitioning frameworks such as ecoregions since that technology did not yet exist.

The narrative classification system consisted of assigning narrative quality ratings such as exceptional (consistent with the Exceptional Warmwater Habitat use), good (Warmwater Habitat use), fair, poor, and very poor, the first two meeting the goals of the Clean Water Act and the latter three failing to attain. The purpose of this narrative classification system was essentially twofold: (1) to provide an objective, systematic basis for assigning aquatic life uses to surface waters; and (2) to provide an objective, standardized approach for determining the magnitude and severity of aquatic life impairments for assessment purposes. Considerable judgment was used in applying these early narrative biological criteria on a site-specific basis. The aggregate impact of these assessments played a major role in setting and evaluating WQS use designations, designing water quality management plans, and developing advanced treatment justifications for municipal sewage treatment plants. These criteria also provided a basis for designating stream and river segments as attaining, partially attaining, or not attaining designated aquatic life uses in the 1982, 1984, and 1986 Ohio EPA 305[b] Reports.

2.3.2 REGIONALLY REFERENCED NUMERICAL BIOLOGICAL CRITERIA

In 1986, a major effort was undertaken to develop regionally referenced and calibrated biological criteria using an array of regional reference sites. The Index of Well-Being was retained in a modified form (Ohio EPA, 1987b) and the Index of Biotic Integrity (IBI; Karr, 1981; Karr et al., 1986) was added. For macroinvertebrates, the Invertebrate Community Index (ICI; Ohio EPA, 1987b; DeShon, 1995) replaced the previously used narrative evaluations for macroinvertebrates. This provides for a more rigorous, ecologically oriented approach to assessing aquatic community health and well-being. The process of deriving the numerical biological criteria is described more extensively by the Ohio EPA (1987a,b; 1989a) and Yoder and Rankin (1995a).

2.3.3 NEW GENERATION BIOLOGICAL COMMUNITY EVALUATION MECHANISMS

New methods and approaches to assessing aquatic community data have been developed in the past 10 to 15 years which have provided a significant advancement in being able to utilize biological community information for aquatic resource characterizations and as arbiters of environmental goal attainment/nonattainment. These include such innovations as the Index of Biotic Integrity (IBI), as originally developed by Karr (1981), and as subsequently modified by many others (Leonard and Orth, 1986; Steedman, 1988; Ohio EPA, 1987b; Miller et al., 1988; Lyons, 1992a; Lyons et al., 1996), the Index of Well-Being (Iwb) developed by Gammon (1976), the Invertebrate Community Index (ICI; Ohio EPA, 1987b; DeShon, 1995), the U.S. EPA Rapid Bioassessment Protocols for fish and macroinvertebrate assemblages (Plafkin et al., 1989), and, more recently, the benthic IBI (BIBI; Kerans and Karr, 1993). These represent what are termed here as new generation community evaluation mechanisms since they represent recent and substantial progress in the seemingly continuous effort to develop improved community indices. While the pursuit of a "better" index has been criticized and even trivialized by some, it is nevertheless an important process in the effort to make biological data a more integral part of the overall environmental and natural resource management process by making it more understandable and usable.

Some have criticized multimetric indices as representing a loss of rich information in the reduction of the data to a single index value (Suter, 1993). However, this presumes that the supporting data is never viewed or examined beyond the calculation of the index itself. The need to examine the basic data is implicit throughout the biological criteria process; thus, such criticisms are without foundation. The need for numerical indices is clearly evident throughout the environmental management process, provided that they emanate from a sound theoretical basis and robust information base. While the need for interpretation by qualified biologists will always be necessary, it is not realistic to expect that judgment alone will be accepted as a substitute for a more empirical process. Considerable biological judgement can be incorporated into biological criteria frameworks.

Some of the early attempts to satisfy the demand for numerical assessment tools produced indices of limited or questionable theoretical rigor. However, the theoretical underpinnings behind the IBI-type indices are much more robust than previous single-dimension indices and community indicators as long as they are used within a framework where the regional faunal considerations, reference condition, and background variability have been adequately accounted for. Steedman (1988) described the IBI as being based on simple, definable ecological relationships that are quantitative as an ordinal, if not linear measure, and that respond in an intuitively correct manner to known environmental gradients. Further, when incorporated with mapping, monitoring, and modeling information, it should be invaluable in determining management and restoration requirements for warmwater streams. As an aggregation of community information, the IBI (and facsimiles thereof) provide a way to organize complex data and reduce it to a scale that is interpretable against communities of a known condition. This is accomplished by an improved stratification and organization of complex ecological information into understandable components or metrics. Simply stated, multimetric indices like the IBI can satisfy the demand for a straightforward numerical evaluation that expresses a relative value of aquatic community health and well-being and allows program managers (who are frequently non-biologists) to, in effect, "visualize" biological integrity. These indices also provide a means to establish numerical biological criteria.

2.3.4 THE INDEX OF BIOTIC INTEGRITY (IBI) MODIFIED BY THE OHIO EPA

Biological criteria are definable end-points that portray the structural and functional integrity of aquatic communities and, by extension, designated aquatic life uses. The Index of Biotic Integrity is used by the Ohio EPA in a modified form and provides an example of using biological judgment in the tailoring of the IBI to local/regional conditions. Several examples of this exist across the U.S., many of which were summarized by Miller et al. (1988). In this process, the Ohio EPA (1987b) derived three different modifications of the original IBI (Table 2.2) that resulted from an extensive process of testing approximately 60 candidate metrics. A wide variety of stream and river sizes occur throughout Ohio that not only contain different fish assemblages, but require the use of different sampling methods. Thus, it was necessary to modify the original IBI for application to these different stream and river sizes and include the selectivity of different sampling gear. The modifications were made in keeping with the guidance provided by Karr et al. (1986).

The modified metrics were selected based on a combination faunal association characteristics (e.g., headwaters vs. large stream assemblages) and sampling gear selectivity considerations (e.g., wading sites vs. boat sites). In each of these modifications, the basic ecological structure and content of the original IBI was preserved. The result was three different modified IBIs (Table 2.2): (1) a headwaters IBI for application to headwater streams (defined as stream locations with a drainage area less than 20 sq. mi.); (2) a wading site IBI for application to stream locations greater than 20 sq. mi. sampled with wading methods; and (3) a boat site IBI for all locations sampled with boat methods. The IBI metrics used to evaluate wading sites most closely approximate those originally proposed by Karr (1981) and refined by Fausch et al. (1984). Wading sites include streams with drainage areas that are generally less than 300 sq. mi. (range 21 to 475 sq. mi.; range of means within the five ecoregions is 44 to 128 sq. mi.). More substantial modifications were made for the metrics used in the boat and headwater sites modified IBIs. Boat sites generally exceed 200 to 300 sq. mi. in drainage area (range 117 to 6479 sq. mi.; range of means for the ecoregions is 225 to 2190 sq. mi.). The modified metrics were selected based on the recognition of the different sampling efficiency and selectivity of the boat electrofishing methods and the different character of larger stream and river fish faunas. Although headwater sites are sampled with wading methods (Ohio EPA, 1989b), these streams have an inherently different faunal composition resulting from the influence of smaller channel and substrate size, temporal flows, and presence or absence of water. The headwater IBI metrics were selected to reflect the character of headwaters fish communities

TABLE 2.2
Modification of Index of Biotic Integrity (IBI) Metrics Used by the Ohio EPA to Evaluate Wading Sites, Headwaters Sites, and Boat Sites

IBI Metric	Headwaters Sites[1,2]	Wading Sites[2]	Boat Sites[3]
1. Number of native species[4]	X	X	X
2. Number of darter species		X	
Number of darter and sculpin species	X		
% Round-bodied suckers[5]			X
3. Number of sunfish species[6]		X	X
Number of headwater species[7]	X		
4. Number of sucker species		X	X
Number of minnow species	X		
5. Number of intolerant species		X	X
Number of sensitive species[8]	X		
6. % Green sunfish			
% Tolerant species	X	X	X
7. % Omnivores	X	X	X
8. % Insectivorous cyprinids			
% Insectivores	X	X	X
9. % Top carnivores		X	X
% Pioneering species[9]	X		
10. Number of individuals			
Number of individuals (less tolerants)[10]	X	X	X
11. % Hybrids			
% Simple lithophils		X	X
Number of simple lithophilic species	X		
12. % Diseased individuals			
% DELT anomalies[11]	X	X	X

[1] Applies to sites with drainage areas <20 sq. mi.
[2] Sampled with wading electrofishing methods.
[3] Sampled with boat electrofishing methods.
[4] Excludes all exotic and introduced species.
[5] Includes all species of the genera *Moxostoma, Hypentelium, Minytrema, Erimyzon, Cycleptus,* and *Ericymba,* and excludes *Catostomus commersoni.*
[6] Includes only *Lepomis* species.
[7] Species designated as permanent residents of headwaters streams.
[8] Includes species designated as intolerant and moderately intolerant (Ohio EPA, 1987b).
[9] Species designated as frequent and predominant inhabitants of temporal habitats in headwaters streams.
[10] Excludes all species designated as tolerant, hybrids, and non-native species.
[11] Includes individuals with deformities, eroded fins or barbels, lesions, and tumors.

in relation to these critical factors. The statistical properties of the Ohio IBIs have been previously examined by Yoder and Rankin (1990) and Fore et al. (1993). Both studies concluded that the Ohio-modified IBIs exhibit acceptably low replicate variability and have the power to discriminate discrete classes of environmental quality (i.e., exceptional, good, fair, etc.).

While the IBI has worked well in Ohio and many parts of the U.S. (Miller et al., 1988), Canada (Steedman, 1988), and Europe (Oberdorff and Hughes, 1992), there are situations in which problems have been encountered. Applications to semi-arid western U.S. drainages (Bramblett and Fausch, 1991) and the coldwater streams of northern states (Lyons, 1992a) have revealed problems with some of the fundamental assumptions of the original IBI. Bramblett and Fausch (1991) encountered difficulties caused by an inherent lack of intolerant species in the highly variable and harsh hydrological conditions of the Arkansas River drainage in Colorado. Lyons (1992b) found that the degradation of coldwater streams in Wisconsin resulted in an increased diversity of fish species, which is a positive occurrence under the basic presumptions of the original IBI. Later work by Lyons et al. (1996) resulted in a modified IBI that was constructed to react in a manner consistent with the inherent ecological characteristics of the coldwater stream types being assessed. This points up the need to consider the inherent character of the regional fauna and the reaction to human-induced environmental changes in developing IBIs for application to the various regions of the U.S.

2.3.5 MODIFIED INDEX OF WELL-BEING (MIwb)

Gammon (1976) originally developed the Index of Well-Being (Iwb) as a multiparameter evaluation of large river fish communities. The Iwb is based on four measures of community diversity, abundance, and biomass, and is an attempt to produce an integrated evaluation of these important and basic fish community attributes. The individual performance of numbers, biomass, and the Shannon index as consistent indicators of the quality of fish communities has historically been disappointing. However, when combined in the Iwb, these individual community attributes work in a more complementary and intuitively predictable manner. For example, an increase in total numbers and/or biomass caused by one or two predominant species is usually offset by a corresponding decline in the Shannon index. In addition, the $\log_e$ transformation of the numbers and biomass components acts to reduce much of the variability inherent to these parameters alone. Gammon (1976) found the variability of each of the four Iwb components as measured by a coefficient of variation to range from 20 to 50%, yet the composite variability reflected by the Iwb was only 7%.

High numbers and/or biomass are commonly, and at times inaccurately, perceived as a positive attribute of a fish assemblage. High numbers and biomass result in a high Iwb score, provided a relative "evenness" is maintained between the abundance of the common species. However, this is not invariable, especially with environmental perturbations that tend to restructure fish communities without corresponding decreases in diversity (e.g., nutrient enrichment, habitat modification). The authors have observed fish communities in habitat-modified streams to exhibit very high numbers, biomass, and moderate species richness. Such communities are usually dominated by tolerant species. Species intolerant of such disturbances either decline in abundance or are eliminated altogether. A net increase in the relative abundance of tolerant species with only modest declines in species richness can yield higher Iwb values. The increased abundance of tolerant species is not always sufficiently offset by decreases in the Shannon indices because species richness is not proportionately influenced. The overall result is an Iwb evaluation that is not reflective of the true response of the fish assemblage to these types of degradation. In fact, Iwb values at some disturbed sites equaled or exceeded those measured at reference or least-impacted sites.

Several modifications of the Iwb were attempted to correct the problem of higher scores at degraded sites. This included the complete elimination of predominant species from the index calculation, selective elimination of species based on their predominance, and a different weighting of the numbers component of the Iwb (Ohio EPA, 1987b). None of these modifications functioned in a consistent manner. The fundamental problem is that the predominance and higher abundance of species that are tolerant to environmental degradation are not sufficiently reflected in the aggregate index. Tolerant species are the last to disappear under the influence of increased environmental degradation, and they may respond favorably (i.e., increase in abundance) to changes in the physical or chemical quality of the environment. Thus, the elimination of the highly tolerant species from the numbers and biomass components of the Iwb was attempted. The Ohio EPA has designated all fish species known to occur in Ohio as highly tolerant, moderately tolerant, intermediate, moderately intolerant, or highly intolerant (Ohio EPA, 1987b) for the purposes of the IBI. This was accomplished by examining a large, statewide database that included nearly 2000 sites and a wide range of environmental quality. While some past attempts to designate species tolerance relied mostly on the existing technical literature and regional fish reference texts, the Ohio EPA method is based on direct observations of species responses in the field. This requires a comprehensive database and should be supplemented by information from the technical literature when necessary.

The modified Iwb retains the same computational formula as the original Iwb developed by Gammon (1976). The major difference is that any of the species designated as highly tolerant, exotic, and hybrid are eliminated from the numbers and biomass components of the Iwb. However, tolerant and exotic species are included in the Shannon index calculations. This modification eliminates the "undesired" effect caused by the high abundance of tolerant species, but retains their "desired" influence in the Shannon indices. To illustrate the effect of this modification, several comparisons were made between key fish community attributes, the modified Iwb, and the conventional Iwb (Ohio EPA, 1987b). In addition, results from different streams and rivers subjected to different types and varying levels of environmental degradation (both chemical and physical) demonstrated the influence that this modification had on an evaluation of fish community health and well-being. These analyses showed that the MIwb can be lower than the original Iwb by as much as one third in highly degraded areas, but by as little as 0.1% in high-quality rivers and streams (Ohio EPA, 1987b). The end result of this modification is an index that is more sensitive to all types of degradation, particularly nutrient enrichment and habitat impacts, which frequently result in an increased abundance and biomass of tolerant species, but which retains the ability to accurately characterize high-quality communities. The MIwb is not applied to headwater sites because our analyses showed a very strong effect of drainage area on biomass, which made its application impractical in these situations (Ohio EPA, 1987b). The computational formula is as follows:

$$\text{Modified Index of Well-Being (MIwb)} = 0.5 \ln N + 0.5 \ln B + \overline{H} \text{ (no.)} + \overline{H} \text{ (wt.)}$$

where

N = CPUE* relative numbers minus species designated highly tolerant (Ohio EPA, 1987b)
B = CPUE relative biomass minus species designated highly tolerant (Ohio EPA, 1987b)
$\overline{H}$ (no.) = Shannon diversity index based on numbers (version that uses $\log_e$)
$\overline{H}$ (wt.) = Shannon diversity index based on numbers (version that uses $\log_e$)
*(CPUE = catch per unit of effort which is per 0.3 km for wading methods and per 1.0 km for boat methods)

2.3.6 IBI/MIwb Relationship

The MIwb is used as a complementary evaluation to the IBI by the Ohio EPA. It is a reflection of overall community production in terms of fish numbers and biomass and usually precedes the IBI in indicating the first stages of fish community recovery from a perturbation. It is especially responsive to severe impacts that result in reduced numbers, biomass, and species richness. The MIwb does not incorporate the more sensitive aspects of the community such as the functional guilds of the IBI. Because the IBI incorporates aspects of the community such as the functional and reproductive guilds, it tends to respond more slowly than the MIwb in terms of exhibiting recovery. Because both indices have different strengths and weaknesses, the authors recommend that both be used, particularly in larger streams (>20 sq. mi.) and rivers. Using the MIwb requires the collection of biomass data (which is not required by current versions of the IBI), but such information is crucial to an accurate assessment of fish community performance.

2.4 USING FISH ASSEMBLAGE INFORMATION IN BIOASSESSMENTS

The primary purpose for using fish assemblage information is to provide a more direct assessment of the status of surface water resources than that offered by surrogate indicators alone. Using this approach focuses more directly on the biological integrity and "fishable" goals of the Clean Water Act. In the water indicators hierarchy (see Figure 2.1), fish assemblage information functions as a response indicator (level 6). This is based on the assertion that the condition of the fish assemblage — as expressed by relative abundance data and further amplified by tools such as the IBI and MIwb — is the end result of environmental processes, including both natural and human-induced changes. Several case study examples of using fish assemblage information in this manner follows.

2.4.1 EVALUATING THE EFFECTIVENESS OF WATER QUALITY MANAGEMENT PROGRAMS

An important use of fish assemblage information in an environmental indicators framework is not only to portray the condition of a water body, but to serve as a composite measure of the effectiveness of the overall water quality management program. As such, fish assemblage information functions within the role of a response indicator (level 6 in Figure 2.1). Using indices such as the IBI and MIwb, a longitudinal portrayal of the sampling results is a commonly used method to visually describe and interpret the magnitude and severity of departures from numerical biological criteria. This is accomplished by plotting the biological index results by river mile for a given study area. Major sources of environmental impact and the applicable numerical biological criteria are indicated on each graph. The results of the fish community assessment in two different Ohio watersheds are used here as an example (Figures 2.9–2.12). This type of analysis is used to demonstrate changes through both space (upstream to downstream) and time (differences between years). Unlike chemical parameters, biological indices integrate chemical, biological, and physical stresses and exposures, and portray the results of water quality management in both aggregate and direct terms. Frequency and duration considerations, which confound most chemical/physical assessments, are integrated by the resident aquatic life in the receiving water body.

2.4.2 SCIOTO RIVER CASE EXAMPLE

The middle segment of the Scioto River, located in central Ohio, is impacted by the city of Columbus, particularly by the two major municipal wastewater treatment plants that can discharge a combined total of nearly 200 million gallons of treated wastewater per day (MGD). During the summer–fall period, this can compose nearly 90% of the total flow of the Scioto River downstream from Columbus. As such, the Scioto River is a wastewater effluent-dominated river during low flow periods. Like most central Ohio watersheds, land use in the upper basin is dominated by row crop agriculture and by urban land use within the major metropolitan areas. Biological sampling has been conducted by the Ohio EPA in a nearly 50-mile reach of the mainstem since 1979. Fish sampling has been conducted each year with the exception of the years 1982, 1983, and 1984, and has preceded and followed major efforts to reduce the pollution effects of the Columbus WWTPs. As shown in Figure 2.9, the fish sampling results revealed a highly degraded fish community within and downstream from Columbus and in proximity to the two WWTPs in 1979. Departures from the biological criteria for the IBI and MIwb indicated poor and very poor quality in lengthy sections of the river. Incomplete recovery was evident in the lower reaches of the study area and was typical of that observed in many similarly impacted rivers of that time period. The next sampling year shown, 1991, revealed extensive improvements in the performance of both the IBI and MIwb, with most sites meeting the applicable biological criteria in the 50-mile study reach. Further incremental improvements were observed in 1994.

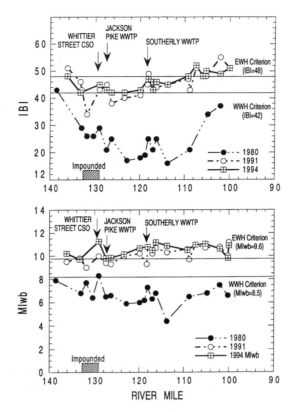

FIGURE 2.9 Results of a fish community assessment conducted in a 50-mile segment of the Scioto River showing IBI (upper) and MIwb (lower) values during 1980, 1991, and 1994 before and after the implementation of water quality-based effluent restrictions at the Columbus, Ohio, Jackson Pike and Southerly wastewater treatment plants (WWTP).

The 1991 sampling followed by three years the construction and operation of advanced wastewater treatment at both Columbus WWTPs, which was designed to meet chemical water quality criteria, particularly for D.O., ammonia-N, and common heavy metals. The result of these improvements resulted in dramatic reductions in the discharge of oxygen-demanding wastes, suspended solids, and ammonia-N by both WWTPs. Figure 2.10 uses the indicators hierarchy to show how administrative actions taken by the Ohio EPA and the U.S. EPA (Level 1: permit issuance, construction grants funding), and the response by the city of Columbus (Level 2: WWTP construction and operation) were followed by reduced stress (Level 3: reduced ammonia-N pollutant loadings), reduced exposure (Level 4: lower ambient ammonia-N concentrations), all of which was followed by the positive response of the fish assemblage (Level 6: IBI, MIwb). In this case, it was the composite condition of the fish community as expressed by the IBI and MIwb that confirmed the effectiveness of the NPDES permitting process, the awarding of grants, the implementation of wastewater treatment improvements, and the relevance of the applicable chemical water quality criteria through the water quality-based permitting process.

2.4.3 Ottawa River Case Example

The Ottawa River, located in northwestern Ohio, is impacted by the city of Lima WWTP, which can discharge a total of nearly 15 MGD of treated wastewater. During the summer–fall period, this can make up nearly 90% of the total flow of the Ottawa River downstream from Lima. Like the Scioto River, the Ottawa River is effluent dominated during low flow periods. There are many other cultural

Using Fish Assemblages in a State Biological Assessment and Criteria Program

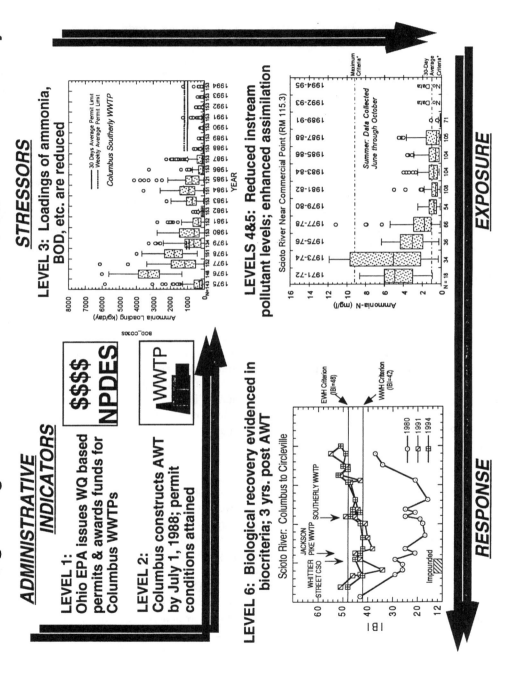

FIGURE 2.10 The progression of indicator results and linkages in the central Scioto River study area showing the effect of permitting, funding, wastewater treatment plant (WWTP) construction, and loading reductions on ambient water quality and fish community condition as portrayed by the IBI.

similarities with the Scioto River case example, such as the sewer system containing combined sewer overflows, a dependency on the Ottawa River as the municipal water supply, and land use in the upper basin being dominated by row crop agriculture. Biological sampling has been conducted by the Ohio EPA in a nearly 50-mile reach of the mainstem since 1977. Fish sampling has been conducted roughly every other year since 1985, and has preceded and followed major efforts to reduce the pollution effects of the Lima WWTP. As shown in Figure 2.11, the fish sampling results revealed a highly degraded fish community within and downstream from Lima and in close proximity to the Lima WWTP in each and every sampling year since 1985. Departures from the biological criteria for the IBI indicated poor and very poor quality in at least a 10-mile long segment immediately downstream from the WWTP. The extremely high incidence of external anomalies (%DELT) on fish is further evidence of the severe biological degradation and, coupled with the very low IBI scores, indicates a response to toxic conditons (Yoder and Rankin, 1995b). No significant recovery was evident in the lower reaches of the study area through 1996 compared to previous years' results (Figure 2.11).

The extensive improvements in the performance of the IBI and MIwb observed in the Scioto River (Figures 2.9 and 2.10) have yet to occur in the Ottawa River. This exists despite the construction and operation of advanced wastewater treatment facilities that were designed to meet chemical water quality criteria, particularly for D.O., ammonia-N, and common heavy metals. These improvements resulted in reduced loadings of oxygen-demanding wastes, suspended solids, and ammonia-N in the Lima WWTP effluent. Figure 2.12 follows the indicators hierarchy in showing how actions taken by the Ohio EPA and the U.S. EPA (Level 1: permit issuance, construction grants funding), and the response by the city of Lima (Level 2: WWTP construction and operation), were followed by reduced stress (Level 3: pollutant loadings) and reduced exposure (Level 4: lower ambient pollutant concentrations), none of which has been corroborated by a positive response of the fish assemblage (Level 6: IBI, MIwb). In this case, the composite condition of the fish community as expressed by the IBI and MIwb failed to confirm the effectiveness of the NPDES permitting process, wastewater treatment, and water quality-based permitting. The reason is that sources other than the Lima WWTP are exerting greater influences on the fish assemblage.

Two major industrial sources — an oil refinery and a chemical manufacturing facility — continue to exert limiting influences on the fish community. Extremely elevated rates of anomalies on fish, coupled with very low IBI and MIwb scores, are a response signature to toxic conditions (Yoder and Rankin, 1995b). Other chemical/physical indicator information not included in Figure 2.12 includes extremely contaminated bottom sediments (heavy metals, hydrocarbons). The oil refinery has been a major source of water quality problems since the turn of the century and many practices at the facility predated the environmental requirements of the 1970s and 1980s. The Ottawa River was literally fishless during large parts of the 20th century. Thus, the residual influence of this facility is still being felt in the Ottawa River even though some fish species now reside in the mainstem. However, compared with the improvements we have observed throughout Ohio during the past 10 years (Ohio EPA, 1997), the Ottawa River has shown very little improvement in response to conventional regulatory and water quality management practices.

The problems inherent to rivers like the Ottawa are the legacy of past practices and the relative capabilities of the conventional regulatory process (i.e., NPDES permitting) to adequately capture all relevant stressors. In the Ottawa River case example, the fish assemblage was uniquely able to demonstrate the magnitude and extent of the impairment, which was not nearly as implicit in commonly used administrative and exposure indicators. State water quality management programs urgently need the type of information provided by biological monitoring to determine when management success has truly been achieved.

2.4.4 Habitat Assessments

Using biological indicators requires the inclusion of habitat quality as a major consideration in addition to chemical/physical water quality. This is especially true of using fish assemblages, the quality of which are strongly influenced by habitat quality. Rankin (1989; 1995) developed the

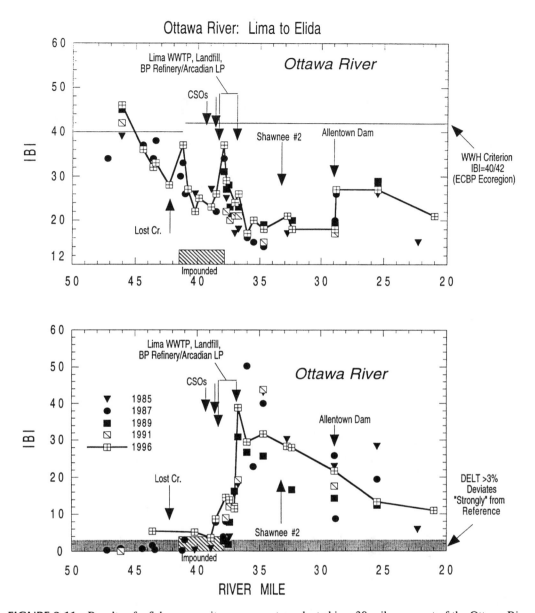

FIGURE 2.11 Results of a fish community assessment conducted in a 30-mile segment of the Ottawa River showing IBI values (upper) and the percentage of fish exhibiting anomalies (%DELT; lower) during 5 years between 1985 and 1996 before and after the implementation of water quality-based effluent restrictions at the Lima, Ohio, wastewater treatment plant (WWTP).

Qualitative Habitat Evaluation Index (QHEI) based on demonstrated relationships with fish assemblage information — specifically, the IBI and the metric components. The IBI is positively correlated with habitat quality as measured by the QHEI, but understanding how this relationship can be used in a management process requires an understanding of the individual component relationships.

Rankin (1989) determined which attributes of the QHEI were both positively and negatively correlated with the IBI. Figure 2.13 shows the relationship between the accumulation of modified habitat attributes (i.e., those attributes that increase with habitat degradation) and the IBI. As the ratio of modified habitat attributes present at a site increase in relation to warmwater attributes (i.e.,

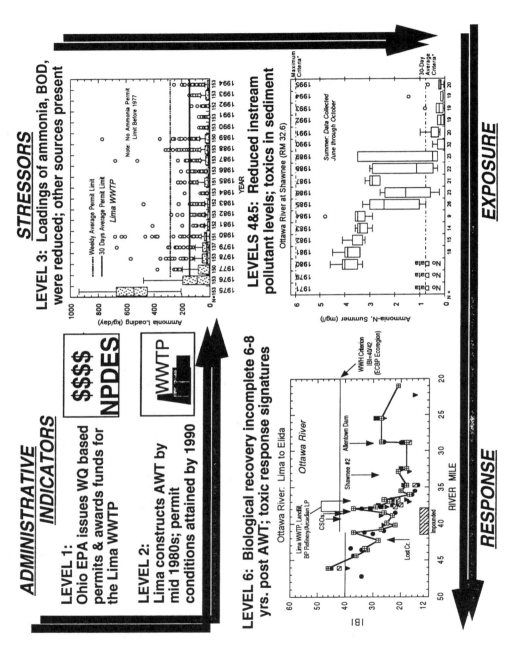

FIGURE 2.12 The progression of indicator results and linkages in the Ottawa River study area showing the effect of permitting, funding, wastewater treatment plant (WWTP) construction, and loading reductions on ambient water quality and fish community condition as portrayed by the IBI.

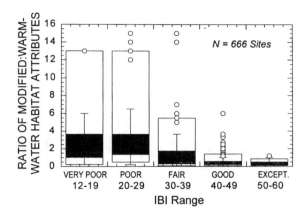

FIGURE 2.13 The ratio of modified habitat to warmwater habitat attributes derived from the Qualitative Habitat Evaluation Index (QHEI) by narrative quality ratings based on IBI scores from 666 Ohio stream sampling locations.

those attributes that increase with improved habitat), the quality of the corresponding fish assemblage decreases. Thus, these types of relationships can be used to predict the potential biological ramifications of proposed projects that may impact habitat before the damage takes place. Ohio EPA has successfully used these relationships to grant or deny the 401 water quality certification of the U.S. Army Corps of Engineers section 404 dredge and fill permits and to review other projects where habitat disturbance is at issue.

2.4.5 STATEWIDE ASSESSMENTS

An important part of state program responsibilities is to periodically report to the U.S. EPA and the Congress on the overall condition of water resources. This is accomplished through reports required by section 305[b] of the Clean Water Act. If the state monitoring program is sufficiently comprehensive in terms of spatial coverage and environmental indicators, then producing a 305[b] report is merely a matter of aggregating information like that presented in the preceding case examples. In Ohio, this is made possible because systematic monitoring of rivers and streams on a statewide basis for nearly 20 years has taken place. The result is an ongoing evaluation of the effectiveness of the overall water quality management process, which in Ohio includes information collected before and after the implementation of major upgrades to municipal WWTPs and the imposition of water quality-based limitations for industries. The results are reported as the aggregate number of river and stream miles that attain aquatic life-based criteria, principally compliance with the biological criteria (Ohio EPA, 1997). Figure 2.14 shows the proportion of assessed river and stream miles that exhibited full attainment of the biological criteria on a biennial basis between 1988 and 1996 and that projected thereafter through the year 2002. The most recent statewide assessment (Ohio EPA, 1997) shows that approximately 50% of the assessed stream and river miles now fully meet designated aquatic life uses. In contrast, if one relied primarily on chemical water quality monitoring alone, the corresponding statistic would be approximately 78% (Ed Rankin, personal communication), which would constitute a serious underestimate of the condition of Ohio's water resources.

National statistics compiled by the U.S. EPA (U.S. EPA, 1996b) show that approximately 70% of rivers and streams meet state-established designated uses. However, this is based on a mix of different assessment approaches, some of which consist of comparatively much less robust indicator frameworks. Efforts to base national estimates on biological information, while handicapped by a lack of information for all states, indicate results more in line with Ohio's experiences (U.S. EPA, 1996a). Thus, the inclusion of biological indicators, which includes fish assemblages in many states, seems essential if we are to more fully protect and restore degraded water resources. This notion

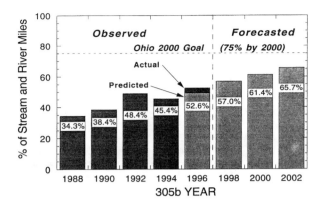

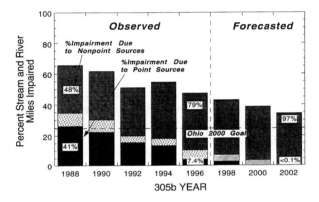

FIGURE 2.14 The aggregate percentage of assessed rivers and stream miles that fully attained aquatic life use criteria (upper) and the proportion that failed to attain aquatic life use criteria (lower) by 305[b] reporting year through 1996 and that projected through reporting year 2000.

is consistent with national compendia that have previously highlighted the significant influence of nonpoint sources and habitat degradation (Judy et al., 1984; Karr, 1991; Benke, 1992). Without a more consistent biological indicators approach, we will continue to overestimate the quality of surface waters and potentially fail to properly address critical and emerging stressors.

The Ohio EPA 305(b) report also supports tracking the Ohio 2001 goals, which include the objective of fully restoring 75% of river and stream miles by the year 2000 (Ohio EPA, 1997). The most recent forecast analysis indicates that at the current rate of improvement, approximately 65% will be fully restored. Furthermore, the lower tier of Figure 2.14 shows that the impairments associated with point sources are being eliminated at a faster rate than those associated with nonpoint sources. The implications of this to the state's water quality management process is that more attention needs to be devoted to nonpoint sources, which includes habitat degradation, if the Ohio 2001 goals are to be met. In this case, fish assemblage information is a key part of the biological criteria process on a statewide basis because it is especially sensitive to the combined effects of habitat degradation and non-point source pollutants such as nutrients. This information has substantially influenced strategic planning, not only within the Ohio EPA, but also within other state and federal agencies involved with non-point sources.

2.4.6 LONG-TERM GOAL ASSESSMENT

One advantage of using numeric biological criteria based on multimetric indices such as the IBI is the ability to quantitatively portray the extent to which a sampling site, reach, or individual river or stream exhibits biological integrity as envisioned by the CWA. The Ohio biological criteria are based on contemporary reference conditions that were revealed by our sampling of reference sites during the 1980s (Yoder and Rankin, 1995a). This provides the basis for establishing short-term objectives based on the goals set forth by the designated aquatic life uses, which are described in the Ohio WQS. Depending on the extent and degree of background landscape disturbance, this contemporary condition may more closely approximate the biological integrity ideal in less disturbed areas than in extensively disturbed regions. In Ohio, landscape disturbance has proceeded unabated for more than two centuries and is most extensive in the Huron/Erie Lake Plain (HELP) ecoregion. Thus, the biocriterion for the IBI that meets today's baseline reference condition (32) in the HELP ecoregion is only 53% of the maximum possible (60). In the much less disturbed Western Allegheny Plateau (WAP) ecoregion, the same threshold (44) is 73% of the maximum. There seems to be an emerging consensus that truly reaching the biological integrity ideal would mean scoring at least 95% (IBI = 57), a level at which less than 5% of river and stream miles in Ohio currently perform. The condition of 99 rivers and streams, as revealed by IBI scores from the 1980s and early 1990s, is illustrated in Figure 2.15. The rank order of streams and rivers is according to the median IBI score of all sites sampled in each. This illustrates a way to comparatively rank the quality of fish assemblages against contemporary reference conditions and the biological integrity ideal simultaneously. As can be seen, this latter ideal is approached by only three or four streams and rivers in Ohio.

Patterns as to why certain streams and rivers harbor exceptional, good, and various degrees (fair, poor, very poor) of degraded fish communities are also apparent. For example, the streams and rivers that rank in the lowest 25% are frequently impacted by severe toxic problems in heavily urbanized and industrial areas. Those ranking in the middle exhibit problems ranging from a combination of agricultural non-point source runoff and habitat modifications to municipal WWTP impacts. Those ranking in the top 10% include most of the exceptional resources, harbor the best populations of rare, threatened, endangered, and declining fish species, and offer the best recreational opportunities. Part of the U.S. EPA environmental indicators process is to report separately on designated aquatic life use attainment (i.e., based on contemporary reference conditions) and biological integrity as separate national indicators (U.S. EPA, 1995b). Figure 2.15 provides a possible approach to reporting on both. The examples in Figures 2.14 and 2.15 illustrate the benefits of the biological assessment and biological criteria process based on a system of tiered designated uses and numerical biological criteria to state programs.

2.4.7 WATERSHED ASSESSMENTS

Using fish assemblages as a key indicator for supporting watershed-based assessment has been demonstrated in numerous instances and includes all of the preceding examples. However, developing specific management approaches on a watershed scale has yet to mature much beyond the past uses of chemical-specific or physical approaches through water quality modeling. Such an approach was very successful in developing water quality-based effluent limitations for toxic substances discharged by point sources under the assumption of steady-state conditions. An important limitation of this approach is that the targeted end-points are indirect surrogates (i.e., substitutes) for the more direct biological indicators. Furthermore, the inherent limitations of water quality modeling are compounded under conditions of variable flow, when natural constituents (e.g., sediment, nutrients) of the environment are considered, and when physical dimensions of the ecosystem (e.g., habitat) must be included. This is known as the total maximum daily loads (TMDL) approach, which attempts to integrate point and nonpoint source inputs.

Two recent studies used the IBI, not only to assess the status of the watershed, but as an end-point to develop watershed-based management goals and directions based on making direct linkages between stressors and the condition of the fish assemblage. Steedman (1988) used the IBI to establish criteria for riparian zone quality and urban land use that would maintain a preferred level of biological performance. Bennet et al. (1993) used the IBI as an end-point for a GIS modeling approach that

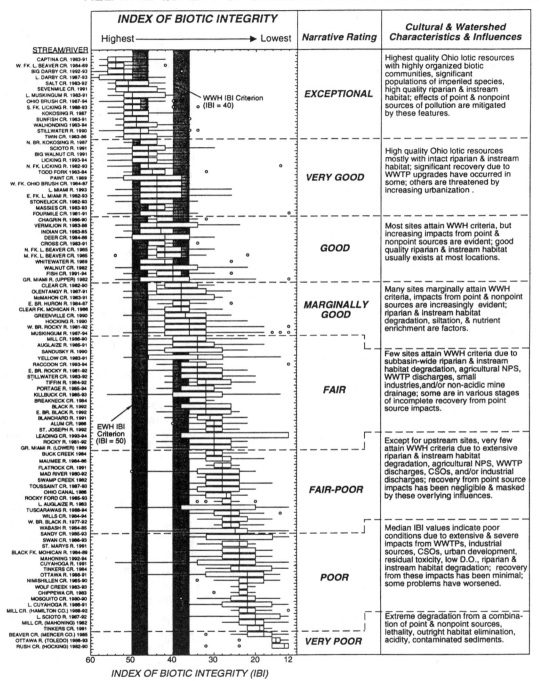

FIGURE 2.15 Frequency distribution of IBI scores derived from electrofishing in 99 Ohio rivers and streams ranked by median IBI score.

included traditional elements of the TMDL approach. They concluded that "... the IBI in a regression equation with land use parameters allows for a method of tying watershed conditions to the biological integrity of a stream without the extremely difficult task of quantifying loading rates for nonpoint source pollutants such as phosphorus and nitrogen." Such an approach offers a method to make a more direct linkage between watershed stressors and biological indicators without the need to risk extrapolating the predictions of chemical/physical modeling to biologically defined designated uses.

2.4.8 OTHER USES OF FISH ASSEMBLAGE INFORMATION

Fish assemblage information is also useful outside the traditional water quality management purview of state WQS and the CWA. One area in which Ohio EPA has been involved is with non-game species, particularly the endangered, threatened, and special-status fish species listed by the Ohio Division of Wildlife. Approximately 30% (53 species) of the Ohio fish fauna is listed as endangered, threatened, special interest, or extinct. This includes 25 of the 41 species designated by the Ohio EPA (1987b) as highly intolerant or sensitive. Based on the fish assemblage data collected by the Ohio EPA since 1978, we estimate that an additional 16 species are in the process of significant declines or fragmentation of their original ranges, some of which are occurring more rapidly than others (Ohio EPA, 1997). This brings to more than 40% the proportion of the Ohio fish fauna that are in a state of imperilment or decline. Tracking the status of an entire statewide fauna is made possible here because of the intensive watershed level of biological assessment that is conducted by the Ohio EPA and supplemented by other agencies. Thus, the "value added" use of the fish assemblage information collected primarily to track trends in water resource quality is exemplified. It also demonstrates the opportunity to utilize the dimensions of the fish assemblage data in ways to refute concerns that such will become collapsed and "lost" in the IBI-based evaluations.

Non-game aquatic communities are not only indicators of acceptable environmental conditions for themselves, but also indicate that the water resource is of an acceptable quality for other wildlife and many human uses. While individual, site-specific watershed and stream disturbances may seem trivial, the aggregate result of these individual impacts emerges in the form of a degraded and declining fish fauna on both watershed and regional scales. Ohio will continue to encounter difficulties in not only highlighting diffuse problems, but in gaining support for action if it does not engage in monitoring and assessment efforts that generate this type of information. Presently, the list of declining fish species is being used alongside that about the imperiled status of other aquatic organism groups (e.g., Unionidae, crayfishes) to implement a revised antidegradation policy in the Ohio WQS.

2.5 CONCLUSION

A monitoring approach, integrating fish assemblage and other biological indicator information that reflects the integrity of the water resource directly, along with water chemistry, physical habitat, toxicity testing, landscape, and other monitoring and source information must be central to accurately defining today's varied and complex problems. Such information is also essential in tracking the progress of management efforts to protect and rehabilitate water resources. The basis for declaring the success of water resource management programs must shift from a reliance on meeting administrative activity measures (e.g., numbers of permits issued, dollars spent, or management practices installed) and a preoccupation with chemical water quality alone, to more integrated and holistic measurements with achieving water resource integrity as an overarching goal. Biological criteria and assessments using fish assemblages can contribute much toward making this shift.

Emphasizing such biological indicator end-points is important because: (1) aquatic life criteria frequently result in the most stringent regulatory and management requirements compared with those for other use categories (i.e., protection for the aquatic life use criteria should assure the protection of other uses); (2) aquatic life uses apply to virtually all water body types and the diverse criteria (i.e., oxygen demanding wastes, nutrients, toxic substances, habitat, physical, and biological factors, etc.) apply to all water resource management issues; and (3) aquatic life uses and the accompanying chemical, physical, and biological criteria provide a comprehensive and accurate ecosystem perspective for guiding water resource management based on promoting the protection of biological integrity.

Biological criteria based on fish assemblages aid greatly in the visualization of aquatic resource goals, values, and attributes. This is a must if we are to change the commonly held view of watersheds and streams as mere catchments and conveyances for receiving municipal and industrial wastes, excess surface and subsurface drainage, and accommodating any number of land uses. The loss of lotic habitat and watershed integrity will continue virtually unabated unless guidelines are set forth under which such activities need to be conducted to conserve biological integrity and biodiversity. Without biological criteria and the accompanying case examples provided by a watershed-based bioassessment effort, such efforts will continue to be seriously handicapped.

ACKNOWLEDGMENTS

This chapter would not have been possible were it not for the many years of effort put into field work, laboratory analysis, and data assessment and interpretation by members, past and present, of the Ohio EPA, Ecological Assessment Unit. Several staff members, including Paul Albeit, Dave Altfater, Brian Alsdorf, Ray Beaumier, Chuck Boucher, Kelly Capuzzi, Bob Miltner, Ed Rankin, Randy Sanders, and Roger Thoma, contributed generously to the development of using fish communities as an integral component of biological criteria. None of this would have been possible without the excellent data management and processing skills supplied by Dennis Mishne. Dan Dudley and Jim Luey contributed at length to the early development and review of the important concepts of biological integrity, ecoregions, reference sites, and biological monitoring in general. Members of the Ohio EPA environmental indicators pilot project team, including Erin Gaskill and Eric Nyhaard, are acknowledged for some of the analyses used herein. Charlie Staudt provided many hours of support in the development of the original computer programs. Finally, Gary Martin and the late Pat Abrams are credited for their solid management support for the concept of biological criteria and biological monitoring in general at the Ohio EPA.

REFERENCES

Angermier, P.L. and I. Schlosser. 1987. Assessing biotic integrity of a fish community in a small Illinois stream, *N. Am. J. Fish. Mgmt.,* 7, 331–338.

Angermier, P.L. and J.R. Karr. 1986. Applying an index of biotic integrity based on stream-fish communities: considerations in sampling and interpretation, *N. Am. J. Fish. Mgmt.,* 6, 418–427.

Barbour, M.T., J. Gerritson, B.D. Snyder, and J.B. Stribling. 1997. Revision to Rapid Bioassessment Protocols for Use in Rivers and Streams: Periphyton, Benthic Macroinvertebrates, and Fish. EPA 841-D-97-002. U.S. EPA, Office of Water, 4503F. Washington, D.C. (Draft).

Barbour, M.T. and et al. 1996. A framework for biological criteria for Florida streams using benthic macro-invertebrates, *J. N. Am. Benth. Soc.,* 15(2), 185–211.

Bayley, P.B., Larimore, R.W., and Dowling, D.C. 1989. Electric seine as a fish-sampling gear in streams, *Trans. Am. Fish. Soc.,* 118, 447–453.

Becker, G.C. 1983. Fishes of Wisconsin. The University Wisconsin Press. Madison, WI.

Benke, A.C. 1990. A perspective on America's vanishing streams, *J. N. Am. Benth. Soc.,* 9(1), 77–88.

Bennet, M.R., J.W. Kleene, and V.O. Shanholtz. 1993. Total Maximum Daily Load Nonpoint Source Allocation Pilot Project. File Report, Dept. of Agricultural Engineering, Blacksburg, VA.

Boccardy, J.A. and E.L. Cooper. 1963. The use of rotenone and electrofishing in surveying small streams, *Trans. Am. Fish. Soc.,* 92, 307–310.

Bramblett, R.G. and K.D. Fausch. 1991. Variable fish communities and the index of biotic integrity in a western Great Plains River, *Trans. Am. Fish. Soc.,* 120, 752.

Clay, W.M. 1975. *The Fishes of Kentucky.* Kentucky Dept. Fish and Wildlife Resources, Frankfort, KY.

Davis, W.S. and T.P. Simon. 1995. *Biological Assessment and Criteria: Tools for Water Resource Planning and Decision Making.* Lewis, Boca Raton, FL.

Denoncourt, R.F. and J.W. Stambaugh, Jr. 1974. An ichthyofaunal survey and discussion of fish species diversity as an indicator of water quality, Codorus Creek drainage, York County, Pennsylvania, *Proc. Pa. Acad. Sci.,* 48, 71–84.

DeShon, J.D. 1995. Development and application of the invertebrate community index (ICI), in W.S. Davis and T.P. Simon (Eds.), *Biological Assessment and Criteria: Tools for Risk-Based Planning and Decision Making.* Lewis, Boca Raton, FL. 217–243.

Doudoroff, P. and C.E. Warren. 1957. Biological indices of water pollution with special reference to fish populations, in *Biological Problems in Water Pollution,* U.S. Public Health Service, Robert A. Taft Sanitary Engineering Center, Cincinnati, OH. 144–163.

Eddy, S. and J.C. Underhill. 1974. *Northern Fishes.* University of Minnesota Press, Minneapolis, MN.

Etnier, D.A. 1993. *The Fishes of Tennessee.* University of Tennessee Press, Knoxville, TN.

Fausch, K.D., Lyons, J., Karr, J.R., and Angermeier, P.L. 1990. Fish communities as indicators of environmental degradation, *American Fisheries Society Symposium,* 8, 123–136.

Fausch, K.D., J.R. Karr, and P.R. Yant. 1984. Regional application of an index of biotic integrity based on stream fish communities, *Trans. Am. Fish. Soc.,* 113, 39–55.

Forbes, S.A. and R.E. Richardson. 1920. The Fishes of Illinois. Ill. Nat. Hist. Surv. Publ. No. 3.

Fore, L.S., J.R. Karr, and L.L. Conquest. 1993. Statistical properties of an index of biotic integrity used to evaluate water resources, *Can. J. Fish. Aquatic Sci.,* 51, 1077–1087.

Funk, J.L. 1954. Movement of stream fishes in Missouri, *Trans. Am. Fish. Soc.,* 85, 39–57.

Funk, J.L. 1958. Relative efficiency and selectivity of gear used in the study of stream fish populations, *23rd N. Am. Wildl. Conf.,* 23, 236–248.

Gammon, J.R., Spacie, A., Hamelink, J.L., and R.L. Kaesler. 1981. Role of Electrofishing in assessing environmental quality of the Wabash River, in *Ecological Assessments of Effluent Impacts on Communities of Indigenous Aquatic Organisms,* Bates, J. M. and Weber, C. I., Eds., ASTM STP 730.

Gammon, J.R. 1980. The use of community parameters derived from electrofishing catches of river fish as indicators of environmental quality, in *Seminar on Water Quality Management Trade-Offs (point source vs. diffuse source pollution).* EPA-905/9-80-009. 335–363.

Gammon, J.R. 1976. The Fish Populations of the Middle 340 km of the Wabash River. Purdue Univ. Water Res. Research Cen. Tech. Rep. 86.

Gammon, J.R. 1973. The Effect of Thermal Inputs on the Populations of Fish and Macroinvertberates in the Wabash River. Purdue Univ. Water Res. Research Cen. Tech. Rep. 32.

Gaufin, A.R. and C.M. Tarzwell. 1953. Discussion of R. Patrick's paper, "aquatic organisms as an aid in solving waste disposal problems," *Sewage Ind. Wastes,* 25(2), 214–217.

Gerking, S.D. 1959. The restricted movement of fish populations, *Biological Review,* 34, 221–242.

Gerking, S.D. 1953. Evidence for the concept of a home range and territory in stream fishes, *Ecology,* 34, 347–365.

Gerking, S.D. 1945. The distribution of the fishes of Indiana. *Investigations of Indiana Lakes and Streams,* 3(1), 1–137.

Hendricks, M.L., C.H. Hocutt, and J.R. Stauffer. 1980. Monitoring of fish in lotic habitats, in C.H. Hocutt and J.R. Stauffer (Eds.), *Biological Monitoring of Fish.* Heath, Lexington, MA. 205–231.

Hill, D.M., E.A. Taylor, and C.F. Saylor. 1975. Status of faunal recovery in the North Fork Holston River, Tennessee and Virginia, *Proc. Ann. Conf. S.E. Assoc. Game Fish. Comm.,* 28, 398–413.

Hubbs, C.L. and K.F. Lagler. 1964. *Fishes of the Great Lakes Region.* University Michigan Press, Ann Arbor, MI.

Hughes, R.M. and J.R. Gammon. 1987. Longitudinal changes in fish assemblages and water quality in the Willamette River, Oregon, *Trans. Am. Fish. Soc.,* 116, 196–209.

ITFM (Intergovernmental Task Force on Monitoring Water Quality). 1995. The Strategy for Improving Water-Quality Monitoring in the United States. Final report of the Intergovernmental Task Force on Monitoring Water Quality. Interagency Advisory Committee on Water Data, Washington, D.C. and Appendices.

ITFM (Intergovernmental Task Force on Monitoring Water Quality). 1992. Ambient Water Quality Monitoring in the United States: First Year Review, Evaluation, and Recommendations. Interagency Advisory Committee on Water Data, Washington, DC.

Jenkins, R.E. and N.M. Burkhead. 1993. *Freshwater Fishes of Virginia.* American Fisheries Society, Bethesda, MD.

Jordan, D.S. 1890. Report of explorations made during summer and autumn of 1888, in the Allegheny region of Virginia, North Carolina, and Tennessee, and in western Indiana, with an account of the fishes found in each of the river basins of those regions, *Bull. U.S. Fish Comm.,* 8(1888), 97–192.

Judy, R.D., Jr., P.N. Seely, T.M. Murray, S.C. Svirsky, M.R. Whitworth, and L.S. Ischinger. 1984. 1982 National Fisheries Survey, Vol. 1. Technical Report Initial Findings. U.S. Fish and Wildlife Service, FWS/OBS-84/06.

Karr, J.R. 1991. Biological integrity: A long-neglected aspect of water resource management, *Ecological Applications,* 1(1), 66–84.

Karr, J.R. 1981. Assessment of biotic integrity using fish communities, *Fisheries,* 6(6), 21–27.

Karr, J.R., K.D. Fausch, P.L. Angermier, P.R. Yant, and I.J. Schlosser. 1986. Assessing Biological Integrity in Running Waters: A Method and Its Rationale. Illinois Natural History Survey Special Publication 5.

Kerans, B.L. and Karr, J.R. 1992. An evaluation of invertebrate attributes and a benthic index of biotic integrity for Tennessee Valley rivers, in *Proc. 1991 Midwest Poll. Biol. Conf.,* EPA 905/R-92/003.

Kirsch, P.H. 1895a. A report on investigations in the Maumee River basin during the summer of 1893, *Bull. U.S. Fish Comm.,* 14(1894), 315–317.

Kirsch, P.H. 1895b. Report on explorations made in Eel River basin in the northeastern part of Indiana in the summer of 1892, *Bull. U.S. Fish Comm.,* 14(1894), 31–41.

Kirtland, J.P. 1838. Report on the zoology of Ohio, *Ann. Rep. Geol. Surv. State of Ohio,* 2, 157–197.

Krumholz, L.A. and W.L. Minckley. 1964. Changes in the fish population in the upper Ohio River following temporary pollution abatement, *Trans. Am. Fish. Soc.,* 93(1), 1–5.

Larimore, R.W. and P.W. Smith. 1963. The fishes of Champaign County, Illinois, as affected by 60 years of stream changes, *Ill. Nat. Hist. Surv. Bull.,* 28(2), 299–382.

Larimore, R.W. 1961. Fish populations and electrofishing success in a warmwater stream, *J. Wildl. Mgmt.,* 25(1), 1–12.

Leonard, P.M. and D.J. Orth. 1986. Application and testing of an index of biotic integrity in small, coolwater streams, *Trans. Am. Fish. Soc.,* 115, 401.

Loftis, J.C., Ward, R.C., and Smillie, G.M. 1983. Statistical models for water quality regulation, *J. Water Poll. Contr. Fed.,* 55, 1098–1106.

Lyons, J., L. Wang, and T. Simonson. 1996. Development and validation of an index of biotic integrity for coldwater streams in Wisconsin, *N. Am. J. Fish. Mgmt.,* 16, 241–256.

Lyons, J. 1992a. Using the Index of Biotic Integrity (IBI) to Measure Environmental Quality in Warmwater Streams of Wisconsin. Gen. Tech. Rep. NC-149. St. Paul, MN: USDA, Forest Serv., N. Central Forest Exp. Sta.

Lyons, J. 1992b. The length of stream to sample with a towed electrofishing unit when fish species richness is estimated, *J. N. Am. Fish. Mgmt.,* 12, 198–203.

Meador, M.R., T.F. Cuffney, and M.E. Gurtz. 1993. Methods for Sampling Fish Communities as Part of the National Water-Quality Assessment Program. U.S. Geol. Surv. Open File Rept. 93-104.

Miller, D. L. and et al. 1988. Regional applications of an index of biotic integrity for use in water resource management, *Fisheries*, 13, 12.

Mills, H.B., W.C. Starrett, and F.C. Bellrose. 1966. Man's Effect on the Fish and Wildlife of the Illinois River. Ill. Nat. Hist. Surv. Biol. Notes 57.

Novotny, D.W. and G.R. Priegel. 1974. Electrofishing Boats, Improved Designs, and Operational Guidelines to Increase the Effectiveness of Boom Shockers. Wisc. DNR Tech. Bull. No. 73, Madison, WI.

Oberdorff, T. and R.M. Hughes. 1992. Modification of an index of biotic integrity based on fish assemblages to characterize rivers of the Seine-Normandie Basin, France, *Hydrobiologia*, 228, 116–132.

Ohio Environmental Protection Agency. 1997. *Ohio Water Resource Inventory. Volume I. Summary, Status, and Trends.* Rankin, E.T., Yoder, C.O., and Mishne, D.A. (Eds.), Ohio EPA Tech. Bull. MAS/1996-7-2-I, Division of Surface Water, Columbus, OH.

Ohio Environmental Protection Agency. 1992. Biological and Habitat Investigation of Greater Cincinnati Area Streams: The Impacts of Interceptor Sewer Line Construction and Maintenance, Hamilton and Clermont Counties, Ohio. OEPA Tech. Rept. EAS/1992-5-1.

Ohio Environmental Protection Agency. 1990. *Ohio Water Resource Inventory. Volume I. Summary, Status, and Trends.* Rankin, E T., Yoder, C.O., and Mishne, D.A. (Eds.), Division of Water Quality Planning and Assessment, Columbus, OH.

Ohio Environmental Protection Agency. 1989a. *Biological Criteria for the Protection of Aquatic Life. Vol. III. Standardized Biological Field Sampling and Laboratory Methods for Assessing Fish and Macroinvertebrate Communities.* Division of Water Quality Monitoring and Assessment, Surface Water Section, Columbus, OH.

Ohio Environmental Protection Agency. 1989b. *Addendum to Biological Criteria for the Protection of Aquatic Life. Vol. II. Users Manual for Biological Field Assessment of Ohio Surface Waters.* Division of Water Quality Planning and Assessment, Surface Water Section, Columbus, OH.

Ohio Environmental Protection Agency. 1988. Ohio Water Quality Inventory — 1988 305(b) Report. Volume I and Executive Summary. Rankin, E.T., Yoder, C.O., and Mishne, D.A. (Eds.), Division of Water Quality Monitoring and Assessment, Columbus, OH.

Ohio Environmental Protection Agency. 1987a. *Biological Criteria for the Protection of Aquatic Life. Vol. I. The Role of Biological Data in Water Quality Assessment.* Division of Water Quality Monitoring and Assessment, Surface Water Section, Columbus, OH.

Ohio Environmental Protection Agency. 1987b. *Biological Criteria for the Protection of Aquatic Life. Vol. II. Users Manual for Biological Field Assessment of Ohio Surface Waters.* Division of Water Quality Monitoring and Assessment, Surface Water Section, Columbus, OH.

Paller, M.H. 1994. Relationships between fish assemblage structure and stream order in South Carolina coastal plain streams, *Trans. Am. Fish. Soc.,* 123, 150–161.

Pflieger, W.L. 1975. *The Fishes of Missouri.* Missouri Dept. of Conservation.

Plafkin, J.L. and et al. 1989. Rapid Bioassessment Protocols for Use in Rivers and Streams: Benthic Macroinvertebrates and Fish. EPA/444/4-89-001. U.S. EPA. Washington, DC.

Rafinesque, C.S. 1820. *Ichthyologia Ohiensis, or Natural History of the Fishes Inhabiting the River and Its Tributary Streams, Preceded by a Physical Description of the Ohio and Its Branches.* W.G. Hunt. Lexington.

Rankin, E.T. 1989. The Qualitative Habitat Evaluation Index (QHEI), Rationale, Methods, and Application. Ohio EPA, Division of Water Quality Planning and Assessment, Ecological Assessment Section, Columbus, OH.

Rankin, E.T. 1995. The use of habitat assessments in water resource management programs, in W. Davis and T. Simon (Eds.), *Biological Assessment and Criteria: Tools for Water Resource Planning and Decision Making.* Lewis, Boca Raton, FL. 181–208.

Rankin, E.T. and Yoder, C.O. 1990. The nature if sampling variability in the Index of Biotic Integrity (IBI) in Ohio streams, in Davis, W.S. (Ed.), *Proceedings of the 1990 Midwest Pollution Control Biologists Conference,* U.S. EPA, Region V, Environmental Sciences Division, Chicago, IL, EPA-905-9-90/005. 9–18.

Robison, H.W. and T.M. Buchanan. 1988. *Fishes of Arkansas.* University of Arkansas Press, Fayetteville, AR.

Sanders, R.S. 1991. Day versus night electrofishing catches from near-shore waters of the Ohio and Muskingum Rivers, *Ohio J. Sci.,* 92, 51–59.

Scott, W.B. And E.J. Crossman. 1973. Freshwater Fishes of Canada. Fisheries Research Board of Canada, *Bull.,* 184.

Simon, T.P. and E.B. Emery. 1995. Modification and assessment of an index of biotic integrity to quantify water resource quality in great rivers, *Regulated Rivers: Research and Management,* 11, 283–298.

Simonson, T. and J. Lyons. 1995. Comparison of catch per effort and removal procedures for sampling stream fish assemblages, *N. Am. J. Fish. Mgmt.,* 15, 419–427.

Smith, C.L. 1985. *The Inland Fishes of New York State.* New York State Department of Environmental Conservation.

Smith, P.W. 1979. *The Fishes of Illinois.* University of Illinois Press, Urbana, IL.

Smith, P.W. 1971. Illinois Streams: A Classification Based on Their Fishes and An Analysis of Factors Responsible for the Disappearance of Native Species. Ill. Nat. Hist. Surv. Biol. Notes 76.

Sparks, R.E. and W.C. Starrett. 1975. An electrofishing survey of the Illinois River, 1959–1974, *Ill. Nat. Hist. Surv. Bull.,* 295–316.

Steedman, R.J. 1988. Modification and assessment of an index of biotic integrity to quantify stream quality in southern Ontario, *Can. J. Fish. Aquatic Sci.,* 45, 492–501.

Suter, G.W., II. 1993. A critique of ecosystem health concepts and indexes, *Environmental Toxicology and Chemistry,* 12, 1533–1539.

Teppen, T.C. and J.R. Gammon. 1976. Distribution and abundance of fish populations in the middle Wabash River, in G.W. Esch and R.W. McFarlane (Eds.), *Thermal Ecology II (ERDA Symposium Series) "CONF-750425."* 272–283.

Tramer, E.J. and P.M. Rogers. 1973. Diversity and longitudinal zonation in fish populations of two streams entering a metropolitan area, *Am. Midl. Nat.,* 90(2), 366–374.

Trautman, M.B. 1957. *The Fishes of Ohio*. Ohio State University Press, Columbus, OH.

Trautman, M.B. 1981. *The Fishes of Ohio*. Ohio State University Press, Columbus, OH.

Tsai, C. 1973. Water quality and fish life below sewage outfalls, *Trans. Am. Fish. Soc.*, 102(2), 281–292.

Tsai, C. 1968. Effects of chlorinated sewage effluents on fishes in upper Patuxent River, Maryland, *Chesapeake Sci.*, 9(2), 83–93.

U.S. Environmental Protection Agency. 1997. Guidelines for Preparation of the Comprehensive State Water Quality Assessments (305(b) Reports) and Electronic Updates. EPA 841-B-97-002A&B. U.S. EPA, Office of Water (4503F), Washington, DC. 20460.

U.S. Environmental Protection Agency. 1996a. Summary of State Biological Assessment Programs for Streams and Rivers. EPA 230-R-96-007. U.S. EPA, Office of Policy, Planning, & Evaluation, Washington, DC. 20460.

U.S. Environmental Protection Agency. 1996b. National Water Quality Inventory: 1994 Report to Congress. EPA 841-R-94-001, U.S. EPA, Office of Water, Washington, DC. 20460.

U.S. Environmental Protection Agency. 1995a. Biological Criteria: Technical Guidance for Streams and Small Rivers. EPA 822-B-94-001, U.S. EPA, Office of Water (WH 4304), Washington, DC. 20460.

U.S. Environmental Protection Agency. 1995b. Environmental Indicators of Water Quality in the United States. EPA 841-R-96-002. Office of Water, Washington, DC. 20460.

U.S. Environmental Protection Agency. 1995c. A Conceptual Framework to Support Development and Use of Environmental Information in Decision-Making. EPA 239-R-95-012. Office of Policy, Planning, and Evaluation, Washington, DC. 20460.

U.S. Environmental Protection Agency. 1991. Environmental Monitoring and Assessment Program. EMAP — Surface Waters Monitoring and Research Strategy — Fiscal Year 1991. EPA/600/3-91/022. Office of Research and Development, Environmental Research Laboratory, Corvallis, OR.

U.S. Environmental Protection Agency. 1990. Biological Criteria, National Program Guidance for Surface Waters. EPA-440/5-90-004. U.S. EPA, Office of Water Regulations and Standards, Washington, DC.

U.S. EPA. 1985. Technical Support Document for Water Quality-Based Toxics Control. EPA/44/4-85/03. U.S. EPA, Office of Water, Washington, DC.

Vincent, R. 1971. River electrofishing and fish population estimates, *Prog. Fish Cult.*, 33(3), 163–169.

Watters, G. T. 1992. Unionids, fishes, and the species-area curve, *Journal of Biogeography*, 19, 481–490.

White, A.M., M.B. Trautman, E.J. Foell, M.P. Kelty, and R. Gaby. 1975. The fishes of the Cleveland metropolitan area including the Lake Erie shoreline, in *Water Quality Baseline Assessment for the Cleveland Area — Lake Erie,* Vol. II. EPA-905/9-75-001.

Yoder, C.O. and E.T. Rankin. 1998. The role of biological indicators in a state water quality management process, *J. Env. Mon. Assess.*, (in press).

Yoder, C.O. and E.T. Rankin. 1995a. Biological criteria program development and implementation in Ohio, in W. Davis and T. Simon (Eds.), *Biological Assessment and Criteria: Tools for Water Resource Planning and Decision Making*. Lewis, Boca Raton, FL. 109–144.

Yoder, C.O. and E.T. Rankin. 1995b. Biological response signatures and the area of degradation value: new tools for interpreting multimetric data, in W. Davis and T. Simon (Eds.), *Biological Assessment and Criteria: Tools for Water Resource Planning and Decision Making*. Lewis, Boca Raton, FL. 263–286.

Yoder, C.O. 1991a. Answering some concerns about biological criteria based on experiences in Ohio, in Gretchin H. Flock (Ed.), *Water Quality Standards for the 21st Century. Proceedings of a National Conference,* U.S. EPA, Office of Water, Washington, DC.

Yoder, C.O. 1991b. The integrated biosurvey as a tool for evaluation of aquatic life use attainment and impairment in Ohio surface waters. *Biological Criteria: Research and Regulation, Proceedings of a Symposium,* December 12–13, 1990, Arlington, Virginia, U.S. EPA, Office of Water, Washington, DC. EPA-440/5-91-005, 110–122.

Yoder, C.O. 1989. The development and use of biological criteria for Ohio rivers and streams, in Gretchin H. Flock (Ed.), *Water Quality Standards for the 21st Century. Proceedings of a National Conference,* U.S. EPA, Office of Water, Washington, DC. 139–146.

3 Collaboration, Compromise, and Conflict: How to Form Partnerships in Environmental Assessment and Monitoring

Thomas P. Simon, Robert M. Goldstein, Patricia A. Bailey, Eric Pearson, Michael Ell, Konrad Schmidt, John Emblom, and Lynn Schlueter

CONTENTS

3.1 Introduction 57
3.2 The Purposes and Challenges of Monitoring 58
3.3 Achieving Common Ground for Collaboration..58
3.4 Strategies for Overcoming Inertia: The Red River of the North Experience 59
3.5 Guidance Based on Our Experience..61
Acknowledgments 62
References 62

3.1 INTRODUCTION

Partnerships between agencies and academicians to develop solutions for environmental problems have faced substantial difficulties. Often, the relationship is based on specific research questions supported by grants. For the agencies, findings from grant-supported research are not always easily translated into assessment objectives. Likewise, the bureaucracy of a governmental agency may present frustrating obstacles to collaboration by academicians. Further, the research opportunities provided through governmental assistance do not always lead to findings publishable in refereed journals.

The need for information in support of resource management strategies has grown in recent decades. The mission of "restoring the chemical, physical, and biological integrity of our nations' surface waters" has resulted in a change in focus and strategy for the U.S. Environmental Protection Agency (EPA) that has occurred slowly over the last 25 years. At the EPA's inception, attention first focused on the most severe point-source problems using biological survey and field monitoring tools. As the agency was required to provide burden of proof, focus shifted from biological to chemical monitoring to document impacts associated with violations of water quality standards and chemical criteria. Emphasis then shifted toward a toxicological monitoring framework to document mortality and sublethal effects from exposure to various effluents with a desire to understand cause-and-effect relationships. Currently, the legal battles provide impetus for the return of biological surveys to document environmental damage, although chemical, physical, and toxicological data are still used to interpret the results of biosurveys. Because of the expertise required to accomplish these types of holistic studies, partnerships are needed to bring together federal and state agencies, academics, local organizations, and others to solve environmental problems.

Federal agencies responsible for environmental monitoring often form partnerships with state and local agencies to accomplish large-scale projects. However, collaboration among agencies with differing objectives often leads to conflict. The success of a monitoring program depends on thorough communication and coordination. Although there may be common ground for initiating a monitoring program, the lack of specific coordinated objectives and protocols often leads to the failure of the program. Recently, these issues that cause conflict in collaborative ventures have been the focus of an intergovernmental task force (ITFM, 1992; 1994; Yoder, 1994).

3.2 THE PURPOSES AND CHALLENGES OF MONITORING

The purposes of monitoring include (1) assessing the condition of the water resource and evaluating change over time, (2) characterizing pollutant problems and ranking water bodies for further investigation or restoration, (3) providing data that support the design, implementation, and mitigation of programs and projects, (4) evaluating the effectiveness of current mitigation or restoration activities, (5) responding to emergencies, such as fish kills or noxious algal blooms, and (6) monitoring for compliance and enforcement activities (ITFM, 1994). Collection of monitoring data without completing the analysis, assessment, and application of results provides little impetus for the continuation and support of monitoring programs. For monitoring programs to be successful, transferral of results to decision-making processes is important. Continuation of monitoring is required for evaluation and validation of decisions taken to manage human activities affecting natural resources.

Monitoring programs are challenged because of: (1) a lack of adequate expertise in government agencies, (2) dwindling financial support to access expertise and conduct necessary studies, and (3) inadequate staffing in both federal and state environmental agencies (USGAO, 1986; ITFM, 1992; Yoder, 1994). Likewise, results from monitoring programs have had an insignificant influence on decisions regarding the protection of natural resources, for several reasons: (1) tardiness of interpretive reports, (2) a lack of cost-effective assessment and monitoring techniques, (3) a lack of coordination and interest among and between agencies, (4) the use of inappropriate methods, and (5) an inability to express results in terms understandable to the public, managers, and those empowered to provide funding (i.e., legislators).

Biologists are often unable or unwilling to transcend the role of scientist to act as an advocate or enter the management hierarchy to develop or influence strategy and planning for their agency. Often, unless opportunities for publication or hands-on field projects are still available in management roles, many biologists are uninterested in making the transition to policy-level management. Thus, a separation or break in communication will occur in many cases within the governmental agencies between current environmental management (often lawyers, engineers, or chemists) and the biologists. When this happens, the ability to influence management decisions with biological data from monitoring programs can be hampered.

3.3 ACHIEVING COMMON GROUND FOR COLLABORATION

Differences in missions or legislative mandates among agencies are often the basis for the inability to collaborate on mutually beneficial projects. For instance, the EPA and state water resource agencies take their definition of biological integrity from Karr and Dudley (1981) as "the ability of an aquatic ecosystem to support and maintain a balanced, integrated, adaptive community of organisms having a species composition, diversity, and functional organization comparable to the best natural habitats within a region." Biological integrity is also a part of the mission of the U.S.

Forest Service (FS), which has interpreted biological integrity as synonymous with "biodiversity." However, Karr (1991) shows that biodiversity "which concentrates only on the structure of the community but not including the functional attributes (e.g., nutrient cycling, trophic interactions, competition, predation, parasitism)" is only one aspect of biological integrity.

In another example, all federal agencies are required by the Endangered Species Act to consult with the U.S. Fish and Wildlife Service (USFWS) concerning permitting and other planning activities that could adversely affect the status of listed species. Habitat modification projects proposed by the U.S. Department of Agriculture, the Natural Resource Conservation Service (NRCS), and the U.S. Army Corps of Engineers (USCOE) must be reviewed by both the EPA and the USFWS. However, the distribution and life cycle requirements of endangered species are often inadequately understood to effectively protect species on the brink of extinction. Inadequate monitoring data reduces the ability to make scientifically valid judgments about environmental damage, which may lead to loss of critical habitat, disjunct distributions, or complete loss of species.

In a final example, the fish and wildlife agency for a particular state is often concerned with managing a water resource for the production of sustainable numbers of large individuals of select species. Therefore, the agency's monitoring targets sport fish species and may group other components of the aquatic community as forage. However, another agency, such as a water-quality resource agency, may need a more comprehensive biological survey and associated habitat assessment. Two agencies may monitor the same river reach, but come to opposite conclusions regarding biological condition. The fish and wildlife agency may consider its objectives met if a stream fish assemblage is dominated by the adult life stage of a particular game species. Conversely, a water-quality agency might conclude the same site is "biologically impaired" because of the dominance of an upper-level carnivore or nonindigenous or exotic species. Both agencies could reach different conclusions and develop different and perhaps contrary management plans because of different objectives.

The development of regionally based biological expectations or "biocriteria" is an important stepping stone in reaching a consensus and consolidating objectives among agencies (McKenzie et al., 1992; Jackson and Davis, 1994; Davis and Simon, 1995). This approach could bridge conflicting monitoring strategies and objectives and provide common ground for assessment and interpretation of the biological data among local, state, and federal agencies (Courtemanch, 1994). By developing and agreeing *a priori* on a single main objective by all the agencies involved, the success of the monitoring or study program is enhanced. Sub-objectives specific to each agency's mission can be met through either ancillary data collection or analysis.

3.4 STRATEGIES FOR OVERCOMING INERTIA: THE RED RIVER OF THE NORTH EXPERIENCE

A multi-agency partnership was initiated to develop biocriteria specific for the Lake Agassiz Plain. This partnership is an example of cooperation among federal, state, and local agencies and academic institutions to conduct a successful monitoring program despite different objectives of each group.

Numerous issues on the concerns for water quality have arisen in the Lake Agassiz Plain, including the effects of over 30 proposed hydroelectric dams, cumulative input sources near Fargo, North Dakota, and the assessment of the condition of the Red River basin for EPA-required reporting. These issues in the Lake Agassiz Plain ecoregion have brought together collaborative efforts among two federal agenies (EPA-Headquarters, Office of Science and Technology; the EPA, Regions 5 and 8; the EPA-Environmental Research Laboratory-Corvallis, and the U.S. Geological Survey [USGS]); four state agencies (North Dakota Department of Health and Consolidated Laboratories [NDHCL]; North Dakota Game and Fish [NDGF]; Minnesota Department of Natural Resources [MDNR]; and Minnesota Pollution Control Agency [MPCA]); one local group, the Red Lake Chippewa (RLC) tribe; and three academic institutions (University of North Dakota; University of Minnesota–Duluth; and University of Minnesota–St. Paul). Additonally, the program was coordinated with the Canadian government through the Red River of the North drainage board and Environment Canada.

The impetus behind the unification of the agencies and local institutions was the development of a standard against which future monitoring projects could be evaluated (Jackson and Davis, 1994; Goldstein et al., 1994). All partners agreed that the primary objective was the evaluation and development of regional reference conditions through the calibration of an index of biotic integrity (Karr et al., 1986). Each agency and group provided expertise and manpower to the project. Selection of representative locations within the Red River Valley allowed use of USGS core monitoring sites from the National Water Quality Assessments (NAWQA) Program and repeat visit sites. Historical monitoring information from MDNR and NDHCL enabled technical outreach and sharing of expertise among the states and EPA regions for the acquisition of new information. Academic institutions provided significant expertise on a variety of study design, organism identification and curation, and assisted in project planning, data analysis, and development of objectives; their most important involvement included the "brainstorming" and development of metrics and analysis of data using mainframe computer applications and Geographic Information Systems (Karr, 1994). Those academic researchers interested in the distribution of fish in each state provided quality assurance checks of vouchered specimens and museum curation of specimens for permanent archival. Data collected through this project supported aspects of several Master of Science projects that resulted in student theses. Minnesota and North Dakota state agencies combined their limited field personnel to train and outfit three field crews. Funding support by EPA headquarters was instrumental in accomplishing the project. The regionalization efforts of the EPA's Environmental Research Laboratory–Corvallis resulted in a spatial framework based on subregions for the states of North Dakota and Minnesota that was an important stride for collaboration for all participating agencies.

Many issues in the Red River basin brought stakeholders together. Foremost was the common goal of developing a standard to assess resource quality in the basin. The steps necessary to develop biological reference conditions (Yoder and Rankin, 1995) included the framework from which the objectives for the project were derived. The main objectives became responsibilities of each group that had efforts targeted at the primary tasks, including review of existing data, collection of new data, and development of regional patterns. Results of the project were summarized through a series of conference calls within the technical group of federal and state biologists and specific writing assignments were made. The project resulted in the formation of biocriteria specific for all size streams in the Lake Agassiz Plain (Pearson et al., 1996; Niemela et al., Chapter 13).

This project united a number of agencies across functional and political boundaries to achieve several goals of mutual benefit. First, the development of regionally based reference conditions across political boundaries provided a common framework for assessment and continued monitoring. This was accomplished through the use of tools and experiences documented by the Ohio Environmental Protection Agency (1989) and Davis and Simon (1995). This not only strengthens biological criteria on a regional basis, but ensures that the point of reference for biological assessment and monitoring is the same for all agencies. Second, the uniting of multiple agencies with different objectives was beneficial in times of dwindling resources. Third, sharing of resources, equipment, expertise, and objectives established an unprecedented relationship among these government agencies and academic institutions. Without this cooperation, none of the agencies involved would have been able to accomplish a project of this magnitude.

The positive history of the Lake Agassiz Plain study became a foundational opportunity for further collaboration and provided additional situations for outreach (Courtemanch, 1994). As program support strengthens and momentum builds, the original individuals involved in the project will, in all likelihood, be replaced by others with a similar vision and desire to work together. The authors have considered whether the success of this project was due to the working relationship of the personnel or the nature of the questions being answered and conclude that it is a combination of both. As individuals work closely together to accomplish task-oriented projects, the expertise (and personality) of each individual may compliment or antagonize another to either build or destroy further collaboration.

3.5 GUIDANCE BASED ON OUR EXPERIENCES

The success of the Lake Agassiz Plain experience was measured by the resulting development of regional expectations and the furtherance of working relationships among the groups. The development of regional and multi-agency involvement has been a "work in progress." This process often requires those groups with closer working relationships (e.g., those in the same state or those with similar responsibilities in monitoring) to mentor or explain decisions and misspoken words to those not familiar with government rhetoric and acronyms. Agencies need to be cautious about using terms and information sources not necessarily recognized or widely known by academic researchers. Also, each member of the group must recognize and respect the capability of the other members so that each person has the ability to voice opinions and feel confident that his/her proposal will be considered. Likewise, personal goals cannot overshadow the needs of the group. Applied research, as demonstrated by this project, is enhanced by including academic expertise. An effective synergism can be obtained through agencies seeking funding support for research institutions and basic researchers seeking opportunities to develop mutually beneficial projects. Agencies often benefit by obtaining information they need, but also achieve better linkages with the scientific community. Often, cooperative relationships can be established to support students in degree projects, internship programs, and other opportunities to provide experience in applied science. Unfortunately, many academic institutions believe that agency support can only be monetary.

Based on experiences in the Red River basin, the following guidance will assist in bridging objectives and enhancing collaboration among most academic institutions, and state and federal agencies.

1. *Employ a basin approach.* In the authors' opinion, to develop effective working relationships, the basin approach may be the best strategy for implementation. The basin approach transcends political boundaries and focuses on providing an overall estimate of the resource. This framework allows agencies and institutions with different monitoring needs to reach a consensus by expanding the scope of the project to meet individual objectives. As watershed data accumulates, further efforts need to compile and assess patterns across ecoregions and larger spatial scales that also transcend political boundaries.
2. *Develop a common objective.* The objective of this project was to develop numerical biocriteria based on a reference condition approach for the Lake Agassiz Plain. This objective was made more difficult because the ecoregion spanned political boundaries. To develop effective reference conditions for the entire resource, the study design needed to recognize spatial factors — including various water body sizes (Davis and Simon, 1995). Crossing political boundaries verifies that the sites typical of the entire region are chosen and used to calibrate the index (Jackson and Davis, 1994; Davis and Simon, 1995). Establishing a common objective ensured participation and support by the various cooperating agencies. The common objective aided in focusing the activities of the project through completion because everyone needed the final product.
3. *Establish lines of communication.* Frequent conversation and open dialogue, including both telephone and face-to-face meetings, will build partnerships and establish trust among cooperators (Yoder, 1994). Goals important to all agencies need to be accentuated while resolving differences of opinion. Discussion of strategies and technical stumbling blocks (i.e., collection methods, analysis techniques, and interpretation of data outputs) should include all partners to the project (Jackson and Davis, 1994). Encourage further

efforts to involve industrial and municipal community scientists who would have an interest in project results (see Snyder et al., Chapter 26). This open sharing of information will assist in diminishing mistrust (Polls, 1994).
4. *Develop plans of action.* A mutually acceptable sampling technique and project plan needs to be the basis for the work. Organization and development of a study design is needed — one that reflects a consensus of the agencies and stakeholders involved and enhances scientific validity. It is not only necessary, but a requirement, that all aspects of the sampling protocols and project plan be completed prior to actual sampling and not changed as the project proceeds.
5. *Procure historical data.* Additional partnerships should be formed within states and regions to collect relevant data among sister agencies. The acquisition of information that addresses the research questions can benefit other agencies. Review needs annually with other agencies and discuss projects that could be of mutual benefit.

ACKNOWLEDGMENTS

The authors appreciate the support of management, especially William Melville, EPA; Mary Knutsen, MPCA; Phil Johnson, EPA, Region 8; and Susan Jackson, EPA-Headquarters. We also acknowledge the efforts of Pat Bailey, who helped bring the diverse interests of all the agencies together. A number of other individuals have helped in this project, including Dave Lorenz, Arthur Lubin, John Peterka, Andy Thompson, B.J. Kratz, Cheryl Ferguson, Chuck Fitz, Paul Seyer, and Steven Kelch. The opinions expressed in this article are those of the authors and do not necessarily represent those of the agencies they represent.

REFERENCES

Courtemanch, D.L. 1994. Bridging the old and new science of biological monitoring, *Journal of the North American Benthological Society,* 13, 117–121.

Davis, W.S. and T.P. Simon. 1995. *Biological Criteria and Assessment: Tools for Water Resource Planning and Decision Making.* Lewis, Boca Raton, FL.

Goldstein, R.M., T.P. Simon, P.A. Bailey, M. Ell, E. Pearson, K. Schmidt, and J.W. Emblom. 1994. Concepts for an index of biotic integrity for the streams of the Red River of the North Basin, in *Proceedings North Dakota Water Quality Symposium.* Fargo, ND, March 30–31, 1994. 169–180.

ITFM (Intergovernmental Task Force on Monitoring Water Quality). 1992. Ambient Water Quality Monitoring in the United States: First Year Review, Evaluation, and Recommendations. Interagency Advisory Committee on Water Data, Washington, DC.

ITFM (Intergovernmental Task Force on Monitoring Water Quality). 1994. The Strategy for Improving Water-Quality Monitoring in the United States. Interagency Advisory Committee on Water Data, Washington, DC.

Jackson, S. and W.S. Davis. 1994. Meeting the goal of biological integrity in water-resource programs in the U.S. Environmental Protection Agency, *Journal of the North American Benthological Society,* 13, 592–597.

Karr, J.R. 1994. Integrating Biology and Statistics in the Development of Biological Criteria. U.S. Environmental Protection Agency, Office of Policy, Planning, and Evaluation, Washington, DC.

Karr, J.R. 1991. Biological integrity: a long-neglected aspect of water resource management, *Ecological Applications,* 1, 66–84.

Karr, J.R. and D. Dudley. 1981. Biological monitoring and environmental assessment, *Environmental Management,* 11, 249–256.

Karr, J.R., K.D. Fausch, P.L. Angermeier, P.R. Yant, and I.J. Schlosser. 1986. Assessing Biological Integrity in Running Waters: A Method and its Rationale. Illinois Natural History Survey, Special Publication 5, Champaign, IL.

McKenzie, D.H., D.E. Hyatt, and V.J. McDonald. 1992. *Ecological Indicators.* Vols. I and II. Elsevier, New York.

Ohio Environmental Protection Agency. 1989. *Biological Criteria for the Protection of Aquatic Life.* Vols. 1, 2, 3. Ohio EPA, Division of Water Quality Planning and Assessment, Ecological Assessment Section, Columbus, OH.

Polls, I. 1994. How people in the regulated community view biological integrity, *Journal of the North American Benthological Society,* 13, 598–604.

U.S. GAO (U.S. General Accounting Office). 1986. The Nation's Water: Key Unanswered Questions About the Quality of Rivers and Streams. GAO/PEMD-86-6. Program Evaluation and Methods Division, Washington, DC.

Yoder, C.O. 1994. Toward improved collaboration among local, State, and Federal agencies engaged in monitoring and assessment, *Journal North American Benthological Society,* 13, 391–398.

Yoder, C.O. and E.T. Rankin. 1995. Biological criteria program development and implementation in Ohio, in W.S. Davis and T.P. Simon (Eds.), *Biological Assessment and Criteria: Tools for water Resource Planning and Decision Making,* Lewis, Boca Raton, FL. 109–144.

4 Historical Biogeography, Ecology, and Fish Distributions: Conceptual Issues for Establishing IBI Criteria

Rex Meade Strange

CONTENTS

4.1 Introduction 65
4.2 Linking Local and Regional Factors 66
    4.2.1 Metapopulation Model 66
    4.2.2 Estimation of Metapopulation Structure in Stream Fishes 67
    4.2.3 Dispersal and Species' Range 68
    4.2.4 Community Diversity as Probability of Regional Movements 69
4.3 Regional and Historical Factors of Diversity 69
    4.3.1 Species and Speciation 69
    4.3.2 Allopatric Speciation...70
    4.3.3 Historical Ecology and Ohio River Basin Fishes 72
    4.3.4 Stream Captures and the Upper Cumberland River 74
4.4 Conclusions 75
References 76

4.1 INTRODUCTION

The Index of Biological Integrity (IBI) has been a useful method for evaluating environmental quality of aquatic habitats through the examination of stream fish communities (Karr, 1981; Fausch et al., 1984; Karr et al., 1986). However, criteria must be based on accurate depictions of community structure that address a hierarchy of scales based on stream, drainage, or ecoregion (Hocutt and Wiley, 1986). Potentially, each regionalization unit may have a different species assemblage for reasons other than human impact. Various methods and techniques have been proposed to establish and evaluate these criteria; these are the focus of the remaining chapters of this book. The purpose of this chapter, however, is to examine the role of history in determining fish distributions and diversity patterns.

    Much of community ecology is based on the concept of the community as a "competitive community" following the philosophies of MacArthur (1972), in which each species competes for some common set of resources (Yodzis, 1993). Consequently, most models of community structure assume homogeneity of species assemblages across a heterogeneous environment, implying that equivalent habitats have the same species assemblages regardless of contingent historical events or geographic position within the region. Trophic guilds are therefore interpreted as the result of species co-evolving or co-adapting themselves as they partition resources and reduce competition. The implication of these assumptions is that communities are fairly rigid structures resistant to change, and local diversity is reducible to biotic and abiotic factors (i.e., energy, temperature, water chemistry). The examination of local processes has been extrapolated to explain both regional and evolutionary processes. However, there is increasing evidence that communities are not so rigid, and there is no connection between local and regional processes (Ricklefs, 1989; Gorman, 1992). A quick survey of diversity patterns among North American fishes (cf. Hocutt and Wiley, 1986) will show that diversity is unevenly distributed across the continent. Similar habitats even in the same biogeographic region may not have similar species assemblages (Osborne and Wiley, 1992). Further, evidence in the form of population genetics and phylogenetic relationships of organisms occurring in similar yet allopatric communities have revealed different histories among those species (Richardson and Gold, 1995; Strange and Burr, 1997; Tibits and Dowling, 1996). This questions many of our assumptions regarding both diversity and environmental assessment. Ricklefs (1989) has argued that a historical approach is necessary to account for diversity.

This chapter discusses patterns of population dynamics and biogeographic patterns relevant to fish distributions and local compositions, and hence, IBI criteria. Emphasis is placed on the importance of the interpenetration of local and regional processes within a historical framework, and that the key to understanding the basis for biodiversity lies in its history. Local processes are the familiar determinant forces of competition, predation, and various stochastic factors that have been used to model community structure. These are considered to operate within a small area of homogeneous habitat on ecological time scales (1 to 100 generations), and act to decrease diversity. Regional processes include dispersal, speciation, and adaptation. These operate on larger spatial and temporal scales (100 to 100,000 generations) and promote diversity. It is the interactions between these processes that determine the nature and distribution of fish faunal diversity.

4.2 LINKING LOCAL AND REGIONAL FACTORS

4.2.1 METAPOPULATION MODEL

It is clear that many fishes are habitat specialists, as each species has a given set of habitats with which it is commonly associated (Kuehne, 1962). Active habitat choice in fishes may be an evolutionary response to fitness maximization, whereby each species enjoys the highest degree of fitness in a given habitat type. Habitat affinities can then be interpreted as adaptations to avoid the harmful effects of natural selection (e.g., competition and predation) in habitats for which individual organisms are not adapted.

Aquatic habitats themselves are distributed nonrandomly across complex environments. Within streams, there is a predictable alternation of riffles and pools, and a given set of relationships between streams of different orders. Ecological processes acted out within habitats typically limit or reduce population growth; therefore, each population has a given (non-zero) probability of becoming extinct if isolated from other populations (Opdam, 1991). In fact, localized habitats and their associated populations are not isolated, but are linked with others by migration driven by active habitat choice. Migration offsets the probability for extinction of a dwindling population and may act to reoccupy a habitat in which the population has already gone extinct. Persistence within any given habitat patch is a balance of the threat of extinction due to ecological processes and migration between habitat patches. This is the model for metapopulation dynamics, whereby local populations are linked by interhabitat migration (Levins, 1969).

The movement of individual fish among habitats and the threat of local extinction is clearly analogous to the balance between colonization and extinction in MacArthur and Wilson's (1967) theory of island biogeography. This can be viewed at different spatial and temporal hierarchical levels. The stochastic back-and-forth movement of individuals between habitats within a species' range in this model is *migration*. The spread of migrants into previously unoccupied habitats outside the species' range is *dispersal*. Note that these two terms are based on the same local phenomenon, yet differ in the scale at which they are perceived. Migration is local and promotes species persistence, whereas dispersal is regional and promotes range expansion. The relationship between these scales may be likened to brownian motion and diffusion, and will play a role in speciation and regional distributions considered later. Metapopulation dynamics link local and regional processes by extending the time frame in which ecological processes act (Ricklefs, 1989).

4.2.2 ESTIMATION OF METAPOPULATION STRUCTURE IN STREAM FISHES

Metapopulation structure within a species can take different forms, depending on the unique ecology and contexts in which it may be found (Harrison and Hastings, 1996). Two different applications of this model to stream fishes using mitochondrial DNA (mtDNA) haplotypes as genetic markers to estimate population movements and connectivity have been described (Strange and Burr, 1995; 1996). The blackside dace *Phoxinus cumberlandensis*, a small schooling minnow occupying pools of headwater streams in the upper Cumberland River of eastern Kentucky, appears to be distributed randomly throughout dendritic stream systems when considered on a local scale. Gene flow and migration estimates indicate no isolation by distance among these populations (an island model of migration), and that approximately one migrant enters the average deme per generation. In other words, blackside dace are panmictic within arrays of local populations (metapopulations) on ecological times scales (1 to 10 generations), with the effect that migration between habitats acts to genetically homogenize metapopulations.

However, a different pattern emerges on a regional level, where metapopulations are organized in a stepping-stone fashion, as would be expected in a stream system. Estimates of gene flow and migration at this level of resolution approximate one individual per metapopulation per generation, indicating a similar local-scale linkage between metapopulations as occurs between demes within metapopulations. This is not enough to overcome the effect of genetic drift because of the greater effective population size of the metapopulation relative to the size of the number of migrants, and an equilibrium between drift and migration maintains a degree of regional genetic heterogeneity.

At an even higher level of spatial and temporal resolution, another factor emerges: *mutation*. Detectable mutation in the form of a new allele occurs within a single individual on the order of 100,000 to 1,000,000 generations. We would expect gradual phylogenetic divergence among metapopulations if they are unconnected by gene flow. Migration and gene flow are frequent enough to spread new alleles throughout the species' range in a geological instant relative to the rate of mutation.

Another pattern of metapopulation structure is exhibited by the Niangua darter *Etheostoma nianguae* (Strange and Burr, 1996). This species occurs in fifth- and sixth-order streams of the Osage River system in central Missouri. Populations are discrete and separated by many river kilometers of higher stream order. Population structure within given streams appears to be essentially panmictic within ecological time frames, but no migration between these metapopulations appears to occur. However, the hypothesis proposed by Mattingly (personal communication) and articulated by Strange and Burr (1996) suggests that stream orders within the Osage system may move downstream during periods of low flow, forcing the darters to move downstream to maintain their stream order position. This would effectively allow the formerly separate populations to coalesce on the scale of every 100 to 1000 generations. Thus, this model specifies that metapopulations maintain homogeneity through migration, diverge because of drift, but ultimately coalesce during periods of low flow. This could account for the current distribution of the species in specific stream systems within the Osage drainage.

Thus, metapopulation structures can take different species-specific forms. The resulting hierarchical structure of migration, drift, and mutation described for these two species (Figure 4.1) are ways in which genetic variability is spatially distributed. Since mutation and recombination are the ultimate source of variability on which selection acts to produce novel adaptations, metapopulation

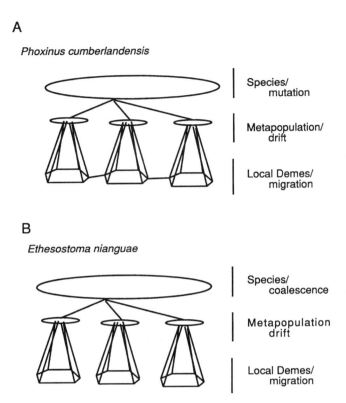

FIGURE 4.1 Hierarchial structure of local migration, drift, mutation, and coalescence in (A) *Phoxinus cumberlandesis* and (B) *Etheostoma nianguae*.

structure links both local extinction-colonization with evolutionary forces. There is no reason to expect that co-occurring species occupying different trophic levels or functional groups will have the same migration patterns (Tibits and Dowling, 1996). Such different patterns can produce a phase-shift between competitor or predator-prey interactions. If Ricklefs' (1989) suggestion regarding history is valid, this is a necessary component of community diversity that must be included in any explanation or utilization of diversity measurements.

4.2.3 Dispersal and Species' Range

A correlate of migration is range limits. All species occupy a restricted range. There are areas in which a given species may be abundant, and others where it does not occur. Species recently introduced into new drainage systems often quickly disperse throughout the system and into the next, until some factor prohibits further expansion. These range boundaries often coincide with a discontinuous habitat distribution, and for fish this could represent drainage divides and waterfalls. In other cases, the cause of the distribution boundary is less clear. If a subtle change in water chemistry, stream gradient, or other factors prohibits a species from expanding its range, why do the populations occurring at the range margin not adapt and spread outward? This regional phenomenon might have a local cause.

There are two interrelated factors that combine to define a species' range (Hoffman and Blows, 1994; Kirkpatric and Barton, 1997). The first is migration or gene flow. Recall that the stochastic movements of individuals between habitat patches are responsible for a species' persistence within its range. Migration offsets local extinction and promotes recolonization of unoccupied sites. Populations well adapted to their environment (higher average fitness) are more likely to contribute to the pool of migrants. Migration is concomitant with gene flow, which in turn promotes genetic homogeneity. Most of the migrants will be from the center if these populations have a higher average fitness than populations at the periphery. Thus, gene flow constantly brings genotypes from the center of a species' range into peripheral populations, thereby lowering their overall fitness and limiting their contribution to the migrant pool. This is the source-sink model of population dynamics.

In turn, niche specificity or its inverse, habitat availability, may limit where a species occurs. Generalists such as the creek chub *Semotilus atromaculatus* are distributed across much of eastern North America in no small part because of their broad habitat requirements. This factor is determined by the species' genotype and extrinsic environmental variables.

Range limitations occur as the result of these two factors, and can be diagrammed as a three-cornered graph with vagility (ability to form migrants), habitat availability, and range size occupying each corner. No one factor can be maximized without minimizing the other two. Thus, migration has the contradictory effect of promoting persistence within the range and negating expansion outside of it. Each species expands its range until it meets hostile environments. If fitness is a measurement of reproductive success through differential selection, this *Lebensraum* is a measurement of the tensions inherent between gene flow and adaptation across a heterogeneous landscape. A quick comparison of species' distributions will reveal that no two species have exactly the same range, another argument in favor of an open community structure.

4.2.4 COMMUNITY DIVERSITY AS PROBABILITY OF REGIONAL MOVEMENTS

With the inclusion of metapopulation models, community structure moves from the product of local species interactions to regional scale processes. This can be stated in strictly descriptive terms. Whittaker (1972) addressed species diversity in terms of alpha, beta and gamma diversities. Alpha diversity represents species richness within individual habitat patches (e.g., how many species occur in riffles of Dog Slaughter Creek). Gamma diversity is the species richness at a regional scale (e.g., how many species occur in the Appalachian highland region). Both of these measurements are simple counts of species present within the respective unit. Beta diversity is a dimensionless number representing species turnover between habitat patches within a region. This can be expressed as a similarity coefficient among species assemblages occurring in any two habitats, and is the result of historical events that define the link between regional and local processes. The greater the turnover between habitats, the greater gamma (regional) diversity, which in turn determines diversity at local scales. The relationship among these diversity measurements can be expressed as:

$$\text{Gamma Diversity} = \text{Alpha diversity} \times \text{Beta diversity}$$

Thus, beta diversity is inversely proportional to the average probability of any species' migration between any two habitat patches. This probability decreases as distances between habitats relative to species vagility increases. Metapopulation dynamics and connectivity between local populations becomes ever more tenuous as dispersal patterns continue and species distributions occupy wider ranges. In short, the probability of extrinsic barriers increases, resulting in allopatric distributions (i.e., vicariance and speciation). The resulting community structure is therefore open, and not fixed by trophic and guild interactions.

4.3 REGIONAL AND HISTORICAL FACTORS OF DIVERSITY

4.3.1 SPECIES AND SPECIATION

The uneven distribution of species diversity across regional scales is attributable to differential speciation and extinction analogous to the distribution of organismal abundances. Speciation is the reproduction of a successful way of life inasmuch as speciation rarely results in unique adaptations

(Gould, 1982). As the sole source for natural diversity, speciation must be considered in any biogeographic model. However, this assumes that biologists are in agreement on the definition and utility of the term "species." Herein, two comments on species — one of definition, the other of pragmatic utility — are offered.

First, the "species question" has received renewed interest over the past 20 years. What is a species and how do we recognize "good species?" Systematists have increasingly rejected the familiar Biological Species Concept (BSC) for other measurements of biodiversity, such as the Phylogenetic Species Concept, or the Evolutionary Species Concept. On the other hand, many non-systematists have retained the BSC for the obvious reason that their interests and problems lie elsewhere and are typically centered on localized scales where the time dimension is not a concern. I would argue that species should be regarded as "individuals" functioning in both ecological and evolutionary processes (Ghiselin, 1974; Hull, 1978) and must be accorded such status (Mayden and Wood, 1995). The alternative is to view species as class constructs, entities that do not function in any meaningful way in natural processes (Hull, 1978).

The majority of fishes encountered by biologists on this continent are sexually reproducing, and it is reasonable to assume that the BSC is applicable to these species. However, it would be a mistake to assume that "reproductive isolation" means absolutely no chance for reproduction. For example, walleye *Stizostedion vitreum* and sauger *S. canadense* occasionally hybridize in the wild, and their hybrids (saugeye) are propagated by fishery management agencies. No one would argue that the two parent species are conspecific; likewise for many species of recognized fishes. Interbreeding is possible or may even occur between good species, but only to the extent that the two remain evolutionarily distinct (Mayr, 1970; Wiley, 1981).

Reproductive isolation must be taken as *de facto* isolation. For example, the frecklebelly darter *Percina stictogaster* is represented by two disjunct populations, one in the Kentucky River and the other in the Green River of Kentucky. Genetic studies (Strange and Burr, 1997) have shown that these populations are independent of each other and represent independent lineages. Thus, they may be regarded as separate species because they are experiencing unique evolutionary events.

Secondly, management personnel possibly feel that the recent spate of new species descriptions and elevations of subspecies (often re-elevations) might in some way hinder management practices or at least add some confusion. There is no justification for this fear because most of these species are allopatric forms that can be regarded as ecological analogs (providing basic ecological research supports this) of their sister species. For example, describing the southern walleye *Stizostedion* sp. of the Mobile basin as a new species (*cf.* Billington and Strange, 1995) will not alter management practices or require the revision of IBI criteria. It is the only *Stizostedion* species in the drainage system. Diversity within the Mobile basin, only the number of fishes occurring in the state of Alabama (a political unit) will not increase.

4.3.2 Allopatric Speciation

Considering the modes in which species are formed, the most frequent mode of speciation among sexually reproducing vertebrates is allopatric (Mayr, 1970; Bush, 1975; Wiley, 1981), and among fish, closest relatives (i.e., sister species) typically occur in adjacent drainage systems (Wiley and Mayden, 1985). Modes of allopatric speciation describe how populations become geographically isolated. Wiley (1981) has characterized three models of allopatric speciation, which are presented here with some revision and examples from the North American fish fauna. Each has its unique set of geographic and phylogenetic predictions.

1. *Vicariance (model 1A sensu Bush 1975; Allopatric speciation model I sensu Wiley 1981).* This is the fragmentation of a previously continuous distribution by geological or climatic barriers. It results in the separation of two comparably sized populations of a single ancestral species. Phylogenetic topologies will conform to the order of vicariance events.

Geographic discontinuity with a relatively high degree of genetic divergence and reciprocal monophylly between populations suggests vicariance (Strange and Burr, 1997). Biogeographic and phylogenetic congruence is expected among broadly sympatric clades (i.e., those that co-occur across much of their joint range), as all should be subject to the same vicariance event regardless of their environmental adaptations (Wiley, 1981).

Vicariance has played a role in the separation and speciation among fishes restricted to the upland habitats of the Ozark and Appalachian highlands of eastern North America (Wiley and Mayden, 1985; Strange and Burr, 1997). These include the species pairs *Etheostoma nianguae* and *E. sagitta*, and *Percina cymatotania* and *P. stictogaster*. Other examples include much of the endemic fauna of the Mobile basin. The southern walleye is a sister species to *S. vitreum* and might be the product of a vicariant event between the Mobile basin and the greater Mississippi River basin (Billington and Strange, 1995).

2. *Speciation through peripheral isolation (e.g., peripatric or founder, sensu Mayr 1963; Model 1B sensu Bush 1975; Model II Allopatric sensu Wiley 1981)*. This is the isolation of a single deme or several closely situated demes marginal to the distribution of the ancestral species. This is the model favored by the punctuated equilibrium theory of evolution described by Eldredge and Gould (1972). Geographically, peripheral isolates form at the margin of the ancestral range and differentiate *in situ*. Character evolution can occur relatively faster in the isolate than the parent population because the isolate is represented by a smaller effective population size. The ancestor persists in this model (in terms of character states), and consequently, phylogenetic reconstructions may not necessarily reflect the pattern of peripheral isolation events.

Peripheral isolates may be formed through microvicariance or dispersal (Frey, 1993). Microvicariance is the formation of a peripheral isolate by the formation by geologic or climatic barriers, and differs from vicariance speciation in the size of the isolated population relative to the parent population (Brooks and McLennan, 1991). This model is independent of the species' vagility. Syntopic (those occurring in the same microhabitats) but not necessarily broadly sympatric species will be affected by a microvicariance event and form peripheral isolates in the same areas.

For fishes, isolation through stream capture is a microvicariant event. These events occur as the result of erosional processes whereby streams cut across drainage divides and divert or "capture" tributary streams from an adjacent drainage. As a result, fishes occurring within these diverted streams are introduced into new drainage systems as peripheral isolates of their parent populations. These events conform to the peripheral isolate model in that they involve one or very few demes. The Cumberland johnny darter *Etheostoma susanae* is a peripheral isolate of *E. nigrum*, having been transferred from the upper Kentucky River into the upper Cumberland River, where it differentiated (Starnes and Starnes, 1979; Strange, 1998). Other examples include the species pair *Phoxinus oreas* (New River drainage) and *P. tennesseensis* (Tennessee River drainage; Starnes and Etnier, 1986), where the implied mechanism for the dispersion of fishes is stream capture (Burr and Warren, 1986). It is argued elsewhere (Strange, 1995) that peripheral isolation through microvicariance is the most frequent mode of speciation among upland stream fishes.

Waif dispersal (*contra* dispersal in the sense of range expansion) is another form of peripheral isolation (Frey, 1993), and is the model for island biogeography (MacArthur and Wilson, 1967). Although long-distance dispersal across geographic barriers has been criticized as unlikely (Wiley and Mayden, 1985; Mayden, 1987), it may be inevitable over geologic time scales. Unlike the above models, speciation through dispersal is dependent on the species' vagility and environmental tolerances. Consequently, phylogenetic and biogeographic patterns among syntopic or broadly sympatric species will not be congruent. Waif dispersal may not have played much of a role in the diversification of North American fishes (Wiley and Mayden, 1985). The best example is the Cuban gar *Atractosteus tristoechus,* which is believed to be sister to the alligator gar *A. spatula* of the North American mainland (Wiley, 1976).

3. *Relictual speciation (Allopatric Model III sensu Wiley 1981).* This involves single demes in which gene flow becomes so restricted that it cannot maintain homogeneity across a species' range. This is typical of rapid changes in climate resulting in isolated and relictual populations. Under these conditions, most populations will share a common ancestor existing immediately prior to range retraction. Consequently, phylogenetic structure will be unresolvable. For example, desert pupfishes *Cyprinodon* spp. possess mitochondrial DNA haplotype assemblages with phylogenetic relationships that are either weakly resolved, incongruent with geographic hypotheses, or contradictory with allozyme data (*cf.* Echelle and Dowling, 1993; Echelle and Echelle, 1993). This is consistent with a broadly distributed species during humid time periods that experienced a range retraction following a climatic shift.

Parapatric speciation differs from allopatric speciation in that divergence between populations occur in spite of a narrow contact zone in which interbreeding continues. Distinguishing between this model and allopatric models is difficult, but can be hypothesized on the basis of the current distributions and geologic history of the region. An example of this model is the species pair *Etheostoma barrense* and *E. rafinesque.* These sister species occur in the Barren River and the upper Green River (plus the lower Barren River), respectively, a distribution consistent with both vicariance and parapatric speciation. However, there is no evidence that the two drainage systems were involved in a geological vicariance or microvicariance event.

The hierarchical metapopulation structures described earlier can be broken at any point in each of these models of speciation. Species are formed by the isolation of single demes, a set of demes, or entire metapopulations. Regions will exhibit speciation patterns and levels of diversity commensurate with historical climatic, geologic, or stream course disturbances. Local species assemblages, in turn, are determined by the pool of species available in any geographic region. Thus, speciation is a historical process typically realized on evolutionary time frames but affecting processes occurring on ecological time frames.

4.3.3 HISTORICAL ECOLOGY AND OHIO RIVER BASIN FISHES

The basic assumption behind historical ecology is that communities contain within them clues to their origin and construction as they represent historical species assemblages (Mayden, 1988; Brooks and McLennan, 1991; Gorman, 1992). One method of examining the historical component in community assembly utilizes a matrix of the presence (1) or absence (0) of a species within each community to estimate a phylogenetic tree among communities using the species as character states. Communities are then organized on the basis of uniquely shared species under the criterion of maximum parsimony. These groups represent long-standing relationships among organisms in terms of local-scale interactions. Thus, we should be able to discriminate between species assemblages that represent different ancestral communities and those that represent more recent origin.

The Ohio River basin is an excellent system in which to model historical components of community structure. It consists of two pre-Pleistocene drainage systems only recently integrated to form the present drainage system (Melhorn and Kempton, 1991; Figure 4.2). These ancestral drainage systems are roughly divided by the position of the Falls of the Ohio at Louisville, Kentucky. Stream systems west (downstream) of this point comprise the lower, or Old Ohio River; streams to the east represent the upper, or Teays River system. Further, the Ohio River basin may be divided into northern and southern components. Northward are those drainage systems of the formerly glaciated areas (glacial till plains); southward are those drainage systems of the unglaciated regions.

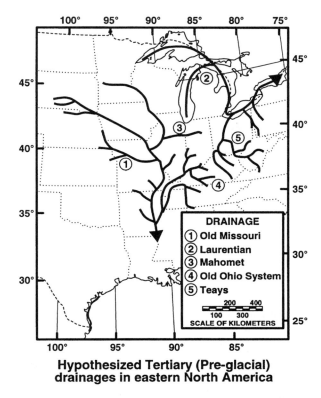

FIGURE 4.2 Pre-Pleistocene drainage patterns of the present Ohio River basin.

Thus, the Ohio River basin contains species assemblages that include pre-Pleistocene and post-Pleistocene communities.

The species considered in an analysis of historical ecology should be restricted to specific habitat types to avoid complications of differential vagilities of the component species. Therefore, scoring of the presence-absence of those fishes typically found in headwater (first- through third-order) streams of the Tennessee, Cumberland, Wabash, Green, Blue, Kentucky, Great Miami, Licking, Scioto, Big Sandy, Kanawha, Muskingum, Allegheny, and Monongahela Rivers was based on the species distribution maps published by Lee et al. (1980). Cumberland River samples were limited to below Cumberland Falls to avoid complications of stream-capture in the analysis (see below). The completed data set was run through PHYLIP's DOLLOP program (Felsenstein, 1995), which minimizes the number of times a species might reinvade a system after extinction. The resulting set of stream-system relationships (Figure 4.3) reflects the historical component in community assembly. Three clades form an unresolved trichotomy at the base, representing the Old Ohio, Teays, and those stream systems of the formerly glaciated regions of the upper Ohio River basin. This can be interpreted as both a conservatism in community structure (Old Ohio and Teays clades) among headwater stream fishes and the plasticity of dispersalist communities occupying the glacial till plains.

An alternative way in which to examine community structure is through the historical biogeography of individual species. The southern redbelly dace *Phoxinus erythrogaster* and johnny darter are broadly distributed in the Ohio River basin, and are not informative in the historical analysis above. However, mtDNA variation within each of these species has been examined and provides evidence for both pre-Pleistocene distributions and post-Pleistocene dispersal (Strange, in preparation).

Phoxinus erythrogaster has a well-resolved geographic structure of its mtDNA haplotypes among lower Ohio River stream systems (Figure 4.4A), whereas little such distinction occurs among

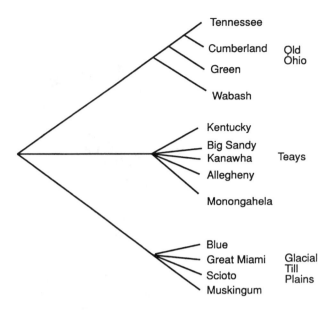

FIGURE 4.3 Cladistic relationships among stream fish communities in the Ohio River basin.

those populations in the upper Ohio River basin. Further, upper Ohio River populations of this species occur only in stream systems originating in the glacial till plains, indicative of a post-Pleistocene dispersal accompanying the integration of the present Ohio River basin. A similar pattern has been reported for the streamline chub *Erimystax dissimilis* (Strange and Burr, 1997). The mtDNA distribution pattern of johnny darter differs from these patterns in that a clear distinction exists between Teays and Old Ohio components of Ohio River basin forms (Figure 4.4B).

Community structure of Ohio River basin headwater stream fishes is the result of many historical events, both pre-Pleistocene and recent dispersal. It is likely that most communities contain both ancient and dispersalist components (Brooks and McLennan, 1991), as has been argued for the regional highland fish fauna (Strange and Burr, 1997). Co-occurring species should have similar biogeographic histories if fish communities represent co-adapted assemblages. Since this is not the case, community diversity must be seen as the result of a regional history that includes climatic shifts, drainage re-arrangements, and range fragmentation. This interpretation of diversity patterns among Ohio River basin fishes represents a combination of two distinct perspectives often contrasted by modern biogeographers. The first is island biogeography (MacArthur and Wilson, 1967), which assumes the stochastic movement of organisms over evolutionary time scales. Contrasting this is vicariance biogeography (e.g., Wiley and Mayden, 1985), which stresses that these movements are often constrained by geologic or climatic factors. Under either model, community structure and diversity are the result of historical factors.

4.3.4 STREAM CAPTURES AND THE UPPER CUMBERLAND RIVER

Stream capture is the major source of fish dispersal across drainage divides (Banarescu, 1990; Burr and Warren, 1986); however, its contribution to community structure has not been adequately demonstrated. Similar upland communities distributed across drainage divides are the result either of stream capture or range retraction following climatic shifts (i.e., relictual distributions). What is required is a model system in which the lower end is blocked to fish dispersal to test the role of stream capture. The upper Cumberland River is the classic example of stream capture in North America (e.g., Kuehne and Bailey, 1961; Burr and Page, 1986; Burr and Warren, 1986; Starnes and Etnier, 1986). It satisfies the requirements needed to test stream captures. Streams of the upper

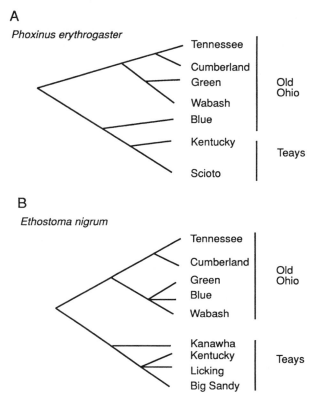

FIGURE 4.4 Phylogenetic relationships among mitochondrial DNA haplotypes of (A) *Phoxinus erythrogaster* and (B) *Etheostoma nigrum* in the Ohio River basin.

Cumberland River are blocked to fish dispersal by Cumberland Falls, a geologic formation 17 m in height (Burr and Warren, 1986). The Cumberland fish fauna is relatively depauperate (39 indigenous species) and includes three endemic forms (*Etheostoma sagitta*, *E. susanae*, and *Phoxinus cumberlandensis*), which is suggestive of an island habitat within the North American highlands. Notable is the lack of large river fishes within the upper Cumberland River system such as *Carpiodes*, *Ictiobus*, *Lepisosteus*, and *Ictalurus* (the latter two introduced in historic times; Burr and Warren, 1986). These species are relatively common below the Falls and in adjacent drainages. Phylogenetic relationships among fishes in the upper Cumberland River indicate that their origin can be traced to three pre-Pleistocene river systems: the Old Ohio River system, the Teays River system, and the Tennessee River (Strange and Burr, 1997; Strange, 1998; Strange, in preparation).

The adjacent Kentucky River system has similar stream morphology and water chemistry, yet has a different fish diversity. A total of 96 native species occur in tributaries of this system (Burr and Warren, 1986). Obviously, IBI criteria for the upper Cumberland River system must differ from the upper Kentucky River system even though the two share many physiological characteristics.

4.4 CONCLUSION

Aquatic communities are assemblages of species united by regional history. This is apparent in both regional historical ecology and comparative historical biogeography of individual fishes. Differing metapopulation structures among co-occurring species promotes (Beta) diversity across regional landscapes. The implication for management agencies in establishing IBI criteria is straightforward. History must be invoked to explain diversity.

Some (Matthews and Robison, 1988; Matthews et al., 1992) have demonstrated that fish assemblages within regions are correlated with water quality, stream type, etc., and management personnel are tempted to use such stream characterizations to establish IBI criteria. However, these particular studies focus on variation within a particular drainage system. Comparisons across drainages are more problematic. It has become clear that drainage systems are historical units that are the proper subjects to consult regarding diversity expectations. Habitat quality may determine what species may persist in a given stream, but regional history determines which and how many species may occupy a habitat. Thus, equivalent habitats in different drainage systems should not be expected to have the same degree of diversity. Some places are just more diverse than others.

REFERENCES

Banarescu, P. 1990. *Zoogeography of Freshwaters.* AULA-Verlag, Wiesbaden.
Billington, N. and R.M. Strange. 1995. Mitochondrial DNA analysis confirms the existence of a genetically divergent walleye population in northeastern Mississippi, *Transactions of the American Fisheries Society,* 124, 770–776.
Brooks, D.R. and D.A. McLennan. 1991. *Phylogeny, Ecology, and Behavior: A Research Program in Comparative Ecology.* University of Chicago Press, Chicago, IL.
Burr, B.M. and L.M. Page. 1986. Zoogeography of the fishes of the lower Ohio-upper Mississippi River basin, in C.H. Hocutt and E.O. Wiley (Eds.), *Zoogeography of North American Freshwater Fishes.* John Wiley & Sons, New York, NY. 287–324.
Burr, B.M. and M.L. Warren. 1986. A distributional atlas of Kentucky fishes, *Kentucky State Nature Preserves Commission Scientific and Technical Series,* 4, 1–398.
Bush, G.L. 1975. Modes of animal speciation, *Annual Review of Ecology and Systematics,* 6, 339–364.
Echelle, A.A. and T.E. Dowling. 1992. Mitochondrial DNA variation and evolution of the Death Valley pupfishes (Cyprinodontidae: *Cyprinodon*), *Evolution,* 46, 193–206.
Echelle, A.A. and A.F. Echelle. 1993. Allozyme perspective on mitochondrial DNA variation and evolution of the Death Valley pupfishes (Cyprinodontidae: *Cyprinodon*), *Copeia,* 1993, 275–287.
Eldredge, N. and S.J. Gould. 1972. Punctuated equilibria: an alternative to phyletic gradualism, in T.J.M. Schopf (Ed.), *Models in Paleobiology.* Freeman Cooper, San Francisco, CA. 82–115.
Fausch, K.D., J.R. Karr, and P.R. Yant. 1984. Regional application of an index of biotic integrity based on stream fish communities, *Transactions of the American Fisheries Society,* 113, 39–55.
Felsenstein, J. 1995. PHYLIP: the phylogeny estimation package manual, version 3.5c. Distributed by the author. University of Washington, Seattle, WA.
Frey, J.K. 1993. Modes of peripheral isolate formation and speciation, *Systematic Biology,* 42, 373–381.
Ghiselin, M.T. 1974. A radical solution to the species problem, *Systematic Zoology,* 23, 536–544.
Gorman, O.T. 1992. Evolutionary ecology and historical ecology: assembly, structure, and organization of stream fish communities, in R.L. Mayden (Ed.), *Systematics, Historical Ecology, and North American Freshwater Fishes.* Stanford University Press, Stanford, CA. 659–688.
Gould, S.J. 1982. Darwinism and the expansion of evolutionary theory, *Science,* 216, 380–387.
Harrison, S. and A. Hastings. 1996. Genetic and evolutionary consequences of metapopulation structure, *Trends in Ecology and Evolution,* 11, 180–183.
Hocutt, C.H. and E.O. Wiley. 1986. *Zoogeography of North American Freshwater Fishes.* John Wiley & Sons, New York.
Hoffman, A.A. and M.W. Blows. 1994. Species borders: ecological and evolutionary perspectives, *Trends in Ecology and Evolution,* 9, 223–227.
Hull, D. 1978. A matter of individuality, *Philosophy of Science,* 45, 335–360.
Karr, J.R. 1981. Assessment of biotic integrity using fish communities, *Fisheries (Bethesda),* 6, 21–27.
Karr, J.R., K.D. Fausch, P.L. Angermeier, P.R. Yant, and I.J. Schlosser. 1986. Assessing Biological Integrity in Running Waters: A Method and Its Rationale. Illinois Natural History Survey Special Publication Number 5.
Kuehne, R.L. 1962. A classification of streams illustrated by fish distribution in an eastern Kentucky stream, *Ecology,* 43, 608–614.

Kuehne, R.A. and R.M. Bailey. 1961. Stream capture and the distribution of the percid fish *Etheostoma sagitta*, with geologic and taxonomic considerations, *Copeia,* 1961, 1–8.

Kirkpatric, M. and N.H. Barton. 1997. Evolution of a species' range, *American Naturalist,* 150, 1–23.

Lee, D.S., C.R. Gilbert, C.H. Hocutt, R.E. Jenkins, D.E. McAllister, and J.R. Stauffer, Jr. 1980. *Atlas of North American Fresh Water Fishes.* North Carolina Biological Survey Number 1980-12.

Levins, R.C. 1969. Some demographic and genetic consequences of environmental heterogeneity for biological control, *Bulletin of the Entomological Society of America,* 15, 237–240.

MacArthur, R.H. 1972. *Geographical Ecology: Patterns in the Distributions of Species.* Harper and Row, New York.

MacArthur, R.H. and E.O. Wilson. 1967. *The Theory of Island Biogeography.* Princeton University Press, Princeton, NJ.

Matthews, W.J. and H.W. Robison. 1988. The distribution of the fishes of Arkansas: a multivariate analysis, *Copeia,* 1988, 358–374.

Matthews, W.J., D.J. Hough, and H.W. Robison. 1992. Similarities in fish distribution and water quality patterns in streams of Arkansas: congruence of multivariate analysis, *Copeia,* 1992, 296–305.

Mayden, R.L. 1988. Vicariance biogeography, parsimony, and evolution in North American freshwater fishes, *Systematic Zoology,* 37, 329–355.

Mayden, R.L. 1992. *Systematics, Historical Ecology, and North American Freshwater Fishes.* Stanford University Press, Stanford, CA.

Mayden, R.L. and R.M. Wood 1995. Systematics, species concepts, and the evolutionary significant unit in biodiversity and conservation biology, *American Fisheries Society Symposium,* 17, 58–113.

Mayr, E. 1970. *Populations, Species and Evolution.* Harvard University Press, Cambridge, MA.

Melhorn, W.N. and J.P. Kempton. 1991. The Teays system: a summary, *Geological Society of America Special Paper,* 258, 125–128.

Opdam, P. 1991. Metapopulation theory and habitat fragmentation: a review of holarctic breeding bird studies, *Landscape Ecology,* 5, 93–106.

Osborne, L.L. and M.J. Wiley. 1992. Influence of tributary spatial position on the structure of warmwater fish communities, *Canadian Journal of Fisheries and Aquatic Sciences,* 49, 671–681.

Ricklefs, R.E. 1989. Speciation and diversity: the integration of local and regional processes, in D. Otte and J.A. Endler (Eds.), *Speciation and Its Consequences.* Sinauer Publishers, Sunderland, MA. 599–622.

Richardson, L.R. and J.R. Gold. 1995. Evolution of the *Cyprinella lutrensis* species group. III. Geographic variation in the mitochondrial DNA of *Cyprinella lutrensis* — the influence of Pleistocene glaciation on population dispersal and divergence, *Molecular Ecology,* 4, 163–171.

Starnes, W.C. and D.A. Etnier. 1986. Drainage evolution and fish biogeography of the Tennessee and Cumberland rivers drainage realm, in C.H. Hocutt and E.O. Wiley (Eds.), *The Zoogeography of North American Freshwater Fishes.* John Wiley & Sons, New York, NY. 325–361.

Starnes, W.C. and L.C. Starnes. 1979. Taxonomic status of the percid fish *Etheostoma nigrum susanae, Copeia,* 1979, 426–430.

Strange, R.M. 1995. Intraspecific Phylogeography of North American Highland Fishes. Masters thesis, Southern Illinois University, Carbondale, IL.

Strange, R.M. 1998. Mitochondrial DNA variation in johnny darters (Pisces: Percidae) from eastern Kentucky supports stream capture for the origin of upper Cumberland River fishes, *American Midland Naturalist,* (in press).

Strange, R.M. and B.M. Burr. 1995. Genetic Variability and Metapopulation Dynamics in the Federally Threatened Blackside Dace, *Phoxinus cumberlandensis* (Pisces: Percidae). Kentucky Fish and Wildlife Resources, Frankfort, KY.

Strange, R.M. and B.M. Burr. 1996. Preliminary Studies of Genetic Variability in the Federally Threatened Niangua Darter, *Etheostoma nianguae.* Missouri Department of Conservation, Camdenton, MO.

Strange, R.M. and B.M. Burr. 1997. Intraspecific phylogeography of North American highland fishes: a test of the Pleistocene vicariance hypothesis, *Evolution,* 51, 885–897.

Tibits, C.A. and T.E. Dowling. 1996. Effects of intrinsic and extrinsic factors on population fragmentation in three species of North American minnows (Teleostei: Cyprinidae), *Evolution,* 50, 1280–1292.

Whittaker, R.H. 1972. Evolution and measurement of species diversity, *Taxon,* 21, 213–251.

Wiley, E.O. 1976. The phylogeny and systematics of fossil and recent gars (Actinopterygii: Lepisosteidae). *Miscellaeneous Publications of the Museum of Natural History, University of Kansas,* 64, 1–111.

Wiley, E.O. 1981. *Phylogenetics: The Theory and Practice of Phylogenetic Systematics.* John Wiley & Sons, New York.

Wiley, E.O. and R.L. Mayden. 1985. Species and speciation in phylogenetic systematics, with examples from the North American fish fauna, *Annals of the Missouri Botanical Garden,* 72, 596–635.

Yodzis, P. 1993. Environment and trophodiversity, in R.E. Ricklefs and D. Schluter (Eds.), *Species Diversity in Ecological Communities: Historical and Geographical Perspectives.* University of Chicago Press, Chicago, IL. 26–38.

5 Applications of IBI Concepts and Metrics to Waters Outside the United States and Canada

Robert M. Hughes and Thierry Oberdorff

CONTENTS

5.1 Introduction 79
5.2 Applications in Countries Outside the United States and Canada 82
    5.2.1 Taxonomic Richness 82
        5.2.1.1 Species Richness 82
    5.2.2 Species Composition (Habitat Guilds) 82
        5.2.2.1 Darter (Benthic) Species Richness 82
        5.2.2.2 Sunfish (Water Column) Species Richness 85
        5.2.2.3 Sucker (Large Long-Lived) Species Richness 86
        5.2.2.4 Intolerant Species Richness 86
        5.2.2.5 Percent Green Sunfish (Tolerant) Individuals 86
    5.2.3 Trophic Composition 87
        5.2.3.1 Percent Omnivorous Individuals 87
        5.2.3.2 Percent Insectivorous Cyprinid (Invertivorous) Individuals 87
        5.2.3.3 Percent Piscivorous (Top Carnivore) Individuals 88
    5.2.4 Individual Abundance and Condition 88
        5.2.4.1 Number of Individuals (Catch Per Unit Effort) 88
        5.2.4.2 Percent Hybrid Individuals (Reproductive Function) 88
        5.2.4.3 Percent Individuals with Anomalies 89
5.3 Conclusions 89
Acknowledgments 90
References 91

5.1 INTRODUCTION

There is a growing need for quantitative, easily implemented, and sensitive biological measures of aquatic ecosystem integrity in both industrialized and developing countries. In highly industrialized countries, water policy is currently changing. Until recently, water quality standards were contingent on defined uses, which were typically human oriented (drinking water, fishing, swimming) or extremely ill defined (aquatic life, fish passage). This kind of policy proved successful in the fight against point source pollution, but appears poorly adapted to integrated management of aquatic ecosystems, which is essential for sustainable conservation of water resources. Increasingly, the goal is not only to preserve ecosystems, but also to rehabilitate them and to restore their natural ecological structures, functions, and integrity. In developing countries, water policies that attempt to anticipate or predict significant economic, social, and ecological impacts rather than react to them, are increas-

TABLE 5.1
IBI Metrics for Midwestern U.S. Streams

Category	Scoring Criteria[a]		
Metric	5	3	1
Species richness			
1. Total number of fish species	b	b	b
2. Number of darter species	b	b	b
3. Number of sunfish species	b	b	b
4. Number of sucker species	b	b	b
Habitat guilds			
5. Number of intolerant species	b	b	b
6. % Individuals as green sunfish	<5	5–20	>20
Trophic guilds			
7. % Individuals as omnivores[c]	<20	20–45	>45
8. % Individuals as insectivorous cyprinids	>45	20–45	<20
9. % Individuals as piscivores[d]	>5	1–5	<1
Abundance			
10. Number of individuals	b	b	b
Reproduction and Condition			
11. % Individuals as hybrids	0	>0–1	>1
12. % Individuals with anomalies[e]	0-2	>2–5	>5

[a] Value approximates (5), deviates somewhat (3), or deviates strongly (1) from the reference condition.
[b] Expected value varies with stream size, region, and basin.
[c] Adult diets typically include ≥25 % plant and ≥25% animal material.
[d] Adult diets usually composed largely of aquatic vertebrates or crayfish.
[e] Disease, eroded fins, lesions, tumors, discoloration, excessive mucous, skeletal abnormalities, missing organs, and other external symptoms.

From Karr et al., 1986; Miller et al., 1988.

ingly necessary for avoiding extreme environmental degradation (Hocutt et al., 1994). In both highly industrialized and less developed countries, it is necessary to develop practical tools using biological community or assemblage-based approaches for monitoring water resource quality.

For aquatic ecosystems, biological indicators can be chosen from many assemblages, but fish are of particular interest because (1) they are present in most water bodies, (2) their taxonomy, ecological requirements, and life histories are generally better known than those of other assemblages, (3) they occupy a variety of trophic levels and habitats, and (4) they have both economic and aesthetic values and thus help raise awareness of the value of conserving aquatic systems.

One integrated tool to quantify the impact of human activities on aquatic ecosystems via fish asssemblages is the Index of Biotic Integrity (IBI, Table 5.1), first formulated by Karr (1981). A similar approach, based on benthic macroinvertebrates, has also been developed (Plafkin et al., 1989; Kerans and Karr, 1994; DeShon, 1995; Fore et al., 1996). The IBI is based on the hypothesis that there are predictable relationships between fish assemblage structure and the physical, chemical, and biological condition of stream systems. The IBI methodology is outlined in Table 5.2.

The IBI was originally developed for midwestern U.S. streams, and integrated 12 attributes of fish assemblages to determine biotic integrity or "health" of the system (Table 5.1). However, Karr (1996) distinguishes between these two concepts, reserving integrity for more-natural ecological conditions, and health for conditions desired by humans but not necessarily natural. Classes of attributes (metrics) in the IBI include species richness, species composition, trophic structure, fish abundance, and fish condition. Each metric reflects the quality of a different aspect of the fish

TABLE 5.2
Principles of Fish Assemblage Assessment with the IBI

1. Select a relatively homogeneous region. A region may be an ecoregion, basin, or fish faunal region that is homogeneous with respect to a combination of environmental characteristics (e.g., climate, physiography, soil, vegetation) and potential fish species.
2. Determine the reference condition(s). References may be based on a set of minimally disturbed reference streams, a disturbance gradient, historical data, paleoecological information, and professional judgment. Expectations will likely differ for water body size, gradient, temperature, or other naturally limiting variables.
3. List candidate metrics and assign species to trophic, tolerance, and habitat guilds. Regional fish texts usually provide this information, at least in developed countries.
4. Sample fish assemblages. This is best done (a) when they are least variable yet most limited by anthropogenic stressors and (b) in a manner yielding a representative collection of species and proportionate abundances, but that (c) is cost-effective.
5. Tabulate numbers of individuals collected by species. Also, determine the total number of individuals collected at each reach.
6. Calculate values for each candidate IBI metric. Typically these are proportions or percents of individuals, or numbers of species in particular categories.
7. Develop scoring criteria. These are based on previously available information from step 2 or from fish data collected at minimally disturbed sites in step 4. Scoring criteria may be continuous (0–1 or 0–10) or based on classes (1, 3, 5 or 0, 5, 10).
8. Calculate metric scores and add these to obtain an IBI score.
9. Evaluate metric and index scores. Consider differences between expected and obtained scores, compare variance results from repeated samples, and assess responsiveness to environmental stressors. Modify or reject metrics that are highly variable or unresponsive, and recalculate if necessary.
10. Interpret IBI score as indicating an acceptable, marginally impaired, or highly impaired fish assemblage; or as excellent, good, fair, poor, or very poor.

assemblage that responds in a different manner to aquatic ecosystem stressors (Margalev, 1963; Fausch et al., 1990; Hughes and Noss, 1992). The combination of metrics reflects insights from individual, population, assemblage, ecosystem, and zoogeographic perspectives (Miller et al., 1988). Furthermore, there is no loss of information as the structure and calculation of IBI preserves both the original data and also provides a metric-by-metric evaluation of stream condition (Harris, 1995). The primary underlying assumptions of the IBI concept are presented in Table 5.3.

Since its introduction, the IBI has been modified almost annually for use in other regions and types of ecosystems throughout the U.S. (Fausch et al., 1984; Leonard and Orth, 1986; Moyle et al., 1986; Thompson and Fitzhugh, 1986; Hughes and Gammon, 1987; Ohio EPA, 1987; Schrader, 1989; Allen, 1989; Fisher, 1989; Langdon, 1989; Crumby et al., 1990; Saylor and Ahlstedt, 1990; Bramblett and Fausch, 1991; Simon, 1991, 1992; Dionne and Karr, 1992; Lyons, 1992; Hoefs and Boyle, 1992; Bailey et al., 1993; Deegan et al., 1993; Goldstein et al., 1994; Lyons et al., 1996) and in Canada (Steedman, 1988; Minns et al., 1994; Richard, 1994). It has also been modified for use outside the U.S. and Canada (Oberdorff and Hughes, 1992; Gutierrez, 1994; Oberdorff and Porcher, 1994; Harris, 1995; Lyons et al., 1995; Didier et al., 1996; Hugueny, 1996; Hocutt et al., 1994; Ganasan and Hughes, 1998). These applications have shown that the IBI concept is widely adaptable, but that metrics must be modified, deleted, or added to reflect regional differences in fish distribution and assemblage structure (Miller et al., 1988; Simon and Lyons, 1995).

All these IBI modifications have retained the same ecological framework, in which fish assemblage data are ordered and interpreted based on their compositional, structural, and functional organization (Fausch et al., 1990; Lyons et al., 1995). The objective in this chapter is to summarize present IBI applications outside the U.S. and Canada, with particular emphasis on the way metrics have been modified while retaining the IBI's theoretical foundations. The goal in doing so is to encourage further applications in other countries.

TABLE 5.3
Assumed Effects of Environmental Degradation on Fish Assemblages

Number of native species, and of those in specialized taxa or guilds, declines[a]
Number of sensitive species declines
Percent of trophic and habitat specialists declines
Total number of individuals declines[a]
Percent of large individuals and the number of size classes decrease
Percent of alien or non-native species or individuals increases
Percent of tolerant individuals increases
Percent of trophic and habitat generalists increases
Percent individuals with anomalies increases

[a] In some waters, especially oligotrophic coldwater systems, increased nutrients and temperatures often result in additional species and individuals.

From Margalef, 1963; Fausch et al., 1990; Hughes and Noss, 1992.

5.2 APPLICATIONS IN COUNTRIES OUTSIDE THE UNITED STATES AND CANADA

Our review located 10 applications of the IBI outside the U.S. and Canada. These applications have occurred in small rivers and wadeable streams on six continents: Europe (Oberdorff and Hughes, 1992; Oberdorff and Porcher, 1994; Oberdorff, 1996, in France; Didier et al., 1996, in Belgium), Africa (Hugueny, 1996, in Guinea; Hocutt et al., 1994, in Namibia), Asia (Ganasan and Hughes, 1998, in India), South America (Gutierrez, 1994, in Venezuela), Australia (Harris, 1995), and North America (Lyons et al., 1995, in Mexico). In the remainder of this section, we address the IBI modifications metric by metric (summarized in Table 5.4).

5.2.1 TAXONOMIC RICHNESS

5.2.1.1 Species Richness

This metric is a common measure of biological diversity that generally declines with environmental degradation (Karr, 1981). It is typically retained in all applications. The major variant was native species richness, which was used by Oberdorff and Porcher (1994), Lyons et al. (1995), Didier et al. (1996), and Hocutt et al. (1994). The only addition was family richness, which was used by Ganasan and Hughes (1998). We believe that restricting this metric to native species is a useful improvement, especially where nonnative species are fairly common or highly invasive. Family richness, or perhaps even order richness in the tropics, seems appropriate for assessing alterations at this level of taxonomic diversity. Oberdorff and Hughes (1992) found that the more primitive families of fishes were the most seriously affected by anthropogenic disturbances in the Seine River basin in France.

5.2.2 SPECIES COMPOSITION (HABITAT GUILDS)

5.2.2.1 Darter (Benthic) Species Richness

Karr (1981) selected this metric because of the sensitivity of darters to the degradation of benthic habitats. Karr et al. (1986) suggested cottid richness or richness of benthic fishes where darters are absent. Benthic species richness was applied by Oberdorff and Hughes (1992), Oberdorff and Porcher (1994), Didier et al. (1996), and Ganasan and Hughes (1998). The number of benthic specialist species was used by Hocutt et al. (1994) and Oberdorff (1996), and percent native benthic

TABLE 5.4
Metric Changes for IBIs Developed in Countries Other than the U.S. and Canada

Original Metric	Substitute	Ref. (Country)
# Fish species	No change	Oberdorff and Hughes (France)
		Oberdorff (France)
		Hugueny et al. (Guinea)
		Ganasan and Hughes (India)[a]
		Harris (Australia)
		Gutierrez (Venezuela)
	# Native fish species	Oberdorff and Porcher (France)
		Didier et al. (Belgium)
		Hocutt et al. (Namibia)
		Lyons et al. (Mexico)
# Darter species	# Benthic species	Oberdorff and Hughes
		Oberdorff and Porcher
		Didier
		Ganasan and Hughes
	# Benthic specialists	Oberdorff
		Hocutt et al.
	# Riffle benthic species	Harris
	% Native benthic individuals	Lyons et al.
	# Mormyrid species	Hugueny
	# Characidiin/parodontin species	Gutierrez
# Centrarchid species	# Water column species	Oberdorff and Hughes
		Ganasan and Hughes
		Lyons et al.
	# Cichlid species	Hugueny
		Hocutt et al.
		Gutierrez
	# Pelagic pool species	Harris
	# Pelagic/rheophilic species	Oberdorff
	Deleted	Oberdorff and Porcher
		Didier et al.
# Sucker species	# Trout/pike age classes	Oberdorff and Hughes
	# Trout/pike/perch ages	Oberdorff
	# Trout age classes	Oberdorff and Porcher
	# Dom./intol. age classes	Didier et al.
	# Large siluriform species	Hugueny et al.
	# Loricariid species	Gutierrez
	# Pool benthic species	Harris
	# Pelagic/rheophilic species	Hocutt et al.
	Deleted	Ganasan and Hughes[b]
		Lyons et al.[b]
# Intolerant species	No change	Oberdorff and Hughes
		Oberdorff
		Didier et al.[c]
		Ganasan and Hughes
		Harris
		Gutierrez
	# Sensitive species	Lyons et al.
	% Sculpin individuals	Oberdorff and Porcher

TABLE 5.4 *(continued)*
Metric Changes for IBIs Developed in Countries Other than the U.S. and Canada

Original Metric	Substitute	Ref. (Country)
	Deleted	Hugueny et al.
		Hocutt et al.
% Green sunfish	% Tolerant individuals	Ganasan and Hughes
		Lyons et al.
	% Roach	Oberdorff and Hughes
		Oberdorff
	% Eel and roach	Oberdorff and Porcher
	% Alien/invasive Individuals	Didier et al.
	% Native individuals	Harris
	% Dominants	Gutierrez
	Deleted	Hugueny et al.
		Hocutt et al.
% Omnivore individuals	No change	Oberdorff and Hughes
		Oberdorff
		Oberdorff and Porcher
		Didier et al.
		Hugueny et al.
		Ganasan and Hughes
		Lyons et al.
	% Omnivorous species	Gutierrez[d]
	% Scavenger individuals	Hocutt et al.[e]
	% Microphagic omnivorous individuals	Harris
% Insectivorous Cyprinid individuals	% Invertivorous individuals	Oberdorff and Hughes
		Oberdorff
		Oberdorff and Porcher
		Hugueny et al.
		Hocutt et al.
	% Microphagic carnivorous individuals	Harris
	% Herbivore individuals	Ganasan and Hughes
	% Insectivore species	Gutierrez
	Deleted	Lyons et al.
		Didier et al.
% Piscivorous individuals	No change	Oberdorff and Hughes
		Hugueny et al.
		Hocutt et al.
		Ganasan and Hughes
	% Piscivorous/invertivorous individuals	Didier et al.
	% Macrophagic carnivorous individuals	Harris
	% Piscivorous species	Gutierrez
	Deleted	Oberdorff
		Oberdorff and Porcher
		Lyons et al.
# Individuals	No change	Hugueny et al.
		Hocutt et al.
		Ganasan and Hughes
		Harris
	Catch/effort	Oberdorff and Hughes
		Lyons et al.

TABLE 5.4 *(continued)*
Metric Changes for IBIs Developed in Countries Other than the U.S. and Canada

Original Metric	Substitute	Ref. (Country)
	# Individuals/100 m^2	Oberdorff
		Oberdorff and Porcher[f]
	Biomass	Didier et al.
	Deleted	Gutierrez
% Hybrid individuals	No change	Hugueny et al.
	% Gravel spawning individuals	Oberdorff and Hughes
	% Generalist spawning individuals	Didier et al.[g]
	% Native livebearing individuals	Lyons et al.[h]
	% Alien/invasive individuals	Hocutt et al.
	% Alien individuals	Ganasan and Hughes
	% Introduced species	Gutierrez
	% Native species	Harris
	Deleted	Oberdorff
		Oberdorff and Porcher
% Anomalies	No change	Oberdorff and Hughes
		Oberdorff
		Oberdorff and Porcher
		Hugueny et al.
		Hocutt et al.
		Ganasan and Hughes
		Lyons et al.
		Harris
		Gutierrez
	Deleted	Didier et al.

[a] Added # of families.
[b] Included in # benthic species metric or % benthic individuals metric.
[c] Added % intolerant individuals.
[d] Added % species as herbivores, % species as detritivores, and % species as parasites.
[e] Added % herbivore/detritivore individuals.
[f] Added total biomass.
[g] Added # of alien/invasive species and % specialist spawner individuals.
[h] Added % alien species.

individuals was proposed by Lyons et al. (1995) because there were too few benthic species in the Mexican streams they studied. Lyons et al. (1995) and Ganasan and Hughes (1998) included both large and small benthic species in this metric, rather than separate them into separate darter and sucker substitute metrics. Harris (1995) proposed number and identity of riffle-dwelling benthic species. Hugueny et al. (1996) introduce mormyrid species richness, while Gutierrez (1994) substituted characidiid and parodontin species richness. All of these substitutes satisfy the objective of assessing the degree to which benthic habitats and the species requiring them are perturbed. Lyons et al. (1995) were perhaps most creative with this metric, using percent benthic individuals instead of benthic species richness because their sites supported only one to three benthic species.

5.2.2.2 Sunfish (Water Column) Species Richness

These fishes are sensitive to the degradation of pool habitats and instream cover (Karr, 1981); other pool-dwelling species were suggested where sunfish diversity is low (Karr et al., 1986). Water column species richness was used by Oberdorff and Hughes (1992), Lyons et al. (1995), and Ganasan and Hughes (1998). Harris (1995) used pelagic pool species. Cichlid species richness was chosen by Gutierrez (1994), Hugueny et al. (1996), and Hocutt et al. (1994) — all working in tropical or subtropical systems. Oberdorff (1996) substituted number of pelagic rheophilic species. On the other hand, Oberdorff and Porcher (1994) and Didier et al. (1996) did not include this metric or a substitute. All four substitutes were appropriate. Despite the frequent occurrence of water column species (salmonids, cyprinids) at their study sites, Oberdorff and Porcher (1994) and Didier et al. (1996) did not explain why they did not use this metric. A water column metric is recommended unless these species are absent, unresponsive to disturbance, or highly variable.

5.2.2.3 Sucker (Large Long-lived) Species Richness

Karr (1981) proposed the sucker (catostomid) metric because these species are sensitive to physical and chemical degradation of waters and they also offer a means to integrate such disturbances over multiple years because of their long lives. Oberdorff and Hughes (1992) used number of trout or pike (*Esox*) age classes; Oberdorff and Porcher (1994) applied number of trout age classes; Didier et al. (1996) applied dominant and intolerant species age classes; and Oberdorff (1996) applied trout, pike, or perch (*Perca*) age classes. Working in tropical waters, Gutierrez (1994) proposed number of loricariids, and Hugueny et al. (1996) substituted number of large siluriform species. Harris (1995) used number of pool-dwelling benthic species, and Hocutt et al. (1994) substituted the number of pelagic rheophilic species. Lyons et al. (1995) included suckers in their percent benthic individuals metric, and Ganasan and Hughes (1998) incorporate all benthic species in their benthic species metric. Both the latter applications occurred with streams in which large, long-lived species were uncommon, but the use of this metric or a substitute such as any of those used above is supported whenever possible.

5.2.2.4 Intolerant Species Richness

This metric was designed to distinguish sites of the highest quality. Only species sensitive to a variety of common and widespread disturbances — particularly sedimentation, turbidity, decreased dissolved oxygen, and warming — were included. Such species are not those particularly sensitive to specific toxic chemicals. Intolerant species are those that first decline with environmental degradation (Karr, 1981; 1986). This metric was used by Oberdorff and Hughes (1992), Gutierrez (1994), Harris (1995), Didier et al. (1996), Oberdorff (1996), and Ganasan and Hughes (1998), while Lyons et al. (1995) proposed number of sensitive species. It was substituted with percent sculpin (*Cottus*) individuals by Oberdorff and Porcher (1994). Didier et al. (1996) also added percent intolerant individuals. The metric was deleted by Hocutt et al. (1994) and Hugueny et al. (1996), who felt that the biology of fishes in Africa was not sufficiently well-known to support designating some species as intolerant. As defined, this metric is extremely useful for detecting the initial signs of ecosystem perturbation and it should be retained if possible. Although Hugueny et al. (1996) found minimal published data on species tolerances, their own data indicated that a percent mormyrid metric would be appropriate, similar to how Oberdorff and Porcher (1994) used percent sculpins. Similar research at reference sites and along a disturbance gradient in southern Africa would likely provide candidates for an intolerant species metric in that region.

5.2.2.5 Percent Green Sunfish (Tolerant) Individuals

Karr (1981) chose the green sunfish (*Lepomis cyanellus*) to represent the degree that a tolerant species increases in relative abundance with disturbance, while Karr et al. (1986) suggested substituting percent tolerant individuals. Lyons et al. (1995) and Ganasan and Hughes (1998) used percent tolerant individuals. Oberdorff and Hughes (1992) and Oberdorff (1996) substituted percent roach (*Rutilus rutilus*), Oberdorff and Porcher (1994) used percent eel (*Anguilla*) and roach, Didier et al. 1996 substituted number of introduced or invasive species and added percent of individuals as introduced or invasive species, while Gutierrez (1994) substituted percent dominant species. Harris (1995) used percent of individuals that are native, which is the complement of percent non-natives. Neither this metric nor a substitute was used by Hocutt et al. (1994) or Hugueny et al. (1996), again because of their uncertainties about the tolerances of African fish species. Both the percent roach and percent eel and roach metrics meet the criteria for this metric; both species can become very abundant in highly polluted waters. Where a single species is not as widely distributed and abundant, the percent tolerant individuals suffices. There is concern with the use of the other substitutes. An introduced/invasive, percent dominant, or percent native metric will only satisfy the requirement for a metric assessing the degraded end of disturbance gradients if those species are also highly tolerant of substantial human perturbations. There are many commonly introduced and occasionally dominant species that are not highly tolerant, and even intolerant, as well as some that are tolerant. If they are tolerant species, it is recommended that they be used directly and not confounded in a metric with other species that are not.

5.2.3 TROPHIC COMPOSITION

5.2.3.1 Percent Omnivorous Individuals

Percent omnivores was proposed by Karr (1981) to assess the degree that the food base is altered to favor species that can digest considerable amounts of both plant and animal foods. It is used in all the non-U.S. IBI adaptations except by Gutierrez (1994), who chose percent species as omnivores; Harris (1995), who prefered percent microphagic omnivores; and Hocutt et al. (1994), who substituted percent opportunistic scavengers. All three substitutes seem appropriate. Our experience elsewhere indicates that this metric is best based on percent individuals vs. percent species, but Gutierrez (1994) lacked abundance data.

5.2.3.2 Percent Insectivorous Cyprinid (Invertivorous) Individuals

This metric was proposed by Karr (1981) as a surrogate for evaluating the degree that the invertebrate assemblage is degraded by environmental changes. In addition, in most U.S. and Canadian streams, invertivores or insectivores are the dominant trophic group of fishes, which makes their assessment important. Oberdorff and Hughes (1992), Oberdorff and Porcher (1994), Oberdorff (1996), Hocutt et al. (1994), and Hugueny et al. (1996) substituted percent invertivores, while Harris (1995) used percent of individuals as microphagic carnivores, also emphasizing that these fishes consume other invertebrates in addition to insects. Gutierrez (1994) used percent insectivorous species. Ganasan and Hughes (1998) substituted percent herbivores for central India rivers, proposing that this guild may be more sensitive than invertivores to disturbances in some tropical rivers. Neither percent insectivorous cyprinids nor a substitute was used by Lyons et al. (1995) or Didier et al. (1996). We agree with the majority here: percent invertivores or microphagic carnivores is preferable and more ecologically accurate than percent insectivores or percent insectivorous cyprinids. A percent herbivore metric as proposed by Gutierrez (1994) and Ganasan and Hughes (1998), or a percent herbivore/detritivore metric (Hocutt et al., 1994), seems promising in tropical and subtropical rivers where such species are important trophic components. Lyons et al. (1995) also reported that herbivores usually occurred at their least degraded sites in Mexico. They also found invertivores absent from sites with naturally depauperate fish assemblages and dropped the metric; it is unclear why Didier et al. (1996) did not include invertivores in Belgium, while Oberdorff and colleagues did so in France. It is recommended that wherever the fauna is rich enough, that invertivores or some substitute group of small organism or specialized feeders be evaluated as metrics.

5.2.3.3 Percent Piscivorous (Top Carnivore) Individuals

These species indicate a trophically diverse assemblage (Karr, 1981); they are susceptible to the bioaccumulation of persistent toxics and, being typically long-lived species, they are affected by long-term physical and chemical habitat alterations. Also, piscivores are often popular game species, and therefore susceptible to exploitation and hatchery stressors. Four authors used the percent piscivore or percent top carnivore metric (Oberdorff and Hughes, 1992; Hugueny et al., 1996; Hocutt et al., 1994; Ganasan and Hughes, 1998). Gutierrez uses percent species as piscivores, and Harris favors percent individuals as macrophagic carnivores. Didier et al. (1996) used percent of individuals as piscivores/invertivores, reflecting that these species typically consume invertebrates as well as small fish. Lyons et al. (1995) and Oberdorff and Porcher (1994) deleted this metric because they dealt with streams largely lacking piscivores; Oberdorff (1996) did not use piscivores because there are too few remaining in western European rivers to produce a sensitive metric. We agree with the use of percent macrophagic carnivores or percent piscivore/invertivores in waters typically lacking true piscivores. In fact, many North American species listed as piscivores in IBI applications also commonly feed on invertebrates. Broadening the metric in this way would also make it as applicable in French rivers (Oberdorff and Porcher, 1994; Oberdorff, 1996) as well as in Belgium (Didier et al., 1996).

Additional trophic metrics have been produced through studies of subtropical or tropical rivers. Like Ganasan and Hughes (submitted), Hocutt et al. (1994) added a percent herbivore/detritivore metric to assess food base quality at the primary production level. Gutierrez (1994), dealing with a rich tropical fauna, included percent detritivore, percent herbivore, and percent parasite metrics. As IBIs are applied to more tropical systems with markedly more diverse fish faunas than occur in the U.S. and Canada, such metrics are likely to be increasingly employed.

5.2.4 INDIVIDUAL ABUNDANCE AND CONDITION

5.2.4.1 Number of Individuals (Catch per Unit Effort)

Individual abundance is a common surrogate for system productivity, and highly disturbed sites are expected to support fewer individuals than high-quality sites (Karr, 1981). This metric, usually as catch-per-unit effort, was retained in all applications, except that of Didier et al. (1996), who substitute total biomass, and Gutierrez (1994), who lacked appropriate quantitative data. Oberdorff and Porcher (1994) also added total biomass. Recognizing the tendency for moderate levels of nutrient and thermal enrichment to elevate fish abundance, Oberdorff and Hughes (1992) and Oberdorff and Porcher (1994) both scored this metric so that very high abundances received lower metric scores than moderate numbers; only very low abundances received the lowest scores. This scoring adaptation is an example of the need to evaluate metric performance along disturbance gradients before applying the IBI in resource assessments (Table 5.2).

5.2.4.2 Percent Hybrid Individuals (Reproductive Function)

Karr (1981) proposed percent hybrids to assess the degree that environmental degradation alters reproductive isolation. Others have broadened the concept to incorporate the degree that alien individuals are favored (Miller et al., 1988) or to evaluate alterations in reproductive guilds (Simon and Lyons, 1995). The percent hybrids metric was retained only by Hugueny et al. (1996), who found none in the rivers they studied, thus making it unresponsive for that set of sites. Oberdorff and Hughes (1992) replaced hybrids with percent gravel spawners, while Didier et al. (1996) used percent of individuals as specialist spawners and percent of individuals as nonspecialist spawners. Lyons et al. (1995) used percent native livebearers and add a percent aliens metric. Gutierrez (1994) used percent introduced species, Harris (1995) substituted percent of species that are native, Hocutt et al. (1994) suggested percent introduced/invasive individuals, and Ganasan and Hughes (1998) proposed percent alien individuals. Oberdorff and Porcher (1994) and Oberdorff (1996) did not use percent hybrids or substitute for it. We agree that the percent hybrids metric has limited usefulness and evaluation of reproductive guilds is favored where this knowledge is available. Furthermore, addition of a percent alien metric in its own right is supported in cases where such species are likely to negatively perturb native species.

5.2.4.3 Percent Individuals with Anomalies

Karr (1981) chose this metric for assessing sites with severe degradation, such as those receiving chronic toxic loads or with toxic sediments. All applications, except that of Didier et al. (1996), retain the percent anomalies metric; these authors do not explain their reasons for dropping the metric, but it should be retained wherever the possibility exists for changes in the incidence of diseased and deformed fish.

5.3 CONCLUSION

All 10 applications of the IBI concept outside the U.S. and Canada have retained the same original ecological framework, including suites of metrics that represent the four major classes of biological attributes. Species richness and the composition metrics evaluate the extent to which key expected elements of biotic diversity are present. Habitat guilds assess physical habitat conditions and occasionally represent composition at the family level. Several trophic composition metrics evaluate integrity associated with functional (food chain) conditions. The fourth set of metrics involves overall assemblage size, reproductive function, the influence of alien individuals, and individual anomalies. The integrity of fish assemblages on all six continents with flowing fresh waters were assessed using these concepts.

Four of Karr's (1981) original 12 metrics were used outside the U.S. and Canada by at least six of ten authors (Table 5.4). Percent individuals with anomalies and percent omnivorous individuals were both used nine and seven times, respectively; the anomaly metric was deleted once, but the omnivore metric was substituted by percent scavengers, percent omnivorous species, and percent microphagic omnivores, and therefore not deleted. Species richness was used by six authors and never deleted; it was replaced with native species richness (a desirable improvement) four times. Seven authors used the intolerant or sensitive species metric and it was only deleted twice; it was replaced by percent sculpin once.

Five original metrics were never used by these authors: darter species richness, sunfish species richness, sucker species richness, percent green sunfish, and percent insectivorous cyprinids (Table 5.4). Darters were replaced with benthic species richness by four authors and by particular benthic subgroups (benthic specialists, riffle species, percent benthic individuals, mormyrids, characidiins/parodontins) in the other cases. Water column or cichlid species richness replaced the sunfish metric three times each, and pelagic/pool or pelagic/rheophilic replaced it twice; it was deleted twice. There was no common substitute for the sucker metric, each replacement only being used once; it was deleted twice. Similarly, the green sunfish metric showed no common alternative; substitute species were used in three cases, percent tolerants were used twice, percent aliens or natives were used once each, and it was deleted twice. The most common replacement for insectivorous cyprinids was percent invertivorous individuals (a more accurate metric), which was used in five cases and deleted twice. Percent hybrids was adopted in only one case and deleted twice; it was replaced by a reproductive guild and an introduced species metric three times each.

The remaining two metrics — percent piscivores and number of individuals — were deleted three times and once, respectively. Piscivores were adopted five times and the major modifications (piscivore/invertivore and macrophagic carnivore) are useful improvements. Number of individuals was used four times, typically with unspecified or nonstandardized sampling methodologies. A measure of catch per unit effort or biomass was employed five times. We believe that measures of total abundance, as well as species composition and proportionate abundance metrics, are highly suspect without standardized and comparable effort. Karr et al. (1986) make several recommendations concerning sampling and data quality. For historical data, they warn that many institutions bias the sampling for particular sport species, do not count individuals accurately, use ineffective sampling methods, or employ unrepresentative sampling designs. Recent work on small streams in the U.S. indicates that lengths 40 times the mean wetted width of the channel (150 m for streams less than 3 m wide) are needed for precise estimates of species richness and proportionate abundances (Lyons, 1992; Angermeier and Smogor, 1995; Paller, 1995).

Three of the papers analyzed, dealing respectively with Australian (Harris, 1995), Belgian (Didier et al., 1996) and Namibian (Hocutt et al., 1994) rivers, only formulated the preliminary (candidate) metrics that could be used for calculating the IBI, but did not test its performance along a stressor gradient. The other applications gave examples of the successful use of the IBI for particular problems: (1) Ganasan and Hughes (1998) found that a modified IBI was responsive to metal and organic pollution and river impoundment in central Indian rivers; (2) in Guinea, Hugueny et al. (1996) documented that the modified IBI was sensitive to effluents from a bauxite plant; (3) Oberdorff and Hughes (1992) and Oberdorff (1996) reported that a modified IBI detected cumulative effects of channelization, agricultural runoff, and urbanization in the Seine River basin, France; (4) Oberdorff and Porcher (1994) indicated that a modified IBI detected effects of salmonid aquaculture effluents on small Brittany streams in France; (5) Gutierrez (1994) reported that a modified IBI revealed the effects of channelization and chemical effluents in small Venezuelan rivers; and (6) Lyons et al. (1995) found close agreement between scores of a modified IBI and independent rankings from an environmental quality index for streams and small rivers of west-central Mexico.

IBI concepts based on assemblage and community ecology can be applied successfully to the markedly differing fish faunas of all six continents inhabited by freshwater fishes. The IBI work outside the U.S. and Canada has produced some promising types of metrics, particularly percent herbivores, percent detritivores, and percent microphagic omnivores, as well as a number of reproductive guilds (percent livebearers, percent specialist spawners, percent nonspecialist spawners) and increasing interest in non-natives (percent native, percent introduced, percent introduced or invasive, percent alien). IBI users should avoid using both complementary metrics (e.g., water column and benthic richness, or percent specialist and nonspecialist spawners). Such metrics are redundant without a third category (e.g., hiders or livebearers). These results support further development and use of this index as an efficient, quantitative biological indicator for water resource assessment and monitoring throughout the world.

Perhaps the most fundamental requirements for further application of the IBI model are increased knowledge of regional fish faunas (biology, ecology and systematics), together with improved sampling methodologies and monitoring programs. In addition, there is a need to evaluate the indices and metrics against multiple stressors to assess their general responsiveness to disturbance, and to conduct repeat sampling at the same sites so that the degree of sampling variance can be related to among-site variance. Only Hugueny et al. (1996) report on both metric responsiveness to disturbance and IBI variance revealed in repeat sampling; however, Hocutt et al. (1994) assessed spatial and temporal variances. These evaluations would benefit from increased regional-scale, government-sponsored research and monitoring by both public and private institutions.

ACKNOWLEDGMENTS

The authors thank Tom Simon for inviting us to prepare this chapter and Jim Karr for a copy of the Gutierrez thesis and continuing advice. John Lyons, Thom Whittier, and Susan Allen-Gil provided helpful and thoughtful reviews of the manuscript. The U.S. Environmental Protection Agency and the French Comité Inter-Agences de l'Eau supported portions of preparation costs through contracts 68-C6-0005 to Dynamac and n° 1302 to the French Conseil Supérieur de la Pêche. The manuscript was subjected to EPA peer and administrative review and cleared for publication.

REFERENCES

Allen, S.M. 1989. *The Adaptation of an Index of Biotic Integrity to Streams of the Coastal Plain Drainage Basins in North Carolina.* M.Sc. Project. School of Forestry and Environmental Studies, Duke University, Durham, NC.

Angermeier, P.L. and R.A. Smogor. 1995. Estimating number of species and relative abundances in stream-fish communities: effects of sampling effort and discontinuous spatial distributions, *Canadian Journal of Fisheries and Aquatic Sciences,* 52, 936–949.

Bailey, P.A., J.W. Enblom, S.R. Hanson, P.A. Renard, and K. Schmidt. 1993. *A Fish Community Analysis of the Minnesota River Basin.* Minnesota Pollution Control Agency, St. Paul, MN.

Bramblett, R.G. and K.D. Fausch. 1991. Variable fish communities and the index of biotic integrity in a western Great Plains river, *Transactions of the American Fisheries Society,* 120, 752–769.

Crumby, W.D., M.A. Webb, and F.J. Bulow. 1990. Changes in biotic integrity of a river in north-central Tennessee, *Transactions of the American Fisheries Society,* 119, 885–893.

Deegan, L.A., J.T. Finn, S.G. Ayvasian, and C. Ryder. 1993. *Feasibility and Application of the Index of Biotic Integrity to Massachusetts Estuaries (EBI).* Massachusetts Department of Environmental Protection, North Grafton, MA.

DeShon, J.E. 1995. Development and application of the invertebrate community index (ICI), in W.S. Davis and T.P. Simon (Eds.), *Biological Assessment and Criteria: Tools for Water Resource Planning and Decision Making.* Lewis, Boca Raton, FL. 217–243.

Didier, J., D. Kestemont, and J.C. Micha. 1996. Indice Biotique d'Intégrité Piscicole pour Evaluer la Qualité Ecologique des Ecosystèmes Aquatiques. Report (n°3) to the Ministère de la Région Wallonne (MRW-DGTRE). Unité de Recherche en Biologie des Organismes, Facultés Universitaires N.D. de la Paix, Namur, Belgium.

Dionne, M. and J.R. Karr. 1992. Ecological monitoring of fish assemblages in Tennessee River reservoirs, in D.H. McKenzie, D.E. Hyatt, and V.J. McDonald (Eds.), *Ecological Indicators.* Vol. 1. Elsevier, New York. 259–281.

Fausch, K.D., J.R. Karr, and P.R. Yant. 1984. Regional application of an index of biotic integrity based on stream fish communities, *Transactions of the American Fisheries Society,* 113, 39–55.

Fausch, K.D., J. Lyons, J.R. Karr, and P.L. Angermeier. 1990. Fish communities as indicators of environmental degradation, in S.M. Adams (Ed.), *Biological Indicators of Stress in Fish.* American Fisheries Society Symposium 8, Bethesda, MD. 123–144.

Fisher, T.R. 1989. *Application and Testing of Indices of Biotic Integrity in Northern and Central Idaho Headwater Streams.* M.Sc. Thesis, University of Idaho, Moscow, ID.

Fore, L.S., J.R. Karr, and R.W. Wisseman. 1996. Assessing invertebrate responses to human activities: evaluating alternative approaches, *Journal of the North American Benthological Society,* 15, 212–231.

Ganasan, V. and R.M. Hughes. 1998. Application of an index of biological integrity (IBI) to fish assemblages of the rivers Khan and Kshipra (Madhya Pradesh), India, *Freshwater Biology,* 40(2).

Goldstein, R.M., T.P. Simon, P.A. Bailey, M. Ell, E. Pearson, K. Schmidt, and J. W. Enblom. 1994. Concepts for an index of biotic integrity for streams of the Red River of the North Basin, in *Proceedings North Dakota Water Quality Symposium,* North Dakota State University, Fargo, ND. 169–179.

Gutierrez, M.A.R. 1994. *Utilizacion de la Ictiofauna como Indicadora de la Integridad Biotica de los Rios Guache y Guanare, Estado Portuguesa, Venezuela.* M.Sc. thesis, Universidad Nacional Experimental de los Llanos Occidentales "Ezequiel Zamora," Guanare, Venezuela.

Harris, J.H. 1995. The use of fish in ecological assessments, *Australian Journal of Ecology,* 20, 65–80.

Hocutt, C.H., P.N. Johnson, C. Hay, and B.J. van Zyl. 1994. Biological basis of water quality assessment: the Kavango River, Namibia, *Reviews Hydrobiologie Tropical,* 27, 361–384.

Hoefs, N.J. and T.P. Boyle. 1992. Contribution of fish community metrics to the index of biotic integrity in two Ozark rivers, in D.H. McKenzie, D.E. Hyatt, and V.J. McDonald (Eds.), *Ecological Indicators.* Vol. 1. Elsevier, New York. 283–303.

Hughes, R.M. and J.R. Gammon. 1987. Longitudinal changes in fish assemblages and water quality in the Willamette River, Oregon, *Transactions of the American Fisheries Society,* 116, 196–209.

Hughes, R.M. and R.F. Noss. 1992. Biological diversity and biological integrity: current concerns for lakes and streams, *Fisheries,* 17(3), 11–19.

Hugueny, B., S. Camara, B. Samoura, and M. Magassouba. 1996. Applying an index of biotic integrity based on fish communities in a west African river, *Hydrobiologia,* 331, 71–78.

Karr, J.R. 1981. Assessment of biotic integrity using fish communities, *Fisheries,* 6(6), 21–27.

Karr, J.R. 1996. Ecological integrity and ecological health are not the same, in P. Schulze (Ed.), *Engineering within Ecological Constraints.* National Academy Press, Washington, DC. 97–109.

Karr, J.R., K.D. Fausch, P.L. Angermeier, P.R. Yant, and I.J. Schlosser. 1986. *Assessing Biological Integrity in Running Waters: A Method and its Rationale.* Illinois Natural History Survey Special Publication 5, Champaign, IL.

Karr, J.R., P.R. Yant, K.D. Fausch, and I.J. Schlosser. 1987. Spatial and temporal variability of the index of biotic integrity in three midwestern streams, *Transactions of the American Fisheries Society,* 116, 1–11.

Kerans, B.L. and J.R. Karr. 1994. A benthic index of biotic integrity (B-IBI) for rivers of the Tennessee Valley, *Ecological Applications,* 4, 768–785.

Langdon, R. 1989. The development of fish population-based biocriteria in Vermont, in T.P. Simon, L.L. Holst, and L.J. Shepard (Eds.), *Proceedings of the First National Workshop on Biological Criteria.* EPA 905-9-89-003. U.S. Environmental Protection Agency, Chicago, IL. 12–25.

Leonard, P.M. and D.J. Orth. 1986. Application and testing of an index of biotic integrity in small, coolwater streams, *Transactions of the American Fisheries Society,* 115, 401–414.

Lyons, J. 1992. Using the Index of Biotic Integrity (IBI) to Measure Environmental Quality in Warmwater Streams of Wisconsin. General Technical Report NC-149. U.S. Department of Agriculture, St. Paul, MN.

Lyons, J. 1992. The length of stream to sample with a towed electrofishing unit when fish species richness is estimated, *North American Journal of Fisheries Management,* 12, 198–203.

Lyons, J., S. Navarro-Perez, P.A. Cochran, E. Santana C., and M. Guzman-Arroyo. 1995. Index of biotic integrity based on fish assemblages for the conservation of streams and rivers in west-central Mexico, *Conservation Biology,* 9, 569–584.

Lyons, J., L. Wang, and T.D. Simonson. 1996. Development and validation of an index of biotic integrity for coldwater streams in Wisconsin. *North American Journal of Fisheries Management,* 16, 241–256.

Margalev, R. 1963. On certain unifying principles in ecology, *American Naturalist,* 97, 357–374.

Miller, D.L., P.M. Leonard, R.M. Hughes, J.R. Karr, P.B. Moyle, L.H. Schrader, B.A. Thompson, R.A. Daniels, K.D. Fausch, G.A. Fitzhugh, J.R. Gammon, D.B. Halliwell, P.L. Angermeier, and D.J. Orth. 1988. Regional applications of an index of biotic integrity for use in water resource management, *Fisheries,* 13(5), 12–20.

Minns, C.K., V.W. Cairns, R.G. Randall, and J.E. Moore. 1994. An index of biotic integrity (IBI) for fish assemblages in the littoral zone of Great Lakes areas of concern, *Canadian Journal of Fisheries and Aquatic Sciences,* 51, 1804–1822.

Moyle, P.B., L.R. Brown, and B. Herbold. 1986. Final Report on Development and Preliminary Tests of Indices of Biotic Integrity for California. U.S. Environmental Protection Agency, Corvallis, OR.

Oberdorff, T. 1996. Réseau Hydrobiologique et Piscicole. Synthèse des Données 1995 sur le Bassin Seine Normandie. Final report to the Agence de l'Eau Seine-Normandie, Conseil Supérieur de la Pêche, Compiègne, France.

Oberdorff, T. and R.M. Hughes. 1992. Modification of an index of biotic integrity based on fish assemblages to characterize rivers of the Seine Basin, France, *Hydrobiologia,* 228, 117–130.

Oberdorff, T. and J.P. Porcher. 1994. An index of biotic integrity to assess biological impacts of salmonid farm effluents on receiving waters, *Aquaculture,* 119, 219–235.

Ohio EPA. 1987. *Biological Criteria for the Protection of Aquatic Life.* Ohio EPA, Division of Water Quality Monitoring and Assessment, Columbus, OH.

Paller, M.H. 1995. Relationships among number of fish species sampled, reach length surveyed, and sampling effort in South Carolina Coastal Plain streams, *North American Journal of Fisheries Management,* 15, 110–120.

Plafkin, J.L., M.T. Barbour, K.D. Porter, S.K. Gross, and R.M. Hughes. 1989. *Rapid Bioassessment Protocols for Use in Streams and Rivers: Benthic Macroinvertebrates and Fish.* EPA/444/4-89-001. U.S. Environmental Protection Agency, Washington, DC.

Richard, Y. 1994. Les Communaute's ichtyologiques du Bassin de la Rivière L'Assomption et 'Intégrité Biotique des Écosystèmes Fluviaux. Envirodoq n° EN940235, Rapport n° QE-89. Direction de Écosystèmes Aquatiques, Ministère de l'Environnement et de la Faune du Québec, Sainte-Foy, Québec, Canada.

Saylor, C.F. and S.A. Ahlstedt. 1990. *Application of Index of Biotic Integrity (IBI) to Fixed Station Water Quality Monitoring Sites.* Tennessee Valley Authority, Aquatic Biology Department, Norris, TN.

Schrader, L.H. 1989. *Use of the Index of Biotic Integrity to Evaluate Fish Communities in Western Great Plains Streams.* M.Sc. Thesis, Colorado State University, Fort Collins, CO.

Simon, T.P. 1991. *Development of Expectations for the Ecoregions of Indiana. I. Central Corn Belt Plain.* EPA 905-9-91-025. U.S. Environmental Protection Agency, Chicago, IL.

Simon, T. P. 1992. *Development of Biological Criteria for Large Rivers with an Emphasis on an Assessment of the White River Drainage, Indiana.* EPA 905-R-92-026. U.S. Environmental Protection Agency, Chicago, IL.

Simon, T.P. and J. Lyons. 1995. Application of the index of biotic integrity to evaluate water resource integrity in freshwater ecosystems, in W.S. Davis and T.P. Simon (Eds.), *Biological Assessment and Criteria: Tools for Water Resource Planning and Decision Making,* Lewis, Boca Raton, FL. 245–262.

Steedman, R.J. 1988. Modification and assessment of an index of biotic integrity to quantify stream quality in southern Ontario, *Canadian Journal of Fisheries and Aquatic Sciences,* 45, 492–501.

Thompson, B.A. and G.R. Fitzhugh. 1986. *A Use Attainability Study: An Evaluation of Fish and Macroinvertebrate Assemblages of the Lower Calcasieu River, Louisiana.* LSU-CFI-29. Center for Wetland Resources, Coastal Fisheries Institute, Louisiana State University, Baton Rouge, LA.

Section II

Guild and Metrics Determination

6 Assessment of Balon's Reproductive Guilds with Application to Midwestern North American Freshwater Fishes

Thomas P. Simon

CONTENTS

6.1 Introduction 97
6.2 Background 98
6.3 Reproductive Guild Placement of Midwestern North American Fishes 98
    6.3.1 A. Nonguarders 98
        6.3.1.1 Open Substratum Spawners (A.1) 98
        6.3.1.2 Brood Hiders (A.2) 109
    6.3.2 B. Guarders 109
        6.3.2.1 Substratum Choosers (B.1) 109
        6.3.2.2 Nest Spawners (B.2) 110
    6.3.3 C. Bearers 111
        6.3.3.1 External Bearers (C.1) 111
        6.3.3.2 Internal Bearers (C.2) 111
6.4 Discussion 111
6.5 Conclusion 112
Acknowledgments 112
References 112

6.1 INTRODUCTION

Similarity in evolutionary peaks and patterns in fishes enabled Balon (1975; 1981; 1985) to develop a framework for classifying reproductive styles. Balon's concept built on Kryzhanovsky's (1949) reasoning that "adaptations of fishes for spawning and development reflect not only the essential ecological factors of the embryonic period, but also the essential factors of all the other intervals of life." This reproductive guild classification was based mainly on form and function in early developmental intervals, on preferred spawning grounds, and on features of reproductive behavior. The difficulty with such a system is that reproductive behavior was well known for less than 2% of the 20,000 known species of fish (Breder and Rosen, 1966), early life history of fishes were imprecisely known representing less than 1% of the known species of fish (Simon, 1986), and, by

98 Assessing the Sustainability and Biological Integrity of Water Resources Using Fish Communities

Balon's own admission, the guild concept was a preliminary proposal reviewing the current knowledge at that time (Balon, 1975).

The present study examines the proper placement of midwestern North American species. The reproductive guild classification is becoming increasingly important as researchers develop alternative characteristics or "metrics" for evaluating North American stream fishes. The Index of Biotic Integrity (Karr et al., 1986) is an important tool that has been developed for determining biological resource quality using multiple aspects of stream fish communities. Alternative metrics include reproductive attributes such as proportion of simple lithophils (Simon, 1991; 1992) or ratio of nest spawning minnows to broadcast spawning minnows (Goldstein et al., 1994). To develop these metrics, species are categorized into appropriate guilds. Inaccurate assessment of biological integrity of a stream reach would result in improper guild placement of species. An increase in reproductive behavior information (Page, 1985; Page and Johnston, 1993; Simon, 1994) and early ontogenetic development data (Hardy, 1978; Auer,1982; Wallus et al., 1990; Simon, 1994) for midwestern North American stream fishes now enable more accurate classification of species.

6.2 BACKGROUND

Subsequent to the original reproductive guild classification by Kryzhanovsky (1949), several other attempts were made to develop a classification based on reproductive styles. Breder and Rosen (1966), in their work on reproductive modes in fishes, developed seven divisions to encompass the known reproductive data. The seven divisions retained a close relationship with Linnean taxonomic arrangement and encompassed various aspects of reproductive behavior and ecology. Nakamura (1969) grouped fish ecologically, based on early development. Page (1985) developed a classification for darters, based on attributes of egg placement, while Page and Johnston (1993) evaluated the same for minnows. The problematic grouping of reproductive behaviors between taxonomic groups makes all but the reproductive guilds a viable concept for application to all species of fishes. Although the reproductive guild framework is acceptable and desirable, further effort is necessary to ensure proper placement of respective species.

The reproductive guild concept has been largely abandoned for several reasons. Balon (1975), in the concept paper, placed species within the framework incorrectly based on spawning placement (Holcik and Hruska, 1966) and insufficient information. His list of taxa was principally limited to Canadian and European species. Insufficient number of guilds (Balon, 1981) or limited ontogenetic information changed guild placement of species (McElman and Balon, 1979; 1980). Although the situation still persists with a lack of reproductive behavior and ontogenetic information for the majority of fishes, a sizable body of information is available for North American species, particularly in the midwestern United States.

6.3 REPRODUCTIVE GUILD PLACEMENT OF MIDWESTERN NORTH AMERICAN FISHES

6.3.1 A. NONGUARDERS

6.3.1.1 Open Substratum Spawners (A.1)

Open substrate spawners are broadcast spawners in which ova and sperm are discharged in large numbers, often in a frenzy of spawning activity. No behavioral provision is made for the zygotes, either before or after spawning. Free embryos and larvae are pelagic or epibenthic, have small yolk provisions (usually a stomodeum), are weakly pigmented, and generally possess an elongated, laterally compressed body (Balon, 1975; 1981; 1985).

Pelagophils (A.1.1) are characterized by numerous buoyant eggs; embryonic respiratory organs are either nonexistent or poorly developed; larvae have little pigmentation; and are nonphotophobic

(Balon, 1981). Important attributes of pelagophils are the possession of an oversized oil globule that acts as a hydrostatic organ. Hydrostatic organs are adaptations to waters well saturated with oxygen that subsequently cause blood vessels to not develop near the body surface. Free embryo stages swim constantly (Balon, 1981). Balon (1975) classified *Coregonus nigripinnis*, *C. reighardi*, *C. zenithicus*, *Notropis atherinoides*, and *Aplodinotus grunniens* in this guild. Only *N. atherinoides* and *A. grunniens* are confirmed members of the guild. Accurate reports for reproductive behavior and ontogenetic information is unavailable for *C. nigripinnis*. Incorrectly placed in the guild are *C. reighardi* and *C. zenithicus,* which reproduce over gravel, rock, and clay substrates and do not have large oil globules (Table 6.1). Additional midwestern North American species that should be added to this guild include *Coregonus artedii* and *Hypophthalmichthys molitrix,* which have pelagic eggs that are released in the water column and develop while maintained in the currents (Table 6.1).

Lithopelagophils (A.1.2) are rock and gravel spawners with pelagic free embryos. Embryos initially have an adhesive chorion that soon become buoyant eggs. Free embryos are pelagic by positive buoyancy or active movement, free embryos have limited embryonic structures, and larvae are unconfirmed nonphotophobic (Balon, 1981). Some species possess a wide fin-fold that enables the embryo to make jumps toward the surface, resulting in a zigzag drift with the current. Balon (1975) indicated the following Canadian species belong in the guild: *Acipenser fulvescens, Dorosoma cepedianum, Coregonus hoyi, Hiodon alosoides, H. tergisus, Lota lota,* and *Morone saxatilis.* All but *M. saxatilis* are confirmed members of the guild (Table 6.1). *Morone saxatilis* is a nonobligatory plant spawner (Hardy, 1978) and is properly classified in the phytolithophil guild (A.1.4). Midwest North American species belonging to this guild include: *Polyodon spatula, Scaphirhynchus platorynchus, Coregonus clupeaformis, C. zenithicus, Osmerus mordax, Couesius plumbeus, Ericymba buccata, Extrarius aestavalis, Hybognathus nuchalis, Lythrurus ardens, Macrhybopsis storeriana, Notropis chalybaeus, N. hudsonius, Phoxinus erythrogaster, Rhinichthys atratulus, R. cataractae, Cycleptus elongatus, Carpiodes carpio, C. cyprinus, C. velifer, Catostomus catostomus, C. commersoni, Erimyzon oblongus, Ictiobus bubalus, I. cyprinellus, I. niger, Minytrema melanops, Stizostedion canadense,* and *S. vitreum.*

Lithophils (A.1.3) are rock and gravel spawners with benthic larvae that hide beneath stones. Free embryos possess moderately developed embryonic respiratory structures, late-appearing pigment, and are photophobic (Balon, 1981). Additional attributes of lithophils include reproduction in lentic and lotic systems, free embryos adapted to well-oxygenated waters and possessing respiratory structures in the ventral finfold and myoseptum, and the appearance of black and yellow pigments late in development. Balon (1975) classified *Coregonus clupeaformis, Osmerus mordax, Clinostomus elongatus, Couesius plumbeus, Notropis anogenus, N. blennius, N. dorsalis, N. rubellus, Rhinichthys atratulus, R. cataractae, Catostomus catostomus, C. commersoni, Hypentilium nigricans, Minytrema melanops, Moxostoma anisurum, M. carinatum, M. duquesnei, M. erythrurum, M. macrolepidotum, M. valenciennesi, Percopsis omiscomaycus, Stizostedion canadense,* and *S. vitreum* as lithophils. Only *H. nigricans, M. anisurum, M. carinatum, M. duquesnei, M. erythrurum, M. macrolepidotum, M. valenciennesi,* and *Percopsis omiscomaycus* are confirmed members of this guild (Table 6.1). Several species removed from this guild and placed in the lithopelagophil guild, since their free larvae are pelagic and not benthic, include: *C. clupeaformis, Osmerus mordax, Couesius plumbeus, Rhinichthys atratulus, R. cataractae, Catostomus catostomus, C. commersoni,* and *Minytrema melanops.* Reproductive or ontogenic information to suggest placement of *Notropis anogenus, N. blennius,* and *N. dorsalis* in this guild is not available. Likewise, *Clinostomus elongatus* is improperly placed in this guild and more accurately reflects reproductive attributes of the brood hiding lithophils (A.2.3; Greeley, 1938; Koster, 1939). Additional midwest North American species in this guild should include *Erimystax x-punctata, Lythrurus umbratilis,* and *E. variatum.*

Phytolithophils (A.1.4) are nonobligatory plant spawners that deposit eggs on submerged items, have late hatching larvae with cement glands in free embryos, have larvae with moderately developed respiratory structures, and have larvae that are photophobic (Balon, 1981). A typical attribute

TABLE 6.1
Reproductive Classifications of Fishes from the Midwestern United States for Computing the Index of Biotic Integrity

Taxa	Reproductive Guild	Literature cited
Petromyzontiformes — lampreys		
Petromyzontidae — lamprey		
Ichthyomyzon bdellium (Jordan), Ohio lamprey	A.2.3	Smith, 1979
I. castaneus Girard, chestnut lamprey	A.2.3	Case, 1970
I. fossor Reighard and Cummins, northern brook lamprey	A.2.3	Reighard and Cummins, 1916
I. unicuspis Hubbs and Trautman, silver lamprey	A.2.3	Trautman, 1957
Lampetra aepyptera (Abbott), least brook lamprey		
L. appendix (DeKay), American brook lamprey	A.2.3	Breder and Rosen, 1966
Petromyzon marinus Linnaeus, sea lamprey	A.2.3	Manion and McLain, 1971
Acipenseriformes — paddlefish, sturgeons		
Polyodontidae — paddlefish		
Polyodon spatula (Walbaum), paddlefish	A.1.2	Parkett, 1961; Friberg, 1972; Wallus et al., 1990
Acipenseridae — sturgeon		
Acipenser fulvescens Rafinesque, lake sturgeon	A.1.2	Scott and Crossman, 1973; Goodyear et al., 1982
Scaphirhynchus platorynchus (Rafinesque), shovelnose sturgeon	A.1.2	Coker, 1929; Pflieger, 1975; Moos, 1978
Lepisosteiformes — gars		
Lepisosteidae — gars		
Atractosteus spatula (Lacepede), alligator gar	A.1.5	Simon and Wallus, 1989
Lepisosteus oculatus Winchell, spotted gar	A.1.5	Riggs and Moore, 1960; Echelle and Riggs, 1972; Tyler and Granger, 1984; Simon and Tybergheim, 1991
L. osseus Linnaeus, longnose gar	A.1.4	Echelle and Riggs, 1972; Yeager and Bryant, 1983; Wallus et al., 1990
L. platostomus Rafinesque, shortnose gar	A.1.5	Riggs and Moore, 1960; Carlander, 1969; Becker, 1983
Amiiformes — bowfin		
Amiidae — bowfin		
Amia calva Linnaeus, bowfin	B.2.5	Balon, 1981; Becker, 1983; Wallus et al., 1990
Anguilliformes — eels		
Anguillidae — eel		
Anguilla rostrata (Lesueur), American eel	N/A	Catadrodromous
Clupeiformes — herring, shad		
Clupeidae — herring		
Alosa chrysochloris (Rafinesque), skipjack herring	A.1.4	Wallus et al., 1990

Assessment of Balon's Reproductive Guilds

Species	Guild	References
A. pseudoharengus (Wilson), alewife	A.1.4	Breder and Rosen, 1966; Jones et al., 1978; Nigro and Ney, 1982
Dorosoma cepedianum (Lesueur), gizzard shad	A.1.2	Bodola, 1966; Shelton, 1972
D. petenense (Gunther), threadfin shad	A.1.5	Shelton and Grinsted, 1972; Wallus et al., 1990
Osteoglossiformes — mooneye		
Hiodontidae — mooneye		
Hiodon alosoides (Rafinesque), goldeye	A.1.2	McPhail and Lindsey, 1970; Wallus et al., 1990
H. tergisus Lesueur, mooneye	A.1.2	Wallus et al., 1990
Salmoniformes — trout, salmon, whitefish		
Salmonidae — salmon and whitefish		
Coregonus artedii Lesueur, cisco or lake herring	A.1.1	Smith, 1954; Dryer and Bail, 1964
C. clupeaformis (Mitchill), lake whitefish	A.1.2	Koelz, 1929; Hart 1930
C. hoyi (Gill), bloater	A.1.2	Koelz, 1929
C. zenithicus (Jordan and Evermann), shortjaw cisco	A.1.2	Van Oosten, 1937; Becker, 1983
Oncorhynchus mykiss Walbaum, rainbow trout	A.2.3	Greeley, 1932; Wallus et al., 1990
O. kisutch (Walbaum), coho salmon	A.2.3	Croot and Margolis, 1992
O. tshawytscha (Walbaum), chinook salmon	A.2.3	Croot and Margolis, 1992
Salvelinus fontinalis (Mitchell), brook trout	A.2.3	Hazzard, 1932; Wallus et al., 1990
S. namaycush (Walbaum), lake trout	A.2.3	Hazzard, 1932
Salmo salar (Walbaum), Atlantic salmon	A.2.3	Scott and Crossman, 1973
S. trutta Linneaus, brown trout	A.2.3	Ottaway et al., 1981; Wallus et al., 1990
Osmeridae — smelt		
Osmerus mordax (Mitchell), rainbow smelt	A.1.2	Langlois, 1954; Rupp, 1959
Umbridae — mudminnows		
Umbra limi (Kirtland), central mudminnnow	B.1.4	Ryder, 1886; Jones, 1973; Wallus et al., 1990
Esocidae — pikes		
Esox americanus Gmelin, grass pickerel	A.1.5	Kleinart and Mraz, 1966; Wallus et al., 1990
E. lucius Linnaeus, northern pike	A.1.5	Frost and Kipling, 1967; McCarraher and Thomas, 1972
E. masquinongy Mitchill, muskellunge	A.1.5	Parsons, 1959; Brewer, 1980
Cypriniformes — carps and minnows		
Cyprinidae — carps and minnows		
Campostoma anomulum (Rafinesque), stoneroller	A.2.3	Hankinson, 1919; Reed, 1958; Miller, 1962
C. oligolepis Hubbs and Greene, largescale stoneroller	A.2.3	Pflieger, 1975
Carassius auratus (Linneaus), goldfish	A.1.5	Webster, 1942; Becker, 1983
Clinostomus elongatus (Kirtland), redside dace	A.2.3	Greeley, 1938; Koster, 1939
Couesius plumbeus (Agassiz), lake chub	A.1.2	Brown et al., 1970
Ctenopharyngodon idella Valenciennes, grass carp	A.1.1	Berg, 1964; Robison and Buchanan, 1988

TABLE 6.1 (continued)
Reproductive Classifications of Fishes from the Midwestern United States for Computing the Index of Biotic Integrity

Taxa	Reproductive Guild	Literature cited
Cyprinella lutrensis (Baird and Girard), red shiner	A.2.4	Minckley, 1973; Pflieger, 1975; Gale, 1986
C. spiloptera Cope, spotfin shiner	A.2.4	Pflieger, 1965
C. whipplei (Girard), steelcolor shiner	A.2.4	Pflieger, 1965
Cyprinus carpio Linneaus, carp	A.1.4	Swee and McCrimmon, 1966; Becker, 1983
Ericymba buccata Cope, silverjaw minnow	A.1.2	Hankinson, 1919; Wallace, 1973
Erimystax dissimilis Kirtland, streamline chub		
E. x-punctata Hubbs and Crowe, gravel chub	A.1.3	Cross, 1967; Becker, 1983
Extrarius aestavalis Girard, speckled chub	A.1.2	Bottrell et al., 1964; Becker, 1983
Hybognathus hankinsoni Hubbs, brassy minnow	A.1.5	Copes, 1975
Hybognathus hayi Jordan, cypress minnow		
H. nuchalis Agassiz, central silvery minnow	A.1.2	Adams and Hankinson, 1928
Hybopsis amblops (Rafinesque), bigeye chub		
H. amnis Hubbs and Greene, pallid shiner		
Hypophthalmichthys molitrix Valenciennes, silver carp	A.1.1	Berry and Low, 1970; Robison and Buchanan, 1988
Luxilus chrysocephalus (Rafinesque), striped shiner	A.2.3	Hankinson, 1932; Raney, 1940b
L. cornutus (Mitchell), common shiner	A.2.3	Adams and Hankinson, 1928; Hankinson, 1932; Raney, 1940a,b; Miller, 1963; 1964
Lythrurus ardens (Cope), rosefin shiner	A.1.2	Rane, 1947a; Yokley, 1974; Steele, 1978
L. fumeus Evermann, ribbon shiner		
L. umbratilis (Girard), redfin shiner	A.1.3	Hunter and Wisby, 1961; Hunter and Hasler, 1965
Macrhybopsis storeriana (Kirtland), silver chub	A.1.2	Kinney, 1954
Nocomis biguttatus (Kirtland), hornyhead chub	A.2.3	Hankinson 1932; Hubbs and Cooper, 1936; Reighard, 1943
N. micropogon (Cope), river chub	A.2.3	Hankinson, 1932; Reighard, 1943
Notemigonus crysoleucus (Mitchell), golden shiner	A.1.5	Cooper, 1935; Latta, 1958; Kramer and Smith, 1960; DeMont, 1982
Notropis anogenus Forbes, pugnose shiner		
N. atherinoides Rafinesque, emerald shiner	A.1.1	Flittner, 1964; Pflieger, 1975
N. ariommus (Cope), popeye shiner		
N. blennius (Girard), river shiner		
N. boops Gilbert, bigeye shiner		
N. buchanani Meek, ghost shiner		

Species	Code	References
N. chalybaeus (Cope), ironcolor shiner	A.1.2	Marshall, 1947
N. dorsalis (Agassiz), bigmouth shiner	A.1.5	Simon, unpublished data
N. heterodon (Cope), blackchin shiner	A.1.5	Simon, unpublished data
N. heterolepis Eigenmann and Eigenmann, blackchin shiner	A.1.2	Greeley and Greene, 1931; Wells and House, 1974
N. hudsonius (Clinton), spottail shiner		
N. ludibundus Cope, sand shiner		
N. photogenis (Cope), silver shiner		
N. rubellus (Agassiz), rosyface shiner	A.2.3	Adams and Hankinson, 1928; Hankinson, 1932; Hubbs and Cooper, 1936; Raney, 1940c; Pflieger, 1955; Reed, 1958; Miller, 1963; 1964
N. texanus (Girard), weed shiner	A.1.5	Black, 1945a; Becker, 1983
N. volucellus (Cope), mimic shiner	B.2.7	Page and Johnson, 1990b
Opsopoeodus emiliae Hay, pugnose minnow		
Phenacobius mirabilis (Girard), suckermouth minnow	A.1.2	Smith, 1908; Raney, 1969; Settles and Hoyt, 1978
Phoxinus erythrogaster (Rafinesque), southern redbelly dace	B.2.7	Hankinson, 1908; Hubbs and Cooper, 1936; Westman, 1938; Gale, 1983
Pimephales notatus (Rafinesque), bluntnose minnow	B.2.7	Forbes and Richardson, 1920; Thomsen and Hasler, 1944; Andrews and Flickinger, 1974; McMillian and Smith, 1974; Cole and Smith, 1987
P. promelas Rafinesque, fathead minnow	B.2.7	Park, 1964; Page and Ceas, 1989
P. vigilax (Baird and Girard), bullhead minnow	A.1.2	Traver, 1929; Raney, 1940d; Schwartz, 1958; Raney, 1969; Bartnik, 1970; Matthews et al., 1982
Rhinichthys atratulus Agassiz, blacknose dace	A.1.2	Greeley and Bishop, 1933; McPhail and Lindsey, 1970
R. cataractae (Valenciennes), longnose dace	A.2.3	Reighard, 1910; Ross, 1975; 1976; 1977; Copes, 1978
Semotilus atromaculatus (Mitchill), creek chub	A.1.4	Berg, 1964; Burkhead and Williams, 1991
Scardinius erythropthalmus (Linneaus), rudd		
Catostomidae — suckers and buffalo		
Cycleptus elongatus (Lesueur), blue sucker	A.1.2	Yeager and Semmens, 1987
Carpiodes carpio (Rafinesque), river carpsucker	A.1.2	Walburg and Nelson, 1966; Cross, 1967
C. cyprinus (Lesueur), quillback	A.1.2	Harlan and Speaker, 1956; Becker, 1983
C. velifer (Rafinesque), highfin carpsucker	A.1.2	Harlan and Speaker, 1956
Catostomus catostomus (Forster), longnose sucker	A.1.2	Geen et al., 1966
C. commersoni Lacepede, white sucker	A.1.2	Reighard, 1920; Breder and Rosen, 1966; McElman and Balon, 1980
Erimyzon oblongus (Mitchill), creek chubsucker	A.1.2	Hankinson, 1919; Becker, 1983; Page and Johnson, 1990a
E. sucetta (Lacepede), lake chubsucker	A.1.4	Cooper, 1936; Becker, 1983
Hypentelium nigricans (Lesueur), northern hogsucker	A.1.3	Raney and Lachner, 1946; Becker, 1983
Ictiobus bubalus (Rafinesque), smallmouth buffalo	A.1.2	Gasaway, 1970; Walburg, 1976

TABLE 6.1 (continued)
Reproductive Classifications of Fishes from the Midwestern United States for Computing the Index of Biotic Integrity

Taxa	Reproductive Guild	Literature cited
I. cyprinellus (Valenciennes), bigmouth buffalo	A.1.2	Walburg, 1964; Becker, 1983
I. niger (Rafinesque), black buffalo	A.1.2	Breder and Rosen, 1966; Becker, 1983
Minytrema melanops (Rafinesque), spotted sucker	A.1.2	McSwain and Gennings, 1972
Moxostoma anisurum (Rafinesque), silver redhorse	A.1.3	Meyer, 1962
M. carinatum (Cope), river redhorse	A.1.3	Becker, 1983
M. duquesnei (Lesueur), black redhorse	A.1.3	Bowman, 1970
M. erythrurum (Rafinesque), golden redhorse	A.1.3	Becker, 1983
M. macrolepidotum (Lesueur), shorthead redhorse	A.1.3	Reighard, 1920; Burr and Morris, 1977
M. valenciennesi Jordan, greater redhorse	A.1.3	Becker, 1983
Siluriformes — bullhead and catfish		
Ictaluridae — bullhead and catfish		
Ameiurus catus (Linnaeus), white catfish	B.2.7	Fowler, 1917
A. melas (Rafinesque), black bullhead	B.2.7	Wallace, 1967
A. natalis (Lesueur), yellow bullhead	B.2.7	Adams and Hankinson, 1926; Cahn, 1927
A. nebulosus (Lesueur), brown bullhead	B.2.7	Smith and Harron, 1903; Breder, 1935; Breder and Rosen, 1966
Ictalurus furcatus (Lesueur), blue catfish	B.2.7	Perry, 1973; Smith, 1979
I. punctatus (Rafinesque), channel catfish	B.2.7	Clemens and Sneed, 1957; Marzolf, 1957; Davis, 1959
Noturus eleutherus Jordan, mountain madtom	B.2.7	Starnes and Starnes, 1985
N. exilis Nelson, slender madtom	B.2.7	Mayden and Burr, 1981
N. flavus Rafinesque, stonecat	B.2.7	Greeley, 1929; Langlois, 1954; Walsh and Burr, 1985
N. gyrinus (Mitchill), tadpole madtom	B.2.7	Menzel and Raney, 1973; Mansueti and Hardy, 1967
N. miurus Jordan, brindled madtom	B.2.7	Taylor, 1969; Scott and Crossman, 1973
N. nocturnus Jordan and Gilbert, freckled madtom	B.2.7	Mayden and Walsh, 1984
N. stigmosus Taylor, northern madtom	B.2.7	Taylor, 1969; Cooper, 1985
Pylodoctis olivaris (Rafinesque), flathead catfish	B.2.7	Fontaine, 1944; Breder and Rosen, 1966
Percopsiformes — cavefish, pirate perch, trout-perch		
Amblyopsidae — cavefish		
Amblyopsis spelaea DeKay, northern cavefish	C.1.4	Poulson, 1961
Typhlichthys subterraneus Girard, southern cavefish	C.1.4	Poulson, 1961

Aphredoderidae — pirate perch		
Aphredoderus sayanus (Gilliams), pirate perch	C.1.4	Martin and Hubbs, 1973
Percopsidae — trout-perch		
Percopsis omniscomaycus (Walbaum), trout-perch	A.1.3	Priegel, 1962; Magnuson and Smith, 1963
Gadiformes — cod		
Gadidae — cod		
Lota lota (Linnaeus), burbot	A.1.2	Cahn, 1936; Breder and Rosen, 1966
Atheriniformes — topminnows, silversides		
Fundulidae — topminnows		
Fundulus catenatus (Storer), northern studfish	A.1.4	Simon, unpublished data
F. diaphanus (Lesueur), banded killifish	A.1.5	Richardson, 1939
F. dispar (Agassiz), northern starhead topminnow	A.1.4	Simon, unpublished data
F. notatus (Rafinesque), blackstripe topminnow	A.1.5	Carranza and Winn, 1954
Poeciliidae — live-bearing fishes		
Gambusia affinis (Baird and Girard), mosquitofish	C.2.1	Simon, unpublished data
Atherinidae — silversides		
Labidesthes sicculus (Cope), brook silverside	A.1.4	Hubbs, 1921; Cahn, 1927
Gasterosteiformes — sticklebacks		
Gasterosteidae — sticklebacks		
Culaea inconstans (Kirtland), brook stickleback	B.2.4	Jacobs, 1948; Winn, 1960
Pungitius pungitius (Linnaeus), ninespine stickleback	B.2.4	McKenzie and Keenleyside, 1970; Scott and Crossman, 1973
Perciformes — basses, sunfish, perch, darters		
Moronidae — temperate basses		
Morone chrysops (Rafinesque), white bass	A.1.4	Riggs, 1955; Chadwick et al., 1966
M. mississippiensis Jordan and Eigenmann, yellow bass	A.1.4	Burnham, 1909
M. saxatilis (Walbaum), striped bass	A.1.4	Hardy, 1978
Centrarchidae — black bass and sunfish		
Ambloplites rupestris (Rafinesque), rock bass	B.2.2	Breder, 1936; Trautman, 1957; Breder and Rosen, 1966
Centrarchus macropterus (Lacepede), flier	B.2.3	Breder, 1936
Lepomis cyanellus Rafinesque, green sunfish	B.2.2	Hunter, 1963; Carlander, 1977
L. gibbosus (Linnaeus), pumpkinseed	B.2.2	Breder and Rosen, 1966; Clark and Keenleyside, 1967
L. gulosus (Cuvier), warmouth	B.2.3	Larimore, 1957
L. humilis (Girard), orangespotted sunfish	B.2.3	Barney and Anson, 1923
L. macrochirus Rafinesque, bluegill	B.2.2	Miller 1963; Becker, 1983
L. megalotis (Rafinesque), longear sunfish	B.2.2	Hankinson, 1908; 1920; Miller, 1963
L. microlophus (Gunther), redear sunfish	B.2.2	Childers, 1967; Gerald, 1970; Hardy, 1978

TABLE 6.1 (continued)
Reproductive Classifications of Fishes from the Midwestern United States for Computing the Index of Biotic Integrity

Taxa	Reproductive Guild	Literature cited
L. punctatus (Valenciennes), spotted sunfish	B.2.2	Carr, 1946; Smith, 1979
Lepomis symmetricus Forbes, bantam sunfish	B.2.2	Burr, 1977
Micropterus dolomieu Lacepede, smallmouth bass	B.2.2	Hubbs and Bailey, 1938; Cleary, 1956; Mraz, 1964
M. punctulatus Rafinesque, spotted bass	B.2.2	Vogele, 1975; Robison and Buchanan, 1988
M. salmoides (Lacepede), largemouth bass	B.2.2	Hubbs and Bailey, 1938; Mraz and Cooper, 1957b; Breder and Rosen, 1966
Pomoxis annularis Rafinesque, white crappie	B.2.5	Pease, 1919; Adams and Hankinson, 1928; Langlois, 1954
P. nigromaculatus (Lesueur), black crappie	B.2.5	Hansen, 1951; Morgan, 1954; Siefert, 1968; Pflieger, 1975
Elassomatidae — pygmy sunfish		
Elassoma zonatum Jordan, banded pygmy sunfish	B.1.4	Mettee, 1974
Percidae — perch and darters		
Ammocrypta clara Jordan and Meek, western sand darter	A.1.6	Simon et al., 1991
A. pellucida (Agassiz), eastern sand darter	A.1.6	Simon et al., 1991
Crystallaria asprella Jordan, crystal darter	A.1.6	Simon et al., 1991
Etheostoma asprigene (Forbes), mud darter	B.1.4	Cummings et al., 1984
E. blennioides (Rafinesque), greenside darter	B.1.4	Lachner et al., 1950; Fahy, 1954
E. caeruleum Storer, rainbow darter	A.2.3	Winn, 1958; Grady and Bart, 1984
E. camurum (Cope), bluebreast darter	B.1.3	Mount, 1959; Page and Simon, 1988
E. chlorosoma (Hay), bluntnose darter	B.1.4	Page et al., 1982; Bart, unpublished data
E. exile (Girard), Iowa darter	B.1.4	Winn, 1958
E. flabellare Rafinesque, fantail darter	B.2.7	Lake, 1936; Simon, 1985; Simon and Layman, 1994
E. gracile (Girard), slough darter	B.1.4	Braasch and Smith, 1967
E. histrio (Jordan and Gilbert), harlequin darter	B.1.4	Simon, 1994
E. maculatum Kirtland, spotted darter	B.1.3	Raney and Lachner, 1939; Page and Simon, 1988

E. microperca Jordan and Gilbert, least darter	B.1.4	Winn, 1958; Burr and Page, 1979
E. nigrum Rafinesque, johnny darter	A.2.7	Winn, 1958; Simon, 1994
E. spectabile (Agassiz), orangethroat darter	A.2.3	Winn, 1958; Simon, unpublished data
E. squamiceps Jordan, spottail darter	B.2.7	Page, 1974b; Simon, 1987
E. tippecanoe Jordan and Evermann, tippecanoe darter	B.1.3	Warren et al., 1986; Page and Simon, 1988
E. variatum Kirtland, variegate darter	B.1.3	Lachner et al., 1950; Simon, pers. observ.
E. zonale (Cope), banded darter	B.1.4	Trautman, 1957; Simon, 1994
Perca flavescens (Mitchill), yellow perch	A.1.4	Thorpe, 1977
Percina caprodes (Rafinesque), logperch	A.2.3	Winn, 1958; Thomas, 1967; 1970
P. copelandi (Jordan), channel darter	A.2.3	Winn, 1953
P. evides (Jordan and Copeland), gilt darter	A.2.3	Hatch, 1982; pers. commun.
P. maculata (Girard), blackside darter	A.2.3	Thomas, 1967; 1970
P. phoxocephala (Nelson), slenderhead darter	A.2.3	Page and Smith, 1971
P. sciera (Swain), dusky darter	A.2.3	Page and Smith, 1970
P. shumardi (Girard), river darter	A.2.3	Simon, 1983; unpublished data
Percina vigil Hay, yellow saddleback darter	A.2.3	Heins and Baker, 1989
Stizostedion canadense (Smith), sauger	A.1.2	Priegel, 1969; Scott and Crossman, 1973
S. vitreum (Mitchill), walleye	A.1.2	Marshall, 1977; McElman and Balon, 1979
Sciaenidae — drum		
Aplodinotus grunniens Rafinesque, freshwater drum	A.1.1	Daiber, 1953; Wirth, 1958
Cottidae — sculpins		
Cottus bairdi Girard, mottled sculpin	B.2.7	Ludwig and Norden, 1964; Becker, 1983
C. carolinae (Gill), banded sculpin	B.2.7	Wallus and Grannemann, 1979
C. cognatus Richardson, slimy sculpin	B.2.7	Breder and Rosen, 1966
Myoxocephalus thompsoni (Girard), deepwater sculpin		

Note: Based on Balon (1981).

of the guild includes reproduction in clear water on submerged plants or, if not available, on other submerged items such as logs, gravel, and rocks. The guild is considered intermediate between the lithophils (A.1.3) and phytophils (A.1.5). Balon (1975) categorized the following Canadian species into this guild: *Alosa pseudoharengus, Hybognathus hankinsoni, H. nuchalis, Macrhybopsis storeriana, Erimystax x-punctata, Cyprinella spiloptera, Lythrurus umbratilis, Notropis volucellus, Labidesthes sicculus, Morone chrysops, Perca flavescens,* and *Etheostoma exile*. Only *Alosa pseudoharengus, Hybognathus hankinsoni, Labidesthes sicculus, Morone chrysops,* and *Perca flavescens* are confirmed as phytolithophils (A.1.4). *Hybognathus nuchalis, Macrhybopsis storeriana,* and *Erimystax x-punctata* free larvae do not possess cement glands and should be placed in the lithopelagophil guild (A.1.2). Additional species are incorrectly classified based on reproductive behavior, such as *Cyprinella spiloptera,* which is a crevice spawning species (Pflieger, 1965); *Lythrurus umbratilis,* a lithophil (Hunter and Wisby, 1961; Hunter and Hasler, 1965); *Notropis volucellus,* a phytophil (Black, 1945a; Becker, 1983); and *Etheostoma exile,* a brood-hiding phytophil (Winn, 1958). Additional midwest North American species in this guild include: *Lepisosteus osseus, Alosa chrysochloris, Cyprinus carpio, Scardinius erythropthalmus, Erimyzon succetta, Fundulus cátenatus, F. dispar,* and *Morone mississippiensis.*

Phytophils (A.1.5) are obligatory plant spawners with adhesive egg envelopes that stick to submerged live or dead plants. Free embryos hatch late and possess cement glands. Larvae also possess extremely well-developed embryonic respiratory structures and have no photophobic reaction (Balon, 1981). Attributes of phytophil species include: adaptations for survival in dense plant growth, muddy bottoms, or low dissolved oxygen; larvae hatch late and are small in size; and respiratory structures are well developed in the anterior yolk and dorsal finfold. Balon (1975) classified *Lepisosteus oculatus, L. osseus, Umbra limi, Esox americanus, E. lucius, E. masquinongy, Notemigonus crysoleucas, Opsopoeodus emiliae, Notropis heterodon, Erimyzon succetta, Ictiobus cyprinellus, Fundulus diaphanus, Etheostoma blennioides,* and *E. microperca* as phytophils. Confirmed phytophils include: *L. oculatus, Umbra limi, Esox americanus, E. lucius, E. masquinongy, Notemigonus crysoleucas, Notropis heterodon,* and *Fundulus diaphanus* (Table 6.1). Improperly classified are *Lepisosteus osseus,* a phytolithophil (A.1.4; Wallus et al., 1990); *Opsopoeodus emiliae,* a guarding, nest-constructing speleophil (B.2.7; Page and Johnson, 1990b); *Erimyzon succetta,* a phytolithophil (A.1.4; Cooper, 1936; Becker, 1983); *Ictiobus cyprinellus,* a broadcast spawning species with pelagic larvae (A.1.2; Walburg, 1964; Becker, 1983); and *Etheostoma blennioides* and *E. microperca,* substrate-choosing phytophils (B.1.4; Fahy, 1964; Burr and Page, 1979). Additional Midwest North American species included in this guild should be *Atractosteus spatula, Lepisosteus platostomus, Dorosoma petenense, Carassius auratus, Hybognathus hankinsoni, Notropis heterolepis, N. volucellus,* and *Fundulus notatus.*

Psammophils (A.1.6) are sand spawners with adhesive eggs, free embryos not possessing cement glands, free embryos with large pectorals, larvae with weakly developed respiratory structures and large neuromasts (cupulae), and phototrophic larvae (Balon, 1981). These species are mainly adapted to life in running waters. Psammophil eggs are small, embryos become active immediately after hatching, and embryos are photophobic, avoiding crevices. Pectoral fins develop early and swimbladders are reduced. Gills function early with wide-open apertures before the mouth is functional. Balon (1975) classified *Notropis heterolepis, N. hudsonius, N. stramineus, Carpiodes cyprinus, Ammocrypta pellucida,* and *Percina caprodes* as members of this guild. Only *Ammocrypta pellucida* is confirmed as a psammophil, based on the above criteria from Balon's list (Table 6.1). Incorrectly classified are *Notropis hudsonius* and *Carpiodes cyprinus,* broadcast spawning lithopelagophils (Greeley and Greene, 1931; Wells and House, 1974; Harlan and Speaker, 1954; Becker, 1983); *N. heterolepis,* a phytophil (Simon, unpublished data); and *Percina caprodes,* a brood-hiding lithophil (A.2.3; Winn, 1958; Thomas, 1967; 1970). No reproductive information exists for *Notropis stramineus,* and its inclusion in this guild must be considered tentative. Additional midwest North American species in this guild should include *Ammocrypta clara* and *Crystallaria asprella* (Simon et al., 1992).

6.3.1.2 Brood Hiders (A.2)

Brood-hiding behavior is similar to broadcasting except that the release of eggs occurs just below, rather than above, the surface of the substrate. Among North American species, brood hiding includes lithophils and speleophils. Lithophils include burying behavior and the construction of nests. Burying behavior of the female includes partially working her body below the surface of the substrate. Either individual or multiple males will mount the female and ova and sperm are expelled. A more liberal definition of speleophils includes crevice spawning behavior that may or may not occur in caves.

Lithophils (A.2.3) are rock and gravel spawners that do not guard their eggs. Large eggs are buried in gravel depressions called redds or in rock interstitial spaces. Free embryos possess large and dense yolk and an extensive respiratory plexus for exogenous and endogenous respiration. Early-emerging larvae are photophobic, while late-hatching larvae are called alevins. Balon (1975) classified *Oncorhynchus kisutch, O. tshawytscha, O. mykiss, Salmo salar, Salvelinus fontinalis, S. namaycush, Nocomis biguttatus, N. micropogon, Semotilus atromaculatus, Etheostoma caeruleum, Percina copelandi, P. maculata,* and *P. shumardi* as brood-hiding lithophils. All of the species listed by Balon are confirmed as brood-hiding lithophils (Table 6.1). Additional midwest North American species in this guild include *Icthyomyzon bdellium, I. casteneus, I. fossor, I. unicuspis, Lampetra appendix, Petromyzon marinus, Salmo trutta, Campostoma anomalum, C. oligolepis, Clinostomus elongatus, Luxilus chrysocephalus, L. cornutus, Notropis rubellus, Etheostoma spectabile, Percina caprodes, P. evides, P. phoxocephala, P. sciera,* and *P. vigil*.

Speleophils (A.2.4) are cave-spawning species that produce few, large, adhesive eggs, hidden in crevices. Free embryos are late hatching and possess extensive respiratory structures (Balon, 1981). Balon did not classify any species to this guild from North America or Europe. North American species should include *Cyprinella lutrensis, C. spiloptera,* and *C. whipplei,* all of which are crevice spawners without parental care. Zygotes are adhesive and large. Free embryos of these species are late-hatching and comparatively larger in size than other cyprinids.

6.3.2 B. GUARDERS

6.3.2.1 Substratum Choosers (B.1)

The substrate choosers exhibit derived behaviors such that the female selects the site of egg deposition (presumably in some species within a male's territory). Typically, the female selects plants or large rocks and a few eggs are expelled and fertilized. Zygotes are specifically placed, abandoned, and receive no direct parental care except territorial behavior provided by the male.

Lithophils (B.1.3) choose rocks for attachment of their eggs, which are usually adhesive, and oval or cylindrical. Free embryos and larvae are pelagic (Balon, 1981). This guild was reclassified (previously guild B.1.1) by Balon (1975); he did not classify any Canadian species in this guild, however. Midwest North American species that should be included are *Etheostoma maculatum, E. tippecanoe,* and *E. camurum*. These three species burrow into the gravel and deposit eggs as individual clusters or clumps. Zygotes of *E. tippecanoe* may be buried 2 to 3 times the body depth of the female below the surface of the gravel. All species maintain a territory defended by the males while spawning; however, the zygotes are not protected after reproduction is complete.

Phytophils (B.1.4) have adhesive eggs that are attached to a variety of plants. Free embryos without cement glands swim instantly after a prolonged embryonic period. This guild, reclassified by Balon (previously B.1.2), is characterized by the male guarding and fanning the eggs. The site of egg placement is sometimes selected and cleaned in advance of spawning. Balon determined that *Pomoxis annularis* was a member of this guild. This author would suggest that *Pomoxis annularis* does not belong in this guild but rather the phytophil (B.2.5) guild. The species constructs a nest that is defended by the male and usually includes submerged plants (Pease, 1919; Adams and Hankinson, 1928; Langlois, 1954) that camouflage the nest and are not building materials from which the nest is constructed. Species that should be included in the guild are: *Elassoma zonatum, Etheostoma asprigene, E. blennioides, E. chlorosoma, E. exile, E. gracile, E. histrio, E. microperca,* and *E. zonale.* Members of *Etheostoma* selectively attach eggs, either individually or in small clusters, on the stems and leaves of submerged and overhanging plants or within the structure of algal mats on rock surfaces. *Etheostoma histrio* is speculated to be a member of this guild based on attributes of free embryos and larvae and the known reproduction locality of adults (T. Simon, unpublished data). Territorial defense has been observed in some species, but no parental care is usually provided.

6.3.2.2 Nest Spawners (B.2)

Nest spawners establish a territory centered within a nest or beneath a cavity. The cavity or substrate is cleared of silt and debris by fin-wagging activities of the male. A ripe female enters the nest or cavity and, following courting by the male, eggs are deposited in the nest or attached beneath the nest stone. Eggs are clustered in proximity to one another and parental guarding is provided by either one or both parents. Cleaning and fanning of the zygotes ensures adequate oxygen. Comparatively fewer numbers of eggs are produced, however; free embryos and larvae have greater yolk provision, greater respiratory structure on the yolk and finfolds, usually hatch later in development, and are more heavily pigmented.

Polyphils (B.2.2) are miscellaneous substrate and material nesters that have adhesive eggs either attached or occur in clusters on any available substrate. Free embryos and larvae have dense yolks with high carotenoid contents, larvae possess well-developed embryonic respiratory structures, and young may feed on the mucus of parents (Balon, 1981). This guild was reorganized (previously B.2.6) from the earlier classification. Balon (1975) indicated that these species are not particular in the selection of nest-building material and substrate. Nests may be circular, usually leaving sticks and rootlets in place. These nests are among or next to plants growing in muddy or sandy shallows of slow rivers or lagoons. This author agrees with Balon's (1975) listing of *Lepomis gibbosus* within this guild. In addition, *Ambloplites rupestris, Lepomis cyanellus, L. macrochirus, L. megalotis, L. microlophus, L. punctatus, L. symmetricus, Micropterus dolomieui, M. punctulatus,* and *M. salmoides* should be included in the guild. These nesting species are not particular in their choice of substrates and may spawn over gravel, sand, clay, or detrital nests (Table 6.1).

Lithophils (B.2.3) are rock and gravel nesters that have spherical or elliptical egg envelopes that are always adhesive. Free embryos are photophobic or possess cement glands and swing tail-up in respiratory motions, larvae possess moderate to well-developed embryonic respiratory structures, and young may first feed on the mucus of parents. Balon (1981) reclassified this guild (previously numbered as B.2.1) and included *Luxilus cornutus, Ameiurus melas, Ambloplites rupestris, Lepomis cyanellus, L. macrochirus, L. megalotis,* and *Micropterus dolomieu.* None of the species identified by Balon (1975) belong in this guild. The centrarchid species listed are properly classified within the polyphils. *Ameiurus melas* is a cavity spawner and does not construct an open nest over gravel and rock (B.2.7; Wallace, 1967). Additional midwest North American species include *Centrarchus macropterus, Lepomis gulosus,* and *L. humilis,* all of which are restricted to spawning over gravel and rock substrates (Table 6.1).

Ariadnophils (B.2.4) are glue-making nesters, in which the male intensively guards eggs deposited in nests that are bound together by a viscid thread spun from a kidney secretion. Eggs and embryos are ventilated by males in spite of well-developed respiratory structures. Nests are constructed from various materials and differ in shape and position. Males may spawn with multiple females, and a single male may sometimes guard several nests. This guild was previously classified as B.2.7 and incorporates the Gasterosteidae.

Phytophils (B.2.5) are plant material nesters that have adhesive eggs and free embryos that hang on plants by cement glands. Larvae possess well-developed respiratory structures in embryos and are assisted in respiration by the fanning of adults (Balon, 1981). Species are adapted to nesting above or on soft muddy bottoms. Balon (1975) classified *Amia calva*, *Micropterus salmoides*, and *Pomoxis nigromaculatus* as members of this guild. *Micropterus salmoides* construct nests over many different substrates and is more appropriately classified as a polyphil (B.2.2). The other species classified in this guild are confirmed members. *Pomoxis annularis* should be included in this guild because it is a species that spawns among plant materials (Table 6.1).

Speleophils (B.2.7) are hole nesters. Two modes of reproductive behavior are included within this guild. Cavity rooftop nesters have moderately developed embryonic repiratory structures, while bottom-burrow nesters have strongly developed respiratory structures (Balon, 1981). The guild is polyphyletic and includes fish guarding their spawn in natural holes and cavities or in specially constructed burrows. The majority of cavity-spawning species deposit eggs on a cleaned area beneath the undersurface of flat stones with nest guarding by the male. Eggs can be laid in either a single or double layer and are well oxygenated. Embryonic structures are only moderately developed. Other members of the guild build elaborate burrows or cavities. The eggs are deposited inside the chamber, often by multiple females. Oxygen concentrations are very low, and free embryos and larvae develop strong respiratory structures. Balon (1975) classified *Pimephales notatus*, *P. promelas*, *Ameiurus natalis*, *A. nebulosus*, *Ictalurus punctatus*, *Noturus flavus*, *N. gyrinus*, *N. miurus*, *Etheostoma flabellare*, *E. nigrum*, *Cottus bairdi*, and *C. cognatus* in this guild. All of the above species are confirmed (Table 6.1). Additional midwest North American freshwater species in this guild include *Opsopoeodus emiliae*, *Pimephales vigilax*, *Etheostoma squamiceps*, and *Cottus carolinae*, which deposit eggs beneath flat stones; while *Ameiurus catus*, *A. melas*, *Ictalurus furcatus*, *Noturus eleutherus*, *N. exilis*, *N. nocturnus*, *N. stigmosus*, and *Pylodictis olivaris* are all burrow or cavity spawners.

6.3.3 C. Bearers

6.3.3.1 External Bearers (C.1)

External bearers carry developing eggs on the surface of their bodies or in externally filled body cavities or special organs (Balon, 1975).

Gill Chamber brooders (C.1.4) include the North American cave fishes that incubate their eggs in gill chambers. The female is thought to take the eggs into her mouth and subsequently pass them into the gill cavity, where they develop until hatching. In addition to the cave fishes, this guild includes *Aphredoderus sayanus* (Martin and Hubbs, 1973).

6.3.3.2 Internal Bearers (C.2)

Eggs are fertilized internally before they are expelled from the body cavity of members of this guild. Special organs or gonopodia are developed to facilitate sperm transfer. Mating does not necessarily coincide with fertilization. After copulation, sperm can be stored for a lifetime. After fertilization, the eggs can be expelled and incubated externally, or retained and incubated in the body cavity of the female, which results in the birth of full-grown juveniles.

Viviparous (C.2.4) fishes have internally fertilized eggs that develop into embryos, alevins, or juveniles. The embryo's partial or entire nutrition and gaseous exchange are supplied by the mother. The exchange between the embryo and mother is via secretory histotrophes, which is ingested or absorbed by the fetus via epithelial absorbtive structures or a yolk sac placenta. The only midwest North American freshwater species is *Gambusia affinis*.

6.4 DISCUSSION

One of the most significant problems with the reproductive guild classification is the definition of the ethnological sections. The original ethnological sections described by Balon (1975) did not adequately clarify the evolutionary progression to more derived reproductive behaviors. The author utilizes the same structure as defined by Balon and clarify several points. Based on evolutionary progression from least to most derived, the broadcast spawners are considered the least specialized. All species that are broadcast spawners are classified in the open substratum spawners (A.1) ethnological section. The polyphyletic brood hiders (A.2) include clustering species that include species making special provisions for the egg placement (e.g., pits or nests without parental care). The guarders reflect a derived condition such that substrate choosers (B.1) may provide parental territoriality during reproduction, but not-long term provision. The most derived guarding behavior, nest spawners (B.2), are nest builders that guard the zygotes and free larvae. The bearers include external and internal modes that reflect the original concepts of Balon (1975).

The movement and reclassification of species within the 33 recognized reproductive guilds occur because of the greater amount of reproductive (Simon, 1995) and ontogentic information available for North American freshwater species (Wallus et al., 1990; Wallus and Kay, 1995). Further efforts need to examine and identify similar groupings for other freshwater and marine faunas. Several species, primarily cyprinids, classified in earlier versions of the reproductive guilds concept (Balon, 1975), lacked reproductive and ontogenetic information. Midwest North American freshwater species lacking sufficient information to enable classification include *Ichthyomyzon aepyptera*, *Erimystax dissimilis*, *Hybognathus hayi*, *Hybopsis amblops*, *H. amnis*, *Notropis ariommus*, *N. blennius*, *N. boops*, *N. buchanani*, *N. dorsalis*, *N. fumeus*, *Phenacobius mirabilis*, and *Myoxocephalus thompsoni*. Tentative classification of these species into guilds is possible based on closely related species. However, identification of spawning requirements and ontogenetic information should be a priority for researchers studying midwest North American freshwater fishes.

6.5 CONCLUSION

Reproductive guild development appears as evolutionary peaks in a landscape of reproductive styles. Balon, in an attempt to expand a previous ecological classification into reproductive styles of all fishes, designated 33 guilds based on form and function in early developmental intervals, on preferred spawning grounds, and on features of reproductive behavior. Problematic classifications in assigning species to correct reproductive guild caused the concept to be abandoned until further behavioral and early ontogentic data became available. Reassignment and confirmation of North American species within the original framework is possible with the increase in ontogenetic and reproductive information for the minnows and darters.

ACKNOWLEDGMENTS

Numerous colleagues have assisted in the development and culture of fishes from which this chapter is based. I appreciate the lasting friendships over the years and extensive assistance from Robert Wallus, Larry Kay, Darrel Snyder, Lee Fuiman, Ed Tyberghein, the late Nancy Garcia, the late Karl Lagler, Joseph Kaskey, Lawrence M. Page, and Robert E. Jenkins. The assistance of these individuals has enabled the spawning of adults, rearing of larvae, and description of early life history stages.

REFERENCES

Adams, C.C and T.L. Hankinson. 1928. The ecology and economics of Oneida Lake. Fish. Bull., New York State College of Forestry 1(4a), *Roosevelt Wildlife Annuals*, 1, 235–548.
Andrews, A.K. and S.A. Flickinger. 1974. Spawning requirements and characteristics of the fathead minnow, *Proceedings Southeastern Association of Game and Fish Commissions*, 27, 759–766.
Auer, N.A. 1982. Identification of Larval Fishes of the Great Lakes Basin with Emphasis on the Lake Michigan Drainage. Great Lakes Fisheries Commission Special Publication 82-3.

Balon, E.K. 1975. Reproductive guilds of fishes: a proposal and definition, *Journal of the Fisheries Research Board of Canada,* 32, 821–864.
Balon, E.K. 1978. Reproductive guilds and the ultimate structure of fish taxocenes: amended contribution to the discussion presented at the mini-symposium, *Environmental Biology of Fishes,* 3, 149–152.
Balon, E.K. 1981. Additions and amendments to the classification of reproductive styles in fishes, *Environmental Biology of Fishes,* 6, 377–389.
Balon, E.K. 1985. *Early Life Histories of Fishes, New Developmental, Ecological, and Evolutionary Perspectives.* Dr. Junk Publ., The Hague, Netherlands.
Barney, R.L. and B.J. Anson. 1923. Life History and Ecology of the Orange-Spotted Sunfish, *Lepomis humilis.* Report U.S. Commission Fisheries 1922, Appendix XV, Bureau of Fisheries Document Number 938, 1–16.
Bartnik, V.G. 1970. Reproductive isolation between two sympatric dace, *Rhinichthys atratulus* and *Rhinichthys cataractae,* in Manitoba, *Journal of the Fisheries Research Board of Canada,* 27, 2125–2141.
Becker, G.C. 1983. *Fishes of Wisconsin.* University of Wisconsin Press, Madison.
Berg, L.S. 1964. Freshwater fishes of the USSR and adjacent countries. 4th ed. Vol. 2. Israel Program for Scientific Translations, Jerusalem.
Berry, P.Y. and M.P. Low. 1970. Comparative studies on some aspects of the morphology of *Ctenopharyngodon idellus, Aristichthys nobilis* and their hybrid (Cyprinidae), *Copeia,* 1970, 708–726.
Black, J.D. 1945a. Natural history of the northern mimic shiner, *Notropis volucellus volucellus* Cope, *Investigation of Indiana Lakes and Streams,* 2, 449–469.
Bodola, A. 1966. Life history of the gizzard shad, *Dorosoma cepedianum* (LeSueur), in western Lake Erie, *U.S. Fish Wildlife Service Fishery Bulletin,* 65, 391–425.
Bottrell, C.E., R.H. Ingersol, and R.W. Jones. 1964. Notes on the embryology, early development, and behavior of *Hybopsis aestivalis tetranemus* (Gilbert), *Transactions of the American Microscopical Society,* 83, 391–399.
Bowman, M.L. 1970. Life history of the black redhorse, *Moxostoma duquesnei* (LeSueur), in Missouri, *Transactions of the American Fisheries Society,* 99, 546–559.
Braasch, M.E. and P.W. Smith. 1967. The life history of the slough darter, *Etheostoma gracile* (Pisces: Percidae). Illinois Natural History Survey Biological Notes 58.
Breder, C.M. 1935. The reproductive habits of the common catfish, *Ameiurus nebulosus* (LeSueur), with a discussion of their significance in ontogeny and phylogeny, *Zoologica,* 19, 143–185.
Breder, C.M. 1936. The reproductive habits of the North American sunfishes (family Centrarchidae), *Zoologica,* 21, 1–48.
Breder, C.M., Jr. and D.E. Rosen. 1966. *Modes of Reproduction in Fishes.* Natural History Press, Garden City, NY.
Brewer, D.L. 1980. A Study of Native Muskellunge Populations in Eastern Kentucky Streams. Kentucky Department Fisheries and Wildlife Research. Fisheries Bulletin Number 64, Frankfort.
Brown, J.H., U.T. Hammer, and G. Koshinsky. 1970. Breeding biology of the lake chub, *Couesius plumbeus,* at Lac la Ronge, Saskatchewan, *Journal of the Fisheries Research Board of Canada,* 27, 1005–1015.
Burkhead, N.M. and J.D. Williams. 1991. An intergeneric hybrid of a native minnow, the golden shiner, and an exotic minnow, the rudd, *Transactions of the American Fisheries Society,* 120, 781–795.
Burnham, C.W. 1909. Notes on the yellow bass, *Transactions of the American Fisheries Society,* 39, 103–108.
Burr, B.M. 1977. The bantam sunfish, *Lepomis symmetricus:* systematics and distribution, and life history in Wolf Lake, Illinois, *Illinois Natural History Survey Bulletin Number,* 31, 437–465.
Burr, B.M. and M.A. Morris. 1977. Spawning behavior of the shorthead redhorse, *Moxostoma macrolepidotum,* in Big Rock Creek, Illinois, *Transactions of the American Fisheries Society,* 106, 80–92.
Burr, B.M. and L.M. Page. 1979. The life history of the least darter, *Etheostoma microperca,* in the Iroquois River, Illinois, Illinois Natural History Survey Biological Notes 112.
Cahn, A.R. 1927. An ecological study of the southern Wisconsin fishes. The brook silverside (*Labidesthes sicculus*) and the cisco (*Leucichthys artedi*) in their relations to the region, *Illinois Biological Monographs,* 11, 1–151.
Cahn, A.R. 1936. Observations on the breeding of the lawyer, *Lota maculosa, Copeia,* 1936, 163–165.
Carlander, K.D. 1969. *Handbook of Freshwater Fishery Biology.* Vol. 1. Iowa State University Press, Ames.
Carlander, K.D. 1977. *Handbook of Freshwater Fishery Biology.* Vol. 2. Iowa State University Press, Ames.

Carr, M.H. 1946. Notes on the breeding habits of the eastern stumpknocker, *Lepomis punctatus punctatus* (Cuvier), *Florida Academy of Science Quarterly Journal,* 9, 101–106.

Carranza, J. and H.E. Winn. 1954. Reproductive behavior of the blackstripe topminnow, *Fundulus notatus, Copeia,* 1954, 273–278.

Case, B. 1970. Spawning behavior of the chestnut lamprey (*Ichthyomyzon casteneus*), *Journal of the Fisheries Research Board of Canada,* 27, 1872–1874.

Chadwick, H.K., C.E. von Geldern, Jr., and M.L. Johnson. 1966. White bass, in A. Calhoun (Ed.), *Inland Fisheries Management.* California Department of Fish and Game, Sacramento. p. 412–422.

Childers, W.F. 1965. Hybridization of Four Species of Sunfishes (Centrarchidae). Unpublished Ph.D. dissertation, University of Illinois, Urbana.

Clark, F.W. and M.H.A. Keenleyside. 1967. Reproductive isolation between the sunfish *Lepomis gibbosus* and *L. macrochirus, Journal of the Fisheries Research Board of Canada,* 24, 495–514.

Cleary, R.E. 1956. Observations affecting smallmouth bass production in Iowa, *Journal of Wildlife Management,* 20, 353–359.

Clemens, H.P and K.E. Sneed. 1957. The Spawning Behavior of the Channel Catfish *Ictalurus punctatus.* U.S. Fish and Wildlife Service Special Science Report, Fisheries Number 219.

Coker, R.E. 1929. Methuselah of the Mississippi, *Scientific Monthly,* 16, 89–103.

Cole, K.S. and R.J. Smith. 1987. Male courting behavior in the fathead minnow, *Pimephales promelas, Environmental Biology of Fishes,* 18, 235–239.

Cooper, G.P. 1935. Some results of forage fish investigations in Michigan, *Transactions of the American Fisheries Society,* 65, 132–142.

Cooper, G.P. 1936. Some results of forage fish investigations in Michigan, *Transactions of the American Fisheries Society,* 65, 132–142.

Copes, F.A. 1975. Ecology of the brassy minnow, *Hybognathus hankinsoni* (Cyprinidae), *Fauna and Flora of Wisconsin Report,* 10, 46–72, University of Wisconsin, Stevens Point, Museum of Natural History.

Copes, F.A. 1978. Ecology of the creek chub, *Semotilus atromaculatus* (Mitchill) in northern waters, *Fauna and Flora of Wisconsin Report,* 12, 1–21, University of Wisconsin, Stevens Point, Museum of Natural History.

Croot, C. and L. Margolis. 1992. *Pacific Salmon Life Histories.* University of British Columbia Press, Vancouver.

Cross, F.B. 1967. *Handbook of Fishes of Kansas.* University of Kansas Museum of Natural History, Miscellaneous Publication 45.

Cummings, K.S., J.M. Grady, and B.M. Burr. 1984. The Life History of the Mud Darter, *Etheostoma asprigene,* in Lake Creek, Illinois. Illinois Natural History Survey Biology Notes 122.

Daiber, F.C. 1953. Notes on the spawning population of the freshwater drum (*Aplodinotus grunniens* Rafinesque) in western Lake Erie, *American Midland Naturalist,* 50, 159–171.

Davis, J. 1959. Management of Channel Catfish in Kansas. University of Kansas Museum of Natural History, Miscellanous Publication Number 21.

DeMont, D.J. 1982. Use of *Lepomis macrochirus* Rafinesque nests by spawning *Notemigonus crysoleucas* (Mitchill) (Pisces: Centrarchidae and Cyprinidae), *Brimleyana,* 8, 61–63.

Echelle, A.A. and C.D. Riggs. 1972. Aspects of the early life history of gars (*Lepisosteus*) in Lake Texoma, *Transactions of the American Fisheries Society,* 101, 106–112.

Fahy, W.E. 1954. The life history of the northern greenside darter, *Etheostoma blennioides blennioides* Rafinesque, *Journal of the Elisha Mitchell Scientific Society,* 70, 139–205.

Flittner, G.A. 1964. Morphology and Life History of the Emerald Shiner *Notropis atherinoides*. Unpublished Ph.D. dissertation, University of Michigan, Ann Arbor.

Fontaine, P.A. 1944. Notes on the spawning of the shovelhead catfish, *Pilodictis olivaris* (Rafinesque), *Copeia,* 1944, 50–51.

Forbes, S.A. and R.E. Richardson. 1920. *The Fishes of Illinois.* 2nd ed. Illinois Natural History Survey, Urbana, IL.

Fowler, H.W. 1917. Some notes on the breeding habits of local catfishes, *Copeia,* 42, 32–36.

Friberg, D.V. 1972. Investigations of Paddlefish Populations in South Dakota and Development and Management Plans. South Dakota Dept. Game, Fish and Parks. Dingell-Johnson Project F-15-R-7, Study 5-9, Pierre, SD.

Frost, W.E. and C. Kipling. 1967. A study of reproduction, early life weight-length relationship and growth of pike, *Esox lucius* L. in Windermere, *Journal of Animal Ecology,* 36, 651–693.

Gale, W.F. 1983. Fecundity and spawning frequency of caged bluntnose minnows — fractional spawners, *Transactions of the American Fisheries Society,* 112, 398–402.

Gale, W.F. 1986. Indeterminate fecundity and spawning behavior of captive red shiners — fractional, crevice spawners, *Transactions of the American Fisheries Society,* 112, 398–402.

Gasaway, C.R. 1970. Changes in the Fish Population in Lake Francis Case in South Dakota in the First 16 Years of Impoundment. U.S. Bureau of Sport Fisheries and Wildlife Technical Paper 56.

Geen, G.H., T.G. Northcote, G.F. Hartman, and C.C. Lindsey. 1966. Life histories of two species of catostomid fishes in Sixteenmile Lake, British Columbia, with particular reference to inlet stream spawning, *Journal of the Fisheries Research Board of Canada,* 23, 1761–1788.

Gerald, J.W. 1970. Species Isolating Mechanisms in the Genus *Lepomis.* Unpublished Ph.D. dissertation, University of Texas, Austin.

Goldstein, R.M, T.P. Simon, P.A. Bailey, M. Ell, K. Schmidt, and J. Emblom. 1994. Proposed metrics for the index of biotic integrity for the streams of the Red River of the North basin, in *Proceedings of the North Dakota Academy of Science.*

Goodyear, C.S., T.A. Edsall, D.M. Ormsby Dempsey, G.D. Moss, and P.E. Polowski. 1982. Atlas of the Spawning and Nursery Areas of Great Lakes Fishes. U.S. Fish and Wildlife Service. FWS/OBS-82/52. Washington, DC.

Grady, J.M. and H.L. Bart, Jr. 1984. Life history of *Etheostoma caeruleum* (Pisces: Percidae) in Bayou Sara, Louisiana and Mississippi, in D.G. Lindquist and L.M. Page (Eds.), *Developments in Environmental Biology of Fishes,* Dr. Junk Publishers, The Hague, Netherlands. 71–81.

Greeley, J.R. 1929. Fishes of the Erie-Niagara watershed, in *A Biological Survey of the Erie-Niagara System.* Supplement for the 18th Annual Report of the New York State Conservation Department for 1935. 150–179.

Greeley, J.R. 1932. The spawning habits of brook, brown, and rainbow trout, and the problem of egg predators, *Transactions of the American Fisheries Society,* 62, 239–247.

Greeley, J.R. 1938. Fishes of the area with annotated list, in *A Biological Survey of the Allegheny and Chemung Watersheds.* Supplement for the 27th Annual Report of the New York State Conservancy Department for 1937. 48–73.

Greeley, J.R. and S.C. Bishop. 1933. Fishes of the upper Hudson watershed with annotated list, in *A Biological Survey of the Upper Hudson Watershed.* Supplement for the 22nd Annual Report for the New York State Conservancy Department for 1932. 64–101.

Greeley, J.R. and C.E. Greene. 1931. Fishes of the area with annotated list, in *A Biological Survey of the St. Lawrence Watershed.* Supplement for the 20th Annual Report for the New York State Conservancy Department for 1930. 44–94.

Hankinson, T.L. 1908. A biological survey of Walnut Lake, Michigan. Report of the State Biological Survey for Michigan 1907, 198–251.

Hankinson, T.L. 1920. Notes on the life histories of Illinois fishes, *Transactions of the Illinois State Academy of Science,* 12, 132–150.

Hankinson, T.L. 1932. Observations on the breeding behavior and habitats of fishes in southern Michigan, *Papers of the Michigan Academy of Science Arts and Letters,* 15, 411–425.

Hardy, J.D., Jr. 1978. *Development of Fishes of the Mid-Atlantic Bight, An Atlas of Egg, Larval and Juvenile Stages. Vol. III. Aphredoderidae Through Rachycentridae.* U.S. Fish Wildlife Service Biological Service Program. FWS/OBS-78/12.

Harlan, J.R. and E.B. Speaker. 1956. *Iowa Fish and Fishing.* Iowa Conservation Commission.

Hatch, J.T. 1982. Life History of the Gilt Darter, *Percina evides* (Jordan and Copeland), in the Sunrise River, Minnesota. Unpublished Ph.D. dissertation, University of Minnesota, Minneapolis.

Hazzard, A.S. 1932. Some phases of the life history of the eastern brook trout, *Salvelinus fontinalis* Mitchell, *Transactions of the American Fisheries Society,* 62, 344–350.

Heins, D.C. and J.A. Baker. 1989. Growth, population structure, and reproduction of the percid fish *Percina vigil, Copeia,* 1989, 727–736.

Holcik and Hruska. 1966. On the spawning substrate of the roach-*Rutilus rutilus* (Linnaeus, 1758) and bream-*Abramis brama* (Linnaeus, 1758) and notes on the ecological characteristics of some European fishes, *Vest. Cesk. Spol. Zool.,* 30, 22–29.

Hubbs, C.L. 1921. An ecological study of the life history of the fresh-water atherine fish *Labidesthes sicculus*, *Ecology*, 2, 262–276.
Hubbs, C.L. and R.M. Bailey. 1938. The Smallmouth Bass. Cranbrook Institute of Science Bulletin, Number 10.
Hubbs, C.L. and G.P. Cooper. 1936. Minnows of Michigan. Cranbrook Institute of Science Bulletin, Number 8.
Hunter, J.R. 1963. The reproductive behavior of the green sunfish, *Lepomis cyanellus*, *Zoologica*, 48, 13–24.
Hunter, J.R. and A.D. Hasler. 1965. Spawning association of the redfin shiner, *Notropis umbratilis*, and the green sunfish, *Lepomis cyanellus*, *Copeia*, 1965, 265–281.
Hunter, J.R. and W.J. Wisby. 1961. Utilization of the nests of green sunfish (*Lepomis cyanellus*) by the redfin shiner (*Notropis umbratilis cyanocephalus*), *Copeia*, 1961, 113–115.
Jacobs, D.L. 1948. Nesting of the brook stickleback. *Proceedings of the Minnesota Academy of Science*, 16, 33–34.
Jones, J.A. 1973. The Ecology of the Mudminnow, *Umbra limi*, in Fish Lake (Anoka County, Minnesota). Ph.D. dissertation, Iowa State University, Ames.
Jones, P.W., F.D. Martin, J.D. Hardy, Jr. 1978. *Development of Fishes of the Mid-Atlantic Bight: An Atlas of Egg, Larval, and Juvenile Stages. Volume 1. Acipenseridae Through Ictaluridae.* U.S. Fish Wildlife Service Biological Services Program. FWS/OBS-78/12, Washington, DC.
Kinney, E.C., Jr. 1954. The life history of the trout perch, *Percopsis omiscomaycus* (Walbaum), in western Lake Erie with notes on associated species. Dissertation Abstracts 206: 1978-1980. Vol. 35. Part II. 30759.
Kleinart, S.J. and D. Mraz. 1966. Life History of the Grass Pickerel (*Esox americanus vermiculatus*) in Southeastern Wisconsin. Wisconsin Department of Natural Resources Technical Bulletin, Number 37, Madison.
Koelz, W. 1929. Coregonid fishes of the Great Lakes, *Bulletin U.S. Bureau of Fisheries*, 43:Part II, 297–643.
Koster, W.J. 1939. Some phases of the life history and relationships of the cyprinid, *Clinostomus elongatus*, *Copeia*, 1939, 201–208.
Kramer, R.H. and L.L. Smith. 1960. Utilization of nests of largemouth bass (*Micropterus salmoides*) by golden shiners (*Notemigonus crysoleucas*), *Copeia*, 1960, 73–74.
Lachner, E.A., E.F. Westlake, and P.S. Handwerk. 1950. Studies on the biology of some percid fishes from western Pennsylvania, *American Midland Naturalist*, 43, 92–111.
Lake, C.T. 1936. The life history of the fantail darter *Catonotus flabellaris flabellaris* (Rafinesque), *American Midland Naturalist*, 17, 816–830.
Langlois, T.H. 1954. *The Western End of Lake Erie and Its Ecology.* J.W. Edward Publishers Inc., Ann Arbor, MI.
Larimore, R.W. 1957. Ecological Life History of the Warmouth (Centrarchidae). Illinois Natural History Survey Bulletin 27.
Latta, W.C. 1958. The ecology of the smallmouth bass, *Micropterus dolomieui* Lacepede, at Waugoshance Point, Lake Michigan, *Dissertation Abstracts*, 18B: 1905.
Ludwig, G.M. and C.R. Norden. 1964. Age, growth, and reproduction of the northern mottled sculpin (*Cottus bairdi bairdi*) in Mt. Vernon Creek, Wisconsin. *Occassional Papers of the Natural History*, Milwaukee Public Museum Number 2.
Karr, J.R., K.D. Fausch, P.L. Angermeier, P.R. Yant, and I.J. Schlosser. 1986. Assessing Biological Integrity in Running Waters: A Method and Its Rationale. Illinois Natural History Survey Special Publication 5.
Kay, L., R. Wallus, B.L. Yeager. 1995. *Reproductive Biology and Early Life History of Fishes in the Ohio River Drainage. Vol. 2. Catostomidae.* Tennessee Valley Authority, Chattanooga, TN.
Kryzhanovsky, S.G. 1949. Eco-morphological principles of development among carps, loaches, and catfishes, *Tr. Inst. Morph. Zhiv. Severtsova*, 1, 5–332. (In Russian) (Part II, Ecological groups of fishes and patterns of their distribution, p. 237–331. Translated from Russian by Fisheries Research Board of Canada Translation Series Number 2945, 1974).
Magnuson, J.J. and L.L. Smith, Jr. 1963. Some phases of the life history of the troutperch, *Percopsis omiscomaycus*, *Ecology*, 44, 83–95.
Manion, P.J. and A.L. McLain. 1971. Biology of Larval Sea Lamprey (*Petromyzon marinus*) of the 1960 Year Class, Isolated in the Big Garlic River, Michigan, 1960–1965. Great Lakes Fisheries Commission Technical Report 16.
Mansueti, A.J. and J.D. Hardy. 1967. Development of fishes of the Chesapeake Bay Region. University of Maryland, Baltimore, Natural Resources Institute, Part I.

Marshall, N. 1947. Studies on the life history and ecology of *Notropis chalybaeus* (Cope), *Journal of the Florida Academy of Science,* 9, 163–188.

Marshall, T.R. 1977. Morphological, physiological, and ethological differences between walleye (*Stizostedion vitreum vitreum*) and pikeperch (*S. lucioperca*), *Journal of the Fisheries Research Board of Canada,* 34, 1515–1523.

Martin, F.D. and C. Hubbs. 1973. Observations on the development of pirate perch *Aphredoderus sayanus* (Pisces: Aphredoderidae) with comments on yolk circulation patterns as a possible taxonomic tool, *Copeia,* 1973, 377–379.

Marzolf, R.C. 1957. The reproduction of channel catfish in Missouri ponds, *Journal of Wildlife Management,* 21, 22–28.

Matthews, W.J., R.E. Jenkins, and J.T. Stryon, Jr. 1982. Systematics of two forms of blacknose dace, *Rhinichnthys atratulus* (Pisces: Cyprinidae) in a zone of syntopy, with a review of the species group, *Copeia,* 1982, 902–920.

Mayden, R.L. and B.M. Burr. 1981. Life history of the slender madtom *Noturus exilis* in southern Illinois (Pisces: Ictaluridae). *Occasional Papers of the Museum of Natural History,* University of Kansas 43.

Mayden, R.L. and S.J. Walsh. 1984. Life history of the least madtom *Noturus hildebrandi* (Siluriformes: Ictaluridae) with comparisons to related species, *American Midland Naturalist,* 112, 349–368.

McCarraher, D.B. and R.E. Thomas. 1972. Ecological significance of vegetation to northern pike *Esox lucius* spawning, *Transactions of the American Fisheries Society,* 101, 560–563.

McElman, J.F. and E.K. Balon. 1979. Early ontogeny of walleye, *Stizostedion vitreum,* with steps of saltatory development, *Environmental Biology of Fishes,* 4, 309–348.

McElman, J.F. and E.K. Balon. 1980. Early ontogeny of white sucker, *Catostomus commersoni,* with steps of saltatory development, *Environmental Biology of Fishes,* 5, 191–224.

McKenzie, J.A. and M.H.A. Keenleyside. 1970. Reproductive behavior of ninespine stickleback (*Pungitius pungitius* L.) in South Bay, Manitoulin Island, Ontario, *Canadian Journal of Zoology,* 48, 55–61.

McMillian, V.E. and R.J.F. Smith. 1974. Agonistic and reproductive behavior of the fathead minnow (*Pimephales promelas* Rafinesque), *Zoologische Tierpsychologic,* 34, 25–58.

McPhail, J.D. and C.C. Lindsey. 1970. Freshwater Fishes of Northwestern Canada and Alaska. Fisheries Research Board of Canada Bulletin. 173.

McSwain, L. and R.M. Gennings. 1972. Spawning behavior of the spotted sucker, *Minytrema melanops* (Rafinesque), *Transactions of the American Fisheries Society,* 101, 738–740.

Menzel, B.W. and E.C. Raney. 1973. Hybrid madtom catfish, *Noturus gyrinus* x *Noturus miurus,* from Cayuga Lake, New York, *American Midland Naturalist,* 90, 165–176.

Mettee, M.F. 1974. A Study of the Reproductive Behavior, Embryology, and Larval Development of the Pygmy Sunfishes of the Genus *Elassoma.* Unpublished Ph.D. dissertation, University of Alabama, Tuscaloosa.

Meyer, W.H. 1962. Life history of three species of redhorse (*Moxostoma*) in the Des Moines River, Iowa, *Transactions of the American Fisheries Society,* 91, 412–419.

Miller, H.C. 1963. The behavior of the pumpkinseed, *Lepomis gibbosus* (Linnaeus), with notes on the behavior of other species of Lepomis and pigmy sunfish, *Elassoma evergladei, Behaviour,* 22, 88–151.

Miller, R.J. 1962. Reproductive behavior of the stoneroller minnow, *Campostoma anomalum pullum, Copeia,* 1962, 407–417.

Miller, R.J. 1963. Comparative morphology of three cyprinid fishes: *Notropis cornutus, Notropis rubellus,* and the hybrid, *Notropis cornutus* x *N. rubellus, Copeia,* 1963, 450–452.

Miller, R.J. 1964. Behavior and ecology of some North American cyprinid fishes, *American Midland Naturalist,* 72, 313–357.

Minckley, W.L. 1973. *Fishes of Arizona.* Arizona Game and Fish Department, Phoenix, AZ.

Moos, R.E. 1978. Movement and Reproduction of Shovelnose Sturgeon, *Scaphirynchus platorynchus* (Rafinesque) in the Missouri River, South Dakota. Ph.D. dissertation, University of South Dakota, Vermillion.

Mount, D.I. 1959. Spawning behavior of the bluebreast darter, *Etheostoma camurum* (Cope), *Copeia,* 1959, 240–243.

Mraz, D. 1964. Observations on Large and Smallmouth Bass Nesting and Early Life History. Wisconsin Conservation Department of Fisheries Research Report Number 11.

Mraz, D. and E.L. Cooper. 1957. Reproduction of carp, largemouth bass, bluegills, and black crappies in small rearing ponds, *Journal of Wildlife Management,* 21, 127–133.

Nakamura, M. 1969. Cyprinid fishes of Japan — studies on the life history of cyprinid fishes of Japan, *Regional Institute Natural Resources Special Publication,* 4, 1–455 (In Japanese).

Nigro, A.A. and J.J. Ney. 1982. Reproduction and early life history of alewife (*Alosa pseudoharengus*) in a Virginia reservoir, *Virginia Journal of Science,* 31, 101.

Ottaway, E.M., P.A. Carling, A. Clarke, and N.A. Reader. 1981. Observations on the structure of brown trout, *Salmo trutta* Linnaeus, redds, *Journal of Fisheries Biology,* 19, 593–607.

Page, L.M. 1974. The Life History of the Spottail Darter, *Etheostoma squamiceps*, in Big Creek, Illinois, and Ferguson Creek, Kentucky. Illinois Natural History Survey Biological Notes 89.

Page, L.M. 1985. Evolution of reproductive behaviors in percid fishes, *Illinois Natural History Survey Bulletin,* 33, 275–295.

Page, L.M. and P.A. Ceas. 1989. Egg attachment in *Pimephales* (Pisces: Cyprinidae), *Copeia,* 1989, 1074–1077.

Page, L.M. and C.E. Johnston. 1990a. Spawning in the creek chubsucker, *Erimyzon oblongus*, with a review of spawning behaviors in suckers (Catostomidae), *Environmental Biology of Fishes,* 27, 265–272.

Page, L.M. and C.E. Johnson. 1990b. The breeding behavior of *Opsopoeodus emilae* (Cyprinidae) and its phylogenetic implications, *Copeia,* 1990, 1176–1180.

Page, L.M. and C.E. Johnston. 1993. The evolution of complex reproductive strategies in North American minnows (Cyprinidae), in R.L. Mayden (Ed.), *Systematics, Historical Ecology, and North American Freshwater Fishes.* Stanford University Press, Stanford, CA. 600–621.

Page, L.M. and T.P. Simon. 1988. Observations on the reproductive behavior and eggs of four species of darters, *Transactions of the Illinois State Academy of Science,* 81, 205–210.

Page, L.M. and P.W. Smith. 1970. The Life History of the Dusky Darter, *Percina sciera*, in the Embarras River, Illinois. Illinois Natural History Survey Biological Notes 69.

Page, L.M. and P.W. Smith. 1971. The Life History of the Slenderhead Darter, *Percina phoxocephala*, in the Embarras River, Illinois. Illinois Natural History Survey Biological Notes 74.

Page, L.M., M.E. Retzer, and R.A. Stiles. 1982. Spawning behavior in seven species of darters (Pisces: Percidae), *Brimleyana,* 8, 135–142.

Parker, H.L. 1964. Natural history of *Pimephales vigilax* (Cyprinidae), *Southwestern Naturalist,* 8, 228–235.

Parsons, J.W. 1959. Muskellunge in Tennessee streams, *Transactions of the American Fisheries Society,* 88, 136–140.

Pease, A.S. 1919. Habits of the Black Crappie in Inland Lakes of Wisconsin. U.S. Commission of Fisheries Report for 1918.

Perry, W.G. 1973. Notes on the spawning of blue and channel catfish in brackish water ponds, *Progressive Fish Culturist,* 35, 164–166.

Pflieger, W.L. 1955. Studies on the life history of the rosyface shiner, *Notropis rubellus* (Agassiz), *Copeia,* 95–104.

Pflieger, W.L. 1965. Reproductive behavior of the minnows *Notropis spilopterus* and *Notropis whipplii, Copeia,* 1965, 1–8.

Pflieger, W.L. 1975. *The Fishes of Missouri.* Missouri Department of Conservation, Jefferson City.

Poulson, T.L. 1961. Cave Adaptations in Amblyopsid Fishes. Unpublished Ph.D. dissertation, University of Michigan, Ann Arbor.

Priegel, G.R. 1962. Plentiful but unknown, *Wisconsin Conservation Bulletin,* 27, 20–21.

Priegel, G.R. 1969. The Lake Winnebago Sauger: Age, Growth, Reproduction, Food Habits and Early Life History. Wisconsin Department of Natural Resources Technical Bulletin Number 43.

Purkett, C.A., Jr. 1961. Reproduction and early development of the paddlefish, *Transactions of the American Fisheries Society,* 90, 125–129.

Raney, E.C. 1940a. Reproductive activities of a hybrid minnow, *Notropis cornutus* x *Nortopis rubellus, Zoologica Society,* 25, 361–367.

Raney, E.C. 1940b. The breeding behavior of the common shiner, *Notropis cornutus* (Mitchill), *Zoologica,* 25, 1–14.

Raney, E.C. 1940c. Nests under the water, *Bulletin of the New York Zoological Society,* 63, 127–135.

Raney, E.C. 1940d. Comparison of the breeding habits of two subspecies of black-nosed dace, *Rhinichthys atratulus* (Hermann), *American Midland Naturalist,* 23, 399–403.

Raney, E.C. 1947a. *Nocomis* nests used by other breeding cyprinid fishes in Virginia, *Zoologica,* 32, 125–132.

Raney, E.C. 1969. Minnows of New York, *The Conservationist,* Pts. I and II. 23, 22–29 and 23, 21–29.

Raney, E.C. and E.A. Lachner. 1939. Observations on the life history of the spotted darter, *Poecilichthys maculatus* (Kirkland), *Copeia*, 1939, 157–165.

Raney, E.C. and E.A. Lachner. 1946. Age, growth, and habits of the hog sucker, *Hypentelium nigricans* (LeSueur), in New York, *American Midland Naturalist*, 36, 76–86.

Reed, R.J. 1958. The early life history of two cyprinids, *Notropis rubellus* and *Campostoma anomalum pullum*, *Copeia*, 1958, 325–327.

Reighard, J. 1910. Methods of studying the habits of fishes, with an account of the breeding habits of the horned dace, *Bulletin of the U.S. Bureau of Fisheries*, 28, 1111–1136.

Reighard, J. 1920. The breeding behavior of suckers and minnows, *Biological Bulletin*, 38, 1–32.

Reighard, J. 1943. The breeding habits of the river chub, *Nocomis micropogon* (Cope), *Papers Michigan Academy of Science, Arts, and Letters*, 29, 397–423.

Reighard, J. and H. Cummins. 1916. Description of a new species of lamprey of the genus *Ichthyomyzon*. Occasional Papers of the Museum of Zoology, University of Michigan, Number 31.

Richardson, L.R. 1939. The spawning behavior of *Fundulus diaphanus* (LeSueur), *Copeia*, 1939, 165–167.

Riggs, C.D. 1955. Reproduction of the white bass, *Morone chrysops*, *Investigations of the Indiana Lakes and Streams*, 4, 87–110.

Riggs, C.D. and G.A. Moore. 1960. Growth of young gar (*Lepisosteus*) in aquaria, *Proceedings of the Oklahoma Academy of Science*, 40, 44–46.

Robison, H.W. and T.M. Buchanan. 1988. Fishes of Arkansas. University of Arkansas Press, Fayetteville.

Ross, M.R. 1975. The Breeding Behavior and Hybridization Potential of the Northern Creek Chub, *Semotilus atromaculatus atromaculatus* (Mitchell). Unpublished Ph.D. dissertation, The Ohio State University, Columbus, OH.

Ross, M.R. 1976. Nest entry behavior of female creek chubs (*Semotilus atromaculatus*) in different habitats, *Copeia*, 1976, 378–389.

Ross, M.R. 1977. Aggression as a social mechanism in the creek chub (*Semotilus atromaculatus*), *Copeia*, 1977, 393–397.

Rupp, R.S. 1959. Variation in the life history of the American smelt in inland waters of Maine, *Transactions of the American Fisheries Society*, 88, 241–252.

Ryder, J.A. 1886. The development of the mudminnow, *American Naturalist*, 20, 823–824.

Schwartz, F.J. 1958. The breeding behavior of the northern blacknose dace, *Rhinichthys atratulus obtusus* Agassiz, *Copeia*, 1958, 141–143.

Scott, W.B. and E.J. Crossman. 1973. Freshwater Fishes of Canada. Fisheries Research Board of Canada Bulletin 184.

Settles, W.H. and R.D. Hoyt. 1978. The reproductive biology of the southern redbelly dace, *Chrosomus erythrogaster* Rafinesque, in a spring-fed stream in Kentucky, *American Midland Naturalist*, 99, 290–298.

Shelton, W.L. 1972. Comparative Reproductive Biology of the Gizzard Shad, *Dorosoma cepedianum* (LeSueur), and the Threadfin Shad, *Dorosoma petenense* (Gunther), in Lake Texoma. Ph.D. dissertation, University of Oklahoma, Norman.

Shelton, W.L. and B.G. Grinsted. 1972. Hybridization between *Dorosoma cepedianum* and *D. petenense* in Lake Texoma, Oklahoma, *Proceedings of the Twenty-Sixth Annual Conference of the Southeastern Association of Game and Fish Commission*, 506–510.

Simon, T.P. 1983. Percidae — perches and darters, in L.E. Holland and M.L. Huston, *A Compilation of Available Literature on the Larvae of Fishes Common to the Upper Mississippi River.* U.S. Fish and Wildlife Service, National Fisheries Research Laboratory, LaCrosse, WI. 248–274.

Simon, T.P. 1985. Descriptions of Larval Percidae Inhabiting the Upper Mississippi River basin (Osteichthyes: Etheostomatini). Unpublished M.S. thesis, University of Wisconsin, LaCrosse.

Simon, T.P. 1986. A listing of regional guides, keys, and selected comparative descriptions of freshwater and marine larval fishes, *Early Life History Section Newsletter*, 7, 10–15.

Simon, T.P. 1987. Descriptions of eggs, larval, and early juveniles of the stripetail darter, *Etheostoma kennicotti* (Putnam) and spottail darter, *E. squamiceps* Jordan (Percidae: Etheostomatini) from tributaries of the Ohio River, *Copeia*, 1987, 433–440.

Simon, T.P. 1990. Development of Index of Biotic Integrity Expectations for the Ecoregions of Indiana. I. Central Corn Belt Plain. U.S. Environmental Protection Agency, Region 5, Chicago, IL. EPA 905/9-91/025.

Simon, T.P. 1992. Development of Biological Criteria for Large Rivers with an Emphasis on an Assessment of the White River Drainage, Indiana. U.S. Environmental Protection Agency, Region 5, Chicago, IL. EPA 905/R-92/006.

Simon, T.P. 1994. Ontogeny and Systematics of Darters (Percidae) with Discussion of Ecological Effects on Larval Morphology. Unpublished Ph.D. dissertation, University of Illinois, Chicago.

Simon, T.P. and S.R. Layman. 1995. Egg and larval development of the striped fantail darter, *Etheostoma flabellare lineolatum* Rafinesque, and duskytail darter, *E.* (*Catonotus*) sp. with comments on the *Etheostoma flabellare* species group, *Transactions of the Kentucky Academy of Science*, 56, 28–40.

Simon, T.P. and E.J. Tyberghein. 1991. Contributions to the early life history of the spotted gar, *Lepisosteus oculatus* Winchell, from Hatchet Creek, Alabama, *Transactions of the Kentucky Academy of Science*, 52, 124–131.

Simon, T.P., E.J. Tyberghein, K.J. Scheidegger, and C.E. Johnston. 1992. Descriptions of protolarvae of the sand darters (Percidae: *Ammocrypta* and *Crystallaria*) with comments on systematic relationships, *Ichthyological Explorations of Freshwaters*, 3, 347–358.

Simon, T.P. and R. Wallus. 1989. Contributions to the early life histories of gar (Actinopterygii: Lepistosteidae) in the Ohio and Teneessee River basins with emphasis on larval development, *Transactions of the Kentucky Academy of Science*, 50, 59–74.

Smith, B.G. 1908. The spawning habits of *Chrosomus erythrogaster* Rafinesque, *Biological Bulletin*, 14, 9–18.

Smith, H.M. and L.G. Harron. 1903. Breeding habits of the yellow catfish, *Bulletin of the U.S. Fisheries Commission*, 22, 149–154.

Smith, P.W. 1979. *Fishes of Illinois*. University of Illinois Press, Champaign.

Starnes, L.B. and W.C. Starnes. 1985. Ecology and life history of the mountain madtom, *Noturus eleutherus* (Pisces: Ictaluridae), *American Midland Naturalist*, 114, 331–341.

Steele, D. 1978. The Reproductive Strategy and Energetics of Rosefin Shiners, *Notropis ardens*, in Barebone Creek, Kentucky. Unpublished Ph.D. dissertation, University of Louisville.

Swee, U.B. and H.R. McCrimmon. 1966. Reproductive biology of the carp, *Cyprinus carpio* L., in Lake St. Lawrence, Ontario, *Transactions of the American Fisheries Society*, 95, 372–380.

Taylor, W.R. 1969. A Revision of the Catfish Genus *Noturus* Rafinesque, with an Analysis of Higher Groups in the Ictaluridae. U.S. National Museum Smithsonian Institute Press, Washington Bulletin 282.

Thomsen, H.P. and A.D. Hasler. 1944. The minnow problem in Wisconsin, *Wisconsin Conservation Bulletin*, 9, 6–8.

Thomas, D.L. 1967. Ecological Relationships of Four *Percina* Darters in the Kaskaskia River. Unpublished Ph.D. dissertation, University of Illinois, Urbana.

Thomas, D.L. 1970. An ecological study of four darters of the genus *Percina* (Percidae) in the Kaskaskia River. Illinois Natural History Survey Biological Notes 70.

Thorpe, J.E. 1977. Morphology, physiology, behavior, and ecology of *Perca fluviatilis* L. and *P. flavescens* Mitchell, *Journal of the Fisheries Research Board of Canada*, 34, 1504–1514.

Trautman, M.B. 1957. *The Fishes of Ohio*. Ohio State University Press, Columbus. (rev. ed. 1981).

Traver, J.R. 1929. The habits of the black-nosed dace, *Rhinichthys atratulus* (Mitchell), *Journal of the Elisha Mitchell Society*, 45, 101–120.

Tyler, J.D. and M.N. Granger. 1984. Notes on food habits, size, and spawning behavior of spotted gar in Lake Lawtonka, Oklahoma, *Proceedings of the Oklahoma Academy of Science*, 64, 8–10.

Van Oosten, J. 1937. The age, growth, and sex ratio of the Lake Superior longjaw, *Leucichthys zenithicus* (Jordan and Evermann), *Papers of the Michigan Academy of Science, Arts, and Letters*, 22, 691–711.

Vogele, L.E. 1975. Reproduction of spotted bass, *Micropterus punctulatus*, in Bull Shoals Reservoir, Arkansas. Technical Paper U.S. Fish and Wildlife Service Number 84.

Walburg, C.H. 1964. Fish Population Studies, Lewis and Clark Lake, Missouri River, 1956–1962. U.S. Fish and Wildlife Service Special Science Report Fisheries Number 482.

Walburg, C.H. 1976. Changes in the Fish Populations of Lewis and Clark Lake, 1956–1962. U.S. Fish and Wildlife Service Research Report 79.

Walburg, C.H. and W.R. Nelson. 1966. Carp, River Carpsucker, Smallmouth Buffalo, and Bigmouth Buffalo in Lewis and Clark Lake, Missouri. U.S. Fish and Wildlife Service Research Report Number 69.

Wallace, C.R. 1967. Observations on the reproductive behavior of the black bullhead (*Ictalurus melas*), *Copeia*, 1967, 852–853.

Wallace, D. 1973. Reproduction of the silverjaw minnow, *Ericymba buccata* Cope, *Transactions of the American Fisheries Society,* 102, 786–793.
Wallus, R. and K.L. Grannemann. 1979. Spawning behavior and early development of the banded sculpin, *Cottus carolinae* (Gill), in R. Wallus and C.W. Voigtlander (Eds.), *Proceedings of a Workshop on Freshwater Larval Fishes.* Tennessee Valley Authority, Knoxville, Tennessee. 199–233.
Wallus, R., T.P. Simon, B.L. Yeager. 1990. *Reproductive Biology and Early Life History of Fishes in the Ohio River Drainage.* Vol. 1. Acipenseridae through Esocidae. Tennessee Valley Authority, Chattanooga, TN.
Walsh, S.J. and B.M. Burr. 1985. Biology of the stonecat *Noturus flavus* (Siluriformes: Ictaluridae), in central Illinois and Missouri streams, and comparisons with Great Lakes populations and congeners, *Ohio Journal of Science,* 85, 85–96.
Warren, M.L., Jr., B.M. Burr, and B.R. Kuhajda. 1986. Aspects of the reproductive biology of *Etheostoma tippecanoe* with comments on egg-burying behavior, *American Midland Naturalist,* 116, 215–218.
Webster, D.A. 1942. The life histories of some Connecticut fishes, in *A Fishery Survey of Some Important Connecticut Lakes.* Bulletin Connecticut State Geological and Natural History Survey 63.
Wells, L. and R. House. 1974. Life history of the spottail shiner (*Notropis hudsonius*) in southeastern Lake Michigan, the Kalamazoo River and western Lake Erie, *U.S. Bureau of Sport Fisheries and Wildlife Research Report,* 78, 1–10.
Westman, J.R. 1938. Studies on the reproduction and growth of the bluntnosed minnow, *Hyborhynchus notatus* (Rafinesque), *Copeia,* 1938, 57–61.
Winn, H.E. 1953. Breeding habits of the percid fish, *Hadropterus copelandi* in Michigan, *Copeia,* 1953, 26–30.
Winn, H.E. 1958. Comparative reproductive behavior and ecology of fourteen species of darters (Pisces: Percidae), *Ecological Monographs,* 28, 155–191.
Winn, H.E. 1960. Biology of the brook stickleback, *Eucalia inconstans, American Midland Naturalist,* 63, 424–440.
Wirth, T.L. 1958. Lake Winnebago freshwater drum, *Wisconsin Conservation Bulletin,* 23, 30–32.
Yeager, B.L. and R.T. Bryant. 1983. Larvae of longnose gar *Lepisosteus osseus* from the Little River in Tennessee, *Journal of the Tennessee Academy of Science,* 58, 20–22.
Yeager, B.L. and K.J. Semmens. 1987. Early development of the blue sucker, *Cycleptus elongatus, Copeia,* 1987, 312–316.
Yokley, P., Jr. 1974. Habitat and reproductive behavior of the rosefin shiner, *Notropis ardens* (Cope), in Lauderdale County, Alabama (Osteichthyes, Cypriniformes, Cyprinidae), *Association of Southeastern Biologists Bulletin,* 21, 93.

7 Toward a United Definition of Guild Structure for Feeding Ecology of North American Freshwater Fishes

Robert M. Goldstein and Thomas P. Simon

CONTENTS

7.1 Introduction 124
7.2 Synthesis of Existing Information for Freshwater Fishes...125
7.3 Proposed Guild Structure...126
    7.3.1 Herbivores 126
        7.3.1.1 Particulate Feeders 127
            7.3.1.1.1 Grazers 127
            7.3.1.1.2 Browsers 127
    7.3.2 Detritivores 128
        7.3.2.1 Filter Feeders...128
            7.3.2.1.1 Suction feeders 128
            7.3.2.1.2 Filterers..128
        7.3.2.2 Particulate Feeders 128
            7.3.2.2.1 Biters 128
            7.3.2.2.2 Scoopers 129
    7.3.3 Planktivores 129
        7.3.3.1 Nondiscriminant Filter Feeders 129
            7.3.3.1.1 Mechanical sieve 129
            7.3.3.1.2 Mucus entrapment...129
            7.3.3.1.3 Ram filtration 129
            7.3.3.1.4 Pump filtration...129
            7.3.3.1.5 Gulping...130
        7.3.3.2 Particulate Feeders 130
            7.3.3.2.1 Size-selective pickers 130
    7.3.4 Invertivores 130
        7.3.4.1 Benthic Predators 130
            7.3.4.1.1 Grazers 130
            7.3.4.1.2 Crushers..131
            7.3.4.1.3 Hunters of mobile benthos..131
            7.3.4.1.4 Lie-in-wait predators...131
            7.3.4.1.5 Tearers 131
            7.3.4.1.6 Diggers 131

                7.3.4.2   Drift Feeders 131
                          7.3.4.2.1 Surface drift feeders 131
                          7.3.4.2.2 Water column drift feeders 132
        7.3.5   Carnivores 132
                7.3.5.1   Whole Body (Piscivore) 132
                          7.3.5.1.1 Stalking 132
                          7.3.5.1.2 Chasing 132
                          7.3.5.1.3 Ambush 132
                          7.3.5.1.4 Protective resemblance 132
                7.3.5.2   Parasites 133
                          7.3.5.2.1 Blood suckers 133
7.4   Discussion 133
        7.4.1   Omnivores, Opportunists, and Generalists 133
        7.4.2   Species with Multiple Trophic States over the Course of Their Lives 134
7.5   Conclusion 135
Acknowledgments 136
References 136
Appendix 7A 139

7.1   INTRODUCTION

The Index of Biotic Integrity (IBI) evaluates the structure and function of fish communities by combining community measures called metrics from three categories: (1) species richness and composition, (2) trophic composition, and (3) abundance and condition (Karr, 1981; Karr et al., 1986). Numerical ratings are assigned to each metric based on the expected values from reference conditions that are as close to pristine as can be found. The sum of the ratings of the individual metrics is the total IBI score. The greater the deviation from the reference community, the greater the change in the sum of the metrics. While species richness and composition metrics are relatively uncomplicated, i.e., most fish can be readily identified to the species level, there are certain difficulties that occur when assigning species to trophic groups. In addition, a nomenclature has been borrowed from evolutionary and population ecology that hinders more than it helps.

Changes in the biotic integrity of the fish community are expressed in the trophic composition of the community. Compared to reference conditions, there are differences in the transport, transformation, and storage of energy in the form of fish biomass. Trophic metrics used in various IBIs include: (1) the proportion of individuals or biomass as omnivores; an increase in omnivores indicates a reduction of biotic integrity; (2) the proportion of individuals or biomass as insectivores; a decrease in the proportion of insectivores indicates a loss of biotic integrity; (3) the proportion of individuals or biomass as top carnivores; an increase in the proportion of top carnivores indicates a decrease in biotic integrity; (4) the proportion of individuals or biomass of benthic insectivores; a decrease is associated with a reduction in biotic integrity; and (5) the proportion or biomass of herbivores; which is usually associated with a decrease in biotic integrity from nutrient input (Goldstein et al., 1994). As trophic metrics have become more specialized with finer resolution, certain problems have arisen that cause confusion and difficulty in the application. One source of confusion results from the difficulty of classifying some species into a trophic group or guild. Another source of confusion is due to terminology. Consider the terms generalist, opportunist, and omnivore. Are they synonymous, or do they denote subtle differences? Unfortunately, the feeding ecology of North American freshwater fishes is, for the most part, poorly known (Gerking, 1994), and recent attempts to identify the nature of fish feeding guilds has not met with satisfactory results (Gerking, 1994). However, the use of guild structure and function are important metric components in evaluating biological integrity — the stability and sustainability of fish communities (Karr, 1981;

Karr et al., 1986; Davis and Simon, 1995; Simon and Emery, 1995). One approach to reduce these difficulties is a single classification system with consistent terminology.

The food chain concept of trophic dynamics as proposed by Lindeman (1942) assumes that species are organized along a stepwise system of trophic levels. Each level classifies organisms that utilize the same or similar food resources. At the base of the trophic pyramid (trophic level I) are the photosynthetic organisms. Trophic level II is made up of species that utilize photosynthetic organisms occurring in trophic level I. For the next steps up, a variety of primary consumers and secondary consumers convert biomass from trophic level 1 into usable energy for the top-level (trophic level IV) carnivores to consume. One of the problems of Lindeman's (1942) trophic scheme is that single species often occupy a variety of trophic levels during their lifetime. O'Neill et al. (1986) suggested that different species can often perform the same or similar trophic function, the same species can perform several functions, and the same species can perform different functions at different times and places. Without sufficient feeding ecology information, it may be impossible to assign a species to a single trophic level.

Most treatments of diet or feeding ecology of fishes are based on anecdotal or qualitative datasets from limited geographic areas. Unfortunately, without quantitative information, the placement of species into correct guilds can be difficult. However, it is the authors' opinion that broad patterns are identifiable and are important indicators of environmental condition. Identification of guilds and the classification of species into those guilds have not received the priority research needed. In order to properly evaluate niche dynamics and breadth, the development of a classification hierarchy is necessary.

Many aspects of feeding ecology can be used to classify the relationships of communities. However, the classification has been limited to what organisms feed upon (food habits), where foraging occurs (foraging habitat), and how food is acquired (foraging habits). Several researchers have evaluated current trophic guild classification schemes (Berkman and Rabeni, 1987; Horowitz, 1978; Lindeman, 1942; Merritt and Cummings, 1984; Moyle and Li, 1979; Schlosser, 1982) and found them inadequate for North American fish species, particularly with regard to fish communities of larger streams and rivers. Thus, this chapter compiles much of the known information on diet, morphology, and behavior for over 400 North American species based on literature and personal investigation to develop a trophic guild classification that might satisfy the needs of other North American investigators for use in biocriteria development (see Appendix).

7.2 SYNTHESIS OF EXISTING INFORMATION FOR FRESHWATER FISHES

Gerking (1994) compiled much of the known information on the feeding ecology of fishes, and found that the field is badly splintered and needs consolidation. He presented a series of definitions that have solidified the field and assisted in the development of our hierarchical classification. There are some differences between what Gerking proposed and the scheme presented here. First, Gerking concluded a classification of fishes could not be possible because the sheer range of variability of feeding ecology was too great to capture. The present scheme, however, limits its scope to concrete examples from a single continent, which may be the most practical way to develop the topic. Second, the often-equivalent classifications of Gerking's hierarchy are placed into a series of classes and modes. It was determined that this would be the best way to deal with incomplete literature without severely compromising the utility of the tool for metric development.

The literature on the feeding ecology of North American freshwater species was evaluated and a database based on published national and state treatments (Scott and Crossman, 1973; Trautman, 1981; Becker, 1983; Robison and Buchanan, 1988; Etnier and Starnes, 1993; Jenkins and Burkhead, 1994) was developed. Information on mouth position, ratio of digestive tract length to total body length, adaptations of the gut, color of the peritoneum, presence of pyloric caeca, teeth presence and

TABLE 7.1
Trophic Classification Scheme for North American Freshwater Fishes

Trophic Class	Trophic Subclass	Trophic Mode
I. Herbivores	Particulate feeder	Grazer
		Browser
II. Detritivores	Filter feeder	Suction feeder
		Filterer
	Particulate feeder	Biters
		Scoopers
III. Planktivores	Filter feeders	Mechanical sieve
		Mucus entrapment
		Ram filtration
		Pump filtration
		Gulping
	Particulate feeder	Size-selective pickers
IV. Invertivores	Benthic predators	Grazers
		Crushers
		Hunters of mobile benthos
		Lie-in-wait predators
		Tearers
		Diggers
	Drift predators	Surface feeders
		Water column feeders
V. Carnivores	Whole body	Stalking
		Chasing
		Ambush
		Protective resemblance
	Parasites	Blood suckers

shape, pharyngeal teeth and plate presence and shape, primary reported diets, trophic level classification (Lindeman, 1942), trophic mode and primary food type, and feeding behavior from these literature sources was also evaluated. Species with either differential or multiple sources of information were kept separate in order to facilitate interpretation. Also evaluated in the laboratory were the anatomical features for a few midwestern species for which there is no published information.

7.3 PROPOSED GUILD STRUCTURE

The classification structure recognizes five feeding guilds and 26 modes of feeding (Table 7.1). Keystone traits of each feeding guild are summarized below. This section describes important attributes of these modes, including representative behavior and alternative feeding strategies or tactics, and provides examples of species illustrating these behaviors.

7.3.1 HERBIVORES

Herbivores are primary consumers that eat green plant biomass made by photosynthesis. Green plants are divided into four broad categories of trophic guild classification: phytoplankton, epilithic algae, epiphytic algae, and macrophytes. Phytoplankton are unicellular or colonial algae that live suspended in the water column. Benthic algae may be epilithic, growing on rocks or hard substrates, or epiphytic, growing on leaves or other soft surfaces. Macrophytes are the aquatic (rooted, emergent, and floating) and terrestrial vascular plants and their reproductive parts (fruits, seeds, flowers, and leaves).

The consumption of algae and higher plants has evolved independently in at least 44 families of fish. Gerking (1994) indicates that 21 freshwater fish families consume plants as a principal part of their diet. No adult herbivorous fish is an obligate plant eater; however, early life history stages may experience ontogenetic niche shifts enabling differential food sources to be used during various life periods (Balon, 1985). Two theories attempt to explain the presence of animal matter in the herbivores' diet. The first suggests that animal matter is inadvertently eaten while grazing. The second suggests that fish will not grow if only macrophytes are eaten (Menzel, 1959). Although phytoplankton might compose the majority of the diet of some species, few if any feed entirely on phytoplankton. During the feeding process, other planktonic organisms typically are consumed. Here, phytoplankton feeders are placed with the more general planktivores (Group III, below).

Herbivores have terminal or subterminal mouths and may have physical adaptations of the maxillary, premaxillary, or supramaxillary structures to aid in feeding. In most cases, the digestive tract is as long or longer than the total length of the individual. The peritoneum is usually dark or colored. Herbivores have two types of alimentary canals; the first lacks a stomach but has an elongated, undifferentiated intestine, while the second type has a well-developed stomach and moderately elongated intestine. In the first type, mastication or chewing is accomplished by the pharyngeal teeth. Mastication is typically a part of the initial tearing or biting of the leaves or fronds. However, opposing pharyngeal teeth act as razors, cutting and shearing ingested plant materials into smaller pieces. The second type grinds food into smaller particles in a thick-walled, muscular stomach. Often, sand grains or other rock particles are responsible for the grinding action, along with the muscular contractions. Herbivore digestive systems exhibit extremes of pH depending on the amount and location of mechanical processing. Enzymes help penetrate the cells of the food or act on the cell contents after mechanical processing.

7.3.1.1 Particulate Feeders

Particulate feeders typically have a thick-walled, muscular stomach and use sand particles for grinding. Two feeding modes within the herbivore particulate feeding subclass are identified. Hiatt and Strasburg (1960) coined the terms "browsers" and "grazers" to describe the two ways of obtaining benthic algal. However, we see this same type of behavior being used to crop aquatic macrophytes. Often, species that fall into this subclass exhibit both modes of feeding behavior.

7.3.1.1.1 Grazers
Grazers crop algae very close to the substrate and may ingest some of the substrate along with the plant material. Grazers feed by pointing the head down and applying their lips and teeth to the substrate. For the most part, they feed on epilithic algae. They obtain food by rasping and suction, and their intestines are invariably packed with calcareous debris and other materials. The central stoneroller *Campostoma anomalum*, largescale stoneroller *C. oligolepis*, chiselmouth *Acrocheilus glutaceus*, and bridgelip sucker *Catostomus columbianus* are examples of grazers.

7.3.1.1.2 Browsers
Browsers bite off pieces of plants above the substrate and feed on epiphytic alga and macrophytes. Browsers point the head upward and use a bite-and-tear technique. These species feed on fruits, seeds, flowers, and leaves. Numerous browser feeding strategies exist, including bite-and-tear, nipping at the tips of filamentous algae (Fryer and Iles, 1972), and picking motions when feeding on filamentous and unicellular algae (Prejs, 1984; Okeyo, 1989; Horn, 1989). Examples of browsers include grass carp *Ctenopharyngodon idella*, ozark minnow *Notropis nubilus*, blackside dace *Phoxinus cumberlandensis*, and Tennessee dace *P. tennessensis*.

7.3.2 Detritivores

Detritus is nonliving, organic matter and its associated microflora in various states of decomposition, which accumulates on the bottom of lakes, ponds, and streams. Detritus is a complex mixture of animal and plant remains that has been broken down chemically and physically by a variety of invertebrate and vertebrate organisms. Detritivores usually possess black peritoneums and extensively coiled guts. The alimentary canal is simple and unspecialized from the esophagus to the rectum. It consists of a sac-like stomach, thin-walled duodenum, and a long (5 to 15 times body length), coiled intestine. Jaw teeth are small, flattened blades adapted for scraping, and the pharyngeal bones grind against each other, reducing the detritus into finer particles. The alimentary tract has two major mechanisms for handling detritus: (1) a simple stomach and long intestine, and (2) a two part stomach and somewhat shorter intestine. Detritivores are placed into two subclasses: filter feeders and particulate feeders.

7.3.2.1 Filter Feeders

The two filter feeding modes are suction feeders and filterers.

7.3.2.1.1 Suction Feeders

Flocculent detritus (ultra-fine organic matter) is sucked into the branchial cavity using lips, some of which possess fine, short teeth. The detritus is sorted in the oral cavity, which is an inverted V-shaped tube in transverse section. Sand and other heavier sediments settle in the apex of the V-shaped tunnel, while the flocculent mixture stays in suspension. Some species may utilize their pectoral fins to stir up the detritus, and then draw the sediments in by suction. Others may position their bodies obliquely downward with their mouths in the bottom ooze. Examples of suction-feeding detritivores include river carpsucker *Carpiodes carpiodes*, highfin carpsucker *C. velifer*, quillback *C. cyprinus*, white sucker *Catostomus commersoni*, and common carp *Cyprinus carpio*.

7.3.2.1.2 Filterers

Detritus is obtained by filtering organic particles from the water. Protraction of the premaxillaries and movement of the head rapidly from side to side while sucking the substrate produces conical depressions in the substrate. When the mouth is opened, the palatine, stenohyoid, and opercular muscles, combined with the muscles of the branchial arches, create a suction action by enlarging the buccal cavity. Solid particles pass through the convex pharyngeal pads projecting from the roof of the mouth and the concave depression of the floor of the mouth. The dorsal pad is coated with a thick, soft membrane studded by minute teeth. Gerking (1994) uses the striped mullet *Mugil cephalus* as a representative of this feeding mode. Here, southern brook lamprey *Ichthyomyzon gagei*, western brook lamprey *Lampetra richardsoni*, sea lamprey ammocete *Petromyzon marinus*, and silver carp *Hypophthalmichthys molitrix* are also classified into this feeding mode.

7.3.2.2 Particulate Feeders

There are two types of particulate feeding modes: biting and scooping.

7.3.2.2.1 Biters

Some species graze on the bottom by biting into the substrate and swallowing the detrital mixture. Others may bite or scrape the thick deposits of periphyton and sedimented detritus off the leaves of macrophytes. The pugnose shiner *Notropis anogenus* is a detritivore that scrapes accumulated detritus from macrophyte leaves. The spotfin shiner *Cyprinella spiloptera* and weed shiner *Notropis texanus* also exhibit this feeding mode.

7.3.2.2.2 Scoopers

Soft substrates are scooped up from the bottom, sorted, and then coarser particles are discarded through the gill covers. The fish opens its mouth and the lower jaw scoops off the top organic layer of the mud. The top organic layer of the mud seems to be the desired diet; however, some underlying sediments are probably ingested as well. Microbranchiospines on the gill arches seem to enable the organism to remove desired particles from the water. Scoopers include species such as bluntnose minnow *Pimephales notatus*, fathead minnow *P. promelas*, pugnose minnow *Opsopoeodus emiliae*, mountain redbelly dace *Phoxinus oreas*, and brassy minnow *Hybognathus hankinsoni*.

7.3.3 PLANKTIVORES

Plankton feeders typically possess elongated pharyngeal teeth, lack a stomach but have an elongated, undifferentiated intestine. Six feeding modes, based on how plankton are filtered from the water column, are proposed. Planktivores are grouped into two trophic classes: nondiscriminant filter feeders and particulate feeders.

7.3.3.1 Nondiscriminant Filter Feeders

Five modes of nondiscriminant filter feeding are recognized: mechanical sieve, mucus entrapment, ram filtration, pump filtration, and gulping.

7.3.3.1.1 Mechanical sieve

This is a passive mode of filtration where feeding is indiscriminate. The gill rakers are the sieve. Smaller particles pass through the interraker spaces, while larger particles are retained. Usually, four pairs of gill rakers are found on gill arches. The rakers are positioned so that they filter water passing through the branchial chamber. The gill arches are moved back and forth by muscular attachments at the base of the arch. The bighead carp *Hypophthalmichthys nobilis* exhibits this feeding mode.

7.3.3.1.2 Mucus entrapment

Greenwood (1953) indicated some cichlid species that feed on phytoplankton utilize a mouth mucus to entangle the algae. The mixture of mucus and algae is then carried posteriorly to the pharyngeal teeth where it is broken into smaller fragments. This may be a secondary feeding mode for some mechanical sieve filter feeders. Mucus cells have been reported for the menhaden *Brevoortia tyrannus*, which uses filtering as its primary feeding mode (Friedland, 1985). River lamprey *Lampetra ayresi* provide an example of this feeding mode.

7.3.3.1.3 Ram filtration

Ram filterers swim with their mouths open and opercles flaring. Water flows into the buccal cavity and out of the opercular openings. The spaces between the gill rakers utilize the mechanical sieve process (see Section 7.3.3.1.1). Ram filtration is not a continuous process. All ram filterers must overcome bow pressure, which is a wave of water pushed ahead of the swimming fish; otherwise, the food item will be pushed to the side rather than ingested (Walters, 1966). Some species overcome bow pressure by opening the mouth and expanding the branchial chamber at the proper instant to avoid bow pressure but not enough to generate suction. Probably, the best-known example of a planktivore ram filtration feeder is the paddlefish *Polyodon spathula*.

7.3.3.1.4 Pump filtration

The fish pumps water into the buccal cavity by a series of rapid, nondirected suctions while the fish is stationary (Holanov and Tash, 1978). The buccal cavity expands quickly and the opercles flare to expel water after the cavity has filled. Pump filterers have differential catch success since some prey have the ability to escape. Examples of pump filtration include the herrings, such as threadfin shad *Dorosoma petenense*, Alabama shad *Alosa alabamae*, American shad *A. sapidissima*, and hickory shad *A. mediocris*.

7.3.3.1.5 Gulping

Gulping is similar to pump filtration except that the mouth is not open as widely and the opercles not as flared as in pump filtering (Gibson and Ezzi, 1985). The suctions are nondirected and relatively large volumes of water are engulfed. Gulping may be a variation or conjunctive mode of feeding with other filter feeding behaviors. Gerking (1994) indicated that this feeding mode is a modification of pump filtration. Friedland (1985) noted gulping in menhaden.

7.3.3.2 Particulate Feeders

Particulate feeding planktivores are size-selective feeders, usually opting for the largest particles.

7.3.3.2.1 Size-selective pickers

Field studies and laboratory experiments have all concluded that many planktivorous species select the largest zooplankton species available. A significant body of work has emerged supporting Optimal Foraging Theory models (Werner, 1974; Werner and Hall, 1974; Savino and Stein, 1982; Mills et al., 1984; Goldstein, 1993). Visual particulate feeding is influenced by reactive distance. Prey size, water clarity, prey behavior, prey translucency, prey shape, color, and other factors affect reactive distance. One of the better-known examples of a particulate planktivore is the blueback herring *Alosa aestivalis* (Brooks and Dodson, 1965). Additional species classified here as size-selective pickers are alewife *A. psuedoharengus*, emerald shiner *Notropis atherinoides*, bridle shiner *N. bifrenatus*, southern redbelly dace *Phoxinus erythrogaster*, shortjaw cisco *Coregonus zenithicus*, fourspine stickleback *Apeltes quadracus*, and ninespine stickleback *Pungitius pungitius*.

7.3.4 INVERTIVORES

Invertivores compose the largest and perhaps the most diverse trophic class. It includes species that feed on the smallest midge, to species that consume large mollusks. Our two subclasses are based on feeding location; they are grouped according to whether they feed primarily on benthos or from the water column.

7.3.4.1 Benthic Predators

Benthic feeders have a wide variety of adaptations to assist in exploiting benthos. Benthos refers to the invertebrate organisms that live on the bottom substrate (e.g., rocks or sand), in the substrate (e.g., sand or mud), or above the substrate (e.g., on vegetation). Benthic predators bite and suck, probe crevices, actively gather, ambush, crush heavy-shelled forms, and crop external invertebrate body parts. Feeding occurs diurnally, nocturnally, or continuously. Morphological adaptations include stationary or protruding jaws; needle-like teeth or crushing molars; ventral sucking mouths or superior mouths; soft vacuum-cleaner lips or bony slashing jaws. Pharyngeal teeth vary from heavy molar-like teeth for crushing to file-like teeth that comb and sort small organisms. Benthic predators are represented by six different feeding modes.

7.3.4.1.1 Grazers

Almost all benthic predators are grazers. This trophic mode can feed on a variety of organisms from habitats ranging from the undersurface of rocks to the tops of vegetation. Food items include the range of invertebrate taxa and may also include periphyton and other portions of the substrate. As such, this category contains the nondiscriminate benthic invertivores. The group is quite large and is useful for separating this general mode of feeding from the more specific groups below.

Examples of grazers include the whitetail shiner *Cyprinella galactura*, red shiner *Cyprinella lutrensis*, fatlips minnow *Phenacobius crassilabrum*, suckermouth minnow *P. mirabilis*, Kanawha minnow *P. teretulus*, and stargazing minnow *P. uranops*.

7.3.4.1.2 Crushers

This group possesses heavy dentine or pharyngeal teeth. Some species have both molariform pharyngeal teeth in addition to jaw teeth to crush prey. Gastropod and mollusk crushers gather shells and feed on the soft parts of the body. Often, crushers possess powerful jaws and massive dentine teeth. Others use a different strategy, including less specialized jaw teeth for picking up gastropods and flat-crowned teeth mouthed on molar-like pharyngeal bones. The redear sunfish *Lepomis microlophus* is known as the shell cracker in some parts of its range because of its feeding mode. Additional species classified in this trophic mode include the extinct harelip sucker *Lagochila lacera*, river redhorse *Moxostoma carinatum*, greater redhorse *M. valenciennesi*, and snail darter *Percina tanasi*.

7.3.4.1.3 Hunters of mobile benthos

Benthic hunters feed either alone or in small groups during either nocturnal or diurnal periods. These groups leave the safety of burrows, crevices, or ledges to stalk prey. Many North American invertivores actively search for benthos. The smallmouth bass *Micropterus dolomieui* is an active feeder on crayfish. Sturgeon and catfish use sensitive barbels to locate benthic invertebrates. Additional species classified as benthic hunters include goldfish *Carassius auratus*, thicklip shiner *Cyprinella labrosa*, slender chub *Erimystax cahni*, tonguetied minnow *Exoglossum laurae*, cutlips minnow *E. maxillingua*, speckled chub *Macrhybopsis aestivalis*, and bigeye chub *Notropis amblops*.

7.3.4.1.4 Lie-in-wait predators

Members of this feeding mode lie in caverns or may hover in mid water waiting for benthic prey to come close enough to capture. Species classified in this trophic mode include Alaska blackfish *Dallia pectoralis*, most darters, and madtoms.

7.3.4.1.5 Tearers

This category is known to rip off tissues of larger bottom organisms. This group includes sponge feeders. The authors are not aware of any freshwater species that exclusively use this trophic mode. With the increase in freshwater sponges in the Great Lakes, one would expect to see some species that utilize sponges as a food item; hence, the inclusion of this mode.

7.3.4.1.6 Diggers

Diggers dig food organisms from the bottom sand or actively move pieces of the bottom substrate to uncover food items. Members of the sculpin genus *Cottus* and darter species, such as logperch *Percina caprodes* and tangerine darter *P. auentiaca*, fit into this feeding mode.

7.3.4.2 Drift Feeders

Drift feeders specialize in taking drifting or swimming invertebrates. Mouths are terminal or supraterminal for taking items either drifting with the current in the water column or on the surface. Most drift in streams occurs during the night (Waters, 1965), while vertical movements of invertebrates in lakes are usually associated with changing light intensity. Therefore, identification of this feeding mode may require confirmation by correlation of nocturnal stomach contents with nocturnal drift sampling or vertical plankton tows.

7.3.4.2.1 Surface drift feeders

Species with terminal or oblique mouth positions approach prey items from below and may sip, gulp, or utilize a buccal pump to engulf prey. The presence of terrestrial invertebrates in the stomach contents is one key to determining this mode. Examples include surface-feeding salmonids, brook silverside *Labidesthes sicculus*, mooneye *Hiodon tergisus*, goldeye *H. alosoides*, bigeye shiner *Notropis boops*, flathead chub *Platygobio gracilis*, many topminnows, and mosquito fish *Gambusia affinis*.

7.3.4.2.2 Water column drift feeders

Mouth position is typically terminal. Fish in this group may position themselves on the bottom, by or behind structure (in many cases, in areas protected from velocity), or suspend in the water column and wait for approaching prey. The common shiner *Luxilus cornutus*, redside dace *Clinostomus elongatus*, blue shiner *Cyprinella caerulea*, steelcolor shiner *Cyprinella whipplei*, whitemouth shiner *Notropis alborus*, and most salmonids, whitefish, and sunfish utilize this feeding mode.

7.3.5 CARNIVORES

Fish that eat or use some part of the body as a primary source of food are not strictly piscivores. The term has been reserved for species that consume fish whole. Although piscivores are often thought to occupy the top of the food chain and are few in number, in certain habitats they may make up a higher percentage (50%) of the assemblage. Simon and Emery (1995) have shown that Great Rivers, such as the Ohio River, possess high percentages of piscivores; Van Oijen (1982) found 86 piscivore species in the Malawi Gulf of Lake Victoria, East Africa.

7.3.5.1 Whole Body (Piscivore)

Here, the term whole body is used, rather than piscivore, to reduce confusion, but piscivore is accurate.

7.3.5.1.1 Stalking

The unobtrusive pursuit of prey, or stalking, is not a widespread strategy of piscivores, but it does occur in a variety of forms. Movement forward and toward the prey item is a result of slow movements until gaining a preferred position, at which time a strike takes place. An alternative strategy utilizes another fish to "shield" the approach. The predator does not mimic the behavior or coloration of the shield, but hides behind it as both swim along the bottom. The burbot *Lota lota* is tentatively classified in this feeding mode.

7.3.5.1.2 Chasing

Large teleosts are known to roam the open seas, sometimes in schools searching for large rations to support their high metabolic needs. They do not often run down individual prey, but commonly seek out schools of favored prey. Some predators will "herd" the schools of prey into large compact groups and then swim through the school thrashing side to side, either stunning or killing prey. The predator then swims beneath the school to consume the sinking prey. The striped bass *Morone saxatilis* may be the only freshwater example in this trophic guild.

7.3.5.1.3 Ambush

Ambush is perhaps the most commonly used strategy of capturing other fish. Predators lie in wait, usually in a location of concealment such as aquatic vegetation, submerged woody debris, or other rock structures. Individuals usually wait in seclusion until the object of choice comes into view; then, with an explosive rush, the predator darts out at the prey. North American species of the Esocidae, Lepisosteidae, and bowfin *Amia calva* utilize the ambush technique in conjunction with protective resemblance.

7.3.5.1.4 Protective resemblance

The mimicry of some part of the natural environment causes the predator to blend in with its habitat. The same coloration that provides predator protection also hides the animal from view while lying-in-wait for its prey to approach. Most species using this approach are benthic inhabitants that feed on invertebrates and fish. Many of these species have lost their swim bladders and remain quietly on the bottom, concealed by cryptic color patterns, or bury themselves beneath the substrate. The walleye *Stizostedion vitreum* and sauger *S. canadense* are classified in this trophic mode even though both species have a swim bladder.

7.3.5.2 Parasites

Although there are numerous parasitic strategies throughout the world that include scale eating, fin tearing, and blood sucking, only the latter occurs in North America.

7.3.5.2.1 Blood suckers

Blood suckers include species that feed on blood from gills, from the back or sides of hosts, and several that burrow under the scales (Kelley and Atz, 1964). Species have either dentary teeth or a large rasping tongue that penetrates the skin or membranes. The parasitic lamprey often attaches to the side of the host and secretes anticoagulants that keep the blood flowing.

7.4 DISCUSSION

The intention here is to provide a framework to classify North American freshwater fishes and provide detailed information to achieve this result (see Appendix 7A for species-specific data). This initial attempt to classify fish into guilds based on a three-level classification scheme may be premature because adequate information is not available for many species. In addition, data gaps exist that require further research to determine more-precise guild membership. It is the authors' intention that as this trophic guild classification scheme is used, it will be refined and improved.

To effectively utilize this classification of North American freshwater fishes, the proposal is that feeding guilds be assigned to the species in a community sample; that the number of individuals and the proportions of the community in each feeding guild be computed for watershed calibration and application; and that the proportions within each subclass and mode (if information is available) be compared with reference conditions or watershed expectations.

Further refinement of the classification to the feeding mode level may provide additional insight. We believe that specific types of perturbations can be identified by the change in feeding mode of certain species. However, the importance of these redefined feeding guilds may be difficult to determine *a priori*. Therefore, when placing taxa into specific subclasses or modes, the practitioner should attempt to limit species membership only to those species actually in the subclass or using the feeding mode in the streams under investigation. Extrapolation from other areas or streams may defeat the purpose of this classification: the comparison of trophic composition between two types of streams.

7.4.1 OMNIVORES, OPPORTUNISTS, AND GENERALISTS

Karr (1971) and Schlosser (1982) defined omnivores as species that have at least 25% each of both plant and animal items in their diet and usually have a dark peritoneum and long gut. Problems with this definition are apparent for nondiscriminate planktivores, such as gizzard shad, that filter feed and consume differential quantities of zooplankton and phytoplankton; for species that may consume both plant and animal food but not in those proportions; and for species that cross trophic classes without consuming plant material (detritivore/invertivore). Although characterization of gizzard shad as an omnivore is appropriate, it does not fully capture the variable ratio of phytoplankton to zooplankton consumption most clupeids have as pelagic filter feeders. Goldstein et al. (1984) suggested a more reduced definition of omnivores as possessing a black peritoneum and a long coiled gut. That definition did not completely identify species such as carp, buffalo, and carpsuckers, which do not have a long coiled gut but seem to be omnivorous. The ultra-fine organic material feeders (UFOM), which are designated in the detritivore guild, are relatively few in number but may account for a large portion of the community biomass. The use of the term "omnivore" is not descriptive nor specific enough to collectively capture UFOM detritivores that generally increase in relative abundance with decreasing environmental quality.

Additional terms such as "generalist" and "opportunist" were created to indicate species that had broad niche breadths and were nonselective consumers (Leonard and Orth, 1986). Smogor and Angermeier (Chapter 9, this volume) identified generalists as indiscriminate feeders that occupied four to five different trophic guilds. Generalists consume wide varieties of specific taxa (e.g., all insects or a variety of plant and animal items). When attempting to evaluate biotic integrity, the ability to discriminate trophic composition is confounded by these definitions. Specific problems have also occurred when foraging encounters include particle size selection or the ability to diet shift with declining resources. In such situations, blacknose dace *Rhinichthys atratulus* and creek chub *Semotilus atromaculatus* can be labeled as generalists. Certainly, the brook trout *Salvelinus fontinalis,* which consumes a randomly encountered drifting mayfly prior to eating a benthic caddisfly after it has eaten a small dace, is an opportunist. Incidental consumption of plant material while targeting invertebrate prey items has confounded trophic characterization of certain species. In the past, qualitative separation criteria, such as minimum diet requirements and specific morphological attributes, were key identifiers of particular terms.

Although there have generally been differences attributed to the terms "opportunist" and "generalist," they both describe a species that may be classified in more than a single trophic guild. The terms reflect survival and trophic strategies and behavior rather than a strict set of food items. Evolutionary development of feeding strategies suggests two diverging strategies: the generalist and the specialist. Most species may be classified into either of these groups depending on morphology (i.e., presence or absence of specialized mouth parts) and behavior. With the objective of making a trophic classification system for evaluating biological integrity, the focus should not be on classifying a species as a generalist, opportunist, or omnivore, but rather classifying species into trophic guilds for comparison to reference conditions or expectations. A classification of omnivores was specifically omitted from this classification. While use of the omnivore classification has been prevalent in IBIs historically, the scope and mixture of the trophic interactions are lost. The classification of a species into multiple trophic classes is indicative of the variability of feeding ecology; and while the terms "omnivore," "generalist," and "opportunist" certainly are indicative of this variability, they do not indicate the breadth of food items nor the types of feeding behaviors. The introduction of another level of resolution, subclass or trophic mode, can reduce the lack of resolution. The objective is to determine the trophic structure of the fish community for comparison with expectations. If the classification is insufficient to detect differences, then either the classification is not sufficiently sensitive or no differences occur. A multiple-level classification system allows for comparison at finer and finer levels of distinction if the information is available. The benefit of additional classification levels becomes apparent when the same species is classified differently between reference conditions and aquatic systems being evaluated, and may indicate not only a change in biotic integrity but also the source. For example, in both reference conditions and the streams being evaluated, common shiners are classified as insectivores and no change in trophic composition is apparent. If the additional classification is applied, then in the reference conditions, common shiners are classified as benthic grazers, water column drift feeders, and surface drift feeders. At the site being evaluated, stomach content analysis indicates they are classified only as surface and water column drift feeders. The loss of the benthic grazing mode is indicative of a reduction in biotic integrity. The reduction in benthos may indicate the source of perturbation.

7.4.2 Species with Multiple Trophic States over the Course of Their Lives

The objective of the classification system described herein is to provide guidance for separation of community members into their various trophic guilds for purposes of metric construction and comparison. The variability in feeding habits of most species involves several changes of guild: (1) during the different life history stages: ontogenetic niche shifts; (2) changes in food items based on availability or selection: diet switching; and (3) changes in food selections based on competition: food partitioning. There is some evidence that feeding mode and prey selection are learned behaviors (Mauck and Coble, 1971; Epinosa and Deacon, 1973; Hansen and Wahl, 1981; Gillen et al., 1981; Colgan et al., 1986), and therefore may change from location to location. In light of the variability inherent in the trophic dynamics of an individual, a species, and an entire community, the question of classification becomes increasingly more difficult.

Information regarding the life stage, size, or level of maturity that must be attained prior to changing trophic guilds is not readily available for most species. Most IBI community analyses ignore individuals less than 25 mm in total length to solve part of this problem. There is an unclear distinction here between a change in guild due to size or age (ontogenetic niche shift) and the addition of feeding modes due to availability (diet switching or partitioning). The distinction might not be necessary for evaluation of biotic integrity. The metrics used in IBIs are designed to determine differences for comparative purposes, but not necessarily the reasons for the differences.

The level of classification used for the trophic metrics determines whether behavioral observations or examination of stomach contents are warranted. Classifications that are limited to the major classes probably do not require stomach content analysis. Species can be assigned to a trophic class, the first level of the hierarchy, based on morphological characteristics and data from the literature. For construction of metrics that require the use of subclass, stomach content examination may be required; and for those metrics that use trophic mode, *in situ* behavioral studies may be required unless mode can be inferred from the stomach contents. This type of resolution is both time consuming and costly. While the proposed classification system is based on morphological features, what occurs in nature will be modified by a host of environmental conditions and species interactions.

Some species may be classified into more than a single guild. Changing food habits (food items) and feeding behavior (feeding mode) are common. Classification into two or more guilds does not negate the classification system nor invalidate the use of trophic metrics for determining biological integrity. The information conveyed by comparison of a species from a reference area, where it may belong in only a single guild, to an area to be evaluated where the species belongs to two guilds is significant. Ontogenetic niche shifts, diet shifting, and food partitioning can be incorporated into trophic metrics. Such metrics can increase the determination of biological integrity by identifying departures from expected behaviors.

In cases where the number of species is low, size or age classes can have different classifications. Multiple trophic classifications for a single species are not unusual. The Ohio EPA (1989) restricts adult classification of carnivores to specific minimum sizes (Table 7.2). The focus of the metrics should be the determination of differences or change. In some cases, it might be sufficient to use a metric that documents only the proportion of species that exhibits a trophic change.

7.5 CONCLUSION

This classification scheme is not intended to be a treatise on trophic ecology; rather, it is a working tool to be used for determining biotic integrity. The proposed trophic classification is based on the concept of "function follows structure." If the species has the morphological and physiological capability to feed on certain types of organisms, then it probably will do so at some instant in time. The plasticity inherent in the feeding modes of fishes is great; but for the purpose of comparing energy flow and biomass through aquatic ecosystems, the classification attempts to balance the level of information needed to define the trophic composition of a community with the effort needed to detect differences in biotic integrity. Workers in the field should compare trophic metrics based on both existing schemes and those proposed to evaluate the capabilities of all methods. In areas with an existing database, known perturbations, and completed IBI evaluations, comparisons of the

TABLE 7.2
Pivotal Size and Weight Categories that Determine Ontogenetic Niche Shifts for Select Carnivores

Species	Size (mm)	Weight (g)
Brown trout	300	550
Grass pickerel	100	7
Chain pickerel	100	7
Northern pike	50	5
Muskellunge	50	5
Burbot	500	1000
White bass	120	20
Striped bass	130	25
Sauger	150	50
Rock bass	120	50
Spotted bass	150	40
Smallmouth bass	50	10
Largemouth bass	80	10
Yellow perch	150	50

Note: Lengths and weights indicated are classification cutoffs for determining minimum sizes when species are classified as carnivores for calculation of an Index of Biotic Integrity. As classified by the Ohio Environmental Protection Agency.

different trophic metrics can be accomplished and tested. This trophic concept is just that — a hypothesis to be tested. The authors believe that this trophic guild system will provide more-accurate determinations of changes in biotic integrity and encourage its testing and modification as needed.

ACKNOWLEDGMENTS

The authors thank the late Shelby Gerking for his encouragement to pursue this subject, and the helpful comments of the reviewers, which greatly improved the paper. The first author received support from the Red River of the North and Upper Mississippi River Study Units of the National Water Quality Assessment Program of the U.S. Geological Survey.

REFERENCES

Balon, E.K. 1985. *Early Life Histories of Fishes: New Developmental, Ecological and Evolutionary Perspectives.* Dr. W. Junk Publishers, Boston.

Becker, G.C. 1983. *Fishes of Wisconsin.* University of Wisconsin Press, Madison.

Berkman, J.E. and C.F. Rabeni. 1987. Effect of siltation in stream fish communities, *Environmental Biology of Fishes,* 18, 285–294.

Brooks, J.L. and S.I. Dodson. 1965. Predation, body size, and the composition of plankton, *Science,* 150, 26–35.

Colgan, P.W., J.A. Brown, and S.D. Orsatti. 1986. Role of diet and experience in the development of feeding behavior in largemouth bass, *Micropterus salmoides, Journal of Fish Biology,* 28, 161–170.

Davis, W.S. and T.P. Simon. 1995. *Biological Assessment and Criteria: Tools for Water Resource Planning and Decision Making.* Lewis, Boca Raton, FL.

Epinosa, F.A. and J.E. Deacon. 1973. The preference of largemouth bass (*Micropterus salmoides* Lacepede) for selected basin species under experimental conditions, *Transactions of the American Fisheries Society*, 102, 335–362.

Etnier, D.A. and W.C. Starnes. 1993. *The Fishes of Tennessee*. University of Tennessee Press, Knoxville, TN.

Friedland, K.D. 1985. Functional morphology of the branchial basket structure associated with feeding in the Atlantic menhaden, *Brevoortia tyrannus*, (Pisces: Clupeidae), *Copeia*, 1985, 1018–1027.

Fryer, G. and T.D. Iles. 1972. *The Cichlid Fishes of the Great Lakes of Africa*. T. F. H. Publications, Neptune, NJ.

Gerking, S.D. 1994. *Feeding Ecology of Fish*. Academic, New York.

Gibson, R.N. and I.A. Ezzi. 1985. Effect of particle concentration on filter- and particulate-feeding in the herring, *Clupea harengus, Marine Biology*, 88, 949–962.

Gillen, A.L., R.A. Stein, and R.F. Carline. 1981. Predation by pellet-reared tiger muskellunge on minnows and bluegills in experimental systems, *Transactions of the American Fisheries Society*, 110, 197–209.

Goldstein, R.M. 1993. Size selection of prey by young largemouth bass, *Proceedings of the Annual Conference of the Southeastern Association of Fish and Wildlife Agencies*, 47, 596–604.

Goldstein, R.M., T.P. Simon, P.A. Bailey, M. Ell, E. Pearson, K. Schmitt, and J.W. Enblom. 1994. Concepts for an index of biotic integrity for streams of the Red River of the North Basin, in B. Seelig (Coordinator), *Proceedings of the North Dakota Water Quality Symposium*, March 30–31, 1994, Fargo. 169–180.

Greenwood, P.H. 1953. Feeding mechanism of the cichlid fish, *Tilapia esculenta* Graham, *Nature (London)*, 172, 207–208.

Haitt, R.W. and D.W. Strasburg. 1960. Ecological relationships of the fish fauna on coral reefs of the Marshall Islands, *Ecological Monographs*, 30, 65–127.

Hansen, M.J. and D.H. Wahl. 1981. Selection of small *Daphnia pulex* by yellow perch in Oneida Lake, New York, *Transactions of the American Fisheries Society*, 110, 64–71.

Holanov, S.H. and J.C. Tash. 1978. Particulate and filter feeding in threadfin shad, *Dorosoma petenense*, at different light intensities, *Journal of Fisheries Biology*, 13, 619–625.

Horn, M. 1989. Biology of marine herbivorous fishes, *Oceanography Marine Biology Annual Review*, 27, 167–272.

Horowitz, R.J. 1978. Temporal variability patterns and the distributional patterns in stream fishes, *Ecological Monographs*, 48, 307–321.

Jenkins, R.E. and N.M. Burkhead. 1994. *Freshwater Fishes of Virginia*. American Fisheries Society, Bethesda, MD.

Karr J.R. 1971. Structure of avian communities in selected Panama and Illinois habitats, *Ecological Monographs*, 41, 207–233.

Karr, J.R. 1981. Assessment of biotic integrity using fish communities, *Fisheries*, 6, 21–27.

Karr, J.R., K.D. Fausch, P.L. Angermeier, P.R. Yant, and I.J. Schlosser. 1986. Assessing Biological Integrity in Running Waters: A Method and Its Rationale. Illinois Natural History Survey Special Publication 5.

Kelley, W.E. and J.W. Atz. 1964. A pygidiid catfish that can suck blood from goldfish, *Copeia*, 1964, 702–704.

Leonard, P.M. and D.J. Orth. 1986. Application and testing of an index of biotic integrity in small, coolwater streams, *Transactions of the American Fisheries Society*, 115, 401–414.

Lindeman, R.L. 1942. The trophic dynamic aspect of ecology, *Ecology*, 23, 399–418.

Mauck, W.L. and D.W. Coble. 1971. Vulnerability of some fishes to northern pike (*Esox lucius*) predation, *Journal of the Fisheries Research Board of Canada*, 28, 957–969.

Menzel, D.W. 1959. Utilization of algae for growth by angelfish, *J. Cons., Cons. Int. Explor. Mer.*, 24, 308–313.

Merritt, R.W. and K.W. Cummins. 1984. *An Introduction to the Aquatic Insects of North America*. Kendell/Hunt Publishing Company, Dubuque, IA.

Mills, E.L., J.L. Confer, and R.C Ready. 1984. Prey selection by young yellow perch: the influence of capture success, visual acuity, and prey choice, *Transactions of the American Fisheries Society*, 113, 579–587.

Moyle, P.B. and H.W. Li. 1979. Community ecology and predator-prey relations in warmwater streams, in R.H. Stroud (Ed.), *Predator-Prey Relationships in Fisheries Management*. American Fisheries Society, Bethesda, MD. 171–178.

Ohio Environmental Protection Agency. 1989. Biological Criteria for the protection of aquatic life. Vol. III. Standardized Biological Field Sampling and Laboratory Methods for Assessing Fish and Macroinvertebrate Communities. Ohio EPA, Division of Water Quality Planning and Assessment, Ecological Assessment Section, Columbus, OH.

Okeyo, D.O. 1989. Herbivory in freshwater fishes: a review, *Israeli Journal of Aquaculture — Bamidgeh,* 41, 79–97.

O'Neill, R.V., D.L. DeAngelis, J.B. Waide, and T.F.A. Allen. 1986. *A Hierarchical Concept of Ecosystems.* Princeton University Press. Princeton, NJ.

Prejs, A. 1984. Herbivory by temperate freshwater fishes and its consequences, *Environmental Biology of Fishes,* 10, 281–296.

Robison, H.W. and T.M. Buchanan. 1988. *Fishes of Arkansas.* The University of Arkansas Press, Fayetteville.

Savino, J.F. and R.A. Stein. 1982. Predator-prey interaction between largemouth bass and bluegill as influenced by simulated submerged vegetation, *Transactions of the American Fisheries Society,* 111, 255–266.

Schlosser, I.J. 1982. Fish community structure and function along two habitat gradients in a headwater stream, *Ecological Monographs,* 52, 395–414.

Scott, W.B. and E.J. Crossman. 1973. Freshwater Fishes of Canada. Fisheries Research Board of Canada Bulletin 184.

Simon, T.P. and E.B. Emery. 1995. Modification and assessment of an Index of Biotic Integrity to quantify water resource quality in Great Rivers, *Regulated Rivers Research and Management,* 11, 283–298.

Trautman, M.B. 1981. *The Fishes of Ohio.* The University of Ohio State Press, Columbus.

Van Oijen, M.J.P. 1982. Ecological differentiation among piscivorous haplochromine cichlids of Lake Victoria (East Africa), *Netherlands Journal of Zoology,* 32, 336–363.

Walters, V. 1966. On the dynamics of filter-feeding by the wavyback skipjack (*Euthynnus affinis*), *Bulletin of Marine Science,* 16, 209–221.

Waters, T.F. 1965. Interpretation of invertebrate drift in streams, *Ecology,* 46, 327–334.

Werner, E.E. 1974. The fish size, prey size, handling time relation in several sunfishes and some implications, *Journal of the Fisheries Research Board of Canada,* 31, 1531–1536.

Werner, E.E. and D.J. Hall. 1974. Optimal foraging and the size selection of prey by the bluegill sunfish (*Lepomis macrochirus*), *Ecology,* 55, 1042–1052.

Appendix 7A

Order	Family	Genus	Species	Common Name	Mouth Position	Ratio of Digestive Tract (DT) to Total Length (TL)[c]	Gut	Color of Peritoneum	Pyloric Caeca	Teeth	Pharyngeal Teeth and Plates	Main Food	First Level Trophic Classification	Second Level Trophic Classification	Trophic Mode	Feeding Behavior
Petromyzon-tiformes	Petromyzon-tidae	Ichthyomyzon	bdellium[d]	Ohio lamprey adult	Subterminal[d]					Yes[d]		Blood of other fish like suckers[d]	Carnivore	Parasite	Blood sucker	
			bdellium[e]	Ohio lamprey ammocoete						No[e]						Filter feeders[e]
			castaneus[a]	Chestnut lamprey ammocoete	Subterminal[a]					Yes		Diatoms and desmids[c]	Herbivore	Filter feeder		Nonparasitic[c]
			castaneus[a]	Chestnut lamprey adult	Subterminal[a]					Yes[e]		Blood of other fish[e]	Carnivore	Parasite	Blood sucker	Nonparasitic[c]
			fossor[a]	Northern brook lamprey ammocoete	Subterminal[a]					No[e]		Microscopic organisms such as diatoms and unicellular algae[c]	Herbivore	Filter feeder		Nonparasitic[c]
			fossor[a]	Northern brook lamprey adult	Subterminal[a]					Yes[e]						Does not feed[e]
			gagei[d]	Southern brook lamprey	Subterminal[d]					Yes[d]		Plankton, particularly diatoms, and organic detrius (Moshin and Galloway, 1997)[f]	Herbivore/detritivore	Filter feeder	Filterer	Filter feeders (Moshin and Galloway, 1997)[f]
			greeleyi[d]	Mountain brook lamprey adult	Subterminal[d]					Yes[d]						Nonparasite[e]
			greeleyi[e]	Mountain brook lamprey ammocoete						No[e]						Do not feed[e]
			unicuspis[a]	Silver lamprey ammocoete	Subterminal[a]					No[e]		Microscopic food drifting downstream; a filter apparatus selects algae, pollen, diatoms, and protozoans (Harlan and Speaker 1956)[f]	Herbivore/detritivore	Filter feeder		Nonparasitic, thrusts heads out of burrow in sandy bottom[c]

Order	Family	Genus	Species	Common Name	Mouth Position	Ratio of Digestive Tract (DT) to Total Length (TL)[c]	Gut	Color of Peritoneum	Pyloric Caeca	Teeth	Pharyngeal Teeth and Plates	Main Food	First Level Trophic Classification	Second Level Trophic Classification	Trophic Mode	Feeding Behavior
			unicuspis[a]	Silver lamprey adult	Subterminal[a]					Yes[a]		Blood of other fish, particularly northern pike, paddlefish, carp, lake sturgeon, suckers, white bass, and catfish[c]	Carnivore	Parasite	Blood sucker	Parasitic, attacking especially at night[f]
			aepyptera[d]	Least brook lamprey adult	Subterminal[d]					Yes[d]						Do not feed[f]
			aepyptera[d]	Least brook lamprey ammocoete						No[e]		Primarily diatoms (Moore and Beamish, 1973)[f]	Herbivore	Filter feeder		
			appendix[a]	American brook lamprey ammocoete	Subterminal[a]							Diatoms and desmids[c]	Herbivore	Filter feeder		Filter feeders[c]
			appendix[a]	American brook lamprey adult	Subterminal[a]					Yes[c]						Do not feed[c]
			ayresi[a]	River lamprey ammocoete	Subterminal[a]							Microscopic plants and animals[b]	Planktivore	Filter feeder	Mucus entrapment	Filter feeders[b]
	Lampetra		ayresi[a]	River lamprey adult	Subterminal[a]					Yes[a]		Blood and fluid of a variety of fishes[b]	Carnivore	Parasite	Blood sucker	Parasitic[b]
			japonica[a]	Arctic lamprey ammocoete	Subterminal[a]							Microscopic plants and animals[b]	Planktivore	Filter feeder		Filter feeders[b]
			japonica[a]	Arctic lamprey adult	Subterminal[a]					Yes[a]		Blood and fluid of fishes[a]	Carnivore	Parasite	Blood sucker	Parasitic[a]
			richardsoni[a]	Western brook lamprey ammocoete	Subterminal[a]							Microscopic plant and animal matter, including desmids, diatoms, algae, and detrius[a]	Herbivore/detritivore	Filter feeder	Filterer	Nonparasitic[a]

Appendix 7A

	richardsoni[a]	Western brook lamprey adult	Subterminal[a]					Do not feed[a]
	tridentata[a]	Pacific lamprey ammocoete	Subterminal[a]		Blood and fluids of fish and other marine vertebrates[a]	Carnivore	Parasite	Blood sucker	Parasitic[a]
	tridentata[a]	Pacific lamprey adult	Subterminal[a]	Yes[a]	Diatoms and other materials that are carried in the water and come to rest on the bottom[c]	Herbivore/detritivore	Filter feeder	Filterer	Filter feeders[c]
Petromyzon	marinus[a]	Sea lamprey ammocoete	Subterminal[a]	No[c]	Blood and tissues of lake trout, large chubs, burbot, and other deep-water species. Then during migration to shallow waters parasitize lake whitefish, lake herring, walleye, yellow perch, round whitefish, sucker, and carp[c]	Carnivore	Parasite	Blood sucker	Parasitic[c]
	marinus[a]	Sea lamprey adult	Subterminal[a]	Yes[a]	Sludgeworms, chironomid larvae, small crustaceans, and plants[a]	Invertivore/herbivore	Benthic	Hunter	Uses sensitive barbels to locate food on bottom[a]
Acipenseridae Acipenser	brevirostrum[a]	Shortnose sturgeon	Subterminal[a]		Food types range widely and composition depends on availability; crayfish, molluscs, insect larvae, fish eggs, fishes (rarely), nematodes, leeches, amphipods, decapods, and a few plants (Harkness, 1923[a])	Invertivore/herbivore	Benthic	Hunter	Uses sensitive barbels to locate food on bottom[a]
Acipenseriformes	fulvescens[a]	Lake sturgeon	Subterminal[a]						

Order	Family	Genus	Species	Common Name	Mouth Position	Ratio of Digestive Tract (DT) to Total Length (TL)[c]	Gut	Color of Peritoneum	Pyloric Caeca	Teeth	Pharyngeal Teeth and Plates	Main Food	First Level Trophic Classification	Second Level Trophic Classification	Trophic Mode	Feeding Behavior
			medirostris[a]	Green sturgeon	Subterminal[a]								Invertivore	Benthic	Hunter	Uses sensitive barbels to locate food on bottom[a]
			oxyrhynchus[a]	Atlantic sturgeon	Subterminal[a]							Wide variety of bottom-dwelling plant and animal material[a]	Invertivore/ herbivore	Benthic	Hunter	Uses sensitive barbels to locate food on bottom[a]
			transmontanus[a]	White sturgeon	Subterminal[a]							For smaller fish; chironomids, mysids, Daphnia and Chaoborus larvae, molluscs, immature mayfly, caddisfly and stonefly, and a few copepods. For larger fish: fish, crayfish, and chironomids[a]	Invertivore/ carnivore	Benthic	Hunter	
		Scaphirhynchus	albus[d]	Pallid sturgeon	Subterminal[d]							Immature aquatic insects, particularly caddisfly larvae, and small fish[d]	Invertivore/ carnivore	Benthic	Hunter	
			platorynchus[e]	Shovelnose sturgeon	Subterminal[e]							68% Potamyia flava larvae, 7% Cheumatopsyche campyla larvae, 17% Hexagenia naiads, and 8% other material, which included immature plecopterans, dipterans, and odonates (Hoopes, 1960)[e]	Invertivore	Benthic	Hunter	Rake bottom with sensitive barbels[e]

Appendix 7A

Order	Family	Genus	species	Common name	Mouth position		Diet	Trophic category		Feeding mode	Notes
	Polyodontidae	Polyodon	spathula[c]	Paddlefish	Subterminal[c]	Yes[c]	Plankton material: small crustaceans, algae, and ephemerid larvae (Wagner, 1908); (Forbes and Richardson, 1920) — mostly entomostracans, larval mayflies, dragonflies, chironomids, aquatic insects, amphipod crustaceans, and leeches[c]	Invertivore/planktivore	Filter feeder	Ram filtration	Primarily plankton feeders; straining plankton with large mouth[c]
Lepisoteiformes	Lepisosteidae	Lepisosteus	oculatus[a]	Spotted gar	Terminal[a]	Yes[a]	Fish — Scott (1967) listed yellow perch and minnows[a]; Fish make up 90% of the diet, the remaining 10% consisting of freshwater shrimp, crayfish, and insects[f]	Carnivore	Whole body	Ambush	Voracious piscivore[a]
			osseus[a]	Longnose gar	Terminal[a]	Yes[a]	Fishes such as carp, silverside, bluegill, gar, largemouth and smallmouth bass, darters, spottail shiner, killifish, black crappie, blackstripe topminnow, sand shiner, cisco and white bass (Haase, 1969; Cahn, 1927)[c]	Carnivore	Whole body	Ambush	
			platostomus[c]	Shortnose gar	Terminal[c]	Yes[c]	Crayfish, perch, sunfish, and bluegills (in Iowa: Potter, 1923); emerging gnats and mayflies (in Illinois: Richardson, 1913); carp (in S. Dakota: Shields, 1957)[c]	Carnivore	Whole body	Ambush	

Order	Family	Genus	Species	Common Name	Mouth Position	Ratio of Digestive Tract (DT) to Total Length (TL)[c]	Gut	Color of Peritoneum	Pyloric Caeca	Teeth	Pharyngeal Teeth and Plates	Main Food	First Level Trophic Classification	Second Level Trophic Classification	Trophic Mode	Feeding Behavior
Amiiformes			spatula[d]	Alligator gar	Terminal[d]					Yes[d]		Blue crabs, turtles, waterfowl, other birds, small mammals, and it also scavenges. Primarily the diet consists of fish, estuarine catfish and drums, and scavenged items such as chicken carcasses[d]	Carnivore	Whole body	Ambush	Very opportunistic predator[d]
	Amiidae	Amia	calva[a]	Bowfin	Terminal[a]		Rudamentary spiral valve		No[a]	Yes[a]	Gular plate[a]	Other fish, crayfish and frogs (Lagler and Hubbs, 1940 - 59% game and pan fish. 17.1% other fish (mainly minnows), and 14.1% crayfish. (Berry, 1955) = 84% gizzard shad[e]	Carnivore	Whole body	Ambush	Clumsy predatory opportunist, generalist, nocturnal[c]
Osteoglossiformes	Hiodontidae	Hiodon	alosoides[a]	Goldeye	Superterminal[a]				Yes[a]	Yes[a]	Yes[a]	No indication of strong food influence, will utilize whatever is most available. Wide variety: aquatic insects and other insects make up the bulk with some fish[a]	Invertivore	Drift	Surface and water column	Often feed at the surface[a]
			tergisus[a]	Mooneye	Superterminal[a]				Yes[a]	Yes[a]	Yes[a]	No indication of strong food influence, will utilize whatever is most available. Wide variety: aquatic insects and other insects make up the bulk with some fish[a]	Invertivore	Drift	Surface and water column	Often feed at the surface[a]

Appendix 7A

Order	Family	Genus	species	Common name	Mouth position	Color/Body	Gill rakers	Teeth	Diet	Trophic category	Feeding mechanism	Feeding mode	Notes
Anguilliformes	Anguillidae	Anguilla	rostrata[a]	American eel	Terminal[a]			Yes[a]	Fishes and invertebrates[a]	Invertivore/carnivore			Voracious piscivore[c]
Clupeiformes	Clupeidae	Alosa	aestivalis[a]	Blueback herring	Terminal[a]; supraterminal[b]		Numerous[a]	Few and small[a]	Plankton, copepods, pelagic shrimp, fish fry, eggs, and insect larvae[a]	Planktivore	Particulate feeder	Size-selective picker	
			alabamae[d]	Alabama shad	Terminal[d]			Weak or absent[d]	Juveniles in freshwater feed on small invertebrates[d]	Planktivore	Filter feeder	Pump filtration	
			chrysochloris[e]	Skipjack herring	Terminal[e]			Yes[e]	Zooplankton, small insect larvae, and small fishes (Eddy and Underhill, 1975)[e]	Planktivore	Filter feeder		
			mediocris[b]	Hickory shad		Somewhat pale, peppered[b]		Tiny on lower jaw[b]	Mainly other fishes; also fish eggs, small crabs, and various pelagic crustaceans (Hildebrand and Schroeder, 1928[b]; Bigelow and Schroeder, 1953)[b]	Planktivore	Filter feeder	Pump filtration	
			pseudoharengus[a]	Alewife	Terminal[a]	Pearly grey[c]	Numerous[a]	Few and small[a]	Zooplankton: 75% cladocerans and copepods (Norden, 1968)[c]	Planktivore	Particulate feeder	Size-selective picker	
		Brevoortia	tyrannus	Menhaden	Terminal[a]			Only on young[g]		Planktivore	Filter feeder	Mucus entrapment/gulping	
			sapidissima[a]	American shad	Terminal[a]	Dark[a] pale or silver[b]	Long	Numerous[a]	Few and small[a]	Small crustaceans, and adult and larval insects (Leim, 1924; Walburg, 1956)[a]	Planktivore	Filter feeder	Pump filtration	
		Dorosoma	cepedianum[e]	Gizzard shad; Eastern gizzard shad[e]	Subterminal[a]		Gizzard, long gut[a]	Numerous[a]	Only on young[g]	Juvenile - plankton; adults - algae[a]; they also graze over the bottom ingesting detritus, sand, and bottom ooze (Baker et al., 1971)[f]	Herbivore	Filter feeder		Completely herbivorous filter feeders

146 Assessing the Sustainability and Biological Integrity of Water Resources Using Fish Communities

Order	Family	Genus	Species	Common Name	Mouth Position	Ratio of Digestive Tract (DT) to Total Length (TL)[c]	Gut	Color of Peritoneum	Pyloric Caeca	Teeth	Pharyngeal Teeth and Plates	Main Food	First Level Trophic Classification	Second Level Trophic Classification	Trophic Mode	Feeding Behavior
Cypriniformes			petenense[a]	Threadfin shad	Terminal[b]				Numerous[b]	No[f]		Algae, protozoans, rotifers, copepods, cladocerans, invert. eggs, organic debris, and fish larvae[b]	Planktivore	Filter feeder	Pump filtration	
	Cyprinidae	Acrocheilus	alutaceus[a]	Chiselmouth	Subterminal[a]		Intestine at least twice body length[a]	Intensely black[a]			Hard cartilaginous sheath with a straight-cutting edge; Yes[a]	Diatoms are the major food for adults[a]	Herbivore	Particulate feeder	Grazer	Specialized: scrapes bottom with chisel-like lower jaw
		Campostoma	anomalum[b]	Central stoneroller	Subterminal[b]		Intestine very long, looped about the gas bladder[b]	Black[b]		Hard ridge on lower jaw[b]	Yes[b]	Algae (Fowler and Taber, 1985), detritus in some settings (Kraatz, 1923; Starrett, 1950b; Burkhead, 1980; Felley and Hill, 1983)[b]	Herbivore	Particulate feeder	Grazer	Grazing minnow
			oligolepis[c]	Largescale stoneroller	Subterminal[c]	DT 3.3–5.2 TL	Long, in transverse loops completely surrounding swim bladder[c]	Black[c]		Hard ridge on lower jaw[c]	Yes[c]	Feeds primarily on algae scraped from submerged objects[c]	Herbivore	Particulate feeder	Grazer	Grazing minnow
		Carassius	auratus[a]	Goldfish	Terminal[f]	DT 1.5 TL	Intestine long, with several loops[f]	Dusky to black[f]		Yes[a]		Larvae and adult aquatic insects, molluscs, crustaceans, aquatic worms, aquatic vegetation[f]	Invertivore/herbivore	Benthic	Hunter	
		Clinostomus	elongatus[a]	Redside dace	Terminal[a]		Intestine short with a single loop[a]	Silvery, speckled[a]			Yes[a]	Food mainly (95%) insects, mainly terrestrial[c]	Invertivore	Benthic and drift	Water column	

Appendix 7A

Genus	Species	Common name	Mouth position	Digestive tract	Color	Schooling	Diet	Feeding mode	Position
	fundaloides[b]	Rosyside dace	Terminal or slightly supraterminal[b]		Silver[b]	Yes[b]	Aquatic and terrestrial insects, and much lesser amounts of worms, arachnids, crayfishes, snails, algae, and detritus (Breder and Crawford, 1922; Flemer and Woolcott, 1966; Gatz, 1979)[b]	Invertivore	Drift	Surface and water column	Drift seeker[b]
Couesius	plumbeus[a]	Lake chub	Terminal[a]	Short, proportional length of digestive tract 0.9-1.2 in total length (Dymond 1926[a])	Silvery, speckled[a]	Yes[a]	Aquatic insect larvae, zooplankton, algae[a]	Invertivore/planktivore		
Ctenopharyngodon	idella[c]	Grass carp	Terminal[c]	Intestine long with several loops[f]		Yes[c]	Macrophytes, when vegetation is unavailable insects and small fishes may be eaten[c]	Herbivore	Particulate feeder	Browser
Cyprinella	analostana[b]	Satinfin shiner	Terminal	DT 1.5-3.0 TL[f]		Yes[b]	Microcrustaceans, terrestrial and aquatic insects, and algae (Flemer and Woolcott, 1966; Gatz, 1979)[b]	Invertivore/planktivore	Drift	Surface and water column	Opportunistic
	caerulea[d]	Blue shiner	Terminal[d]			Yes[d]	Terrestrial insects supplemented with occasional mayfly and caddisfly immatures[d]	Invertivore	Drift	Surface and water column	Surface and midwater feeder[d]
	callistia[d]	Alabama shiner	Subterminal[d]		Dark[d]	Yes[d]	Primarily midge, blackfly, and caddisfly larvae and a few mayfly nymphs[d]	Invertivore	Benthic	
	camura[d]	Bluntface shiner	Terminal[d]			Yes[d]				

Order	Family	Genus	Species	Common Name	Mouth Position	Ratio of Digestive Tract (DT) to Total Length (TL)[c]	Gut	Color of Peritoneum	Pyloric Caeca	Teeth	Pharyngeal Teeth and Plates	Main Food	First Level Trophic Classification	Second Level Trophic Classification	Trophic Mode	Feeding Behavior
			galactura[b]	Whitetail shiner	Subterminal[b], terminal[f]		Short[f]	Silvery with speckles[f]			Yes[b]	Diverse allochthonous and benthic organisms, including worms, mites, insects, larval fishes, and plant material (Outten, 1958)[b]; Crayfish and clams; terrestrial and aquatic insects and some small fish (Outten, 1958)[f]	Invertivore	Benthic and drift	Grazer	
			labrosa[b]	Thicklip shiner	Subterminal[b]						Yes[b]	In decreasing importance: water mites and larvae of midges, blackflies, mayflies, stoneflies, and beetles (M.G. Pfleger)[b]	Invertivore	Benthic	Hunter	
			lutrensis[d]	Red shiner	Terminal[d]	DT 0.7-0.8 TL	S-shaped[d]	Silvery with numerous large, dark chromatophores[d]			Yes[d]	Both terrestrial and aquatic insects, and algae (Lewis and Gunning, 1959, as C. whipplei; Laser and Carlander, 1971)[d]	Invertivore/ herbivore	Benthic	Grazer	
			spiloptera[a]	Spotfin shiner	Subterminal[a]	DT 0.6TL	Short, single-S-shaped loop[c]	Silvery[a]			Yes[a]	Terrestrial and aquatic insects; microcrustaceans, decapods, water mites, plant material including seeds, and detritus (Starrett, 1950b; Minckley, 1963; Mendelson, 1975; Hess, 1983; Angermeier 1985; Vadas, 1990)[b]	Invertivore/ detritivore	Particulate feeder	Biter	

Species	Common name	Mouth	DT/TL	Gut	Color		Pharyngeal teeth	Food	Trophic	Habitat	Feeding mode
trichroistia[d]	Tricolor shiner	Terminal					Yes[d]	Half terrestrial insects, half aquatic immature dominated by mayfly nymphs[d]	Invertivore	Benthic and drift	
venusta[d]	Blacktail shiner	Terminal[d]		Short[f]	Silver with dark speckles[f]		Yes[d]	Surface insects (Hambrick and Hibbs, 1977), and benthic invertebrates (Hale, 1963)[d]	Invertivore	Benthic and drift	
whipplei[f]	Steelcolor shiner	Terminal[f]		Short[f]	Silvery, heavily sprinkled with black pigment[f]		Yes[f]	Terrestrial insects found near the surface and small invertebrates that are found at varying depths[f]	Invertivore	Drift	Surface and water column
Cyprinus carpio[a]	Common carp	Subterminal[a]	DT 1.6TL		Dusky[f]	No[a]	Yes, molar like[a]	Plant tissue, aquatic insects, crustaceans, annelids, and molluscs[a]	Invertivore/detritivore	Benthic/filter feeder	Grazer/suction feeder
Erimystax cahni[b]	Slender chub	Subterminal[b]					Yes[b]	Insect larvae, small snails[c]	Invertivore	Benthic	Hunter
dissimilis[c]	Streamline chub	Subterminal[c]		Intestine long[f]	black[f]	Yes[e]	Yes[f]				
insignis[b]	Blotched chub	Subterminal[b]					Yes[b]	A variety of immature insects, mainly midge and blackfly larvae, a few other invertebrates, and large amounts of microscopic plants and detritus (Harris, 1986)[b]	Invertivore/planktivore		
x-punctata	Gravel chub	Subterminal[a]	DT > TL	Several loops, longer than TL[c]	Uniformly dark brown[a]		Yes[a]	Detailed studies not available, but most likely aquatic insect larvae[c]	Invertivore	Benthic	
Exoglossum laurae[b]	Tonguetied minnow	Subterminal[b]		Gut S-shaped[b]	Silver[b]		Yes[b]	Small crustaceans and insect larvae (Greeley, 1927)[b]	Invertivore	Benthic	Hunter
maxillingua[a]	Cutlips minnow				Silvery white[a]	No[a]	Yes[a]	Aquatic insect larvae and molluscs[a]	Invertivore	Benthic	Hunter

Order	Family	Genus	Species	Common Name	Mouth Position	Ratio of Digestive Tract (DT) to Total Length (TL)[c]	Gut	Color of Peritoneum	Pyloric Caeca	Teeth	Pharyngeal Teeth and Plates	Main Food	First Level Trophic Classification	Second Level Trophic Classification	Trophic Mode	Feeding Behavior
		Hemitremia	flammea[d]	Flame chub	Terminal[d]						Yes[d]	Primarily midge larvae supplemented with isopods, oligochaetes, hemipterans, and snails (Sossamon, 1990)[d]	Invertivore	Benthic and drift		
		Hybognathus	hankinsoni[a]	Brassy minnow	Subterminal[a]	DT 3.8-4.1 TL	Intestine elongate, distinctly coiled on right side[a]			No[a]	Yes[a]	Phytoplankton and other algae, zooplankton, some aquatic insects. Long, looped intestine suggests feeding on plant material (Copes, 1975) 94% algae, 5% organic debris, 0.5% animal matter[c]	Planktivore/ detritivore	Particulate feeder	Scooper	
			hayi[f]	Cypress minnow	Terminal[f]	DT over twice the TL[f]	Intestine long and coiled[f]	Black[f]			Yes[f]		Detritivore			Detritivore
			nuchalis[c]	Mississippi silvery minnow	Subterminal[b]	DT 5.4-5.8 TL	Intestine elongate, distinctly coiled on right side[a]	Uniformly black[a]			Yes[a]	Bottom ooze and algae have been reported from the stomachs[c]	Detritivore	Particulate feeder		Known as "mudeating minnows"
			placitus[f]	Plains minnow	Subterminal[f]	DT over twice the TL[f]	Long, coiled[f]	Black[f]			Yes[f]	Miller and Robison (1973) speculated that it feeds on benthic microflora such as algae and diatoms[f]	Herbivore			Herbivorous[f]
			regius[b]	Eastern silvery minnow	Subterminal[b]	DT over twice the TL[f]	Gut very long, spirally coiled[f]	Black[f]			Yes[b]	Ooze, diatoms, and other algae (Raney, 1939a)[b]	Herbivore/ detritivore			

Appendix 7A

Genus	Species	Common name	Mouth	Gut length	Intestine	Color	Stomach	Food	Trophic guild	Feeding location	Feeding mode	Feeding group
Hypophthal-michthys	molitrix[f]	Silver carp	Supraterminal[f]	DT 3-6 × TL[f]	Several loops, 3-6 times longer than TL[f]		Yes[f]	Phytoplankton and detritus; the filtering capability makes tiny green and blue-green algal cells available as food, along with some larger zooplankters[d]	Herbivore/detritivore	Filter feeder	Filterer	Pelagic filter feeders
	nobilis[f]	Bighead carp	Supraterminal[f]	DT 3-5 × TL[f]	Several loops, 3-5 times longer than TL[f]		Yes[f]	Zooplankton, clumps of algae, and insect larvae; capable of switching to phytoplankton when zooplankton and detritus are scarce (Cremer and Smitherman, 1980)[d]	Invertivore/planktivore	Filter feeder	Sieve	Filter feeders
Luxilus	albeolus[b]	White shiner	Terminal				Yes[b]	Aquatic and terrestrial insects (Surat et al., 1982; Hess, 1983)[b]	Invertivore			
	cardinalis[f]	Cardinal shiner	Terminal[f]		Short[f]	Black[f]	Yes[f]					
	cerasinus[b]	Crescent shiner	Terminal[b]				Yes[b]	Aquatic insects and at times terrestrial insects; worms, zooplankton, snails, larval fishes, and plant material are eaten in small amounts (Schwartz and Dutcher, 1962; Gatz, 1979; Mauney, 1979; Surat et al., 1982)[b]	Invertivore	Benthic and drift		
	chrysoceph-alus[b]	Striped shiner	Terminal	DT 0.7-0.8 TL	S-shaped[e]	Dark brown overlying a silvery background	Yes[c]	Aquatic and terrestrial insects are dominant; other foods include small crayfishes, fish eggs, small fishes, algae, and detritus (Gillen and Hart, 1980; Angermeier, 1985)[b]	Invertivore	Benthic and drift		

Order	Family	Genus	Species	Common Name	Mouth Position	Ratio of Digestive Tract (DT) to Total Length (TL)[c]	Gut	Color of Peritoneum	Pyloric Caeca	Teeth	Pharyngeal Teeth and Plates	Main Food	First Level Trophic Classification	Second Level Trophic Classification	Trophic Mode	Feeding Behavior
			coccogenis[b]	Warpaint shiner	Terminal[b]						Yes[b]	Aquatic and terrestrial insects; worms and spiders are minor food items (Outten, 1957)[b]	Invertivore	Benthic and drift		
			cornutus[a]	Common shiner	Terminal[a]	DT 0.8 TL	Intestine short S-shaped[c]	Black			Yes[a]	Aquatic insects, adult and larvae as well as algae and other aquatic plants[a]	Invertivore	Benthic and drift	Watercolumn and surface	Versatile
			pilsbryi	Duskystripe shiner	Terminal[f]		Short[f]	Black[f]			Yes[f]	Aquatic insects (chironomid larvae and pupae), terrestrial insects, and algae[f]	Invertivore	Benthic and drift		
			zonatus[f]	Bleeding shiner	Terminal[f]		Short[f]	Black[f]			Yes[f]	Insects and other small invertebrates found floating on the surface of the water or drifting in the current (Pflieger, 1975)[f]	Invertivore	Drift		
		Lythurus	ardens[b]	Rosefin shiner	Terminal						Yes[b]	Terrestrial insects in warmer times of the year (Meredith and Schwartz, 1959; Lotrich, 1973; Small, 1975)[b]; benthic aquatic insects, algae, and detritus (Surat et al., 1982)[b]	Invertivore	Benthic and drift		
			fumeus[f]	Ribbon shiner	Terminal[f]		Short[f]	Silvery with dark speckles[f]			Yes[f]					Often feeds from the surface[f]
			lirus[d]	Mountain shiner	Terminal[d]						Yes[d]					

Appendix 7A

Genus	species	Common name	Mouth	Eye/DT	Intestine	Coloration	Sex dim	Diet	Trophic	Location	Notes
	snelsoni[f]	Ouachita shiner	Terminal[f]		Short[f]	Silvery, sprinkled with dark speckles[f]	Yes[f]	Simulids, chironomids, and ephemeropterans[f]	Invertivore	Benthic and drift	
	umbratilis[e]	Redfin shiner	Terminal[a]	DT 0.5-0.6 TL	Short, S-shaped loop[e]	Silvery with speckles	Yes[a]				
Machybopsis	aestivalis[c]	Speckled chub	Subterminal[c]	DT 0.5-0.7 TL	Simple S-shaped duct[c]	Silvery	Yes[c]	Immature aquatic insects[c]	Invertivore	Benthic	Hunter
	gelida[f]	Sturgeon chub	Subterminal[f]		Short[f]	Silvery w/ scattered speckles[f]	Yes[f]				
	meeki[f]	Sicklefin chub	Terminal[f]		Short[f]	Silvery w/ scattered speckles[f]	Yes[f]				Benthic taste feeder[f]
	storeriana[a]	Silver chub	Subterminal[a]	DT 0.6-0.7 TL	Short single S-shaped loopc	Silvery[a]	Yes[a]	Young: chadocerans, copepods, chironomids[c] Adults: mayflies, chironomids[a]	Planktivore/ invertivore		
Margariscus	Margarita[a]	Pearl dace	Terminal[b]	DT 0.7-0.9 TL	Intestine w/single loop[a]	Silvery w/ speckles on dorsal part[a]	Yes[b]	Microcrustaceans and immature insects including fingernail clams, snails, small fishes, and plant matter[b]	Invertivore/ carnivore		
Mylocheilus	caurinus[a]	Peamouth	Slightly subterminal[a]			Dusky[a]	Yes[a]	Aquatic and terrestrial insects, planktonic crustaceans, molluscs, occasionally small fishes[b]	Invertivore/ carnivore		
Nocomis	asper[f]	Redspot chub	Terminal[f]		Short[f]	Black[f]	Yes[f]	Insect larvae and adults, crustaceans, other invertebrates, and plant material (Miller and Robison, 1973)[f]	Invertivore/ herbivore		Surface or midwater feeder[f]
	biguttatus[a]	Hornyhead chub	Terminal[a]	DT 0.9 TL	Short single S-shaped duct[c]	Uniformly brown[a]	Yes[a]	Filamentous algae, diatoms, cladocerans, insect larvae for young; snails, insect larvae, annelids, crayfish in older fish[f]	Invertivore/ herbivore		

Order	Family	Genus	Species	Common Name	Mouth Position	Ratio of Digestive Tract (DT) to Total Length (TL)[c]	Gut	Color of Peritoneum	Pyloric Caeca	Teeth	Pharyngeal Teeth and Plates	Main Food	First Level Trophic Classification	Second Level Trophic Classification	Trophic Mode	Feeding Behavior
			effusus[d]	Redtailed chub	Terminal[d]						Yes[d]					
			leptocephalus[b]	Bluehead chub	Subterminal[b]		Intestine whorled[b]	Dusky or dark[b]			Yes[b]	Aquatic insects and plant material, particularly algae (Flemer and Woolcott, 1966; Gatz, 1979)[b]	Invertivore/herbivore			
			micropogen[a]	River chub	Nearly terminal[a]		Intestine short[a]	Uniformly brown or black, with darker speckles[a]			Yes[a]	Algae and zooplankton when young; aquatic insect larvae, snails, crayfish, possibly some fish when older[c]	Planktivore/invertivore			
			playrhynchus[b]	Bigmouth chub	Subterminal[b]		Intestine usually S-shaped, occasionally with a short anterior accessory loop[b]	All dark or partly silver[b]			Yes[b]	Immature mayflies, caddisflies, midges, and blackflies (Hess, 1983); probably consumes other invertebrates and occasional fishes[b]	Invertivore	Benthic		
			raneyi[b]	Bull chub	Terminal		Intestine unwhorled[b]				Yes[b]	Benthic and drifting insects, snails, and occasionally crayfishes and fishes; large amounts of filamentous algae sometimes taken perhaps incidentally while searching for animal food (Jenkins and Lachner in Lee et al., 1980)[b]	Invertivore/herbivore			

Genus	species	common name	mouth	DT	Intestine	Peritoneum		Diet	Trophic	Feeding	Position	Notes
Notemigonus	crysoleucas[a]	Golden shiner	Oblique[a]	DT 0.6-0.7 TL	Intestine short[a]	Dusky[a]	Yes[a]	Keast and Webb (1966) found main foods as follows: Cladocera 90% by volume, flying insects 20%, chironomid pupae 30%, and filamentous algae[b]	Invertivore/herbivore	Particulate feeder	Surface and water column	Midwater and surface feeder
Notropis	alborus[b]	Whitemouth shiner	Subterminal[b]				Yes[b]	Microcrustaceans, mites, diatoms, and detritus (Gatz 1979; R. K. Law, in litt.)[b]	Planktivore	Filter feeder		
	altipinnis[b]	Highfin shiner	Terminal[b]				Yes[b]	Terrestrial insects and lesser amounts of filamentous algae (Gatz, 1979[b]); microcrustaceans and terrestrial insects, some aquatic insects (A.H. Campbell)[b]	Invertivore/herbivore			
	ambiops	Bigeye chub	Subterminal[f]		Short[f]	Silvery[f]	Yes[f]	Microcrustaceans and midge larvae[c]	Invertivore			Hunter
	ammophilus[d]	Orangefin shiner	Subterminal[d]				Yes[d]	Primarily midge larvae[d]	Invertivore			Benthic
	amnis[c]	Pallid shiner	Subterminal[c]	DT 0.5 TL	Short, S-shaped loop[c,H103]	Silvery, sometimes with faint speckles[f]	Yes					
	amoenus	Comely shiner	Subterminal				Yes					
	anogenus[a]	Pugnose shiner	Terminal[a]	DT 0.6-0.7 TL	S-shaped[c]	Dark[a]	Yes[a]	Minute plant and animal organisms, and organic detritus[a]	Detritivore	Particulate feeder	Biter	Grazes on plants
	ariommus	Popeye shiner	Terminal				Yes					
	asperifrons[d]	Burhead shiner	Terminal[d]				Yes[d]					

156 Assessing the Sustainability and Biological Integrity of Water Resources Using Fish Communities

Order	Family	Genus	Species	Common Name	Mouth Position	Ratio of Digestive Tract (DT) to Total Length (TL)[c]	Gut	Color of Peritoneum	Pyloric Caeca	Teeth	Pharyngeal Teeth and Plates	Main Food	First Level Trophic Classification	Second Level Trophic Classification	Trophic Mode	Feeding Behavior
			atherinoides[a]	Emerald shiner	Terminal[a]	DT 0.5–0.6 TL	Short, S-shaped loop[c]	Silvery and speckled[a]			Yes[a]	Microcrustaceans, some midge larvae, and algae[a]	Planktivore	Particulate feeder	Size-selective picker	Move with planktonic food source up at dusk, back down at dawn
			atrocaudalis[f]	Blackspot shiner	Slightly subterminal[f]						Yes[f]					
			bairdi[f]	Red River shiner							Yes[f]					
			bifrenatus[a]	Bridle shiner	Terminal[a]			Silvery, lightly speckled[a]			Yes[a]	Small planktonic animals of various kinds including chironomids, entomostacans, cladocerans, copepods (Harrington, 1948)[b]	Planktivore	Particulate feeder	Size-selective picker	Predaceous
			blennius[a]	River shiner	Subterminal[a]	DT 0.6–0.7 TL	Intestine short S-shaped loop[c]	Silvery[a]			Yes[a]	Insect feeder (Becker, 1983); Daphnia, caddisflies, mayflies, corixids, tendipedid larvae, algae, and other plant material[f]	Invertivore	Benthic and drift		
			boops[f]	Bigeye shiner	Terminal[f]		Short[f]	Black[f]			Yes[f]	Surface insects, occasionally leaping from water to capture hovering insects (Trautman, 1957)[d]	Invertivore	Drift	Surface and water column	
			buccatus[b]	Silverjaw minnow	Subterminal[b]			Silver[b]			Yes[b]	Worms, microcrustaceans, insects, fingernail clams, snails, algae, and detritus (Hoyt, 1970; Lotrich, 1973; Wallace, 1976); midge larvae are consistantly important in the diet[b]	Invertivore	Benthic and drift		

Appendix 7A

buchanani	Ghost shiner	Subterminal[a]	DT 0.6 TL	Single S-shaped loop[c]	Silvery[c]	Yes[c]			
chalybaeus	Ironcolor shiner	Terminal or slightly subterminal[b]	DT 0.6 TL	Single S-shaped loop[c]	Silvery with numerous dark speckles[c]	Yes[c]	minute insect larvae (Cope, 1869); small animals (Marshall, 1947)	Invertivore	
chiliticus	Redlip shiner	Subterminal[b]				Yes	Aquatic and terrestrial insects; occasionally worms, centipedes, plant material, detritus (L.A. Goodwin)[b]	Invertivore/detritivore	Benthic and drift
chrosomus	Rainbow shiner	Terminal[d]				Yes[d]			
dorsalis	Bigmouth shiner	Subterminal[a]	DT 0.4-0.5 TL	Short, S-shaped[c]	Silvery[a]	Yes[a]	Calopsectra (60%), Dicronomini (8.3%), Chironomid larvae of the tribe Chironomini (7.2%), and chironomid larvae, 70% of food was of benthic origin (Mendelson, 1972)[c]	Insectivore	Benthic
girardi	Arkansas River shiner	Subterminal[f]		Short[f]	Silver[f]	Yes[f]		Invertivore	
greenei	Wedgespot shiner	Subterminal[f]		Short[f]	Silvery with dark speckles[f]	Yes[f]			Feed on organisms that are exposed by movement of the sand or are washed downstream[f]
heterodon	Blackchin shiner	Terminal[a]	DT 0.7-0.8 TL	S-shaped[c]	Silvery[a]	Yes[a]	Small crustaceans and small insects[a]	Invertivore	
heterolepis	Blacknose shiner	Subterminal[a]	DT 0.4-0.6 TL	Short S-shaped[c]	Silvery[a]	Yes[a]	Cladocera, insects, green algae[a]	Invertivore/herbivore	
hubbsi	Bluehead shiner	Terminal[f]				Yes[f]			

158 Assessing the Sustainability and Biological Integrity of Water Resources Using Fish Communities

Order	Family	Genus	Species	Common Name	Mouth Position	Ratio of Digestive Tract (DT) to Total Length (TL)[c]	Gut	Color of Peritoneum	Pyloric Caeca	Teeth	Pharyngeal Teeth and Plates	Main Food	First Level Trophic Classification	Second Level Trophic Classification	Trophic Mode	Feeding Behavior
			hudsonius[a]	Spottail shiner	Subterminal[a]	DT 0.6–0.7 TL	Intestine short S-shaped[c]	Silvery[a]			Yes[a]	Dymond (1926) noted plankton feeding such as Daphnia and Bosmina as well as aquatic insect larvae such as Chironomidae and Ephemeroptera[a]	Invertivore/ planktivore			
			hypsinotus	Highback chub	Subterminal						Yes	Microcrustaceans and aquatic and terrestrial insects	Invertivore/ planktivore			
			leuciodus[b]	Tennessee shiner	Terminal or very slightly subterminal[b]						Yes[b]					
			lineapuncta-tus	Lined chub							Yes	Primarily chirono-mid larvae and pupae, along with some larger aquatic insects such as Hydopsyche and Glossosoma trichopteran larvae, heptageniid and baetid mayflies; terrestrial spiders, snails, beetles, and Hemipterans are also taken	Invertivore			
			maculatus[d]	Tailight shiner	Terminal[f]		Short[f]	Silvery, usually with a few scattered dark speckles[f]			Yes[f]	Microcrustaceans, rotifers, unicellular algae, and small dipteran larvae (Beach, 1974; Cowell and Barnett, 1974)[d]	Planktivore			

Appendix 7A

									Particulate feeder	Browser	
nubilus[c]	Ozark minnow	Terminal sub-terminal[c]	DT 1.9–2.4 TL	Elongate with regular coils[c]	Black[c]	Yes[c]	Green algae, blue-green algae, diatoms, other plant material, and sand	Herbivore			
ortenburgeri	Kiamichi shiner	Extremely oblique[f]				Yes[f]					
ozarcanus[f]	Ozark shiner	Subterminal[f]		Short[f]	Silvery[f]	Yes[f]					
perpallidus	Peppered shiner			Short[f]	Silvery with scattered dark speckles[f]	Yes[f]	Squatic and terrestrial insects, predominantly dipterans[f]	Invertivore			
photogenis[b]	Silver shiner	Terminal[b]		Short[f]	Silvery with a few scattered, large melanophores midventrally[f]	Yes[b]	Terrestrial insects, aquatic insect larvae, microcrustaceans, worms, and algae (Gruchy et al., 1973; Smith et al., 1981; Hess, 1983; Parker and McKee, 1984a)[b]	Invertivore/planktivore			Benthic, midwater, and surface feeder
potteri	Chub shiner					Yes[f]	60% benthic invertebrates, 13% fish, 8% open-water invertebrates (mostly Cladocera), and 16% substrate particles (Felley, 1984)[f]	Invertivore/carnivore			
procne[b]	Swallowtail shiner	Subterminal[b]				Yes[b]	Worms, mites, microcrustaceans, aquatic and terrestrial insects, diatoms, and filamentous algae[b]	Invertivore/herbivore			

Order	Family	Genus	Species	Common Name	Mouth Position	Ratio of Digestive Tract (DT) to Total Length (TL)[c]	Gut	Color of Peritoneum	Pyloric Caeca	Teeth	Pharyngeal Teeth and Plates	Main Food	First Level Trophic Classification	Second Level Trophic Classification	Trophic Mode	Feeding Behavior
			rubellus[a]	Rosyface shiner	Terminal[a]	DT 0.5–0.6 TL	Short, S-shaped loop[a]	Silvery[a]			Yes[a]	Reed (1975) found aquatic insects (71.9%), algae diatoms (18.1%), and inorganic material (10%)[a]	Invertivore/ detritivore/ herbivore			
			rubricroceus	Saffron shiner	Subterminal						Yes	Aquatic and terrestrial insects, and also takes worms, millipedes, spiders, algae, and vascular plants (Outten, 1958)	Invertivore/ herbivore			
			rupestris[d]	Bedrock shiner						Yes[d]		Primarily midge larvae and chydorid cladocerans with a few water mites, copepods, and small mayflies[d]	Invertivore/ planktivore			
			sabinae[f]	Sabine shine	Subterminal[f]		Short[f]	Silvery with a few faint speckles[f]			Yes[f]					
			scabriceps[b]	New River shiner	Subterminal[b]						Yes[b]	A Virginia collection showed caddisfly larvae and leeches[b]	Invertivore			
			semperasper[b]	Roughhead shiner	Terminal[b]						Yes[b]	Immature aquatic insects (S.W. Hipple)[b]	Invertivore			
			shumardi[d]	Silverband shiner	Terminal[d]		Short[f]	Silvery[f]			Yes[d]					
			spectrunculus[d]	Mirror shiner	Subterminal[d]						Yes[d]					
			stilbius[d]	Silverstripe shiner	Terminal[d]						Yes[d]					
			stramineus[a]	Sand shiner	Terminal[a]	DT 0.7–0.9 TL	Single S-shaped loop[c]	Silvery lightly speckled[c]			Yes[c]	Terrestrial and aquatic insects along with bottom ooze diatoms[a]	Invertivore/ detritivore			

Appendix 7A

Species	Common name	Mouth	Gut length	Intestine	Peritoneum	Yes/No	Diet	Trophic guild	Feeding location	Feeding mode
telescopus	Telescope shiner	Terminal[f]			Short[f]	Black[f]	Yes[f]	Blackflies and midges (Pflieger, 1975; Hess, 1983); immature benthic insects and adult midges (K.M. Sullivan)[b]	Invertivore		Biter
texanus[c]	Weed shiner	Terminal[c]	DT 0.6–0.7 TL	Single S-shaped loop[c]	Silvery, with a few dark speckles, to dusky[c]		Yes[c]	Plant debris, including filamentous algae and animal material[f]	Detritivore	Particulate feeder	
topeka volucellus[a]	Topeka shiner Mimic shiner	Subterminal Terminal[a]	DT 0.6–0.7 TL	Single S-shaped loop[c]	Silvery lightly speckled[a]		Yes[a]	Entomostracans (especially *Daphnia*), insects (particularly Chironomidae), and green and blue-green algae[a]	Invertivore/herbivore		
wickliffi[d]	Channel shiner	Terminal					Yes[d]				
xaenocephalus[d]	Coosa shiner	Terminal					Yes[d]				
Opsopoeodus emiliae[a]	Pugnose minnow	Terminal[a]	DT 0.5 TL	Intestine short S-shaped[c]	Silvery, heavily speckled with chromatophores[a]		Yes[a]	Filamentous algae, unidentified debris, plant fibers, eggs, crustaceans, and sand[a]	Detritivore	Particulate feeder	Scooper
Phenacobius catostomus[d]	Riffle minnow	Subterminal[d]					Yes[d]				
crassilabrum[b]	Fatlips minnow	Subterminal[b]					Yes[b]	Larvae of midges and the cranefly *Antocha*[b]	Invertivore	Benthic	Grazer
mirabilis[b]	Suckermouth minnow	Subterminal[b]	DT 0.6–0.7 TL				Yes[b]	Insect larvae, mainly midges and caddisflies[b]	Invertivore	Benthic	Grazer
teretulus[b]	Kanawha minnow	Subterminal[b]					Yes[b]	Immature mayflies, caddisflies, and dipterans[b]	Invertivore	Benthic	Grazer
uranops[b]	Stargazing minnow	Subterminal[b]					Yes[b]	Mayfly and caddisfly larvae[b]	Invertivore	Benthic	Grazer
Phoxinus cumberlandensis[d]	Blackside dace	Terminal[d]				Dark[d]	Yes[d]	Attached algae; immature insects are important in winter (Starnes and Starnes, 1981)[d]	Invertivore/herbivore	Particulate feeder	Browser

162 Assessing the Sustainability and Biological Integrity of Water Resources Using Fish Communities

Order	Family	Genus	Species	Common Name	Mouth Position	Ratio of Digestive Tract (DT) to Total Length (TL)[c]	Gut	Color of Peritoneum	Pyloric Caeca	Teeth	Pharyngeal Teeth and Plates	Main Food	First Level Trophic Classification	Second Level Trophic Classification	Trophic Mode	Feeding Behavior
			eos[a]	Northern redbelly dace		DT 1.1-1.6 TL	Intestine longer than body, coiled[e]	Uniformly black[a]			Yes[a]	Algae such as diatoms and filamentous algae, zooplankton, aquatic insects[e]	Invertivore/ planktivore			Omnivorous[e]
			erythrogaster[d]	Southern redbelly dace	Terminal	DT 1.6-1.7 TL	Long and Looped[d]	Dark[d]			Yes[f]	Invertebrates and diatoms (Philips, 1969)[d]	Planktivore	Particulate feeder	Size-selective picker	
			neogaeus[a]	Finescale dace	Terminal[a]	DT 0.6-0.8 TL	Intestine short; a single S-curve[c]	Uniformly black[a]			Yes[a]	Insects, crustaceans, plankton[c]	Invertivore/ planktivore			Omnivorous[c]
			oreas[b]	Mountain redbelly dace	Subterminal[b]		Long gut[b]					Algae, detritus (Flemer and Woolcott, 1966; W.J. Matthews)[b]	Detritivore	Particulate feeder	Scooper	Herbivorous/ detritivorous[b]
			tennessensis[b]	Tennessee dace	Subterminal[b]		Intestine long, whirled[b]	Dark[b]			Yes[b]	Living and decaying plants (Starnes and Jenkins, 1988)[b]	Herbivore	Particulate feeder	Browser	Herbivorous[b]
	Pimephales		notatus[a]	Bluntnose minnow	Inferior[c]	DT 1.0-1.2 TL	Intestine elongated, coiled[f]	Uniformly black[a]			Yes[a]	Kesrad and Webb (1966) - 20-50% bottom ooze, 2-30% chironomid larvae, and 10-75% Cladocera[b]	Detritivore	Particulate feeder	Scooper	Bottom feeder of organic detritus[b]
			promelas[a]	Fathead minnow	Nearly terminal[a]	DT 1.4-1.7 TL	Intestine long but variable[c]	Uniformly black[a]			Yes[a]	Algae, bottom mud, organic detritus, aquatic insect larvae, and zooplankton[c]	Detritivore/ invertivore	Particulate feeder	Scooper	High vegetative diet; opportunistic[c]
			tenellus[f]	Slim minnow	Terminal, slightly oblique[f]		Intestine short with single S-shaped loop[f]	Silvery[f]			Yes[f]					
			vigilax[c]	Bullhead minnow	Terminal[c]	DT 0.6-0.7 TL	Short intestine; S-shaped no coils[c]	Silvery[c]			Yes[c]	Algae, other vegetation, and small snails and other small bottom dwelling animals[c]	Herbivore/ invertivore			Omnivorous[c]

Appendix 7A

Genus	Species	Common name	Mouth	DT/TL	Intestine	Coloration		Diet	Feeding type		Water column	Notes
Platygobio	*gracilis*[a]	Flathead chub	Subterminal[a]			Silvery[a]	Yes[a]	Olund and Cross (1961) - 35% Corixidae, and terrestrial insects 21% ants, 30% beetles, and dipterous flies 9%[c]	Invertivore	Drift	Surface and water column	Predaceous, using both sight and gustatory or taste buds associated with the barbels
Ptychocheilus	*oregonensis*[a]	Northern squawfish	Terminal[a]		Intestine short[a]	Speckled[a]	Yes[a]	Various fish: shiners, sticklebacks; terrestrial insects and some plankton[a]	Carnivore			
Rhinichthys	*atratulus*[a]	Blacknose dace	Subterminal[b]	DT 0.8-0.9 TL		Uniformly brown[a]	Yes[a]	Aquatic insect larvae[a]	Invertivore			
	cataractae[a]	Longnose dace	Subterminal[a]	DT 0.6-0.8 TL	Simple, S-shaped[c]	Silvery speckled with brown[a]	Yes[a]	Reed (1959) - 90% adult or immature stages of blackflies, midges, and mayflies[a]	Invertivore			Bottom feeder
	falcatus[a]	Leopard dace	Subterminal[a]				Yes[a]	Aquatic insect larvae and terrestrial insects[a]	Invertivore			
	osculus[a]	Speckled dace	Subterminal[a]			Uniformly dark brown[a]	Yes[a]					
Richardsonius	*balteatus*[a]	Redside shiner	Terminal[a]		Long and coiled[a]	Pale, silvery, lightly speckled w/ black[a]	Yes[a]	Fry feed on diatoms, copepods, ostracods; larger fish are mainly insectivorous, immature forms of most aquatic insects[a]	Invertivore/ planktivore			
Semotilus	*atromaculatus*[a]	Creek chub	Terminal[c]	DT 0.6-0.7 TL	Intestine short, a single loop[a]	Silvery w/ light speckling[a]	Yes[a]	Young feed on planktonic organisms; adults feed on insect larvae such as beetles, mayflies, caddisflies, and chironomids; large males will also eat crayfish and small fishes[a]	Invertivore/ carnivore			

Order	Family	Genus	Species	Common Name	Mouth Position	Gut	Color of Peritoneum	Pyloric Caeca	Teeth	Pharyngeal Teeth and Plates	Main Food	First Level Trophic Classification	Second Level Trophic Classification	Trophic Mode	Feeding Behavior
			corporalis[a]	Fallfish	Slight subterminal[a]	Intestine short, single main loop[a]	Silvery, scattered speckles[a]			Yes[a]	Aquatic insect larvae, terrestrial insects take from surface, crustaceans, and fishes[a]	Invertivore/carnivore			
		Tinca	tinca[a]	Tench	Terminal[a]					Yes[a]	Aquatic insect larvae and molluscs[a]	Invertivore			
	Catostomidae	Carpiodes	carpio[c]	River carpsucker	Subterminal[c]	Much coiled[c]				Yes[c]	Periphyton, small planktonic plants and animals; 60% organic matter and 40% plants, etc.[c]	Planktivore/detritivore	Filter feeder	Suction feeder	
			cyprinus[a]	Quillback	Subterminal[a]	Intestine long w/ 6-9 coiled loops[a]			No[a]	Yes[a]	Immature insects, other invertebrates, and organic material[a]	Invertivore/detritivore	Benthic/filter feeder	Grazer/suction feeder	
			velifer[c]	Highfin carpsucker	Subterminal[c]	Much coiled[c]				Yes[c]	93% bottom ooze and algae, and 7% insects (Harrison, 1950)[c]	Detritivore	Filter feeder	Suction feeder	
		Catostomus	catostomus[a]	Longnose sucker	Subterminal[a]	Intestine long, undifferentiated, 2 or 3 coils; stomach scarcely differentiated[a]	Variable, silver to shiny black[a]	No[a]	No[a]	Yes[a]	A food list in order of frequency: Amphipods, Trichoptera, chironomids larvae and pupae, Ephemeroptera, ostracods, gastropods, Coleptera, Pelecypods, Copepods, Cladocerans, and plants[a]	Invertivore			
			columbianus[a]	Bridgelip sucker	Subterminal[a]	Intestine long, 6-14 loops anterior to liver, little differentiated[a]	Black[a]	No[a]	No[a]	Yes[a]	Algae scraped from rocks and invertebrates[a]	Herbivore	Particulate feeder	Grazer	

Genus	species	Common name	Mouth	Intestine	Color	Benthic/filter feeder	Grazer/suction feeder	Diet	Trophic category
	commersoni[a]	White sucker	Subterminal[a]	Intestine long w/ 4-5 coils, simple[a]	Pale or lightly speckled[a]	No[a]	Yes[a]	Invertebrates such as chironomids, trichopterans, and molluscs[a]	Invertivore/detritivore
	macrocheilus[a]	Largescale sucker	Subterminal[a]	Intestine long and coiled[a]	Dark[a]	No[a]	Yes[a]	Carl (1936) noted; ostracods, copepods, amphipods, Trichoptera, Chironomidae, larvae of other aquatic insects, molluscs, fish eggs, diatoms, and detritus[a]	Invertivore
Cycleptus	platyrhynchus[a]	Mountain sucker	Subterminal[a]	Intestine long w/ 6-10 coils anterior to liver[a]	Black or dusky[a]	No[a]	Yes[a]	Diatoms, other algae, higher plants, dipterous larvae and pupae[a]	Herbivore/invertivore
	elongatus[c]	Blue sucker	Subterminal[c]				Yes[c]	Insects, crustaceans, and plant material including algae[a]	Invertivore/herbivore
Erimyzon	oblongus[c]	Creek chubsucker	Subterminal[c]			No[c]	Yes[c]	Copepods, cladocerans, chironomid larvae and other bottom organisms[b]	Invertivore
	sucetta[d]	Lake chubsucker	Slightly subterminal[a]	Intestine long w/ several coils[a]	Silvery[a]	No[a]	Yes[a]	Copepods, cladocerans, chironomid larvae and other bottom organisms 70% vegetable matter (Cahn 1927)[c]	Invertivore/herbivore
Hypentelium	etowanum[d]	Alabama hogsucker	Subterminal[d]						
	nigricans[a]	Northern hog sucker	Subterminal[a]	Long, little differentiated, 5 coils anteriorly[a]	Dusky w/ some black pigment[a]	No[a]	Yes[a]	Insect larvae, crustaceans, diatoms, and other minute vegetation[a]	Invertivore/herbivore
	roanokense	Roanoke hog sucker	Subterminal					Insect larvae (mostly Diptera), algae, detritus (W.J. Matthews et al.)[b]	Invertivore/detritivore

Order	Family	Genus	Species	Common Name	Mouth Position	Ratio of Digestive Tract (DT) to Total Length (TL)[c]	Gut	Color of Peritoneum	Pyloric Caeca	Teeth	Pharyngeal Teeth and Plates	Main Food	First Level Trophic Classification	Second Level Trophic Classification	Trophic Mode	Feeding Behavior
		Ictiobus	bubalus[c]	Smallmouth buffalo	Subterminal[c]		Much elongated, with loops running parallel to body axis[c]		No[c]	No[c]	Yes[c]	Zooplankton and attached algae, Chironomidae, Baetidae, and Trichoptera[a]	Invertivore/ herbivore			
			cyprinellus[a]	Bigmouth buffalo	Terminal[a]		Intestine very long w/ at least 4 loop.-	Black[a]	No[a]	No[a]	Yes[a]	For young, copepods, cladocerans, chironomid larvae, diatoms; for older fish entomostraca, aquatic beetles, molluscs, and amphipods[a]	Invertivore			
			niger[c]	Black buffalo	Subterminal[c]		Much elongated, with loops running parallel to body axis[c]				Yes[c]					
		Lagochila[d]	lacera[d]	Harelip sucker[d]	Subterminal[d]		Short intestine[d]					Snails[d]	Invertivore			
		Minytrema	melanops[a]	Spotted sucker	Subterminal[a]		Intestine long, little differentiated[a]	Colorless to white[a]	No[a]	No[a]	Yes[a]	Molluscs and other invertebrates, mainly immature insects 19.4% copepods and ostracods, 77.2% cladocerans (Bur, 1976)[c]	Invertivore			
		Moxostoma	anisurum[a]	Silver redhorse	Subterminal[a]		Intestine long, little differentiated, 5 rounded coils obvious ventrally[a]	Colorless to silvery[a]	No[a]	No[a]	Yes[a]	(Meyer, 1962) 91% Chironomidae, 62% Ephemeroptera, 18% Trichoptera; (Harrison, 1950) 64% insects, 21% plants, 8% organic material, 4% fish, and 3% crustaceans[c]	Invertivore			

Appendix 7A

ariommum	Bigeye jumprock	Subterminal				Immature mayflies, and caddisflies, and particularly midges, blackflies, cranefly (*Antocha*) and detritus[b]	Invertivore		
atripinne[d]	Blackfin sucker	Subterminal[d]				Midge larvae, microcrustacea, and occasional larger insect larvae (Timmons et al., 1983)[d]	Invertivore		
carinatum[a]	River redhorse	Subterminal[a]	Intestine long, 5-6 rounded coils obvious ventrally[a]	Colorless to silvery[a]	No[a]	Yes[a]	Invertebrates such as immature insects and molluscs[a]	Invertivore	Benthic	Crusher
cervinum	Black jumprock	Subterminal				Insect larvae (mainly midges), water mites, small amounts of algae and detritus[b]	Invertivore		
duquesnei[a]	Black redhorse	Subterminal[a]	Intestine long, 5-6 rounded coils[a]	Silvery[a]	No[a]	Yes[a]	Dipterans, ephemeropterans, cladocerans, copepods, ostracods, amphipods[a]	Invertivore		
erythrurum[a]	Golden redhorse	Subterminal[a]	Intestine long, little differentiated, 2-3 long coils[a]	Colorless to silvery[a]	No[a]	Yes[a]	Immature insects, worms, molluscs; 46.5% Tricoptera, 27.1% Tendipedidae, 13.6% Ephemeroptera, 3.4% Sphaeriidae, 2.1% Copepoda (Bur, 1976)[c]	Invertivore		
hamiltoni[b]	Rustyside sucker	Subterminal[b]					Detritus, small numbers of mayfly, caddisfly, and true fly larvae[b]	Invertivore/ detritivore		
hubbsi[a]	Copper redhorse	Subterminal[a]	Intestine long, 4 long coils[a]	Black[a]	No[a]	Yes[a]	Molluscs and immature insects[a]	Invertivore		

168 Assessing the Sustainability and Biological Integrity of Water Resources Using Fish Communities

Order	Family	Genus	Species	Common Name	Mouth Position	Ratio of Digestive Tract (DT) to Total Length (TL)[c]	Gut	Color of Peritoneum	Pyloric Caeca	Teeth	Pharyngeal Teeth and Plates	Main Food	First Level Trophic Classification	Second Level Trophic Classification	Trophic Mode	Feeding Behavior
			macroiepi-dotum[a]	Shorthead redhorse	Subterminal[a]		Intestine long, little differentiated, 6 rounded coils slightly to right side[a]	Colorless to silvery[a]	No[a]	No[a]	Yes[a]	Immature forms of ephemeroterans, trichopterans, chironomids, tipulids, ostracods, molluscs, oligochaetes, various crustaceans, and diatoms[a] (Bur, 1976) 41.2% Tendipedidae, 39.3% Trichoptera, 12.5% Simuliidae, 1.7% Sphaeriidae[c]	Invertivore			
			pappillosum[b]	V-lip redhorse	Subterminal[b]						Yes[b]	Ostracods, vascular plants, silt (Gatz, 1979); aquatic insects almost certainly are consumed[b]	Invertivore/ detritivore			
			poecilurum[d]	Blacktail redhorse	Subterminal[d]						Yes[f]	Detritus, diatoms, and a variety of small invertebrates such as microcrustacea, rotifers, caddisfly larvae, and phantom midge (Chaoborus) larvae[d]	Detritivore/ invertivore			
			rhothoecum[d]	Torrent sucker	Subterminal[b]							Algae, other plant material, and detritus; less amounts of insect larvae, mainly midges (Flemer and Woolcott, 1966; W.J. Matthews et al.)[b]	Detritivore/ herbivore			
			robustum[b]	Smallfin redhorse	Subterminal[b]							Aquatic insects, detritus[b]	Invertivore/ detritivore			

Appendix 7A

Order	Family	Genus	Species	Common name	Mouth position	Gut	Peritoneum	Barbels?	Spines?	Diet	Trophic category	Habitat/food size	Crusher
			valenciennesi[a]	Greater redhorse	Subterminal[a]	Intestine long, little differentiated, 5-7 rounded coils[a]	Black[a]	No[a]	No[a]	Yes[a]	Invertebrates, immature insects, worms, and molluscs[a] (Rimsky-Korsakoff, 1929) 5% midge larvae, 25% molluscs, 60% crustaceans, and 10% plants[c]	Invertivore	Benthic	Crusher
Siluriformes	Ictaluridae	Ameiurus	brunneus[b]	Snail bullhead	Subterminal[b]		Silver[b]		Yes[b]	Yes[b]	Insect larvae, snails, shiners, filamentous algae, and aquatic macrophytes (Yerger and Relyea, 1968)[b]	Invertivore/herbivore/carnivore	Benthic/particulate/whole body	
			catus	White catfish	Slightly subterminal						Aquatic insects, other invertebrates, fishes, plants (Menzel, 1945; Carlander, 1969)[a]	Invertivore/herbivore/carnivore	Benthic/particulate/whole body	
			melas[a]	Black bullhead	Slightly subterminal[b]; terminal[a]	DT 0.8-1.5 TL	Intestine well differentiated, coiled[a]	Heavily speckled w/ black[a]	No[a]	Yes[a]	Yes[b]	Variety of invertebrates and fishes (Carlander, 1969)[a] 54.1% Hyallela azteca, 19.2% insect larvae, 15.1% organic detritus, 4.5% insect adults, 3.4% fungi and algae, 2.8% small crayfish, 0.9% misc. (Darnell and Meierotto, 1962)[c]	Invertivore/carnivore	Benthic/whole body	
			natalis[a]	Yellow bullhead	Slightly subterminal[b]; terminal[a]	DT 1.3 TL	Intestine well differentiated[a]	Gray[a]	No[a]	Yes[a]	Yes[a]	Various aquatic invertebrates and fishes (Flemer and Woolcott, 1966; Russell, 1976; Becker, 1983)[b] crayfish, amphipods, cladocerans, insect larvae, copepods, snails, and algae (Pearse, 1921a)[c]	Invertivore/carnivore	Benthic/whole body	

Order	Family	Genus	Species	Common Name	Mouth Position	Ratio of Digestive Tract (DT) to Total Length (TL)[c]	Gut	Color of Peritoneum	Pyloric Caeca	Teeth	Pharyngeal Teeth and Plates	Main Food	First Level Trophic Classification	Second Level Trophic Classification	Trophic Mode	Feeding Behavior
		Ictalurus	nebulosus[a]	Brown bullhead	Slightly subterminal[a]; terminal[a]	DT 1.1–1.4 TL	Intestine well differentiated[a]	Silvery to grey[a]	No[a]	Yes[a]	Yes[a]	Insects, fish eggs, mollusks, and plants (Carlander, 1969)[c]	Invertivore/ herbivore/ carnivore	Benthic/ particulate/ whole body		
			platycephalus[b]	Flat bullhead	Slightly subterminal					Yes[b]	Yes[b]	Aquatic invertebrates and fishes (Olmsted and Cloutman, 1979)[b]	Invertivore/ carnivore	Benthic/ whole body		
			furcatus[f]	Blue catfish	Subterminal[f]					Yes[f]		An array of invertebrates, fishes, and occasionally frogs[f]	Invertivore/ carnivore	Benthic/ whole body		
			punctatus[a]	Channel catfish	Subterminal[a]		Intestine well differentiated, coiled[a]	Speckled w/ black[a]	No[a]	Yes[a]	Yes[a]	Mayflies, caddisflies, chironomids, molluscs, crayfish, crabs, green algae, larger plants, tree seeds (Bailey and Harrison, 1948)[a]	Invertivore/ carnivore	Benthic/ whole body		
		Noturus	albater[f]	Ozark madtom	Subterminal[f]					Yes[f]		Aquatic insects, notably dipterans[f]	Invertivore	Benthic	Lie-in-wait/ ambush	
			baileyi[d]	Smoky madtom	Subterminal[d]											
			elegans[d]	Elegant madtom	Subterminal[d]											
			eleutherus	Mountain madtom	Subterminal					Yes[e]		Immature aquatic insects (Starnes and Starnes, 1985)[b]	Invertivore	Benthic	Lie-in-wait/ ambush	
			exilis[e]	Slender madtom	Terminal or subterminal[e]	DT 1.1–1.3 TL	Coiled[e]			Yes[e]		Mostly caddisflies, a trace of midgeflies, unidentifiable insect parts, filamentous algae, and debris[e]	Invertivore/ detritivore	Benthic/ particulate feeder	Lie-in-wait/ ambush	
			flavater[f]	Checkered madtom	Subterminal[f]					Yes[f]						

Appendix 7A

flavipinnis	Yellowfin madtom	Subterminal[a]			No[a]	Yes[a]	Exclusively eats immature forms of all major groups of benthic aquatic insects (Jenkins, 1975b; P. Shute, 1984)[b]	Invertivore	Benthic
flavus[a]	Stonecat	Subterminal[a]	DT 1.3 TL	Intestine coiled and well differentiated[a]	Colorless[a]	Yes[a]	Yes[a]	64% aquatic riffle insects, 14% fish, including spotfin shiners, common shiners, and a bullhead minnow, 9% crayfish and earthworms, 7% filamentous algae and weed seeds of terrestrial origin, 5% undetermined organic matter (Harrison, 1950)[c]	Invertivore/ carnivore	Benthic/whole body
gilberti	Orangefin madtom	Subterminal						Immature aquatic insects, particularly mayflies, hydropsychid caddisflies, and midges[b]	Invertivore	Benthic
gyrinus[a]	Tadpole madtom	Terminal[a]	DT 1.0-1.4 TL	Intestine well differentiated[a]	Colorless[a]	No[a]	Yes[a]	Cladocera, ostracods, Hyalella, chironomids, and other immature insects[a]; 44% insects, 18.3% oligochaetes, 28.3% small crustaceans, 5.9% plants, 0.1% snails, 0.1% algae, 3% silt and debris (Pearse, 1918)[c]	Invertivore/ planktivore	Benthic/ particulate feeder
hildebrandi[d]	Least madtom	Subterminal[d]						Primarily midge and caddis larvae, supplemented with mayfly and stonefly nymphs[d]	Invertivore	Benthic	Lie-in-wait/ ambush

Order	Family	Genus	Species	Common Name	Mouth Position	Ratio of Digestive Tract (DT) to Total Length (TL)[c]	Gut	Color of Peritoneum	Pyloric Caeca	Teeth	Pharyngeal Teeth and Plates	Main Food	First Level Trophic Classification	Second Level Trophic Classification	Trophic Mode	Feeding Behavior
			insignis	Margined madtom	Subterminal							Variety of aquatic invertebrates, mostly aquatic insect larvae; fishes and terrestrial insects taken occasionally (Bowman, 1932; 1936; Flemer and Woolcott, 1966)[b]	Invertivore	Benthic	Lie-in-wait/ ambush	
			lachneri[f]	Ouachita madtom	Terminal[f]					Yes[f]		Ephemeropterans, dipterans (Chironomidae), coleopterans, trichopterans, isopods, copepods, and gastropods[f]	Invertivore	Benthic		
			leptacanthus[d]	Speckled madtom	Subterminal[d]							Primarily midge larvae[d]	Invertivore	Benthic		
			miurus[a]	Brindled madtom	Slightly subterminal[a]				No[a]	Yes[a]		Immature insects, other invertebrates, and plants[b]	Invertivore	Benthic		
			munitus[d]	Frecklebelly madtom	Subterminal		Intestine well differentiated[a]	Colorless[a]				Diet consisted primarily of hydropsychid caddisfly larvae, ephemerellid mayfly nymphs, and blackfly and midge larvae (G. Miller, 1984)[d]	Invertivore	Benthic		
			nocturnus[d]	Freckled madtom	Subterminal[d]					Yes[f]		Aquatic insect immatures dominated by mayflies, caddisflies, midges, and blackflies[d]	Invertivore	Benthic		Lie-in-wait/ ambush
			phaeus[d]	Brown madtom	Terminal to slightly subterminal											
			stanauli[d]	Pygmy madtom	Subterminal[d]					Yes[f]						

Appendix 7A

				Mouth position		Intestine			Diet	Trophic category	Feeding location	Feeding mode
		stigmosus[d]	Northern madtom	Subterminal				Yes[f]	Snails, isopods, mayflies, dragonflies, caddisflies, stoneflies, aquatic lepidopterans, aquatic beetles, and dipterans[f]	Invertivore	Benthic	Lie-in-wait/ambush
		taylori[f]	Caddo madtom									
		trautmani[e]	Scioto madtom	Subterminal								
	Pylodictis	olivaris[c]	Flathead catfish	Terminal[c]	DT 1.0 TL			Yes[c]	Young: microcrustaceans and insect larvae; adults; crayfishes, clams, and particularly fishes (Minckley and Deacon, 1959; Guier et al., 1984)[b]	Invertivore/carnivore	Benthic and drift/whole body	Passive predator[c]
Salmoniformes												
Esocidae	Esox	americanus	Redfin pickerel	Terminal[a]		Intestine long and undifferentiated[a]	No[a]	Yes[a]	Young-of-the-year: entirely inverts, such as cladocerans, amphipods, isopods, and immature stages of aquatic insects.[a] Adults: other fish[a]	Carnivore	Whole body	Ambush
		americanus vermiculatus[a]	Grass pickerel	Terminal[a]		Intestine long and undifferentiated[a]	No[a]	Yes[a]	Young cladocerans, amphipods, ostero cods, isopods, immature or adult insects from orders; Diptera, Plecoptera, Hemiptera[a]; Adults: other fish, crayfish[a]	Invertivore carnivore	Benthic and drift/whole body	Lie-in-wait-ambush
		lucius[a]	Northern pike	Terminal[a]		Intestine long and undifferentiated[a]	No[a]	Yes[a]	Young: larger zooplankton, some immature insects;[a] Adults: other fish, frogs, crayfish[a]	Carnivore	Whole body	Ambush

Order	Family	Genus	Species	Common Name	Mouth Position	Ratio of Digestive Tract (DT) to Total Length (TL)c	Gut	Color of Peritoneum	Pyloric Caeca	Teeth	Pharyngeal Teeth and Plates	Main Food	First Level Trophic Classification	Second Level Trophic Classification	Trophic Mode	Feeding Behavior
			masquinon-gya	Muskellunge	Terminala		Long intestinea		Noa	Yesa	Yesa	Young: larger zooplankton;a Adults: other fish.a	Carnivore	Whole body	Ambush	
			nigera	Chain pickerel	Terminala		Intestine long and undifferentiateda		Noa	Yesa	Yesa	Young: plankton, immature aquatic insects;a Adults: other fish, crayfisha	Carnivore	Whole body	Ambush	
	Umbridae	Dallia	pectoralisa	Alaska blackfish	Terminala				Noa	Yesa	Yesa	Largely dipteran insect l.-a.a, ostracods, cladocerans, and snails fish usually not part of dieta	Invertivore	Benthic and drift	Lie in wait/ambush	Moves up slowly on food items, which it takes with a quick dart
		Umbra	limia	Central mudminnow	Terminala				Noa	Yesa	Yesa	Young: ostracods, newly hatched snails;a After 20 mm: insect larvae and adults, molluscs, amphipods, isopods, and arachnidsa	Invertivore	Benthic and drift		
			pygmaeab	Eastern mudminnow	Terminalb					Yes, smallb		Midge larvae, small crustaceans, small crayfish, other insects, and rarely fish (Flemer and Woolcott, 1966; Gatz, 1979; G.B. Pardue and M.T. Huish)b	Invertivore	Benthic and drift		
	Osmeridae	Hypomesus	olidusa	Pond smelt	Supertermi-nala				Yesa	Smalla	Yesa	Rotifers for young-of-the-year;a Rotifers, algae, insects, and crustaceans for adultsa	Invertivore	Benthic and drift		
		Osmerus	mordaxa	Rainbow smelt	Terminala		Shorta	Silvery with dark specklesa	Yesa	Yes, large on tonguea	Yesa	Mysis relicta, various other invertebrates, and to a small extent other fishesa	Invertivore/carnivore	Benthic and drift/whole body		

Appendix 7A

Family	Genus	Species	Common name	Mouth position				Diet	Trophic	Feeding mode	Habitat
	Spirinchus	thaleichthys[a]	Longfin smelt	Supraterminal[a]	Physostomous[a]	Yes[a]	Yes[a]	As young in freshwater - Neomysis mercedis[a]	Invertivore	Benthic and drift	Water column
Salmonidae	Coregonus	artedi[a]	Cisco, Lake herring	Terminal[a]		Yes[a]	Yes[a]	Plankton, algae, copepods, cladocera; various other insects, eggs, and other small fish[a]	Planktivore/ invertivore	Particulate feeder/drift	
		autumnalis[a]	Arctic cisco	Terminal[a]		Yes[a]	On tongue[a]	Crustaceans and small fish[a]	Planktivore/ carnivore	Particulate feeder/whole body	
		clupeaformis[a]	Lake whitefish	Terminal[a]			Small and weak[a]	Young: plankton; Adults: insect larvae, molluscs, amphipods, fish and fish eggs, etc.[a]	Invertivore/ carnivore	Particulate feeder/whole body	
		hoyi[a]	Bloater	Terminal[a]				Mysis relicta 48.2%, Pontoporeia affinis 60.7%, copepods, eggs and small molluscs (Wells and Beeton, 1963)[a]	Planktivore/ invertivore	Particulate feeder/drift	
		huntsmani[a]	Atlantic whitefish	Terminal[a]			Small[a]	Amphipods, periwinkles, and other invertebrates[a]	Invertivore	Benthic and drift	
		kiyi[a]	Kiyi	Terminal[a]				Mysis relicta and Pontoporeia sp.; 69.7% Pontoporeia sp. and 30.3% Mysis sp. (Barsamin, 1967)[a]	Planktivore	Particulate feeder	
		laurettae[a]	Bering cisco	Terminal[a]			On tongue[a]	Crustaceans[a]	Planktivore		
		nasus[a]	Broad whitefish	Subterminal[a]		Yes[a]	Weak[a]	Insect larvae, molluscs, crustaceans[a]	Invertivore	Benthic and drift	
		reighardi[a]	Shortnose cisco	Terminal[a]				Mysis relicta and Pontoporeia hoyi[a]	Planktivore	Particulate feeder	
		sardinella[a]	Least cisco	Terminal[a]			On tongue[a]	Planktonic crustaceans, aquatic and terrestrial insects[a]	Planktivore/ invertivore	Particulate feeder/drift	

Order	Family	Genus	Species	Common Name	Mouth Position	Ratio of Digestive Tract (DT) to Total Length (TL)[c]	Gut	Color of Peritoneum	Pyloric Caeca	Teeth	Pharyngeal Teeth and Plates	Main Food	First Level Trophic Classification	Second Level Trophic Classification	Trophic Mode	Feeding Behavior
			zenithicus[a]	Shortjaw cisco	Terminal[a]							Pontoporeia hoyi, Mysis relicta, planktonic Crustacea, and aquatic insect larvae[a]	Planktivore	Particulate feeder	Size-selective picker	
		Oncorhynchus	clarki[a]	Cutthroat trout	Terminal[a]				Yes[a]	Yes[a]	Basibranchial plate[a]	Insects and small fishes[a]	Invertivore/ carnivore	Benthic and drift/whole body		
			gorbuscha[a]	Pink salmon	Terminal[a]				Yes[a]	Yes[a]	Basibranchial plate[a]	Some nymphal and larval insects[a]	Invertivore			
			keta[a]	Chum salmon	Terminal[a]				Yes[a]	Yes[a]	Basibranchial plate[a]	Variety of mature and immature insects, as well as copepods and nematodes[a]	Invertivore/ planktivore			
			kisutch[a]	Coho salmon	Terminal[a]				Yes[a]	Yes[a]	Basibranchial plate[a]	Young: larvae of dipterans, tricoptera, plecoptera, and coleoptera, as well as sockeye salmon fry.[a] Adults: (Great Lakes) rainbow smelt and alewife[a]	Invertivore/ carnivore			
			mykiss[a]	Rainbow trout	Terminal[a]				Yes[a]	Yes[a]	Basibranchial plate[a]	Invertebrates, other fish, and fish eggs[a]	Invertivore/ carnivore			
			nerka[a]	Kokanee and sockeye salmon	Terminal[a]				Yes[a]	Yes, but small[a]	Basibranchial plate[a]	Plankton, Chironomid pupae (70% volume), terrestrial insects, and water mites (Northcote and Lorz, 1966)[a]	Invertivore/ planktivore			
			tshawytscha[a]	Chinnok salmon	Terminal[a]				Yes[a]	Yes[a]	Basibranchial plate[a]	Terrestrial insects, Crustacea, chironomid larvae, pupae, and adults, corixids, caddisflies, mites, spiders, aphids, Corethra larvae, and ants[a]	Invertivore/ carnivore			

Prosopium	*coulteri*[a]	Pygmy whitefish	Subterminal[a]	Yes (simple)[a]	Small and on the tongue[a]	Crustaceans, aquatic insect larvae, chironomids, and eggs[a]	Invertivore
	cylindraceum[a]	Round whitefish	Subterminal	Yes[a]	Small and on the tongue[a]	Mayfly larvae and pupae, various other invertebrates, small fish, and fish eggs[a]	Invertivore/carnivore
	williamsoni[a]	Mountain whitefish	Subterminal[a]	Yes[a]	Small and on the tongue[a]	Insect larvae, occasionally small fishes, and fish eggs[a]	Invertivore/carnivore
Salmo	*salar*[a]	Atlantic salmon	Terminal[a]	Yes[a]	Basibranchial plate[a]	Various aquatic and terrestrial invertebrates; most importantly larvae and nymphs of chironomids, mayflies, caddisflies, blackflies, and stoneflies[a]	Invertivore/carnivore
	trutta[a]	Brown trout	Terminal[a]	Yes[a]	Basibranchial plate[a]	Aquatic and terrestrial invertebrates, crustaceans, molluscs, frogs, and fishes[a]	Invertivore/carnivore
Salvelinus	*alpinus*[a]	Arctic char	Terminal[a]	Yes[a]	Basibranchial plate[a]	Plankton, various invertebrates, fishes, and algae smelt can make up to 60% of diet of larger adult fish[a]	Invertivore/carnivore
	fontinalis[a]	Brook trout	Terminal[a]	Yes[a]	Basibranchial plate[a]	Any living creature its mouth can accomodate from insects, to fish, to small mammals, field mice, etc.[a]	Invertivore/carnivore
	malma[a]	Dolly varden	Terminal[a]	Yes[a]	Basibranchial plate[a]	Invertebrates, eggs, and fish[a]	Invertivore/carnivore

Order	Family	Genus	Species	Common Name	Mouth Position	Ratio of Digestive Tract (DT) to Total Length (TL)[c]	Gut	Color of Peritoneum	Pyloric Caeca	Teeth	Pharyngeal Teeth and Plates	Main Food	First Level Trophic Classification	Second Level Trophic Classification	Trophic Mode	Feeding Behavior
			namaycush[a]	Lake trout	Terminal[a]				Yes[a]	Yes[a]	Basibranchial plate[a]	Freshwater sponges, aquatic and terrestrial invertebrates, many fishes (primarily ciscoes, alewife, whitefish, smelt, and sculpins for adults), and small mammals[a]	Invertivore/ carnivore			
		Stenodus	leucichthys[a]	Inconnu	Terminal[a]				Yes[a]	Fine and small[a]		Invertebrates as young and fish as adults[a]	Invertivore/ carnivore			
		Thymallus	arcticus[a]	Arctic grayling	Terminal[a]				Yes[a]	Small[a]		Various terrestrial invertebrates, various aquatic invertebrates, fishes, fish eggs, lemmings, and plankton crustaceans[a]	Invertivore/ planktivore/ carnivore			
Percopsi-formes	Percopsidae	Percopsis	omiscomaycus[a]	Trout-perch	Subterminal[a]	DT 0.5 TL			Yes, unique in number and arrangement[a]	Yes[a]		Insect larvae, fish, and amphipods[a]; 80% chironomid larvae, 48% amphipods, 39% cladocerans, 29% copepods, fish eggs observed in only two stomachs (Tomlinson and Junde, 1977)[c]	Invertivore/ carnivore	Benthic		Bottom feeder
	Aphredoderidae	Aphredoderus	sayanus[c]	Pirate perch	Terminal[c]	DT 1.2-1.6 TL				Yes[c]		Small aquatic insects and small fish[a]	Invertivore/ carnivore			

Appendix 7A

Gadiformes	Gadidae	*Lota*	*lota*[a]	Burbot	Subterminal (slightly)[a]		Yes[a]		74% fish, 26% invertebrates; dominant items 76% sculpins, 51% coregonid chubs, 37% *Pontoporeia*, 34% trout-perch, 26% *Mysis*. Note: study done before advent of alewife (Van Oosten and Deason, 1938)[c]	Invertivore/ carnivore			Voracious nocturnal predator
		Microgadus	*tomcod*[a]	Atlantic tomcod									
Atheriniformes	Cyprinodontidae	*Fundulus*	*catenatus*[d]	Northern studfish	Terminal[a]			Yes[d]	Aquatic insect larvae and snails (McCaskill et al., 1972; fisher, 1981)[d]	Invertivore			
			chrysotus[d]	Golden topminnow	Supraterminal				Primarily insects taken at surface (Hunt, 1953)[d]	Invertivore	Drift	Surface feeder	Primarily a surface feeder
			diaphanus[a]	Banded killifish	Terminal[a]	DT 0.5 TL	Intestine w/ single loop, about one half body length[a]	Silvery and lightly speckled[a]		Yes[a]	For young: chironomid larvae, ostacods, cladocerans, copopods, amphipods, flying insects; For older fish: same as young but also newly hatched Odonata and Ephemeroptera nymphs, molluscs, and turbellarians[c]	Invertivore/ planktivore		
			dispar[d]	Northern starhead topminnow	Supraterminal				Terrestrial insects, snails, small crustaceans, and some algae (Forbes and Richardson, 1920; Gunning and Lewis, 1955)[d]	Invertivore			

180 Assessing the Sustainability and Biological Integrity of Water Resources Using Fish Communities

Order	Family	Genus	Species	Common Name	Mouth Position	Ratio of Digestive Tract (DT) to Total Length (TL)[c]	Gut	Color of Peritoneum	Pyloric Caeca	Teeth	Pharyngeal Teeth and Plates	Main Food	First Level Trophic Classification	Second Level Trophic Classification	Trophic Mode	Feeding Behavior
			heteroclitus[a]	Mummichog	Terminal[a]			Uniformly black[a]		Yes[a]		Diatoms, amphipods, other crustaceans, molluscs, fish eggs, and small fish and also vegetation, such as eel grass[a]	Planktivore/ invertivore/ herbivore/ carnivore			
			julisia[d]	Barrens topminnow	Supraterminal							Microcrustacea, midges, amphipods and isopods, small mayflies, terrestrial insects and snails[d]	Planktivore/ invertivore			
			lineolatus[b]	Lined topminnow	Terminal[b]							Insects (aquatic and terrestrial), algae, seeds[b]	Invertivore/ herbivore			
			notatus[c]	Blackstripe topminnow			Short and straight or with single forward loop[c]					75% of food filamentous algae, Spirogyra and Zygnema predominating, the remainder of its diet was insects picked from surface (Cahn, 1927)[f]	Herbivore/ invertivore	Particulate feeder/ drift	Browser/ surface feeder	
			notti[e]	Bayou topminnow	Supraterminal	DT .75 TL	Short[e]			Yes[e]		Half the food ingested consisted of insects, half of which were terrestrial; molluscs, crustaceans, and delicate aquatic vegetation constituted the remainder of the diet (Forbes and Richardson, 1920)[e]	Invertivore/ herbivore			

Appendix 7A

Order	Family	Genus	species	Common name	Mouth position			Diet	Diet category	Feeding mode 1	Feeding mode 2
			olivaceus[d]	Blackspotted topminnow	Slightly supraterminal			Terrestrial arthropods, aquatic insects, small crustaceans, and diatoms (Rice, 1942; Thomerson and Wooldridge, 1970)[d]	Invertivore/ herbivore		
			rathbuni[b]	Speckled killifish	Terminal[b]			Midge larvae, other aquatic insects, and a small shiner (Lee et al., 1980)[b]	Invertivore		
			stellifer[d]	Southern studfish	Terminal[d]		Yes[d]				
	Poeciliidae	Gambusia	affinis	Western mosquitofish	Terminal				Invertivore	Drift	Surface feeder
			holbrooki[b]	Eastern mosquitofish	Supraterminal[b]		Yes, small[b]	Mosquito larvae and pupae, worms, microcrustaceans, mites, terrestrial and other aquatic insects, snails, and algae (Hildebrand and Schroeder, 1928; Krumholz, 1948; Harrington and Harrington, 1961; Gatz, 1979)[b]	Invertivore		
	Atherinidae	Labidesthes	sicculus[a]	Brook silverside	Terminal[a]	DT 0.5 TL	Short, S-shaped[c]	No[a]	Keast and Webb (1966) in a: 80% Cladocera, 40% small flying insects, and 50% Chaoborus	Planktivore/ Invertivore	Particulate feeder/drift	Surface feeder
		Menidia[d]	beryllina[d]	Inland silverside	Terminal[d]				Zooplankton (Saunders, 1959[c]; midge larvae, mayfly larvae, and fallen terrestrial insects[d]	Planktivore/ invertivore	Particulate feeder/drift	Surface feeder
Gasterosteiformes	Gasterosteidae	Apeltes	quadracus[a]	Fourspine stickleback	Terminal[a]			Yes[a]	Planktonic plants and animals[b]	Planktivore	Particulate feeder	Size-selective picker

Order	Family	Genus	Species	Common Name	Mouth Position	Ratio of Digestive Tract (DT) to Total Length (TL)[c]	Gut	Color of Peritoneum	Pyloric Caeca	Teeth	Pharyngeal Teeth and Plates	Main Food	First Level Trophic Classification	Second Level Trophic Classification	Trophic Mode	Feeding Behavior
		Culaea	*inconstans*[a]	Brook stickleback	Terminal[a]	DT 0.5 TL				Yes[a]		Larvae of a wide variety of aquatic insects and crustaceans, and also eggs and larvae of other species, as well as snails, oligochaetes, and algae[a]; 47% insects, 38% entomostracans (Pearse, 1918)[c]	Planktivore/invertivore	Particulate feeder		
		Gasterosteus	*aculeatus*[a]	Threespine stickleback	Terminal[a]					Yes[a]		Various worms, small crustaceans, aquatic insects, and larvae, drowned aerial insects, eggs, and fry of fish including their own[b]	Invertivore			
			wheatlandi[a]	Blackspotted stickleback												
		Pungitius	*pungitius*[a]	Ninespine stickleback	Oblique[a]	DT 0.4 TL	Short[c]			Yes[a]		61% *Pontoporeia*, 21% *Mysis*, 8% copepods, 10% other misc. items (Griswold and Smith, 1973)[c]	Planktivore	Particulate feeder	Size-selective picker	
Scorpaeni-formes[d]	Cottidae	*Cottus*	*aleuticus*[a]	Coastrange sculpin	Terminal[a]					Yes[a]		Aquatic insects and other benthic invertebrates, especially molluscs; in autumn, salmon eggs may be eaten extensively[b]	Invertivore	Benthic	Digger	Rolls over stones
			asper[a]	Prickly sculpin	Terminal[a]					Yes[a]		Chironomid and trichopteran larvae and other bottom invertebrates such as molluscs[b]	Invertivore	Benthic	Digger	Rolls over stones

Appendix 7A

baileyi[b]	Black sculpin	Terminal[b]	Yes[b]	Mayfly, midge, blackfly, and cranefly larvae (Novak and Estes, 1974)[b]	Invertivore	Benthic	Digger	Rolls over stones
bairdi[a]	Mottled sculpin	Terminal[a]	Yes[a]	Chironomid larvae and mayfly nymphs for smaller fish; less chironomids and more larger mayfly and stonefly nymphs with caddisfly larvae and crayfish[c]	Invertivore	Benthic	Digger	Rolls over stones
carolinae[b]	Banded sculpin	Terminal[b]	Yes[b]	Small crustaceans and insects; larger sculpins feed on crayfish and fishes more than small one (Minckley et al., 1963)[b]	Invertivore	Benthic	Digger	Rolls over stones
cognatus[a]	Slimy sculpin	Terminal[a]	Yes[a]	Koster (1937), aquatic insect larvae and nymphs made up 50–85% of diet, the more important insect groups were mayflies, caddisflies, dipterous larvae, stoneflies, and dragonflies[a]	Invertivore	Benthic	Digger	Rolls over stones
confusus[a]	Shorthead sculpin	Terminal	Yes[a]		Invertivore	Benthic	Digger	Rolls over stones
girardi[b]	Potomac sculpin	Supraterminal[b]	Yes[b]	Benthic insect larvae, mostly mayflies, caddisflies, and midges, amphipods, crayfish, snails, other invertebrates and fantail darters taken rarely (Matheson, 1979)[b]	Invertivore	Benthic	Digger	Rolls over stones

184 Assessing the Sustainability and Biological Integrity of Water Resources Using Fish Communities

Order	Family	Genus	Species	Common Name	Mouth Position	Ratio of Digestive Tract (DT) to Total Length (TL)[c]	Gut	Color of Peritoneum	Pyloric Caeca	Teeth	Pharyngeal Teeth and Plates	Main Food	First Level Trophic Classification	Second Level Trophic Classification	Trophic Mode	Feeding Behavior
			hypselurus[f]	Ozark sculpin						Yes[f]		Crayfish (Orconectes), trichoptera larvae (Hydropsychidae), plecoptera, ephemeroptera, diptera (Chironomus and Simulium), as well as gastropods, elmid beetles, and miscellaneous animal material (Cooper, 1975)[f]	Invertivore	Benthic	Digger	
			rhotheus[a]	Torrent sculpin	Subterminal[a]					Yes[a]		Planktonic crustaceans, midge and mayfly larvae for young; fish are almost the exclusive food for sculpins larger than 70 mm[b]	Invertivore	Benthic	Digger	Rolls over stones
			ricei[a]	Spoonhead sculpin	Subterminal[a]					Yes[a]		Planktonic crustaceans in deep lakes; aquatic insect larvae in inshore regions[a]	Invertivore	Benthic	Digger	Rolls over stones
		Myoxocephalus	quadricornis	Fourhorn sculpin	Terminal					Yes[a]		Mysis, Pontoporeia, and chironomid larvae (Dymond, 1926; McPhail and Lindsey, 1970; McAllister, 1961)[c]	Invertivore	Benthic	Digger	Rolls over stones
Perciformes	Percichthyidae	Morone	americana[a]	White perch	Terminal[a]					Yes[a]		Microcrustaceans, grass shrimp, crayfishes, insect larvae, and fishes (Hildebrand and Schroeder, 1928)[c]	Invertivore/ carnivore			

Appendix 7A

Family	Genus	species	Common name	Mouth position	DT	Diet	Eye color	Piscivore	Trophic category	Body part	Feeding behavior/location
		chrysops[a]	White bass	Terminal[a]		Alewife, black crappie, and crayfishes (Boaze, 1972; Kohler, 1980)[b]		Yes[a]	Invertivore/carnivore		
		mississippiensis[c]	Yellow bass	Terminal[c]	DT 1.5 TL	92% Cladocera, 92% Copepoda, 68% Chironomidae larvae, 52% Chironomidae pupae, 16% Chaoborinae, 20% fish remains, 28% Ostracoda; Ephemeroptera, *Hydracarina*, and Corixidae occurred less frequently (Helm, 1964)[c]		Yes[c]	Invertivore/carnivore		
		saxatilis[a]	Striped bass	Terminal[a]		Fishes, Squids, clams, lobsters, crabs, shrimps, and other invertebrates (Smith, 1907)[b]		Yes[a]	Invertivores/carnivore	Whole body	Chasing
Centrarchidae	Acantharchus	pomotis[a]	Mud sunfish	Terminal[b]		Crustaceans and insects[b]		Yes[b]	Invertivore		
	Ambloplites	ariommus[d]	Shadow bass	Terminal[d]		Small fish, crayfish, and larger insects, including dragonfly, mayfly, and stonefly nymphs[d]			Invertivore/carnivore		
		cavifrons[d]	Roanoke bass	Terminal or supraterminal							
		constellatus[f]	Ozark bass	Terminal or supraterminal		Fishes, insects, particularly crayfish[b]			Invertivore/carnivore		
		rupestris[a]	Rock bass	Terminal slightly oblique[a]			Colorless[a]	Yes[a]	Invertivore/carnivore		
	Centrarchus	macropterus[b]	Flier	Supraterminal		Aquatic insects, crayfish, and fish[b]		Yes[a]	Invertivore		
						Arachnids, minute crustaceans, and insects (Flemer and Woolcott, 1966)[b]			Invertivore	Drift	Water column

Order	Family	Genus	Species	Common Name	Mouth Position	Ratio of Digestive Tract (DT) to Total Length (TL)[c]	Gut	Color of Peritoneum	Pyloric Caeca	Teeth	Pharyngeal Teeth and Plates	Main Food	First Level Trophic Classification	Second Level Trophic Classification	Trophic Mode	Feeding Behavior
		Elassoma	zonatum[d]	Banded pygmy sunfish	Terminal[d]					Yes[d]		Microcrustacea supplemented with midge larvae, large crustacea (amphipods and isopods), mayfly nymphs, and small snails and clams[d]	Invertivore	Drift	Water column	
		Enneacanthus	chaetodon[b]	Blackbanded sunfish	Terminal[b]					Yes[b]		Small invertebrates associated w; aquatic macrophytes (Seal, 1914; Schwartz, 1961; Wujewicz, 1982)[b]				
			gloriosus[b]	Bluespotted sunfish	Terminal or supraterminal[b]					Yes[b]		Insects and other small invertebrates[b]	Invertivore	Drift	Water column	
			obesus[b]	Banded sunfish	Supraterminal[b]					Yes[b]		Microcrustaceans and insects (Cohen, 1977)[b]	Invertivore	Drift	Water column	
		Lepomis	auritus[a]	Redbreast sunfish	Terminal, somewhat oblique[a]		Intestine well differentiated[a]	Colorless[a]	Yes[a]	Yes[a]	Yes[a]	Immature aquatic insects[b]	Invertivore	Drift	Water column	
			cyanellus[a]	Green sunfish	Terminal oblique[a]		Intestine long and well differentiated[a]	White to silvery[a]	Yes[a]	Yes[a]	Yes[a]	Insects, molluscs, and small fishes[a]	Invertivore/carnivore	Drift/wholebody	Water column/ambush	
			gibbosus[a]	Pumpkinseed	Terminal, slightly oblique[a]		Intestine well differentiated[a]	Silvery[a]	Yes[a]	Yes[a]	Yes[a]	Variety of insects, small fishes, and other invertebrates[a]	Invertivore/carnivore	Drift/wholebody	Water column/chasing	
			gulosus[c]	Warmouth	Terminal[c]					Yes[c]	Yes[c]	Insects, snails, crayfishes, and fishes (Larimore, 1957; Flemer and Woolcott, 1966; Gatz, 1979)[b]	Invertivore/carnivore	Drift/wholebody	Water column/chasing	

humilis[d]	Orangespotted sunfish	Terminal[d]					Yes[d]	Mainly microcrustaceans and aquatic insect larvae (Barney and Anson, 1922); Tennessee specimens contained both chironomid larvae and terrestrial insects[d]	Invertivore	Drift	Water column/ surface	Benthic and surface feeders[d]
macrochirus[d]	Bluegill	Terminal, slightly oblique[a]		Intestine well differentiated[a]	Silvery[a]	Yes[a]	Yes[a]	Insects, crustaceans, and plant material Keast and Webb (1966): 50% of the food volume was chironomid larvae; Moffett and Hunt (1943); 22% of the diet was plant material[a]	Invertivore	Drift	Water column/ surface	
marginatus[d]	Dollar sunfish	Terminal[d]					Yes[a]	Midge larvae and microcrustaceans (McLane, 1955); Tennessee specimens contained much detritus and filamentous algae with a few terrestrial insects[d]	Invertivore	Drift	Water column/ surface	Both benthic and surface feeders[d]
megalotis[a]	Longear sunfish	Terminal, slightly oblique[a]		Intestine well differentiated[a]	Silvery[a]	Yes[a]	Yes[a]	Insects and other invertebrates[a]	Invertivores	Drift	Surface feeder	Feeds extensively at the surface
microlophus[b]	Redear sunfish	Terminal oblique[b]					Yes[b]	Snails and small mussels, aquatic insects also (Emig, 1966b; Wilbur, 1969)[b]	Invertivore	Benthic and drift	Crusher	

Order	Family	Genus	Species	Common Name	Mouth Position	Ratio of Digestive Tract (DT) to Total Length (TL)[c]	Gut	Color of Peritoneum	Pyloric Caeca	Teeth	Pharyngeal Teeth and Plates	Main Food	First Level Trophic Classification	Second Level Trophic Classification	Trophic Mode	Feeding Behavior
			punctatus[d]	Spotted sunfish	Terminal[d]							Midge larvae and other immature insects; also microcrustaceans such as amphipods and cladocerans (McLane, 1955)[d]	Invertivore	Drift	Water column/surface	
			symmetricus[d]	Bantam sunfish	Terminal[d]						Yes[d]	Small crustaceans, midge larvae, snails, supplemented with surface feeding on hemipterans and terrestrial insects[d]	Invertivore	Drift	Water column/surface	
		Micropterus	coosae[d]	Redeye bass	Terminal							Terrestrial insects, crayfish, small fishes, salamanders, and aquatic insect larvae[d]	Invertivore/carnivore			
			dolomieu[a]	Smallmouth bass	Terminal, slightly oblique[a]		Intestine well differentiated[a]	Silvery[a]	Yes[a]	Yes[a]	Yes[a]	Insects, crayfish, and fishes Tester (1932a): 60–90% of food volume was crayfish, fishes 10–30%, and aquatic and terrestrial insects 0–10%[a]	Invertivore/carnivore	Benthic/whole body	Hunter/ambush	
			punctulatus[b]	Spotted bass	Terminal or slightly supraterminal[b]					Yes[b]		Insects, crayfishes, and fishes (Vogele, 1975; Hess, 1983)[b]	Invertivore/carnivore	Wholebody		
			salmoides[a]	Largemouth bass	Terminal, little oblique[a]		Intestine well differentiated[a]	Silvery[a]	Yes[a]	Yes[a]	Yes[a]	45% fish, 14.2% algae, 13.4% amphipods, 12.1% crayfish, 10.7% insect adults, 3.6% plants (Pearse, 1921a)[c]	Invertivore/carnivore	Wholebody		Sight feeder

Appendix 7A

Family	Genus	species	Common name	Mouth position	Other				Food notes	Trophic	Habitat
	Pomoxis	*annularis*[a]	White crappie	Terminal somewhat oblique[a]	Intestine well differentiated[a]	Silvery[a]	Yes[a]	Yes[a]	Yes[a]	Aquatic insects, crustaceans, and fish; Hansen (1951), fishes 57.8% of volume, aquatic insects 34.9% of volume[a]	Invertivore/ carnivore	Whole-body
		nigromaculatus[a]	Black crappie	Terminal oblique[a]	Intestine well differentiated[a]	Silvery[a]	Yes[a]	Yes[a]	Yes[a]	Keast and Webb (1966), Keast and Welch (1968), and Keast (1968a,b); food volume up to: 70% Chaoborus larvae, 50% cladocerans, 20% copepods, 25% fishes, 15% flying insects, 25% chironomid pupae and larvae, etc.[a]	Invertivore/ carnivore	
Percidae	*Crystallaria*	*asprella*[d]	Crystal darter	Slightly subterminal				Yes[d]		Specimens from Tombigbee River, TN contained only heptageniid mayfly nymphs[d]	Invertivore	Benthic
	Ammocrypta	*beani*[d]	Naked sand darter	Slightly subterminal						Midge larvae[d]	Invertivore	Benthic
		clara[c]	Western sand darter	Terminal[c]						Aquatic insects (Forbes and Richardson, 1920)[b]	Invertivore	Benthic
		pellucida[a]	Eastern sand darter	Subterminal[a]						Midge larvae[d]		
		vivax[d]	Scaly sand darter	Terminal or slightly subterminal				Yes[a]		Midge larvae, entomostracans[b]	Invertivore	Benthic

Order	Family	Genus	Species	Common Name	Mouth Position	Ratio of Digestive Tract (DT) to Total Length (TL)[c]	Gut	Color of Peritoneum	Pyloric Caeca	Teeth	Pharyngeal Teeth and Plates	Main Food	First Level Trophic Classification	Second Level Trophic Classification	Trophic Mode	Feeding Behavior
		Etheostoma	acuticeps[d]	Sharphead darter	Terminal[d]							Mainly mayfly, midge, and blackfly larvae; caddisfly and other larvae are also taken[b]; in the Nolichucky River, TN food consisted of 31% blackfly larvae, 30% mayfly nymphs, and 26% midge larvae[d]	Invertivore	Benthic		
			aquali[d]	Coppercheek darter	Terminal							Stomachs of 44 specimens contained 39% hydropsychid caddis larvae, 20% midge larvae, 16% mayfly nymphs, 7% blackfly larvae, and a wide variety of other aquatic immatures and a few snails[d]	Invertivore	Benthic		
			asprigene[d]	Mud darter	Terminal							Midge and caddisfly larvae, mayfly nymphs, and isopods[d]	Invertivore	Benthic		
			baileyi[d]	Emerald darter	Subterminal[d]							Midge larvae and pupae with some caddisfly and mayfly immatures and microcrustaceans[d]	Invertivore	Benthic		
			barbouri[d]	Teardrop darter	Terminal							Midge and blackfly larvae, copepods and cladocerans, and immature mayflies and caddisflies[d]	Invertivore	Benthic		

Appendix 7A

barrenense[d]	Splendid darter	Subterminal		Dipteran larvae (midges and blackflies), baetid mayfly nymphs and water mites[d]	Invertivore	Benthic
bellum[d]	Orangefin darter	Terminal[d]				
blennioides[b]	Greenside darter	Subterminal[b]	Yes[b]	Midge larvae, cladocerans, copepods, and various other insect larvae (Turner, 1921)[b]	Invertivore	Benthic
blennius[d]	Blenny darter	Subterminal[d]		Midge, blackfly, mayfly, and caddisfly larvae, in that order of importance[d]	Invertivore	Benthic
boschungi[d]	Slackwater darter	Terminal to slightly subterminal		Aquatic isopods, amphipods, mayfly nymphs, midge larvae, and limpets[d]	Invertivore	Benthic
caeruleum[f]	Rainbow darter	Subterminal[a]	Yes[a]	Midge and mayfly larvae, along with various other invertebrates (Turner, 1921)[b]	Invertivore	Benthic
camurum[b]	Bluebreast darter	Terminal[b]		Dipteran, mayfly, and other insect larvae (Stiles, 1972; Bryant, 1979)[b]	Invertivore	Benthic
chlorobranchium	Greenfin darter	Terminal				
chlorosomum[c]	Bluntnose darter	Terminal	Yes[c]	Similar to Johnny darter: midge and mayfly larvae (Pflieger)[c]	Invertivore	Benthic
cinereum[b]	Ashy darter	Subterminal		Midge and burrowing mayfly larvae and oligochaete worms (Shepard and Burr, 1984[b])	Invertivore	Benthic
collettei[f]	Creole darter			Aquatic insects, mainly mayflies[f]	Invertivore	Benthic

192 Assessing the Sustainability and Biological Integrity of Water Resources Using Fish Communities

Order	Family	Genus	Species	Common Name	Mouth Position	Ratio of Digestive Tract (DT) to Total Length (TL)[c]	Gut	Color of Peritoneum	Pyloric Caeca	Teeth	Pharyngeal Teeth and Plates	Main Food	First Level Trophic Classification	Second Level Trophic Classification	Trophic Mode	Feeding Behavior
			collis[b]	Carolina darter	Terminal[b]							Microcrustaceans, true flies (mostly midges), and occasionally mayfly larvae (E.H. Knicely)[b]	Invertivore	Benthic		
			coosae[d]	Coosa darter	Terminal to slightly subterminal[d]							Midge and blackfly larvae (78%), supplemented with cladocera, copepods, mayfly nymphs, and caddisfly larvae[d]	Invertivore	Benthic		
			cragini[f]	Arkansas darter								Snails are a large part of diet, isopods 58%, ephemeropterans 12%, chironmids 8% (Taber et al., 1986)[f]	Invertivore	Benthic		
			crossopterum[d]	Fringed darter	Terminal[d]							Aquatic insect immatures, including midge and caddisfly, larvae and mayfly nymphs, and crustaceans such as isopods, amphipods, and small crayfishes[d]	Invertivore	Benthic		
			ditrema[d]	Coldwater darter	Terminal[d]							Midge larvae, amphipods, isopods and copepods[d]	Invertivore	Benthic		
			duryi[d]	Black darter	Subterminal											
			emieri[d]	Cherry darter	Terminal to slightly subterminal											
			enzonum[f]	Arkansas saddled darter												
			exile[e]	Iowa darter	Subterminal[e]					Yes[e]		Midge larvae, mayfly larvae, and amphipods[e]	Invertivore	Benthic		

Appendix 7A

flabellare[a]	Fantail darter	Terminal[a]		Mayfly larvae, beetle larvae, caddisfly larvae, midge larvae, and corixids[a,b]	Invertivore	Benthic
flavum[d]	Saffron darter	Subterminal				
fusiforme[b]	Swamp darter	Terminal	Yes[a]	Microcrustaceans and aquatic insect larvae, particularly midges (Flemer and Woolcott, 1966; Gatz, 1979, Schmidt and Whiteworth, 1979)[b]	Invertivore	Benthic
gracile[d]	Slough darter	Terminal[d]		Primarily midge larvae, mayfly nymphs, and microcrustacea[d]	Invertivore	Benthic
histrio[d]	Harlequin darter	Slightly subterminal		Midge, blackfly, and caddisfly larvae, and mayfly nymphs[d]	Invertivore	Benthic
jordani[d]	Greenbreast darter	Terminal[d]		58% midge and other dipteran larvae, 24% mayfly nymphs, and smaller numbers of water mites and caddisfly larvae (O'Neil, 1980)[d]	Invertivore	Benthic
juliae[d]	Yoke darter			Larvae and naiads of aquatic insects, with chironomid larvae the major food item[f]	Invertivore	Benthic
kanawhae[b]	Kanawha darter	Subterminal		Immature insects, mainly mayflies, caddisflies, midges, and blackflies (R.D. Bumgarner)[b]	Invertivore	Benthic
kennicotti[d]	Stripetail darter	Terminal[d]		Midge larvae, mayfly and stonefly nymphs, and crustaceans[d]	Invertivore	Benthic

194 Assessing the Sustainability and Biological Integrity of Water Resources Using Fish Communities

Order	Family	Genus	Species	Common Name	Mouth Position	Ratio of Digestive Tract (DT) to Total Length (TL)[c]	Gut	Color of Peritoneum	Pyloric Caeca	Teeth	Pharyngeal Teeth and Plates	Main Food	First Level Trophic Classification	Second Level Trophic Classification	Trophic Mode	Feeding Behavior
			longimanum[b]	Longfin darter	Subterminal							Mayfly, caddisfly, and midge larvae, and water mites[b]	Invertivore	Benthic		
			luteovinctum[d]	Redband darter	Terminal to slightly subterminal							Midge larvae[d]	Invertivore	Benthic		
			lynceum[d]	Brighteye darter	Subterminal							Midge and blackfly larvae and mayfly nymphs[d]	Invertivore	Benthic		
			maculatum[e]	Spotted darter	Terminal[e]											
			microlepidum[d]	Smallscale darter	Terminal[d]											
			microperca[d]	Least darter	Terminal[a]					Yes[a]		Crustaceans and suitably small benthic organisms[a]	Invertivore	Benthic		
			moorei[j]	Yellowcheek darter								Aquatic dipteran larvae (Chironomidae and Simuliidae); stoneflies, mayflies and caddisflies[f]	Invertivore	Benthic		
			neopterum[d]	Lollypop darter	Terminal[d]											
			nigripinne[d]	Blackfin darter	Terminal[d]											
			nigrum[a]	Johnny darter	Subterminal					Yes[a]		Midge and mayfly larvae and some copepods[a]	Invertivore	Benthic		
			obeyense[d]	Barcheek darter	Terminal to slightly subterminal							Midge larvae and mayfly nymphs; young feed heavily on copepods[d]	Invertivore	Benthic		
			olivaceum[d]	Dirty darter	Terminal[d]							Midge larvae, mayfly, caddisfly, and stonefly immatures, isopods and amphipods[d]	Invertivore	Benthic		

Appendix 7A

olmstedi[b]	Tessellated darter	Subterminal	Immature insects and other small invertebrates; midge larvae and frequently eaten[b]	Invertivore	Benthic
osburni[b]	Candy darter	Subterminal	Larvae of mayflies, caddisflies, and true flies (T.E. Inman)[b]	Invertivore	Benthic
pallididor-sum[d]	Paleback darter		Small crustaceans, mayfly larvae, and other immature aquatic insects (Hamrick and Robinson, 1979)[f]	Invertivore	Benthic
parvipinne[d]	Goldstripe darter	Terminal	Midge larve, dipteran pupae, caddisfly larvae, dytiscid beetle larvae, and small crayfish[d]	Invertivore	Benthic
podoste-mone[b]	Riverweed darter	Subterminal[b]	Immature benthic insects including midge and caddisfly larvae, immature mayflies, beetles, and other dipterans, water mites, snails, and fish eggs are rarely eaten (Matthews et al., 1982; Haxo et al., 1985)[b]	Invertivore	Benthic
proeliare[d]	Cypress darter	Terminal to slightly subterminal	Primarily midge larvae and microcrustacea, with isopods, amphipods, and mayfly nymphs (Rice, 1942; Burr and Page, 1978)[d]	Invertivore	Benthic

Order	Family	Genus	Species	Common Name	Mouth Position	Ratio of Digestive Tract (DT) to Total Length (TL)[c]	Gut	Color of Peritoneum	Pyloric Caeca	Teeth	Pharyngeal Teeth and Plates	Main Food	First Level Trophic Classification	Second Level Trophic Classification	Trophic Mode	Feeding Behavior
			punctulatum[d]	Stippled darter	Terminal[f]							Isopods 66.6%, mayfly nymphs 12.4%, and caddisfly larvae 8%; amphipods, crayfish, and earthworms[f]	Invertivore	Benthic		
			pyrrhogaster[d]	Firebelly darter	Terminal to slightly subterminal							Primarily midge larvae[d]	Invertivore	Benthic		
			radiosum[f]	Orangebelly darter								Crustaceans, mayfly and other insect larvae[f]	Invertivore	Benthic		
			rufilineatum[b]	Redline darter	Terminal[b]							Truefly larvae (particularly midges); other insects and invertebrates (Fisher and Pearson, 1987)[b]	Invertivore	Benthic		
			rupestre[d]	Rock darter	Slightly subterminal											
			sagitta[d]	Arrow darter	Terminal[d]							Mayfly nymphs, blackfly and midge larvae, and lesser numbers of caddisfly, stonefly, and beetle larvae[d]	Invertivore	Benthic		
			sanguifluum[d]	Bloodfin darter	Terminal[d]											
			serrifer[b]	Sawcheek darter	Terminal[b]							Virginia specimens took amphipods (31% of total items) and isopods (5%); stonefly (1%), mayfly (22%), beetle (7%), and midge (31%) larvae; and snails (3%) (K.A. Tyler)[b]	Invertivore	Benthic		

Appendix 7A

simoterum[b]	Snubnose darter	Subterminal[b]	Midge larvae and other aquatic insect larvae, microcrustaceans water, mites, and snails (Page and Mayden, 1981)[b]	Invertivore	Benthic
smithi[d]	Slabrock darter	Subterminal[d]	Primarily midge, mayfly, and caddisfly immatures, with copepods and other microcrustacea more prevalent in young fish[d]	Invertivore	Benthic
spectabile[d]	Orangethroat darter	Terminal[d]	Midge and blackfly larvae, mayfly nymphs, isopods, amphipods, and caddisfly larvae (Cross, 1967; Small, 1975)[d]	Invertivore	Benthic
squamiceps[d]	Spottail darter	Terminal[d]			
stigmaeum	Speckled darter	Subterminal	Midge larvae and macrocrustacea[d]	Invertivore	Benthic
striatulum[d]	Striated darter	Terminal[d]			
swaini[d]	Gulf darter	Terminal[d]	Midge larvae supplemented with isopods and immature blackflies, mayflies, caddisflies, and dragonflies (Ruple et al., 1983, 1984)[d]	Invertivore	Benthic
swannanoa[b]	Swannanoa darter	Subterminal[b]	Immature mayflies, caddisflies, and midges (R. L. Steele)[b]	Invertivore	Benthic
tippecanoe[b]	Tippecanoe darter	Terminal[b]	Mayfly, caddisfly, and midge larvae (A. R. Clarke)[b]	Invertivore	Benthic

Order	Family	Genus	Species	Common Name	Mouth Position	Ratio of Digestive Tract (DT) to Total Length (TL)[c]	Gut	Color of Peritoneum	Pyloric Caeca	Teeth	Pharyngeal Teeth and Plates	Main Food	First Level Trophic Classification	Second Level Trophic Classification	Trophic Mode	Feeding Behavior
			trisella[d]	Trispot darter	Subterminal[d]							Wide range of invertebrates dominated by midge larvae and mayfly nymphs; stonefly nymphs, caddisfly larvae, and copepods are seasonally important[d]	Invertivore	Benthic		
			tuscumbia[d]	Tuscumbia darter	Terminal[d]							Amphipods, physid snails, and predominantly midge larvae[d]	Invertivore	Benthic		
			variatum[b]	Variegate darter	Subterminal[b]							Midge larvae, other immature insects, and mites (Turner, 1921; Wehnes, 1973; in Page, 1983; Nemecek, 1978)[b]	Invertivore	Benthic		
			virgatum[d]	Striped darter	Terminal[d]							Midge larvae[b]	Invertivore	Benthic		
			vitreum[b]	Glassy darter	Terminal							Midge larvae supplemented by other insect larvae and other invertebrates (Stiles, 1972; Bryant, 1979)[b]	Invertivore	Benthic		
			vulneratum[b]	Wounded darter	Terminal[b]											
			wapiti[d]	Boulder darter	Terminal[d]							Microinvertebrates[f]				
			whipplei	Redfin darter												
			zonale[b]	Banded darter	Terminal							Midge and blackfly larvae (Wynes and Wissing, 1982)[b]	Invertivore	Benthic		
			zonistium[d]	Bandfin darter	Subterminal[d]							Midge larvae[d]	Invertivore	Benthic		

Appendix 7A

				Intestine well differentiated[a]	Silvery[a]	Yes, thick[a]	Yes[a]			
Perca	*flavescens*[a]	Yellow perch	Terminal slightly oblique[a]					Immature insects, larger invertebrates, fishes, and fish eggs[a]	Invertivore/ carnivore		
Percina	*antesella*[d]	Amber darter	Terminal to slightly subterminal[d]					Snails and limpets, immature aquatic insects, particularly caddisfly larvae and mayfly nymphs (Freeman, 1977)[d]	Invertivore	Benthic	
	aurantiaca[b]	Tangerine darter	Subterminal					Immature midges, mayflies, and caddisflies[b]	Invertivore	Benthic	Digger
	burtoni[b]	Blotchside darter	Subterminal[b]					Benthic insects[b]	Invertivore	Benthic	
	caprodes[c]	Logperch	Subterminal[c]				Yes[a]	Aquatic insect larvae, particularly mayfly and midge larvae (Turner, 1921)[a]	Invertivore	Benthic	Digger
	copelandi[a]	Channel darter	Subterminal[a]				Yes[a]	Mayfly and midge larvae and large amounts of algae and bottom debris and microcrustaceans (Turner, 1921)[a]	Invertivore/ herbivore	Benthic	
	crassa[b]	Piedmont darter	Terminal[b]					Aquatic insects (Gatz, 1979)[b]	Invertivore	Benthic	
	evides[c]	Gilt darter	Terminal[c]					Immature caddisflies, blackflies, and midges (Hickman and Fitz, 1978; Becker, 1983; Hatch, 1983)[b]	Invertivore	Benthic	
	gymnocephala jenkinsi[d]	Appalachia darter Conasauga logperch	Terminal Subterminal[d]								
	macrocephala[b]	Longhead darter	Terminal[b]					Mayfly larvae and crayfishes (Page, 1978)[b]	Invertivore	Benthic	

Order	Family	Genus	Species	Common Name	Mouth Position	Ratio of Digestive Tract (DT) to Total Length (TL)[c]	Gut	Color of Peritoneum	Pyloric Caeca	Teeth	Pharyngeal Teeth and Plates	Main Food	First Level Trophic Classification	Second Level Trophic Classification	Trophic Mode	Feeding Behavior
			maculata[a]	Blackside darter	Terminal[a]					Yes[a]		Mayfly and midge larvae, corixid nymphs, copepods, and fish[a]	Invertivore	Benthic		
			nasuta[]	Longnose darter												
			nigrofasci-ata[d]	Blackbanded darter	Terminal[d]							Midge, blackfly, mayfly, and caddisfly larvae (Mathur, 1973b)[d]	Invertivore	Benthic		
			notogramma[b]	Stripeback darter	Terminal[b]							Insects and other invertebrates[b]	Invertivore	Benthic		
			oxyrhynchus[b]	Sharpnose darter	Terminal[b]							Mayfly and caddisfly larvae; midge and blackfly larvae also taken (Hess, 1983)[b]	Invertivore	Benthic		
			palmaris[d]	Bronze darter	Terminal							Blackfly larvae and pupae, and hydropsychid caddis larvae[d]; blackfly, mayfly, and midge larvae (Wieland, 1982)[d]	Invertivore	Benthic		
			pantherina[]	Leopard darter								Small insect larvae (Robison, 1978c)[f]	Invertivore	Benthic		
			peltata[b]	Shield darter	Terminal[b]							Insects[b]	Invertivore	Benthic		
			phoxoceph-ala[d]	Slenderhead darter	Terminal[d]							Mayfly nymphs, midge larvae, caddisfly larvae (Karr, 1963; Thomas, 1970)[d]	Invertivore	Benthic		
			rex[a]	Roanoke logperch	Subterminal[b]							Immature insects; midges, caddisflies (Burkhead, 1983)[b]	Invertivore	Benthic		
			roanoka[b]	Roanoke darter	Terminal[b]							Mayfly, blackfly, and midge larvae (Hobson 1979; Matthews et al. 1982; Hess 1983)[b]	Invertivore	Benthic		

Appendix 7A

Genus	species	Common name	Mouth position	Intestine			Diet	Trophic guild	Position	Defense	Behavior
	sciera[b]	Dusky darter	Terminal[b]				Midge and blackfly larvae, seasonally supplemented with caddisfly and mayfly larvae (Page and P. Smith, 1970; Miller, 1983)[b]	Invertivore	Benthic		
	shumardi[a]	River darter	Terminal[a]			Yes[a]	Larvae of mayflies, midges, caddisflies, and blackflies[a]	Invertivore	Benthic		
	squamata[d]	Olive darter	Terminal[d]								
	tanasi[d]	Snail darter	Subterminal[d]				Specializes on small pleurocerid river snails as well as some physid snails and limpets; snails made up 60% of the diet with caddisfly larvae, midge and blackfly larvae, and a few mayfly nymphs also eaten[d]	Invertivore	Benthic	Crusher	
	uranidea[f]	Stargazing darter					Snails and Limpets (Thompson, 1974)[f]	Invertivore	Benthic	Crusher	
	vigil[M]	Saddleback darter	Terminal to slightly subterminal			Yes[a]	Snails when available, caddisfly larvae, midge larvae, and small mayfly nymphs (Thompson, 1974; Miller, 1983)[d]	Invertivore	Benthic		
Stizostedion	canadense[a]	Sauger	Terminal very slightly oblique[a]	Intestine well differentiated[a]	White[a]	Yes[a]	Fishes and various invertebrates, depending on size of and age of fish[a]	Invertivore/carnivore	Wholebody	Protective resemblance	Sight predators and negatively phototrophic

Order	Family	Genus	Species	Common Name	Mouth Position	Ratio of Digestive Tract (DT) to Total Length (TL)[c]	Gut	Color of Peritoneum	Pyloric Caeca	Teeth	Pharyngeal Teeth and Plates	Main Food	First Level Trophic Classification	Second Level Trophic Classification	Trophic Mode	Feeding Behavior
		Stizostedion	vitreum[a]	Walleye	Terminal[a]		Short intestine, well differentiated[a]	White[a]	Yes[a]	Yes[a]	Yes[a]	Fishes and various invertebrates, depending on size and are of fish, as well as availability of food sources[e]	Invertivore/carnivore	Wholebody	Protective resemblance	Sight predators and negatively photo-trophic
	Sciaenidae	Aplodinotus	grunniens[a]	Freshwater drum	Subterminal[a]					Yes[a]	Yes[a]	Mayflies and amphipods; in larger fish, fish and crayfish appear more frequent in diet (Daiber, 1952)[b]	Invertivore/carnivore	Wholebody		Rolls over stones
	Mugilidae	Mugil	cephalus[d]	Striped mullet	Terminal[d]		Long[f]					Adults and subadults are almost exclusively detritivores, but young feed on small invertebrates such as copepods and mosquito larvae[d]	Detritivore/invertivore	Filter feeder	Filterer	

[a] Referenced out of *Freshwater Fishes of Canada*: W. B. Scott and E. J. Crossman 1973. Fisheries Research Board of Canada, Ottawa.
[b] Referenced out of *Freshwater Fishes of Virginia*: Robert E. Jenkins and Noel M. Burkhead 1994. American Fisheries Society Bethesda, Maryland.
[c] References out of *Fishes of Wisconsin*: George C. Becker 1983. The University of Wisconsin Press.
[d] References out of *The Fishes of Tennessee*: David A. Etnier and Wayne C. Starnes 1993. The University of Tennessee Press/Knoxville.
[e] References out of *The Fishes of Ohio*: Milton B. Trautman 1981. Ohio State University Press.
[f] References out of *Fishes of Arkansas*: Henry W. Robison and Thomas M. Buchanan 1988. The University of Arkansas Press, Fayetteville, Arkansas.

8 Influence of the Family Catostomidae on the Metrics Developed for a Great River Index of Biotic Integrity

Erich B. Emery, Thomas P. Simon, and Robert Ovies

CONTENTS

8.1 Introduction 203
8.2 Materials and Methods 204
    8.2.1 Study Area 204
    8.2.2 Sample Methods 204
8.3 Results and Discussion 206
    8.3.1 Species Composition 206
    8.3.2 Longitudinal Trends 207
    8.3.3 Temporal Trends 212
        8.3.3.1 Pike Island Lock Chamber 212
        8.3.3.2 Gallipolis Lock Chamber 214
        8.3.3.3 McAlpine Lock Chamber 214
        8.3.3.4 Uniontown Lock Chamber 214
    8.3.4 Multimetric Applications 214
        8.3.4.1 Species Richness Metrics 215
        8.3.4.2 Proportional Metrics 215
        8.3.4.3 Relative Number of Individuals 218
    8.3.5 Influence of Suckers on Other Metrics 218
        8.3.5.1 Total Number of Species 218
        8.3.5.2 Percent Insectivores 218
        8.3.5.3 Percent Omnivores 222
        8.3.5.4 Catch per Unit Effort 222
        8.3.5.5 Percent Simple Lithophils 222
        8.3.5.6 Percent DELT Anomalies 222
8.4 Conclusion 222
References 223

8.1 INTRODUCTION

The 15 sucker species composing the genera *Carpiodes*, *Catostomus*, *Cycleptus*, *Hypentelium*, *Ictiobus*, *Minytrema*, and *Moxostoma* are common inhabitants of the Ohio River (Pearson and Krumholz, 1984). Catostomids are primarily insectivores that utilize their subterminal mouth for suction feeding (Etnier and Starnes, 1993; see Goldstein and Simon, Chapter 6). Most species are simple lithophilous spawners, scattering eggs among the gravel substrates without parental care, and frequently utilize smaller tributary streams for spawning (Trautman, 1981; Etnier and Starnes, 1993). Species intolerance to pollution varies widely; however, the round-bodied suckers are generally considered responsive to changes in water quality and habitat composition.

Catostomids are a major component of the Ohio River community and, because of their importance, they are considered reliable environmental indicators (Ohio EPA, 1989a; Simon and Emery, 1995). Suckers compose 12% of the total number and 14% of all the species collected. The family is a significant component of a Great River fish community index being developed for the Ohio River (Ohio EPA, 1989a; Simon and Emery, 1995). The sucker family will be a single metric; however, it will ultimately influence several of the other metrics. Sucker abundance will influence several of the species composition metrics (e.g., total number of species, number of sucker species, and number of round-bodied sucker species), and several of the proportional metrics (e.g., percent round-bodied suckers, percent omnivores, percent insectivores, and percent simple lithophils).

Historical information for the Catostomidae enables long-term temporal trend assessment; and the overall knowledge for each of the species defines community function (Pearson and Krumholz, 1984; Pearson and Pearson, 1989). Reproductive and feeding patterns, tolerances to impairment, and other ecological parameters have been adequately described for each of the species occurring in the Ohio River (Kay et al., 1996).

The purpose of this study is to evaluate longitudinal and ecological patterns, and the influence of the family on individual Index of Biotic Integrity (IBI) metrics. Although the Catostomidae are well studied in streams, little is known of the community structure and function in the Ohio River. The intention here is to explore the influence of suckers on the structure and function of a Great River fish community.

8.2 MATERIALS AND METHODS

8.2.1 STUDY AREA

The Ohio River begins at the confluence of the Monongahela and Allegheny Rivers, flowing southwesterly to the Mississippi River near Cairo, Illinois (Figure 8.1). The Ohio River is 981 linear miles and is impounded by 20 navigation dams that provide a 9-foot minimum depth for river commerce and navigation. The study area crosses four ecoregions: Western Allegheny Plateau, Interior Plateau, Interior River Lowland, and the Mississippi Alluvial Plain (Omernik, 1987; Omernik and Gallant, 1989).

The Ohio River fish community was sampled using boat electrofishing methods at 339 locations along the mainstem between 1990 and 1996. Intensive surveys of five pools that provide spatial coverage of a location every two river miles were conducted.

8.2.2 SAMPLE METHODS

Fish community surveys were conducted using night, nearshore boat electrofishing methods and by introducing rotenone into lock chambers of the navigation dams. Lock chambers were sampled by opening the downstream gates for 8 h prior to sampling the previous night. The following morning, the chamber gate was closed and a concentration of 2.5% rotenone was introduced into the lock chamber. The rotenone caused the fish to rise to the surface where they were netted, placed into containers, and returned to the shoreline for processing. Fish were sorted according to species into 3-cm size classes, and total batch weights were measured for each size class. Lock chambers were sampled on a rotating basis so that all navigation lock chambers were covered every two or three years. Four lock chambers were selected for temporal analysis for this chapter because of the longitudinal position along the river: Pike Island, Gallipolis, McAlpine, and Uniontown Lock

FIGURE 8.1 Map of the Ohio River showing the study area and four lock chambers sampled between 1978 and 1993.

chambers were selected to represent upper, upper-mid, mid-lower, and lower sections of the river and were sampled between 1978 to 1995.

Night electrofishing boat samples were collected between July and October 1991 to 1996. A total of 339 stations were collected using standardized field, laboratory, and data processing methods (Ohio EPA, 1989 a,b). Fish were collected using conventional Great River methods (Sanders, 1988;

Simon and Sanders, Chapter 17). A two- or three-person crew operated an 18-ft aluminum jon boat equipped with a 5000-watt generator and a Smith Root VI-A pulsed DC electrofishing unit. Sampling at each zone began no sooner than 30 minutes after sunset. The gear had an effective depth of 10 to 15 ft, which was most effective in the shallow nearshore zone within 75 m of shore. Each 500-m zone was sampled for 2000 to 3000 s, depending on the complexity of the habitat. All observed fish were netted, placed into an aerated holding tank, weighed, measured, and returned to the river. Batch weights based on 3-cm size classes were measured for each species. All fish were identified to species using Gerking (1955), Pflieger (1973), and Trautman (1981). All collected fish were identified for deformities, eroded fins, lesions, and tumors (DELT; Sanders et al., Chapter 8).

8.3 RESULTS AND DISCUSSION

8.3.1 Species Composition

Fifteen species of sucker have been collected from the Ohio River (Table 8.1). Suckers can be categorized into two groups based on body morphology: deep-bodied and round-bodied. Deep-bodied suckers include members of the genera *Ictiobus* and *Carpiodes*. The round-bodied group includes the genera *Catostomus*, *Cycleptus*, *Hypentelium*, *Minytrema*, and *Moxostoma*.

The most frequently collected sucker in the Ohio River is the smallmouth buffalo *Ictiobus bubalus*, while the white sucker *Catostomus commersoni* is the least collected species (Figure 8.2). Both the white sucker and blue sucker *Cycleptus elongatus* distribution patterns have changed. The blue sucker has rarely been collected since 1978; however, lock chamber data from before 1978 shows that the species was present in the lower Ohio River. The white sucker is a tolerant species and was once more prevalent in the upper Ohio River, but appears to have been displaced by other species as the water quality of the Ohio River improved with the closing of the steel mills in Pittsburgh (Figure 8.3).

TABLE 8.1
Sucker Species Collected from the Ohio River During Lock Chamber Rotentone Sampling Between 1978 and 1993, and During Nightboat Electrofishing Between 1990 and 1995

Common Name/Scientific Name	Trophic Guild		Reproductive Guild		
	Insectivore	Omnivore	Tolerance	Simple Lithophil	Other
Blue sucker, *Cycleptus elongatus*	X		Sensitive	X	
White sucker, *Catostomus commersoni*		X	Tolerant	X	
River carpsucker, *Carpiodes carpio*		X	Tolerant		X
Quillback, *C. cyprinus*		X	Tolerant		X
Highfin carpsucker, *C. velifer*	X		Tolerant		X
Northern hogsucker, *Hypentelium nigricans*	X		Sensitive	X	
Smallmouth buffalo, *Ictiobus bubalus*		X	Tolerant		X
Bigmouth buffalo, *I. cyprinellus*		X	Tolerant		X
Black buffalo, *I. niger*		X	Tolerant		X
Spotted sucker, *Minytrema melanops*	X		—	X	
Silver redhorse, *Moxostoma anisurum*	X		Sensitive	X	
River redhorse, *M. carinatum*	X		Sensitive	X	
Black redhorse, *M. duquesnei*	X		Sensitive	X	
Golden redhorse, *M. erythrurum*	X		Sensitive	X	
Shorthead redhorse, *M. macrolepidotum*	X		Sensitive	X	

Note: Characteristics of sucker species include trophic guild, tolerance, and reproductive guild classification based on Kay et al. (1994).

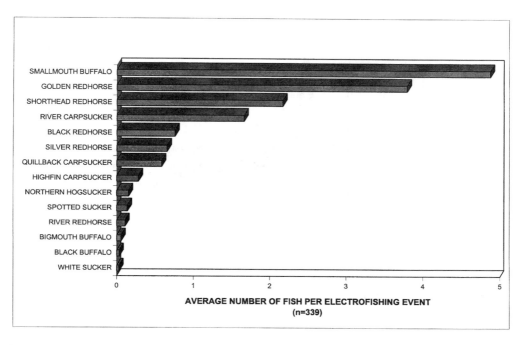

FIGURE 8.2 Average number of sucker species collected per night electrofishing sample event. The number of collection attempts is based on 339 sample events.

8.3.2 Longitudinal Trends

Longitudinal patterns of Ohio River suckers show that specific species are more prevalent along certain stretches of the Ohio River (Figure 8.3). Pearson and Krumholz (1984) graphed distribution patterns for all Ohio River species based on historical data. The distribution of sucker species along longitudinal gradients is important for showing patterns for accurate portrayal of calibrated metric expectations. For example, various species are often limited to specific reaches of the Ohio River.

The round-bodied suckers (Figure 8.3) are most frequently collected in the upper 600 miles of the Ohio River, while deep-bodied suckers are more typical in the lower 300 miles but are found riverwide because of characteristics of the lacustrine zone of navigation pools. Specific species patterns show that the redhorse species reaches its highest abundance in the upper 450 miles of the Ohio River, rarely occurring below the McAlpine Dam (Ohio River Mile 605.0) and former Falls of the Ohio River.

Five species of redhorse occur in the Ohio River (Figure 8.3A). The golden redhorse *Moxostoma erythrurum* is most dominant in the middle Ohio River, while population numbers decline in the lower and upper portions of the river. The shorthead redhorse *M. macrolepidotum* occurs in the upper Ohio River and maintains a high relative abundance until the McAlpine Pool. Black redhorse *M. duquesnei* is most common in the upper 300 miles of the Ohio River, a decline in the population density occurring below the Meldahl Pool. Silver redhorse *M. anisurum* is found in the upper Ohio River pools; population densities decline rapidly with distance downstream from the Allegheny, Monongahela, and Ohio River confluence. The river redhorse *M. carinatum* is so rarely collected that distribution patterns cannot be discerned.

Three additional round-bodied suckers — spotted sucker *Minytrema melanops*, white sucker, and northern hogsucker *Hypentelium nigricans* — occur in the Ohio River (Figure 8.3B). The spotted sucker is frequently collected from the middle Ohio River, while the northern hogsucker and white sucker are commonly collected in the upper third of the river.

208 Assessing the Sustainability and Biological Integrity of Water Resources Using Fish Communities

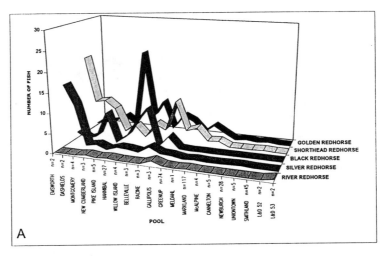

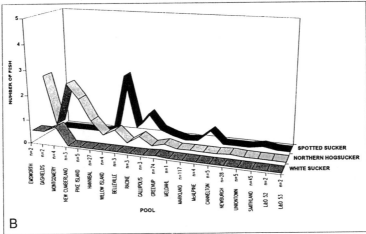

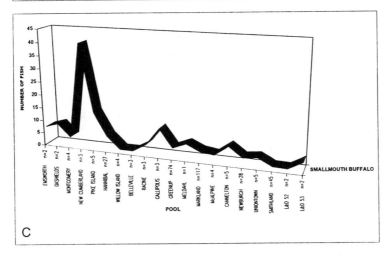

FIGURE 8.3 Average number of sucker species collected along the length of the Ohio River per electrofishing event. (A) redhorse species; (B) other round-bodied suckers; (C) smallmouth buffalo; (D) buffalo species; and (E) carpsucker species.

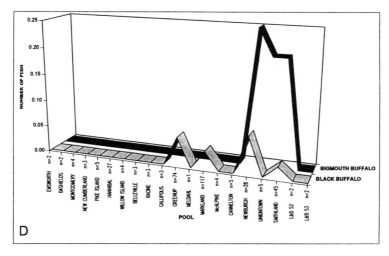

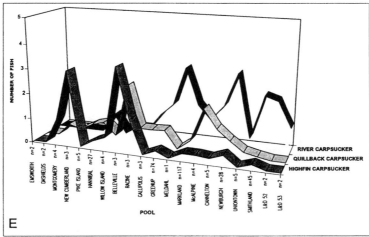

FIGURE 8.3 *Continued.*

The deep-bodied suckers are typically associated with the lower half of the Ohio River (Figures 8.3C, D, E). The three species of *Ictiobus* are generally associated with ultra-fine particulate organic matter. The smallmouth buffalo occurs predominantly in the upper 300 miles of the Ohio River, but is frequently collected riverwide. The black buffalo *I. niger* and bigmouth buffalo *I. cyprinellus* are generally limited to the lower half of the Ohio River, reaching their highest population densities in the lower 300 miles (Figure 8.3D). The quillback *Carpiodes cyprinus* is limited to the middle Ohio River (Figure 8.3E); this distribution pattern was also observed by Pearson and Krumholz (1984).

The distribution patterns of suckers are a function of a variety of factors, including habitat requirements, feeding guild trophic dynamics, and reproductive requirements. The changing nature of the Ohio River from the upper river above "The Falls of the Ohio" to downstream reflects a natural break in river gradient. The influence of the navigation pools also contributes to the decline in gradient within each individual navigation pool. The round-bodied sucker populations are most dominant in the upper half of the Ohio River above the falls. Increased flow causes the sorting of substrate particle sizes, enabling coarse, silt-free substrates to accumulate in the upper portion of the river and below each navigation dam in the tailwater or riverine zones. The transition in the lower portions of a pool to a lacustrine zone causes ultra-fine particulate organic matter, the preferred

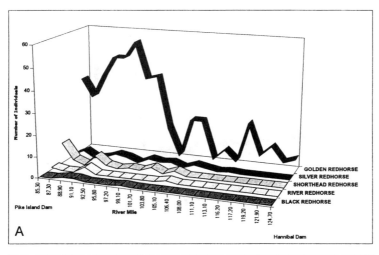

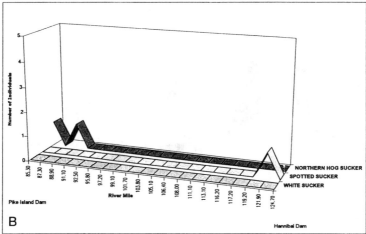

FIGURE 8.4 Average number of sucker individuals collected at 2-mile intervals in the Hannibal Pool per electrofishing event. (A) redhorse species; (B) other round-bodied suckers; (C) carpsucker species; and (D) buffalo species.

food of the deep-bodied suckers, to accumulate. The trophic dynamics of the deep-bodied suckers are mostly detrital and, morphologically, the species possesses a black peritoneum and longer coiled gut (Becker, 1983; Etnier and Starnes, 1993). Goldstein and Simon (Chapter 6) have classified these deep-bodied suckers as detrital feeders.

The deep-bodied suckers are generally found in the transition and lacustrine zones of a navigation pool (Figures 8.4C, D). River carpsucker and quillback are the most abundant in the lacustrine zones of a navigation pool, while highfin carpsucker *Carpiodes velifer* is common in the transition zone of the navigation pool. The presence of smallmouth buffalo throughout the navigation pool shows that the species may be a habitat generalist and not influenced by habitat changes. In Hannibal Pool, the round-bodied suckers, including species of redhorses, hogsucker, spotted sucker, and white sucker, are typically found in the upper riverine zone of the navigation pool (Figures 8.4A, B).

Longitudinal patterns of sucker distribution may be a result of the geomorphology of the Ohio River. As the gradient declines from upstream to downstream in the Ohio River, substrate changes occur from coarse substrates in the upper reaches to sand-clay substrates in the lower reaches.

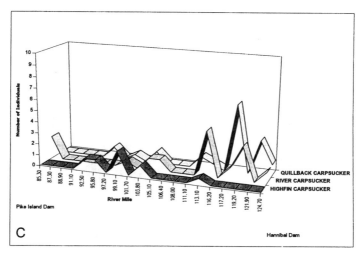

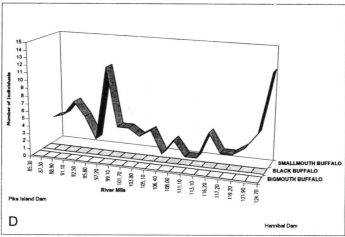

FIGURE 8.4 *Continued.*

These changes affect the reproductive and feeding habitat available to round-bodied suckers. Kay et al. (1994) report that most species of redhorse require coarse, silt-free substrates at the head of riffles for reproduction. Often, these species are known to return to smaller tributaries for spawning (Curry and Spacie, 1984; Moss et al., 1983; Page and Johnston, 1990; Matheny and Rabeni, 1995). Since it is not known whether round-bodied suckers utilize the mainstem Ohio River for reproduction, the reduced number of tributary streams in the lower river may preclude the occurrence of some round-bodied species (E. Emery, unpublished data).

Feeding requirements of suckers are adapted for specialized feeding (Goldstein and Simon, Chapter 6). Round-bodied suckers utilize coarse sand and gravel areas where they search for benthic invertebrates. All species root around in the substrate, using the papillose lips and subterminal mouth for removing macroinvertebrates from the substrate (Kay et al., 1994). Substrate composition changes from gravel to clay in the lower river; thus, a change in substrate composition may not support an adequate macroinvertebrate food base. The presence of round-bodied suckers in the lower river may be limited to the areas beneath the navigation dams where the riverine zones of the navigation pools are most similar to preimpoundment conditions.

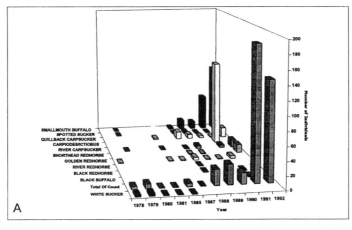

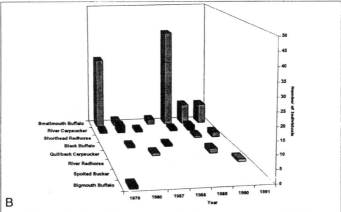

FIGURE 8.5 Species abundance of suckers collected in the Ohio River at various lock chambers between 1978 and 1993. (A) Pike Island Lock and Dam; (B) Gallipolis Lock and Dam; (C) McAlpine Lock and Dam; and (D) Uniontown Lock and Dam.

8.3.3 TEMPORAL TRENDS

The biological integrity of the fish community of the Ohio River has shown improvement based on correlated water quality improvements (Reash and Van Hassel, 1988). Krumholz and Minckley (1964) reported that, during the steel mill strike of 1959 that reduced pollution loadings to the Ohio River, white sucker invaded the mainstem from tributaries and backwater areas. Lock chamber collections conducted between 1978 and 1993 have shown an increased species richness, a lower number of tolerant and omnivorous species, and increased percentages of insectivores and carnivores, and few DELT anomalies.

Temporal trends of catostomids in four representative lock chambers sampled between 1978 and 1993 were examined (Figure 8.5). These four lock chambers show trends in sucker abundance and species richness during the recovery of the Ohio River after the industrialized city of Pittsburgh began to close steel mills.

8.3.3.1 Pike Island Lock Chamber

The Pike Island lock chamber occurs in the upper Ohio River at ORM 84.2. This section of the river was heavily polluted during the period when the steel mills and numerous other related

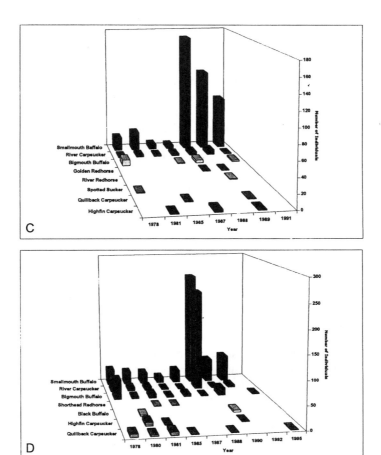

FIGURE 8.5 *Continued.*

industries thrived in the upper Ohio Valley (Figure 8.5A). Increases in species richness and relative abundance of round- and deep-bodied suckers increased with reductions in pollutant loadings. The only species showing a decline with increased water quality improvements was the white sucker (paired t-test, $p < 0.01$). Prior to 1987, white sucker was the most abundant species and occurred at nearly all sampling events. This species is considered tolerant of a wide range of water quality conditions and can be found in highly degraded areas. We observed a decline in the relative abundance of white sucker as other sucker species returned to the Pike Island Pool. Since 1988, not a single white sucker has been collected; but eight other sucker species have recolonized the area. Between 1988 and 1992, sucker species increased in relative abundance and tolerant sucker species declined. Recolonization of the Pike Island Pool by eight sucker species considered sensitive suggests that improvements in the water quality of the Ohio River during this time have been a result of anthropogenic influences. Smallmouth buffalo relative abundance increased significantly (paired t-test, $p < 0.05$) during 1991 and 1992. The increase in smallmouth buffalo is likely a result of the species' sustained reproductive period between May and September (Kay et al., 1994). Spawning success for this species is improved when water levels inundate adjacent fields and low areas (Becker, 1983). Increased water levels during the spring would have also benefited river carpsucker; however, the sustained water levels throughout the year increased smallmouth buffalo abundance (U.S. Geological Survey, unpublished Ohio River stage data).

8.3.3.2 Gallipolis Lock Chamber

The Gallipolis lock chamber is located at ORM 279.2. The dramatic changes in species richness and relative abundance at Pike Island Pool were not observed at the Gallipolis lock chamber (Figure 8.5B). Black buffalo and river carpsucker were collected throughout the study period and did not reveal any temporal trend (paired t-test, $p < 0.10$). Shorthead redhorse, black buffalo, and quillback were collected on only two occasions. The bigmouth buffalo was collected only during 1979 and has not been collected since. The river redhorse, collected in 1991, is a mollusk sight feeder. The presence of river redhorse may be the result of increased mussel populations. The appearance of shorthead and river redhorse may be an early indication of improving water quality conditions in the upper Ohio River. Smallmouth buffalo relative abundance was statistically significant during 1988 (paired t-test, $p < 0.01$), and was also significant during 1979 (paired t-test, $p < 0.01$).

8.3.3.3 McAlpine Lock Chamber

The McAlpine lock chamber is located at ORM 604.0 near Louisville, Kentucky. The relative abundance of the sucker community near the McAlpine lock chamber suggests that the community was possibly suppressed prior to 1987. Few individuals of any species were typically collected during any sample event, and six species ($\bar{x} = 3.5$ species per sample event) were collected between 1978 and 1987 (Figure 8.5D). Smallmouth buffalo numbers increased statistically (pairwise t-test, $p > 0.05$) between 1987 and 1988, and seven species of suckers have been collected. Five species of deep-bodied suckers have been collected from the McAlpine lock chamber, compared with three species of round-bodied suckers. The sucker community in the McAlpine lock chamber has increased species richness and relative number in response to changing environmental conditions. However, unlike the Pike Island lock chamber, it is not obvious what the stresses were that have been moderated. An increased species richness of round-bodied suckers during 1989 could be the result of reduced runoff during the drought of 1988. Another plausible explanation could be that, in the lower 300 miles of the Ohio River, the sucker community lacks the species richness of round-bodied suckers observed above the Falls of the Ohio.

8.3.3.4 Uniontown Lock Chamber

The Uniontown lock and dam is located near Uniontown, Indiana, at ORM 846.0. No temporal trend in species richness was observed before 1992 (paired t-test, $p < 0.10$); however, a reduction in species richness since 1992 (paired t-test, $p < 0.01$) has been a result of declining round-bodied suckers (Figure 8.5D). Species collected from the Uniontown lock and dam are dominated by deep-bodied suckers. No significant community changes occurred for most species, with the exception of smallmouth buffalo and river carpsucker. During the drought of 1988, these two species increased dramatically in relative abundance (paired t-test, $p < 0.01$). A possible reason for increased relative abundance of these two species may be the result of improved spawning success in years when water levels rise in the spring to flood marshes or low-lying areas (Walburg, 1976; Becker, 1983). Such was the case in the spring of 1988. In subsequent years, the relative abundance of smallmouth buffalo and river carpsucker returned to expected ranges.

8.3.4 MULTIMETRIC APPLICATIONS

Ecologically significant species groups are typically included in a multimetric index for assessing ecosystem integrity (Karr, 1981; Ohio EPA, 1989a; Simon and Emery, 1995). Karr (1981) suggested that the family Catostomidae is an important measure of flowing rivers in the midwest; suckers as a component of a freshwater stream and small river community were used when developing the Index of Biotic Integrity (IBI). The Ohio EPA (1989a) modified the IBI for large inland rivers of Ohio and used the proportion of round-bodied suckers and number of sucker species in the boat method IBI. Sanders (1991) applied the inland large river metrics to an evaluation of the Ohio River along the Ohio shoreline. Simon and Emery (1995) modified the IBI for use in Great Rivers. They evaluated several species richness metrics, number and proportion of round-bodied sucker species and individuals for evaluating the Western Allegheny Plateau, which is the upper 300 miles of the Ohio River.

The suckers are capable of influencing several of the IBI metrics currently being evaluated for consideration in an index for Great Rivers. Suckers influence species richness metrics, trophic and proportional metrics, catch and relative abundance measures, and DELT anomaly metrics.

8.3.4.1 Species Richness Metrics

Metrics typically considered for development of the IBI include: total number of species, total number of sucker species, number of round-bodied sucker species, and number of deep-bodied sucker species (Figure 8.6). The riverwide expectations for the total number of sucker species show a decreasing trend in species richness with increasing drainage area (Figure 8.6A). A lower total number of sucker species richness for the lower river is expected.

The decline in total number of sucker species is a result of the declining number of round-bodied sucker species. The number of round-bodied sucker species along the Ohio River shows declining trends with increasing drainage area (Figure 6B). The occurrence of round-bodied suckers below ORM 600 is erratic. Use of the total number of round-bodied suckers as a metric in the Ohio River fish index is likely, although it may not be excluded for use below ORM 600. Round-bodied suckers are important indicators of a high quality resource, and increasing number of species would reflect increasing biological integrity.

Deep-bodied suckers are uniformly distributed in the Ohio River (Figure 8.6C). No drainage area relationship was observed; however, the increase of deep-bodied suckers may be a reflection of declining biological integrity. The deep-bodied suckers are detritivores and are tolerant of thermal increase (Gammon, 1983; Simon, 1992). The deep-bodied suckers become more important in the lower river because they typically are the only representatives. This group may replace the round-bodied suckers as a metric for the lower portion of the river, but would reflect declining biological integrity.

8.3.4.2 Proportional Metrics

As sucker species increase and dominate the relative catch, they influence various community function characteristics. These functional characteristics of Great River fish communities are often utilized as proportion metrics (Karr et al., 1986; Simon and Emery, 1995). Metrics considered for a Great River index included: percent round-bodied sucker individuals, percent deep-bodied sucker individuals, percent round-bodied sucker biomass, and percent deep-bodied sucker biomass.

In the Ohio River the percent of round-bodied sucker individuals declines with increasing drainage area (Figure 8.7A). Although the round-bodied suckers are an important community member in the upper 500 river miles, in the lower 400 river miles round-bodied sucker individuals rarely contribute more than 5% of the community. Thus, the use of round-bodied suckers should be considered an indicator of high biological integrity for the upper Ohio River (above ORM 520). Deep-bodied suckers attain their highest relative abundance in the middle Ohio River (between ORM 400 and 500), but the variance in relative abundance is not statistically different (MANOVA, $p < 0.10$) for the entire Ohio River (Figure 8.7B). If one considers an increase in percent of deep-bodied sucker individuals to reflect a decline in biological integrity, this metric should reflect lower integrity characteristics when increasing.

Percent community composition expressed as biomass has not been routinely used for development of an IBI (Goldstein et al. 1994; Niemela et al., Chapter 12). In the Red River of the North drainage systems of northwestern Minnesota and northeastern North Dakota, few species are

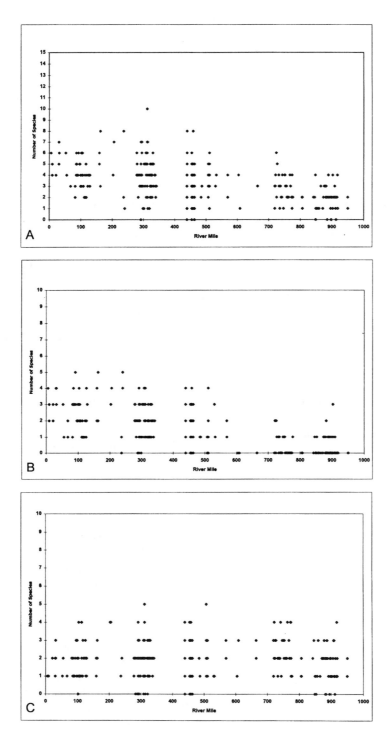

FIGURE 8.6 Influence of the number of suckers species individuals on various species composition metrics considered for use in an index of biotic integrity. (A) total number of sucker species; (B) round-bodied sucker species; and (C) deep-bodied sucker species.

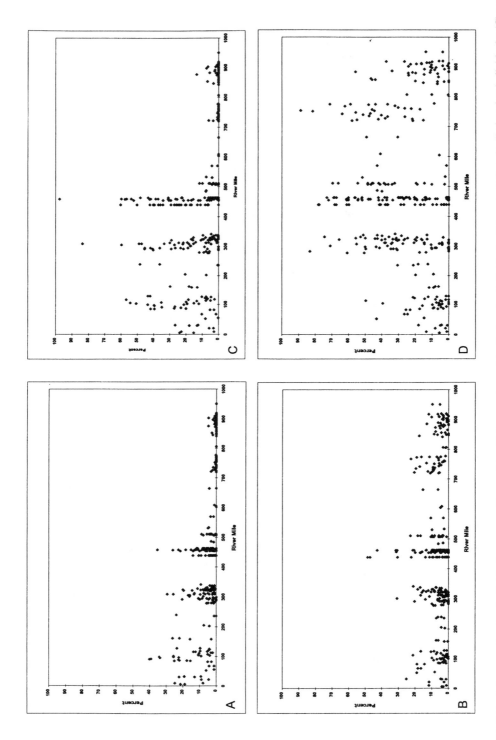

FIGURE 8.7 Influence of sucker individuals on proportion metrics. (A) percent round-bodied sucker individuals; (B) percent deep-bodied sucker individuals; (C) percent round-bodied sucker biomass; and (D) percent deep-bodied sucker biomass.

naturally present. For the Red River, trophic dynamics are better measured using biomass than percentage of individuals. Generally, reasons that many indices have not used biomass are simply because either the information does not exist or the information is used in another index (Index of Well-Being, Gammon, 1973). Examination of biomass for Great River fish communities can more adequately depict the roles that suckers contribute to the entire community. The percent biomass of round-bodied suckers shows how significant this family is in the upper Ohio River (Figure 8.7C). The dramatic decline of round-bodied sucker biomass near ORM 520 shows ranges from nearly 98% of the community biomass above the Falls of the Ohio, to seldom above 10% of the community below the Falls. The proportion of the total catch biomass that is deep-bodied suckers is greater than proportions based on numbers of fish caught. Typically, individual deep-bodied suckers weigh 2.0 kg and thus contribute significantly to the total biomass. The biomass of the deep-bodied suckers is variable (Figure 8.7D); however, they contribute significantly (MANOVA, $p < 0.05$) in the middle and lower Ohio River. Pools of the middle and lower Ohio River possess large lacustrine zones that may enhance the competitive advantage of deep-bodied suckers. However, regardless of where sampling is conducted in the Ohio River, the fish community biomass is clearly dominated by suckers, representing as much as 90% of the total catch biomass at any site.

8.3.4.3 Relative Number of Individuals

Sanders (1991) used the relative number of round-bodied suckers in the catch as a metric for the upper Ohio River. We evaluated the relative number of total suckers, round-bodied suckers, and deep-bodied suckers (Figure 8.8). The total number of suckers declined just below river mile 500 (Figure 8.8A), which is certainly a reflection of the loss of round-bodied suckers (Figure 8.8B), since the deep-bodied suckers are more evenly distributed throughout the Ohio River (Figure 8.8C).

8.3.5 INFLUENCES OF SUCKERS ON OTHER METRICS

The dominant presence of suckers will indirectly affect other aspects of IBI metrics. Suckers will indirectly affect select metrics within the species composition and richness, trophic composition, reproductive guild, abundance, and individual health and condition metrics (Karr, 1981; Karr et al., 1986; Simon and Emery, 1995). Specific metrics affected by the presence of suckers include: total number of species, the percent trophic composition of the catch as insectivores, the percent trophic composition of the catch as omnivores, catch per unit effort (CPUE), percent simple lithophils, and frequency of deformities, eroded fins, lesions, and tumors (DELT) (Figure 8.9).

8.3.5.1 Total Number of Species

Our collections show that catostomids represent 14% of all species collected by the night boat electrofishing method between 1991 and 1996 (Figure 8.9A). The suckers are represented by 14 species, while the other 15 families of the Ohio River contain 88 species. Simon and Emery (1995) showed that the total number of species metric for the Western Allegheny Plateau ecoregion portion of the Ohio River would need an increase of 10 species to change integrity scores; however, changes in as few as three species could be enough to change integrity classes.

8.3.5.2 Percent Insectivores

The trophic composition of insectivores in the Ohio River would be directly affected by the round-bodied suckers. Our collections show that suckers compose over 32% of all insectivorous fish collected (Figure 8.9B). The round-bodied suckers are all categorized as insectivores, while the deep-bodied suckers are omnivores (Ohio EPA, 1989a; Simon and Emery, 1995). Simon and Emery (1995) show that differences between integrity scores for insectivores are separated by 20% total

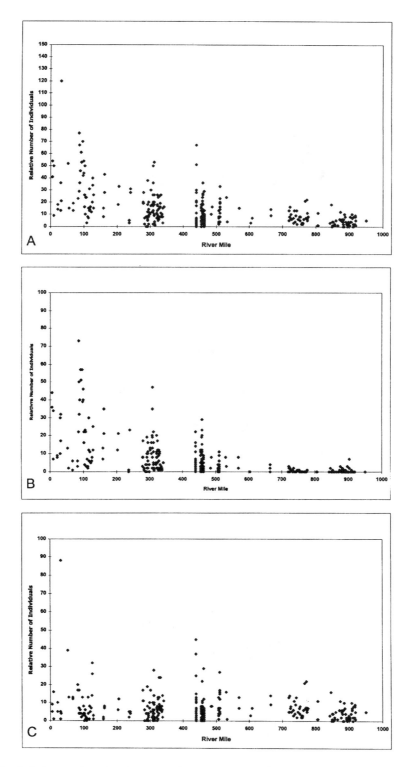

FIGURE 8.8 Relative number of suckers on various metrics considered for use in an index of biotic integrity. (A) total number of suckers; (B) round-bodied suckers; and (C) deep-bodied suckers.

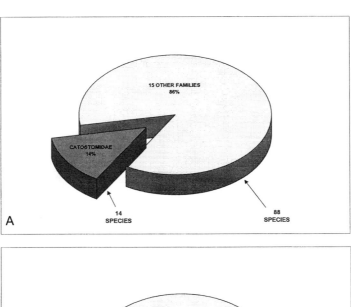
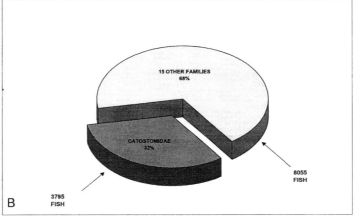
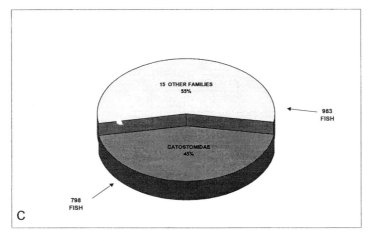

FIGURE 8.9 Contributions of the sucker family to the total catch for specific characteristics of the Ohio River. (A) total number of species; (B) percent of catch as insectivore individuals; (C) percent of catch as omnivore individuals; (D) catch per unit of effort; (E) percent of catch as simple lithophils; and (F) percent of DELT anomalies.

Influence of the Family Catostomidae on the Metrics Developed for a Great River IBI

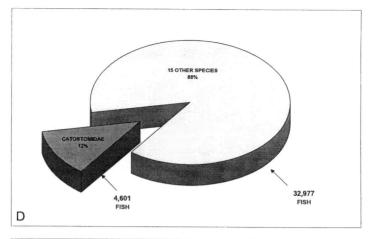

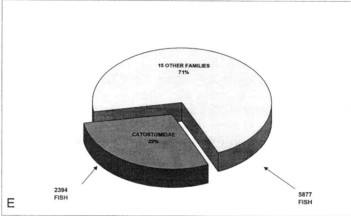

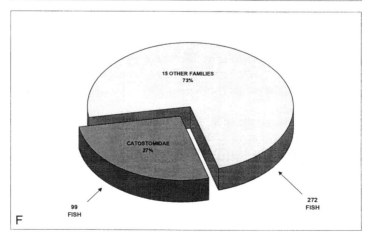

FIGURE 8.9 *Continued.*

catch proportions. If either an increase or decrease in deep-bodied suckers occurred, a reduction in percent insectivores would cause a decline in site biological integrity.

8.3.5.3 Percent Omnivores

The trophic composition of omnivores would be directly affected by an increase in deep-bodied sucker individuals. The deep-bodied suckers compose nearly half (45%) of all omnivores collected (Figure 8.9C). Simon and Emery (1995) showed that a change in about 7% of omnivore abundance could change integrity scores. As deep-bodied suckers increase at a site, biological integrity declines.

8.3.5.4 Catch per Unit Effort

High sucker abundance indirectly affects the catch per unit effort. Our collections in the Ohio River resulted in a total catch of 37,578 fish. Of the total catch, 4601 were suckers — contributing 12% of the catch (Figure 8.9D). Simon and Emery (1995) showed that the total catch would need to increase by 250 fish to change integrity scores at a site in the upper Ohio River.

8.3.5.5 Percent Simple Lithophils

The proportion of simple lithophilous spawning fish are considered a measure of habitat disturbance since these species spawn in the interstitial spaces of cobble and gravel substrates (Ohio EPA, 1989a). Simon and Emery (1995) classified all round-bodied suckers as simple lithophilous spawning species. Author collections show catostomids make up nearly 30% of all simple lithophils sampled in the Ohio River (Figure 8.9E). Simon and Emery (1995) showed that changes of about 20% simple lithophils would change integrity scores. Simple lithophils are important components of the Ohio River and require high-quality, silt-free substrates for spawning.

8.3.5.6 Percent DELT Anomalies

The occurrence of DELT anomalies is an indicator of community health at the lowest extremes of biological integrity (Karr, 1981; Karr et al., 1986). Increases in DELT anomalies are a result of increased contaminant levels (Sanders et al., Chapter 8). Since contaminants are often found in the sediments, benthic species that either feed, reproduce, or are in proximity to the sediments are the most vulnerable to exposure. In addition, long-lived species are the most important indicators since they will generally be exposed for longer durations. Simon and Emery (1995) suggested that proportions of DELT anomalies should be limited to species attaining lengths of 200 mm TL. The occurrence of specific DELT anomalies are important response signatures that can categorize impact type.

As DELT anomalies increase, stresses on individual fish are reflected especially in sucker species. Our collections show that nearly 27% of all anomalies observed in the Ohio River occurred in suckers (Figure 8.9F). Simon and Emery (1995) show that increases in DELT between 1 to 3% of the total catch would reflect differences in biological integrity between reference conditions and the most degraded site in the Ohio River.

8.4 CONCLUSION

The family Catostomidae composes a significant amount of the fish fauna biomass of the Ohio River. The species of this family are important environmental indicators of river health and are used as an important attribute of Great River fish communities in a multimetric index. The fish community of the Ohio River was sampled by electrofishing at 339 locations and 4 navigation lock and dams. Spatial distribution patterns associating community structure with drainage area riverwide show that, within each navigation pool, community structure is related to distance downstream from the navigation dam. The Catostomidae of the Ohio River has demonstrated an increase in community diversity and complexity spatially as the water quality of the system has improved. However, the displacement of redhorses by buffalo and carpsuckers in the lower 300 miles of the Ohio River might be a result of differing habitat characteristics, food base alterations, and reproductive limitations imposed by fewer tributaries.

As water pollution control measures have been incorporated into the management scheme for the Ohio River, the quality of the water and the resources it supports have improved. The most significant improvement was observed at the Pike Island lock and dam. This dam is in the upper portion of the river where round-bodied suckers dominate the fish community and relative abundance is greatest. Suckers become a smaller component of the community in the middle and lower stretches of the Ohio River. Changes in sucker community structure and function as a result of water quality improvements are not as obvious. Since the sucker community of the upper Ohio River was the most impacted by pollution it was the most able to show improvement. The increase in round-bodied sucker species richness and relative abundance in the upper river is an important environmental indicator since this species is sensitive to water quality impact and is more likely to show response to improved conditions.

The deep-bodied suckers of the lower Ohio River may not have been as affected historically by the polluted conditions resulting from industrialization. Therefore, sucker community temporal trends observed at Gallipolis, McAlpine, and Uniontown Pools may not be a result of impact and recovery, but actually of a stable community with natural fluctuations from water levels. Improved species richness and relative abundance in the lower Ohio River would require extensive habitat restoration of backwater areas and inundated gravel bars.

Distribution trends of suckers longitudinally along the river and within navigational pools must be accounted for in the development of reference condition expectations. The influence of suckers on metrics in a Great River IBI must be evaluated to determine which component of this family best describes the structure and function of the Ohio River fish community. Sucker metrics could be developed to reflect species richness, proportion within the community, or simply relative numbers. The metric development described here highlights the round-bodied suckers or focuses on the deep-bodied group, or could include all suckers occurring in the Ohio River. The conclusion is that the round-bodied suckers should be featured as a metric to characterize the upper Ohio River above the Falls of the Ohio, while all sucker species should be used to evaluate the Great River community below the falls. If the deep-bodied suckers are used, they should reflect a loss of biological integrity. The proportion of total catch as biomass shows the importance of the suckers: round-bodied suckers in the upper river and deep-bodied suckers in the lower river. The influence of the round-bodied suckers should be an indication of high biological integrity, while the increase in deep-bodied suckers are a reflection of declining biological integrity.

The catostomids could greatly influence six other metrics being considered for inclusion in numeric biocriteria for the Ohio River. Sucker influence must be considered in the development of the fish index component of biocriteria. The development of biological criteria must take sucker presence into consideration so as not to allow the catostomids to determine the integrity scores of too many of the metrics.

REFERENCES

Becker, G.C. 1983. *The Fishes of Wisconsin.* University of Wisconsin Press, Madison.
Curry, K.D. and A. Spacie. 1984. Differential use of stream habitat by spawning catostomids, *American Midland Naturalist,* 111, 267–279.
Etnier, D.A. and W. Starnes. 1993. *The Fishes of Tennessee.* University of Tennessee Press, Knoxville.
Gammon, J.R. 1973. The Effect of Thermal Inputs on the Populations of Fish and Macroinvertebrates in the Wabash River. Purdue University Water Resources Center Technical Report 32, West Lafayette.

Gammon, J.R. 1983. Changes in the fish community of the Wabash River following power plant start-up: projected and observed, *Aquatic Toxicology and Hazard Assessment: Sixth Symposium,* 802, 350–366. ASTM STP, Philadelphia.

Gerking, S.D. 1955. Key to the fishes of Indiana, *Investigations of Indiana Lakes and Streams,* 4, 49–86.

Goldstein, R.M., T.P. Simon, P.A. Bailey, M. Ell, K. Schmidt, and J.W. Emblom. 1994. Proposed metrics for the index of biotic integrity for the streams of the Red River of the North basin, *Proceedings of the North Dakota Water Quality Symposium,* March 30–31, Fargo, ND.

Karr, J.R. 1981. Assessment of biotic integrity using fish communities, *Fisheries,* 6, 21–27.

Karr, J.R., K.D. Fausch, P.L. Angermeier, P.R. Yant, and I.J. Schlosser. 1986. Assessment of Biological Integrity in Running Waters: A Method and Its Rationale. Illinois Natural History Survey Special Publication 5.

Kay, L.A., R. Wallus, and B.L. Yeager. 1996. *Reproductive Biology and Early Life History of Fishes of the Ohio River Drainage. Vol. 2. Catostomidae.* Tennessee Valley Authority, Knoxville.

Krumholz, L.A. and W.L. Minkley. 1964. Changes in the fish population in the upper Ohio River following temporary pollution abatement, *Transactions of the American Fisheries Society,* 93, 1–5.

Matheny, M.P. and C.F. Rabeni. 1995. Patterns and movement and habitat use by northern hogsuckers in an Ozark stream, *Transactions of the American Fisheries Society,* 124, 886–897.

Moss, R.E., J.W. Scanlan, and C.S. Anderson. 1983. Observations on the natural history of the blue sucker (*Cycleptus elongatus* LeSueur) in the Neosho River, *American Midland Naturalist,* 109, 15–22.

Ohio Environmental Protection Agency (OEPA). 1989a. *Biological Criteria for the Protection of Aquatic Life. Vol. III. Standardized Biological Field and Laboratory Methods for Assessing Fish and Macroinvertebrate Communities.* Ohio EPA, Division of Water Quality Planning and Assessment, Ecological Assessment Section, Columbus, OH.

Ohio Environmental Protection Agency (OEPA). 1989b. *Biological Criteria for the Protection of Aquatic Life. Vol. II. Users Manual for Biological Field Assessment of Ohio Surface Water.* Ohio EPA, Division of Water Quality Planning and Assessment, Ecological Assessment Section, Columbus, OH.

Omernik, J.M. 1987. Ecoregions of the conterminous United States, *Annals of the Association of American Geographers,* 73, 133–136.

Omernik, J.M. and A.L. Gallant. 1988. Ecoregions of the Upper Midwest States. EPA 600-3-88-037. U.S. Environmental Protection Agency, Corvallis, OR.

Page, L.M. and C.E. Johnston. 1990. Spawning in the creek chubsucker, *Erimyzon oblongus,* with a review of spawning behavior in suckers (Catostomidae), *Environmental Biology of Fishes,* 27, 265–272.

Pearson, W.D. and L.A. Krumholz. 1984. *Distribution and Status of Ohio River Fishes.* Oak Ridge National Laboratory. ORNL/Sub/79-7831/1.

Pearson, W.D. and B.J. Pearson. 1989. Fishes of the Ohio River, *Ohio Journal of Science,* 89, 181–187.

Pflieger, W.L. 1973. *The Fishes of Missouri.* Missouri Department of Conservation, Columbia.

Reash, R.J. and J.H. Van Hassel. 1988. Distribution of upper and middle Ohio River fishes. II. Influence of zoogeographic and physiochemical tolerance factors, *Journal of Freshwater Ecology,* 4, 459–476.

Sanders, R.E. 1989. Comparison of day and night electrofishing catches along the shores of navigation dam pools in the Muskingum and Ohio Rivers, *Ohio Journal of Science.*

Sanders, R. 1991. A 1990 Night Electrofishing Survey of the Upper Ohio River Mainstem (RM 40.5 to 270.8) and Recommendations for a Long-Term Monitoring Program. Ohio Department of Natural Resources, Columbus, OH.

Simon, T.P. 1992. Development of Biological Criteria for Large Rivers with an Emphasis on an Assessment of the White River Drainage, Indiana. EPA 905-R-92-026. U.S. Environmental Protection Agency, Region 5, Chicago, IL.

Simon, T.P. and E.B. Emery. 1995. Modification and assessment of an index of biotic integrity to quantify water resource quality in Great Rivers, *Regulated Rivers Research and Management,* 11, 283–298.

Trautman, M.B. 1981. *The Fishes of Ohio.* The Ohio State University Press, Columbus.

Walburg, C.H. 1976. Changes in the Fish Populations of Lewis and Clark Lake, 1956–1962, and Their Relation to Water Management and the Environment. U.S. Fish and Wildlife Service Research Report 79.

9 The Use of External Deformities, Erosion, Lesions, and Tumors (DELT Anomalies) in Fish Assemblages for Characterizing Aquatic Resources: A Case Study of Seven Ohio Streams

Randall E. Sanders, Robert J. Miltner, Chris O. Yoder, and Edward T. Rankin.

CONTENTS

9.1 Introduction 225
    9.1.1 Use of External DELT Anomalies in Bioassessments 226
    9.1.2 Study Area 227
9.2 Methods 229
    9.2.1 Examination of Fish for External DELT Anomalies 229
    9.2.2 Statistical Analyses of Stream Quality 230
9.3 Results and Discussion 231
    9.3.1 Predominant Types of DELT Anomalies 231
    9.3.2 Predominant Species with DELT Anomalies 236
    9.3.3 DELT Anomalies vs. Stream Quality 237
    9.3.4 Severity of DELT Anomalies (Severe vs. Mild) 239
9.4 Conclusion 243
Acknowledgments 244
References 244

9.1 INTRODUCTION

Since 1966, an increasing number of studies have reported the occurrence of external fish abnormalities (e.g., deformities, fin erosion, open sores, and tumors) in a variety of aquatic habitats (streams, lakes, estuaries, and marine) and discussed possible relationships between anomalies and environmental quality. Consistently, these studies have reported either a low number or percentage of anomalies at nonpolluted sites or a high number or percentage at polluted sites affected by industrial and sewage discharges, or both (Mills et al., 1966; Shotts et al., 1972; Komada, 1980; Berra and Au, 1981; Murchelano and Ziskowski, 1982; Sherwood, 1982; Cross, 1985; Bengtsson

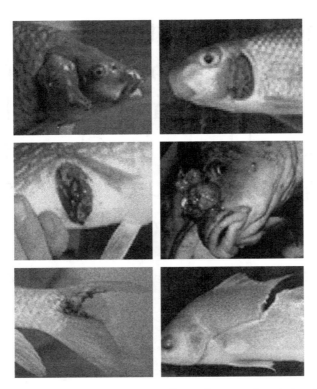

FIGURE 9.1 Five examples of external DELT anomalies and one predator-induced injury. Upper left: A common carp with a severely deformed head (knothead). Upper right: A golden redhorse with a severely eroded gill cover. Middle left: A northern pike with a large lesion/ulceration. Middle right: A common carp with large nasal tumors. Lower left: A silver redhorse with multiple DELT anomalies (severely eroded fin and tumors). Lower right: Predator injuries, such as the one a heron made on this freshwater drum, are not coded as a DELT anomaly by the Ohio EPA.

et al., 1985; Bengtsson and Larsson, 1986; Baumann et al., 1987; Malins et al., 1988; Reash and Berra, 1989; Lindesjoo and Thulin, 1990; McCain et al., 1992; Fournie and Summers, 1996). As the result of these and similar studies, the occurrence of fish anomalies has become an increasingly accepted indicator of environmental quality for water resources and fish health (Karr, 1986; Plumb, 1994; Yoder and Rankin, 1995).

9.1.1 Use of External DELT Anomalies in Bioassessments

Fish with external deformities, erosion, lesions (defined by the Ohio EPA as open sores or exposed tissue), and tumors (neoplasms) (DELT anomalies, Figure 9.1) were first observed by Ohio Environmental Protection Agency (Ohio EPA) biologists in 1979 while sampling a large urban stream, the Scioto River in downtown Columbus. The Scioto River receives a combination of pollutants from industrial and municipal effluent, combined sewer overflows, and urban runoff. Stream flows are also dominated by municipal effluent downstream from Columbus during periods of low flow (Yoder and Rankin, 1995). After a literature review showed a positive relationship between degraded water quality and anomalies (Sniezko, 1962; Mills et al., 1966; Pippy and Hare, 1969; Shotts et al., 1972; Hickey, 1972; Sniezko, 1974), DELT anomalies were incorporated into the Ohio EPA's stream monitoring program. As part of the development of biological criteria in 1987, the percentage of fish with DELT anomalies was then used as a modification of Karr's original metric for the percentage of diseased fish in the Index of Biotic Integrity (IBI; Karr, 1981; Ohio EPA, 1987a;

TABLE 9.1
Index of Biotic Integrity (IBI) Scoring Criteria for the Percentage of Fish with Deformities, Erosion, Lesions/Ulcers, and Tumors Captured at Headwater, Wading, and Boatable Sites

	Ohio EPA DELT Scoring Criteria		
	"5"	"3"	"1"
Headwater and wading sites	≤0.1%	>0.1 ≤ 1.3%	>1.3%
Boat sites	<0.5%	0.5 ≤ 3.0%	>3.0%

Note: Criteria are based on data from more than 300 reference sites throughout Ohio. If DELT is present on >1 individual at sites with ≤1000 total fish, then a score of "5" is assigned; or if 2 DELT individuals are collected at a site with ≤1000 total fish, then a score of "3" is assigned.

1987b; Table 9.1). Blackspot *(Uvulifer ambloplitis* and *Crassiphiala bulboglossa)*, anchor worm *(Lernaea cyprinacea)*, and other parasites were excluded from the metric (but are still recorded) due to the lack of a consistent, inverse relationship to environmental quality (Allison et al., 1977; Berra and Au, 1981; Whittier et al., 1987). Biocriteria were subsequently adopted into Ohio Water Quality Standards regulations in 1990 (Ohio EPA, 1992a). After realizing a high degree of variability in the severity of DELT anomalies in Ohio streams (e.g., slight fin erosion vs. severely eroded fins and skeletal and multiple anomalies), field criteria for classifying the severity of DELT anomalies were developed by Ohio EPA biologists in 1991.

In addition to IBI scores, longitudinal graphs of the percentage of DELT anomalies and photographs are frequently used in agency reports, presentations, and news releases along with other results to effectively portray the biological and chemical quality of Ohio streams (Ohio EPA, 1992b; 1995; 1996). While many audiences do not easily relate to numerical values that reflect nonattainment of Clean Water Act goals, examples of fish with DELT anomalies are clearly understood by the regulated community, resource managers, and the general public as indicators of a polluted water resource. Through the years, however, it has become obvious that not all fish species are equally susceptible, and a number of questions have arisen as to the Ohio EPA's use of these types of abnormalities. The objectives of this study were to try to answer some of these questions by determining: (1) What are the predominant types of DELT anomalies? (2) Are some fish species more susceptible to DELT anomalies? (3) Is there a statistical relationship between the percentage of DELT anomalies and the two biological indices (IBI and Modified Index of Well-Being [MIwb]) used by the Ohio EPA to assess overall stream quality? (4) Is the severity of DELT anomalies useful in assessing environmental quality?

9.1.2 STUDY AREA

Bioassessment results from seven streams sampled by the Ohio EPA between 1991 and 1995 were selected for this study to encompass a wide range of environmental quality (exceptional to very poor), stream sizes (drainage areas from 6.4 to 75,350 square miles), and land uses (woodland, agricultural, industrial, and urban). The database consisted of 360 electrofishing collections from 139 sampling locations spanning 746 river miles (1194 km). Approximately 181 hours of electrofishing was conducted over a total cumulative distance of 114 miles (184 km). Brief descriptions of the seven streams used in the present study are provided in Table 9.2.

TABLE 9.2
Descriptions of the Seven Ohio Streams Used in the DELT Study

1. *Ottawa River:* Located in northwestern Ohio, the Ottawa River is a medium-size tributary of the Auglaize River (Lake Erie basin). Land use within the 365-square-mile watershed is predominantly agricultural, with the exception of the urban/industrial complex within Lima. Major point source discharges to the river within and near Lima consist of the municipal sewerage system, composed of five large combined sewer overflow structures (CSOs) and a 001, an oil refinery, and a chemical complex. The stream also receives landfill leachate and chemical spills. During periods of low flow, more than 80% of the stream flow is effluent downstream from the three major wastewater discharges. Historically, the Ottawa River has been one of Ohio's most severely impacted streams downstream from Lima. A total of 23 samples from 8 locations (RM 46.1–1.2) were collected (between 27 June and 23 August 1991) and used in this study (Ohio EPA, 1992).
2. *Mahoning River:* Located in northeastern Ohio and western Pennsylvania, the Mahoning River watershed drains a total of 1140 square miles of the upper Ohio River basin. Land use within the watershed consists of a mixture of agriculture, heavy industry, and urbanization. The mainstem receives a multitude of municipal and industrial discharges, CSOs, and stormwater runoff. Historically, the Mahoning River has been severely impacted, but has shown some recovery. In 1994, the mainstem was in predominantly nonattainment of its Warmwater Habitat (WWH) aquatic life use designation throughout the lower 90 miles. Stream quality ranged from exceptional in its headwaters to very poor downstream from Youngstown (Ohio EPA, 1996). A total of 101 samples from 39 locations (RM 100.6–0.2) were collected (between 6 July and 21 September 1994) and used in this study.
3. *Ohio River:* Forming most of Ohio's eastern and southern boundaries, near-shore waters of the Ohio River now occupy Ohio due to impoundment by 10 high-lift navigational locks and dams. The mainstem receives a multitude of point source discharges from sewage treatment plants, heavy industries (e.g., steel mills, chemical plants), coal-fired power plants, and CSOs (ORSANCO, 1994). The mainstem also receives a considerable amount of barge traffic; however, a considerable amount of the watershed (drainage areas ranged from 23,450 to 75,350 square miles) is comprised of agricultural land uses and woodlands. A total of 102 fall night samples from 35 locations (RM 48.8–487.2) were collected (between 23 September 1991 and 31 October 1995) and used in this study (Sanders, 1995).
4. *Salt Creek:* Located in southcentral Ohio, Salt Creek is a medium-size tributary of the Scioto River. Land use within the 555-square-mile watershed is dominated by woodlands and agriculture. The mainstem receives effluent from one small municipal sewage plant in the upper half of the watershed and a second moderate size plant in the headwaters of a major tributary. A total of 23 samples from 13 locations (RM 41.2–0.3) were collected (between 29 June and 22 October 1992) and used in this study; 75% of the mainstem sites were in FULL attainment of Exceptional Warmwater Habitat (EWH) in 1992 (Ohio EPA, unpublished data).
5. *Little Miami River:* Located in southwestern Ohio, the Little Miami River flows in a southwesterly direction to its confluence with the Ohio River. Land use within the 1757-square-mile watershed is predominantly agricultural, followed by forests and suburbanization. The mainstem and tributaries received a cumulative total of approximately 50 million gallons per day (MGD) of municipal and county sewage treatment plant effluent in 1993. Portions of the watershed are developing rapidly, however, and the quantity of effluent is expected to increase. With the exception of elevated total phosphorus and fecal bacteria concentrations, water quality was generally good throughout the mainstem in 1993. Mainstem sediment quality was also good, with few elevated metals or organic compounds; 41% of the mainstem was in FULL attainment of EWH in 1993 (Ohio EPA, 1995). A total of 87 samples from 32 locations (RM 102.1–0.2) were collected (between 29 July and 14 October 1993) and used in this study (Ohio EPA, 1995).
6. *East Fork Little Miami River:* Located in southwestern Ohio, the East Fork is the largest tributary of the Little Miami River. Land use within its 499-square-mile watershed is largely agricultural, but is becoming increasing suburbanized in the lower third, which receives effluent from four WWTPs. Approximately 50% of the lower 15 miles of the mainstem was in FULL attainment of EWH in 1993 (Ohio EPA, 1995). A total of 16 samples from eight locations (RM 15.5–1.4) were collected (between 20 July and 8 September 1993) and used in this study.
7. *Little Beaver Creek:* Located on the east side of Dayton, Ohio, Little Beaver Creek is a small tributary (26.4-square-mile drainage area) of Beaver Creek, which drains into the Little Miami River near Xenia. The stream contains one of the most developed watersheds of the seven streams selected, and receives more than 8 MGD of effluent (the largest discharge in the LMR watershed). Approximately 93% of the lower 4.7 miles of the tributary was in nonattainment of WWH in 1993 (Ohio EPA, 1995). A total of eight samples from four locations (RM 4.7–0.1) were collected (between 7 July and 18 August 1993) and used in this study.

9.2 METHODS

Standardized field, laboratory, and data processing methods and procedures were used in this study (Ohio EPA, 1987a,b; 1989a,b,c). All fish sampling in this study was conducted with two principal types of gasoline-powered, pulsed DC electrofishing gear: (1) 1750-watt pulsator/generator combination (T&J Manufacturers) designed for smaller wadeable streams, and (2) boat-mounted, 3500-watt generator and pulsator combinations (Smith-Root Type 3.5 or 5.0 GPP units) with a straight electrode configuration for wider and deeper boatable streams. Sampling was conducted during the day except in the Ohio River, where night electrofishing was used for improved catches of most species (Sanders, 1992). Each sampling site consisted of a fixed distance ranging from approximately 150 to 200 meters for wading methods and 500 meters for boat-sampled sites. The time required to sample and process the catch at each site ranged from 1 to 3 hours.

Gross external anomalies are defined by the Ohio EPA (Ohio EPA, 1989c) as external skin, fin, or subcutaneous disorders visible to the naked eye during normal sampling procedures (i.e., when the fish are captured, identified, sorted, weighed, and counted). Deformities, erosion (of fins, barbels, and gill covers), skin lesions (i.e., open sores, ulcerations, exposed tissue), and tumors are 4 of the 15 types of anomalies (other types include common parasites and other abnormalities) recorded by Ohio EPA biologists during a fish survey. The DELT anomalies include many of the most obvious external clinical signs of infectious diseases and parasites (Plumb, 1994), and the use of such a "super group" reduces the significance of a misidentification. While every effort is made to correctly identify anomalies, the interchangeability of terms such as eroded and lesion by pathologists makes it less critical what to call an abnormality than whether the anomaly is present or absent. Ohio EPA staff have learned that with minimal training and established criteria, the field identification of DELT anomalies is relatively easy. General definitions and characteristics of the 15 different anomalies recorded are described by the Ohio EPA (1989c). Field biologists are urged to refer to textbooks on fish health for further information and pictures of specific anomalies (e.g., Post, 1987; Plumb, 1994). Questionable specimens were preserved for lab verification.

9.2.1 EXAMINATION OF FISH FOR EXTERNAL DELT ANOMALIES

Although all fish captured are identified and counted during Ohio EPA stream surveys, only fish that are weighed (individually, in aggregate, or by subsample) are examined for external anomalies at most sites (i.e., sites with drainage areas greater than 20 square miles). Typically, this represents most of the fish captured; however, subsampling is recommended when large catches of certain species (e.g., gizzard shad) occur in order to save time on processing in the field. Subsamples by species typically included at least 15 individuals for larger species such as common carp, suckers, and sunfishes, and at least 50 individuals for smaller species such as minnows and darters. A conscious effort was made at all sites to "randomly" select fish for weighing and not skew the results by looking for fish with or without DELTs for weighing. At headwater sites (drainage areas less than or equal to 20 square miles) where fish are not weighed (because the MIwb is not applicable), all fish counted were examined for external DELT anomalies. Once detected, the severity of all DELT anomalies was determined as mild or severe using the standardized criteria (Table 9.3). The time used for examining specimens in the field typically consisted of less than 10 seconds (i.e., long enough to determine the presence or absence of anomalies on one side and classify the severity). For most species, this consisted of looking at one side of the fish, including all visible fins. Ictalurids (catfish) were also examined ventrally for barbel anomalies and parasites.

Anomaly data from each sampling location was then entered into the Ohio EPA fish community database (Ohio ECOS) along with numbers and weight(s) by species. Since the Ohio EPA does not currently calculate the percentage of DELT anomalies for each species (in report form), the percentage by species in this study was based only on fish that were examined for DELTs. The

TABLE 9.3
Ohio EPA's Field Criteria for Determining the Severity (Mild or Severe) of External Deformities, Erosion, Lesions, and Tumors (DELT Anomalies)

Type of DELT Anomaly (FINS Code)	Severity Criteria
Deformed fin, head, vertebrae, barbel, and other body parts	
Mild	One deformed fin or branched barbel.
Severe	Two or more deformed fins or barbels; or any body (head, vertebrae, abdomen, or other body part) deformity.
Eroded fin, gill cover, or barbel	
Mild	One or two barbels eroded less than half the barbel length, or a fin ray not eroded past ray fork.
Severe	Three or more eroded barbels; or a barbel eroded more than half its total length; >2 fins eroded or fin eroded past a single ray fork or if gill cover is eroded showing exposed gill(s).
Lesion (open sore, exposed tissue, or ulceration)	
Mild	<2 lesions < the size of the largest scale.
Severe	>3 lesions or a lesion > the size of the largest scale or raw tissue.
Tumor	
Mild	<2 tumors < the diameter of the eye (count patches of Lymphocystis as one tumor).
Severe	> 3 tumors or one tumor larger than the diameter of the eye.

total percentage of all species was calculated for each site, however, by Ohio ECOS, which computes (and sums) each type of anomaly for each species in each sample as a weighted number based on percent occurrence among weighed fish times the total number of that fish species in the sample. A fish with two or more different types of DELT anomalies is coded as an M (for multiple DELT anomalies) to avoid inflating the true percentage. Obvious injuries (e.g., fish-eating-bird or hooking injuries) are not included.

9.2.2 STATISTICAL ANALYSES OF STREAM QUALITY

The two fish indices incorporated into Ohio's biological criteria, the IBI and MIwb, were used to determine if a statistically significant relationship existed between stream quality and the percentage of DELT anomalies. The IBI, first introduced by Karr (1981), consists of 12 metrics that assess fish assemblages based on species richness and composition, trophic composition, abundance, and health. Ohio EPA's modified versions for headwater, wading, and boatable streams were used. The MIwb, a modified version of the Index of Well-Being (Gammon, 1976), is a measure of the fish community based on a calculation using relative number, biomass, and the Shannon Diversity Index (based on numbers and weight) from which highly tolerant and exotic fishes are removed from the numbers and biomass calculations. Higher scores for both indices typically reflect improving quality of fish assemblages or stream quality in general.

The percentage of DELT anomalies from the electrofishing samples collected from the seven streams used in this study were regressed against the MIwb and IBI (n = 346 and 360, respectively) scores computed for those samples. Before computing regression functions, percentage of DELTs were transformed to fit model assumptions with a Ln (y + 1) transformation (Neter et al., 1990). For the regression of Ln(DELT + 1) on IBI, plots of residuals against estimated values demonstrated nonconstancy of error variance over the range of IBI scores ≥35; therefore, only data for IBI scores <35 were considered using the logarithmic transformation (Table 9.4). Percentage of DELTs for IBI scores ≥35 were subsequently transformed using 1/Y + 1 (Neter et al., 1990), resulting in a normal distribution and constancy of error variance as demonstrated by a whisker plot of the

TABLE 9.4
Parameters and Significance Tests for Regressions of Percentage DELT Anomalies Against MIwb, IBI ≥ 35, and IBI < 35

Variable	Coefficient	Standard Error	Student's t	p
Constant	3.9735	0.1771	22.433	<0.0001
MIwb	−0.3149	0.0205	15.344	<0.0001
Constant	−0.0863	0.1620	−0.53	0.5947
IBI ≥35	0.0132	0.0036	3.66	<0.0001
Constant	4.8979	0.2732	17.93	<0.0001
IBI <35	−0.1185	0.1063	−11..5	<0.0001

Note: The coefficient for IBI ≥ 35 is positive due to an inverse transformation.

transformed variable and a plot of residuals against estimated values. Residual and normal probability plots constructed for the regression of Ln(DELT + 1) on MIwb did not indicate a strong departure from normality or equality of error variance over the range of observed values.

9.3 RESULTS AND DISCUSSION

Of the 102,164 fish examined for external anomalies, one or more DELT anomalies was observed on 2,657 fish (2.6%). Of the 109 total species and 8 hybrids examined, one or more DELT anomalies was observed on 62 of the species (56.9%) and 5 hybrids (Table 9.5). By stream, Salt Creek (one of Ohio's highest quality rural streams) had the lowest overall percentage of DELT anomalies for both the total number of individuals examined (0.4%) and number of species afflicted (23.1%, Table 9.6). Conversely, the highest percentages of total individuals and species with anomalies occurred in two of Ohio's most biologically and chemically impacted streams, the Ottawa (8.1 and 56.5%, respectively) and Mahoning (7.1 and 60.0%, respectively) rivers. Similarly, the maximum and median values of DELT anomalies by stream were also lowest in Salt Creek (3.0 and 0.3%, respectively) and highest in the Ottawa (57.5 and 13.1%, respectively) and Mahoning (37.9 and 8.2%) rivers (Table 9.6). All but one of the streams had at least one site with no observed DELT anomalies. The exception, Little Beaver Creek, also had a relatively low minimum value of 0.3%.

9.3.1 PREDOMINANT TYPES OF DELT ANOMALIES

Of the 2657 fish with DELT anomalies (all data pooled), the predominant type of anomaly was erosion (56.0%), followed by deformities (30.1%), lesions (7.0%), and tumors (1.2%). Fish with multiple DELT anomalies accounted for 5.6% of all fish with anomalies. Similar patterns in frequency by type were exhibited in six of the streams (Table 9.6). Little Beaver Creek, however, had more deformities than erosion. The overall percentage of DELT anomalies represented by deformities ranged from a minimum of 13.6 in the Ottawa River to a maximum of 44.4 in Little Beaver Creek. The total percentage of deformities was also high and only slightly less than erosion in the Mahoning River, which has multiple complex toxic problems. The overall percentage of erosion was between 40.0 and 60.7 in six of the streams, but reached a maximum of 77.9 in the Ottawa River. Skin lesions were between 0.7 and 12.7%, and tumors between 0.0 and 11.9%. The highest overall percentage of tumors occurred in Salt Creek where all five fish were observed with tumor-like growths apparently caused by the lymphocystis virus. The high rate of tumors in Salt

TABLE 9.5
Summary of the Fishes Collected in Seven Ohio Streams and the Incidence of DELT Anomalies by Taxa

FAMILY Common Name (Scientific Name)	No. of Streams Collected in	Total No. of Fish Examined	No. of Streams with a DELT	Total No. with DELT Anomalies	% with DELT Anomalies
ESOCIDAE (pikes)					
muskellunge x northern pike hybrid	2	5	1	2	40.0
muskellunge (Esox masquinongy)	1	15	1	2	13.3
grass pickerel (Esox americanus vermiculatus)	3	22	1	2	9.1
ICTALURIDAE (catfishes)					
black bullhead (Ameiurur melas)	3	15	1	5	33.3
channel catfish (Ictalurus punctatus)	6	1643	6	236	14.4
yellow bullhead (Ameiurur natalis)	6	231	5	30	13.0
brown bullhead (Ameiurur nebulosus)	1	55	1	6	10.9
flathead catfish (Pylodictis olivaris)	4	326	2	14	4.3
stonecat (Noturus flavus)	4	97	3	4	4.1
mountain madtom (Noturus eleutherus)	2	50	0	0	0.0
brindled madtom (Noturus miurus)	1	20	0	0	0.0
tadpole madtom (Noturus gyrinus)	1	1	0	0	0.0
CYPRINIDAE (carps and minnows)					
common carp x goldfish hybrid	6	235	6	71	30.2
common carp (Cyprinus carpio)	7	2624	7	749	28.5
goldfish (Carassius auratus)	4	324	3	77	23.8
golden shiner (Notemigonus crysoleucas)	4	350	2	11	3.1
river chub (Nocomis micropogon)	4	768	2	19	2.5
creek chub (Semotilus atromaculatus)	6	2545	5	32	1.3
redfin shiner (Lythrurus umbratilis)	1	93	1	1	1.1
rosyface shiner (Notropis rubellus)	4	215	2	2	0.9
steelcolor shiner (Cyprinella whipplei)	3	287	1	1	0.3
spotfin shiner (Cyprinella spiloptera)	7	3083	4	9	0.3
central stoneroller (Campostoma anomalum)	7	5615	4	6	0.1
emerald shiner (Notropis atherinoides)	5	5330	3	4	<0.1
striped shiner (Luxilus chrysocephalus)	7	1407	1	1	<0.1
fathead minnow (Pimephales promelas)	4	1073	1	1	<0.1
bluntnose minnow (Pimephales notatus)	7	5769	2	4	<0.1
sand shiner (Notropis stramineus)	6	1190	1	1	<0.1
silver chub (Macrhybopsis storeriana)	3	1245	0	0	0.0
gravel chub (Erimystax x-punctatus)	3	20	0	0	0.0
blacknose dace (Rhinichthys atratulus)	5	854	0	0	0.0
tonguetied minnow (Exoglossum laurae)	1	1	0	0	0.0
suckermouth minnow (Phenacobius mirabilis)	6	159	0	0	0.0
southern redbelly dace (Phoxinus erythrogaster)	1	2	0	0	0.0
silver shiner (Notropis photogenis)	5	431	0	0	0.0

TABLE 9.5 (continued)
Summary of the Fishes Collected in Seven Ohio Streams and the Incidence of DELT Anomalies by Taxa

FAMILY Common Name (Scientific Name)	No. of Streams Collected in	Total No. of Fish Examined	No. of Streams with a DELT	Total No. with DELT Anomalies	% with DELT Anomalies
rosefin shiner (Lythrurus ardens)	3	44	0	0	0.0
river shiner (Notropis blennius)	2	89	0	0	0.0
spottail shiner (Notropis hudsonius)	1	74	0	0	0.0
whitetail shiner (Cyprinella galactura)	1	1	0	0	0.0
mimic shiner (Notropis volucellus)	2	129	0	0	0.0
channel shiner (Notropis wickliffi)	1	321	0	0	0.0
silverjaw minnow (Notropis buccatus)	5	291	0	0	0.0
bullhead minnow (Pimephales vigilax)	4	138	0	0	0.0
other minnow hybrids	2	2	0	0	0.0
CATOSTOMIDAE (suckers)					
bigmouth buffalo (Ictiobus cyprinellus)	4	23	2	5	21.7
black buffalo (Ictiobus niger)	3	142	3	24	16.9
blue sucker (Cycleptus elongatus)	1	7	1	1	14.3
river redhorse (Moxostoma carinatum)	4	80	3	10	12.5
spotted sucker (Minytrema melanops)	4	147	2	14	9.5
silver redhorse (Moxostoma anisurum)	6	641	6	57	8.9
white sucker (Catostomus commersoni)	5	2459	4	193	7.8
quillback (Carpiodes cyprinus)	5	582	5	44	7.6
highfin carpsucker (Carpiodes velifer)	4	64	2	4	6.2
river carpsucker (Carpiodes carpio)	4	299	3	13	4.3
golden redhorse (Moxostoma erythrurum)	6	2890	6	107	3.7
smallmouth buffalo (Ictiobus bubalus)	4	1054	3	38	3.6
black redhorse (Moxostoma duquesnei)	6	1080	6	38	3.5
shorthead redhorse (Moxostoma macrolepidotum)	4	1319	3	41	3.1
northern hog sucker (Hypentelium nigricans)	7	2381	4	62	2.6
creek chubsucker (Erimyzon oblongus)	1	14	0	0	0.0
river carpsucker x quillback hybrid	2	3	0	0	0.0
UMBRIDAE (mudminnows)					
central mudminnow (Umbra limi)	1	8	1	1	12.5
PERCIDAE (perches)					
walleye (Stizostedion vitreum)	3	202	2	17	8.4
sauger x walleye (S. canadense x S.vitreum)	5	61	2	2	3.3
yellow perch (Perca flavescens)	2	415	1	12	2.9
sauger (Stizostedion canadense)	4	3100	3	22	0.7
johnny darter (Etheostoma nigrum)	5	225	1	1	0.4
Logperch (Percina caprodes)	5	757	1	1	0.1
greenside darter (Etheostoma blennioides)	7	1371	1	1	0.1
dusky darter (Percina sciera)	2	48	0	0	0.0
blackside darter (Percina maculata)	3	19	0	0	0.0
slenderhead darter (Percina phoxocephala)	4	60	0	0	0.0

TABLE 9.5 *(continued)*
Summary of the Fishes Collected in Seven Ohio Streams and the Incidence of DELT Anomalies by Taxa

FAMILY Common Name *(Scientific Name)*	No. of Streams Collected in	Total No. of Fish Examined	No. of Streams with a DELT	Total No. with DELT Anomalies	% with DELT Anomalies
river darter *(Percina shumardi)*	1	127	0	0	0.0
channel darter *(Percina copelandi)*	1	62	0	0	0.0
eastern sand darter *(Ammocrypta pellucida)*	2	24	0	0	0.0
banded darter *(Etheostoma zonale)*	5	552	0	0	0.0
variegate darter *(Etheostoma variatum)*	2	159	0	0	0.0
bluebreast darter *(Etheostoma camurum)*	2	2	0	0	0.0
rainbow darter *(Etheostoma caeruleum)*	6	808	0	0	0.0
orangethroat darter *(Etheostoma spectabile)*	4	60	0	0	0.0
rainbow x orangethroat darter hybrid	1	1	0	0	0.0
fantail darter *(Etheostoma flabellare)*	6	982	0	0	0.0
CENTRARCHIDAE (sunfishes)					
pumpkinseed *(Lepomis gibbosus)*	4	340	2	22	6.5
green sunfish *(Lepomis cyanellus)*	7	2734	4	182	6.7
largemouth bass *(Micropterus salmoides)*	7	1239	7	79	6.4
hybrid sunfish	7	169	2	6	3.6
rock bass *(Ambloplites rupestris)*	6	1040	4	30	2.9
warmouth *(Lepomis gulosus)*	4	37	1	1	2.7
spotted bass *(Micropterus punctulatus)*	5	1576	4	39	2.5
white crappie *(Pomoxis annularis)*	6	632	3	13	2.1
smallmouth bass *(Micropterus dolomieu)*	6	2222	4	30	1.4
bluegill *(Lepomis macrochirus)*	7	4429	5	42	0.9
black crappie *(Pomoxis nigromaculatus)*	6	342	2	2	0.6
longear sunfish *(Lepomis megalotis)*	7	2091	3	11	0.5
orangespotted sunfish *(Lepomis humilis)*	4	10	0	0	0.0
redear sunfish *(Lepomis microlophus)*	3	7	0	0	0.0
PERCICHTHYIDAE (temperate basses)					
striped bass *(Morone saxatilis)*	1	116	1	2	1.7
white bass *(Morone chrysops)*	5	1188	2	6	0.5
white x striped bass hybrid	2	231	1	1	0.4
white perch *(Morone americanus)*	2	5	0	0	0.0
LEPISOSTEIDAE (gars)					
longnose gar *(Lepisosteus osseus)*	3	169	2	3	1.8
shortnose gar *(Lepisosteus platostomus)*	1	2	0	0	0.0
HIODONTIDAE (mooneyes)					
mooneye *(Hiodon tergisus)*	2	84	1	1	1.2
CLUPEIDAE (herrings)					
gizzard shad *(Dorosoma cepedianum)*	7	16,497	5	171	1.0

TABLE 9.5 *(continued)*
Summary of the Fishes Collected in Seven Ohio Streams and the Incidence of DELT Anomalies by Taxa

FAMILY Common Name *(Scientific Name)*	No. of Streams Collected in	Total No. of Fish Examined	No. of Streams with a DELT	Total No. with DELT Anomalies	% with DELT Anomalies
skipjack herring *(Alosa chrysochloris)*	2	368	0	0	0.0
threadfin shad *(Dorosoma petenense)*	1	5	0	0	0.0
SCIAENIDAE (drums)					
freshwater drum *(Aplodinotus grunniens)*	5	6864	4	18	0.3
PETROMYZONTIDAE (lampreys)					
silver lamprey *(Ichthyomyzon unicuspis)*	1	19	0	0	0.0
ohio lamprey *(Ichthyomyzon bdellium)*	1	2	0	0	0.0
least brook lamprey *(Lampetra aepyptera)*	1	6	0	0	0.0
american brook Lamprey *(Lampetra appendix)*	1	1	0	0	0.0
POLYODONTIDAE (paddlefish)					
paddlefish *(Polyodon spathula)*	1	1	0	0	0.0
ANGUILLIDAE (freshwater eels)					
american eel *(Anguilla rostrata)*	1	1	0	0	0.0
AMIIDAE (bowfins)					
bowfin *(Amia calva)*	1	6	0	0	0.0
CYPINODONTIDAE (killifishes)					
blackstripe topminnow *(Fundulus notatus)*	3	43	0	0	0.0
ATHERINIDAE (silversides)					
brook silverside *(Labidesthes sicculus)*	5	107	0	0	0.0
COTTIDAE (sculpins)					
mottled sculpin *(Cottus bairdi)*	4	164	0	0	0.0
TOTALS		102,164	—	2657	—

Note: The percentage of DELT anomalies is listed in descending order by family (based on the species with the highest value) and taxa. No species were deleted due to low numbers. Nomenclature follow Robins et al. (1991).

Creek is surprising, however, because it was the only stream that did not have a fish with two or more (multiple) DELT anomalies. In the other six streams, fish with multiple anomalies represented 3.4 to 7.5% of the total anomalies.

In contrast to the present study results, two similar Ohio studies reported deformities as the predominant anomaly type in three streams surveyed near Mansfield, Ohio. Berra and Au (1981) found spinal curvature followed by deformed fins the most common types of anomalies in Cedar Fork (a small headwater tributary), while Reash and Berra (1989) also reported that deformities were also the most common type of anomalies observed on fishes in the Clear Fork and Rocky Fork (two larger streams within the same watershed). Fournie and Summers (1996) reported skin lesions (mostly fin erosion) as the most common type of anomaly in the Virginian and Louisianian provinces. Gill erosion, however, was apparently included in a category of branchial and gill abnormalities.

Many previous studies have focused on single types of external anomalies (e.g., deformities or eroded fins), which makes their results difficult to compare to the present study (Sherwood, 1982;

TABLE 9.6
Summary Statistics for the Seven Study Area Streams

	Ottawa River	Mahoning River	Ohio River	Salt Creek	Little Miami River	East Fork	Little Beaver Creek
Total no. of fish examined	7052	14,460	41,267	9571	23,456	3674	2684
Total % fish with DELT(s)	8.1	7.1	1.0	0.4	2.1	2.4	1.7
% of total species afflicted	56.5	60.0	37.2	23.1	50.0	38.2	45.8
Minimum % DELT	0.0	0.0	0.0	0.0	0.0	0.0	0.3
Maximum % DELT	57.5	37.9	9.2	3.0	11.7	6.9	5.1
Median % DELT	13.1	8.2	0.9	0.3	1.9	2.9	1.8
Total no. of DELTs + M	574	1025	401	42	481	89	45
% Deformed (of total no. DELTs)	13.6	39.3	28.4	28.6	31.0	27.0	44.4
% Eroded fins and gill covers	77.9	46.6	56.4	50.0	50.9	60.7	40.0
% Lesions (skin)	0.7	8.0	6.7	9.5	12.7	9.0	2.2
% Tumors	0.3	0.8	2.0	11.9	1.2	0.0	6.7
% Multiple DELTs (M)	7.5	5.3	6.5	0.0	4.2	3.4	6.7
DELT severity ratio	0.87	0.75	0.84	0.45	0.46	0.37	1.81

Note: "M" denotes fish with multiple DELT anomalies. The DELT Severity Ratio (based on criteria presented in Table 9.3) is calculated by summing the number of fish with severe and multiple DELTs divided by the number of fish with mild DELTs.

Murchelano and Ziskowski, 1982; Cross, 1985; Bengtsson et al., 1985; Bengtsson and Larsson, 1986; Baumann et al., 1987; Lindesjoo and Thulin, 1990). Eroded fins are, however, a clinical sign for at least two of the most common bacterial infections (Plumb, 1994). Columnaris *(Flavobacterium columnare)* is one of the most common fish diseases frequently associated with fin rot. Another, motile *Aeromonas septicemia* (MAS) has been one of the most frequently diagnosed bacterial diseases of fish based on data compiled by the Fish Disease Committee of the Southern Division of the American Fisheries Society (Plumb, 1994). Post (1987) reports that all freshwater fishes are susceptible to both of these diseases.

9.3.2 PREDOMINANT SPECIES WITH DELT ANOMALIES

The results showed a wide range in the overall percentage of DELT anomalies by species for the 117 taxa collected from the 7 streams (Table 9.5). The taxa with the highest overall percentages of DELTs were muskellunge x northern pike (40.0), black bullhead (33.3), common carp x goldfish hybrid (30.2), carp (28.5), goldfish (23.8), and bigmouth buffalo (21.7). The species with the majority of the total DELT anomalies observed in the study (number of each species with DELTs/total number of fish with DELTs [n = 2657]), however, were common carp (28.2%), followed by channel catfish (8.9%), white sucker (7.3%), green sunfish (6.8%), and gizzard shad (6.4%). These percentages of DELTs for the first four species greatly exceeded their relative abundance based on the total catch (e.g., common carp had 28.2% of the total DELT anomalies, but represented only 2.6% of the catch). The relative abundance of gizzard shad, however, was more than double the percentage of the total number of DELTs (16.1% vs. 6.4).

The most frequently occurring species with DELT anomalies were common carp and largemouth bass, the only two species collected in all seven streams with DELT anomalies. Six other species collected in two or more streams (black buffalo, silver redhorse, quillback, golden redhorse, black redhorse, and channel catfish) and one hybrid (common carp x goldfish) had at least one individual with a DELT anomaly in all streams from which collected.

By family, the highest percentages of afflicted species (with more than one species) were Esocidae (pikes, 100), Catostomidae (suckers, 88.2%), Centrachidae (sunfishes, 85.7), Percichthyidae (temperate basses, 75), Ictaluridae (catfishes, 66.7), and Cyprinidae (carps and minnows, 50%). Based on the mean percent of species within families, families with the highest occurrences were Esocidae (20.8), Ictaluridae (8.9), and Catostomidae (7.4), followed by Cyprinidae (2.9) and Centrarchidae (2.6).

Conversely, no DELT anomalies were observed on 50 taxa, 29 of which were minnows and darters. Other families of fish with more than 10 individuals collected and no observed anomalies included Petrmyzondidae (lampreys), Cyprinodontidae (killifishes), Atherinidae (silversides), and Cottidae (sculpins). In general, the data show a higher percentage of anomalies for the larger, longer-lived, pollution-tolerant taxa than for the smaller, shorter-lived, pollution-sensitive taxa. Many of the highest percentages of anomalies were detected on medium- to large-size bottom-feeding taxa (carp, suckers, and catfish). Many of the smaller benthic species (e.g., darters), however, rarely had an anomaly. Recent field observations support the present study's results that some minnow species are less susceptible to DELT-type anomalies. During the summer of 1996, extra time was used in the examination of DELT anomalies on minnows from one of the Ottawa River's most severely impacted sites (overall percentage of DELT anomalies was 29.2). Results showed four (fathead minnow, spotfin shiner, central stoneroller, and redfin shiner) of the six minnow species collected did not have any DELT anomalies, while all three of the sunfish species collected had at least one DELT anomaly. Additionally, the percentage of DELTs on creek chubs and bluntnose minnows was lower (20.0 and 21.4, respectively) than on bluegills, largemouth bass, and green sunfish (100, 100, and 41.2, respectively). Age may also contribute to the presence of DELT anomalies on various species (e.g., younger fish may have less anomalies because of a shorter exposure time to various stressors). However, observations in the Ottawa River have revealed high percentages of eroded fins on juvenile bluegills.

A previous Ohio study of anomalies on stream fishes reported that white suckers in Cedar Fork had the highest percentage (0.9) of the six species afflicted by anomalies (Berra and Au, 1981). That study also supports the results shown in this study, that even the most susceptible species do not have high percentages everywhere. Also in Ohio, Berra and Au (1981) reported that two abundant species, rainbow darter and creek chub, had no deformities. Elsewhere, Plumb (1994) reports that common carp, channel catfish, and goldfish are particularly susceptible to columnaris disease. Carp and goldfish are also commonly afflicted by *Aeromonas salmonicida achromogens* (also referred to as "atypical nonmotile *Aeromonas*"), which can result in DELT-type anomalies (Plumb, 1994). Fournie and Summers (1996) reported higher rates of abnormalities in demersal fish species (bottom dwelling) than in pelagic or piscivorous fishes.

9.3.3 DELT Anomalies vs. Stream Quality

Regressions of the percentage of DELT anomalies against both the IBI and MIwb showed significant inverse relationships. High percentages of DELTs were associated with poor or very poor quality fish assemblages, while consistently low levels of the anomalies were correlated with very good to exceptional assemblages (Figure 9.2). Measured against the IBI, the percentage of DELT anomalies increased linearly with decreasing fish community performance across a range of scores indicating normal to exceptionally good fish assemblages. However, the number of DELTs increased exponentially at degraded sites (i.e., sites with IBI scores <35). Although the percentage of DELT anomalies is used in the calculation of IBI scores, the percentage can only influence scores by a total of four units. Elevated percentages of DELTs often occur in conjunction with less than full attainment of aquatic life use designations, which requires multiple IBI metrics to deviate from the reference condition and results in deductions of more than 12 units. An IBI analysis (by metric) of an impaired segment of the upper Little Miami River shows impacted fish assemblages scored poorly due to a shift from top carnivores to omnivorous and lower than expected

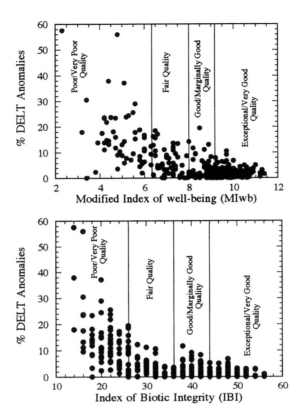

FIGURE 9.2 Scatter plots of the percent DELT anomalies by Modified Index of Well-Being (MIwb, top graph) and Index of Biotic Integrity (IBI, bottom graph) showing the narrative evaluations for stream quality based on Ohio EPA biological criteria.

numbers of intolerant species along with elevated DELT percentages (Figure 9.3). Measured against the MIwb, which is not influenced by anomalies, the percentage of DELTs tended to increase exponentially across the full range scores. The differing nature of regression functions between the IBI and MIwb could be due to the narrow interval of scores representing normal to exceptional communities for the MIwb compared to that for the IBI, and not to differing sensitivities to environmental perturbations.

Previous Ohio EPA reports on study area streams (Ohio EPA, 1992b; 1995; 1996) have shown longitudinal trends in the percentage of DELT anomalies and the IBI and MIwb that also support an inverse relationship. Similar longitudinal trends of low to no DELT anomalies at upstream control or background sites, but increased percentages of DELTs downstream from point source discharges and other pollution sources, are also shown by individual species in different streams (Figure 9.4) and multiple species in a single stream (Figure 9.5). Box plots of the percentages of DELT anomalies from the seven study area streams and 13 other Ohio streams are shown in Figure 9.6. These plots show the percentage of DELT fish is predominantly less than 3.0 in least-impacted streams (and upstream control sites in impacted streams) and greater than 3.0 in streams with multiple point-source discharges and low flows dominated by effluent.

At least one other Ohio study and two coastal studies have reported similar trends between point source discharges and the prevalence of DELT-type anomalies. Reash and Berra (1989) found that unpolluted sites (Clear Fork and Rocky Fork) had a lower percentage of the total catch afflicted by fin erosion and deformities (0.9 and 0.7, respectively) than polluted Rocky Fork sites (10.4) within and downstream from Mansfield. Reash and Berra (1989) also found similar anomaly

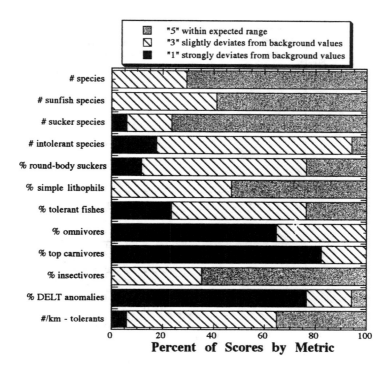

FIGURE 9.3 Metric analysis for Index of Biotic Integrity scores from an impaired segment of the Little Miami River downstream from Xenia, Ohio. Percent of metric values (by metric) scoring a "1", "3", and "5" are shown. Values that score a "1" strongly deviate from the expected are shown in solid black. A predominance of solid black indicates a rather severely degraded component of a fish assemblage. Values that score a "3" slightly deviate from the expected, and a "5" value is within the expected range for the biotic integrity of water resources.

trends for individual species. The percentage of individuals with erosion and deformities for three species markedly increased at the polluted sites within and downstream from Mansfield (creek chub from 0.8 to 7.5, white sucker from 7.9 to 41.1, and green sunfish from 3.6 to 28.3). Lindesjoo and Thulin (1990) also found a clear correlation between an industrial Swedish bleach pulp mill effluent and fin erosion on perch and ruffe with decreasing frequencies at increasing distance from the discharge point. Cross (1985) also reported the highest percentage of fin erosion close to a southern California municipal wastewater outfall and declining rates with increasing distance from the point source.

Compared to upstream background sites, the percent of species afflicted with DELT anomalies also markedly increased downstream from multiple point source discharges in the study area streams. In the Ottawa River, only 13.1% of species had anomalies upstream from Lima (River Mile 46.1) compared to 61.1% downstream from Lima (River Mile 34.7). Table 9.6 shows similar values in the percentage of species afflicted in a rural stream (23.1 in Salt Creek) and a markedly higher level in a highly urbanized industrial stream (60.0% in the Mahoning River). Similar to Salt Creek, Berra and Au (1981) also reported a low percent of total species afflicted (17.6) and overall rate of anomalies (0.26%) in Cedar Fork, a small rural tributary of Clear Fork with a predominantly agricultural watershed.

9.3.4 SEVERITY OF DELT ANOMALIES (SEVERE VS. MILD)

The severity of DELT anomalies also appeared positively related to the degree of impact. Box plots of common carp in four of the streams show markedly lower percentages of severe DELT

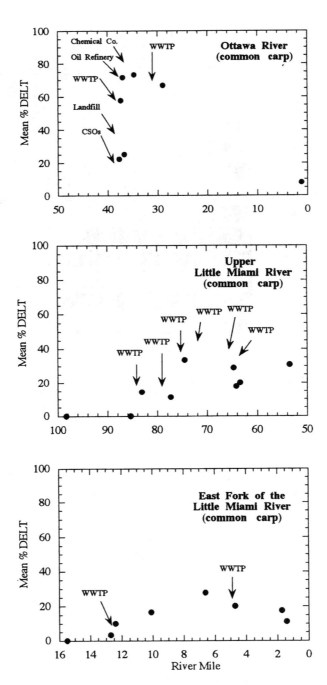

FIGURE 9.4 Longitudinal scatter plots (upstream to downstream) of the mean percentages of common carp with DELT anomalies in the Ottawa River (top), upper Little Miami River (middle), and lower East Fork of the Little Miami River (bottom). Only point source discharges (effluent) of more than 0.75 million gallons per day (MGD) are shown.

anomalies in two streams predominated by municipal WWTP discharges (Little Miami River and its East Fork) than two streams that receive (and historically received) heavy industrial effluent, municipal effluent, and larger quantities of untreated sewage overflows (Mahoning and Ottawa

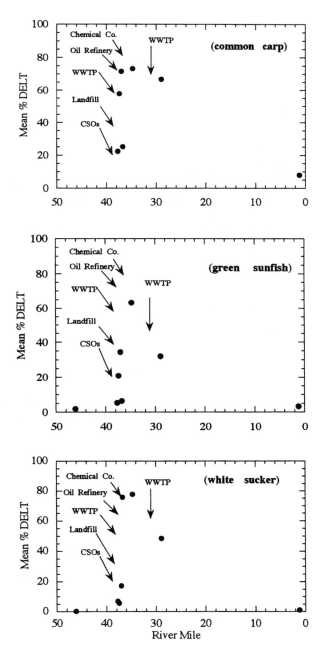

FIGURE 9.5 Longitudinal scatter plots of the mean percentage of common carp (top), green sunfish (middle), and white sucker (bottom) with DELT anomalies in the Ottawa River. Point source dischargers shown had volumes of more than 0.75 million gallons per day (MGD).

Rivers; Figure 9.7). The 75th percentile values in both nonindustrial streams were equal to or less than the 25th percentile values for the two industrial streams. The ratio of fish with severe plus multiple DELT anomalies to mild DELT anomalies also differs (Table 9.6). The least impacted streams primarily influenced by agricultural runoff and municipal wastewater had values < 0.5, while the more severely impacted streams with complex, multiple causes and sources of pollutant-related stresses (e.g., municipal and industrial discharges, combined sewer overflows, urban and

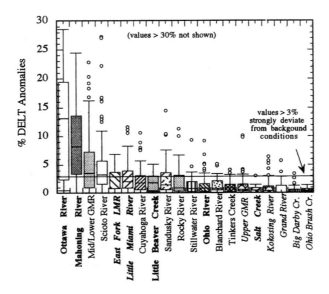

FIGURE 9.6 Box plots (upper, middle, and lower horizontal lines show 75th, 50th, and 25th percentile values, respectively) of the percentages of DELT anomalies in 20 Ohio streams ranked by median values. The seven study area streams are shown in bold. Streams with Exceptional Warmwater Habitat aquatic life use designations are shown in italics.

agricultural runoff, contaminated sediments, and toxics) contained values > 0.5. Yoder and Rankin (1995) developed patterns of response between the IBI, MIwb, and DELT anomalies, termed "biological response signatures," which are combinations of fish community attributes that consistently indicate a general type of environmental or pollution stress. High occurrences of DELT

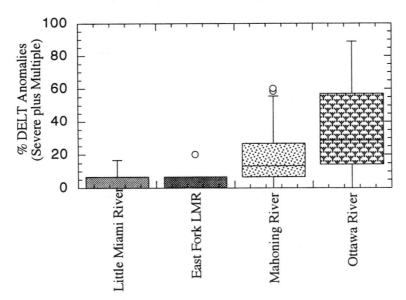

FIGURE 9.7 Box plot (upper, middle, and lower horizontal lines show 75th, 50th, and 25th percentile values, respectively) comparisons of the percentages of fish with severe and multiple DELT anomalies in two nonindustrial streams (Little Miami River and East Fork of the Little Miami River) with multiple sewage plant discharges and two streams with large industrial discharges (Mahoning and Ottawa Rivers).

anomalies (>10–15%) in combination with low IBI and MIwb scores reflecting poor to very poor community condition consistently occurred in stream segments stressed by complex combinations of municipal and industrial sources that frequently exhibited toxicity to bioassay test organisms or exceedences of known chemical toxicity thresholds both in the water column and bottom sediments. A direct relationship between common carp with "knothead" and the severity of pollution in the Illinois River was first reported in 1928 (Thompson, 1928; Mills et al., 1966). In Ohio, common carp with severe knothead (Figure 9.1) have been consistently captured downstream from large industrial effluents (e.g., Mahoning and Ottawa rivers), but have rarely been observed in streams receiving only municipal sewage effluent (e.g., Little Miami River and East Fork of the Little Miami River). The total percent of fish with anomalies in the Ohio River and Beaver Creek was relatively low; however, the severity ratio was high and indicative of impacts from complex industrial and municipal effluents.

Reash and Berra (1989) also found that the severity of fin erosion (based on number of fins afflicted) and the prevalence of fin erosion and deformities were greater at polluted sites than at nonpolluted sites. Additionally, the most common type of anomaly at nonpolluted sites in Clear Fork were fin deformities (usually considered a mild deformity by the Ohio EPA if afflicting only one fin) compared to spinal curvatures (always considered a heavy deformity by the Ohio EPA) at the polluted sites in Rocky Fork. In addition to the severity of individual anomaly types, the presence of multiple types of DELT anomalies on individual fish also appears to be indicative of more severe impacts. Salt Creek was the only stream in the present study with no multiple DELTs recorded. Similarly, Berra and Au (1981) found no individual fish in Cedar Fork to have more than one type of anomaly.

9.4 CONCLUSION

Overall, these results from seven Ohio streams support the findings of previous studies, which have shown or suggested an inverse relationship between the prevalence or percentage of DELT-type anomalies and environmental quality. As used by the Ohio EPA, the percentage of external DELT anomalies in fish assemblages has proven to be an effective IBI metric and accurate indicator of chemical water and sediment quality in streams over a wide range of drainage areas and geographically different basins throughout Ohio. When compared with background levels, elevated occurrences of DELT anomalies in Ohio have been most often found in association with point source discharges, particularly those associated with industrial and municipal wastewater effluents. The incidents of DELTs in wild Ohio fish assemblages has also been useful because conclusions can be made based on both elevated and nonelevated levels. Increased or elevated rates are typically indicative of fish assemblages stressed by chemical pollutants. High bacterial levels (e.g., motile members of the genus *Aeromonas*) may also play an important role in association with sewage-related effluent and combined sewer overflow releases. Low or no anomalies in Ohio typically suggest good chemical water and sediment quality. Biological impairment has been detected in Ohio streams in conjunction with low or no anomalies; however, the cause of impairment was typically due to physical factors (e.g., physical habitat) as opposed to chemical factors. This also corresponds to the findings of Yoder and Rankin (1995).

Two factors that contribute to the usefulness of DELT anomalies as a reliable and accurate indicator of stream quality include the susceptibility of many fish species and a consistently low natural rate at background or reference sites. The results of this study, based on similar longitudinal trends, also suggest that the detection of anomalies could be limited to the more susceptible species (e.g., common carp, white suckers, redhorse species) to reduce the amount of time spent on processing fish and eliminate the possibility of potentially skewing the results by capturing large numbers of species that are not as susceptible (e.g., most minnow species). This would only be recommended, however, if one or more of the selected species were present at all sampling locations. The selection should also include several species from each family to account for different assemblages due to changes in stream size (i.e., headwaters to large rivers). Based on the results of this study, good indicator taxa include the common carp, goldfish, common carp x goldfish hybrid, channel catfish, bullheads, white sucker, redhorses and most other sucker species, green sunfish, largemouth bass, and pike species. Methods may also be improved by not including juvenile fish (e.g., set a minimum length limit) in the DELT rate. The severity of DELT anomalies based on the Ohio EPA's criteria for severe and mild cases also appears important in assessing the complexity and severity of impacted fish assemblages.

After more than 10 years of use in the Ohio EPA's surface water program, the percentage of DELT anomalies has proven to be a reliable indicator of fish community condition. This indicator has responded in an intuitively correct and predictable manner and has been informative across a wide gradient of environmental conditions and stresses. It has been most helpful in identifying sites degraded by multiple and cumulative stresses. Preserved specimens or photographs of fish with DELT anomalies have also proven to be an effective communicator of degraded stream quality to resource managers, the regulated community, and the general public. It is important to realize, however, that elevated occurrences of DELT anomalies may be part of the recovery process for many of the most historically impaired streams. As pollution levels have been reduced in many of Ohio's larger streams, pollution-sensitive fish species have returned; however, many have locally elevated levels of anomalies. The return of these species (even with a DELT anomaly) should be construed as a positive sign of recovery, as well as an indication of continued toxicity or other pollution-related stress(es) and the need for additional pollution abatement measures. With millions of dollars being spent to monitor and restore stream quality throughout the United States, accurate environmental indicators such as DELT anomalies may also prove useful in other states, so that future abatement measures are well directed and result in measurable improvements. Biologists interested in duplicating the fish sampling methods used in this study are encouraged to contact the Ohio EPA.

ACKNOWLEDGMENTS

This study would not have been possible without the staff of the Ecological Assessment Section of the Ohio EPA. The authors would like to thank Dennis Mishne and Charlie Staudt for database management and computer programming. We also thank two anonymous reviewers for providing helpful comments and improving the quality of this manuscript.

REFERENCES

Allison, L.N., J.G. Hnath, and W.G. Yoder. 1977. Manual of Common Diseases, Parasites, and Anomalies of Michigan Fishes. Michigan Department of Natural Resources, Lansing. Fish Management Report Number 8.

Baumann, P.C., W.D. Smith, and W.K. Parland. 1987. Tumor frequencies and contaminant concentrations in brown bullhead from an industrialized river and a recreational lake, *Transactions of the American Fisheries Society,* 116, 79–86.

Bengtsson, B.-E., A. Bengtsson, and M. Himberg. 1985. Fish deformities and pollution in some Swedish waters, *AMBIO,* 14, 32–35.

Bengtsson, B.-E. and A. Larsson. 1986. Vertebral deformities and physiological effects in fourhorn sculpin *(Myoxocephalus quadricornis)* after long-term exposure to a simulated heavy metal-containing effluent, *Aquatic Toxicology,* 9, 215–229.

Berra, T.M. and R. Au. 1978. Incidence of blackspot disease in fishes in Cedar Fork Creek, Ohio, *Ohio Journal of Science,* 78, 318–322.

Berra, T.M. and R. Au. 1981. Incidence of teratological fishes from Cedar Fork Creek, Ohio, *Ohio Journal of Science,* 81, 225–229.

Cross, J. N. 1985. Fin erosion among fishes collected near a southern California municipal wastewater outfall (1972–82), *Fisheries Bulletin,* 83, 195–206.

Esch, G.W. and T.C. Hazen. 1980. Stress and body condition in a population of largemouth bass: implications for red-sore disease, *Transactions of the American Fisheries Society,* 109, 532–536.

Fournie, J.W. and J.K. Summers. 1996. Prevalence of gross pathological abnormalities in estuarine fishes, *Transactions of the American Fisheries Society,* 125, 581–590.

Gammon, J.R. 1976. The Fish Populations of the Middle 340 km of the Wabash River. Purdue University Water Resources Research Center Technical Report 86.

Hickey, C.R. 1972. Common Abnormalities in Fishes, Their Causes and Effects. New York Ocean Science Laboratory, Technical Report 0013.

Karr, J.R. 1981. Assessment of biotic integrity using fish communities, *Fisheries,* 6, 21–27.

Karr, J.R., K.D. Fausch, P.L. Angermier, P.R. Yant, and I.J. Schlosser. 1986. Assessing Biological Integrity in Running Waters: A Method and Its Rationale. Illinois Natural History Survey Special Publication 5.

Komanda, N. 1980. Incidence of gross malformations and vertebral anomalies of natural and hatchery *Plecoglossus altivelis, Copeia,* 1980, 29–35.

Lindesjoo, E. and J. Thulin. 1990. Fin erosion of perch *(Perca fluviatilis)* and ruffe *(Gymnocephalus cernua)* in a pulp mill effluent area, *Diseases of Aquatic Organisms,* 8, 119–126.

Malins, D.C., B.B. McCain, J.T. Landahl, M.S. Myers, M.M. Krahn, D.W. Brown, S.-L. Chan, and W.T. Roubal. 1988. Neoplastic and other diseases in fish in relation to toxic chemicals: an overview, *Aquatic Toxicology,* 11, 43–67.

McCain, B.B., S.-L. Chan, M.M. Krahn, D.W. Brown, M.S. Myers, J.T. Landahl, S. Pierce, R. C. Clark, Jr., and U. Varanasi. 1992. Chemical contamination and associated fish diseases in San Diego bay, *Environmental Science Technology,* 26, 725–733.

Mills, H.B., W.C. Starrett, and F.C. Bellrose. 1966. Man's Effect on the Fish and Wildlife of the Illinois River. Illinois Natural History Survey Biological Notes 57.

Murchelano, R.A. and J. Ziskowski. 1982. Fin rot disease in the New York Bight (1973–1977), in G.F. Mayer (Ed.), *Ecological Stress and the New York Bight: Science and Management.* Estuarine Research Federation, Columbia, SC. p. 347–358.

Neter, J., W. Wasserman, and M.H. Kutner. 1990. *Applied Linear Statistical Models: Regression, Analysis of Variance, and Experimental Designs.* Third ed. Irwin, Inc. Homewood, IL.

Ohio Environmental Protection Agency. 1987a. *Biological Criteria for the Protection of Aquatic Life. Vol. I. The Role of Biological Data in Water Quality Assessment.* Division of Water Quality Monitoring and Assessment, Surface Water Section, Columbus, OH.

Ohio Environmental Protection Agency. 1987b. *Biological Criteria for the Protection of Aquatic Life. Vol. II. Users Manual for Biological Field Assessment of Ohio Surface Waters.* Division of Water Quality Monitoring and Assessment, Surface Water Section, Columbus, OH.

Ohio Environmental Protection Agency. 1989a. Ohio EPA Manual of Surveillance Methods and Quality Assurance Practices, Updated Edition. Division of Environmental Services, Columbus, OH.

Ohio Environmental Protection Agency. 1989b. Addendum to Biological Criteria for the Protection of Aquatic Life. Users Manual for Biological Field Assessment of Ohio Surface Waters. Division of Water Quality Planning and Assessment, Surface Water Section, Columbus, OH.

Ohio Environmental Protection Agency. 1989c. *Biological Criteria for the Protection of Aquatic Life. Vol. III. Standardized Biological Field Sampling and Laboratory Methods for Assessing Fish and Macroinvertebrate Communities.* Division of Water Quality Planning and Assessment, Columbus, OH.

Ohio Environmental Protection Agency. 1992a. State of Ohio Water Quality Standards. Chapter 3745-1 of the Administrative Code. Division of Water Quality Planning and Assessment, Columbus, OH.

Ohio Environmental Protection Agency. 1992b. Biological and Water Quality Study of the Ottawa River, Hog Creek, Little Hog Creek, and Pike Run (Hardin, Allen, and Putnam Counties, Ohio). Division of Water Quality Planning & Assessment, Ecological Assessment Section, Columbus, OH.

Ohio Environmental Protection Agency. 1995. *Biological and Water Quality Study of the Little Miami River and Selected Tributaries (Clark, Greene, Montgomery, Warren, Clermont, and Hamilton Counties, Ohio).* Vol. 1. Division of Surface Water, Monitoring and Assessment Section, Columbus, OH.

Ohio Environmental Protection Agency. 1996. *Biological and Water Quality Study of the Mahoning River Basin (Ashtabula, Columbiana, Portage, Mahoning, Stark, and Trumbull Counties, Ohio; Lawrence and Mercer Counties, Pennsylvania).* Vol. 1. Division of Surface Water, Monitoring and Assessment Section, Columbus, OH.

Ohio River Valley Water Sanitation Commission. 1994. *Ohio River Water Quality Fact Book: A Compendium of Information for Use in Water Quality Analysis of the Ohio River.* ORSANCO, Cincinnati OH.

Pippy, J.H. and G.M. Hare. 1969. Relationship of river pollution to bacterial infection in salmon and suckers, *Transactions of the American Fisheries Society,* 1969, 685–690.

Plumb, J.A., 1994. *Health Maintenance of Cultured Fishes: Principal Microbial Diseases.* CRC Press, Boca Raton, FL.

Post, G. 1987. *Textbook of Fish Health.* T.F.H. Publications, Inc., Neptune City, NJ.

Reash R.J. and T.M. Berra. 1989. Incidence of fin erosion and anomalous fishes in a polluted stream and a nearby clean stream, *Water, Air, and Soil Pollution,* 47, 47–63.

Robins, C.R., C.E. Bond, J.R. Brooker, E.A. Lachner, R.N. Lea, and W.B. Scott. 1991. Common and scientific names of fishes from the United States and Canada, 5th ed. American Fisheries Society Special Publication 20, Bethesda, MD.

Sanders, R. E. 1992. Day versus night electrofishing catches from near-shore waters of the Ohio and Muskingum rivers, *Ohio Journal of Science,* 92, 51–59.

Sanders, R.E. 1995. Ohio's Near-Shore Fishes of the Ohio River: 1991 to 2000 (Year Four: 1994 Results). Division of Surface Water, Monitoring and Assessment Section, Columbus, OH.

Sherwood, M.J. 1982. Fin erosion, liver condition, and trace contaminant exposure in fishes from three costal regions, in G.F. Mayer (Ed.), *Ecological Stress and the New York Bight: Science and Management.* Estuarine Research Federation, Columbia, SC. p. 359–377.

Shotts, E.B., J.L. Gaines, Jr., L. Martin, and A.K. Prestwood. 1972. *Aeromonas*-induced deaths among fish and reptiles in an eutrophic inland lake, *Journal of the American Veterinary Medical Association,* 161, 603–607.

Sniezko, S.F. 1962. The Control of Bacterial and Virus Diseases of Fishes. Biological Problems in Water Pollution, 3rd seminar. U.S. Public Health Service Publication Number 999-WP-25:281-282.

Sniezko, S.F. 1974. The effects of environmental stress on outbreaks of infectious diseases of fishes, *Journal of Fish Biology,* 6, 197–208.

Thompson, D.H. 1928. The "knothead" carp of the Illinois River, *Illinois Natural History Survey Bulletin,* 17, 285–320.

Whittier, T.R., D.P. Larsen, R.M. Hughes, C.M. Rohm, A.L. Gallant, and J.M. Omernik. 1987. The Ohio Stream Regionalization Project: A Compendium of Results. U.S. Environmental Protection Agency, Freshwater Research Laboratory, Corvallis, OR EPA/600/3-87/025.

Yoder, C.O. and E.T. Rankin. 1995. Biological response signatures and the area of degradation value: new tools for interpreting multimetric data, in W. Davis and T. Simon (Eds.), *Biological Assessment and Criteria: Tools for Water Resource Planning and Decision Making.* Lewis, Boca Raton, FL. p. 263–286.

Section III

Regional Applications of the IBI

10 Effects of Drainage Basin and Anthropogenic Disturbance on Relations Between Stream Size and IBI Metrics in Virginia

Roy A. Smogor and Paul L. Angermeier

CONTENTS

10.1 Introduction 249
10.2 Methods 251
    10.2.1 Data Source 251
    10.2.2 Fish Sampling 252
    10.2.3 Habitat Variables and Disturbance Categories 253
    10.2.4 Physiography, Drainage, and Stream Size 254
    10.2.5 Potential IBI Metrics 254
    10.2.6 Statistical Tests and Considerations 256
10.3 Results 260
    10.3.1 Relations at Least-Disturbed Sites 260
    10.3.2 Disturbance Effects on Stream Size-vs.-Metric Relations 261
10.4 Discussion 261
    10.4.1 Interpretation and General Recommendations 261
    10.4.2 Taxonomic Metrics in Virginia 268
    10.4.3 Functional Metrics in Virginia 269
    10.4.4 Further Considerations 269
Acknowledgments 270
References 270

10.1 INTRODUCTION

The Index of Biotic Integrity (IBI) incorporates fish-assemblage attributes (called metrics) that reflect predominant anthropogenic effects on streams (Karr et al., 1986). Each IBI metric describes a particular taxonomic, trophic, reproductive, or tolerance feature of the assemblage (e.g., number of darter species, proportion of individuals as top carnivores, proportion as lithophilous spawners, proportion as members of tolerant species). An IBI score represents comparisons between metric values at a sample site and those expected under conditions least affected by anthropogenic disturbance. These expectations serve as predetermined criteria that are used as standards of comparison for scoring individual IBI metrics; hereafter, these standards are referred to as "metric criteria." If an observed metric value closely matches its criterion value, then the metric is assigned an arbitrary numeric score (typically 5). If the observed value differs moderately from its criterion, then the metric is assigned a lower score (typically 3). If the observed value differs greatly from its criterion (a condition reflecting high anthropogenic disturbance), then the metric is assigned the lowest score (typically 1). The IBI score for a site is simply the sum of these metric scores; a high score represents fish-assemblage attributes similar to those of a least-disturbed assemblage, i.e., high biotic integrity.

Because least-disturbed fish assemblages differ naturally across geographic regions and across environmental gradients within regions (e.g., stream size, temperature; Ohio EPA, 1988; Lyons, 1992; Smogor, 1996), setting appropriate metric criteria (i.e., metric values representative of typical least-disturbed conditions) requires that IBI users understand how fish assemblages vary at the spatial scales pertinent to bioassessment objectives. For example, several IBI programs in the U.S. operate within state boundaries (e.g., Ohio EPA, 1988; Lyons, 1992). At this scale, to ensure accurate IBIs, developers and users need to understand how metric criteria vary among physiographies, ecoregions, and major river drainages, and across stream sizes ranging from small creeks to large rivers.

Users of the IBI typically adjust taxonomic-metric criteria on the basis of empirical relations between stream size and taxa richness (Fausch et al., 1984; Karr et al., 1986; Ohio EPA, 1988; Lyons, 1992); such adjustments seem justifiable given that many studies have documented increases in fish-species richness with increasing stream size (Kuehne, 1962; Sheldon, 1968; Whiteside and McNatt, 1972; Lotrich, 1973; Horwitz, 1978; Evans and Noble, 1979; Platts, 1979). For example, the criterion value for assigning a score of 5 to the metric, total number of species, typically increases with increasing stream size. However, automatic incorporation of such stream-size adjustments in an IBI is unwarranted. Most of the studies just cited examined only one possible IBI metric, total species richness, across a limited range of stream sizes (e.g., first- through fifth-order) in a *single* region, watershed, or drainage. Moreover, many did not account for other sources of variation that potentially obscure, confound, or otherwise affect the expression of stream size-vs.-richness relations, such as differing levels of anthropogenic degradation at sample sites. Accounting for these potential effects can enhance the IBI's intended function, i.e., to solely reflect anthropogenic disturbance effects.

To set appropriate metric criteria for a geologically and hydrologically complex area (e.g., the state of Virginia), one should compare how fish metrics vary with stream size relative to how they differ across regions, drainages, and relevant environmental gradients within each. Recent studies suggest that inter- and intraregional differences in geomorphology, hydrology, or local habitat can obscure or even preclude the expression of stream size-vs.-richness relations (Matthews, 1986; Maurakis et al., 1987; Beecher et al., 1988; Ohio EPA, 1988; Morin and Naiman, 1990; Lyons, 1992). Despite these findings, we know of no IBI study that has explicitly addressed the influence of region, drainage group, and level or type of anthropogenic disturbance on the expression of stream size-vs.-metric relations. Detecting such relations apparently depends on the range of stream sizes being investigated. For example, the Ohio EPA (1988) evidenced strong stream size-vs.-metric relations, but only across stream-size ranges smaller than the entire range considered.

Contrary to taxon-richness patterns, very few published uses of an IBI have explicitly addressed potential stream-size effects on nontaxonomic metrics, e.g., proportional abundance of omnivores, of top carnivores, of simple, lithophilous spawners (hereafter called *functional metrics*). Consequently, very few IBIs have functional metric criteria that differ by stream size (but see Ohio EPA, 1988). Criteria for various functional metrics — typically invariant across stream sizes — commonly are borrowed from previous IBI applications, but rarely are justified further for their appropriateness. Ignoring potential relations between stream size and functional metrics seems contrary to some well-known hypotheses regarding aquatic community organization and contrary to evidence that such relations exist.

The River Continuum Concept (Vannote et al., 1980) explicitly predicts changes in fish trophic structure from headwater streams (typically first- to third-order) to large rivers (typically ≥ seventh-order). Similarly, Horwitz (1978) and Schlosser (1982; 1987) hypothesized a trophic progression from a predominance of generalized-feeding, small fishes in headwaters (first- to third-order), to one of specialized-feeding, large fishes in midsize streams (fourth- to sixth-order), to one of detritivorous/herbivorous fishes in large rivers (seventh- to eighth-order). They postulated that increasing environmental stability with increasing stream size effects increased food availability and biotic interactions, and consequently results in a predictable trophic progression. Paller (1994) provided limited empirical support for this hypothesis: at first- through fourth-order coastal plain streams in South Carolina, he found trophic differences between second-order and larger streams. Although this trophic progression hypothesis remains untested for most stream sizes and most spatial scales broader than within-stream, it seems prudent to explicitly consider these potential effects of stream size when setting IBI trophic-metric criteria.

Based on reasoning similar to that of Horwitz (1978) and Schlosser (1982; 1987), Townsend and Hildrew's (1994) habitat-templet model predicts decreased age at reproduction, body size, potential lifespan, and incidence of parental care in species occurring in spatially and temporally heterogeneous lotic habitats versus those in more homogeneous ones. Therefore, one can reasonably expect that some reproductive or life-history attributes of fish assemblages would differ between smaller, upstream (i.e., more dynamic) and larger, downstream (i.e., more stable) lotic habitats. Consistent with this expectation, Schlosser (1990) found that lifespan, maximum body size, and age at maturity were greatest for large-river (seventh- to twelfth-order) fishes and least for headwater (first- to third-order) fishes of the Illinois River basin.

To date, few reproductive or life-history attributes have been considered for use as IBI metrics, despite their potential for reflecting anthropogenic disturbance (Simon, Chapter 6; Goldstein and Simon, Chapter 7). One commonly used and useful metric — proportional abundance of "tolerant" individuals — broadly incorporates species' reproductive features in the sense that species typically classified as tolerant of anthropogenic disturbance tend to have more generalized life-history or reproductive strategies. Despite the conceptual bases for expecting reproductive, life-history, or tolerance metrics to vary with stream size, we know of only one IBI study (Ohio EPA, 1988) that explicitly addressed such variation. Examining how reproductive and tolerance metrics vary with stream size allows setting appropriate metric criteria, a necessity for an accurate IBI.

This chapter expands on the results of Smogor (1996), who examined how potential IBI metrics varied at least-disturbed sites among physiographies, ecoregions, and drainage groups to determine the most appropriate regional framework for a statewide IBI in Virginia. Based on analyses of interregion variation in metrics, Smogor recommended that an IBI program for Virginia comprise three distinct versions of the index, each adapted for one of three physiographic regions (hereafter, also called "IBI regions"): Coastal Plain, Piedmont, and Mountain (Figure 10.1). For sites in each IBI region, this chapter examines: (1) how individual taxonomic, trophic, reproductive, and tolerance metrics vary with stream size at least-disturbed sites, and (2) how these relations with stream size vary among the major drainage basins in each IBI region. These two steps help reveal the expected natural relations between stream size and metrics within selected regions. Further examination reveals: (3) whether these expected relations differ between least- and most-disturbed sites in each relevant region. Accounting for differences in stream size-vs.-metric relations among regions and disturbance levels, general recommendations are offered regarding which Virginia IBI metrics would require criteria adjusted for stream size.

10.2 METHODS

10.2.1 Data Source

The data used comes from a 1987 to 1990 survey of Virginia stream fishes (Angermeier and Smogor, 1992). Survey data included catch-per-effort of fish species and estimates of selected instream and riparian habitat measures at each of 189 sampling sites. Sites occurred in third- through sixth-order streams across most of the physiographic regions and major river drainages of the state (Figure

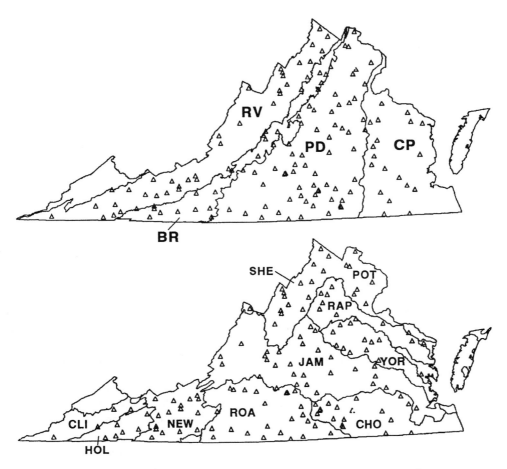

FIGURE 10.1 Maps of Virginia showing the locations of sites sampled in a survey of warmwater streams 1987–1990. Upper map shows physiographic regions: CP = Coastal Plain, PD = Piedmont, BR = Blue Ridge, RV = Ridge and Valley; for analyses, Mountain sites = BR + RV sites. Lower map shows major river drainages: SHE/POT = Shenandoah/Potomac; RAP = Rappahannock; YOR = York; JAM = James; CHO = Nottoway, Meherrin, and Blackwater subdrainages of the Chowan River drainage; ROA = Roanoke; NEW = New; HOL = Holston; CLI = Clinch. A subset, i.e., 114 least- or most-disturbed sites, of these sites was used in analyses.

10.1). Sites ranged from about 50 to 250 m long and drained areas 7 to 454 km²; sampling effort (i.e., stream-surface area sampled) was approximately proportional to stream size.

10.2.2 Fish Sampling

Stream sites that were at least 1 km from major confluences were chosen to limit potential downstream-size "effects" on IBI metrics (e.g., Osborne et al., 1992). Subsequent analysis showed that the stream size of the nearest downstream confluence was related minimally to metric values (Smogor, 1996). At each site, an electric seine (Angermeier et al., 1991) was employed to collect fish in a series of representative channel units (e.g., riffles, runs, pools). For most sites, each habitat unit was blocked bank-to-bank with upstream and downstream nets (0.64-cm mesh). Workers made two seine passes in an upstream direction in each blocked habitat unit. Dipnets (0.64-cm mesh) were used to capture all stunned fish, and the fish were identified as to species. Individuals greater than 100 mm (total length) usually were returned to the stream; smaller individuals and those difficult to identify on-site were preserved and identified in the laboratory.

10.2.3 HABITAT VARIABLES AND DISTURBANCE CATEGORIES

Smogor (1996) ranked 189 statewide sites from least to most disturbed, as part of an initial investigation of regional variation in fish-assemblage attributes. Site rankings were based on 10 selected land-use, riparian, and instream (collectively, "habitat") variables that presumably reflected common anthropogenic disturbance: watershed deforestation, watershed urbanization, watershed mining, and local depletion of well-vegetated riparian zones and of instream cover for fish (see Table 10.1). Hereafter, the term "disturbance" is used to refer to anthropogenic effects on stream-fish assemblages and stream habitat.

Watershed land use and degradation of riparian areas and instream habitat structure can alter fish assemblages via effects on flow regime, energy source, water quality, physical habitat, or biotic interactions (Larimore and Smith, 1963; Smith, 1971; Karr and Schlosser, 1978; Karr and Dudley, 1981; Angermeier and Karr, 1984; Karr et al., 1985; Berkman and Rabeni, 1987; Karr, 1991; Detenbeck et al., 1992; Weaver and Garman, 1994). For example, deforestation and mass removal of riparian woody vegetation increases sunlight, nutrients, and sediments to streams. Increases in temperature, nutrient enrichment, and excessive siltation can alter fish richness and abundance via changes in water quality, altered flows, food availability, spawning substrate, and cover. Moreover, because streams occur in drainage hierarchies, localized habitat disturbances can have much broader, cumulative upstream or downstream effects (e.g., Karr et al., 1985; Pringle, 1997). Although the habitat variables used did not represent all possible anthropogenic effects on fish assemblages in Virginia, they do reflect many of the most-documented and pervasive effects (Jenkins and Burkhead, 1994; Virginia Department of Environmental Quality, 1994).

To rank sites, 10 variables were used in three watershed-scale classes and in one on-site class (Table 10.1). Percent barren, urban, and forest were the proportions of land, as each, in the watershed upstream of each site. Number of mining or urban point sources (of pollution) was the number of coal-mine dumps, tailings ponds, coal tipples, or water or sewage treatment plants in the watershed upstream of each site. All on-site variables were visually estimated or measured coincident with fish sampling. Further sampling details are described by Smogor (1996).

TABLE 10.1
Habitat Variables Used to Determine 63 Least- and 51 Most-Disturbed Sites

Variable	Range
Mining variables	
Percent barren	0–24
Number of mining point-sources	0–13
Urban variables	
Percent urban	0–49
Number of urban point-sources	0–6
Forest variables	
Percent forest	11–100
On-site variables	
Bank erosion	slight–high
Riparian width (in meters)	1–50+
Percent riparian as forest	0–100
Percent bankside as trees or shrubs (i.e., woody cover)	5–100
Percent instream woody cover	0–92

Note: Habitat variables are arranged in three watershed-scale and one on-site classes. The range of each habitat variable across all sites is shown.

For habitat variables in each of the four classes, sites were ranked by each variable; higher ranks indicated less-disturbed conditions. Then, for each site, ranks were summed across variables to yield a sum for each class. The entire range of class sums was divided into thirds, and each site was assigned a standardized class score (of 1, 2, or 3) that represented the site's membership in the least-, moderately, or most-disturbed third of the class. Finally, for each site, the four standardized class scores were summed to yield a total score, which ranged potentially from 4 to 12. By standardizing class scores, equal weight was given to each disturbance class because there was little *a priori* evidence to suggest otherwise. The highest-scoring 66 sites were categorized as least-disturbed, and the lowest-scoring 53 sites as most-disturbed. The sites were screened for differences in fish-sampling effort, and five sites that were relatively undersampled were omitted; the five lay farthest below a regression line fitted to a plot of mean site width versus site length sampled. The final sample comprised 63 least- and 51 most-disturbed sites.

No single class of habitat variables dominated the ranking scheme; there were no strong correlations between any possible pair of standardized class scores (all absolute values of Spearman's rho [r_s] < 0.45). Furthermore, individual habitat variables were not excessively redundant. For example, among least-disturbed sites, only one correlation between habitat variables had an absolute value of $r_s > 0.50$.

10.2.4 PHYSIOGRAPHY, DRAINAGE, AND STREAM SIZE

The physiographic region of each site was determined using a coverage (ArcInfo) of physiographic boundaries that were digitized from 1:24,000 U.S. Geological Survey topographic maps. A few sites lay on or very near physiographic boundaries; a physiography was assigned to them, based on personal on-site observations of surrounding landforms and vegetation, stream gradient, and instream mineral substrates. The major river drainage of each site was determined by using a U.S. Geological Survey Hydrologic Unit Map of Virginia (1:500,000; see Figure 10.1). Table 10.2 shows numbers of sites in selected combinations of region, disturbance category, and major river drainage. Because sites were unevenly distributed across these categories, drainage-basin groups (hereafter called "basins") were formed to allow adequate sample sizes for analyses (Table 10.2). Watershed area upstream of each site served as a measure of stream size. Watershed area was calculated by computer program (ArcInfo), from watershed boundaries digitized from U.S. Geological Survey topographic maps (1:24,000).

10.2.5 POTENTIAL IBI METRICS

We examined metrics that presumably reflect effects of typical anthropogenic disturbance on fish communities *and* are relatively easy to determine from field data (Karr et al., 1986; Miller et al., 1988; Fausch et al., 1990; Simon and Lyons, 1995). Some of the metrics examined have been used widely in IBI analyses; others were judged potentially useful, although they had not been considered previously. For metrics that were counts of species (e.g., number of native darters; see Table 10.3), square-root transformed values ($[X+0.5]^{0.5}$) were employed for analyses. For metrics that were proportions, arcsine transformed values (arcsine $X^{0.5}$) were used. Only those metrics that varied enough in each physiography and that had approximately normal distributions (after transformation) were incorporated. For example, not included in the analyses was "number of intolerant species" (see below, in this Section 10.2.5), a commonly used IBI metric, because too few sites had intolerant species. Similarly excluded was "number of native sucker species" because few sucker species occur in Virginia's Coastal Plain (Jenkins and Burkhead, 1994) and because distributions of this metric across Piedmont and Mountain sites were asymmetric with extreme outliers.

Taxonomic, trophic, reproductive, and tolerance classifications of species (Table 10.4) were based on various regional texts (e.g., Pflieger, 1975; Jenkins and Burkhead, 1994) and on personal

TABLE 10.2
Number of Sites, by Physiographic Region, in Each Disturbance Category (Least- vs. Most-Disturbed) on Each Major River Drainage or Basin

	Group	CP	PD	MT	Piedmont Least-	Piedmont Most-	Mountain Least-	Mountain Most-
Disturbance category								
Least-disturbed		9	29	25				
Most-disturbed		6	17	28				
Major river drainage								
Shenandoah/Potomac	CHE	1	6	15	0	6	8	7
Rappahannock/York	CHE	5	8	3	5	3	3	0
James	CHE	3	10	10	6	4	7	3
CHE total		9	24	28	11	13	18	10
Roanoke/Nottoway/Blackwater	ALB	6	22	3	18	4	1	2
New	OHIO	0	0	12	0	0	4	8
Holston/Clinch	OHIO	0	0	10	0	0	2	8
OHIO total		0	0	22	0	0	6	16

Note: CP = Coastal Plain; PD = Piedmont; MT = Mountain. Drainage-basin groups (Group) used for analyses are: CHE = Chesapeake Bay-basin streams, ALB = Albemarle Sound-basin streams, and OHIO = Ohio River-basin streams. The three Mountain sites in the Albemarle basin were excluded from one-way ANCOVAs.

TABLE 10.3
Fish Metrics Used in Analyses of Relations with Stream Size

Metrics	Code
Taxonomic metrics	
Number of native species	NATSP
Number of native minnow species	MINSP
Number of sunfish species	SUNSP
Number of native darter species	DARSP
Number of nonnative species	NONNATSP
Trophic metrics	
Proportion as generalist feeders	GENPRP
Proportion as benthic, specialist invertivores	BINVPRP
Proportion as specialist carnivores	CARNPRP
Reproductive metrics	
Proportion as mineral-substrate, simple spawners, i.e., simple lithophils	SLITHPRP
Proportion as various-substrate, manipulative spawners, i.e., generalist spawners	VMANPRP
Number of late-maturing (>2 yr) species	AGE3SP
Tolerance metric	
Proportion as members of tolerant species	TOLPRP

Note: Metrics are arranged in four classes: taxonomic, trophic, reproductive, and tolerance. "Proportion" is proportion of individuals. Subsequent tables refer to metrics by their codes.

experience. Native versus non-native status (by major river drainage) follows Jenkins and Burkhead (1994). Number of native minnows comprised all native species in the family Cyprinidae, and number of sunfish comprised all species in Centrarchidae. Number of native sunfish was not used as a metric because many sites had no natives. Number of native darter species comprised all native *Percina* spp. or *Etheostoma* spp.

"Tolerant" species were classified as those generally known to be affected least detrimentally by typical anthropogenic disturbances to streams and watersheds; many of these species' historical ranges or local proportional abundances have increased with increases in anthropogenic disturbance (e.g., common carp, *Cyprinus carpio*; bluntnose minnow *Pimephales notatus*; green sunfish, *Lepomis cyanellus*). Each species' tolerance rating was determined before ranking sites from least- to most-disturbed in order to minimize bias in classifications (Smogor, 1996). Karr et al. (1986) recommended that less than 10% of the species in an IBI region be classified as "intolerant." Similarly, it is suggested that any IBI "tolerance" classification be limited to a small percentage of the included species; this ensures that the "tolerance" metric reflects exclusively the lowest end of the biotic integrity continuum, i.e., only those severely disturbed sites dominated by tolerant species or individuals. Only 12% of all Virginia species sampled were classified as tolerant.

For trophic metrics, classification of each species was based on three factors: number of food types typically eaten, feeding behavior, and feeding group. Four food types were designated: (1) detritus, (2) algae or vascular plants, (3) invertebrates, and (4) fish (including fish blood) or crayfish. "Generalist feeders" were species in which adults eat three or four food types; "specialists" eat one or two types. Two mutually exclusive feeding behaviors, benthic and non-benthic, were designated. Benthic species feed, as adults, mostly along the stream bottom and have specialized morphology (e.g., inferior mouth) for doing so. Fish species were assigned to one of five feeding groups based on the primary food type(s) of subadults and adults: (1) detritivore/algivore/herbivore, (2) algivore/herbivore/invertivore, (3) invertivore, (4) invertivore/piscivore, (5) piscivore or fish parasite. Group 4 comprised species in which subadults eat primarily invertebrates, but adults eat primarily fish or crayfish (e.g., American eel, *Anguilla rostrata*; crappies, *Pomoxis spp.*; yellow perch, *Perca flavescens*). "Carnivores" were species in groups 4 or 5 (Table 10.3).

For reproductive metrics, one life-history trait and two classification factors were considered: species' age at maturity, spawning substrate, and spawning behavior. "Late-maturing species" are defined as those in which females typically do not spawn before 3 years of age. Four spawning-substrate categories are used: (1) uses no substrate or has pelagic eggs, (2) typically uses vegetation or organic debris, (3) known to use various substrates and not restricted to unsilted mineral substrates, (4) obligately uses unsilted mineral substrates, i.e., lithophilic. Also designated were "manipulative" vs. "simple" (nonmanipulative) spawners (Table 10.3). "Manipulative spawners" build nests, depressions, or cavities, *or* actively guard eggs or young (e.g., lampreys, catfishes, sunfishes). "Simple spawners" exhibit relatively little nest preparation or parental care.

10.2.6 STATISTICAL TESTS AND CONSIDERATIONS

For least-disturbed sites in Piedmont and in Mountain, we used one-way analysis of covariance (ANCOVA, SAS 1990) to examine how each metric varied between two basins, simultaneously with stream size (i.e., watershed area as covariate). Small sample sizes precluded adequate multivariate analyses. Small sample sizes also precluded adequate parametric analyses for Coastal Plain sites. Therefore, for least-disturbed Coastal Plain sites, bivariate correlations (Spearman's rho) between each metric and stream size, in each of two basins, were examined.

Based on results from least-disturbed sites in each physiography, we then examined how relations between stream size and selected metrics differed between least- and most-disturbed sites. Hereafter, variation in metrics is referred to as the "effects" of stream size, of basin, or of disturbance on each metric.

TABLE 10.4
Taxonomic, Trophic, and Reproductive Classifications of Fish Species That Were Sampled at 143 Wadeable Stream Sites in Virginia, 1987–1990

Species	TRO	NUM	BEN	AGE	SUB	MANIP	TOL
Petromyzontidae							
Ichthyomyzon greeleyi	DAH	2	+	6	MIN	+	I
Lampetra aepyptera	DAH	2	+	6	MIN	+	
Lampetra appendix	DAH	2	+	5	MIN	+	
Petromyzon marinus	PIS	1		9	MIN	+	
Amiidae							
Amia calva	PIS	1		4	VEG	+	
Clupeidae							
Alosa aestivalis	INV	1		4	NON		
Dorosoma cepedianum	AHI	2		2	VAR		T
Salmonidae							
Oncorhynchus mykiss	IP	2		1	MIN	+	
Salmo trutta	IP	2		1	MIN	+	
Salvelinus fontinalis	IP	2		2	MIN	+	I
Anguillidae							
Anguilla rostrata	IP	2		5	NON		
Esocidae							
Esox americanus	PIS	1		2	VEG		
Esox lucius	PIS	1		2	VEG		
Esox niger	PIS	1		2	VEG		
Umbridae							
Umbra pygmaea	INV	1		1	VAR	+	
Cyprinidae							
Campostoma anomalum	DAH	2	+	2	MIN	+	
Clinostomus funduloides	INV	1		2	MIN		
Cyprinus carpio	AHI	4		3	VAR		T
Cyprinella analostana	INV	2		1	VAR		
Cyprinella galactura	INV	2		2	VAR		
Cyprinella spiloptera	INV	3		1	VAR		
Erimystax insignis	AHI	3	+	1	MIN		
Exoglossum laurae	INV	1		2	MIN	+	I
Exoglossum maxillingua	INV	1		2	MIN	+	
Hybognathus regius	DAH	2		2	VAR		
Luxilus albeolus	INV	1		1	MIN		
Luxilus cerasinus	INV	2		2	MIN		
Luxilus chrysocephalus	INV	4		2	MIN		
Luxilus coccogenis	INV	1		2	MIN		
Luxilus cornutus	INV	4		2	MIN		
Lythrurus ardens	INV	3		1	MIN		
Lythrurus lirus	INV	1		1	MIN		
Margariscus margarita	INV	3		1	MIN	+	
Nocomis leptocephalus	AHI	3		3	MIN	+	
Nocomis micropogon	INV	3		3	MIN	+	
Nocomis platyrhynchus	INV	3		3	MIN	+	
Nocomis raneyi	INV	3		3	MIN	+	
Notemigonus crysoleucas	AHI	2		2	VAR		T
Notropis alborus	INV	2		1	MIN		

TABLE 10.4 *(continued)*
Taxonomic, Trophic, and Reproductive Classifications of Fish Species That Were Sampled at 143 Wadeable Stream Sites in Virginia, 1987–1990

Species	TRO	NUM	BEN	AGE	SUB	MANIP	TOL
Notropis altipinnis	INV	2		1	VAR		
Notropis amblops	INV	1	+	1	MIN		I
Notropis amoenus	INV	1		1	MIN		
Notropis buccatus	AHI	3	+	1	MIN		T
Notropis chalybaeus	INV	2		1	VAR		
Notropis chiliticus	INV	2		1	MIN		
Notropis hudsonius	INV	2		2	VAR		
Notropis leuciodus	INV	1		1	MIN		
Notropis photogenis	INV	2		1	MIN		
Notropis procne	INV	2		2	MIN		
Notropis rubricroceus	INV	2		1	MIN		
Notropis rubellus	INV	1		1	MIN		
Notropis scabriceps	INV	1		2	MIN		
Notropis semperasper	INV	1		2	MIN		
Notropis spectrunculus	INV	1		1	MIN		
Notropis stramineus	INV	3		1	MIN		
Notropis telescopus	INV	1		2	MIN		
Notropis volucellus	INV	3		1	VAR		
Phenacobius teretulus	INV	1	+	2	MIN		
Phoxinus oreas	DAH	3		1	MIN		
Pimephales notatus	AHI	3		1	VAR	+	T
Pimephales promelas	AHI	3		1	VAR	+	T
Rhinichthys atratulus	INV	3		2	MIN		T
Rhinichthys cataractae	INV	2		2	MIN		
Semotilus atromaculatus	IP	4		1	MIN	+	T
Semotilus corporalis	IP	4		2	MIN	+	
Catostomidae							
Catostomus commersoni	AHI	3	+	3	MIN		T
Erimyzon oblongus	INV	3	+	2	VAR		
Hypentelium nigricans	INV	2	+	3	MIN		
Hypentelium roanokense	INV	3	+	2	MIN		
Moxostoma anisurum	INV	3	+	5	MIN		
Moxostoma ariommum	INV	2	+	3	MIN		I
Moxostoma cervinum	INV	3	+	2	MIN		
Moxostoma duquesnei	INV	3	+	3	MIN		
Moxostoma erythrurum	INV	3	+	4	MIN		
Moxostoma hamiltoni	AHI	3	+	3	MIN		
Moxostoma macrolepidotum	INV	3	+	4	MIN		
Moxostoma pappillosum	INV	3	+	4	MIN		
Moxostoma rhothoecum	AHI	3	+	3	MIN		I
Ictaluridae							
Ameiurus brunneus	IP	3		3	VAR	+	
Ameiurus catus	IP	3		3	MIN	+	
Ameiurus melas	IP	3		2	MIN	+	T
Ameiurus natalis	IP	3		2	VAR	+	
Ameiurus nebulosus	IP	3		3	VAR	+	
Ameiurus platycephalus	IP	3		3	VAR	+	

TABLE 10.4 (continued)
Taxonomic, Trophic, and Reproductive Classifications of Fish Species That Were Sampled at 143 Wadeable Stream Sites in Virginia, 1987–1990

Species	TRO	NUM	BEN	AGE	SUB	MANIP	TOL
Ictalurus punctatus	IP	3		3	VAR	+	
Noturus flavus	INV	2		3	MIN	+	
Noturus gyrinus	INV	1	+	2	VAR	+	
Noturus insignis	INV	2		3	MIN	+	
Aphredoderidae							
Aphredoderus sayanus	INV	2		1	VAR	+	
Cyprinodontidae							
Fundulus catenatus	INV	1		1	MIN		
Fundulus diaphanus	INV	1		1	VAR		
Fundulus heteroclitus	INV	2		1	VAR		
Fundulus rathbuni	INV	1		1	VAR		
Poeciliidae							
Gambusia affinis	INV	1		0	NON		T
Cottidae							
Cottus bairdi	INV	1	+	2	VAR	+	
Cottus baileyi	INV	1	+	2	VAR	+	
Cottus carolinae	INV	1	+	2	VAR	+	
Cottus cognatus	INV	1	+	2	VAR	+	
Cottus girardi	INV	1	+	2	VAR	+	
Cottus sp.	INV	1	+	2	VAR	+	
Cottus sp.	INV	1	+	2	VAR	+	
Moronidae							
Morone americana	IP	2		2	MIN		
Centrarchidae							
Acantharchus pomotis	INV	2		2	VAR	+	
Ambloplites cavifrons	IP	2		2	MIN	+	I
Ambloplites rupestris	IP	2		2	MIN	+	
Centrarchus macropterus	INV	2		2	VAR	+	
Enneacanthus gloriosus	INV	1		2	VAR	+	
Enneacanthus obesus	INV	1		2	VAR	+	
Lepomis auritus	IP	2		2	MIN	+	
Lepomis cyanellus	IP	2		1	VAR	+	T
Lepomis gibbosus	INV	1		1	VAR	+	
Lepomis gulosus	IP	2		1	VAR	+	
Lepomis macrochirus	INV	1		1	VAR	+	T
Lepomis megalotis	INV	1		2	MIN	+	
Lepomis microlophus	INV	1		2	VAR	+	
Micropterus dolomieu	IP	2		2	MIN	+	
Micropterus punctulatus	IP	2		2	VAR	+	
Micropterus salmoides	PIS	1		2	VAR	+	
Pomoxis annularis	IP	2		2	VAR	+	
Pomoxis nigromaculatus	IP	2		2	VAR	+	
Percidae							
Etheostoma blennioides	INV	1	+	2	VAR		
Etheostoma caeruleum	INV	1	+	1	MIN		
Etheostoma collis	INV	1	+	1	VAR		
Etheostoma flabellare	INV	1	+	2	MIN	+	T

TABLE 10.4 *(continued)*
Taxonomic, Trophic, and Reproductive Classifications of Fish Species That Were Sampled at 143 Wadeable Stream Sites in Virginia, 1987–1990

Species	TRO	NUM	BEN	AGE	SUB	MANIP	TOL
Etheostoma fusiforme	INV	1	+	1	VAR		T
Etheostoma kanawhae	INV	1	+	2	MIN		
Etheostoma longimanum	INV	1	+	1	MIN	+	
Etheostoma nigrum	INV	1	+	1	VAR	+	T
Etheostoma olmstedi	INV	1	+	1	VAR	+	T
Etheostoma podostemone	INV	1	+	1	MIN	+	
Etheostoma rufilineatum	INV	1	+	1	MIN		
Etheostoma serrifer	INV	1	+	1	VAR		
Etheostoma simoterum	INV	1	+	1	VAR		
Etheostoma vitreum	INV	1	+	1	VAR		
Etheostoma zonale	INV	1	+	1	VAR		
Percina caprodes	INV	1	+	2	MIN		
Percina gymnocephala	INV	1	+	2	MIN		
Percina notogramma	INV	1	+	2	MIN		
Percina oxyrhynchus	INV	1	+	1	MIN		
Percina peltata	INV	1	+	2	MIN		
Percina roanoka	INV	1	+	2	MIN		
Perca flavescens	IP	2		3	VAR		

Note: Trophic groups (TRO) are: DAH = detritivore/algivore/herbivore, AHI = algivore/herbivore/invertivore, INV = invertivore, IP = invertivore/piscivore, and PIS = piscivore or parasite. Number of food types (NUM) shows number of types typically eaten: (a) detritus, (b) algae or vascular plants, (c) invertebrates, or (d) fish or blood. Benthic feeders (BEN) are shown with a "+." Female age at reproduction (in years) is shown as "AGE." Spawning substrates (SUB) are: NON = none or pelagic, VEG = vegetation or organic debris, VAR = not restricted to particular substrates, and MIN = restricted to unsilted mineral substrates from sand to boulder. Nest preparers or parental-care givers (MANIP) are shown with a "+." Simple spawners are species that exhibit no nest preparation or parental guarding/care (i.e., MANIP not a "+"). For example, simple lithophils are species with "MIN" for SUB and a blank for "MANIP." Tolerant species are shown as "T" and intolerant species as "I" for variable, TOL. See Section 10.2 for further explanation.

Overall, distributions of metrics and their residuals were nearly symmetric and had no extreme outliers; however, small sample sizes of some groups increased the potential for nonhomogeneity of variances. Moreover, pronounced interaction (e.g., watershed area × basin) occurred in several of the ANCOVAs, which indicated possible violation of the assumption of equal stream size-vs.-metric slopes among statistical groups. For these reasons, we also examined stream size-vs.-metric relations by using bivariate correlations (Spearman's rho); ANCOVAs were supplemented by examining plots of metrics vs. watershed area: (1) for sites in each physiography and (2) within physiographies, for sites in each basin, disturbance category, or disturbance category within basin.

10.3 RESULTS

10.3.1 RELATIONS AT LEAST-DISTURBED SITES

At least-disturbed sites in each physiography, there were few consistencies in how metrics varied with stream size and between basins. Stream-size effects (including interactions with basin) largely predominated over sole basin effects. Sole stream-size effects were most pronounced at Piedmont and at Mountain sites: more native species occurred at larger sites and vice versa (Table 10.5; Figures 10.2 and 10.3), consistent with much previous justification for adjusting this metric for stream size (e.g., Fausch et al., 1984; Karr et al., 1986). Relations between stream size and the other taxonomic metrics were mostly inconsistent across physiographies or across basins in each physiography (Table 10.5; Figures 10.2 and 10.3). For example, number of native darters was greater at larger Piedmont and larger Coastal Plain sites in Chesapeake basin (Figures 10.2 and 10.4); however, the opposite relation held for Chesapeake-basin sites in Mountain (Figure 10.3) and for Albemarle-basin sites in Coastal Plain (Figure 10.4).

Relations between stream size and taxonomic metrics predominated only at Mountain sites, where only one functional metric, proportion as specialist carnivores, was related with stream size (Table 10.5; Figure 10.3). In contrast, in Piedmont and in Coastal Plain, relations between stream size and functional metrics predominated (Table 10.5; Figures 10.2 and 10.4), a result inconsistent with the prevailing IBI emphasis to use invariant (with stream size) functional-metric criteria.

10.3.2 Disturbance Effects on Stream Size-vs.-Metric Relations

Stream size relations for the two taxonomic metrics that varied most with stream size in Piedmont, number of native species and of native darter species, held for sites considered across the entire range of disturbance (Table 10.6; Figure 10.2). However, for three of the four functional metrics that varied most with stream size in Piedmont (i.e., VMANPRP, AGE3SP, TOLPRP in Table 10.5), the relations were most pronounced at least-disturbed sites. Similarly, for Mountain sites, stream size-vs.-metric relations held, across disturbance categories, for all four taxonomic metrics exhibiting such relations (i.e., NATSP and SUNSP in both basins; DARSP and NONNATSP each in a single basin); whereas, the single functional metric (CARNPRP) was related to stream size only at least-disturbed sites (Table 10.6; Figure 10.3). The somewhat unexpected negative relation for darter species at least-disturbed Chesapeake basin sites (lower left plot in Figure 10.3) resulted from the four largest, least-disturbed sites being in the Shenandoah River drainage, a drainage notably depauperate in native darter species compared to most other Chesapeake basin drainages (Jenkins and Burkhead, 1994). For Coastal Plain sites, the four strongest relations between stream size and functional metrics occurred at least-disturbed sites (Figure 10.4). In summary, in each physiography, stream-size relations with functional metrics were evident almost exclusively at least-disturbed sites; whereas, most relations with taxonomic metrics held across disturbance categories.

10.4 DISCUSSION

10.4.1 Interpretation and General Recommendations

Very few IBI studies have explicitly examined the appropriateness of each metric and its criteria with respect to how the metric varies across spatial scales, environmental gradients (e.g., stream size, temperature, stream gradient), and disturbance levels relevant to the intended bioassessment. Contrary to prior IBI emphases, our results showed that only a few taxonomic but several functional metrics varied with stream size and that most of these relations differed among as well as within IBI regions. Despite conceptual arguments and some prior evidence to the contrary, we found few generally applicable patterns in the way taxonomic or functional metrics varied with stream size. Therefore, until well-supported, generally applicable patterns emerge, an empirical approach to developing and using metric criteria and the IBI is recommended. First, IBI regions and relevant environmental gradients in each region should be explicitly defined and justified. Then, for each IBI region, metric criteria and their adjustments should be determined by examining empirical relations between each metric and each environmental gradient. Adjusting IBI metrics to account for natural variation in relations with stream size requires careful consideration, and one should not

TABLE 10.5
Summaries of One-Way ANCOVAs of Each Fish Metric, At Least-Disturbed Sites, Between Two Basins in Piedmont or in Mountain

	R²	F	P	EFFECT	SS	F	P
			Piedmont (N = 29)				
NATSP	0.24	2.62	0.0728	WAREA	1.60	7.82	0.0098
				BASIN	0.05	0.22	0.6426
				WAR × BAS	0.04	0.20	0.6562
MINSP	0.17	1.77	0.1792				
SUNSP	0.09	0.84	0.4858				
DARSP	0.33	4.10	0.0170	WAREA	0.52	8.16	0.0085
				BASIN	0.31	4.76	0.0387
				WAR × BAS	0.25	3.85	0.0609
NONNATSP	0.09	0.87	0.4682				
GENPRP	0.22	2.30	0.1023	WAREA	0.02	0.85	0.3667
				WAR × BAS	0.02	0.79	0.3839
				BASIN	0.01	0.22	0.6413
BINVPRP	0.05	0.47	0.7034				
CARNPRP	0.10	0.90	0.4559				
SLITHPRP	0.02	0.19	0.9006				
VMANPRP	0.31	3.74	0.0240	WAREA	0.21	5.45	0.0279
				BASIN	0.21	5.46	0.0278
				WAR × BAS	0.16	4.22	0.0506
AGE3SP	0.42	6.07	0.0030	WAREA	36.5	17.71	0.0003
				BASIN	1.16	0.56	0.4604
				WAR × BAS	0.64	0.31	0.5820
TOLPRP	0.24	2.56	0.0777	WAREA	0.15	7.61	0.0107
				BASIN	0.01	0.34	0.5657
				WAR × BAS	0.01	0.29	0.5978
			Mountain (N = 25)				
NATSP	0.28	2.56	0.0838	WAREA	1.09	6.45	0.0195
				BASIN	0.15	0.91	0.3515
				WAR × BAS	0.14	0.81	0.3793
MINSP	0.19	1.54	0.2341				
SUNSP	0.35	3.65	0.0302	BASIN	0.53	4.56	0.0453
				WAR × BAS	0.45	3.83	0.0644
				WAREA	0.17	1.47	0.2399
DARSP	0.31	2.98	0.0560	WAR × BAS	0.43	5.01	0.0368
				BASIN	0.36	4.21	0.0535
				WAREA	0.00	0.02	0.8835
NONNATSP	0.24	2.16	0.1251	BASIN	0.43	2.27	0.1471
				WAR × BAS	0.25	1.33	0.2620
				WAREA	0.04	0.23	0.6373
GENPRP	0.23	1.97	0.1517	WAREA	0.06	2.36	0.1401
				WAR × BAS	0.05	1.73	0.2030
				BASIN	0.02	0.91	0.3519
BINVPRP	0.03	0.23	0.8725				
CARNPRP	0.45	5.54	0.0062	WAREA	0.16	5.88	0.0249
				WAR × BAS	0.04	1.52	0.2316
				BASIN	0.03	1.23	0.2813
SLITHPRP	0.07	0.51	0.6801				
VMANPRP	0.07	0.49	0.6944				

TABLE 10.5 (continued)
Summaries of One-Way ANCOVAs of Each Fish Metric, At Least-Disturbed Sites, Between Two Basins in Piedmont or in Mountain

	R^2	F	P	EFFECT	SS	F	P
AGE3SP	0.16	1.25	0.3190				
TOLPRP	0.03	0.23	0.8710				

Note: Watershed area (WAREA or WAR) is the covariate. For each ANCOVA, R^2 is the amount of variance in each metric that was accounted for simultaneously by basin (BASIN or BAS) and stream-size effects. For each ANCOVA with $R^2 > 0.20$, all possible effects are shown in order of strength (based on Type III sums of squares [SS]). Also included are F-test value (F) and statistical probability (P) for each test. Degrees of freedom for each omnibus test were 3 for model, 25 for error for PD; and 3,20 for MT. See Table 10.3 for codes of metrics.

expect that such adjustments can be universally applied. Notwithstanding the potential advantages of adjusting some metrics for stream size, one should be cautious in applying such adjustments without adequate justification. Many previous IBI applications have simply borrowed metric criteria and adjustments that were based on relations for streams outside the regions intended for assessment. The results presented in this chapter suggest that such practice can unnecessarily confound an IBI-based assessment by failing to account for variation specific to each metric in each region considered.

A major advantage of adjusting metric criteria for stream size is that adjusted criteria increase the metric's ability to reflect anthropogenic rather than stream-size effects. For example, relations for Piedmont sites in Virginia (see Figure 10.2) show that proportions of tolerants or of generalist spawners as high as 0.50 may nonetheless represent relatively undisturbed conditions in smaller streams (i.e., <25 km^2); whereas, such proportions in larger streams (i.e., >150 km^2) would probably indicate anthropogenic degradation. Adjusting the criteria of these two functional metrics would enhance the ability of a Piedmont IBI to reflect conditions solely attributable to disturbance.

When examining stream size-vs.-metric relations to determine appropriate metric criteria, one should try to include data from the entire range of stream sizes likely to be assessed; relations may not be constant across this range. Our results for Piedmont (Figure 10.2) and for Mountain (Figure 10.3) sites showed that species richness tended to increase with stream size, but then level off in the largest streams. The Ohio EPA (1988) and Lyons (1992) evidenced similar patterns. The stream size at which this leveling-off occurs can vary by metric among geographic regions (e.g., Lyons, 1992; Paller, 1994), emphasizing the need to account not only for region-specific variation, but also for that region-specific variation evident across the range of stream sizes relevant to bioassessment objectives. We judge that our results and interpretations would apply to Virginia warmwater wadeable streams that range up to about 400 km^2, and caution against extrapolation to larger streams that would require sampling with non-wading gears. We also recommend examining more large (i.e., >250 km^2) streams in PD and in MT, and more small (i.e., first - and second-order) streams in CP, to help determine definitive metric criteria for Virginia IBIs.

When examining stream size-vs.-metric relations to determine appropriate metric criteria, one should first consider only those relations that represent a typical, least-disturbed fish assemblage. Because little is known about how disturbance effects on metrics may vary across stream sizes, we believe that stream size-vs.-metric relations representative of least-disturbed conditions are the most reasonable standard for adjusting metric criteria. Aggregating data from sites representing different levels of disturbance can confound attempts to discern the truly "least-disturbed" relations between stream size and metrics. For example, our results showed that stream size-vs.-metric relations, especially for functional metrics, were most evident at least-disturbed sites. Lack of previous

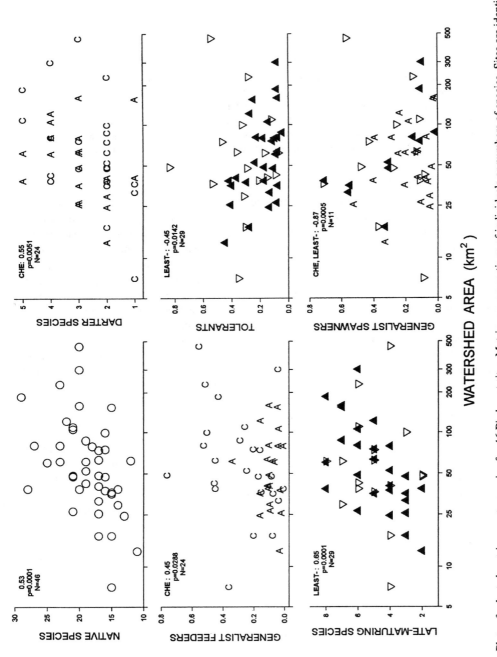

FIGURE 10.2 Plots of selected metrics vs. stream size for 46 Piedmont sites. Metrics are proportions of individuals or numbers of species. Sites are identified by basin (C or CHE = Chesapeake, A or ALB = Albemarle) or by disturbance category (dark, upward-pointing triangles = least-disturbed sites; open, downward triangles = most-disturbed sites). For lower right plot, triangles represent Chesapeake sites only. Strongest relations in each plot are summarized by showing site group (e.g., LEAST- = least-disturbed sites), Spearman correlation coefficient, statistical probability (p), and sample size (N).

Effects of Drainage Basin and Anthropogenic Disturbance

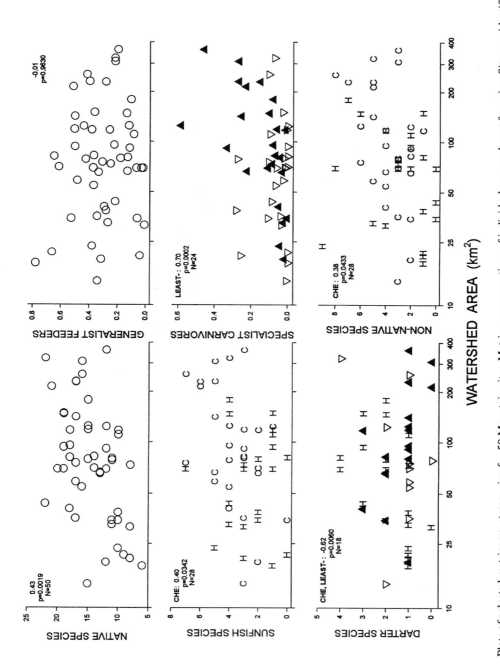

FIGURE 10.3 Plots of selected metrics vs. stream size for 50 Mountain sites. Metrics are proportions of individuals or numbers of species. Sites are identified by basin (C or CHE = Chesapeake, H or OHI = Ohio) or by disturbance category (dark, upward-pointing triangles = least-disturbed sites; open, downward triangles = most-disturbed sites). For lower left plot, triangles represent Chesapeake sites only. Strongest relations in each plot are summarized by showing site group (e.g., LEAST- = least-disturbed sites), Spearman correlation coefficient, statistical probability (p), and sample size (N).

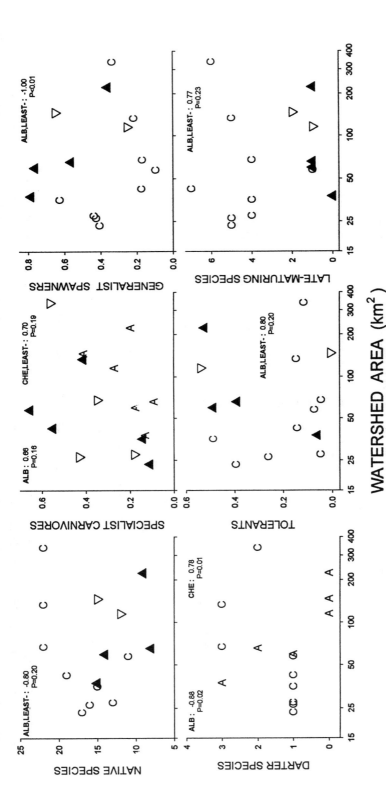

FIGURE 10.4 Plots of selected metrics vs. stream size for 15 Coastal Plain sites. Metrics are proportions of individuals or numbers of species. Sites are identified by basin (C or CHE = Chesapeake, A or ALB = Albemarle) or by disturbance category in a particular basin (dark, upward-pointing triangles = least-disturbed sites; open, downward triangles = most-disturbed sites). Strongest relations in each plot are summarized by showing site group (e.g., ALB, LEAST- = least-disturbed sites in Albemarle), Spearman correlation coefficient, and statistical probability (p).

TABLE 10.6
Summaries of One-Way ANCOVAs of Differences in Selected Fish Metrics Between Least- and Most-Disturbed Sites in Piedmont, in Basin within Piedmont, in Mountain, or in Basin within Mountain

	R^2	F	P	EFFECT	SS	F	P
All Piedmont sites (N = 46)							
NATSP	0.25	4.58	0.0073	WAREA	1.99	12.19	0.0011
				DISTRB	0.06	0.40	0.5320
				WAR × DIS	0.10	0.64	0.4281
AGE3SP	0.25	4.77	0.0060	WAREA	19.52	8.09	0.0054
				WAR × DIS	13.71	5.68	0.0217
				DISTRB	13.16	5.45	0.0244
TOLPRP	0.24	4.44	0.0085	WAR × DIS	0.10	3.20	0.0808
				WAREA	0.05	1.59	0.2149
				DISTRB	0.04	1.38	0.2461
Piedmont sites in Chesapeake basin (N = 24)							
DARSP	0.35	3.59	0.0317	WAREA	0.84	10.49	0.0041
				WAR × DIS	0.18	2.31	0.1446
				DISTRB	0.17	2.11	0.1616
GENPRP	0.30	2.84	0.0641	WAREA	0.10	2.08	0.1652
				DISTRB	0.01	0.26	0.6158
				WAR × DIS	0.00	0.00	0.9870
VMANPRP	0.27	2.42	0.0960	WAR × DIS	0.32	6.46	0.0194
				DISTRB	0.31	6.14	0.0223
				WAREA	0.14	2.84	0.1074
Piedmont sites in Albemarle basin (N = 22)							
DARSP	0.20	1.47	0.2567				
GENPRP	0.21	1.56	0.2343	WAREA	0.02	1.39	0.2533
				WAR × DIS	0.02	1.36	0.2592
				DISTRB	0.01	1.00	0.3306
VMANPRP	0.17	1.21	0.3352				
All Mountain sites (N = 50)							
NATSP	0.24	4.82	0.0053	WAREA	3.26	13.87	0.0005
				WAR × DIS	0.08	0.36	0.5535
				DISTRB	0.05	0.23	0.6351
CARNPRP	0.34	7.88	0.0002	WAREA	0.21	7.46	0.0089
				WAR × DIS	0.20	7.30	0.0096
				DISTRB	0.14	5.11	0.0286
Mountain sites in Chesapeake basin (N = 28)							
SUNSP	0.25	2.70	0.0684	WAREA	0.93	5.49	0.0278
				DISTRB	0.28	1.65	0.2118
				WAR × DIS	0.21	1.24	0.2772
DARSP	0.21	2.08	0.1289	WAR × DIS	0.49	5.04	0.0343
				DISTRB	0.48	4.93	0.0361
				WAREA	0.05	0.51	0.4805
NONNATSP	0.23	2.43	0.0901	WAREA	0.83	5.92	0.0228
				DISTRB	0.02	0.12	0.7371

TABLE 10.6 *(continued)*
Summaries of One-Way ANCOVAs of Differences in Selected Fish Metrics Between Least- and Most-Disturbed Sites in Piedmont, in Basin within Piedmont, in Mountain, or in Basin within Mountain

	R^2	F	P	EFFECT	SS	F	P
				WAR × DIS	0.00	0.00	0.9508
GENPRP	0.13	1.16	0.3464				
				Mountain sites in Ohio basin (N = 22)			
SUNSP	0.27	2.18	0.1255	WAREA	0.06	0.24	0.6273
				DISTRB	0.04	0.19	0.6645
				WAR × DIS	0.00	0.00	0.9967
DARSP	0.23	1.76	0.1910	WAREA	0.50	3.99	0.0610
				WAR × DIS	0.02	0.14	0.7141
				DISTRB	0.01	0.08	0.7780
NONNATSP	0.38	3.64	0.0328	DISTRB	0.66	1.83	0.1926
				WAR × DIS	0.27	0.74	0.4004
				WAREA	0.06	0.16	0.6934
GENPRP	0.07	0.43	0.7334				

Note: Watershed area (WAREA or WAR) is the covariate. For each ANCOVA, R^2 is the amount of variance in each metric that was accounted for simultaneously by disturbance (DISTRB or DIS) and stream-size effects. For each ANCOVA with $R^2 > 0.20$, all possible effects are shown in order of strength (based on Type III sums of squares [SS]). Also included are F-test value (F) and statistical probability (P) for each test. Degrees of freedom for each omnibus test were 3 for model, 25 for error for PD; and 3,20 for MT. See Table 10.3 for codes of metrics.

evidence of relations between stream size and functional metrics may have resulted from the lack of accounting for varying levels of disturbance among sample sites.

10.4.2 Taxonomic Metrics in Virginia

For Virginia, Piedmont and Mountain versions of the IBI would require adjustments (for stream size) in number of native species, a result consistent with many previous IBI applications. Numbers of sunfish species and of non-native species (both intercorrelated: $r_s = 0.75$ at 50 Mountain sites and $r_s = 0.86$ at 28 Chesapeake-basin sites in Mountain) would require adjustment in a Mountain IBI, whereas number of darter species would require adjustment in a Piedmont IBI. Adjustments for these three metrics are especially warranted for sites in the Chesapeake basin (see Figures 10.2 and 10.3).

For the Coastal Plain, the strongest relations between stream size and taxonomic metrics differed between basins and were largely contrary to well-evidenced relations: numbers of native species and of darter species were least at largest sites in the Albemarle basin, whereas number of darter species was highest at largest sites in the Chesapeake basin. These apparent basin differences may have resulted from local variation unaccounted for by these analyses.

The largest two sites in the Albemarle basin were swamp-like and naturally less species-rich; whereas, their Chesapeake-basin counterparts were more speciose, flowing streams. This type of natural local variation, common at sites throughout the Coastal Plain in Virginia and North Carolina (Smock and Gilinsky, 1992; personal observation) can confound attempts to relate metrics with stream size or anthropogenic disturbance, especially attempts that address large spatial scales while ignoring more localized variation. Moreover, many aspects of Coastal Plain fish communities, including effects of anthropogenic disturbance, are less understood than those of more-upland stream-fish communities. Local variation remains problematic until comprehensive studies in the Coastal Plain can distinguish and attribute variation in metrics to natural versus anthropogenic causes. Our experience suggests that consideration of intraregion differences (in fish metrics) across previously less-emphasized environmental gradients (e.g., pH, productivity) may be necessary to adequately characterize metric variation in Coastal Plain streams.

10.4.3 Functional Metrics in Virginia

Relations between stream size and several functional metrics were evident at Piedmont and at Coastal Plain sites, but few such relations were found at Mountain sites. Consequently, for Virginia, Piedmont and Coastal Plain IBIs would require adjustments for some functional metrics, contrary to previous IBI emphases. Inconsistencies in relations (discussed earlier in Section 10.4.1) suggest that few such adjustments would be universally applicable with respect to metric, IBI region, or basin within region.

For both Piedmont or Coastal Plain IBIs, adjustments in two metrics (proportions as generalist spawners and as tolerants) seem especially warranted, although adjustments would differ between basin and in their direction of effect in each physiography. For example, relations were most evident at Albemarle sites in Coastal Plain, but most evident at Chesapeake sites in Piedmont. Proportion as tolerants increased with stream size in Coastal Plain, but decreased with stream size in Piedmont (Figures 10.2 and 10.4). Some of these relations may have resulted from correlations between the two metrics: at 4 least-disturbed Albemarle sites in Coastal Plain, $r_s = -0.80$, and at 11 least-disturbed Chesapeake sites in Piedmont, $r_s = 0.87$. For each physiography, larger sample sizes of least-disturbed sites in each basin or relevant subregion would allow the use of multivariate analyses to help reveal apparent regional effects that could be due to interrelations of metrics.

10.4.4 Further Considerations

We recommend that future research should attempt to determine and account for relevant sources of intraregional variation in IBI metrics. Use of multivariate statistical techniques would be especially informative. By addressing interrelations within and among sets of disturbance measures and biotic-response measures (e.g., IBI metrics), a properly applied multivariate analysis can provide a more concise, more comprehensive, and more realistic analysis than that provided by uni- and bivariate analyses alone (Gittins, 1985; Huberty and Morris, 1989; e.g., Smogor and Angermeier, Chapter 23). Moreover, a properly applied multivariate analysis necessarily includes statistical and graphical examinations of lesser-order relations and therefore subsumes a bivariate- or univariate-based approach.

Developers and users of the IBI could benefit from a more advanced, mechanism-based understanding of how anthropogenic disturbance affects fish assemblages and of the spatial scales at which one can best distinguish such effects from natural variation. However, investing in the pursuit of this "advanced" understanding has disadvantages. Although one could apparently improve an IBI by examining and incorporating the many possible sources of intraregion variation in metrics (e.g., stream size, basin, stream gradient, water chemistry, interactions of these factors), attempting to account for each source, for each metric, via an adjustment of the IBI would be too cumbersome, data-demanding, and impractical (if not impossible) for many users of the index. In practice, the potential advantages of incorporating adjustments to metric criteria might not always outweigh the disadvantages of the added complexity and difficulty in metric scoring and interpretation. While we advocate careful planning and empirical justification in the development of an IBI, we also recognize the "real-world" constraints on developing and using the resultant index. Presently, Virginia versions of the IBI based on our findings would be incomplete and fallible, but often adequate and useful. Present implementation of a practical and reasonably accurate way to assess ongoing stream degradation offers greater potential for saving streams than does waiting for the development of a "perfect" index.

ACKNOWLEDGMENTS

The authors thank P. Braaten, L. Calliott, G. Galbreath, T. Hash, M. Kosala, P. Lookabaugh, D. Magoulick, B. Mueller, T. Richards, S. Smith, J. Tousignant, and C. Weiking for fieldwork assistance, and D. Orth, E. Smith, A. Dolloff, and two anonymous reviewers for helpful criticism. Fieldwork and some data analysis were funded by Federal Aid in Sport Fish Restoration grant F-86-R from the Virginia Department of Game and Inland Fisheries. Additional analysis was supported by funding from the U.S. Environmental Protection Agency.

REFERENCES

Angermeier, P. L. and J. R. Karr. 1984. Relationships between woody debris and fish habitat in a small warmwater stream, *Transactions of the American Fisheries Society*, 113, 716–726.

Angermeier, P. L. and R. A. Smogor. 1992. Development of a Geographic Information System for Stream and Fisheries Data in Virginia. Final report of Federal Aid Project F-86 to Virginia Department of Game and Inland Fisheries, Richmond, VA.

Angermeier, P. L., R. A. Smogor, and S. D. Steele. 1991. An electric seine for collecting fish in streams, *North American Journal of Fisheries Management*, 11, 352–357.

Beecher, H. A., E. R. Dott, and R. F. Fernau. 1988. Fish species richness and stream order in Washington State streams, *Environmental Biology of Fishes*, 22, 193–209.

Berkman, H. E. and C. F. Rabeni. 1987. Effects of siltation on stream fish communities, *Environmental Biology of Fishes*, 18, 285–294.

Detenbeck, N. E., P. W. DeVore, G. J. Niemi, and A. Lima. 1992. Recovery of temperate-stream fish communities from disturbance: a review of case studies and synthesis of theory, *Environmental Management*, 16, 33–53.

Evans, J. W. and R. L. Noble. 1979. The longitudinal distribution of fishes in an east Texas stream, *American Midland Naturalist*, 101, 333–334.

Fausch, K. D., J. R. Karr, and P. R. Yant. 1984. Regional application of an index of biotic integrity based on stream fish communities, *Transactions of the American Fisheries Society*, 113, 39–55.

Fausch, K. D., J. Lyons, J. R. Karr, and P. L. Angermeier. 1990. Fish communities as indicators of environmental degradation, *American Fisheries Society Symposium*, 8, 123–144.

Fore, L. S., J. R. Karr, and R. W. Wisseman. 1996. Assessing invertebrate responses to human activities: evaluating alternative approaches, *Journal of the North American Benthological Society*, 15, 212–231.

Gittins, R. 1985. *Canonical Analysis. A Review with Applications in Ecology*. Springer-Verlag, New York.

Horwitz, R. J. 1978. Temporal variability patterns and the distributional patterns of stream fishes, *Ecological Monographs*, 48, 307–321.

Huberty, C. J. and J. D. Morris. 1989. Multivariate analysis versus multiple univariate analyses, *Psychological Bulletin*, 105, 302–308.

Jenkins, R. E. and N. M. Burkhead. 1994. *The Freshwater Fishes of Virginia*. American Fisheries Society, Bethesda, MD.

Karr, J. R. 1991. Biological integrity: a long-neglected aspect of water resource management, *Ecological Applications*, 1, 66–84.

Karr, J. R. and I. J. Schlosser. 1978. Water resources and the land-water interface, *Science*, 201, 229–234.

Karr, J. R. and D. R. Dudley. 1981. Ecological perspective on water quality goals, *Environmental Management*, 5, 55–68.

Karr, J. R., L. A. Toth, and D. R. Dudley. 1985. Fish communities of midwestern rivers: a history of degradation, *BioScience*, 35, 90–95.

Karr, J. R., K. D. Fausch, P. L. Angermeier, P. R. Yant, and I. J. Schlosser. 1986. Assessing Biological Integrity in Running Waters: A Method and Its Rationale. Illinois Natural History Survey Special Publication 5, Urbana, IL.

Kuehne, R. A. 1962. A classification of streams illustrated by fish distribution in an eastern Kentucky creek, *Ecology*, 43, 608–614.

Larimore, R. W. and P. W. Smith. 1963. The fishes of Champaign County, Illinois, as affected by 60 years of stream changes, *Illinois Natural History Survey Bulletin*, 28, 299–382.

Lotrich, V. A. 1973. Growth, production, and community composition of fishes inhabiting a first-, second-, and third-order stream of eastern Kentucky, *Ecological Monographs*, 43, 377–397.

Lyons, J. 1992. Using the Index of Biotic Integrity (IBI) to Measure Environmental Quality in Warmwater Streams of Wisconsin. General Technical Report NC-149, U.S. Department of Agriculture, Forest Service. North Central Forest Experiment Station, St. Paul, MN.

Matthews, W. J. 1986. Fish faunal "breaks" and stream order in the eastern and central United States, *Environmental Biology of Fishes*, 17, 81–92.

Maurakis, E. G., W. S. Woolcott, and R. E. Jenkins. 1987. Physiographic analyses of the longitudinal distribution of fishes in the Rappahannock River, Virginia, *Association of Southeastern Biologists Bulletin*, 34, 1–14.

Miller, D. L., P. M. Leonard, R. M. Hughes, J. R. Karr, P. B. Moyle, L. H. Schrader, B. A. Thompson, R. A. Daniels, K. D. Fausch, G. A. Fitzhugh, J. R. Gammon, D. B. Haliwell, P. L. Angermeier, and D. J. Orth. 1988. Regional applications of an index of biotic integrity for use in water resource management, *Fisheries (Bethesda)*, 13(5), 12–20.

Morin, R. and R. J. Naiman. 1990. The relation of stream order to fish community dynamics in boreal forest watersheds, *Polskie Archiwum Hydrobiologii*, 37, 135–150.

Ohio Environmental Protection Agency (EPA). 1988. Biological Criteria for the Protection of Aquatic Life. Vol. II. Users Manual for Biological Field Assessment of Ohio Surface Waters. Ohio Environmental Protection Agency, Division of Water Quality Monitoring and Assessment, Surface Water Section, Columbus, Ohio.

Osborne, L. L., S. L. Kohler, P. B. Bayley, D. M. Day, W. A. Bertrand, M. J. Wiley, and R. Sauer. 1992. Influence of stream location in a drainage network on the Index of Biotic Integrity, *Transactions of the American Fisheries Society*, 121, 635–643.

Paller, M. H. 1994. Relationships between fish assemblage structure and stream order in South Carolina Coastal Plain streams, *Transactions of the American Fisheries Society*, 123, 150–161.

Pflieger, W. L. 1975. *The Fishes of Missouri*. Missouri Department of Conservation, Columbia, MO.

Platts, W. S. 1979. Relationships among stream order, fish populations, and aquatic geomorphology in an Idaho river drainage, *Fisheries (Bethesda)*, 4(2), 5–9.

Pringle, C. M. 1997. Exploring how disturbance is transmitted upstream: going against the flow, *Journal of the North American Benthological Society*, 16, 425–438.

SAS. 1990. *SAS/STAT User's Guide*. Vols. 1 and 2. SAS Institute, Cary, NC.

Schlosser, I. J. 1982. Fish community structure and function along two habitat gradients in a headwater stream, *Ecological Monographs*, 52, 395–414.

Schlosser, I. J. 1987. A conceptual framework for fish communities in small warmwater streams, in W. J. Matthews and D. C. Heins (Eds.), *Community and Evolutionary Ecology of North American Stream Fishes*. University of Oklahoma Press, Norman, OK. 17–24.

Schlosser, I. J. 1990. Environmental variation, life history attributes, and community structure in stream fishes: implications for environmental management and assessment, *Environmental Management*, 14, 621–628.

Sheldon, A. 1968. Species diversity and longitudinal succession in stream fishes, *Ecology*, 49, 193–197.

Simon, T. P. and J. Lyons. 1995. Application of the index of biotic integrity to evaluate water resource integrity in freshwater ecosystems, in W. S. Davis and T. P. Simon (Eds.). *Biological Assessment and Criteria: Tools for Water Resource Planning and Decision Making*. Lewis Publishers, Boca Raton, FL. 245–262.

Smith, P. W. 1971. Illinois Streams: A Classification Based on Their Fishes and An Analysis of Factors Responsible for Disappearance of Native Species. Illinois Natural History Survey Biological Notes, Number 76. Urbana, IL.

Smock, L. A. and E. Gilinsky. 1992. Coastal Plain blackwater streams, in C. T. Hackney, S. M. Adams, and W. H. Martin (Eds.), *Biodiversity of the Southeastern United States: Aquatic Communities*. Wiley, New York. 271–313.

Smogor, R. A. 1996. Developing an index of biotic integrity for warmwater wadable streams in Virginia. M.S. Thesis, Department of Fisheries and Wildlife Sciences, Virginia Polytechnic Institute and State University, Blacksburg, VA.

Townsend, C. R. and A. G. Hildrew. 1994. Species traits in relation to a habitat templet for river systems, *Freshwater Biology,* 31, 265–275.

Vannote, R. L., G. W. Minshall, K. W. Cummins, J. R. Sedell, and C. E. Cushing. 1980. The river continuum concept, *Canadian Journal of Fisheries and Aquatic Sciences,* 37, 130–137.

Virginia Department of Environmental Quality. 1994. Virginia water quality assessment for 1994, 305(b) Report to EPA and Congress. Virginia Department of Environmental Quality Information Bulletin Number 597. Richmond, VA.

Weaver, L. A. and G. C. Garman. 1994. Urbanization of a watershed and historical changes in a stream fish assemblage, *Transactions of the American Fisheries Society,* 123, 162–172.

Whiteside, B. G. and R. M. McNatt. 1972. Fish species diversity in relation to stream order and physicochemical conditions in the Plum Creek drainage basin, *American Midland Naturalist,* 88, 90–101.

11 Characteristics of Fish Assemblages and Environmental Conditions in Streams of the Upper Snake River Basin in Eastern Idaho and Western Wyoming

Terry R. Maret

CONTENTS

11.1 Introduction 273
    11.1.1 Environmental Setting 275
    11.1.2 Previous Fishery Studies 277
11.2 Characterization of Reference Streams: Case Study 279
    11.2.1 Methods 279
    11.2.2 Results and Discussion 280
11.3 Characterizing Major Stream Types: Case Study 284
    11.3.1 Methods 284
    11.3.2 Results and Discussion 286
11.4 Important Components of an Index of Biotic Integrity 295
11.5 Future Needs 296
Acknowledgments 297
References 297

11.1 INTRODUCTION

The U.S. Geological Survey (USGS) National Water Quality Assessment (NAWQA) Program was initiated to define the status and trends in the quality of the nation's surface- and ground-water resources (Leahy et al., 1990). The upper Snake River Basin (USNK) in eastern Idaho and western Wyoming was 1 of 20 NAWQA Program study units to begin full implementation in 1991. The surface-water component of the NAWQA Program included the collection of biological information to aid in the interpretation and assessment of changes in stream quality (Gurtz, 1994). This biological information consisted of ecological surveys that characterized fish, macroinvertebrates, algae, and associated riparian and instream habitats. One important aspect of this program addresses the relations of physical and chemical characteristics of streams and associated fish assemblages. The analyses of these relations in this chapter are part of the multiple lines of evidence the NAWQA

Program uses to assess stream quality. This chapter primarily summarizes a number of NAWQA reports containing historical and current (1990 to 1995) case studies on fish and associated stream habitats in the USNK. More specific information regarding the various reports or case studies can be found in the references cited within each section.

Specific objectives of this chapter are to: (1) describe fish assemblages and their spatial patterns within the USNK; (2) identify and characterize some of the predominant environmental variables that affect fish assemblages; (3) identify attributes, or metrics, of fish assemblages, which will be useful in evaluating biotic integrity; and (4) discuss future data needs related to using fish assemblages to assess conditions of western streams. This chapter will provide a framework for using fish assemblages to develop an Index of Biotic Integrity (IBI) for streams in the USNK.

Human activities can alter physical, chemical, or biological conditions of surface water. Many rivers and streams in the conterminous United States have been degraded as a result of nonpoint source pollutants, fragmentation by dams and diversions, habitat alteration, and introduction of non-native fish species (Moyle, 1986; Heede and Rinne, 1990; Allan and Flecker, 1993; Doppelt et al., 1993; Dynesius and Nilsson, 1994). Human alterations of physical, chemical, or biological conditions in lotic systems usually result in changes in the distribution and structure of fish assemblages. In fact, many endemic fish species of the western United States are endangered, threatened, or of special concern as a result of human activities (Warren and Burr, 1994).

Fish assemblages, which are groups of species that co-occur in the same area, are structured by local, regional, and historical processes operating at various spatial and temporal scales (Tonn, 1990). The habitat structure of a stream is determined by climate, geology, vegetation, and other features of the surrounding watershed (Frissell et al., 1986), and stream classification schemes have been developed that are based on measures of stream morphology (Rosgen, 1994). Fish assemblages are most directly influenced by local physical and chemical characteristics of the stream habitat. The depauperate fish fauna of the western United States have been attributed, in part, to natural geological barriers like waterfalls and mountain ranges (Smith, 1981). Thus, comparisons of fish assemblages in different ecoregions having similar land surface form, potential natural vegetation, land use, and soils (Omernik and Gallant, 1986) within a geographic region can enhance understanding of the relative importance of environmental factors influencing the distribution of stream fish (Jackson and Harvey, 1989). Comparisons of historical and recent fish distributions also can provide information on whether changes in occurrence patterns of various species are the result of human activities or natural processes.

Several ecologists have used multivariate analyses to identify and interpret patterns in assemblage structure as they relate to environmental conditions (Gauch, 1982). These multivariate analyses summarize patterns of association within a species-by-sample data matrix for purposes of classification. Ordination frequently is used to summarize patterns within this matrix by defining a series of axes that express the major environmental gradients in assemblage structure. Multivariate analyses are effective for identifying similarities among sites with respect to various physical, chemical, and biological characteristics, and for depicting relations between assemblage patterns and environmental gradients. Hypotheses can be formulated from these exploratory analyses about relations between fish assemblages and environmental variables.

Documenting spatial and temporal changes in fish assemblages among streams can provide important information on stream quality and the biotic integrity of freshwater ecosystems. Karr and Dudley (1981) defined biotic integrity as the "ability to support and maintain a balanced, integrated, adaptive community of organisms having a species composition, diversity, and functional organization comparable to that of natural habitat of the region." Because aquatic assemblages integrate the characteristics of their environment, they provide useful measures for evaluating the effects of human activities in a river basin (Karr, 1991). However, before the effects of human alterations to streams can be evaluated, biological criteria are required for least-disturbed, or "reference," streams or are formulated from historical data (Hughes et al., 1986). Least-disturbed streams are defined as being relatively unaltered by humans; thus, observed distribution and assemblage structure should approximate the historical arrangement of fish species in a basin.

Information about the kinds of species and their relative abundances provides a direct measure of beneficial uses of surface water for coldwater aquatic life and salmonid spawning, helps detect problems that other monitoring methods may overlook or underestimate, and provides the basis for systematically measuring the progress of pollution abatement programs (U.S. EPA, 1990). One approach to evaluating biotic integrity is the IBI, which is a multimetric rating based on structure, composition, and functional attributes of a fish assemblage (Karr et al., 1986). This index depends on regional reference information to score individual fish metrics, and different assemblages associated with specific regions may require different metrics for evaluating biotic integrity (Miller et al., 1988). The IBI has been modified successfully for use in many different types of streams throughout North America (Simon and Lyons, 1995). However, more data are needed on the response of entire fish assemblages in coldwater streams to environmental change (Lyons et al., 1996).

A number of independent studies have documented correlations between biological and environmental variables and ecoregions (Hughes and Larsen, 1988; Hughes et al., 1994). Consequently, many state-level monitoring programs have been relatively successful in using an IBI and ecoregion approach to implement aquatic biological assessment programs (Fausch et al., 1984; Gallant et al., 1989). Specific instream biological monitoring protocols for fish have been developed for wadeable streams of the Pacific Northwest (Hayslip, 1993; Chandler et al., 1993). Fisher (1989) developed a modified fish IBI for small headwater streams of northern Idaho and found index scores significantly correlated with timber harvest, road density, and cobble embeddedness. Few studies have examined relations between entire fish assemblages and measured environmental factors for the major environmental settings of the USNK.

11.1.1 Environmental Setting

The 35,800-sq. mi. USNK extends about 450 river miles from its headwaters in southern Yellowstone National Park to King Hill in south-central Idaho (Figure 11.1). Land surface elevation above sea level ranges from 13,770 ft. for mountain peaks in the headwaters of the Snake River to 2500 ft. at King Hill. Most streams in the basin originate in foothill or montane regions (6000 to 10,000 ft. in elevation). Maupin (1995) provided a detailed discussion of the geology, climate, hydrology, and land use in the basin.

The geology of the basin is characterized largely by basalt flows in the lowlands of the central and southern parts, and by intrusive volcanic, sedimentary, and metamorphic rocks in the uplands and mountains to the north, south, and east (Maupin, 1995). Basalt flows in the northern part of the basin prevent northern streams such as the Big Lost and Little Lost Rivers from reaching the Snake River.

Climate in most of the basin is semiarid, and annual precipitation ranges from 10 to 20 inches. At higher elevations in the eastern part of the basin, annual precipitation can average more than 20 inches. Precipitation occurs primarily as snow, and peak flows in streams result from spring snowmelt.

The basin contains about 8460 mi. of streams (Maret, 1995). Streamflow in the Snake River and its major tributaries is highly regulated by dams and diversions, primarily for agricultural use and hydroelectric power generation. Irrigation projects have resulted in about 5700 mi. of canals and about 1300 mi. of drains in the basin (U.S. Water and Power Resources Service, 1981), and water transfer from one river basin to irrigate crops in another is common practice. Ecological consequences of interbasin transfer of water include changes in streamflow, introduction of exotic species, and alteration of habitat (Meador, 1992).

Clark (1994) described in detail the characteristics of surface-water quality and hydrology of the basin. Thurow et al. (1988) reported that surface water is generally high in alkalinity (greater

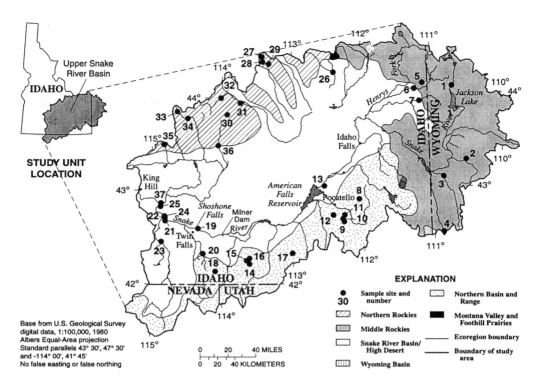

FIGURE 11.1 Ecoregions and sample sites in the upper Snake River Basin. (Ecoregions modified from Omernik and Gallant, 1986).

than 150 mg/L as $CaCO_3$), contains large concentrations of various ions, and generally supports productive aquatic assemblages. Upland streams in forested watersheds and lowland streams in rangeland (desert shrubland) watersheds are typified by coarse substrates (gravel and cobbles), high gradients (greater than 1.0%), well-defined riffle-pool habitats, and sparse macrophyte growth. Springs are typified by a wide variety of substrates (sand to large basalt boulders), low gradients (1.0% or less), and abundant macrophyte growth. Large rivers and streams in agricultural watersheds are typified by fine-grained substrates, low gradients, and abundant macrophyte growth.

Water years 1988 to 1992 were extremely dry years in the USNK, and streamflows were smaller than historical averages throughout the basin. Streamflows were variable during water years 1993 to 1995 and continued to be smaller than historical averages for many parts of the USNK; however, some streamflows actually exceeded historical averages during water years 1993 and 1995. Streamflows at most gaging stations on the mainstem of the Snake River were smaller than average during the sampling period, 1993 to 1995 (G.M. Clark, U.S. Geological Survey, written communication, 1996).

Springs along the Snake River between Milner Dam and King Hill provide more than 50% of the discharge measured at King Hill on the Snake River. Many of the springs along the Snake River between Twin Falls and Hagerman are used for commercial trout production. More than 80% of the nation's trout supply is produced in this area (Brockway and Robinson, 1992).

Designated beneficial uses of streams in the basin are agriculture, industry, public water supply, recreation, and propagation of fish and wildlife, and criteria have been developed in state water-quality standards to protect these beneficial uses from impairment (Idaho Department of Health and Welfare, 1989). In addition, coldwater aquatic life is a designated use for most streams in the basin, and these streams are suitable for protection and maintenance of viable assemblages of aquatic organisms whose optimal growing temperature is below 18°C (Idaho Department of Health and Welfare, 1990).

Nonpoint-source pollution and water diversions are the predominant influences on surface-water quality in the basin. Pollutants of greatest concern that have been associated with habitat degradation of streams include nutrients, sediment, bacteria, organic waste, and elevated water temperature (Idaho Department of Health and Welfare, 1989). Beneficial uses of streams most impaired by pollutants include sustaining coldwater biota, salmonid spawning, and water-contact recreation (Maret, 1995).

Water quality of the middle reach of the Snake River (between Milner Dam and King Hill; Figure 11.1) is affected by irrigation drainage, fish-farm effluent, municipal effluent, hydrologic modification, and dams (Brockway and Robinson, 1992). As a result of these activities, segments of this river were listed as "water-quality limited" in 1990 because nuisance weed growth had exceeded water-quality criteria, and standards established for protection of coldwater biota and salmonid spawning (Idaho Department of Health and Welfare, 1995).

Land use in the basin comprises 50% rangeland, 23% forest land, and 21% agricultural land; the remaining area, classified as "other," comprises barren soil or rock with little vegetation, urban areas, waterbodies, wetlands, and tundra (Maupin, 1995). Most agricultural lands are adjacent to the Snake River because of irrigation needs. Livestock grazing is common throughout the basin. Logging, mining, and recreation also are predominant land uses. Population in the basin is about 435,000. The largest cities are Idaho Falls, Pocatello, and Twin Falls.

Four ecoregions compose more than 99% of the land area in the basin: Snake River Basin/High Desert (SRB), 50%; Middle Rockies (MR), 23%; Northern Basin and Range (NBR), 18%; and Northern Rockies (NR), 9% (Figure 11.1). The Wyoming Basin and Montana Valley and Foothill Prairies ecoregions compose less than 1% of the land area in the basin. Vegetation in the upper elevations consists of coniferous forests, whereas sagebrush (*Artemisia* sp.) communities dominate the lowlands. Typical woody vegetation in riparian areas consists of river birch (*Betula occidentalis*), alder (*Alnus* sp.), dogwood (*Cornus stolinifera*), willow (*Salix* sp.), and poplar (*Populus* sp.).

11.1.2 Previous Fishery Studies

Fish assemblages in the Snake River have been investigated since the late 1800s. Gilbert and Evermann (1895) and Evermann (1896) described fish distribution in the middle reach of the Snake River and tributaries before hydroelectric-power development. Anadromous chinook salmon (*Oncorhynchus tshawytscha*) and Pacific lamprey (*Lampetra tridentata*) have been eliminated from the basin downstream from Shoshone Falls since the construction of hydroelectric-power facilities on the mainstem Snake River.

Fish species in the USNK (see Table 11.1 for common and scientific names) are adapted primarily for coldwater habitats and are predominantly represented by the families Salmonidae (trout), Cottidae (sculpin), Cyprinidae (minnows), and Catostomidae (suckers). The fish fauna in this basin are represented by 25 native species belonging to 5 families, and an additional 14 species introduced primarily to enhance sport fishery (Maret, 1995).

Shoshone Falls, a large waterfall on the Snake River near the city of Twin Falls, prevents upstream migration of fish. Native species living only in the Snake River and its tributaries downstream from the falls include the bridgelip sucker, largescale sucker, chiselmouth, leopard dace, northern squawfish, peamouth, white sturgeon, Wood River sculpin, and Shoshone sculpin (Maret, 1995).

The Wyoming Game and Fish Department (WGFD) and Idaho Department of Fish and Game (IDFG) have performed many fishery studies to assess sport-fishery populations and associated habitats. Simpson and Wallace (1982), Thurow et al. (1988), and Baxter and Stone (1995) described fish species distributions in the USNK. Maret (1995) summarized in detail the fish species in the basin and land uses affecting their habitat.

Eight native fish species are recognized as Species of Special Concern in the basin (Table 11.1). Six of these fish species currently are listed as Species of Special Concern by IDFG: the white

TABLE 11.1
Fish Origin, Trophic Group, and Tolerance to Organic Compounds, Sediment, and Warmwater Pollution in Streams in the Upper Snake River Basin

Family	Common Name	Species	Origin	Trophic Group of Adults	Tolerance to Pollution
Acipenseridae	White sturgeon (SC)	*Acipenser transmontanus*	Native	Omnivore	Intolerant
Catostomidae	Bluehead sucker (SC)	*Catostomus discobolus*	Native	Herbivore	Tolerant
	Bridgelip sucker	*Catostomus columbianus*	Native	Herbivore	Intermediate
	Largescale sucker	*Catostomus macrocheilus*	Native	Omnivore	Tolerant
	Mountain sucker	*Catostomus platyrhynchus*	Native	Herbivore	Intermediate
	Utah sucker	*Catostomus ardens*	Native	Omnivore	Tolerant
Centrarchidae	Black crappie	*Pomoxis nigromaculatus*	Introduced	Insectivore	Tolerant
	Bluegill	*Lepomis macrochirus*	Introduced	Insectivore	Tolerant
	Largemouth bass	*Micropterus salmoides*	Introduced	Piscivore	Tolerant
	Smallmouth bass	*Micropterus dolomieui*	Introduced	Piscivore	Intermediate
Cichlidae	Tilapia	*Tilapia* sp.	Introduced	Insectivore	Tolerant
Cottidae	Mottled sculpin	*Cottus bairdi*	Native	Insectivore	Intermediate
	Paiute sculpin	*Cottus beldingi*	Native	Insectivore	Intolerant
	Shorthead sculpin	*Cottus confusus*	Native	Insectivore	Intolerant
	Shoshone sculpin (SC)	*Cottus greenei*	Native	Insectivore	Intolerant
	Wood River sculpin (SC)	*Cottus leiopomus*	Native	Insectivore	Intolerant
Cyprinidae	Carp	*Cyprinus carpio*	Introduced	Omnivore	Tolerant
	Chiselmouth	*Acrocheilus alutaceus*	Native	Herbivore	Intermediate
	Fathead Minnow	*Pimephales promelas*	Introduced	Omnivore	Tolerant
	Hornyhead chub (SC)	*Nocomis biguttatus*	Native	Omnivore	Intermediate
	Leatherside chub (SC)	*Gila copei*	Native	Insectivore	Intermediate
	Leopard dace	*Rhinichthys falcatus*	Native	Insectivore	Intermediate
	Longnose dace	*Rhinichthys cataractae*	Native	Insectivore	Intermediate
	Northern squawfish	*Ptychocheilus oregonensis*	Native	Piscivore	Tolerant
	Peamouth	*Mylocheilus caurinus*	Native	Insectivore	Intermediate
	Redside shiner	*Richardsonius balteatus*	Native	Insectivore	Intermediate
	Speckled dace	*Rhinichthys osculus*	Native	Insectivore	Intermediate
	Utah chub	*Gila atraria*	Native	Omnivore	Tolerant
Ictaluridae	Black bullhead	*Ameiurus melas*	Introduced	Omnivore	Tolerant
	Brown bullhead	*Ameiurus nebulosus*	Introduced	Omnivore	Tolerant
	Channel catfish	*Ictalurus punctatus*	Introduced	Omnivore	Tolerant
Percidae	Walleye	*Stizostedion vitreum*	Introduced	Piscivore	Intermediate
	Yellow perch	*Perca flavescens*	Introduced	Insectivore	Intermediate
Petromyzonidae	Pacific lamprey (E)	*Lampetra tridentata*	Native	Parasite	Intermediate
Salmonidae	Brook trout	*Salvelinus fontinalis*	Introduced	Insectivore	Intolerant
	Brown trout	*Salmo trutta*	Introduced	Insectivore	Intolerant
	Bull trout (SC)	*Salvelinus confluentus*	Native	Piscivore	Intolerant
	Chinook salmon (E)	*Oncorhynchus tshawytscha*	Native	Insectivore	Intolerant
	Chutthroat trout (SC)	*Oncorhynchus clarki* sp.	Native	Insectivore	Intolerant
	Mountain whitefish	*Prosopium williamsoni*	Native	Insectivore	Intolerant
	Rainbow trout[a]	*Oncorhynchus mykiss* sp.	Native	Insectivore	Intolerant

Note: Data from Scott and Crossman (1973); Simpson and Wallace (1982); Sigler and Sigler (1987). Common and scientific names according to Robins and others (1991). SC, Species of Special Concern; E, eliminated from the upper Snake River Basin (Maret, 1995).

[a] Native in the Snake River and tributaries downstream from Shoshone Falls.

sturgeon, Shoshone sculpin, Wood River sculpin, bull trout, cutthroat trout, and leatherside chub. These species, and the redband trout (*Oncorhynchus mykiss gibbsi*), are candidates for threatened and endangered listing by the U.S. Fish and Wildlife Service (Idaho Conservation Data Center, 1994). The WGFD has listed four species as Species of Special Concern in the Wyoming part of the basin: cutthroat trout, leatherside chub, bluehead sucker, and hornyhead chub (Robin Jones, Wyoming Natural Diversity Database, written communication, 1992). A more complete discussion of the distribution and habitat needs of these Species of Special Concern is given in a report by Maret (1995).

Maret (1995) grouped five discrete drainages on the basis of cluster analysis of fish species presence or absence (Figure 11.2): (1) Part of a large river fishery in the Snake River between Shoshone Falls and King Hill contained a large number of species, many of which were introduced and adapted to warmwater habitats; (2) the upper Snake River and South Fork Snake River upstream from Shoshone Falls contained a high-quality cutthroat trout fishery; (3) Henrys Fork, Teton River, Salt River, Portneuf River, Willow Creek, and Blackfoot River upstream from Shoshone Falls contained a cutthroat trout fishery with introduced species; (4) Snake River tributaries downstream from Shoshone Falls, including Rock Creek and Big Wood River, contained a trout fishery dominated by introduced trout species consisting primarily of brown and rainbow trout; and (5) Big Lost River contained a trout fishery with few native species.

Fish assemblages in headwater first- and second-order streams in the basin typically comprise few species and low abundances (Robinson and Minshall, 1994; Maret et al., 1997). Trout and/or sculpins typically make up entire collections from many of the streams sampled. These investigators also noted a shift from an intolerant assemblage composed predominantly of coldwater species such as trout to a more tolerant warmwater assemblage in streams affected by human disturbances.

11.2 CHARACTERIZATION OF REFERENCE STREAMS: CASE STUDY

Before the effects of human alterations to streams can be evaluated, biological criteria for water-quality monitoring require data based on reference streams or other suitable historical data (Hughes et al., 1986; Hayslip, 1993). For example, the application of an IBI (Index of Biotic Integrity) is dependent on regional reference site information to score individual fish metrics (Karr, 1991). The assessment of fish assemblages in relation to reference site environmental variables such as landscape and instream habitat characteristics is important for watershed management and stream restoration activities.

This case study was designed specifically to describe and evaluate least-disturbed wadeable streams and springs in the USNK. The study was a cooperative effort between the USGS and Idaho State University in Pocatello.

11.2.1 METHODS

Maret et al. (1997) evaluated fish assemblages and environmental variables in 37 least-disturbed, first- through sixth-order streams and springs in the USNK (see Figure 11.1 for site locations). Least-disturbed sites were selected using criteria established by Hughes et al. (1986). Stream sites also were selected to represent each of the four major ecoregions of the basin. Fish sampling and habitat characterization were conducted during base-flow conditions in the summer and autumn of 1990 through 1994. Fish sampling procedures followed methods outlined by Meador et al. (1993a). Environmental variables were measured using procedures described by Platts et al. (1983) and Meador et al. (1993b). Multivariate and multimetric analyses were used to evaluate the fish and environmental data.

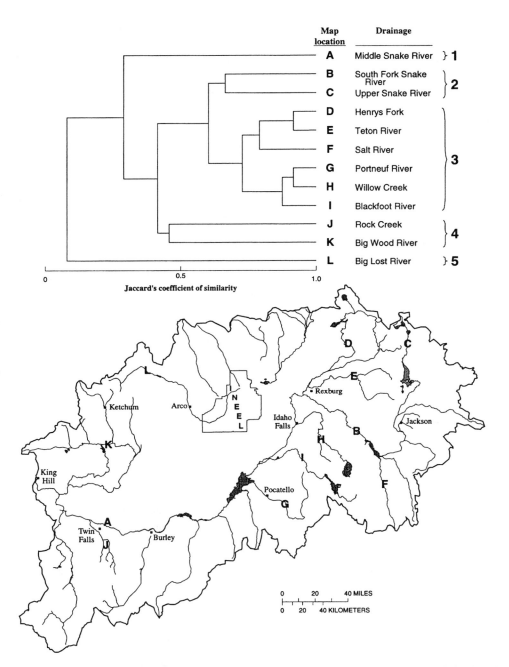

FIGURE 11.2 Dendrogram showing relative similarities in fish species in 12 drainages and relative locations of each drainage in the upper Snake River Basin.

11.2.2 Results and Discussion

Overall, 19 fish species were collected in the families Salmonidae, Cottidae, Cyprinidae, and Catostomidae (Figure 11.3). The number of species collected at a site ranged from 1 to 11. The greatest number of species (14) was collected in the SRB ecoregion and the fewest (8) in the NR ecoregion. However, only six species were collected from springs in the SRB ecoregion. Species common to the four ecoregions were the mottled sculpin, Paiute sculpin, speckled dace, and rainbow

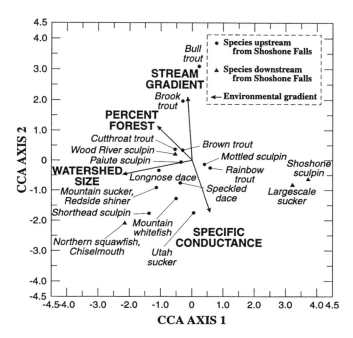

FIGURE 11.3 Canonical correspondence analysis (CCA) ordination plots showing 19 fish species collected at 37 sample sites in the upper Snake River Basin (species optimum in a sample defines the sample location along the environmental gradients of specific conductance, watershed size, percent forest, and stream gradient).

trout. Rainbow trout were collected at 54% of the sites. Paiute sculpin, brook trout, mottled sculpin, and cutthroat trout also were commonly collected at 27 to 38% of the sites. Rainbow trout are considered native in the USNK, although their historical range is limited to the Snake River and its tributaries downstream from Shoshone Falls (Simpson and Wallace, 1982). Consequently, brown trout and brook trout were the only introduced species collected (35% of the sites). Chiselmouth, largescale sucker, and northern squawfish, all tolerant species, were collected at only two sites downstream from Shoshone Falls on the Snake River.

In general, fish metrics showed high variation among sample sites (Table 11.2). Salmonids and cottids were predominant among study streams. Juvenile salmonids were collected at 89% of the sites and typically represented 57% of the assemblage at a site. The MR ecoregion had the greatest number of total species, native species, intolerant species, and percent cottids, but the lowest percent salmonids. In contrast, the NBR ecoregion had the lowest percent cottids and the highest percent salmonids. On the basis of a principal components analysis (PCA), six of these fish metrics were determined to be important in describing fish assemblages in the USNK: percent cottids, number of native species, total number of species, percent salmonids, number of salmonid species, and percent introduced species.

One or more fish metrics was significantly correlated with the environmental variables of watershed size, specific conductance, and stream gradient. Differences in these environmental variables were most apparent between the SRB and the other three ecoregions. The presence/absence of specific species for assessing changes in specific areas of the USNK was also useful. For example, Shoshone sculpin, Wood River sculpin, and bull trout are common only in isolated basins or springs that represent relatively different aquatic habitats in the basin. The percent juvenile salmonids appears to be a useful metric to indicate biotic integrity because almost 90% of the study sites contained juvenile salmonids. This metric also would be a direct measure of a recognized beneficial use of coldwater streams in Idaho (Chandler et al., 1993). Whether the percent juveniles decreases in degraded streams of the basin has yet to be determined. Other metrics, such as fish condition

TABLE 11.2
Summary of 11 Fish Metrics and Corresponding Means, Standard Deviations, and Ranges, by Ecoregion, from 37 Sample Sites in the Upper Snake River Basin

Fish Metric		Middle Rockies N = 7	Northern Basin and Range N = 11	Snake River Basin/High Desert N = 9	Northern Rockies N = 10
Number of total species*	Mean	4.7	1.9	3.4	2.0
	SD	2.8	1.3	2.1	1.3
	Range	2–11	1–4	1–8	1–5
Number of native species*	Mean	3.7	1.0	2.6	1.0
	SD	2.4	1.0	2.1	0.9
	Range	1–9	0–3	0–7	0–3
Number of introduced species	Mean	1.0	0.9	0.9	1.0
	SD	1.1	0.7	0.3	0.6
	Range	0–3	0–2	0–1	0–2
Number of intolerant species*	Mean	2.3	0.6	0.6	1.2
	SD	0.9	0.6	1.0	0.4
	Range	1–4	0–2	0–3	1–2
Number of salmonid species	Mean	2.0	1.3	1.1	1.4
	SD	1.2	0.6	0.7	0.7
	Range	1–4	1–3	0–3	1–3
Percent introduced species	Mean	12.0	57.0	34.1	47.2
	SD	19.9	43.3	30.4	42.2
	Range	0–59	0–100	0–100	0–100
Percent cottids*	Mean	66.0	7.4	49.3	31.8
	SD	20.3	14.5	28.0	37.3
	Range	38–94	0–46	0–97	0–83
Percent salmonids*	Mean	26.9	84.5	36.7	68.0
	SD	17.8	27.3	29.5	37.5
	Range	6–59	11–100	0–100	7–100
Percent invertivores	Mean	84.4	92.0	87.3	79.8
	SD	21.7	20.5	15.5	39.9
	Range	50–100	29–100	56–100	0–100
Percent intolerant species	Mean	93.1	94.2	86.2	99.8
	SD	13.2	18.4	17.9	0.6
	Range	62–100	36–100	46–100	98–100
Percent juvenile salmonids	Mean	56.2	65.3	47.0	57.9
	SD	20.0	22.7	30.1	37.7
	Range	33–84	36–100	0–87	0–100

Note: Metrics with an asterisk differ significantly ($p < 0.05$) among ecoregions; N, number of samples; SD, standard deviation.

and health, might also prove useful as indicators of biotic integrity, but this cannot be determined without information from degraded sites.

Results of the canonical correspondence analysis (CCA) (ter Braak, 1988) of fish species composition and environmental variables for the 37 sample sites are illustrated in Figure 11.4. Axes 1 and 2 of the CCA accounted for 66% of the explained joint variance in the fish species and environmental variables. Eight relatively unrelated but biologically relevant variables were used in the CCA: percent forest, watershed size, stream gradient, percent canopy opening, substrate size, percent embeddedness, pH, and specific conductance. Percent forest, watershed size, stream gra-

Characteristics of Fish Assemblages and Environmental Conditions

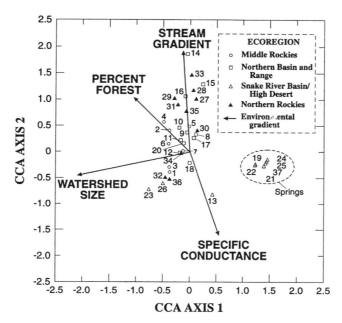

FIGURE 11.4 Canonical correspondence analysis (CCA) ordination plots showing 37 sample sites in the upper Snake River Basin based on fish species occurrence and the environmental variables specific conductance, watershed size, percent forest, and stream gradient. (Sites also are identified by their respective ecoregion.)

dient, and specific conductance were identified in the forward selection process by CCA as significant in the ordination of species data ($p < 0.05$).

The CCA of species presence and environmental gradients indicates the influence of multiple environmental variables on the distribution of species assemblages (Figure 11.3). Here, species locations in the ordination diagram represent the measured environmental optima in relation to a particular species. In addition, species that plot near the center of Figure 11.3 are more likely to occur in a wide range of habitat types or ecoregions. Rainbow trout, mottled sculpin, Paiute sculpin, and speckled dace are examples of these species that were collected in all four ecoregions.

The CCA also revealed some distributional differences in fish assemblages as a result of a natural barrier, land use, and watershed size. Four of the five fish species collected only downstream from Shoshone Falls (northern squawfish, chiselmouth, Shoshone sculpin, and largescale sucker) were located on the perimeter of the ordination diagram (Figure 11.3). These species, in addition to Utah sucker, were collected only from streams in predominantly rangeland watersheds in the lower part of the basin. These streams were characterized by relatively high specific conductance and low gradients. In contrast, bull trout and brook trout were collected from streams characterized by low specific conductance and high gradients. These salmonids, in addition to cutthroat trout, brown trout, and the Wood River sculpin, also were collected from streams in forested watersheds. Shoshone sculpin and bull trout were collected from springs or streams in smaller watersheds than those where northern squawfish and chiselmouth were collected.

Patterns in fish species distributions and assemblages, whether using taxonomic or functional descriptors, did not correspond well with accepted ecoregion boundaries. However, the findings do suggest that fish assemblages and habitat were distinctly different between spring sites in the SRB ecoregion and the other sites sampled (Figure 11.4). Ecoregion differences were evident only between the SRB and other ecoregions, primarily due to spring sites. The three stream sites in the SRB ecoregion did tend to plot together near the lower half of Figure 11.4; they had low gradients, were below 5200 ft. in elevation, and represented primarily rangeland watersheds. In contrast, stream sites in the upper half of Figure 11.4 had higher gradients, were at higher elevation, and represented primarily forested watersheds. Whittier et al. (1988) also showed relations between ecoregions containing distinct landscapes (montane and desert) and fish assemblages in Oregon streams.

The lack of correspondence between fish assemblage patterns and ecoregions in the USNK might be, in part, because many of the sampling sites were located near the boundary of adjacent ecoregions and, hence, may represent conditions characteristic of one or both ecoregions. The lack of correspondence between fish assemblages and the montane ecoregions may be the result of introduced fish species. Even though many of the sites sampled could be regarded as relatively undisturbed and remote, more than half (56%) of these, including sites upstream from Shoshone Falls that contained rainbow trout, contained introduced salmonids. These introductions effectively homogenize the fish assemblage in such depauperate systems. Correspondence between fish assemblages and ecoregions in the western United States probably would be more definitive if introduced species and stocked salmonids were not included in the data analyses. Additional study is needed to characterize fish assemblages of large streams and rivers in basins that encompass more than one ecoregion.

The data suggest some difficulties in the development of biotic indices using fish metrics at the ecoregion level. Results showed that least-disturbed streams in the basin are historically species depauperate. For example, streams in the NR ecoregion typically contained only one salmonid species. In all likelihood, some of the headwater montane sites probably lacked native fish species until introductions were made. Fisher (1989) also reported low fish species richness in northern Idaho headwater streams.

A common perception about aquatic ecosystems is that higher diversity equates with higher biotic integrity and, thus, better environmental quality. For example, the number of fish species in warmwater streams generally declines with decreasing environmental quality and is reflected in the scoring of metrics used in an IBI (Karr et al., 1986). This assumption of a positive relation between species richness and biotic integrity does not hold for coldwater streams of the USNK. Coldwater least-disturbed streams of the USNK tend to contain fewer fish species than do sites affected by human activities where tolerant species have been introduced. Therefore, efforts to maximize fish diversity in coldwater streams of the USNK might not result in higher biotic integrity.

11.3 CHARACTERIZING MAJOR STREAM TYPES: CASE STUDY

Data for this case study were collected as part of the NAWQA Program to characterize aquatic biota and associated habitats in surface water. This study included the sampling of fish assemblages and measurement of environmental variables for some of the major stream types in the USNK — large rivers, agricultural streams, and least-disturbed streams and springs in forested and/or rangeland watersheds.

11.3.1 METHODS

Fish assemblages and environmental variables were evaluated for 30 (first- through seventh-order) streams in the USNK (Figure 11.5). Fish surveys were conducted on stream reaches during base-flow conditions in summer and autumn, 1993 through 1995. Representative reaches were selected on the basis of criteria outlined by Meador (1993b). Reach length usually depended on the presence of at least two repeating geomorphic channel units (riffle, run, pool) per site. Site descriptions and specific information describing reach characteristics can be found in a report by Maret (1997).

Sites were selected for sampling by spatially stratifying the basin by ecoregion, land use, and site type. Five sites represented large rivers and included nonwadable tributaries of the Snake River (large river site type); 5 sites represented streams characterized by a direct association with irrigated

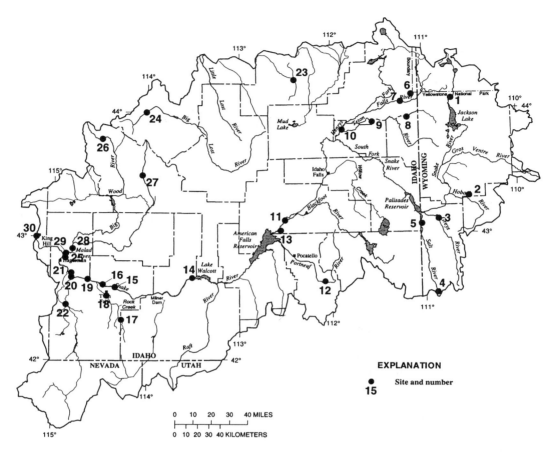

FIGURE 11.5 Site locations for case study on characterization of stream types in the upper Snake River Basin.

agriculture, row crop production, and livestock grazing (agricultural site type); and 20 sites represented reference streams and springs; 14 of these 20 sites were streams located primarily in forested and (or) rangeland watersheds (reference stream site type), and five were springs located in rangeland watersheds (reference spring site type). All the large river and agricultural sites had some form of water regulation (diversions or dams) upstream. Only four of the 20 reference sites had any form of water regulation upstream.

Six sites were selected to evaluate year-to-year variability in fish assemblages for various site types and environmental settings. These included Snake River at Flagg Ranch (site 1), Salt River near Etna (site 5), Portneuf River at Topaz (site 12), Rock Creek at Twin Falls (site 18), Big Lost River near Chilly (site 24), and Snake River at King Hill (site 30). Multiple reaches were also sampled at Rock Creek at Twin Falls and Big Lost River near Chilly to evaluate reach variability in fish assemblages. Fish sampling procedures followed methods outlined by Meador et al. (1993a). The collected fish were identified to species level, measured for total length and weight, examined for anomalies, and returned to the stream.

Twenty-four environmental variables consisting of watershed, hydrologic, and habitat characteristics were evaluated for each site (see Maret, 1997, for complete list). Watershed size, stream order, and land use were determined using a geographic information system (GIS). The following physical, hydrologic, and physicochemical habitat characteristics were determined at three to six transects within each stream reach sampled: reach length, width, depth, width/depth ratio, velocity, discharge, discharge as percent coefficient of variation (CV), specific conductance, water temperature, pH, dissolved oxygen, percent dissolved oxygen saturation, substrate size, percent embeddedness, percent substrate fines, percent cover, and percent open canopy (Platts et al., 1983; Meador et al., 1993b).

Fish metrics were used to compare taxonomically dissimilar communities over large spatial scales and to provide direct measures of biotic integrity and aquatic life beneficial uses (Karr, 1991). Fourteen fish metrics were determined: total number of fish collected, total number of fish per minute of electrofishing, total number of species, number of native species, percent anomalies, percent introduced species, percent common carp, percent cottids, percent salmonids, percent juvenile salmonids (<4 in. total length), percent adult salmonids (>8 in. total length), number of intolerant species, percent omnivores, and percent coldwater adapted. These fish metrics are based on ecological principles; many have predictable responses to human activities (Karr, 1991). Each species was categorized according to geographic origin (native or introduced); trophic group; and tolerance to organic pollution, warm water, and sediment pollution using protocols developed by Chandler et al. (1993). Water temperature preferences (cold- or warmwater adapted) were assigned using data compiled by Idaho Department of Health and Welfare (Don Zaroban, written communication, 1995).

A variety of analytical methods were used to describe and evaluate fish assemblages and environmental variables. Initially, the characteristics of the fish species collected were described and spatially displayed to evaluate patterns in the data. Fish collected at selected sites during multiple years or among multiple reaches were compared to determine spatial and temporal variability. Patterns were evaluated on the basis of the two *a priori* classification schemes: ecoregions and site types. Multivariate and multimetric analyses then were used to evaluate the fish and environmental data. Multivariate analyses are based on statistical algorithms, whereas multimetric analyses incorporate more descriptive ecological information. As a result of multivariate analyses, selected fish metrics and environmental variables were examined further using regression analysis and box plots, and medians were statistically tested among site types. These latter analyses helped evaluate specific relations between fish metrics and environmental variables and provided a visual description of variability among site types. Finally, selected fish metrics for the mainstem Snake River and its major tributaries were compared to identify longitudinal changes (from upstream to downstream) in fish assemblages.

11.3.2 Results and Discussion

A total of 5295 fish were collected during this study. The number of fish collected from any one site ranged from 11 to 666, and the mean for all sites was 115. Twenty-six fish species were collected (Figure 11.6) representing the families Catostomidae, Centrarchidae, Cottidae, Cyprinidae, Ictaluridae, Percidae, and Salmonidae. About 73% of the species collected from all sites were native species. The number of species collected from any one site ranged from 2 to 11, and the mean for all sites was 6.

The greatest number of species (19) were collected from large rivers and the fewest (8) were collected from the springs (Figure 11.6). Species common to all site types included the mottled sculpin, rainbow trout, redside shiner, speckled dace, and Utah sucker. Rainbow trout, mottled sculpin, and speckled dace were the most frequently collected species at 73, 60, and 50% of the sites, respectively. Shorthead sculpin, fathead minnow, and black bullhead were collected from only one site. Shoshone sculpin were also rare, collected only from a few spring sites near Hagerman. Species collected only from reference stream sites were brook trout, shorthead sculpin, and Wood River sculpin.

The total number of species and number of native species collected from all sites are geographically summarized in Figure 11.7. Most collections contained fewer than 200 individuals and comprised fewer than eight native species. Fewer than 100 individuals and fewer than five native

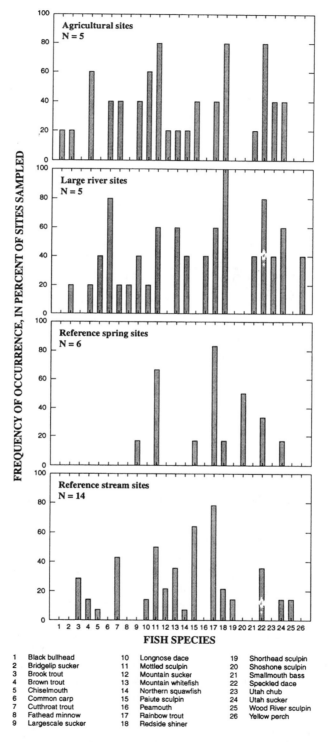

FIGURE 11.6 Frequency of occurrence of fish species by site types, upper Snake River Basin.

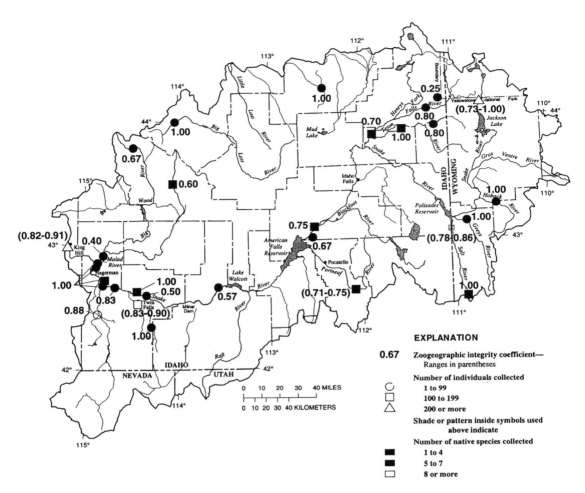

FIGURE 11.7 Summary of fish abundances, native species, and range of zoogeographic integrity index values for all sites in the upper Snake River Basin. (Locations and site numbers are the same as shown in Figure 11.5.)

species typically were collected from reference stream sites located at high elevations in the drainage basins. Maret et al. (1997) reported similar findings for small, first- and second-order reference streams throughout the USNK. The greatest number of individuals and native species were collected from the Snake River at Flagg Ranch and Snake River at King Hill sites. Although 10 species were collected at King Hill (site 30), this is only about one-half the number of native species that historically have been collected from this large river downstream from Shoshone Falls (Maret, 1995). The loss of the anadromous species from the basin accounts for some of this difference.

In general, native fish species are most abundant and diverse in relatively undisturbed environments (Moyle, 1986). Therefore, the occurrence of native fish species in relation to non-native species can indicate the extent of habitat degradation, which may include physical and (or) chemical changes detrimental to native fish species. Major faunal shifts in many streams in the western United States are the result of introduced fish species. Often, introduced fish species are better adapted than native species to thrive in altered habitats (Moyle, 1994). The status of biotic integrity is related to the extent of habitat disturbance and the occurrence of native versus introduced species. The zoogeographic integrity coefficient (ZIC), an index derived from the ratio of the number of native species to the total number of species, was used to evaluate the degree of habitat disturbance, whereby a value of 1 indicated an undisturbed environment and a value of 0 indicated a highly disturbed environment (Elvira, 1995). The range of ZIC values is spatially displayed for each sample site in Figure 11.7.

The basin-wide introduction of intolerant salmonid species, including brook, brown, and rainbow trout, confounds the use of introduced species as a metric for measuring habitat degradation. Introduced species were collected from a number of the reference streams during this study (Figure 11.7). The lowest ZIC index value of 0.25 was for the reference stream site, Robinson Creek at Warm River (site 6), where three of the four species collected were introduced salmonids. The native cutthroat trout was not collected from this site. Therefore, the ZIC index may not be a useful indicator of environmental disturbance where intolerant fish species such as salmonids commonly have been introduced and have become part of the resident fishery.

Eight fish metrics were selected to best illustrate the relations among fish metrics and site types (Figure 11.8, A–H). Number of species and number of native species were significantly greater for agricultural and large river site types than for reference stream and spring site types (Figures 11.8A and B). Fewer introduced species were collected from many of the reference stream and spring sites than from agricultural sites (Figure 11.8E). Springs typically contained only native species, with the exception of Devils Washbowl (site 15) upstream from Shoshone Falls, where rainbow trout were collected. Reference stream and spring sites also were generally smaller than agricultural and large river sites, which typically contain more fish species than do smaller streams (Fausch et al., 1984).

Percent omnivores and percent common carp are metrics that typically increase with increasing habitat degradation or other environmental disturbance. Percent omnivores was higher for agricultural and large river sites than for reference stream and spring sites, and common carp were collected only from the agricultural and large river sites (Figures 11.8C and D).

The number of intolerant species (Figure 11.8F) varied greatly for all site types except springs. The median was highest for reference stream sites and was significantly different from that for large river and spring sites.

Percent salmonid and percent cottid metric values also varied greatly; values ranged from 0 to more than 80% for some site types (Figures 11.8G and H). The median percent salmonids was highest for agricultural sites, primarily because of large percentages of salmonids collected from Salt River (site 5) and Rock Creek (site 18). According to the IDFG, Rock Creek is heavily stocked with rainbow and brown trout upstream and downstream from site 18 (Fred Partridge, Idaho Department of Fish and Game, written communication, 1993), which would likely inflate the median percent salmonids. The large river site type typically had few salmonids, and median values were significantly lower than those for agricultural and reference stream site types. The percent cottids was highest for reference stream and spring sites. Median percent cottids for agricultural and large river sites was significantly lower than for reference stream sites.

A CCA (canonical correspondence analysis) ordination is shown for all sites in Figure 11.9, excluding site 9 due to missing data. This analysis was designed to detect patterns of variation in the species assemblages that can best be explained by the measured environmental variables. Environmental variables with long vectors were more strongly correlated with the ordination axes than were those with short vectors. In other words, long vectors depict greater influence of that environmental variable in structuring the fish assemblage. This ordination analysis was constrained by the environmental variables shown in the figures and directly relates the gradients of the environmental variables to the fish assemblages.

Sites did not appear to correspond to ecoregions. CCA axis 1 (Figure 11.9) appeared to separate sites on the basis of watershed size. Large river and agricultural site types generally were spread along axis 1 (lower right), and the reference stream and spring site types were spread along axis 2 (upper left). There was high overlap in the kinds of species among site types in the CCA ordination.

Multiple-year and multiple-reach sites (1, 5, 12, 18, 24, and 30) generally grouped near each other (Figure 11.9), indicating that fish assemblages and environmental variables for these sites

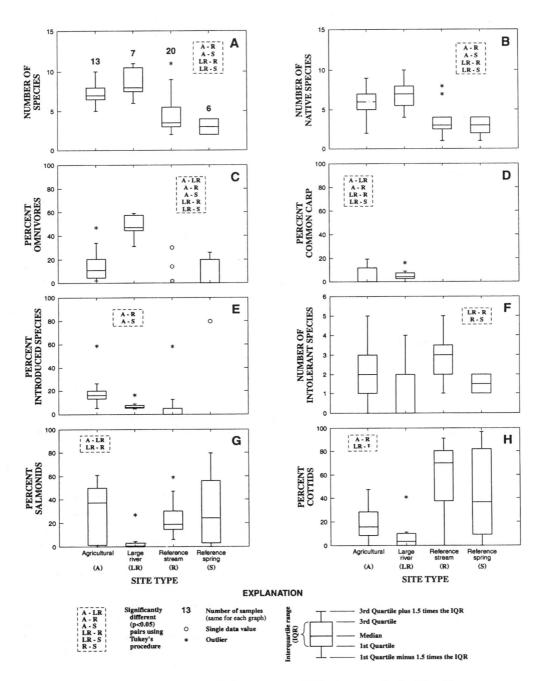

FIGURE 11.8 Selected fish metrics for all site types sampled in the upper Snake River Basin.

were similar. This similarity supports the premise that fish assemblages from a representative reach were indicative of local conditions.

CCA axes 1 and 2 accounted for 51 and 29%, respectively, of the explained joint variance in the fish assemblages and environmental variables. Most of the variability was accounted for by elevation, watershed size, percent agricultural land, and discharge CV, with eigenvalues of 0.40, 0.38, 0.28, and 0.24, respectively. Six of the eight variables were statistically significant in the

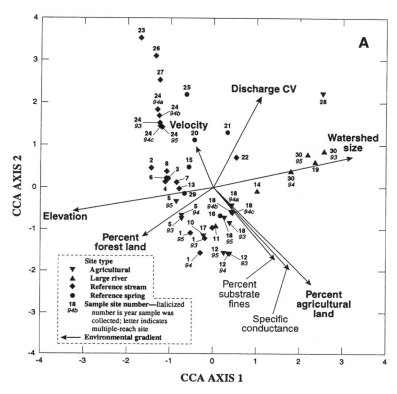

FIGURE 11.9 Canonical correspondence analyses (CCA) ordination plots of site scores by site type along selected environmental gradients, upper Snake River Basin. (Environmental variables in bold letters were significant ($p < 0.05$) along one or both axes.)

ordination for one or both axes, indicating that these six variables were correlated with the ordination axes. These six variables were elevation, watershed size, percent agricultural land, discharge CV, percent forest land, and velocity. Greater absolute values of canonical coefficients indicated stronger correlation between a variable and the axis tested. The variables with the strongest correlations along each axis had the greatest influence on the species composition of the samples. Most sites were categorized along axis 1 by elevation and watershed size, with canonical coefficients of –0.61 and 0.39, respectively. Percent forest land, percent agricultural land, and discharge CV, with canonical coefficients of –0.93, –0.61, and 0.39, respectively, were strongly correlated along axis 2. Even though specific conductance and percent substrate fines were not significant in the ordination for either axis, the strong correlation with percent agricultural land was evident with the alignment of the environmental vectors (Figure 11.9). Land use can be an important, large-scale factor affecting composition and structure of fish assemblages. Studies have shown the effects of siltation on salmonid fisheries (Chapman, 1988) and relations of fish assemblages to specific conductance (Matthews et al., 1992).

According to Jongman et al. (1995), sites that differ by a score of 4 standard deviations of CCA axis units tend to have few, if any, species in common. Sites approached this amount of separation on axis 1 and exceeded this amount of separation on axis 2. For example, the amount of separation between sites 12 and 23 was more than 5 for axis 2, which suggests that these sites have no species in common. These two sites are quite different. Medicine Lodge Creek (site 23) is a reference stream in a rangeland watershed and is a disjunct drainage where only two intolerant species were collected (rainbow trout and shorthead sculpin). The Portneuf River at Topaz (site 12) is severely affected by agricultural land use, as indicated by excessive siltation (83 to 92%

substrate fines and 88 to 99% embeddedness) and the presence of relatively large numbers of common carp.

The CCA analysis indicates that most sites are separated by stream size, which was inversely correlated with elevation. The development of an IBI using fish assemblages will need to account for this in deriving scoring criteria. Land use is another important factor to consider in development of an IBI. Streams in agricultural areas of the basin generally have fish assemblages different from those of other stream types sampled; therefore, metrics selection should account for this difference. CCA analysis also suggests that percentage of agricultural land use can be used as a indicator of human disturbance when validating and testing an IBI.

The results of this study indicate that GIS and multivariate analyses can be useful tools in characterizing fish assemblages and patterns relating to environmental variables at various landscape scales. Once similar patterns in fish assemblages are identified using these tools, a multimetric analysis, with indices such as the IBI, can be developed for various geographic regions and/or site types.

Habitats in the mainstem Snake River and major tributary sites of the USNK have been degraded as a result of agricultural land use and water regulation by diversions and reservoirs (Maret, 1995). These habitat changes are reflected in the type of fish assemblage inhabiting the Snake River and its major tributaries. Ten sites were selected (Figure 11.10A–F) to illustrate longitudinal changes in six fish metrics and differences in fish assemblages on the mainstem Snake River and major tributaries. All sites were located in watersheds where agricultural land use constituted 15 to 35% of the watershed, and where streamflow was highly regulated by upstream diversions and reservoirs. The exception was Snake River at Flagg Ranch (site 1), which is located in a primarily forested watershed where streamflow is not highly regulated. In a downstream direction, corresponding to river mile and elevation, sites 1, 11, 14, 19, and 30 represented the mainstem Snake River, and sites 5, 10, 12, 18, and 28 generally represented agricultural site types. The river mile for tributary sites represented the tributary's confluence with the Snake River. The range of elevations for all sites was about 6800 ft. at Snake River at Flagg Ranch to about 2500 ft. at Snake River at King Hill. The six fish metrics were number of native species, percent introduced species, percent omnivores, percent common carp, percent salmonids, and percent coldwater-adapted species. Maximum metric values were used to characterize multiple-year and multiple-reach sites.

The number of native species would be expected to decrease with habitat degradation, which may also allow for the invasion of introduced species. The number of native species (Figure 11.10A) ranged from 2 for Malad River near Gooding (site 28) to 10 for Snake River at King Hill (site 30). The number of native species collected from Malad River near Gooding was distinctly lower than from other sites. The high variability of discharge for this site is at least a partial explanation for the low number of species. A low of four native species was recorded for the Snake River near Minidoka (site 14). Only 1% of the fish collected from site 14 were coldwater species (Figure 11.10F). A possible explanation for the absence of coldwater species may be that Lake Walcott, a relatively shallow impoundment immediately upstream from the collection site, may be causing increased water temperatures at the site.

Introduced species (Figure 11.10B) were collected from all mainstem and tributary sites. Metric values would be expected to increase with habitat degradation. Percent introduced species was relatively constant and ranged from 5 to 26 for all sites except Malad River near Gooding (site 28), where percent introduced species was about 60. Moreover, three introduced salmonid species, likely the result of past stockings, were collected from the most upstream reference site, Snake River at Flagg Ranch (site 1).

Percent omnivores (Figure 11.10C) for all sites ranged widely from 6 to 92. Metric values would be expected to increase with habitat degradation. Rock Creek at Twin Falls (site 18), a tributary to the Snake River, had the lowest percent omnivores. Salmonid and cottid species, all of which are invertivores and indicators of higher quality habitats (Table 11.3), were predominant at

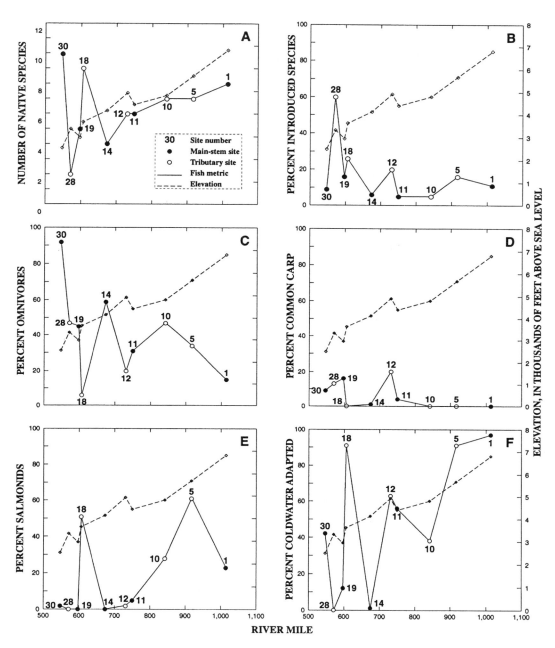

FIGURE 11.10 Selected fish metrics for five mainstem Snake River sites and five major tributary sites from Snake River at Flagg Ranch, Wyoming, to Snake River at King Hill, Idaho, influenced by agricultural land use. (River Mile for tributaries represents the confluence with the Snake River.)

this site. The Snake River near Minidoka (site 14) and at King Hill (site 30) had the highest percent omnivores (59 and 92, respectively), due primarily to the abundance of catostomids. For comparison, Hughes and Gammon (1987) used percent omnivores as a metric for evaluating trophic structure of fish assemblages in the Willamette River, a large, coldwater river in Oregon. They determined that fish assemblages exceeding 50% omnivores indicated habitat degradation, compared with less degraded reference habitats of the upper mainstem Willamette. Using their criteria, the predomi-

TABLE 11.3
Fish Metrics Used in Assessing Biotic Integrity of Coldwater Streams in Idaho

Metrics	Fisher (1989)	Robinson and Minshall (1992)	Chandler et al. (1993)
Species richness and composition			
Number of intolerant species	X	X	X
Number of introduced species	X	X	X
Percentage introduced species		X	
Number of native species		X	
Number of salmonid species	X	X	X
Number of species	X	X	
Percentage of salmonids	X		
Number of tolerant species		X	
Abundance of tolerant species		X	
Tolerant species biomass		X	
Trophic composition			
Number of benthic insectivores		X	X
Percentage of insectivores	X	X	X
Percentage of omnivores		X	X
Percentage of salmonids	X	X	
Percentage of carnivores		X	
Abundance, condition, and age structure			
Average salmonid length	X		
Average salmonid weight	X		
Salmonid condition index		X	
Percentage of anomalies	X		X
Percentage of hybrids	X		
Salmonid biomass	X	X	
Salmonid abundance		X	
Total fish abundance		X	X
Total fish biomass	X	X	
Percentage of young-of-year salmonids		X	X
Comparative index			
Jaccard's coefficient of similarity			X

Note: X denotes metric was used or recommended in assessing biotic integrity.

nance of omnivores at these Snake River sites suggests an unbalanced trophic structure, possibly resulting from habitat degradation. In addition, percent omnivores ranged from 0 to 15 for the Snake River at Flagg Ranch, suggesting that trophic structures for the Snake River near Minidoka and at King Hill sites are unbalanced, compared with this Snake River reference site.

Percent common carp (Figure 11.10D), a metric representing an introduced and highly tolerant species, also was used by Hughes and Gammon (1987). They determined that 10% common carp was a metric criteria representing conditions strongly deviating from reference conditions and indicating habitat degradation. Carp were collected from four mainstem and two tributary sites. Percent common carp was higher than 10 for Portneuf River at Topaz (site 12), Snake River near Buhl (site 19), and Malad River near Gooding (site 28). The presence of this species in these coldwater streams is a strong indication of habitat degradation, and this metric appears to be consistently responsive to environmental disturbance.

Percent salmonids (Figure 11.10E), a metric representing coldwater, intolerant species that are commonly present in the basin, ranged from 0 to 57. All mainstem and tributary sites were expected to support salmonid species on the basis of historical records and recent studies (Maret, 1995). Salmonids were not collected from Snake River near Minidoka (site 14), Snake River near Buhl (site 19), or Malad River near Gooding (site 28). The absence of salmonids at these locations may be indicative of habitat degradation. In contrast, percent salmonids collected from Snake River at Flagg Ranch (site 1), Salt River near Etna (site 5), and Rock Creek at Twin Falls (site 18) ranged from 23 to 61, percentages that would be expected at these sites.

Because most streams in the basin have been designated for beneficial use of coldwater biota (Maret, 1995), the fish assemblages would be expected to consist primarily of coldwater-adapted species (Figure 11.10F) in the absence of habitat degradation. Some mainstem Snake River sites near the lower end of the basin were not supporting a viable coldwater fishery, especially sites 14 and 19 (Minidoka and Buhl). Even though the fish assemblage for Snake River at King Hill (site 30) appeared to recover slightly, evidenced by collection of some coldwater species, salmonids were noticeably lacking in the fish assemblage, composing only 0 to 2% of the total number of fish collected. In addition, some of the tributaries to the middle reach of the Snake River, such as Malad River near Gooding (site 28), no longer supported a coldwater fishery. Analysis of the preceding metrics generally supports the water-quality-limited designation for the middle reach of the Snake River between Milner Dam and King Hill.

11.4 IMPORTANT COMPONENTS OF AN INDEX OF BIOTIC INTEGRITY

Fish metrics used to assess biotic integrity in Idaho coldwater streams are summarized in Table 11.3; 26 fish metrics, including species richness and composition, trophic composition, abundance, condition, age structure, and comparative index, have been tested in Idaho streams. Most of these metrics are associated with salmonid fishery attributes. The Idaho Division of Environmental Quality has published protocols for the use of fish information to assess Idaho streams (Chandler et al., 1993).

The NAWQA Program findings indicated that an IBI for the coldwater streams of the USNK will be simpler than versions previously developed for warmwater streams, which comprised 12 metrics (Karr et al., 1986). At least six fish metrics — number of native species, percent introduced species, percent omnivores, percent common carp, percent salmonids, and percent coldwater-adapted species — appeared to respond to varying degrees of human influence on the Snake River and its major tributaries. Using these metrics in an IBI will increase the probability of an accurate assessment (Fore et al., 1996).

In warmwater streams, the number of fish species generally declines with decreasing environmental quality and is reflected in the scoring of metrics used in an IBI (Karr et al., 1986). This assumption of a positive relation between species richness and biotic integrity does not extend to coldwater streams of the USNK. Coldwater reference streams of the USNK tend to contain fewer fish species than do sites affected by human activities, such as agricultural sites where tolerant species have been introduced. Lyons and others (1996) also noted that coldwater streams in Wisconsin contained fewer species than did warmwater streams, and attributed this difference to thermal preferences of fish species. Many more species are adapted to a warmwater habitat than to a coldwater habitat. Only when the degradation becomes so severe that even warmwater introduced species are lost would species richness decline.

Generally, most types of watershed degradation increase summer water temperatures in streams (Doppelt et al., 1993). Percent coldwater-adapted species is therefore a particularly useful metric because warming of a stream provides a less favorable habitat for native coldwater species and a more favorable habitat for a larger number of relatively tolerant introduced warmwater species. Moyle (1986) also reported this shift from a few native coldwater species to many warmwater introduced species for degraded streams in California. Where colonization from downstream reaches is hindered by waterfalls or disjunct drainages such as the Big Lost River drainage, warmwater species may not replace coldwater species, and species richness may remain relatively low or may decline after degradation. In addition, many reservoirs in the USNK have been stocked with warmwater species and provide a refugia for warmwater species to colonize favorable habitats.

Ideally, stocked salmonids should not be included in the calculation of an IBI because their presence does not directly reflect biotic integrity. A stream regularly stocked with trout may or may not have high biotic integrity, and the presence of these salmonids would likely inflate final IBI scores because some of the metrics in a coldwater IBI would be influenced by salmonid abundance. During the USNK studies, differentiation between natural and stocked fish at some sites was extremely difficult; therefore, all stocked fish were included in all metric calculations. Perhaps IDFG stocking records, along with physical appearance and size distribution, can be used in future studies to identify stocked fish.

Stream size was inversely related to elevation in the USNK studies, and both variables were strongly related to fish assemblages. Consequently, selection and scoring of metrics for an IBI would be influenced by stream size and elevation. In addition, reference spring sites appeared to support a fish assemblage different from that of other site types, which may require some specific adjustments in an IBI.

The data collected during the NAWQA Program, in addition to data from other studies, particularly for small first- and second-order streams, will provide the framework for developing an IBI for the various streams of the USNK. Ultimately, development of an IBI will provide an understanding of probable fish assemblages in the absence of significant human disturbance. Perhaps the most important result of developing an IBI is that it will provide resource managers with a formalized approach for establishing water-quality goals and evaluating the status of aquatic life beneficial uses.

11.5 FUTURE NEEDS

Much remains to be done in the West before an IBI approach is widely accepted as a monitoring tool. Descriptions of entire fish assemblages, particularly nongame species identifications of cottids and cyprinids, are needed to better understand relations between fish species and environmental conditions. The alarming disappearance of amphibians in many of the nation's mountainous streams emphasizes the need for inclusion of these species with fishery collections. Museum vouchering of hard-to-identify specimens as a common practice would provide better documentation of species occurrence and taxonomic characteristics. A peer-reviewed checklist of fish species in the Northwest, including guilds of pollution tolerances, trophic group of adults, species of concern, and temperature preferences, would provide a summary of the current level of professional knowledge regarding autecology. This formalized checklist would provide consistency and credibility to regional indices developed for water-quality assessment programs. More information is needed on fish assemblages and associated habitats of large rivers in the western United States. Most of these large rivers have been hydrologically modified by dams and reservoirs and affected by cumulative pollutants, but they are perhaps the least understood of all stream types. Last, before metrics are selected for inclusion in an IBI, more validation of the metrics against known levels of environmental disturbance is needed. Metric responsiveness to varying degrees of human disturbance in different environmental settings is paramount to the development of water-quality assessment tools designed to evaluate aquatic life beneficial uses of surface water.

This chapter provides a better understanding of the fish species and major environmental factors controlling their distribution and abundance in the USNK. The next step will be to put these concepts and principles into practice.

ACKNOWLEDGMENTS

Numerous USGS employees assisted in collecting and processing data during the high-intensity phase of the USNK NAWQA Program: Michael A. Beckwith, Gregory M. Clark, Steven L. Goodbred, Stephanie J. Eisenbarth, Molly A. Maupin, William H. Mullins, Douglas S. Ott, Michael G. Rupert, Terry M. Short, Kenneth D. Skinner, Marc A. Sylvester, Ian R. Waite, and Catherine A. Wolf, Marc S. Gamblin, Fred E. Partridge, Richard Scully, and Charles D. Warren, IDFG, contributed valuable advice and field assistance. John W. Kiefling, WGFD, also provided valuable assistance with the collection of field data. G. Wayne Minshall and Christopher T. Robinson, Idaho State University, supplied useful information and technical assistance with data analyses.

REFERENCES

Allan, J.D. and A.S. Flecker. 1993. Biodiversity conservation in running waters, *Bioscience*, 43, 32–43.

Baxter, G.T. and M.D. Stone. 1995. *Fishes of Wyoming*. Wyoming Game and Fish Department, Cheyenne.

Brockway, C.E. and C.W. Robinson. 1992. Middle Snake River Water Quality Study. Phase I: Final Report. University of Idaho, Idaho Water Resources Research Institute, Kimberly Research and Extension Center, Kimberly.

Chandler, G.L., T.R. Maret, and D.W. Zaroban. 1993. Protocols for Assessment of Biotic Integrity (Fish) in Idaho Streams. Idaho Department of Health and Welfare, Division of Environmental Quality, Water Quality Monitoring Protocols, Boise. Report Number 6.

Chapman, D.W. 1988. Critical review of variables used to define effects of fines in redds of large salmonids, *Transactions of the American Fisheries Society*, 117, 1–21.

Clark, G.M. 1994. Assessment of Selected Constituents in Surface Water of the Upper Snake River Basin, Idaho and Western Wyoming, Water Years 1975–89. U.S. Geological Survey Water Resources Investigations Report 93-4229.

Doppelt, B., M. Scurlock, C.A. Frissell, and J.R. Karr. 1993. *Entering the Watershed — A New Approach to Save America's River Ecosystems*. Island Press, Washington, DC.

Dynesius, M. and C. Nilsson. 1994. Fragmentation and flow regulation of river systems in the northern third of the world, *Science*, 266, 753–762.

Elvira, B. 1995. Native and exotic freshwater fishes in Spanish river basins, *Freshwater Biology*, 33, 103–108.

Evermann, B.W. 1896. A preliminary report upon salmon investigations in Idaho in 1894, *Bulletin of the United States Fish Commission for 1895*, 15, 253–284.

Fausch, K.D., J.R. Karr, and P.R. Yant. 1984. Regional application of an index of biotic integrity based on stream fish communities, *Transactions of the American Fisheries Society*, 113, 39–55.

Fisher, T.R., 1989. Application and Testing of Indices of Biotic Integrity in Northern and Central Idaho Headwater Streams. M.S. thesis, University of Idaho, Moscow.

Fore, L.S., J.R. Karr, and R.W. Wisseman. 1996. Assessing invertebrate responses to human activities — evaluating alternative approaches, *Journal of the North American Benthological Society*, 15, 212–231.

Frissell, C.A., W.J. Liss, C.E. Warren, and M.D. Hurley. 1986. A hierarchical framework for stream habitat classification — viewing streams in a watershed context, *Environmental Management*, 10, 199–214.

Gallant, A.L., T.R. Whittier, D.P. Larsen, J.M. Omernik, and R.M. Hughes. 1989. Regionalization as a Tool for Managing Environmental Resources. EPA 600/3-89-060. U.S. Environmental Protection Agency, Corvallis, OR.

Gauch, H.G., Jr. 1982. *Multivariate Analysis in Community Ecology*. Cambridge University Press, New York.

Gilbert, C.H. and B.W. Evermann. 1895. A report upon investigations in the Columbia River basin, with descriptions of four new species of fishes, *Bulletin of the United States Fish Commission for 1894*, 14, 169–179.

Gurtz, M.E. 1994. Design of biological components of the National Water-Quality Assessment (NAWQA) Program, in S.L. Loeb and A. Spacie (Eds.), *Biological Monitoring of Aquatic Systems*. Lewis, Boca Raton, FL. 323–354.

Hayslip, G.A. (Ed.). 1993. Region 10 In-Stream Biological Monitoring Handbook for Wadeable Streams in the Pacific Northwest. U.S. Environmental Protection Agency, Environmental Services Division, Seattle, WA.

Heede, B.H. and J.N. Rinne. 1990. Hydrodynamic and fluvial morphologic processes — implications for fisheries management and research, *North American Journal of Fisheries Management,* 10, 249–268.

Hughes, R.M. and J.R. Gammon. 1987. Longitudinal changes in fish assemblages and water quality in the Willamette River, Oregon, *Transactions of the American Fisheries Society,* 116, 196–209.

Hughes, R.M., S.A. Heiskary, W.J. Matthews, and C.O. Yoder. 1994. Use of ecoregions in biological monitoring, in S.L. Loeb and A. Spacie (Eds.), *Biological Monitoring of Aquatic Systems.* Lewis, Boca Raton, FL. 125–151.

Hughes, R.M. and D.P. Larsen. 1988. Ecoregions — an approach to surface water protection, *Journal of the Water Pollution Control Federation,* 60, 486–493.

Hughes, R.M., D.P. Larsen, and J.M. Omernik. 1986. Regional reference sites — a method for assessing stream potentials, *Environmental Management,* 10, 629–635.

Idaho Conservation Data Center. 1994. Rare, Threatened and Endangered Plants and Animals of Idaho. Idaho Department of Fish and Game, Boise.

Idaho Department of Health and Welfare. 1989. Idaho Water Quality Status Report and Nonpoint Source Assessment, 1988. Idaho Department of Health and Welfare, Division of Environmental Quality, Boise.

Idaho Department of Health and Welfare. 1990. Idaho Water Quality Standards and Wastewater Treatment Requirements, Title 1, Chapter 2. Idaho Department of Health and Welfare, Division of Environmental Quality, Boise.

Idaho Department of Health and Welfare. 1995. The Middle Snake River Nutrient Management Plan. Division of Environmental Quality, South Central Idaho Regional Office, Twin Falls.

Jackson, D.A. and H.H. Harvey. 1989. Biogeographic associations in fish assemblages — local vs. regional processes, *Ecology,* 70, 1472–1484.

Jongman, R.H.G., C.J.F. Ter Braak, and O.F.R. Van Tongeren (Eds.), 1995. *Data Analysis in Community and Landscape Ecology.* Cambridge University Press, New York.

Karr, J.R. 1991. Biological integrity — a long-neglected aspect of water resource management, *Ecological Applications,* 1, 66–84.

Karr, J.R. and D.R. Dudley. 1981. Ecological perspective on water quality goals, *Environmental Management,* 5, 55–68.

Karr, J.R., K.D. Fausch, P.L. Angermeier, P.R. Yant, and I.J. Schlosser. 1986. Assessing Biological Integrity in Running Waters — A Method and Its Rationale. Illinois Natural History Survey, Special Publication 5, Champaign.

Leahy, P.P., J.S. Rosenshein, and D.S. Knopman. 1990. Implementation plan for the National Water-Quality Assessment Program. U.S. Geological Survey Open-File Report 90-174.

Lyons, J., L. Wang, and T.D. Simonson. 1996. Development and validation of an index of biotic integrity for coldwater streams in Wisconsin, *North American Journal of Fisheries Management,* 16, 241–256.

Maret, T.R. 1995. Water-Quality Assessment of the Upper Snake River Basin, Idaho and Western Wyoming — Summary of Aquatic Biological Data for Surface Water Through 1992. U.S. Geological Survey Water-Resources Investigations Report 95-4006.

Maret, T.R. 1997. Characteristics of Fish Assemblages and Related Environmental Variables for Streams of the Upper Snake River Basin, Idaho and Western Wyoming, 1993–95. U.S. Geological Survey Water-Resources Investigations Report 97-4087.

Maret, T.R., C.T. Robinson, and G.W. Minshall. 1997. Fish assemblages and associated environmental correlates in least-disturbed streams of the upper Snake River Basin, western USA, *Transactions of the American Fisheries Society,* 126, 200–216.

Matthews, W.J., D.J. Hough, and H.W. Robison. 1992. Similarities in fish distribution and water quality patterns in streams of Arkansas — congruence of multivariate analyses, *Copeia,* 1992, 296–305.

Maupin, M.A. 1995. Water-Quality Assessment of the Upper Snake River Basin, Idaho and Western Wyoming — Environmental Setting, 1980–92. U.S. Geological Survey Water-Resources Investigations Report 94-4221.

Meador, M.R. 1992. Inter-basin water transfer — ecological concerns, *Fisheries,* 17(2), 17–22.

Meador, M.R., T.F. Cuffney, and M.E. Gurtz. 1993a. Methods for Sampling Fish Communities as Part of the National Water-Quality Assessment Program. U.S. Geological Survey Open-File Report 93-104.

Meador, M.R., C.R. Hupp, T.F. Cuffney, and M.E. Gurtz. 1993b. Methods for Characterizing Stream Habitat as Part of the National Water-Quality Assessment Program. U.S. Geological Survey Open-File Report 93-408.

Miller, D.L., P.M. Leonard, R.M. Hughes, J.R. Karr, P.B. Moyle, L.H. Schrader, B.A. Thompson, R.A. Daniels, K.D. Fausch, G.A. Fitzhugh, J.R. Gammon, D.B. Halliwell, P.L. Angermeier, and D.J. Orth. 1988. Regional applications of an index of biotic integrity for use in water resource management, *Fisheries,* 13(5), 12–20.

Moyle, P.B. 1986. Fish introductions into North America — patterns and ecological impacts, in H.A. Mooney and J.A. Drake (Eds.), *Ecology of Biological Invasions of North America and Hawaii.* Springer-Verlag, New York. 27–43.

Moyle, P.B. 1994. Biodiversity, biomonitoring, and the structure of stream fish communities, in S.L. Loeb and A. Spacie (Eds.), *Biological Monitoring of Aquatic Systems.* Lewis, Boca Raton, FL. 171–186.

Omernik, J.M. and A.L. Gallant. 1986. Ecoregions of the Pacific Northwest. EPA 600/3-86/033. U.S. Environmental Protection Agency, Corvallis, OR.

Platts, W.S., W.F. Megahan, and G.W. Minshall. 1983. Methods for Evaluating Stream, Riparian, and Biotic Conditions. U.S. Forest Service, General Technical Report INT-138, Ogden, UT.

Robins, C.R., R.M. Bailey, C.E. Bond, J.R. Brooker, E.A. Lachner, R.N. Lea, and W.B. Scott. 1991. Common and Scientific Names of Fishes from the United States and Canada. American Fisheries Society Special Publication 20, Bethesda, MD.

Robinson, C.T. and G.W. Minshall. 1992. Refinement of Biological Metrics in the Development of Biological Criteria for Regional Biomonitoring and Assessment of Small Streams in Idaho, 1991–1992. Idaho State University, Department of Biological Sciences, Pocatello.

Robinson, C.T. and G.W. Minshall. 1994. Biological Metrics for Regional Biomonitoring and Assessment of Small Streams in Idaho. Idaho State University, Stream Ecology Center, Pocatello.

Rosgen, D.L. 1994. A classification of natural rivers, *Catena,* 22, 169–199.

Scott, W.B. and E.J. Crossman. 1973. Freshwater Fishes of Canada. Fisheries Research Board of Canada Bulletin 184.

Sigler, W.F. and J.W. Sigler. 1987. *Fishes of the Great Basin.* University of Nevada Press, Reno.

Simon, T.P. and J. Lyons. 1995. Application of the index of biotic integrity to evaluate water resource integrity in freshwater ecosystems, in W.S. Davis and T.P. Simon (Eds.), *Biological Assessment and Criteria: Tools for Water Resource Planning and Decision Making.* Lewis, Boca Raton, FL. 245–262.

Simpson, J.C. and R.L. Wallace. 1982. *Fishes of Idaho.* The University Press of Idaho, Moscow.

Smith, G.R. 1981. Late Cenozoic freshwater fishes of North America, *Annual Reviews of Ecology and Systematics,* 12, 163–193.

Ter Braak, C.J.F., 1988. CANOCO — A FORTRAN Program for Canonical Community Ordination by Partial Detrended Canonical Correspondence Analysis, Principal Components Analysis and Redundancy Analysis (Version 2.1). Ministry of Agriculture and Fisheries, Agricultural Mathematics Group, Technical Report LWA-88-02, Wageningen, The Netherlands.

Thurow, R.F., C.E. Corsi, and V.K. Moore. 1988. Status, ecology, and management of Yellowstone cutthroat trout in the upper Snake River drainage, Idaho, *American Fisheries Society Symposium 4,* 25–36.

Tonn, W.M. 1990. Climate change and fish communities — a conceptual framework, *Transactions of the American Fisheries Society,* 119, 337–352.

U.S. Environmental Protection Agency. 1990. Biological Criteria — National Program Guidance for Surface Waters. EPA 440/5-90-004. U.S. Environmental Protection Agency, Office of Water Regulations and Standards, Washington, DC.

U.S. Water and Power Resources Service. 1981. Water and Power Resources Service Project Data. U.S. Water and Power Resources Service [Bureau of Reclamation], Boise, ID. 583–1212.

Warren, M.L., Jr. and B.M. Burr. 1994. Status of freshwater fishes of the United States — overview of an imperiled fauna, *Fisheries,* 19(1), 6–18.

Whittier, T.R., R.M. Hughes, and D.P. Larsen. 1988. Correspondence between ecoregions and spatial patterns in stream ecosystems in Oregon, *Canadian Journal of Fisheries and Aquatic Sciences,* 45, 1264–1278.

12 Classification of Freshwater Fish Species of the Northeastern United States for Use in the Development of Indices of Biological Integrity, with Regional Applications

David B. Halliwell, Richard W. Langdon, Robert A. Daniels, James P. Kurtenbach, and Richard A. Jacobson

CONTENTS

12.1	Introduction	302
	12.1.1 Overview of the Index of Biotic Integrity	303
12.2	Methodology	303
	12.2.1 Information Sources	304
12.3	Fish Species Distribution and Diversity	304
	12.3.1 Glaciation Effects on Species Richness and Composition	304
	12.3.2 Historical Environmental Perturbations	311
	12.3.3 Resident Fish Species Distribution	311
12.4	Aquatic Ecoregions	313
	12.4.1 Aquatic Reference Condition	313
12.5	Current Northeast Fish Species Distribution	313
	12.5.1 New Jersey	313
	12.5.2 New York	313
	12.5.3 Vermont-West	314
	12.5.4 New England	314
	12.5.5 Native Fish Species	315
	12.5.6 Introduced Fish Species	316
	12.5.7 Naturalized Fish Species	316
12.6	Resident Fish Species and Local Origins	316
	12.6.1 Resident Fish Species Definition and Groups	316
12.7	IBI and Trophic Categories	317
	12.7.1 Northeast Trophic Groups	317
	12.7.2 Generalist Feeders	317
	12.7.3 Water Column Insectivores	318
	12.7.4 Benthic Insectivores	318

12.7.5 Top Carnivores 318
  12.7.6 Minor Trophic Categories 318
12.8 IBI and Aquatic Habitat Classification 319
  12.8.1 Macrohabitat Class 319
12.9 IBI and Fish Species Tolerance Classification 319
  12.9.1 Disjunct Fish Species Populations 319
12.10 Considerations for Northeast IBI Formulation 320
12.11 Northeast IBI Regional (State) Applications 320
  12.11.1 Fish Sampling Methods 320
  12.11.2 Northeast IBI Metric Summary 321
  12.11.3 Species Richness and Composition Metrics 322
  12.11.4 Trophic Guild Structure Metrics 325
  12.11.5 Fish Abundance and Condition Metrics 325
12.12 Validation of Northeastern Lotic IBI Versions 326
  12.12.1 Constraints of Existing Northeastern IBI Versions 326
Acknowledgments 327
References 327
Appendix 12A 335
Appendix 12B 337

12.1 INTRODUCTION

The Index of Biotic Integrity (IBI) is based on the premise that biological communities (e.g., fish assemblages) respond to and are modified by human activities that may alter the physical, chemical, and biological processes of aquatic ecosystems (Karr, 1981; 1991; Karr et al., 1986). The IBI comprises 12 metrics used to assess: (1) fish species richness and composition; (2) trophic guild structure; and (3) relative abundance and condition. Each metric is scored using a standard that is generated from minimally impacted regional reference sites (Hughes et al., 1986; Hughes, 1995).

Extensive background information is required to develop an IBI. To use an IBI, it is necessary to classify fishes according to their origins, trophic or feeding status, aquatic habitats, and tolerance to environmental degradation (Fausch et al., 1990; Schlosser, 1990; Bramblett and Fausch, 1991). This information is required for the development of component IBI metrics, such as native fish species richness, proportion of fish as introduced species, number of benthic insectivores, and proportion of intolerant or tolerant fish species. This type of information is widely diffused throughout the peer-reviewed and nonreviewed literature. Researchers who have modified original indices of biological integrity for the Northeast (Miller et al., 1988) have relied on information from midwestern sources (Karr et al., 1986; Whittier et al., 1987; Plafkin et al., 1989) and local accounts.

The intent of this chapter is to summarize information on the origin, current distributional pattern, characteristic macrohabitat, trophic guild, and tolerance to environmental stress for 150 fish species found in freshwaters in the northeastern United States (New York, New Jersey, and the six New England states). Recognizing the relatively depauperate nature of fish assemblages in the northeastern United States, differences in fish species richness and distribution among northeastern U.S. subregions are documented. Fish species macrohabitat is classified by size and type of waterbody, and temperature regime. Fish species trophic classification, based on dominant food type and habitat from which food is taken, include categories such as: generalist feeder; water column insectivore; benthic insectivore; and top carnivore. This chapter also provides general classes of fish species tolerance to environmental perturbations.

Any IBI developed to bioassess freshwater environments in the northeastern U.S. needs to be modified according to aquatic habitat type and associated resident fish assemblages. When developing an IBI for the Northeast, it is particularly important to account for: (1) low native and resident fish species richness; (2) species origins and status of introduced fish species; and (3) the high proportion of generalist feeders within both regional reference and degraded sites.

This classification of freshwater fishes provides a synthesis of current information to facilitate further IBI development in the northeastern U.S. In addition, this chapter provides and compares recent fish assemblage IBI applications from the northeastern states of Vermont (Langdon, 1989), New Jersey (Kurtenbach, 1994), and Connecticut (Jacobson, 1994).

12.1.1 Overview of the Index of Biotic Integrity

Biotic integrity is best monitored at the community or assemblage level (Hughes et al., in revision). Similar to Simon and Lyons (1995), the IBI is viewed not strictly as a community analysis, but as an analysis of several hierarchical levels of biology. A representative sample of the fish assemblage is used to assess the relative condition of the resident biota within an aquatic community.[1]

Karr (1981) originally developed the IBI for resident fish assemblages to assess the relative quality of small to medium-sized, warmwater streams in the midwestern U.S. (Fausch et al., 1984; Karr et al., 1986). The IBI has since undergone a variety of modifications for use in other parts of the country (Miller et al., 1988; Southerland and Stribling, 1995), and has been applied in over 35 states (Karr and Dionne, 1991). Outside the U.S. (see Hughes and Oberdoff, Chapter 5), the IBI has been modified for streams in southern Ontario (Steedman 1988), France (Oberdorff and Hughes, 1992), Mexico (Lyons et al., 1995), and Australia (Harris, 1995). Other modifications of the IBI have been applied to estuarine fish assemblages in the Chesapeake Bay (Jordan et al., 1991) and Massachusetts (Deegan et al., 1993). The development of lentic IBI metrics are also under investigation (Karr and Dionne, 1991; Larsen et al., 1991; Whittier and Paulsen, 1992; Jennings et al., Chapter 21; Whittier, Chapter 22). An IBI has recently been developed specifically for Great Lakes littoral fish assemblages (Minns et al., 1994; Thoma, Chapter 16).

Past IBI development using fish assemblages in northeastern streams includes work in New York and Massachusetts (Miller et al., 1988), Vermont (Langdon, 1989; Burnham et al., 1991), Connecticut (Jacobson, 1994), and New Jersey (NJ-DEP, 1993; Kurtenbach, 1994).

Simon and Lyons (1995) provide a comprehensive overview of IBI applications to date, both lotic and lentic, including responses to criticisms of IBI concepts and use (Suter, 1993). Statistical properties of the IBI were reviewed by Fore et al. (1994) and Hughes et al. (in revision). Fish sampling protocols were studied by Angermeier and Karr (1986), Angermeier and Schlosser (1989), Lyons (1992), Simonson and others (1994), Angermeier and Smogor (1995), Paller (1995), and Simonson and Lyons (1995).

12.2 METHODOLOGY

Fish distribution, ecology, and tolerance information includes data from New York, New Jersey, and the New England states. Fish occurrence patterns are not structured along political boundaries, hence, distribution information is presented on the basis of state-specific as well as biogeographic boundaries. Presenting fish distribution information based on regional similarity supports current biomonitoring programs in the Northeast, including the U.S. Environmental Protection Agency's Environmental Monitoring and Assessment Program (EMAP: Larsen et al., 1991; Whittier and Paulsen, 1992) and the U.S. Geological Survey's National Water Quality Assessment Program

---

[1] The terms "assemblage" and "community" should not be used synonymously (Fauth et al., 1996). Accordingly, the term "assemblage" is used to denote a subset of a biological community composed of a single phylogenetic group (e.g., fish, aquatic insects, zooplankton, etc.).

(NAWQA: Gurtz, 1994; Gilliom et al., 1995). This broad scale information also supports regionally based bioassessments (Larsen and Christie, 1993; Whittier et al., 1997).

12.2.1 INFORMATION SOURCES

Natural history and distribution information for 150 freshwater fish species from a variety of historical and unpublished records from museums, academic institutions, and natural resource agencies is compiled in Table 12.1 (see Appendix A). Fish macrohabitat information, trophic classifications, and environmental tolerance information are derived from regional and state fish texts (Bigelow and Schroeder, 1953; Carlander, 1969; 1977; Scott and Crossman, 1973; Hubbs and Lagler, 1974; Lee et al., 1980; Trautman, 1981; Smith, 1985; Page and Burr, 1991) and numerous species-specific references (e.g., Keast, 1966; 1985; Robinson et al., 1996; Whittier et al., 1997).

Table 12.1 lists all fishes reproducing in freshwater environments in the Northeast, including landlocked forms of marine origin, such as introduced (e.g., rainbow smelt, alewife, white perch) and naturally occurring forms (e.g., sticklebacks and killifish). Fish species are cited by common names only in the text, after Hunter (1990), and scientific nomenclature is listed in Table 12.1, according to Robins et al. (1991).

Numerous fish species have been historically reported from Northeast freshwaters, whose occurrence is not recent, episodic, or of short duration (e.g., extirpated, exotic, or estuarine). Thirty of these rarely encountered fish species are listed in Appendix B and omitted from further consideration.

12.3 FISH SPECIES DISTRIBUTION AND DIVERSITY

The zoogeography of northeastern freshwater fish species has been determined by the post-glacial dispersal characteristics of the individual species, hydrological and physical conditions, and by climatic and human-induced events (Gilbert, 1980; Hocutt and Wiley, 1986). Hydrologic and physical incidents include major, system-wide effects, such as glaciation; and those that initially have local effects, such as headwater stream capture and lowland swamping (Halliwell, 1989). Anthropogenic impacts include a variety of mechanisms, such as intentional or accidental species introductions, the construction of canals that join drainages, and dams that alter fish habitats and local dispersal routes.

12.3.1 GLACIATION EFFECTS ON SPECIES RICHNESS AND COMPOSITION

Native fishes in the northeastern U.S. are almost exclusively a post-glacial fauna (Smith, 1985). Limited freshwater fish habitats existed in the region during the most recent glacial event (Wisconsinan stage) and native northeastern fishes have reached their present distributions in the past 8000 to 16,000 years. The depauperate nature of the native fish fauna is a direct result of the relatively short time available for fish to recolonize and the rapid formation of physical barriers following glaciation (Gilbert, 1976; 1980; Smith, 1985). Also, the postulated Atlantic Slope refugia, particularly for the New England region, were relatively depauperate (Schmidt, 1986; Stemberger, 1995). In contrast, the native freshwater fish fauna of the nonglaciated midwestern (Figure 12.1) and southeastern U.S. is comparatively rich (Moyle and Herbold, 1987; Warren and Burr, 1994).

Individual lakes and ponds, exclusive of the Great Lakes, Finger Lakes, and Lake Champlain, generally contain between three and 20 fish species. Wadeable streams and rivers generally have one to 15 species and, more typically, only six to eight species. Commonly encountered fish species from representative northeastern lotic and lentic datasets are listed in Table 12.2. The lake fish fauna of northwestern Maine and southcentral New Jersey (i.e., Pinelands) differ considerably from the typical fish assemblages generally found in northeastern lentic waters.

TABLE 12.1
Classification of Freshwater Fish Species of the Northeastern United States

	Native Distribution	Ecoregion	Occur	Water Class	Temp.	Trophic Class	Tolerance
PETROMYZONTIDAE (lampreys)							
Lampetra appendix (American brook lamprey)	NE [VT-E + ME]	BCG	L-R	B	C	NF	I
Ichthyomyzon fossor (northern brook lamprey)	NY and VT-W	CD	R	B	C	NF	I
Ichthyomyzon greeleyi (mountain brook lamprey)	New York	C	R	B	C	NF	I
Ichthyomyzon bdellium (Ohio lamprey)	New York	C	R	S	C-W	PF	M
Ichthyomyzon unicuspis (silver lamprey)	NY and VT-W	CD	L	R-L	C-W	PF	I
Petromyzon marinus (sea lamprey)	Northeast	ABD-G	L	R-L	C-W	PF	M
ACIPENSERIDAE (sturgeons)							
Acipenser brevirostrum (shortnose sturgeon)	NE [VT-E-W]	BCF	L-R	R	W	BV	I
Acipenser fulvescens (lake sturgeon)	NY and VT-W	CD	R	R-L	W	BV	I
Acipenser oxyrinchus (Atlantic sturgeon)	NE [VT-E-W]	BCF	L-R	R	W	BV	I
LEPISOSTEIDAE (gars)							
Lepisosteus osseus (longnose gar)	NY and VT-W	CD	L	S-L	W	TC	M
AMIIDAE (bowfins)							
Amia calva (bowfin)	1-2 Int 3-4 6 and 8	CD	L	S-L	W	TC	T
ANGUILLIDAE (freshwater eels)							
Anguilla rostrata (American eel)	Northeast	A-G	C	S-L	W	TC	T
CLUPEIDAE (herrings)							
Alosa aestivalis (blueback herring)	NE Int VT-W	BCFG	L	R-L	W	PI	M
Alosa pseudoharengus (alewife)	NE [VT-E]	BCFG	L-C	S-L	C-W	PI	M
Alosa sapidissima (American shad)	NE [VT-W]	BCFG	L	R	W	PI	M
Dorosoma cepedianum (gizzard shad)	1 3-4 6-7 Int 2	BCFG	L	R-L	W	PH	T
CYPRINIDAE (carps and minnows)							
Campostoma anomalum (central stoneroller)	New York	CD	C	S	C-W	BH	T
Carassius auratus (goldfish)	Int Northeast	Exotic	C	R-L	W	GF	T
Clinostomus elongatus (redside dace)	New York	ACD	C	B	W	WC	I
Couesius plumbeus (lake chub)	NE [NJ-CT-RI]	AC-E	L-R	R-L	C	GF	M
Cyprinella analostana (satinfin shiner)	NY and NJ	ACEG	C	S	W	WC	T
Cyprinella spiloptera (spotfin shiner)	NY-NJ and VT-W	CDE	C	S-L	W	WC	T
Cyprinus carpio (common carp)	Int Northeast	Exotic	C	R-L	W	GF	T

TABLE 12.1 (continued)
Classification of Freshwater Fish Species of the Northeastern United States

	Native Distribution	Ecoregion	Occur	Water Class	Temp.	Trophic Class	Tolerance
Erimystax dissimilis (streamline chub)	New York	C	R	S	W	BI	I
Erimystax x-punctatus (gravel chub)	New York	C	R	S	W	BI	I
Exoglossum laurae (tonguetied minnow)	New York	C	L	S	W	BI	M
Exoglossum maxillingua (cutlips minnow)	I-4 Int 6	CDFG	C	S-L	W	BI	I
Hybognathus hankinsoni (brassy minnow)	NY and VT-W	AD	L	S	W-B	BH	M
Hybognathus regius (eastern silvery minnow)	NE [4-5] Int 9	B-DFG	L-R	R-L	W	BH	I
Luxilus chrysocephalus (striped shiner)	New York	CD	L	S	W	WC	T
Luxilus cornutus (common shiner)	Northeast	A-EG	L-C	S-L	C-W	GF	M
Lythrurus umbratilis (redfin shiner)	New York	D	U	S	W	WC	T
Macrhybopsis storeriana (silver chub)	New York	D	R	S	W	BI	I
Margariscus margarita (pearl dace)	NY VT-W ME	ACD	L-C	S-L	C-B	GF	M
Nocomis biguttatus (hornyhead chub)	New York	CD	L	S	W	GF	M
Nocomis micropogon (river chub)	New York	CD	L	S-R	W	GF	M
Notemigonus crysoleucas (golden shiner)	Northeast	A-G	C	S-L	W	GF	T
Notropis amblops (bigeye chub)	New York	CD	L	S	W	WC	M
Notropis amoenus (comely shiner)	NY and NJ	EG	L	S-R	W	WC	T
Notropis anogenus (pugnose shiner)	New York	D	R	S-R	W	WC	I
Notropis atherinoides (emerald shiner)	1 and 2 Int 6 and 9	CD	L	R-L	W	WC	M
Notropis bifrenatus (bridle shiner)	NE [VT-E]	A-EG	L	S-L	W	WC	I
Notropis buccatus (silverjaw minnow)	New York	C	R	B	W	WC	T
Notropis chalybaeus (ironcolor shiner)	NY and NJ	EFG	R	S	W-B	WC	M
Notropis dorsalis (bigmouth shiner)	New York	C	L	B	W	WC	M
Notropis heterodon (blackchin shiner)	NY and VT-W	ACD	R	R-L	W	WC	I
Notropis heterolepis (blacknose shiner)	1 and 2 and 7-9	ACD	U	B-L	C-W	BI	I
Notropis husdonius (spottail shiner)	NE Int Maine	BCDG	L-C	R-L	W	WC	M
Notropis photogenis (silver shiner)	New York	C	L	S	W	WC	T
Notropis procne (swallowtail shiner)	NY and NJ	CG	L	S	W	WC	M
Notropis rubellus (rosyface shiner)	NY and VT-W	CD	C	S-R	W	WC	I
Notropis stramineus (sand shiner)	NY and VT-W	CD	L	S-L	W	GF	M
Notropis volucellus (mimic shiner)	1 and 2 Int 6 and 7	A-D	L	S-L	W	GF	M

Phoxinus eos (northern redbelly dace)	NE [NJ-CT-RI]	ACD	L-R	S-L	C-B	GF	M
Phoxinus neogaeus (finescale dace)	1 and 2 and 7-9	AC	L	S-L	C-B	GF	M
Pimephales notatus (bluntnose minnow)	1 and 2 Int 3-8	B-EG	C-U	S-L	W	GF	T
Pimephales promelas (fathead minnow)	1 and 2+7-9 Int 3-6	B-EG	C-U	S-L	W	GF	T
Rhinichthys atratulus (blacknose dace)	Northeast	A-G	C	B-S	C-W	GF	T
Rhinichthys cataractae (longnose dace)	Northeast	A-EG	C	B-S	C-W	BI	M
Scardinius erythrophthalmus (rudd)	Int 1-2, 6 and 9	Exotic	U	S-L	W	WC	T
Semotilus atromaculatus (creek chub)	Northeast [RI]	AC-EG	C	S-L	C-W	GF	T
Semotilus corporalis (fallfish)	Northeast	A-EG	C	S-L	C-W	GF	M
CATOSTOMIDAE (suckers)							
Carpiodes cyprinus (quillback)	NY and VT-W	CD	R	R-L	W	GF	T
Catostomus catostomus (longnose sucker)	NE [NJ-CT-RI]	AC-E	U-R	S-L	C	BI	I
Catostomus commersoni (white sucker)	Northeast	A-EG	C	S-L	C-W	GF	T
Erimyzon oblongus (creek chubsucker)	NE [VT-E-W]	B-G	L	S-L	W	GF	I
Hypentelium nigricans (northern hog sucker)	New York	ACD	C	S	W	GF	M
Moxostoma anisurum (silver redhorse)	NY and VT-W	CD	L-R	S-R	W	BI	M
Moxostoma carinatum (river redhorse)	New York	CD	R	S-R	W	BI	I
Moxostoma duquesnei (black redhorse)	New York	C	R	R	W	GF	I
Moxostoma erythrurum (golden redhorse)	New York	CD	L	R	W	BI	I
Moxostoma macrolepidotum (shorthead redhorse)	NY and VT-W	CD	L-R	S-R	W	BI	M
Moxostoma valenciennesi (greater redhorse)	NY and VT-W	CD	L-R	R-L	W	BI	I
ICTALURIDAE (bullhead catfishes)							
Ameiurus catus (white catfish)	1 and 3 Int 4-9	CFG	L	R-L	W	TC	M
Ameiurus natalis (yellow bullhead)	1 and 3 Int 6-8	BCE-G	L	S-L	W-B	GF	T
Ameiurus nebulosus (brown bullhead)	Northeast	A-G	C	S-L	W	GF	T
Ictalurus punctatus (channel catfish)	1 and 2 Int 3-4+6-8	AD	L	R-L	W	TC	M
Noturus flavus (stonecat)	NY and VT-W	CD	L-R	S-L	W	BI	M
Noturus gyrinus (tadpole madtom)	1 and 3 Int 6+8	C-F	L-R	S	W	BI	M
Noturus insignis (margined madtom)	1 and 3 Int 6+8	ACDG	L-R	S	W	BI	M
Noturus miurus (brindled madtom)	New York	CD	L	S	W	BI	M
ESOCIDAE (pikes)							
Esox americanus americanus (redfin pickerel)	NE [VT-E]	BCFG	L-C	S-L	W-B	TC	M
Esox americanus vermiculatus (grass pickerel)	New York	CD	L	S-L	W	TC	M
Esox lucius (northern pike)	1 and 2 Int 3-9	ACD	L-C	R-L	C-W	TC	I
Esox masquinongy (muskellunge)	1 and 2 Int 3 and 9	CD	U	R-L	C-W	TC	I
Esox niger (chain pickerel)	Northeast	A-G	C	S-L	W	TC	M

TABLE 12.1 (continued)
Classification of Freshwater Fish Species of the Northeastern United States

	Native Distribution	Ecoregion	Occur	Water Class	Temp.	Trophic Class	Tolerance
UMBRIDAE (mudminnows)							
Umbra limi (central mudminnow)	1 and 2 Int 4 and 6	CD	L	S-L	W-B	GF	T
Umbra pygmaea (eastern mudminnow)	NY and NJ	CFG	L	S-L	W-B	GF	T
OSMERIDAE (smelts)							
Osmerus mordax (rainbow smelt)	NE Int VT-E-W	B-F	L	S-L	C	GF	I
SALMONIDAE (trouts)							
Coregonus artedi (cisco or lake herring)	NY and VT-W	ACD	L	L	C	PL	I
Coregonus clupeaformis (lake whitefish)	1 and 2 and 7-9	ACD	L-U	R-L	C	BI	I
Prosopium cylindraceum (round whitefish)	1 and 2 and 7-9	AC	U-R	L	C	BI	I
Oncorhynchus mykiss (rainbow trout)	Int Northeast	Exotic	L	S	C	TC	I
Salmo salar (Atlantic salmon)	1 and 9 Int 4-8	A-E	L	S-L	C	TC	I
Salmo trutta (brown trout)	Int Northeast	Exotic	C	S	C	TC	I
Salvelinus alpinus (Arctic char)	ME Ext. 7 and 8	A	R	L	C	TC	I
Salvelinus fontinalis (brook trout)	Northeast	A-G	C	B-L	C	TC	I
Salvelinus namaycush (lake trout)	1-2+7-9 Int 3 and 6	AC-E	L	L	C	TC	I
PERCOPSIDAE (trout-perches)							
Percopsis omiscomaycus (Trout-perch)	NY and VT-W	CD	L-R	S-L	C-W	WC	M
APHREDODERIDAE (pirate perches)							
Aphredoderus sayanus (pirate perch)	NY and NJ	DF	L	S-L	W-B	WC	M
GADIDAE (cods)							
Lota lota (burbot)	NE [NJ and RI]	ACD	L-R	S-L	C	TC	M
FUNDULIDAE (killifishes)							
Fundulus diaphanus (banded killifish)	Northeast	A-G	C	S-L	W	WC	T
Fundulus heteroclitus (mummichog)	NE [VT-E-W]	BCF	L	S-L	C-W	GF	T
POECILIIDAE (livebearers)							
Gambusia holbrooki (eastern mosquitofish)	New Jersey	F	U	S-L	W	WC	T
Gambusia affinis (western mosquitofish)	Int NY and NJ	Exotic	U	S-L	W	WC	T
ATHERINIDAE (silversides)							
Labidesthes sicculus (brook silverside)	New York	CD	L	S-L	W	WC	I

Classification of Freshwater Fish Species of the Northeastern United States

Species	Region						
GASTEROSTEIDAE (sticklebacks)							
Culaea inconstans (brook stickleback)	1 and 2 and Maine	ACD	U	S-L	C-W	WC	I
Gasterosteus aculeatus (threespine stickleback)	NE [VT-E-W]	B-DF	R	S-L	C-W	WC	M
Apeltes quadracus (fourspine stickleback)	NE [VT-E-W]	BCF	U	S	C-W	WC	M
Pungitius pungitius (ninespine stickleback)	NE [VT-E-W]	A-DF	U	S-L	C-W	WC	M
COTTIDAE (sculpins)							
Cottus bairdi (mottled sculpin)	NY and VT-W	CD	L	B-L	C-W	BI	M
Cottus cognatus (slimy sculpin)	Northeast [RI]	AC-EG	L-U	B-L	C	BI	I
MORONIDAE (temperate basses)							
Morone americana (white perch)	NE Int VT-E-W	B-DFG	L	R-L	C-W	TC	M
Morone chrysops (white bass)	New York	CD	L	R-L	W	TC	T
Morone saxatilis (striped bass)	NE [VT-W]	BCFG	L	R	W	TC	?
CENTRARCHIDAE (sunfishes)							
Acantharchus pomotis (mud sunfish)	NY and NJ	CF	L	S-L	W-B	WC	M
Ambloplites rupestris (rock bass)	1 and 2 Int 3-8	ACDE	C	S-L	C-W	TC	M
Enneacanthus chaetodon (blackbanded sunfish)	New Jersey	F	L	S-L	W-B	WC	I
Enneacanthus gloriosus (bluespotted sunfish)	NY and NJ	CDFG	L	S-L	W-B	WC	I
Enneacanthus obesus (banded sunfish)	NE [2+7+9]	BF	L-R	S-L	W-B	WC	I
Lepomis auritus (redbreast sunfish)	Northeast	A-CEG	C	S-L	W	GF	M
Lepomis cyanellus (green sunfish)	NY Int 3-4+6	D	U	S-L	W	GF	T
Lepomis gibbosus (pumpkinseed)	Northeast	A-G	C	S-L	W	GF	M
Lepomis macrochirus (bluegill)	NY Int 2-8	CDE	C	S-L	W	GF	T
Micropterus dolomieu (smallmouth bass)	1-2 Int 3-9	CD	C	R-L	C-W	TC	M
Micropterus salmoides (largemouth bass)	NY Int 2-9	CD	C	S-L	W	TC	M
Pomoxis annularis (white crappie)	NE Int 2-4+6	CD	U	R-L	W	TC	T
Pomoxis nigromaculatus (black crappie)	NY Int 2-9	CD	C	S-L	W	TC	M
PERCIDAE (perches)							
Ammocrypta pellucida (eastern sand darter)	NY and VT-W	C	R	S	W	BI	I
Etheostoma blennioides (greenside darter)	New York	CD	L	S	W	BI	I
Etheostoma caeruleum (rainbow darter)	New York	CD	L	B-S	W	BI	I
Etheostoma camurum (bluebreast darter)	New York	C	R	S	W	BI	I
Etheostoma exile (Iowa darter)	NY and VT-W	CD	L-R	S-L	W	BI	M
Etheostoma flabellare (fantail darter)	NY and VT-W	ACD	L	B-S	W	BI	M
Etheostoma fusiforme (swamp darter)	NE [VT-E-W]	BF	L-U	S-L	W-B	BI	I
Etheostoma maculatum (spotted darter)	New York	C	R	S	W	BI	I
Etheostoma nigrum (johnny darter)	New York	CD	L	S-L	W	BI	M

TABLE 12.1 (continued)
Classification of Freshwater Fish Species of the Northeastern United States

	Native Distribution	Ecoregion	Occur	Water Class	Temp.	Trophic Class	Tolerance
Etheostoma olmstedi (tessellated darter)	Northeast [ME]	A-EG	C	S-L	C-W	BI	M
Etheostoma variatum (variegate darter)	New York	C	L	S	W	BI	M
Etheostoma zonale (banded darter)	New York	C	L	S-R	W	BI	I
Perca flavescens (yellow perch)	Northeast	A-G	C	S-L	C-W	TC	M
Percina caprodes (logperch)	NY and VT-W	CD	L	S-L	W	BI	M
Percina copelandi (channel darter)	NY and VT-W	CD	L-R	R-L	W	BI	I
Percina evides (gilt darter)	New York	C	R	S-R	W	BI	I
Percina macrocephala (longhead darter)	New York	C	R	S-R	W	BI	I
Percina maculata (blackside darter)	New York	CD	L	S	W	BI	M
Percina peltata (shield darter)	NY and NJ	CG	U	S	C-W	BI	M
Stizostedion vitreum (walleye)	1 and 2 Int 3-4+6-8	CD	L-C	R-L	C-W	TC	M
SCIAENIDAE (drums)							
Aplodinotus grunniens (freshwater drum)	NY and VT-W	CD	L	R-L	W	BV	M
SOLEIDAE (soles)							
Trinectes maculatus (hogchoker)	NE [VT-E-W]	BF	C	S-R	W	GF	I

Notes: **Key — Fish Assemblage Classes:** Native Distribution [brackets] = ABSENCE; Int = introduced; NE = Northeast; 1 = New York; 2 = Vermont-West; 3 = New Jersey; 4 = Connecticut; 5 = Rhode Island; 6 = Massachusetts; 7 = Vermont-East; 8 = New Hampshire; 9 = Maine.
Ecoregion (Omernik, 1987, showing state inclusion): A = Northeastern highlands (ME, NH, VT, MA, CT, NY, NJ); B = Northeastern coastal zone (New England, NY); C = Northern Appalachian plateau and uplands (NY, VT-W); D = Erie/Ontario Lake plain (NY); E = Northcentral Appalachians (NY, NJ); F = Middle Atlantic coastal plain (NJ); G = Northern piedmont (NJ).
Frequency of Occurrence: C = Common, widespread; U = Uncommon, sporadic; L = Local occurrence only; R = Rare.
Water Class: B = Brooks (smaller flowing waters: <5 meters wide); S = Streams (intermediate flowing waters: 5-10 meters wide); R = Rivers (larger flower waters: >10 meters wide); L = Lakes (inclusive of ponds and reservoirs).
Water Type — Thermal Regime: C = Coldwater; C-W = Inhabits both types/coolwaters; W = Warmwater; C-B = Cold-Bogs; W-B = Warm-Bogs.
Trophic Class (* primary classes): * GF = Generalist feeder (e.g., omnivorous fishes); * WC = Water column insectivore (e.g., numerous cyprinids); * BI = Benthic insectivore (e.g., sculpins, darters); * TC = Top carnivore (e.g., salmonids, esocids, black basses); NF = Nonparasitic filterers (e.g., brook lampreys); PF = Parasitic filterers (e.g., sea lamprey); BV = Benthic invertivores (e.g., sturgeons); PI = Planktivorous invertivores (e.g., clupeids); PH = Planktivorous herbivores (e.g., gizzard shad); BH = Benthic herbivores (e.g., *Hybognathus* spp.).
Tolerance (to environmental perturbations) Class (Tol): I = Intolerant; M = Intermediate; T = Tolerant.

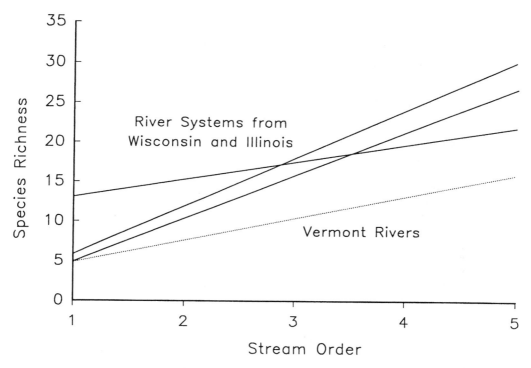

FIGURE 12.1 Comparison of maximum species richness lines between three midwestern river drainages (Karr et al., 1986) and stream drainages in Vermont (Langdon, 1989).

12.3.2 HISTORICAL ENVIRONMENTAL PERTURBATIONS

Human disturbance has affected fish populations in the northeastern U.S. for over 200 years. European settlement in the late 1700s and early 1800s brought widespread change in land use (Thomson, 1958). Most of the region was deforested by the mid-1800s, leading to degraded water quality by siltation and increased water temperature, which reduced the abundance and distribution of coldwater fishes (Titcomb, 1926; Waters, 1995). By the late 1800s, aquatic habitats were further altered and migratory patterns disrupted by the construction of small mill-dams throughout the region, many of which survive today (Schmidt, 1986). By the turn of the century, industrial pollution had adversely affected water quality, fish abundance, and distribution (Gordon, 1937; Bailey, 1938; Bailey and Oliver, 1939). Since the late 1980s, point-source water pollution has been reduced in the Northeast, while nonpoint sources of pollution (e.g., erosion, sedimentation, and leaching) have increased and continue to negatively impact water quality.

It is probable that few waterbodies currently support truly pristine fish assemblages; even remote lakes and streams are not unaffected. Airborne pollutants, associated with acid deposition and transport of metals and organic pollutants (e.g., mercury and PCBs) can be detected throughout remote regions in the northeastern U.S. (Haines, 1981; Haines and Baker, 1986; Gallagher and Baker, 1990).

12.3.3 RESIDENT FISH SPECIES DISTRIBUTION

Human activities have altered the distribution of most resident fishes so that it is difficult to delineate their "natural" ranges. By the mid-1800s, many currently popular game fishes (e.g., largemouth and smallmouth bass, brook and brown trout, northern pike, chain pickerel, bluegill, yellow perch, and brown bullhead), as well as goldfish and common carp, were extensively moved throughout

TABLE 12.2
Rank Order of Occurrence and Trophic Status of the 20 Most Commonly Encountered Fish Species in Northeastern Wadeable Streams and Lakes and Ponds

Wadeable Streams		Rank	Lakes and Ponds	
Species	Trophic Status		Species	Trophic Status
Blacknose dace	GF	1	Pumpkinseed	GF
White sucker	GF	2	Yellow perch	TC
Longnose dace	BI	3	Golden shiner	GF
Brook trout	TC	4	Brown bullhead	GF
Common shiner	GF	5	Largemouth bass	TC
Tessellated darter	BI	6	Chain pickerel	TC
Creek chub	GF	7	Bluegill	GF
Brown trout	TC	8	White sucker	GF
Fallfish	GF	9	Black crappie	TC
Slimy sculpin	BI	10	American eel	TC
Pumpkinseed	GF	11	Brook trout	TC
Largemouth bass	TC	12	Banded killifish	WC
American eel	TC	13	Smallmouth bass	TC
Brown bullhead	GF	14	White perch	TC
Bluegill	GF	15	Yellow bullhead	GF
Northern redbelly dace	GF	16	Redbreast sunfish	GF
Longnose sucker	BI	17	Creek chub	GF
Golden shiner	GF	18	Fallfish	GF
Redbreast sunfish	GF	19	Common shiner	GF
Chain pickerel	TC	20	Tessellated darter	BI

Data Source Summary — Sample Sizes

Lotic			Lentic		
(Halliwell, 1989)	MA-DFW	1430	ME-DIFW	2039	(Fenderson)
(Jacobson, 1994)	CT-DEP	1090	EMAP-SW	195	(Whittier)
(Langdon, 1989)	VT-DEC	314	CT-DEP	87	(Jacobson)
(Kurtenbach, 1994)	NJ-EPA	164	NJ-DFW	85	(Kurtenbach)

Note: Generally includes species occurring at frequencies >10%. Regional differences exist. In Maine (Northwestern) lakes: rainbow smelt, northern redbelly dace, lake chub, blacknose dace, and landlocked salmon are among the top twenty most commonly encountered fishes. In New Jersey (Pineland) lakes: bluespotted sunfish, mud sunfish, swamp darter, eastern mudminnow, and banded sunfish are among the top twenty most commonly encountered fishes.

the Northeast (Cardoza et al., 1993). Similarly, baitfish species, such as golden shiner and fathead minnow, have been widely distributed by official and illegal introductions into numerous waterbodies, including northern coldwater lakes in Maine and the Adirondacks of New York. Historically, yellow perch and golden shiner rarely occurred in Adirondack lakes except for a few of the larger lakes (Mather, 1886). In the last century, human intervention has enabled these species to become prominent members of the fish assemblage in most larger Adirondack lakes, as well as many smaller upland northeastern lakes (George, 1981; Gallagher and Baker, 1990).

The prevalence of introduced species, particularly top carnivores (gamefish) and cyprinids (baitfish), generally increases with human use and population density (Whittier et al., 1997). The presence of introduced species can be considered a perturbation of the aquatic ecosystem (deviation from the natural state fish assemblage), locally reducing biological integrity (Karr et al., 1986; Bramblett and Fausch, 1991; Lyons et al., 1996; Mundahl and Simon, Chapter 15).

12.4 AQUATIC ECOREGIONS

Omernik (1987) classified ecoregions in the U.S. by evaluating regional patterns of land use, vegetation, soils, and land surface. It is hypothesized that the flora and fauna among ecoregions are more similar than between adjacent ecoregions (Hughes et al., 1987). This correspondence between stream fish assemblages and aquatic ecoregions has been confirmed in several states including Ohio (Larsen et al., 1986), Kansas (Hawkes et al., 1986), Oregon (Hughes et al., 1987), Arkansas (Rohm et al., 1987), and Wisconsin (Lyons, 1989). Ecoregion mapping has been evaluated for use in establishing numerical biocriteria for aquatic life in only two northeastern states (i.e., Vermont, Langdon, 1989; and Massachusetts, Griffith, 1994).

12.4.1 AQUATIC REFERENCE CONDITION

The evaluation of any aquatic assemblage is based on a comparison between the observed condition and the expected condition. The expected condition represents the "biological potential" for that particular site, as defined by a regional standard or reference condition (Hughes et al., 1986; Hughes, 1995). An important preliminary step in developing numerical biocriteria is the identification of regional reference sites (U.S. EPA, 1990). Reference sites support fish assemblages that reflect the unique qualities of that ecoregion and are minimally impacted. It is inappropriate to select sites in regulated streams or fluctuating reservoirs as references when evaluating the effect of these types of alterations (Karr and Dionne, 1991).

Generally, reference sites in lotic waters support fish assemblages dominated by top carnivores (i.e., trout, pickerel, black bass) and benthic insectivores (e.g., sculpin and darters). Generalist feeders (e.g., white sucker, blacknose dace, creek chub) may occur at low densities in reference stream fish assemblages (Lyons et al., 1996). Comparatively little has been done to establish reference conditions for lentic systems (Baker et al., 1997).

12.5 CURRENT NORTHEAST FISH SPECIES DISTRIBUTION

Some 150 fish species inhabit freshwater environments in the northeastern states (Table 12.1). Only 27 (18%) of these species occur throughout the entire northeast region.

12.5.1 NEW JERSEY

The distribution of freshwater fishes in New Jersey generally corresponds with Omernik's (1987) Middle Atlantic Coastal Plain and Northern Piedmont ecoregions (Figure 12.2); 78 fish species occur in New Jersey freshwaters. Eastern mosquitofish and blackbanded sunfish occur only in New Jersey within the northeast region. New Jersey shares 10 additional species in common with only New York (Table 12.1: satinfin, comely, ironcolor, and swallowtail shiners; eastern mudminnow; pirate perch; western mosquitofish; mud sunfish; bluespotted sunfish, and shield darter).

12.5.2 NEW YORK

Nearly all of the fishes listed for the Northeast inhabit the waters of New York (98%). South-flowing rivers possess 34 fish species (i.e., Allegheny, Susquehanna, and Delaware Rivers) generally corresponding with the Northern Appalachian Plateau and Uplands and the North Central Appalachian ecoregions (Figure 12.2). Three species (i.e., redfin shiner, silver chub, and pugnose shiner) are

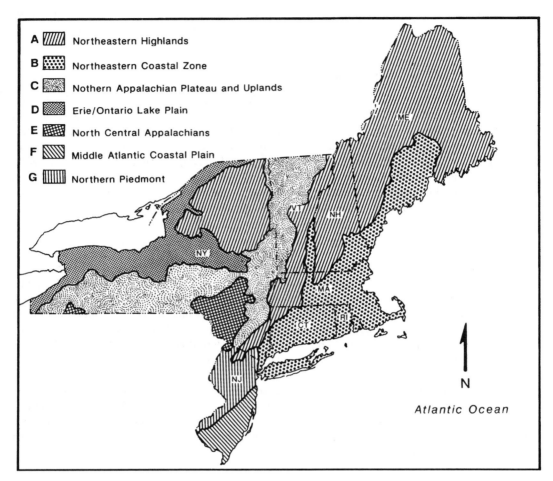

FIGURE 12.2 Map delineating Omernick's (1987) ecoregions for the northeastern U.S.

found only in Lakes Erie and Ontario and their tributaries, in partial correspondence with the Lake Erie and Ontario Lake Plain ecoregion.

12.5.3 VERMONT-WEST

In Vermont, fish species richness patterns differ considerably between western or Lake Champlain drainages (VT-W) and the eastern Connecticut River drainage (VT-E). The Lake Champlain drainage has 90 fish species, whose distribution generally corresponds to the Northern Appalachian Plateau and Uplands ecoregion. In common with New York, the Lake Champlain drainage has 37 species that do not occur in the New England drainages east of the Green Mountains. These relatively species-rich VT-W drainages are excluded from our fish faunal characterization of the "New England" fish region.

12.5.4 NEW ENGLAND

The New England freshwater fish fauna comprises 80 species. New York and New England share 13 species in common that are absent from New Jersey waters, including: lake chub, pearl dace, blacknose shiner, northern redbelly dace, finescale dace, brook stickleback, and lake, and round whitefish. These fish are primarily northern coolwater species (Lyons et al., 1996) with distributions generally corresponding with the Northeastern Highlands ecoregion of northern New York (i.e., Adirondacks), Vermont, New Hampshire, and Maine. Several northern species, (e.g., lake chub and northern redbelly dace) occur as disjunct relict populations in Massachusetts along the southern periphery of their distributions (Halliwell, 1989). The only native fish species present in New England, and absent from New York and New Jersey, is the Arctic char, which is native to Maine and has been extirpated from New Hampshire and Vermont (Kircheis, 1980).

12.5.5 Native Fish Species

The New York fish fauna has 141 native fish species, which is about twice as rich as that of the Lake Champlain drainage (VT-W) and nearly three times greater than the native fish fauna of New Jersey or the combined New England states (Figure 12.3). The large number of native species found in New York reflects the five major drainage systems, including the southwestern Allegheny River watershed draining into the species-rich Ohio River, a tributary to the Mississippi River.

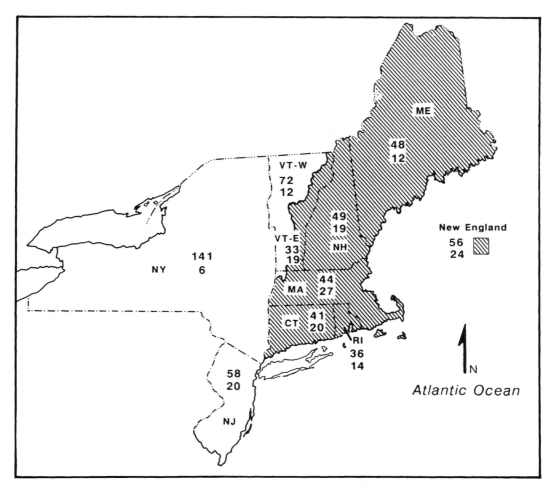

FIGURE 12.3 Number of established fish species, native (upper number) and introduced (lower number), by northeastern state and New England combined (cross-hatched area).

12.5.6 Introduced Fish Species

Introduced fish species found in northeastern freshwaters include exotic oriental and European species (e.g., goldfish, common carp, brown trout, and rudd). Rudd were introduced by the baitfish trade and are known to populate several waters in Maine and Massachusetts (Hartel et al., in press). In New York, a population of rudd has been present in a single stream system (Roeliff-Jansen Kill) since 1936 (Smith, 1985). Rudd are capable of hybridizing with the native golden shiner under laboratory conditions (Burkhead and Williams, 1991) and are currently prohibited from sale in most northeastern states.

Rainbow trout (salmon), from Pacific coast origins, have been extensively stocked in northeastern waters to support a recreational trout fishery. Unlike the well-established brown trout, self-sustaining populations of rainbow trout are rare in the Northeast. Several native estuarine and typically anadromous species (e.g., white perch, rainbow smelt, and alewife) have been introduced into numerous inland lakes and reservoirs throughout the Northeast.

12.5.7 Naturalized Fish Species

Several fish species were introduced as sportfish throughout the Northeast over a century ago (Smith, 1985; Schmidt, 1986). In Maine, Kendall (1914) deemed "the introduced smallmouth bass as sufficiently established to be admitted as a state fish." Largemouth and rock bass, bluegill, black crappie, yellow bullhead, and the exotic brown trout are all widespread and often common to abundant in southern New England waters, where they may be considered "naturalized" fish species.

More recently introduced fish species do not warrant such consideration, including white catfish, native to New York and New Jersey waters and established in several southern New England waters. White crappie and green sunfish are represented by incidental local stockings only (Whitworth et al., 1968) and have expanded their range over the past 25 years in the New Jersey Northern Piedmont ecoregion. Other native New York sportfish introduced on a local basis into New Jersey and southern New England waters include lake trout, northern pike, muskellunge, channel catfish, and walleye.

12.6 RESIDENT FISH SPECIES AND LOCAL ORIGINS

Two basic assumptions for applying an IBI are: (1) the fish sample is representative of the entire resident fish assemblage (Karr et al., 1986); and (2) that metric formulation and site classifications are conducted by trained biologists familiar with the local fish fauna (Fausch et al., 1990). An awareness of local species origins is required to ensure that these basic IBI assumptions are met. Also, appropriate selection of IBI metrics requires an understanding of the ecological and evolutionary mechanisms forming resident fish assemblages (Bramblett and Fausch, 1991; Strange, Chapter 4).

12.6.1 Resident Fish Species Definition and Groups

Development of an IBI requires clear definition of fish species that actually reflect ambient environmental conditions based on residency. The term "resident" species is used to distinguish a natural or adaptive assemblage of fish (Karr and Dudley, 1981) represented by several age classes that populate similar aquatic environments (Ricklefs, 1990). In most cases, fish species may be classified into one of the following four groups:

1. *Resident indigenous:* naturally occurring native species populating suitable aquatic habitats: e.g., brook trout, slimy sculpin, and longnose sucker in coldwaters; brown bullhead, pumpkinseed, and golden shiner in warmwaters.
2. *Resident naturalized:* well-established nonnative species populating suitable aquatic habitats: e.g., exotic brown trout in coldwaters; introduced smallmouth bass in coolwaters.

3. *Nonresident transient:* nonpopulating fish species found to occur in unsuitable aquatic habitats: e.g., individual lentic fish species sampled above or downstream of impoundments (e.g., yellow perch, bluegill, and chain pickerel).
4. *Nonresident stocked:* nonpopulating fish species introduced for a recreational fishery only: e.g., salmonid stocks, inclusive of catchable-sized "native" brook trout of hatchery origins.

In formulating an IBI for the northeastern U.S. it is strongly recommended that only resident fish species be used to estimate taxonomic richness response metrics. The occurrence of nonresident fishes in a sample are often unpredictable, not replicable, and do not necessarily indicate site-specific environmental conditions. Simon and Lyons (1995) have also recommended the exclusion of migratory and stocked fish (Lyons et al., 1996), as well as transient reservoir or lentic species (Karr et al., 1986; Bramblett and Fausch, 1991).

12.7 IBI AND TROPHIC CATEGORIES

To develop indices of biological integrity, it is necessary to assign each fish species to a trophic category (Keast, 1966; Austen et al., 1994). Karr et al. (1986) originally recognized fish species from four feeding categories: herbivore, omnivore, insectivore, and piscivore. Each of these trophic categories are well represented in midwestern streams (Whittier et al., 1987; Bramblett and Fausch, 1991). However, fish assemblages in the Northeast are comparatively depauperate, with most trophic groups represented by few species.

Tailoring trophic categories to specific geographical areas is appropriate for developing regional indices of biological integrity (Leonard and Orth, 1986; Miller et al., 1988). For example, Minns et al. (1994) used three fish trophic groups (i.e., generalist, specialist, and piscivore) to characterize littoral zone fish assemblages in selected areas of the Great Lakes.

Since most northeastern freshwater fish may be considered carnivorous, classification of species is based on a combination of dominant food items and the habitat type from which the food items are taken. Possible combinations of prey type and feeding habitats can be viewed as a continuum (Keast, 1966; Gerking, 1994; Goldstein and Simon, Chapter 7). At opposite ends of the continuum are feeding habitat specialists and generalists (Leonard and Orth, 1986). Specialists tend to forage in restricted habitats and exhibit a relatively narrow range of preferred prey types. In contrast, generalists exploit a wider range of habitats in search of a greater variety of prey types.

12.7.1 NORTHEAST TROPHIC GROUPS

Four trophic categories for northeastern U.S. freshwaters are recommended, a generalist feeding group (including omnivores) and three specialist feeding groups: water column insectivore; benthic insectivore; and top carnivore (including piscivores). A herbivorous category was not deemed useful for northeastern waters, where only four, locally distributed fish species are classified as strict herbivores (i.e., gizzard shad, central stoneroller, brassy minnow, and eastern silvery minnow).

When assigning fish species to a particular trophic category, it is also necessary to consider life-history stage. Certain species, such as parasitic lampreys, change diets and feeding ecology dramatically as they mature. Although ontogenetic diet niche shifts in other species are more gradual (e.g., black bass), they are equally important to consider (Olson, 1996). Species that show ontogenetic diet niche shifts can be classed as water column insectivores as young fish and top carnivores as older and larger individuals.

12.7.2 GENERALIST FEEDERS

The term "generalist" is used to describe fish species that consume a wide variety of food types from a wide variety of habitats. In the Northeast, this is a relatively large group composed primarily of omnivores. Common omnivorous fish species include white sucker, golden and common shiner, bluntnose and fathead minnow, and brown and yellow bullhead. Here, several species listed as insectivores by other authors (Karr et al., 1986; Whittier et al., 1987; Miller et al., 1988) are classified as generalist feeders (e.g., pumpkinseed, bluegill, fallfish, and creek chub). These species forage across many habitat types and consume a diversity of insect prey. Species in the generalist group consume both aquatic and terrestrial invertebrates, small fish, and vegetation from bottom, mid-water, and surface locations. Generalist feeders are able to adapt to an unstable forage base, which is characteristic of degraded habitat conditions (Fausch et al., 1990). More than 50% of the fish species commonly encountered in northeastern freshwater habitats can be classified as generalist feeders (Table 12.2).

12.7.3 WATER COLUMN INSECTIVORES

Water column insectivores are species that primarily consume insects taken from the water surface or the upper half of the water column (Keast, 1966). Such fish species may occasionally consume other invertebrates as well. Most water column feeders are cyprinids (e.g., spottail and bridle shiner). This group can also include juvenile stages of larger piscivorous species, such as trout, bass, pickerel, and perch. Invertebrates, particularly insects, are the dominant prey item of younger fish; whereas older, larger fish prey predominantly on fish (Keast, 1985).

12.7.4 BENTHIC INSECTIVORES

Benthic insectivorous fish species are "specialists" that primarily consume insects taken from the bottom substrate. This group is represented by species such as sculpins, darters, and several minnow species (e.g., cutlips minnow). Specialists are morphologically and behaviorally constrained to more specific prey types and habitats and are generally less tolerant of environmental degradation (Karr et al., 1986).

12.7.5 TOP CARNIVORES

Top carnivores are large, mobile fish species capable of capturing active, mobile prey, inclusive of larger invertebrates (Hughes et al., in revision; Goldstein and Simon, Chapter 7). This group primarily includes adults of piscivorous species such as black bass, trout, pickerel, and perch. In cold headwater streams, where trout are the only top carnivore present, they are classified as top carnivores, regardless of age or size.

12.7.6 MINOR TROPHIC CATEGORIES

Several northeastern freshwater fishes are difficult to place into any single major trophic category. Such species occupy a position near the specialist end of the trophic continuum and have restricted distributions in the Northeast. With the exception of freshwater brook lampreys (nonfeeding as adults) and silver and Ohio lampreys (parasitic), all of these fishes are anadromous or landlocked, including sturgeons (benthic invertivores) that feed essentially on mollusks and crustaceans; alewife, blueback herring, and American shad (planktivores); lamprey ammocoetes (filter feeders); and adult sea lampreys (parasitic carnivores). With the exception of lake sturgeon, lamprey ammocoetes, and introduced lacustrine alewife and blueback herring populations, most of these species are limited to larger coastal riverine habitats in the Northeast. The impacts of dams on fish migration tend to complicate the evaluation of biological integrity when assessing these anadromous fish species populations. These largely migratory fish species should be considered as a separate grouping in the formulation of indices of biological integrity for northeastern aquatic environments.

12.8 IBI AND AQUATIC HABITAT CLASSIFICATION

The development of IBI metrics and assignment of scoring criteria are dependent on the type of aquatic habitat studied. Lotic environments can be classified on the basis of size, water chemistry, elevation, gradient, substrate, and water temperature (Pennak, 1971; Platts, 1980; Platts et al., 1983). In contrast, lentic waters have traditionally been classified according to degree of trophic development (e.g., oligotrophic, mesotrophic, eutrophic, and dystrophic) (Sly and Busch, 1992).

12.8.1 MACROHABITAT CLASS

Fish species are classified along two macrohabitat variables (Table 12.1) based on waterbody type (lotic, lentic, bog) and temperature regime (cold, cool, and warmwater) (Table 12.1). Some species are not restricted to a single macrohabitat class; for example, yellow perch populate both flowing and standing waters that range from cold- to warmwater. All species tend to prefer an optimum range of environmental conditions within this spectrum of aquatic habitats (Eaton et al., 1995). As a result, macrohabitats are listed as those aquatic environments where a species is capable of reproducing. Species classified as lotic and lentic sometimes require both of these environments to complete their life cycles (e.g., landlocked forms of anadromous species such as Atlantic salmon, rainbow smelt, alewife, and lamprey).

Flowing water habitat for fish species are classified according to the summer base-flow wetted width of the stream channel, from small to large: "brooks," <5 m wide; "streams," 5 to 10 m wide; and "rivers," >10 m wide. The macrohabitat class "bog" is used for fish species associated with stained waters that are typically acid and brown in color, following the classification of Graham (1993) for naturally acidic lakes in New Jersey.

12.9 IBI AND FISH SPECIES TOLERANCE CLASSIFICATION

The recognition of habitat-induced stress factors and the use of fish assemblages as indicators of environmental degradation play an integral role in the development of Indices of Biological Integrity (Fausch et al., 1990). Three general classes of fish species tolerance to environmental perturbations are used: intolerant, intermediate or moderate, and tolerant. To date, only midwestern U.S. fish tolerance classifications are available in the literature (Karr et al., 1986; Whittier et al., 1987; Plafkin et al., 1989; Poff and Allan, 1995; Wichert, 1995). A number of fishes listed as intolerant in midwestern classifications appear to be more tolerant in northeastern waters (e.g., mottled sculpin, rock bass, and fantail darter). This variation may be due to clinal variation in a fish species' environmental requirements over their geographic range as well as varying responses to different degrees of environmental degradation (Leonard and Orth, 1986; Simon and Lyons, 1995; Whittier and Hughes, 1998).

12.9.1 DISJUNCT FISH SPECIES POPULATIONS

The environmental tolerance of a disjunct fish population may differ considerably on the periphery of its range when contrasted to populations within its primary geographical range (Karr, 1991). Disjunct Massachusetts relict populations of lake chub and northern redbelly dace occur along the extreme southern fringe of their primarily northern distribution (Halliwell, 1989). Similarly, disjunct populations of lake chub and redside dace occur in New York waters (Smith, 1985). These limited fish populations may be considered intolerant, based on a precarious existence where populations have low genetic viability (Soule, 1987). Conversely, where these species are commonly found, they are generally considered to be moderately tolerant (Table 12.1).

12.10 CONSIDERATIONS FOR IBI FORMULATION

The formulation of Indices of Biological Integrity for aquatic environments in the northeastern U.S. can be complicated for several reasons, including: (1) low regional and site-specific native and resident fish species richness; (2) widespread abundance of introduced fish species; and (3) a large proportion of the resident fish assemblage at both reference and impacted sites are generalist feeders. Low species richness may pose a problem in IBI formulation since the theoretical framework of the index is dependent on at least moderate levels of species richness (Fausch et al., 1990) and diverse trophic guilds (Miller et al., 1988; Bramblett and Fausch, 1991). A fish assemblage of at least four resident species is recommended for IBI application in Vermont's wadeable streams (Langdon, 1989). Species richness metrics require a distinction between native and introduced fish species, since an abundance of introduced fish are indicative of degraded aquatic conditions (Hughes and Gammon, 1987). As discussed, it is also important to clearly define those species that reflect ambient environmental conditions by virtue of their sustained residency in the aquatic habitat being assessed. Not surprisingly, several different IBIs are needed to adequately assess northeastern aquatic habitats. An IBI should use only validated metrics and provide: criteria for metric scoring; a well-defined list of species included in each metric; the rationale used to define expected metric performance; and constraints of use based on geographic region and waterbody type.

12.11 NORTHEAST IBI REGIONAL (STATE) APPLICATIONS

An important assumption of the IBI is that individual metrics may be modified based on regional differences in fish assemblage structure and distribution, yet still retain the theoretical framework of the index.(Fausch et al., 1984). Simon and Lyons (1995) recognized that generating an acceptable set of expectations is perhaps the most difficult part in developing modifications of the IBI and effectively applying an existing version to a new geographic area.

As reported by Miller et al. (1988), IBI modifications were initially developed for northeastern streams based on pilot studies in Massachusetts and New York. These early studies were largely based in large, cool- and warmwater stream drainages (Simon and Lyons, 1995). Additional modifications of the IBI have been developed for fish assemblages in wadeable streams in Vermont (Langdon, 1989), northern New Jersey (Kurtenbach, 1994), and Connecticut (Jacobson, 1994). Many of these streams are perennially coldwater streams populated by wild trout. Within both impacted and reference sites, drainage areas ranged from 13 to 259 sq. km in New Jersey, 5 to 300 sq. km in Vermont, and 36 to 42 sq. km in Connecticut. A separate, unpublished IBI (NJ-DEP, 1993) was also developed for wadeable New Jersey streams in the Northern Piedmont ecoregion having drainage areas less than 78 sq. km. Hughes and Omernik (1981; 1983) found watershed area to be a more useful and accurate definition of relative stream size than the widely used ordering system originally developed by Horton (1945) and modified by Strahler (1957).

12.11.1 Fish Sampling Methods

Fish assemblages were sampled by fishery biologists during summer base-flow conditions using backpack, float-boat, or stream-side electrofishing gear, without block nets. Generally, a single upstream pass was made, including several pool-riffle sequences (Angermeier and Karr, 1986) within at least a 100-meter stream reach. A smaller range of sample reach lengths (40 to 80 meters) were used in Connecticut studies (Jacobson, 1994). Paller (1995) found that it was more efficient to sample a larger area with a single electrofishing pass than to sample a smaller area with multiple passes. For wadeable streams less than 5 to 6 meters wide, a single, multihabitat electrofishing pass is generally acceptable if an adequate length of 100 to 200 meters is sampled. This sampling design approximates currently recommended specifications of 30 to 40 stream widths to obtain representative fish samples in small streams (Lyons, 1992; Angermeier and Smogor, 1995; Paller, 1995; Simonson and Lyons, 1995; Barbour et al., in revision). All fish greater than 25 mm TL (1 year or older in the CT-SI) were identified to species, counted, checked for disease and anomalies, and, in most cases, voucher specimens collected.

12.11.2 NORTHEAST IBI METRIC SUMMARY

New Jersey and Vermont stream researchers have attempted to maintain the theoretical IBI framework as developed by Karr et al. (1986) for warmwater streams in the Midwest (Table 12.3). The New Jersey IBI (Table 12.4) uses 10 of the original IBI metrics, while the Vermont IBI (Table 12.5) uses 9, with omission of the number of water column species (i.e., sunfish and trout species). In contrast, the Connecticut Specific Index (SI) uses only eight metrics (Table 12.6), including several replacement metrics that take into account the prevalence of lentic fish species in Connecticut headwater streams. The metrics and scoring criteria of the Connecticut SI are tailored to the depauperate, coldwater fish assemblage structure characteristic of small, headwater streams in Connecticut. A reduction in the number of metrics (8–10 vs. 10–12) reflects, in part, the simplified structure and function of coldwater fish assemblages relative to warmwater fish assemblages (Simon and Lyons, 1995; Lyons et al., 1996).

TABLE 12.3
Comparison of IBI Metrics Developed in New Jersey, Vermont, and Connecticut for Headwater Streams

No.	IBI Metric	New Jersey	Vermont	Connecticut
	Species Richness and Composition			
1	# Fish species present (native/wild trout)	Resident – Trout	Lotic Fish + Trout	Species Only + Trout
2	# Benthic insectivores (BI)	Number	Number	Percent
3	# Water column species (WC)	Trout	Deleted	Deleted
4	# Sucker species	Deleted	Deleted	Deleted
5	# Intolerant species	Number	Number	% Lotic[a]
6	% Individual indicators	WS[b]	WS (WS + CRC)[c]	Dace[d]
	Trophic Guild Structure			
7	% Individual omnivores	Generalists	Generalists	Deleted
8	% Individual insectivores	Cyprinids	WC and BI	WC/BI[e]
9	% Individual top carnivores	+ Trout	+ Trout	Lentic
	Fish Abundance and Condition			
10	# Individuals/unit area	Number of resident fish (Lotic)		
++	# Individuals/unit area	—	—	Lentic
11	% Individuals as hybrids	Deleted	Deleted	Deleted
12	% Individuals with DELT[f]	Retained	Retained	Deleted
Total Number of Metrics:		10	9	8

Note: Based on the original 12-metric IBI configuration of Karr et al. (1986).

[a] Percent of individuals as lotic species.
[b] WS = white sucker.
[c] Originally white sucker (WS); currently creek chub (CC) + WS.
[d] Blacknose and longnose dace (*Rhinichthys* spp.).
[e] Ratio of water column (WC) to benthic insectivores (BI).
[f] Disease and deformities: eroded fins, lesions, tumors.

TABLE 12.4
New Jersey (Northern) Index of Biological Integrity

IBI Category and Metric	Scoring Criteria		
	5	3	1
Species Richness and Composition			
# Resident lotic species[a] (− trout)	Varies with stream size		
# Benthic (insectivorous) species	Varies with stream size		
# Water column species[b]	Varies with stream size		
# Intolerant/sensitive species	Varies with stream size		
% Individuals as white sucker	<10	10–30	>30
Trophic Guild Structure			
% Individuals as generalist feeders[c]	<20	20–45	>45
% Individuals as insectivorous cyprinids	>45	20–45	<20
% Individuals as top carnivores[d] (or trout)	>5	1–5	<1
Fish Abundance and Condition			
# Individuals as lotic residents	>250	75–250	<75
% Individuals with disease and anomalies[e]	<2	2–5	>5

IBI Score Class
  Excellent   45–50
  Good   37–44
  Fair   29–36
  Poor   10–28

[a] Excluding resident trout species.
[b] Includes resident trout and sunfish species.
[c] Carp, goldfish, creek chub, fathead minnow, banded killifish.
[d] Excluding American eel in non-coastal streams.
[e] Excluding low-level incidence of blackspot disease.

Two of the original twelve IBI metrics (Karr et al., 1986) are deleted from all northeastern IBI applications (i.e., number of sucker species and percent as hybrids). Only "total resident lotic species number" and "resident lotic fish abundance" metrics are retained in a generally unmodified form (Table 12.3). The nine metrics common between the New Jersey IBI (Table 12.4) and Vermont IBI (Table 12.5) differ primarily in scoring criteria. Metric adjustments in scoring criteria are made to account for stream size and physical-chemical reach conditions, including thermal regime, elevation, and measures of stream fertility. These measures are generally indicative of stream gradient differences encountered in northeastern wadeable-sized streams. As expected, high-elevation headwater streams are generally smaller, colder, and relatively infertile, in contrast to low gradient headwater stream systems.

12.11.3 SPECIES RICHNESS AND COMPOSITION METRICS

Total number of fish species: This metric is used in all current northeastern IBI applications (NJ-VT-CT), however, only "resident" (native and naturalized) lotic species are included. The New Jersey IBI excludes salmonid species from this metric to avoid redundancy with the "number of sunfish and trout species" metric. In the Vermont IBI, all resident trout populations count as a single "species." Generally, the more degraded the stream, the lower the number of resident fish species. Site species richness varies with stream size; hence, maximum (resident) species richness lines (MSRL) are used for metric scoring. The Vermont IBI uses separate MSRLs for cold (elevation

TABLE 12.5
Vermont (Statewide) Index of Biological Integrity

IBI Category and Metric	Scoring Criteria		
	5	3	1
Species Richness and Composition			
# Resident lotic species (+ trout)	Varies with stream size		
# Benthic (insectivorous) species	>1	1	0
# Intolerant/sensitive resident (lotic) species			
Elevation >125 m	>1	1	0
Elevation <125 m	1	0	0
% Individuals as white sucker and creek chub	<20	20–40	>40
Trophic Guild Structure			
% Individuals as generalist feeders			
Elevation >210 m	<20	20–45	>45
Elevation <210 m	<30	30–60	>60
% Individuals as benthic + water column insectivores			
Elevation >210 m	>65	30–65	<30
Elevation <210 m	>55	20–55	<20
% Individuals as top carnivores (+ trout)			
Coldwater sites	>20	3–20	<3
Warmwater sites	>10	3–10	<3
Fish Abundance and Condition			
# Individuals as lotic residents			
Elevation >210 m	—	—	—
Alkalinity >9 ppm	>10	7–10	<7
Alkalinity <9 ppm	<6	3–6	<3
Elevation <210 m	>20	10–20	<10
% Individuals with disease and anomalies	<2	2–5	>5

IBI Score Class		Conditions for Use and Scoring
Excellent	41–45	1. For wadeable streams only.
Good	33–37	2. At least (4) resident species, including (1) generalist feeder.
Fair[a]	27–29	
Poor[a]	<27	3. Only individuals more than 25 mm TL.
		4. Only resident stream fish species.

[a] Site fails to meet VT Class B/C water quality standards.

>210 m) and warmwater streams (elevation <210 m). The Connecticut SI does not use MSRLs, as that index was specifically designed to assess small coldwater streams characterized by low rates of variability (Jacobson, 1994).

Number of darter species: Naturally low regional diversity in darter species warrants the modification of this metric in all northeastern IBI versions to include all benthic fish species such as darters, sculpins, madtoms, and numerous cyprinids (Miller et al., 1988; Langdon, 1989). All of these species are also classified as insectivores (Table 12.1). This metric is further modified in the Connecticut SI as "percent of individuals" that are benthic (insectivorous) species. Benthic fishes are dependent on aquatic insects as a food source and are susceptible to low-level stream perturbations (Karr et al., 1986).

Number of sunfish species: This metric is retained only in the New Jersey IBI, and has been modified to include both sunfish and resident (nonstocked) salmonid species, equivocal to water

TABLE 12.6
Connecticut (Coldwater) Specific Index (CT-SI)

IBI Category and Metric	Scoring Criteria		
	5	3	1
Species Richness and Composition			
# Resident lotic species (+ trout)	>6	4–6	<4
% Benthic (insectivorous) species[a]	>40	30–40	<30
	<60	60–70	>70
% Individuals as resident lotic species	>90	60–90	<60
% Individuals as dace (*Rhinichthys* species)	>40	20–40	<20
	<60	60–80	>80
Trophic Guild Structure			
% Ratio of benthic–water column[b] insectivores	>0.8	0.5–0.8	<0.5
	<2	2–3	>3
% Individuals as lentic piscivores[c]	<2	2–5	>5
Fish Abundance and Condition			
# Individuals as lotic residents	>250	100–250	<100
# Individuals as lentic fishes[d]	<5	5–20	>20

IBI Score Class*
- Excellent   34–40
- Good   28–32
- Fair   22–26
- Poor   16–20
- Very Poor   8–14

[a] Cutlips minnow, blacknose and longnose dace, creek chubsucker, white sucker, slimy sculpin, tessellated darter.
[b] Atlantic salmon, brook and brown trout, creek chub, fallfish, common shiner, and redbreast sunfish.
[c] American eel, chain and redfin pickerel, largemouth bass.
[d] Golden shiner, rock bass, pumpkinseed, bluegill, yellow perch.

Note: **Conditions for Use and Scoring:** "No Fish" class omitted, as fish were collected at all sites sampled. Brown bullhead only representative "lentic benthic insectivorous" fish.

column species (as in Steedman, 1988). This metric has been deleted in the Vermont IBI and Connecticut SI versions due to low regional expectations for sunfish (Miller et al., 1988; Bramblett and Fausch, 1991). Generally, pool or water column fish species decline with increased stream degradation and loss of in-stream cover (e.g., pool depth and large woody debris).

Number of sucker species: This metric has been deleted from all Northeast IBI applications due to naturally low taxonomic diversity (Leonard and Orth, 1986; Bramblett and Fausch, 1991). With the exception of western New York, only three species of suckers occur in the Northeast. White sucker are ubiquitous and are sympatric with either longnose sucker or creek chubsucker. Longnose sucker and creek chubsucker are both ecologically and distributionally separated in northeastern waters. In place of the sucker metric, New Jersey (NJ-DEP, 1993) proposed substituting "number of cyprinid species," excluding exotic carp and goldfish, as an alternate metric that has similar sensitivity to physical and chemical habitat degradation.

Number of intolerant species: New Jersey and Vermont have retained this metric (following Miller, 1988), while Connecticut has replaced "number of intolerant species" with "percent of sample individuals that are classified as lotic species." The simple presence of resident lotic fish species are apparently indicative of higher quality stream conditions in Connecticut.

Percent of individuals as green sunfish: Green sunfish are not native to the Northeast, with the exception of the Great Lakes drainages in western New York (Smith, 1985). This metric has been modified in every Northeast IBI application by replacing green sunfish with tolerant species surrogates. The New Jersey IBI uses white sucker (Miller et al., 1988; NJ-DEP, 1993), and the Connecticut SI uses both blacknose and longnose dace (*Rhinichthys* spp.), after Steedman (1988). The Vermont IBI originally used white sucker (Langdon, 1989), but has changed to using creek chub (Leonard and Orth, 1986) in combination with white sucker. An increased improvement in the range of performance for this metric has been observed since this change. Several authors (Karr et al., 1986; Miller et al., 1988; Paller et al., 1996) suggest using "percent of fish that are tolerant" in place of green sunfish, and others suggest using the "number or percent of exotic species" (Hughes and Gammon, 1987; Crumby et al., 1990; Bramblett and Fausch, 1991). A promising, related metric proposed by Jacobson (1994) in Connecticut streams is a measure of nonresident fish presence or habitat transients (e.g., lentic fishes associated with impounded sections of streams and rivers).

12.11.4 TROPHIC GUILD STRUCTURE METRICS

Percent of individuals as omnivores: New Jersey and Vermont modified this metric to include "generalist feeders." Generalist feeders are species (e.g., blacknose dace and creek chub) that consume a wide assortment of food types from a wide variety of habitats. Generalist species may increase in abundance in disturbed habitats and degraded waters. The omnivore or generalist trophic class was not used in the Connecticut SI application.

Percent of individuals as insectivorous cyprinids: This metric is retained in the New Jersey IBI and modified in the Vermont IBI to include all insectivorous fish species (Miller, 1988; Bramblett and Fausch, 1991), including those classified as benthic and water column fishes (Table 12.1). A trophic shift from insectivore to generalist feeder generally occurs with increasing stream perturbation. In the Connecticut SI, this metric is replaced by "ratio of benthic to water column (insectivorous) fish species."

Percent of individuals as piscivores or top carnivores: This metric is retained in the New Jersey and Vermont IBI as "top carnivores," including resident trout as predators of large invertebrates as well as fish (Karr et al., 1986; Hughes et al., in press). Within warmwater IBI applications, larger size classes of yellow perch and white perch are classified as top carnivores (Table 12.1). Connecticut modified this metric as "lentic piscivores," including chain pickerel and largemouth bass (Jacobson, 1994). American eel are considered a transient species in noncoastal streams and are generally excluded from this metric in northeastern IBI applications. The presence of resident top carnivore species often indicates a healthy and trophically diverse fish assemblage (Karr et al., 1986).

12.11.5 FISH ABUNDANCE AND CONDITION METRICS

Number of individuals per unit area or catch per effort: All Northeast IBI versions retain this metric. Only individuals considered stream (lotic) residents enter into the scoring of this metric. In addition, the Connecticut SI includes "the number of lentic individuals" (negative response metric) as one of its eight metrics (Table 12.6). In general, degraded sites would be expected to have reduced numbers of lotic residents and increased numbers of lentic individuals.

Percent of hybrid individuals: This metric has been deleted from all northeastern IBI applications. Hybridization in fish is difficult to recognize during field studies (Leonard and Orth, 1986; Miller et al., 1988) and does not appear to be a common occurrence in fish assemblages inhabiting wadeable northeastern streams (Plafkin et al., 1989).

Percent of individuals displaying DELT (Deformity, Eroded fins, Lesions, or Tumors): This metric is used in the New Jersey and Vermont IBI versions and deleted from the Connecticut SI.

The occurrence of blackspot disease (cysts of the parasite *Neascus*), at minor levels of infestation, is not included. The simple presence of parasites, at low levels of incidence and infection, is not necessarily indicative of habitat degradation (Steedman, 1991; Sanders et al. Chapter 9).

12.12 VALIDATION OF NORTHEASTERN LOTIC IBI VERSIONS

Northeastern IBIs are currently being tested and validated through correlation of index scores with site-specific, physical-chemical variables and land-use practices. A review of Vermont IBI scores and subjective site evaluations of perturbation variables from 181 sites on 95 wadeable streams indicated that the Vermont IBI consistently discriminated between three levels of disturbance: severe, moderate, and minimal (reference condition). Similarly, in New Jersey, the IBI is currently used as a screening tool to determine the existence and relative severity of biological impairment in larger, wadeable streams. Unpublished studies (USGS-NAWQA) have shown a close relationship between IBI and land-use patterns in New Jersey watersheds.

A stepwise linear regression analysis of Vermont IBI scores indicated that four of the nine IBI metrics explained 66% of the total dataset variation from wadeable streams. In order of correlation, these metrics were: number of intolerant species; percent benthic species; percent generalists; and resident lotic species richness. Jacobson (1994) found the Connecticut SI to be both sensitive and stable, as well as an excellent measure of the integrity of wadeable-sized coldwater streams in Connecticut. Application of the Connecticut SI was unbiased by differences in fish assemblage structure among drainage basins and mean basin SI scores were significantly different, decreasing from west to east (i.e., Housatonic 27.7, Farmington 24.9, and Thames 21.1).

Both Vermont and New Jersey fish IBI versions show general correspondence with benthic macroinvertebrate measures of biological integrity (see Snyder et al., Chapter 26). At Vermont sites, where both assemblages were sampled, the IBI and a modified Hilsenhoff Biotic Index were significantly correlated ($p < 0.05$, n = 65). Separate fish and benthic macroinvertebrate determinations of compliance with water quality standards were in general agreement in 70 to 85% of the evaluations where both groups were sampled in New Jersey and Vermont streams.

12.12.1 Constraints of Existing Northeastern IBI Versions

Current IBI versions in the Northeast are shaped by both stream size (see Smogor and Angermeier, Chapter 10) and fish assemblage structure relative to habitat thermal regimes. Except for earlier efforts (see Miller et al., 1988), all northeastern IBI applications to date were developed on the basis of sampling wadeable-sized streams. New Jersey and Vermont IBI versions are applied to cold-, cool-, and warmwater habitats, but are limited in New Jersey to streams with drainage areas of 13 sq. km or more, and in Vermont to streams supporting four or more resident fish species. Smaller (first- and some second-order) headwater trout production waters are generally excluded from IBI application due to their naturally depauperate resident fish species richness. Insufficient fish assemblage information is provided from assessments of small streams in the Northeast to effectively apply conventional Indices of Biological Integrity.

Similar to Lyons et al. (1996), the Connecticut Specific Index (CT-SI) was designed to assess the biological integrity of only smaller, wadeable-sized coldwater streams. A major difference between the Wisconsin and Connecticut "IBI" versions is that a portion of Connecticut's streams studied were moderate to high gradient, while only low-gradient (≤10 m/km or 1%) streams were studied in Wisconsin.

Future IBI development in the Northeast will further explore the potential for applying separate IBI versions to assess different habitats (e.g., upland vs. lowland, coldwater vs. warmwater), as demonstrated for Vermont streams and more recently by Lyons and others (1996) for Wisconsin's typically low-gradient wadeable coldwater streams.

ACKNOWLEDGMENTS

The authors thank Thom Whittier, Mark Bain, C. Lavett Smith, Karsten Hartel, and two anonymous reviewers for their helpful comments on earlier drafts of this chapter. We thank state fish biologists Ken Cox (VT), Neal Hagstrom (CT), and Kendall Warner (ME) for assistance in reviewing fish species classifications. Owen Fenderson kindly provided the Maine lakes fish species occurrence summary data, and Thom Whittier summarized the EMAP-SW Northeast Lakes fish species data. We also appreciate the support of the New England Association of Environmental Biologists (NEAEB) for sponsoring several IBI-related sessions and workshops at their annual meetings in recent years (1992–1994).

REFERENCES

Angermeier, P.L. and J.R. Karr. 1986. Applying an index of biotic integrity based on stream-fish communities: considerations in sampling and interpretation, *North American Journal of Fisheries Management,* 6, 418–429.

Angermeier, P.L. and I.J. Schlosser. 1989. Species-area relationships for stream fishes, *Ecology,* 70, 1450–1462.

Angermeier, P.L. and R.A. Smogor. 1995. Estimating number of species and relative abundances in stream-fish communities: effects of sampling effort and discontinuous spatial distributions, *Canadian Journal of Fisheries and Aquatic Sciences,* 52, 936–949.

Austen, D.J., P.B. Bayley, and B.W. Menzel. 1994. Importance of the guild concept to fisheries research and management, *Fisheries,* 19(6), 12–20.

Baker, J.R., D.V. Peck, and D.W. Sutton (Eds.). 1997. Environmental Monitoring and Assessment Program — Surface Waters: Field Operations Manual for Lakes. EPA/620/R-97/001. U.S. Environmental Protection Agency, Washington, DC.

Bailey, J.R. and J.A. Oliver. 1939. The fishes of the Connecticut Watershed, in H.E. Warfel, (Ed.), *Biological Survey of the Connecticut Watershed.* New Hampshire Fish and Game Department Survey Report No. 4. 152–189.

Bailey, R.M. 1938. The fishes of the Merrimack Watershed, in E.E. Hoover, (Ed.), *Biological Survey of the Merrimack Watershed.* New Hampshire Fish and Game Department Survey Report No. 3. 149–185.

Barbour, M.T., J. Gerritsen, B.D. Snyder, and J.B. Stribling (in press). Rapid Bioassessment Protocols for Use in Streams and Rivers: Periphyton, Benthic Macroinvertebrates, and Fish. U.S. Environmental Protection Agency, Office of Water, Washington, DC, (in revision).

Bean, T.H. 1903. Catalogue of the Fishes of New York. Bulletin of the New York State Museum 60, Zoology 9, Albany, NY.

Bigelow, H.B. and W.C. Schroeder. 1953. *Fishes of the Gulf of Maine.* USDI-FWS Fishery Bulletin 74, Vol. 53.

Bouton, D.M. 1994. Strategies and Near Term Operational Plan for the Management of Endangered, Threatened, and Special Concern Fishes of New York. New York Department of Environmental Conservation, Division of Fish and Wildlife, Bureau of Fisheries, Albany, NY.

Bramblett, R. and K.D. Fausch. 1991. Variable fish communities and the index of biotic integrity in a western Great Plains river, *Transactions of the American Fisheries Society,* 120, 752–769.

Burkhead, N.M. and J.D. Williams. 1991. An intergeneric hybrid of a native minnow, the golden shiner, and an exotic minnow, the rudd, *Transactions of the American Fisheries Society,* 120, 781–795.

Burnham, D., S. Fiske, and R.W. Langdon. 1991. Compliance monitoring of the aquatic biota in Vermont, in *Biological Criteria: Research and Regulation, Proceedings of a Symposium,* U.S. EPA, Office of Water, Washington, DC. 135–137.

Cardoza, J.E., G.S. Jones, T.W. French, and D.B. Halliwell. 1993. *Exotic and Translocated Vertebrates of Massachusetts.* Massachusetts Division of Fisheries and Wildlife, Westboro, MA, Fauna of Massachusetts Series No. 6, 2nd ed.

Carlander, K.D. 1969. *Handbook of Freshwater Fishery Biology,* Vol. 1, Iowa State University Press.

Carlander, K.D. 1977. *Handbook of Freshwater Fishery Biology,* Vol. 2, Iowa State University Press.

Clayton, G., C. Cole, S. Murawski, and J. Parrish. 1976. Common Marine Fishes of the Massachusetts Coastal Zone: A Literature Review. Institute for Man and Environment, Publication No. R-76-16, University of Massachusetts, Amherst, MA.

Collette, B.B. and K.E. Hartel. 1988. An Annotated List of the Fishes of Massachusetts Bay. National Oceanic and Atmospheric Administration Technical Memoir, National Marine Fisheries Service, NEC-51.

Cooper, G.P. 1939a. A Biological Survey of the Waters of York County and the Southern Part of Cumberland County, Maine, *Maine Department Inland Fisheries and Game Report*, 1, 1–58.

Cooper, G.P. 1939b. A biological survey of thirty-one lakes and ponds of the upper Saco River and Sebago Lake drainage systems in Maine, *Maine Department Inland Fisheries and Game Report*, 2, 1–147.

Cooper, G.P. 1940. A biological survey of the Rangeley Lakes, with special reference to the trout and salmon, *Maine Department Inland Fisheries and Game Report*, 3, 1–182.

Cooper, G.P. 1941. A biological survey of lakes and ponds of the Androscoggin and Kennebec River drainage systems in Maine, *Maine Department Inland Fisheries and Game Report*, 4, 1–238.

Cooper, G.P. 1942. A biological survey of lakes and ponds of the central coastal area of Maine. *Maine Department Inland Fisheries and Game Report*, 5, 1–184.

Crumby, W.D., M.A. Webb, F.J. Bulow, and H.J. Cathey. 1990. Changes in biotic integrity of a river in north-central Tennessee, *Transactions of the American Fisheries Society*, 119, 885–893.

Deegan, L.A., J.T. Finn, S.G. Ayvazian, and C. Ryder. 1993. Feasibility and Application of the Index of Biotic Integrity to Massachusetts Estuaries (EBI). Final Report to MA-EOEA, Massachusetts Department Environmental Protection, Worcester, MA.

Deevey, E.S. and J.S. Bishop. 1941. A Fishery Survey of Important Connecticut Lakes. Connecticut Geological and Natural History Survey, Bulletin Number 63.

Eaton, J.G., J.H. McCormick, B.E. Goodno, D.G. O'Brien, H.G. Stefany, M. Hondzo, and R.M. Scheller. 1995. A field information-based system for estimating fish temperature tolerances, *Fisheries*, 20(4), 10–18.

Everhart, W.H. 1976. *Fishes of Maine*. Maine Department of Inland Fish and Game, 4th ed.

Evermann, B.W. and W.C. Kendall. 1902a. An annotated list of the fishes known to occur in Lake Champlain and its tributary waters, *Report of the U.S. Commission on Fish and Fisheries*, (1901), 217–225.

Evermann, B.W. and W.C. Kendall. 1902b. An annotated list of the fishes known to occur in the St. Lawrence River, *Report of the U.S. Commission on Fish and Fisheries*, (1901), 227–240.

Facey, D.E. 1991. *Survey of Fishes of the Lake Champlain Basin*. Vermont Department of Fish and Wildlife Non-game and Natural Heritage Program.

Facey, D.E. and G.W. LaBar. 1990. *A Survey of the Fishes of Vermont Waters of the Southern Lake Champlain Basin*. Vermont Department of Fish and Wildlife Non-game and Natural Heritage Program.

Fausch, K.D., J.R. Karr, and P.R. Yant. 1984. Regional application of an index of biotic integrity based on stream fish communities, *Transactions of the American Fisheries Society*, 113, 39–55.

Fausch, K.D., J. Lyons, J.R. Karr, and P.L. Angermeier. 1990. Fish communities as indicators of environmental degradation, in S.M. Adams (Ed.), *Biological Indicators of Stress in Fish*. American Fisheries Society Symposium Number 8, Bethesda, MD. 123–144.

Fauth, J.E., J. Bernardo, M. Camara, W.J. Resetarits, Jr., J.V. Buskirk, and S.A. McCollum. 1996. Simplifying the jargon of community ecology: a conceptual approach, *American Naturalist*, 147, 282–286.

Fore, L.S., J.R. Karr, and L.L. Conquest. 1994. Statistical properties of an index of biological integrity used to evaluate water resources, *Canadian Journal of Fisheries and Aquatic Sciences*, 51, 1077–1087.

Fowler, H.W. 1906. The Fishes of New Jersey, in *Annual Report New Jersey State Museum, Part II*, Trenton, NJ. 35–477.

Fowler, H.W. 1952. A list of fishes of New Jersey with offshore species, *Proceedings of the Philadelphia Academy of Natural Sciences*, 104, 89–151.

French, T.W. and J.E. Cardoza. (In press). *Rare and Endangered Vertebrate Species in Massachusetts*. Massachusetts Division of Fisheries and Wildlife and Massachusetts Audubon Society Press.

Gallagher, J. and J.P. Baker. 1990. Current status of fish communities in Adirondack lakes, in *An Interpretive Analysis of Fish Communities and Water Chemistry, 1984–87*. Adirondack Lakes Survey Corporation, Ray Brook, NY. 3-11-3-44.

George, C.J. 1981. *The Fishes of the Adirondack Park*. New York State Department of Environmental Conservation, Lake Monograph Program.

Gerking, S.D. 1994. *Feeding Ecology of Fish*. Academic Press, San Diego, CA.

Gilbert, C.R. 1976. Composition and derivation of the North American freshwater fish fauna, *Florida Scientist,* 39, 104–111.

Gilbert, C.R. 1980. Zoogeographic factors in relation to biological monitoring of fish, in C.H. Hocutt and J.R. Stauffer, Jr. (Eds.), *Biological Monitoring of Fish,* Lexington Books, Lexington, MA. 309–355.

Gilliom, R.J., W.M. Alley, and M.E. Gurtz. 1995. *Design of the National Water-Quality Assessment Program: Occurrence and Distribution of Water-Quality Conditions.* U.S. Geological Survey Circular 1112.

Gordon, B.L. 1974. *The Marine Fishes of Rhode Island.* Watch Hill, Rhode Island.

Gordon, M. 1937. The fishes of eastern New Hampshire, in E.E. Hoover, (Ed.), *Biological Survey of the Androscoggin, Saco, and Coastal Watersheds.* New Hampshire Fish and Game Department, Survey Report No. 2, 101–118.

Graham, J.H. 1993. Species diversity of fishes in naturally acidic lakes in New Jersey, *Transactions of the American Fisheries Society,* 122, 1043–1057.

Greeley, J.R. 1927. Fishes of the Genesee region with annotated list, in *A Biological Survey of the Genesee River System.* Supplement to the 16th Annual Report of the New York State Conservation Department (1926). 47–66.

Greeley, J.R. 1928. Fishes of the Oswego watershed, in *A Biological Survey of the Oswego River System.* Supplement to the Seventeenth Annual Report of the New York State Conservation Department (1927). 84–107.

Greeley, J.R. 1929. Fishes of the Erie-Niagara watershed, in *A Biological Survey of the Erie-Niagara System.* Supplement to the 18th Annual Report of the New York State Conservation Department (1928). 150–179.

Greeley, J.R. 1930. Fishes of the Champlain watershed., in *A Biological Survey of the Champlain Watershed.* Supplement to the 19th Annual Report of the New York State Conservation Department (1929). 44–87.

Greeley, J.R. 1934. Fishes of the Raquette watershed with an annotated list, in *A Biological Survey of the Raquette Watershed.* Supplement to the 23rd Annual Report of the New York State Conservation Department (1923). 53–108.

Greeley, J.R. 1935. Fishes of the watershed with an annotated list, in *A Biological Survey of the Mohawk-Hudson Watershed.* Supplement to the 24th Annual Report of the New York State Conservation Department (1934). 63–101.

Greeley, J.R. 1936. Fishes of the area with an annotated list, in *A Biological Survey of the Delaware and Susquehanna Watersheds.* Supplement to the 25th Annual Report of the New York State Conservation Department (1935). 45–88.

Greeley, J.R. 1937. Fishes of the area with an annotated list, in *A Biological Survey of the Lower Hudson Watershed.* Supplement to the 26th Annual Report of the New York State Conservation Department (1936). 45–103.

Greeley, J.R. 1938. Fishes of the area with an annotated list, in *A Biological Survey of the Allegheny and Chemung Watersheds.* Supplement to the 27th Annual Report of the New York State Conservation Department (1937). 48–73.

Greeley, J.R. 1939. The fresh-water fishes of Long Island and Staten Island with annotated list, in *A Biological Survey of the Fresh Waters of Long Island.* Supplement to the 28th Annual Report of the New York State Conservation Department (1938). 29–44.

Greeley, J.R. 1940. Fishes of the watershed with annotated list, in *A Biological Survey of the Lake Ontario Watershed.* Supplement to the 29th Annual Report of the New York State Conservation Department (1939). 42–81.

Greeley, J.R. and S.C. Bishop. 1932. Fishes of the watershed with annotated list, in *A Biological Survey of the Oswegatchie and Black River Systems.* Supplement to the 21st Annual Report of the New York State Conservation Department (1931). 54–92.

Greeley, J.R. and S.C. Bishop. 1933. Fishes of the upper Hudson watershed with annotated list, in *A Biological Survey of the Upper Hudson Watershed.* Supplement to the 22nd Annual Report of the New York State Conservation Department (1932). 64–101.

Greeley, J.R. and C.W. Greene. 1931. Fishes of the area with annotated list, in *A Biological Survey of the Saint Lawrence Watershed.* Supplement to the 20th Annual Report of the New York State Conservation Department (1930). 44–94.

Griffith, G.E., J.M. Omernik, S.M. Pierson, and C.W. Kiilsgaard. 1994. The Massachusetts Ecological Regions Project. U.S.-EPA/ERL, Corvallis, OR, for MA-DEP/DWPC.

Gurtz, M.E. 1994. Design of biological components of the National Water-Quality Assessment (NAWQA) Program, in Loeb, S.L. and A. Spacie (Eds.), *Biological Monitoring of Aquatic Systems*, Lewis, Boca Raton, FL. 323–354.

Guthrie, R.C., J.A. Stolgitis, and W.L. Bridges. 1973. Pawcatuck River Watershed — Fisheries Management Survey. Rhode Island Division of Fish and Wildlife, Fisheries Report Number 1.

Guthrie, R.C. and J.A. Stolgitis. 1977. Fisheries Investigations and Management in Rhode Island Lakes and Ponds. Rhode Island Division of Fish and Wildlife, Fisheries Report Number 3.

Haines, T.A. 1981. Acid precipitation and its consequences for aquatic ecosystems: a review, *Transactions of the American Fisheries Society*, 110, 669–707.

Haines, T.A and J.P. Baker. 1986. Evidence of fish population responses to acidification in the eastern United States, *Water, Air, and Soil Pollution*, 31, 605–629.

Halliwell, D.B. 1984. *A List of the Freshwater Fishes of Massachusetts*. Massachusetts Division of Fisheries and Wildlife, Westboro, MA, Faunal Series, No. 4, 2nd ed..

Halliwell, D.B. 1989. *A Classification of Streams in Massachusetts — To Be Used as a Fisheries Management Tool*. Ph.D. dissertation, University of Massachusetts, Amherst.

Harris, J.H. 1995. The use of fish in ecological assessments, *Australian Journal of Ecology*, 20, 65–80.

Hartel, K.E. 1992. Non-native Fishes Known from Massachusetts Freshwaters. Occasional Report No. 2, Museum of Comparative Zoology, Harvard University, Cambridge, MA.

Hartel, K.E., D.B. Halliwell, and A.E. Launer. (in press). *The Inland Fishes of Massachusetts*. Harvard University, Museum of Comparative Zoology and Massachusetts Audubon Society Press.

Hastings, R.W. 1979. Fish of the Pine Barrens, in R.T.T. Forman (Ed.), *Pine Barrens: Ecosystem and Landscape*, Academic, New York. 489–504.

Hawkes, C.L., D.L. Miller, and W.G. Layher. 1986. Fish ecoregions of Kansas: stream fish assemblage patterns and associated environmental correlates, *Environmental Biology of Fishes*, 17, 267–279.

Hocutt, C.H. and E.O. Wiley. 1986. *The Zoogeography of North American Freshwater Fishes*, John Wiley & Sons, New York.

Horton, R.E. 1945. Erosional development of streams and their drainage basins — hydrophysical approach to quantitative morphology, *Geological Society of America Bulletin*, 56, 275–370.

Hubbs, C.L. and K.F. Lagler. 1974. *Fishes of the Great Lakes Region*. University of Michigan Press, Ann Arbor.

Hughes, R.M. 1995. Defining acceptable biological status by comparing with reference conditions, in W.S. Davis and T.P. Simon (Eds.), *Biological Assessment and Criteria: Tools for Water Resource Planning and Decision Making*. Lewis, Boca Raton, FL. 31–47.

Hughes, R.M., D.P. Larsen, and J.M. Omernik. 1986. Regional reference sites: a method for assessing stream potentials, *Environmental Management*, 10, 629–635.

Hughes, R.M. and J.R. Gammon. 1987. Longitudinal changes in fish assemblages and water quality in the Willamette River, Oregon, *Transactions of the American Fisheries Society*, 116, 196–209.

Hughes, R.M., E. Rexstad, and C.E. Bond. 1987. The relationship of aquatic ecoregions, river basins, and physiographic provinces to the ichthyogeographic regions of Oregon, *Copeia*, 1987, 423–432.

Hughes, R.M. and R.F. Noss. 1992. Biological diversity and biological integrity: current concerns for lakes and streams, *Fisheries*, 17(3), 11–20.

Hughes, R.M., L. Reynolds, P.R. Kaufmann, A.T. Herlihy, T. Kincaid, and D.P. Larsen. (in revision). Development and application of an index of fish assemblage integrity for wadeable streams in the Willamette Valley, Oregon, USA, *Canadian Journal of Fisheries and Aquatic Sciences*.

Hunter, J. (Ed.). 1990. *Writing for Fishery Journals*. American Fishery Society, Bethesda, MD.

Jacobson, R.A. 1994. Application of the Index of Biotic Integrity to Small Connecticut Streams. Master's Thesis, University of Connecticut, Storrs.

Jennings, M.J., L.S. Fore, and J.R. Karr. 1995. Biological monitoring of fish assemblages in Tennessee Valley Reservoirs, *Regulated Rivers: Research and Management*, 11, 263–274.

Jordan, S.J., P.A. Vaas, and J. Uphoff. 1991. Fish assemblages as indicators of environmental quality in Chesapeake Bay, in *Biological Criteria: Research and Regulation — Proceedings of a Symposium*. EPA-440/5-91-005. U.S. Environmental Protection Agency, Office of Water, Washington, DC. 73–80.

Karr, J.R. 1981. Assessment of biotic integrity using fish communities, *Fisheries*, 6(6), 28–30.

Karr, J.R. 1991. Biological integrity: a long-neglected aspect of water resource management, *Ecological Applications*, 1, 66–84.

Karr, J.R. and D.R. Dudley. 1981. Ecological perspective on water quality goals, *Environmental Management*, 5, 55–68.

Karr, J.R., K.D. Fausch, P.L. Angermeier, P.R. Yant, and I.J. Schlosser. 1986. Assessing Biological Integrity in Running Waters: A Method and Its Rationale. Illinois Natural History Survey Special Publication No. 5.

Karr, J.R. and M. Dionne. 1991. Designing surveys to assess biological integrity in lakes and reservoirs, in *Biological Criteria: Research and Regulation — Proceedings of a Symposium*. U.S.-EPA, Office of Water, Washington, DC, EPA-440/5-91-005. 62–72.

Keast, A. 1966. Trophic Interrelationships in the Fish Fauna of a Small Stream. Great Lakes Research Division, University of Michigan, Publication No. 15, 51–79.

Keast, A. 1985. The piscivore feeding guild of fishes in small freshwater ecosystems, *Environmental Biology of Fishes*, 12, 119–129.

Kendall, W.C. 1914. The Fishes of Maine. *Proceedings of the Portland (Maine) Society of Natural History*, Vol. 3(1).

Kircheis, F.W. 1980. The landlocked charrs of Maine: the sunapee and the blueback, in *Charrs Salmonid Fishes of the Genus Salvelinus*, E.K. Balon (Ed.), D.W. Junk, The Netherlands. 749–755.

Kircheis, F.W. 1994. Update on freshwater fish species reproducing in Maine, *Maine Naturalist*, 2(1), 25–28.

Kurtenbach, J.P. 1994. Index of Biotic Integrity Study of Northern New Jersey Drainages. U.S. EPA, Region 2, Division of Environmental Science and Assessment, Edison, NJ.

Langdon, R.W. 1989. The development of fish population-based biocriteria in Vermont, in T.P. Simon, L.L. Holst, and L.J. Shephard (Eds.), *Proceedings of the First National Workshop on Biocriteria*, EPA 905-9-89-003. U.S. Environmental Protection Agency, Region 5, Chicago, IL. 12–25.

Langdon, R.W. 1994. A Fish Survey of Lake Champlain Tributaries in Addison County, Vermont. Vermont Department of Environmental Conservation. Report prepared for the Vermont Department of Fish and Wildlife Non-game and Natural Heritage Program.

Larsen, D.P., J.M. Omernik, R.M. Hughes, C.M. Rohm, T.R. Whittier, A.J. Kinney, A.L. Gallant, and D.R. Dudley. 1986. Correspondence between spatial patterns in fish assemblages in Ohio streams and aquatic ecoregions, *Environmental Management*, 10, 815–828.

Larsen, D.P., D.L. Stevens, A.R. Selle, and S.G. Paulsen. 1991. Environmental Monitoring and Assessment Program, EMAP-Surface Waters, a Northeast Lakes Pilot, *Lake and Reservoir Management*, 7, 1–11.

Larsen, D.P. and S.J. Christie (Eds.). 1993. EMAP-Surface Waters 1991 Pilot Report. EPA/620/R-93/003. U.S. Environmental Protection Agency, Corvallis, OR.

Lee, D.S., C.R. Gilbert, C.H. Hocutt, R.E. Jenkins, D.E. McAllister, and J.P. Stauffer, Jr. 1980. *Atlas of North American Freshwater Fishes*. North Carolina State Museum of Natural History, Raleigh, NC.

Leonard, P.M. and D.J. Orth. 1986. Application and testing of an index of biotic integrity in small, coolwater streams, *Transactions of the American Fisheries Society*, 115, 401–414.

Lyons, J. 1989. Correspondence between the distribution of fish assemblages in Wisconsin streams and Omernik's ecoregions, *American Midland Naturalist*, 122, 163–182.

Lyons, J. 1992. The length of stream to sample with a towed electrofishing unit when fish species richness is estimated, *North American Journal of Fisheries Management*, 12, 198–203.

Lyons, J., S. Navarro-Perez, P.A. Cochran, E. Santana C., and M. Gusman-Arroyo. 1995. Index of biotic integrity based on fish assemblages for the conservation of streams and rivers in west-central Mexico, *Conservation Biology*, 9, 569–584.

Lyons, J., L. Wang, and T.D. Simonson. 1996. Development and validation of an index of biotic integrity for coldwater streams in Wisconsin, *North American Journal of Fisheries Management*, 16, 241–256.

MacMartin, J.M. 1962. *Vermont Stream Survey 1952–1960*. Vermont Fish and Game Department, Montpelier, VT.

Mather, F. 1886. Memoranda relating to Adirondack Fishes with Descriptions of New Species from Surveys Made in 1882. State of New York Adirondack Survey. Appendix to the 12th Report.

Miller, D.L., P.M. Leonard, R.M. Hughes, J.R. Karr, P.B. Moyle, L.H. Schrader, B.A. Thompson, R.A. Daniels, K.D. Fausch, G.A. Fitzhugh, J.R. Gammon, D.B. Halliwell, P.L. Angermeier, and D.J. Orth. 1988. Regional applications of an index of biotic integrity for use in water resource management, *Fisheries*, 13(5), 12–20.

Minns, C.K., V.W. Cairns, R.G. Randall, and J.E. Moore. 1994. An index of biotic integrity (IBI) for fish assemblages in the littoral zone of Great Lakes' areas of concern, *Canadian Journal of Fisheries and Aquatic Sciences*, 51, 1804–1822.

Moyle, P.B. and B. Herbold. 1987. Life history patterns and community structure in stream fishes of western North America: comparisons with eastern North America and Europe, in W. J. Matthews and D.C. Heins (Eds.), *Community and Evolutionary Ecology of North American Stream Fishes,* University of Oklahoma Press. 25–32.

New Jersey Department Environmental Protection (NJ-DEP). 1993. *Development and Modification of an Index of Biotic Integrity for New Jersey Lotic Waters.* NJ-DEP, Division of Science and Research, Trenton, NJ.

Oberdorff, T. and R.M. Hughes. 1992. Modification of an index of biotic integrity based on fish assemblages to characterize rivers of the Seine-Normandie Basin, France, *Hydrobiologia,* 228, 117–130.

Olson, M.H. 1996. Ontogenetic niche shifts in largemouth bass: variability and consequences for first-year growth, *Ecology,* 77, 179–190.

Omernik, J.M. 1987. Ecoregions of the conterminous United States, *Annals of the Association of American Geographers,* 77, 118–125.

Page, L.M. and B.M. Burr. 1991. *A Field Guide to Freshwater Fishes: North America North of Mexico.* Peterson Field Guide Series No. 42.

Paller, M.H. 1995. Relationships among number of fish sampled, reach length surveyed, and sampling effort in South Carolina coastal plain streams, *North American Journal of Fisheries Management,* 15, 110–120.

Paller, M.H., M.J. Reichert, and J.M. Dean. 1996. Use of fish communities to assess environmental impacts in South Carolina coastal plain streams, *Transactions of the American Fisheries Society,* 125, 633–644.

Pennak, R.W. 1971. Towards a classification of biotic habitats, *Hydrobiologia,* 38, 321–334.

Plafkin, J.L., M.T. Barbour, K.D. Porter, S.K. Gross, and R.M. Hughes. 1989. *Rapid Bioassessment Protocols for Use in Streams and Rivers: Benthic Macroinvertebrates and Fish.* U.S. Environmental Protection Agency, Washington, DC.

Platts, W.S. 1980. A plea for fishery habitat classification, *Fisheries,* 5(1), 2–6.

Platts, W.S., W.F. Megahan, and G.W. Minshall. 1983. Methods for Evaluating Stream, Riparian, and Biotic Conditions. U.S. Forest Service, General Technical Report INT-138.

Poff, N.L., and J.D. Allan. 1995. Functional organization of stream fish assemblages in relation to hydrological variability, *Ecology,* 76, 606–627.

Ricklefs, R. 1990. *Ecology.* 3rd ed. W. H. Freeman, New York.

Robins, C.R., R.M. Bailey, C.E. Bond, J.R. Brooker, E.A. Lachner, R.N. Lea, and W.B. Scott. 1991. *Common and Scientific Names of Fishes from the U.S. and Canada.* American Fisheries Society Special Publication 20, 5th ed.

Robinson, B.W., D.S. Wilson, and G.O. Shea. 1996. Trade-offs of ecological specialization: an intraspecific comparison of pumpkinseed sunfish phenotypes, *Ecology,* 77, 170–178.

Rohm, C.M., J.W. Giese, and C.C. Bennett. 1987. Evaluation of an aquatic ecoregion classification of streams in Arkansas, *Journal of Freshwater Ecology,* 4, 127–140.

Saila, S.B. and D. Horton. 1957. *Fisheries Investigations and Management in Rhode Island Lakes and Ponds.* Rhode Island Division of Fish and Game, Fisheries Publication No. 3.

Scarola, J.F. 1973. *Freshwater Fishes of New Hampshire.* New Hampshire Fish and Game Department, Division of Inland and Marine Fisheries.

Schlosser, I.J. 1990. Environmental variation, life history attributes, and community structure in stream fishes: implications for environmental management and assessment, *Environmental Management,* 14, 621–628.

Schmidt, R.E. 1986. Zoogeography of the northern Appalachians, in C.H. Hocutt and E.O. Wiley (Eds.), *The Zoogeography of North American Freshwater Fishes.* John Wiley & Sons, New York. 137–159.

Scott, W.B. and E.J. Crossman. 1973. *Freshwater Fishes of Canada.* Fisheries Research Board of Canada Bulletin 184.

Simon, T.P. and J. Lyons. 1995. Application of the index of biotic integrity to evaluate water resource integrity in freshwater ecosystems, in W.S. Davis and T.P. Simon (Eds.), *Biological Assessment and Criteria: Tools for Water Resource Planning and Decision Making.* Lewis, Boca Raton, FL. 245–262.

Simonson, T.D., J. Lyons, and P.D. Kanehl. 1994. Quantifying fish habitat in streams: transect spacing, sample size, and a proposed framework, *North American Journal of Fisheries Management,* 14, 607–615.

Simonson, T.D. and J. Lyons. 1995. Comparison of catch per effort and removal procedures for sampling stream fish assemblages, *North American Journal of Fisheries Management,* 15, 419–427.

Sly, P.G. and W.D.N. Busch. 1992. A system for aquatic habitat classification of lakes, in *The Development of an Aquatic Habitat Classification System for Lakes,* CRC Press, Boca Raton, FL. 15–26.

Smith, C.L. 1985. *The Inland Fishes of New York State.* New York State Department of Environmental Conservation, Albany.

Soule, M.E. (Ed.). 1987. *Viable Populations for Conservation.* Cambridge University Press, New York.

Southerland, M.T. and J.B. Stribling. 1995. Status of biological criteria development and implementation, in W.S. Davis and T.P. Simon (Eds.), *Biological Assessment and Criteria: Tools for Water Resource Planning and Decision Making,* Lewis, Boca Raton, FL. 81–96.

Steedman, R.J. 1988. Modification and assessment of an index of biotic integrity to quantify stream quality in southern Ontario, *Canadian Journal of Fisheries and Aquatic Sciences,* 45, 492–501.

Steedman, R.J. 1991. Occurrence and environmental correlates of black spot disease in stream fishes near Toronto, Ontario, *Transactions of the American Fisheries Society,* 120, 494–499.

Stemberger, R.S. 1995. Pleistocene refuge areas and postglacial dispersal of copepods of the northeastern U.S., *Canadian Journal of Fisheries and Aquatic Sciences,* 52, 2197–2210.

Stiles, E.W. 1978. *Vertebrates of New Jersey.* E.W. Stiles Publishers, Somerset, NJ.

Strahler, A.N. 1957. Quantitative analysis of watershed geomorphology, *Transactions of the American Geophysical Union,* 38, 913–920.

Suter, G.W. 1993. A critique of ecosystem health concepts and indexes, *Environmental Toxicology and Chemistry,* 12, 1533–1539.

Thomson, B.F. 1958. *The Changing Face of New England.* Houghton Mifflin, Boston.

Titcomb, J.W. 1926. Forests in relation to freshwater fishes, *Transactions of the American Fisheries Society,* 56, 122–129.

Trautman, M.B. 1981. *The Fishes of Ohio.* Ohio State University Press, Columbus.

U.S. Environmental Protection Agency. 1990. *Biological Criteria: National Program Guidance for Surface Waters.* Office of Water Regulations and Standards, Washington, DC.

Warren, M.L. and B.M. Burr. 1994. Status of freshwater fishes of the U.S.: overview of an imperiled fauna, *Fisheries,* 19(1), 6–18.

Waters, T.F. 1995. *Sediment in Streams: Sources, Biological Effects, and Control.* American Fisheries Society, Monograph Number 7.

Whittier, T.R., D.P. Larsen, R.M. Hughes, C.M. Rohm, A.L. Gallant, and J.M. Omernik. 1987. The Ohio Stream Regionalization Project: A Compendium of Results. EPA/600/3-87/025. U.S. Environmental Protection Agency, Environmental Research Laboratory, Corvallis, OR.

Whittier, T.R., and S.G. Paulsen. 1992. The surface waters component of the Environmental Monitoring and Assessment Program (EMAP): an overview, *Journal of Aquatic Ecosystem Health,* 1, 13–20.

Whittier, T.R., D.B. Halliwell, and S.G. Paulsen. 1997. Cyprinid distributions in northeast U.S.A. lakes: evidence of regional-scale minnow biodiversity losses, *Canadian Journal of Fisheries and Aquatic Sciences,* 54, 1593–1607.

Whittier, T.R. and K.E. Hartel. 1997. First records of redear sunfish (*Lepomis microlophus*) in New England. *Northeastern Naturalist,* 4: 237–240.

Whittier, T.R. and R.M. Hughes. 1998. Evaluation of fish species tolerances to environmental stressors in northeast USA lakes. *North American Journal of Fisheries Management,* (in press).

Whitworth, W.R. 1996. *Freshwater Fishes of Connecticut.* State Geological and Natural History Survey, Connecticut Department of Environmental Protection, Bulletin Number 114.

Whitworth, W.R., P.L. Berrien, and W.T. Keller. 1968. *Freshwater Fishes of Connecticut.* State Geological and Natural History Survey, Connecticut Department of Environmental Protection, Bulletin No. 1.

Wichert, G.A. 1995. Effects of improved sewage effluent management and urbanization on fish associations of Toronto streams, *North American Journal of Fisheries Management,* 15, 440–456.

Appendix 12A
Publications Providing Information on the Historical and Current Distribution of Northeastern United States Freshwater Fish Species

New York	Mather, 1886; Evermann and Kendall, 1902a; 1902b; Bean, 1903; Greeley, 1927 to 1940; Greeley and Bishop, 1932; 1933; Greeley and Greene, 1931; George, 1981; Smith, 1985; Bouton, 1994; Whittier et al., 1997.
New Jersey	Fowler, 1906; 1952; Stiles, 1978; Hastings, 1979; New Jersey DEP, 1993; Graham, 1993; Kurtenbach, 1994; Whittier et al., 1997.
Maine	Kendall, 1914; Cooper, 1939a; 1939b; 1940; 1941; 1942; Everhart, 1976; Kircheis, 1994; Whittier et al., 1997; Whittier and Hartel, 1997.
Vermont	Everman and Kendall, 1902a; 1902b; MacMartin, 1962; Facey and LaBar, 1990; Facey, 1991; Langdon, 1994; Whittier et al., 1997.
New Hampshire	Gordon, 1937; Bailey, 1938; Bailey and Oliver, 1939; Scarola, 1973; Whittier et al., 1997.
Rhode Island	Saila and Horton, 1957; Guthrie et al., 1973; Gordon, 1974; Guthrie and Stolgitis, 1977; Whittier et al., 1997.
Massachusetts	Halliwell, 1984; 1989; Whittier et al., 1997; French and Cardoza, (in press); Hartel et al., (in press).
Connecticut	Deevey and Bishop, 1941; Whitworth et al., 1968; Jacobson, 1994; Whitworth, 1996; Whittier et al., 1997.

Appendix 12B
Rarely Encountered Fish Species of Inland Waters in the Northeastern United States

Type	Examples
Marine strays or Estuarine species Clayton et al., 1976 Collette and Hartel, 1988 Hartel et al., (in revision) Smith, 1985 Whitworth, 1996	Hickory shad (*Alosa mediocris*) Atlantic tomcod (*Microgadus tomcod*) Atlantic needlefish (*Strongylura marina*) Sheepshead minnow (*Cyprinodon variegatus*) Striped killifish (*Fundulus majalis*) Rainwater killifish (*Lucania parva*) Inland silverside (*Menidia beryllina*) Northern pipefish (*Syngnathus fuscus*)
Exotics with unique records/populations Cardoza et al., 1993 Hartel, 1992 Smith, 1985 Whitworth, 1996	Walking catfish (*Clarias batrachus*) MA Oscar (*Astronotus ocellatus*) MA, RI Snakehead (*Channa* sp.) MA, RI Red piranha (*Pygocentrus nattereri*) MA Warmouth (*Lepomis gulosus*) NY Redear sunfish (*Lepomis microlophus*) VT Bitterling (*Rhodeus sericeus*) NY Tench (*Tinca tinca*) CT
Introduced sterile Triploids or hybrids	Grass carp (*Ctenopharyngodon idella*) Tiger muskellunge (*Esox lucius* x *masquinongy*)
Extirpated fishes (w/o recent records) Smith, 1985	Paddlefish (*Polyodon spathula*) Shortnose gar (*Lepisosteus platostomus*) Lake chubsucker (*Erimyzon sucetta*) Longear sunfish (*Lepomis megalotis*) Black bullhead (*Ameiurus melas*)
Fishes reported from only Lake Champlain or Lake Ontario Smith, 1985	Mooneye (*Hiodon tergisus*) Sauger (*Stizostedion canadense*) Spoonhead sculpin (*Cottus ricei*) Deepwater sculpin (*Myoxocephalus thompsoni*)
Pacific salmon (*Oncorhynchus* spp.) Scott and Crossman, 1973 Smith, 1985	Chinook salmon (*O. tshawytscha*) Coho salmon (*O. kisutch*) Pink salmon (*O. gorbuscha*) Sockeye (Kokanee) (*O. nerka*)

13 Development of an Index of Biotic Integrity for the Species-Depauperate Lake Agassiz Plain Ecoregion, North Dakota and Minnesota

Scott Niemela, Eric Pearson, Thomas P. Simon, Robert M. Goldstein, and Patricia A. Bailey

CONTENTS

13.1 Introduction 340
13.2 Study Area 341
    13.2.1 Drainage Features of the Red River of the North Basin 341
    13.2.2 Lake Agassiz Plain Ecoregion...341
    13.2.3 Historical Red River Basin Data...342
13.3 Materials and Methods 343
    13.3.1 Establishing the Reference Condition...343
    13.3.2 Criteria for Selecting Reference Sites...344
    13.3.3 Fish Community Sampling Procedures 344
13.4 Results..345
    13.4.1 Metrics 345
    13.4.2 Species Composition and Richness Metrics...345
        13.4.2.1 Metric 1: Total Number of Fish Species (All Stream Sizes)...............345
        13.4.2.2 Metric 2: Proportion of Headwater Species (Headwater Streams), Number of Benthic Insectivores (Moderate-Sized Streams), and Proportion of Round-Bodied Suckers (Large Rivers) 345
            13.4.2.2.1 Proportion of headwater species.........................347
            13.4.2.2.2 Number of benthic insectivore species............................347
            13.4.2.2.3 Proportion of round-bodied suckers 347
        13.4.2.3 Metric 3: Number of Minnow Species (Headwater and Moderate-Sized Streams), Proportion of Large River Individuals (Large Rivers) 352
            13.4.2.3.1 Number of minnow species...............................352
            13.4.2.3.2 Proportion of large river individuals 352
        13.4.2.4 Metric 4: Evenness (All Stream Sizes) 352
        13.4.2.5 Metric 5: Number of Sensitive Species (All Stream Sizes).................353
        13.4.2.6 Metric 6: Proportion of Tolerant Individuals (All Stream Sizes) 353

	13.4.3	Trophic Guilds 353
		13.4.3.1 Metric 7: Proportion of Omnivore Biomass (All Stream Sizes) 353
		13.4.3.2 Metric 8: Proportion of Insectivore Biomass (All Stream Sizes) 354
		13.4.3.3 Metric 9: Proportion of Pioneer Species (Headwater Streams) and Proportion of Piscivore Biomass (Moderate and Large Rivers) 354
		13.4.3.3.1 Proportion of piscivore biomass 354
		13.4.3.3.2 Proportion of pioneer species 354
	13.4.4	Individual Condition, Reproductive Guilds, and Abundance 356
		13.4.4.1 Metric 10: Number of Individuals Per Meter (All Stream Sizes) 356
		13.4.4.2 Metric 11: Proportion of Individuals as Simple Lithophilic Spawners (All Stream Sizes) 356
		13.4.4.3 Metric 12: Proportion of Individuals with Deformities, Eroded Fins, Lesions and Tumors (DELT) (All Stream Sizes) 356
	13.4.5	Alternative Metrics 357
	13.4.6	Scoring Modifications 357
		13.4.6.1 Headwater Streams 358
		13.4.6.2 Moderate-Sized Streams and Large Rivers 358
13.5	Discussion 358
	13.5.1	Lake Agassiz Plain 358
		13.5.1.1 Species Composition 358
		13.5.1.2 Trends in IBI Scoring 358
	13.5.2	Headwater Streams, Moderate-Sized Streams, and Large Rivers 360
		13.5.2.1 Species Composition 360
		13.5.2.2 Trends in IBI Scoring 360
	13.5.3	Minnesota and North Dakota 361
		13.5.3.1 Species Composition 361
		13.5.3.2 Trends in IBI Scoring 361
	13.5.4	The Red River 362
		13.5.4.1 Species Composition 362
		13.5.4.2 Trends in IBI Scoring 363
	13.5.5	Variability 363
Acknowledgments 363
References 364

13.1 INTRODUCTION

One of the most common and widely accepted analytical tools used to measure the integrity of rivers and streams is the Index of Biotic Integrity (IBI). Biological integrity is defined as "the ability to support and maintain a balanced, integrated, adaptive community of organisms having a species composition, diversity, and functional organization comparable to that of the natural habitat of the region" (Karr and Dudley, 1981). The IBI relies on multiple parameters (termed "metrics") based on fish community structure and function, to evaluate a complex biotic system. It incorporates professional judgment in a systematic and sound manner, but sets quantitative criteria that enable determination of a continuum between very poor and excellent conditions based on species richness and composition, trophic and reproductive constituents, and fish abundance and condition. Since the metrics are differentially sensitive to various perturbations (e.g., siltation or toxic chemicals), as well as various degrees or levels of change within the range of integrity, conditions at a site can be determined with considerable accuracy.

However, a single IBI cannot be applied universally. The IBI was developed in midwestern wadeable streams and rivers (Karr, 1981; Karr et al., 1986) and modifications are needed when applying the IBI in other regions of the country to account for regional variation in fish community composition (Miller et al., 1988; Simon and Lyons, 1995). Various permutations of the original IBI have been developed for use in warmwater streams in northern sections of the Midwest (Simon and Lyons, 1995), Wisconsin (Lyons, 1992), southern Minnesota (Bailey et al., 1993), and southern Ontario (Steedman, 1988). However, species-depauperate systems of the Hudson Bay drainage have not been evaluated.

The purpose of this study is to develop potential IBI metrics to evaluate the Lake Agassiz Plain ecoregion (formerly the Red River Valley ecoregion) within the Red River Basin. We propose specific Index of Biotic Integrity criteria including the development of metrics and maximum species richness lines to delineate areas of high quality within the Lake Agassiz Plain ecoregion of the Red River Basin and to evaluate rivers and streams in the Lake Agassiz Plain ecoregion.

13.2 STUDY AREA

13.2.1 Drainage Features of the Red River of the North Basin

The Red River Basin drains 17,500 sq. mi. in northwestern Minnesota, 21,000 sq. mi. in eastern North Dakota, and 800 sq. mi. in northeastern South Dakota (Renard et al., 1986). The Red River is the major drainage unit in the basin (Figure 13.1). The river meanders northward for 394 miles to the United States–Canadian border (IRRPB, 1995). The mean annual flow of the Red River increases from 519 cfs (cubic feet per second) at its source at the confluence of the Bois de Sioux and Otter Tail Rivers (Renard et al., 1986), to 3386 cfs at the United States–Canadian border (IRRPB, 1995). Most of the flow occurs during the spring and early summer months when snow melt and heavy rains can cause severe flooding. The gradient gradually declines from 1.3 ft./mi. at Wahpeton/Breckenridge to 0.2 ft./mi. at the U.S.–Canadian border (Stoner et al., 1993). About 75% of the Red River flow comes from Minnesota tributaries (Tornes and Brigham, 1994). The largest tributary to the Red River in Minnesota is the Red Lake River, with a mean annual flow of 1110 cfs; while the largest North Dakota tributary of the Red River is the Sheyenne River, with a mean annual flow of 184 cfs.

13.2.2 Lake Agassiz Plain Ecoregion

The present study area was once entirely inundated by an immense glacial lake known as Lake Agassiz. Lake Agassiz once covered over 22,000 sq. mi. of land in Minnesota, South Dakota, and North Dakota (Waters, 1977). During the 12,000 years since the last glacial retreat, the waters of the basin have warmed, erosion has cut new and deeper stream channels, drainage patterns have changed, and vegetation has changed. Many of these changes were accelerated by human activities. The rich glacial lake sediments left behind by Lake Agassiz were highly desired for agriculture. Currently, over 90% of the land in Minnesota counties bordering the Red River has been cultivated (MPCA, 1994). Soils of the Red River Basin are by nature, poorly drained. A massive system of drainage ditches, channelized streams, and flood control impoundments have been built to bring the land into production (USGS, 1974).

Omernik (1987) and Omernik and Gallant (1988) defined the ecoregions of the U.S. from maps of land-surface form, soil types, potential natural vegetation, and land use. The Lake Agassiz Plain ecoregion is the only ecoregion that is entirely included in the Red River Basin (Figure 13.1). Four additional ecoregions occur within the Red River watershed along the outer edges of the basin:

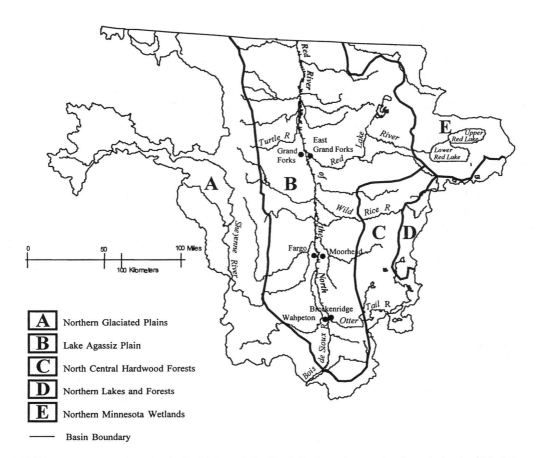

FIGURE 13.1 Major rivers in the Red River of the North Basin and ecoregion boundaries (modified from Omernik, 1987).

North Central Hardwood Forests, Northern Minnesota Wetlands, Northern Glaciated Plains, and Northern Lakes and Forests.

Headwater reaches of most major tributaries to the Red River originate in these four outer ecoregions. The majority of these tributary streams are higher in gradient, have definitive geomorphological units (i.e., riffles, pools, and runs), and are dominated by course gravel or cobble substrates. As tributary streams enter the Lake Agassiz Plain ecoregion, their gradient decreases and stream morphology begins to slow, forming meandering runs. The fine alluvial deposits of Lake Agassiz are reflected in the substrate of the rivers and streams. These physical changes associated with Lake Agassiz affect water chemistry characteristics by contributing to an increase in turbidity, conductivity, and total suspended solids.

13.2.3 HISTORICAL RED RIVER BASIN DATA

Historical fish community data exist for the Red River and some of its larger tributaries (Enblom, 1982; Renard et al., 1983; Renard et al., 1986; Hanson et al., 1984; Neel, 1985; Peterka, 1978; Peterka, 1991), but the smaller streams have generally been neglected (Figure 13.2). Renard et al. (1986) conducted fish community assessments at 14 locations on the Red River; a total of 36 species were collected, representing 12 families. Carp (*Cyprinus carpio*) composed over 50% of the total catch by weight. Game fish were not common in the collection. The most common game fish species were channel catfish (*Ictalurus punctatus*) and walleye (*Stizostedion vitreum*), comprising

Development of an IBI for the Species-Depauperate Lake Agassiz Plain Ecoregion

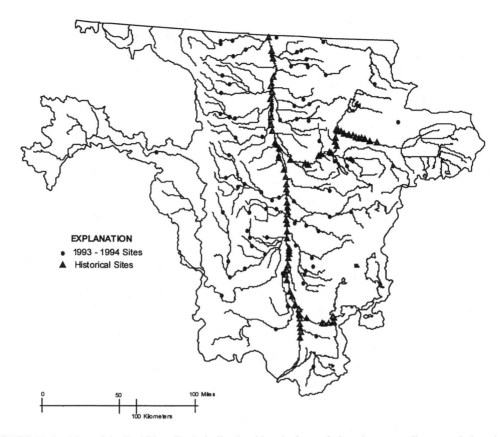

FIGURE 13.2 Map of the Red River Basin indicating historical sample locations as well as sample locations during the 1993–1994 study period.

8.2 and 3.2% of the total catch, respectively. Renard et al. (1983) surveyed the fish community at 8 sites on the Red Lake River in Minnesota, the largest tributary of the Red River; 38 fish species representing 13 families were collected during the survey. Of the six rivers in North Dakota that Peterka (1991) surveyed, Park River had the greatest number (18) of fish species.

Goldstein (1995), summarizing historical fish community data from the Red River Basin, reported that western tributaries of the Red River supported fewer species than those in the east. Of the 75 species found in the major rivers of the basin, 51 were present in the western half; whereas, 71 were found in the east. Species richness was reported to be influenced by drainage area (DA) and the number of ecoregions through which a river flows.

13.3 MATERIALS AND METHODS

13.3.1 Establishing the Reference Condition

Reference conditions define the physical, chemical, and biological expectations of minimally impacted rivers and streams (Hughes et al., 1982; Whittier et al., 1987; Hughes, 1995). Properly defined reference conditions provide a reasonable benchmark for comparison to measure the degree of water quality degradation (Hughes and Omernik, 1981; Hughes et al., 1986).

Reference conditions for the Lake Agassiz Plain ecoregion were based on 111 locations surveyed throughout the ecoregion during June through August 1993-1994 (Figure 13.2). Sites representing "least impacted" areas of the basin were considered to be candidate reference sites.

The selection of candidate reference sites was based on information gathered from consultation with local resource managers, historic fish collections by the Minnesota Department of Natural Resources and the North Dakota Department of Health, and on-site reconnaissance. Also, obtaining an adequate spatial coverage of the ecoregion determined the location of sites. In addition to the candidate reference sites, select data from past MDNR fisheries surveys were reviewed, and 102 sites with equivalent data collection objectives were also added to the database.

13.3.2 Criteria for Selecting Reference Sites

Reference sites are defined as possessing the following characteristics: (1) natural stream geomorphology (i.e., the stream channel has not been altered by dredging or channelization); (2) continuous riparian area along the reach extending laterally three times the channel width (e.g., land use is consistent laterally, soils and vegetation are undisturbed); (3) vegetation is undisturbed (i.e., woody vegetation has not been logged or brushed out and grasses have not been cut, burned, or treated with herbicides); (4) stream segment has not been stocked with forage or game fish species; (5) no point sources or in flowing springs, ditches, or drainage canals; (6) reach geomorphology is consistent with segment geomorphology; (7) no stream habitat improvements (e.g., wing dams, rip rap); (8) river sites are not snagged (e.g., all woody debris remains in the stream until moved by natural processes); (9) no dams or diversions upstream or downstream within two meander cycles or two replications of major geomorphological units; and (10) no bridges upstream within two meander cycles or two replications of major geomorphological units.

Few sites in the Lake Agassiz Plain met all criteria because natural, undisturbed land cover in the ecoregion is almost nonexistent. Therefore, candidate reference sites could not be considered pristine or undisturbed. Rather, these sites represent the best conditions given the anthropogenic impacts, channelization, and cultural eutrophication that have occurred in the Lake Agassiz Plain ecoregion.

13.3.3 Fish Community Sampling Procedures

All sites were rigorously sampled to get representative, distance-specific, quantitative estimates of species richness and biomass (Angermeier and Karr, 1986). Sampling was conducted in all-size river reaches in the Lake Agassiz Plain ecoregion from the headwater streams (drainage area [DA] <200 sq. mi., to the large rivers with DA > 1,500 sq. mi.). The reach length of each site was determined by multiplying the mean stream width by 15, with a maximum reach length of 500 m. Therefore, stream reach length varied with stream size. Sites ranged in length from 50 m at some headwater sites to 500 m at large river sites.

Sampling was conducted during the summer months to take advantage of low and stable flow conditions (Karr et al., 1986). Unfortunately, 1993 and 1994 were somewhat atypical because water levels remained high in the Lake Agassiz Plain throughout the summer months. Gear selection was dependent on stream size, velocity, substrate, and depth. However, only one electrofishing gear type was needed at each site to collect a representative sample (Jung and Libosvarsky, 1965; Ohio EPA, 1989). A generator-powered backpack electrofishing unit or a T&J pulsed-DC generator (300 V, 1750 watts) mounted in a Coleman Sport-canoe, or attached to a long-line was used to sample fish at headwater and moderately sized wadeable streams (1–1500 sq. mi. DA). Large river sites (>1500 sq. mi. DA) were sampled using a Coleman Sport canoe or boat electrofisher (Ohio EPA, 1989; U.S. EPA, 1988). Sampling occurred along both shorelines in streams wider than 5 m or followed a serpentine pattern on both shores for streams less than 5 m wide.

At each site, attempts were made to collect all fish encountered. Adult and juvenile specimens from each site were counted and identified to species utilizing the taxonomic keys of Becker (1983), Gerking (1955), and Trautman (1981). Smaller and more difficult to identify taxa were preserved for later examination and identification in the laboratory. The young-of-the-year fish less than 20 mm in length were not included in the analysis. Early life stages exhibit high initial mortality (Simon and Lyons, 1995) and were difficult to collect with gear designed for larger fish (Angermeier and Karr, 1986). Specimens greater than 20 mm total length (TL) were easily collected using our gear. Juvenile specimens greater than 20 mm TL function in distinct trophic guilds and reflect mature species attributes.

All fish were examined for the presence of gross external anomalies. Incidence of anomalies was defined as the presence of externally visible morphological anomalies (i.e., deformities, fin or gill erosion, lesions/ulcers, and tumors (DELT)). Specific anomalies include: fin rot; pugheadedness; Aeromonas (causes ulcers, lesions, and skin growth, and formation of pus-producing surface lesions accompanied by scale erosion); dropsy (puffy body); swollen eyes; fungus; ich; curved spine; and swollen-bleeding mandible or opercle. Incidence is expressed as percent of anomalous fish among all fish collected. Hybrid species encountered in the field (e.g., hybrid centrarchids, cyprinids) were recorded on the data sheet and, when possible, potential parental combinations were recorded.

13.4 RESULTS

13.4.1 METRICS

Goldstein et al. (1994) proposed a series of metrics for the Red River Basin that addressed the unique environmental attributes of the region. The current study expands on the conceptual framework laid out by Goldstein et al. (1994) by developing an IBI for the rivers and streams of the Lake Agassiz Plain ecoregion of Minnesota and North Dakota.

Twelve metrics are presented for each of three stream size classifications: headwater streams, moderate-sized streams, and large rivers (Table 13.1). Differences between headwater, moderate-sized, and large river sites were determined by searching for bimodal patterns in the metric specific data set plots. The transition between headwater and moderate sized streams occurred at 200 sq. mi. DA and between moderate-sized streams and large rivers at 1500 sq. mi. DA. Maximum scoring lines were drawn following the procedure of Fausch et al. (1984) and the Ohio EPA (1987). Several metrics are drainage size dependent. Drainage size effects were determined by evaluating trends in species or proportions of individuals with increasing (log transformed) DA. Narrative category descriptions (excellent, good, etc.) were assigned by comparing the IBI scores and fish community attributes with the narrative category descriptions provided by Karr (1981) (Table 13.2).

13.4.2 SPECIES COMPOSITION AND RICHNESS METRICS

13.4.2.1 Metric 1: Total Number of Fish Species (All Streams Sizes)

This metric is considered to be one of the most powerful in determining stream condition because a direct correlation exists between high-quality resources and the number of fish species in warm-water assemblages (Ohio EPA, 1987; Davis and Lubin, 1989; Plafkin et al., 1989; Simon, 1991). Moreover, highly diverse fish communities often contain specialized species that are typically less able to cope with changes in water quality.

This metric is strongly correlated with DA at headwater and moderate-sized stream and river sites up to 1500 sq. mi. DA. Large river sites (>1500 sq. mi. DA) did not show a DA relationship, but instead reached an asymptote (Figure 13.3).

13.4.2.2 Metric 2: Proportion of Headwater Species (Headwater Streams), Number of Benthic Insectivores (Moderate-Sized Streams), and Proportion of Round-Bodied Suckers (Large Rivers)

Karr's (1981) original IBI metrics included the number of darter species. Darters are insectivorous, habitat specialists, and sensitive to physical and chemical environmental disturbances (Page, 1983;

TABLE 13.1
Index of Biotic Integrity Metrics Used to Evaluate Headwater (<200 sq. mi. DA), Moderate (200–1500 sq. mi. DA), and Large River (>1500 sq. mi. DA) Sites in the Lake Agassiz Plain Ecoregion

Category	Metric	Headwater Scoring Classification			Moderate Stream Scoring Classification			Large River Scoring Classification		
		5	3	1	5	3	1	5	3	1
Species richness and composition	Total number of species	Varies with drainage area			Varies with drainage area			Varies with drainage area		
	Evenness	>0.8	>0.6 and ≤0.8	≤0.6	>0.8	>0.6 and ≤0.8	≤0.6	>0.8	>0.6 and ≤0.8	≤0.6
	Number of minnow species	Varies with drainage area								
	% Pioneer individuals[a]	Varies with drainage area			>7	>4 and ≤7	≤4	>16	>8 and ≤16	≤8
	% Headwater individuals	>50	>25 and ≤50	≤25						
	Number of benthic insectivore species				>7	>4 and ≤7	≤4			
	% Large river individuals							>40	>20 and ≤40	≤20
	Round-bodied suckers									
Trophic composition	% Piscivore biomass[a]				>20 and <30	>10 and ≤20, ≥30 and <40	≤10 and ≥40	>20 and <30	>10 and ≤20, ≥30 and <40	≤10 and ≥40
	% Omnivore biomass[a]	<33	≤33 and <60	≥66	<33	≥33 and <60	≥66	<33	≥33 and <60	≥66
	% Insectivore biomass[a]	>60	>30 and ≤60	≤30	>60	<30 and <60	≤30	>60	>30 and ≤60	≤30
Reproductive guild	% Simple lithophilic spawners[a]	>60	>30 and ≤60	≤30	>60	>30 and ≤60	≤30	>60	>30 and ≤60	≤30
Functional guild	% Tolerant individuals	<31	≥31 and <62	≥62	<31	≥31 and <62	≥62			
	Number of sensitive species	Varies with drainage area			Varies with drainage area			Varies with drainage area		
Abundance and condition	Number of individuals per meter	Varies with drainage area			Varies with drainage area			>60	>3 and ≤6	≤3
	% DELT	<1	≥1 and <4	≥4	<1	≥1 and <4	≥4	<1	≥1 and <4	≥4
Alternative metrics	% Headwater individuals				>50	>25 and ≤50	≤25			
	% Pioneer individuals				Varies with drainage area			Varies with drainage area		
	Number of minnow species							<7	>4 and ≤7	≤4
	% Subterminal mouth minnows				>20	>10 and ≤20	≤10			
	Number of sucker species				>4	>2 and ≤4	≤2			

[a] Special scoring procedures are required when the number of species metric is scored a "1" or when the catch is extremely low (<25 individuals at headwater sites, <50 individuals at moderate sized streams and large rivers).

TABLE 13.2
Attributes of Index of Biotic Integrity (IBI) Classification, Total IBI Scores, and Integrity Classes from Karr et al. (1986), and for the Lake Agassiz Plain

Karr IBI Score	Lake Agassiz Plain	Integrity Class	Attributes
58–60	51–60	Excellent	Comparable to the best situation without human disturbance; all regionally expected species for the habitat and stream size, including the most intolerant forms, are present with a full array of age (size) classes; balance trophic structure.
48–52	41–50	Good	Species richness somewhat below expectations, especially due to the loss of the most intolerant forms; some species are present with less than optimal abundances or size distributions; trophic structure shows some signs of stress.
40–44	31–40	Fair	Signs of additional deterioration include loss of intolerant forms, fewer species, highly skewed trophic structure (e.g., increasing frequency of omnivores and other tolerant species); older age classes of top predators may be rare.
28–34	21–30	Poor	Dominated by omnivores, tolerant forms, and habitat generalists; few top carnivores; growth rates and condition factors commonly depressed; hybrids and diseased fish often present.
12–22	12–20	Very Poor	Few fish present, mostly introduced or tolerant forms; hybrids common; disease, parasites, fin damage, and other anomalies regular.
0			No Fish, repeated sampling finds no fish.

Kuehne and Barbour, 1983). However, few darter species occur in streams of the Lake Agassiz Plain ecoregion. For this reason, the number of darter species metric is replaced by three separate metrics.

13.4.2.2.1 Proportion of headwater species

Headwater fish species are associated with stable flow conditions, permanent habitat, low environmental stress, and higher biological integrity (Ohio EPA, 1987; Simon, 1991). Five species that are found in the Lake Agassiz Plain are designated as headwater taxa (Table 13.3). No drainage area relationship was observed for this metric in headwater streams. (Figure 13.3).

13.4.2.2.2 Number of benthic insectivore species

Suckers, madtoms, and some minnows are benthic insectivores and functionally occupy the same niche type as darters (Ohio EPA, 1987; Simon, 1991). Nineteen species are designated as benthic insectivores (Table 13.3). With the exception of blacknose dace (*Rhinichthys atratulus*), which may be behaviorally plastic (Leonard and Orth, 1986), the remainder of the species are found in riffle habitats, usually over clean gravel substrates. The benthic insectivore metric was applied to moderate-sized streams because most of these species were absent at a significant number of headwater sites. This metric asymptotes at moderate stream sizes and declines at the largest DAs (Figure 13.3).

13.4.2.2.3 Proportion of round-bodied suckers

Karr (1981) included the number of sucker species in his original metrics. Round-bodied suckers include members of the sucker genera *Minytrema, Hypentelium, Moxostoma, Cycleptus,* and *Erimyzon* (see Emery et al., Chapter 7). Here, members of the sucker genus *Catostomus* are not included because members of this genus are able to tolerate a wide variety of environmental perturbations. Due to their long life cycles (10–20 years), suckers provide a long-term assessment of past environmental conditions. Sucker species, other than those in the genus *Catostomus,* are intolerant to habitat and water quality degradation (Phillips and Underhill, 1971; Karr et al., 1986; Trautman, 1981; Becker, 1983). Unlike smaller benthic insectivores that are often difficult to collect in large rivers, round-bodied suckers are effectively sampled with electrofishing gear and comprise

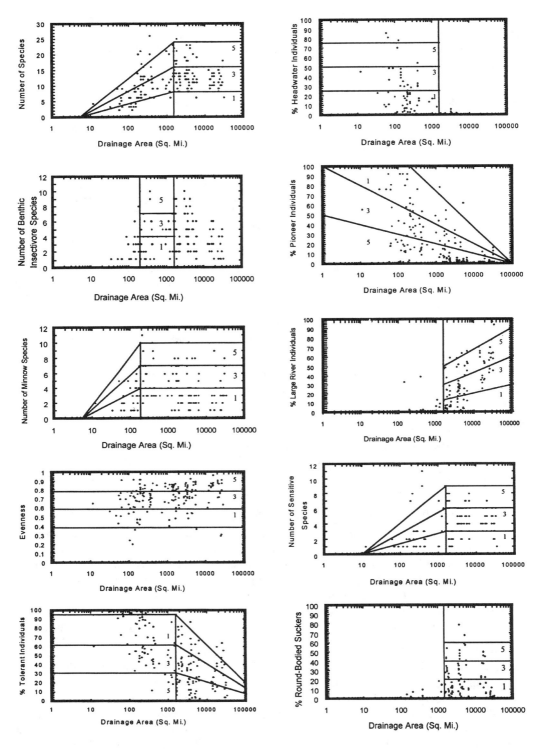

FIGURE 13.3 Maximum scoring lines for determining trends in species composition and richness metrics with increasing drainage area for the Lake Agassiz Plain ecoregion.

TABLE 13.3
Metric Classifications for all Documented Species (Current and Historic) Found in the Red River Basin

Scientific Name	Common Name	Classification
ACIPENSERIDAE	STURGEONS	
Acipenser fulvescens	lake sturgeon	LR BI IN SL
AMIIDAE	BOWFINS	
Amia calva	bowfin	PI
COTTIDAE	SCULPINS	
Cottus bairdi	mottled sculpin	HW BI IN SN
CATOSTOMIDAE	SUCKERS	
Carpiodes carpio	river carpsucker	LR OM SU
Carpiodes cyprinus	quillback	LR TL OM SU
Catostomus commersoni	white sucker	TL OM SL SU
Hypentelium nigricans	northern hog sucker	BI INRB SL SN SU
Ictiobus cyprinellus	bigmouth buffalo	LR TL OM SU
Moxostoma anisurum	silver redhorse	BI RB IN SL SN SU
Moxostoma erythrurum	golden redhorse	BI RB IN SL SN SU
Moxostoma macrolepidotum	shorthead redhorse	BI RB IN SL SN SU
Moxostoma valenciennesi	greater redhorse	BI RB IN SL SN SU
CENTRARCHIDAE	SUNFISH	
Ambloplites rupestris	rock bass	PI SN
Lepomis cyanellus	green sunfish	TL IN PN
Lepomis gibbosus	pumpkinseed	IN
Lepomis humilis	orangespotted sunfish	IN
Lepomis macrochirus	bluegill	IN
Micropterus dolomieu	smallmouth bass	PI SN
Micropterus salmoides	largemouth bass	PI
Pomoxis annularis	white crappie	PI
Pomoxis nigromaculatus	black crappie	PI
CYPRINIDAE	MINNOWS	
Campostoma anomalum	central stoneroller	MN PN SM
Campostoma oligolepis	largescale stoneroller	MN PN SM
Cyprinella spiloptera	spotfin shiner	MN IN
Cyprinus carpio	common carp	MN TL OM
Hybognathus hankinsoni	brassy minnow	MN
Luxilus cornutus	common shiner	MN IN
Macrhybopsis storeriana	silver chub	BI MN LR IN SM
Margariscus margarita	pearl dace	HW MN IN
Nocomis biguttatus	hornyhead chub	MN IN SN
Notemigonus crysoleucas	golden shiner	MN TL IN
Notropis anogenus	pugnose shiner	MN OM SN
Notropis atherinoides	emerald shiner	MN LR IN

TABLE 13.3 *(continued)*
Metric Classifications for all Documented Species (Current and Historic) Found in the Red River Basin

Scientific Name	Common Name	Classification
Notropis blennius	river shiner	MN IN SL SM
Notropis dorsalis	bigmouth shiner	BI MN IN SM
Notropis heterodon	blackchin shiner	MN IN SN
Notropis heterolepis	blacknose shiner	MN IN SN
Notropis hudsonius	spottail shiner	MN LR IN
Notropis rubellus	rosyface shiner	MN IN SL SN
Notropis stramineus	sand shiner	BI MN IN SN SM
Notropis texanus	weed shiner	MN IN
Notropis volucellus	mimic shiner	MN IN SN
Phoxinus eos	northern redbelly dace	HW MN IN SL SN
Phoxinus neogaeus	finescale dace	HW MN IN SN
Pimephales notatus	bluntnose minnow	MN TL OM PN SM
Pimephales promelas	fathead minnow	MN TL OM PN
Platygobio gracilis	flathead chub	MN IN
Rhinichthys atratulus	blacknose dace	BI HW MN TL IN SL
Rhinichthys cataractae	longnose dace	BI MN IN SL SN SM
Semotilus atromaculatus	creek chub	MN TL IN PN
ESOCIDAE	PIKES	
Esox lucius	northern pike	PI
Esox masquinongy	muskellunge	PI
FUNDULIDAE	KILLIFISHES	
Fundulus diaphanus	banded killifish	IN
GADIDAE	CODFISHES	
Lota lota	burbot	LR PI SL
GASTEROSTEIDAE	STICKLEBACKES	
Culaea inconstans	brook stickleback	HW TL IN
HIODONTIDAE	MOONEYES	
Hiodon alosoides	goldeye	LR IN SN
Hiodon tergisus	mooneye	LR IN SN
ICTALURIDAE	CATFISHES	
Ameiurus melas	black bullhead	TL OM
Ameiurus natalis	yellow bullhead	OM
Ameiurus nebulosus	brown bullhead	OM
Ictalurus punctatus	channel catfish	LR TL PI
Noturus flavus	stonecat	BI IN SN
Noturus gyrinus	tadpole madtom	BI IN

TABLE 13.3 *(continued)*
Metric Classifications for all Documented Species (Current and Historic) Found in the Red River Basin

Scientific Name	Common Name	Classification
LEPISOSTEIDAE	**GARS**	
Lepisosteus osseus	longnose gar	LR PI
MORONIDAE	**TEMPERATE BASSES**	
Morone chrysops	white bass	LR PI
PERCIDAE	**DARTERS**	
Etheostoma caeruleum	rainbow darter	BI IN SL SN
Etheostoma exile	iowa darter	BI IN
Etheostoma microperca	least darter	BI IN
Etheostoma nigrum	johnny darter	BI IN PN
Perca flavescens	yellow perch	IN
Percina caprodes	logperch	BI IN SL SN
Percina maculata	blackside darter	BI IN SL
Percina shumardi	river darter	LR BI IN SL
Stizostedion canadense	sauger	LR PI SL
Stizostedion vitreum	walleye	LR PI SL
PERCOPSIDAE	**TROUT PERCHES**	
Percopsis omiscomaycus	trout-perch	BI SN IN
PETROMYZONTIDAE	**LAMPREYS**	
Ichthyomyzon castaneus	chestnut lamprey	LR PI
Ichthyomyzon unicuspis	silver lamprey	LR PI
SALMONIDAE	**TROUTS**	
Coregonus artedi	cisco	
Coregonus clupeaformis	lake whitefish	IN
Oncorhynchus mykiss	rainbow trout	PI
Salmo trutta	brown trout	PI
Salvelinus fontinalis	brook trout	PI SN
SCIANIDAE	**DRUMS**	
Aplodinotus grunniens	freshwater drum	LR TL IN
UMBRIDAE	**MUDMINNOWS**	
Umbra limi	central mudminnow	TL IN

Note: HW = headwater species, BI = benthic insectivore species, RB = round-bodied sucker species, MN = minnow species, LR = larger river species, TL = tolerant species, OM = omnivore species, IN = insectivore species, PN = pioneer species, PI = piscivore species, SL = simple lithophil species, SN = sensitive species, SM = subterminal mouth minnow species, SU = sucker species.

a significant component of large riverine fish faunas. For these reasons, we applied the proportion of round-bodied suckers for DA > 1500 sq. mi. Of the nine sucker species present in the Red River Basin, five are considered round-bodied (Table 13.3). The proportion of round-bodied suckers did not show any relationship with DA for large rivers (Figure 13.3).

13.4.2.3 Metric 3: Number of Minnow Species (Headwater and Moderate-Sized Streams), Proportion of Large River Individuals (Large Rivers)

Karr's (1981) original Index of Biotic Integrity metrics utilized the presence of sunfish species to evaluate the quality of pool habitats. Pools often act as "sinks" for the accumulation of toxins and suspended sediment. Thus, this metric evaluates the quality of pool substrates (i.e., gravel and boulder), instream cover (Pflieger, 1975; Trautman, 1981), and the associated aquatic macroinvertebrate community (Forbes and Richardson, 1920; Becker, 1983).

13.4.2.3.1 Number of minnow species

The use of minnows to evaluate pools in the Lake Agassiz Plain is suggested because in this ecoregion minnows, rather than sunfish, are the dominant inhabitants of stream pool habitats. In addition, minnows are wide-ranging; they are susceptible to collection gear in most streams and rivers of the ecoregion, and their numbers tend to increase with higher biological integrity. As many as 11 different minnow species occur at locations <1500 sq. mi. DA (Table 13.3). This metric asymptotes at moderate-sized streams.

13.4.2.3.2 Proportion of large river individuals

Since minnows are often difficult to collect in large rivers, another metric was substituted for the original "sunfish" metric proposed by Karr (1981). Although not a complete replacement for the evaluation of pool habitat, the lack of sunfish species in the Lake Agassiz Plain facilitated the substitution of the proportion of individuals that are large river species.

Just as certain species are characteristic of headwater streams, others are commonly found in large river habitats. Pflieger (1975), Burr and Warren (1986), Simon (1992), and Simon and Emery (1995) found that a characteristic fish faunal assemblage was apparent in large and great river habitats. Simon (1992) found that certain fish species appeared at approximately 2000 sq. mi. DA. Herein, 19 species that occur in the Lake Agassiz Plain ecoregion are designated as large river species (Table 13.3). A lower proportion of large river taxa suggests a loss of biological integrity in large river habitats. An increasing drainage relationship for large rivers in the Lake Agassiz Plain ecoregion was observed (Figure 13.3).

13.4.2.4 Metric 4: Evenness (All Stream Sizes)

Evenness describes the distribution and abundance of individuals among species by comparing the observed diversity to a theoretical maximum diversity (Pielou, 1975). If all species are equally abundant, the evenness value equals a maximum value of 1. The greater the differences in abundance, the smaller the evenness (value approaches 0).

Oftentimes, in degraded environments, tolerant species dominate the fish community at the expense of other less-dominant species. As their relative abundance increases, evenness is reduced such that even without a loss of species (species richness), total diversity declines due to the reduction in evenness. Using evenness as a metric compliments the tolerant species metric by providing a measure of the degree to which tolerant species dominate a particular environment. Thus, reduced evenness indicates a loss of biotic integrity. No drainage relationship was observed (Figure 13.3); however, no relationship was expected since this community attribute is not influenced by stream size categories, but disruptions in community stability.

13.4.2.5 Metric 5: Number of Sensitive Species (All Stream Sizes)

Karr (1981) included only those species considered highly intolerant to a variety of disturbances in the number of intolerant species metric. Karr et al. (1986) further defined intolerant taxa as those that decline with decreasing environmental quality, and disappear as viable populations when the aquatic environment degrades to the "fair" category (Karr et al., 1986). Because very few species at headwater sites are classified as highly intolerant, the Ohio EPA (1987) modified the intolerant species metric to include some species classified as moderately intolerant. This modified version of Karr's original intolerant metric was called the "sensitive species metric." The number of sensitive species distinguishes between streams of highest quality. An absence of these species would indicate an anthropogenic stress or loss of habitat.

The criteria for determining intolerance for the Lake Agassiz Plain were based on the Ohio EPA (1989) and Underhill's (1989) documentation of historical changes in the distribution of Minnesota species, and supplemental information from regional ichthyofaunal texts (Pflieger, 1975; Smith, 1979; Trautman, 1981; Becker, 1983; Burr and Warren, 1986). Designation of too many species as sensitive will prevent this metric from discriminating among the highest quality resources. Some 25 species that occur in the Lake Agassiz Plain ecoregion were classified as sensitive (Table 13.3). The number of sensitive species increases with DA among headwater and moderate-sized streams and asymptotes in large rivers (Figure 13.3).

13.4.2.6 Metric 6: Proportion of Tolerant Individuals (All Stream Sizes)

The green sunfish (*Lepomis cyanellus*) is tolerant to degraded environments where it often successfully competes against less-tolerant species and becomes a dominant member of the fish community. Karr (1981) proposed a metric to measure the proportion of green sunfish in the fish community in midwestern streams. Although the green sunfish is widely distributed in the Midwest, it is most commonly collected in small streams and is not commonly found in the Lake Agassiz Plain ecoregion. Karr et al. (1986) suggested additional tolerant species could be substituted for the green sunfish if they responded in a similar manner. By increasing the number of tolerant species in this metric, the sensitivity for various sized streams and rivers is improved.

Several species in the Lake Agassiz Plain are known to increase in abundance with increasing degradation of stream quality. These species exhibit tolerance to a wide range of water quality problems, including thermal loadings, siltation, habitat degradation, and certain toxins (Gammon, 1983; Ohio EPA, 1989). The list of tolerant species for the Lake Agassiz Plain is based on Ohio EPA data (1989) and on expert consensus opinion between Minnesota and North Dakota ichthyologists. 15 species that occur in the Lake Agassiz Plain ecoregion are considered tolerant (Table 13.3). No DA relationship was evident for DA < 1500. However, a negative DA relationship was observed in large rivers (>1500 sq. mi.) (Figure 13.3).

13.4.3 Trophic Guilds

13.4.3.1 Metric 7: Proportion of Omnivore Biomass (All Stream Sizes)

Goldstein et al. (1994) defined an omnivore as a species that consumes significant quantities of both plant and animal materials (including detritus) and has the physiological ability (usually indicated by the presence of a long coiled gut and dark peritoneum) to utilize both. Fishes that do not feed on plants but on a variety of animal material are not considered omnivores. Dominance of omnivores suggests specific components of the food base are less reliable, increasing the success of more opportunistic species. Although bullheads (*Ameiurus* sp.) do not possess a dark peritoneum and therefore do not fit the definition of an omnivore, the professional opinion of local biologists is that the bullheads function as omnivorous members of the fish community in the Lake Agassiz Plain ecoregion. This metric evaluates the intermediate to low categories of environmental quality.

Here, the use of the proportion of omnivore biomass rather than percentage of individuals since biomass more appropriately reflects the utilization of energy within the fish community; 11 species that occur in the Lake Agassiz Plain ecoregion were classified as omnivores (Table 13.3). No relationship with DA was found for headwater, moderate streams, or large river sites (Figure 13.4). The lack of a DA pattern was expected since degraded habitats are not exclusive to any particular size waterbody.

13.4.3.2 Metric 8: Proportion of Insectivore Biomass (All Stream Sizes)

The proportion of insectivores is a modification of the Karr et al. (1986) original IBI metric, the proportion of insectivorous Cyprinidae. The inclusion of all insectivorous species was based on the observation that all areas of the Lake Agassiz Plain ecoregion do not possess high proportions of insectivorous cyprinids in high-quality streams. Insectivorous species are an important link in transferring energy between lower trophic levels to keystone predator species. This metric is intended to respond to a depletion of the benthic macroinvertebrate community, which comprises the primary food base for most insectivorous fishes. As disturbance increases, the diversity of insect larvae decreases, triggering an increase in the omnivorous trophic level. Thus, this metric varies inversely with the omnivore metric with increased environmental degradation. There are 51 insectivorous fish species that occur in the Lake Agassiz Plain ecoregion (Table 13.3). No relationship existed between DA and proportion of insectivorous fishes in the Lake Agassiz Plain ecoregion (Figure 13.4).

13.4.3.3 Metric 9: Proportion of Pioneer Species (Headwater Streams) and Proportion of Piscivore Biomass (Moderate-Sized Streams and Large Rivers)

Karr (1981) developed the carnivore metric to measure community integrity at upper trophic levels of the fish community. Carnivores depend on lower trophic levels for survival and are therefore dependent on a well-balanced, stable fish community structure indicative of high-quality environments. Here, this metric is retained for moderate-sized and large river sites in the Red River Basin; however, proportion of pioneer species was substituted for headwater streams, which typically are too small to possess carnivores.

13.4.3.3.1 Proportion of piscivore biomass
This metric includes individuals of species in which the adults are predominantly piscivores, although some may feed on other vertebrates or invertebrates such as crayfish (Karr et al., 1986). Species that are opportunistic do not fit into this metric, e.g., creek chub (*Semotilis atromaculatus*) (Karr et al., 1986; Ohio EPA, 1987). There are 19 piscivore species found in the Lake Agassiz Plain ecoregion (Table 13.3).

Because most piscivores found in this region are managed for sport fishing, the natural balance or proportion of piscivores can sometimes be higher than expected. Therefore, an upper limit of 30% piscivore biomass is placed on this metric. Karr (1981) suggested that the proportion of piscivores should be a reflection of DA. Such a correlation in streams greater than 200 sq. mi. DA was not found in the Lake Agassiz Plain ecoregion (Figure 13.4).

13.4.3.3.2 Proportion of pioneer species
Piscivores are generally not abundant in headwater streams; thus, an alternate metric was developed by the Ohio EPA (1987) to determine the permanence of the stream habitat. Smith (1971) identified a signature assemblage of small stream species, termed "pioneer species." These species are the first to colonize sections of headwater streams after desiccation and they predominate in unstable environments affected by anthropogenic stresses. Thus, a high proportion of pioneer species indi-

Development of an IBI for the Species-Depauperate Lake Agassiz Plain Ecoregion

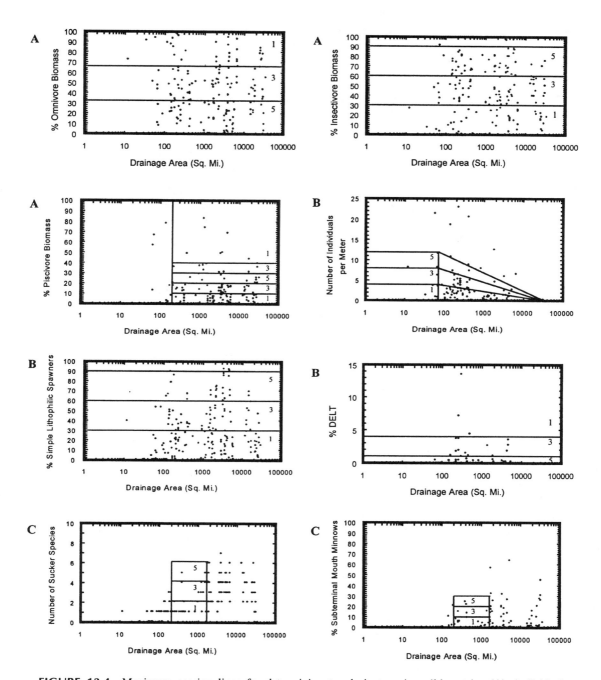

FIGURE 13.4 Maximum scoring lines for determining trends in trophic guild metrics (A), individual condition, reproductive guilds, and abundance metrics (B), and alternative metrics (C), with increasing drainage area for the Lake Agassiz Plain ecoregion.

cates an environment that is unstable or stressed. There are seven species that occur in the Lake Agassiz Plain ecoregion that are considered pioneer species (Table 13.3). The number of pioneer species decreases with DA (Figure 13.4).

13.4.4 INDIVIDUAL CONDITION, REPRODUCTIVE GUILDS, AND ABUNDANCE

13.4.4.1 Metric 10: Number of Individuals per Meter (All Stream Sizes)

This metric evaluates population density and is expressed as catch-per-unit effort. Catch per unit effort for the Lake Agassiz Plain ecoregion was calculated as the total number of individuals collected per meter of stream reach sampled. (See Section 13.3.3). Low catch per unit effort values may indicate that biotic integrity is being compromised and that the normal trophic relationships of fish communities are being altered. Although biotic integrity is generally considered to be positively correlated with catch per unit effort, numerous caveats exist. For example, as biotic integrity increases, the catch per unit effort often becomes more variable because numerous physical and chemical characteristics of the stream begin to limit species abundance. Also, under certain circumstances (e.g., channelization), reduced canopy cover may increase stream productivity by allowing light to penetrate the stream surface. This may lead to an increase in the abundance of certain tolerant fishes (Ohio EPA, 1987). Lyons (1992) found that abundance, excluding tolerant species, was greatest at fair quality sites in Wisconsin warmwater streams and lower at sites classified as excellent. Karr et al. (1986) suggest that this metric is most sensitive at intermediate to low ends of sensitivity. A DA-dependent relationship was observed for the Lake Agassiz Plain ecoregion (Figure 13.4).

13.4.4.2 Metric 11: Proportion of Individuals as Simple Lithophilic Spawners (All Stream Sizes)

The Ohio EPA (1987) replaced the proportion of hybrids (Karr et al., 1986), with the proportion of simple lithophilic spawners. The original hybrid metric's intent was to assess the extent to which degradation has altered reproductive isolation among species. However, difficulties in identification, lack of occurrence in headwater and impacted streams, and presence of hybrids in high-quality streams among certain taxa (e.g., cyprinids and centrarchids) caused a lack of sensitivity for the hybrid metric.

Spawning guilds are known to be affected by habitat quality (Balon, 1975; Berkman and Rabeni, 1987). Simple lithophilic spawners broadcast eggs that come into contact with the substrate. Eggs develop in the interstitial spaces between sand, gravel, and cobble without parental care. Simple lithophilic spawners are particularly sensitive to siltation since they require clean gravel or cobble substrates for reproductive success. Berkman and Rabeni (1987) observed an inverse correlation between simple lithophilic spawners and the proportion of silt in streams. Historically, some simple lithophilic spawners have experienced significant range reductions due to increased silt loads. Eighteen species that occur in the Lake Agassiz Plain ecoregion are considered simple lithophils (Table 13.3). No relationship with DA was observed at stream and river sites for the proportion of lithophilic species (Figure 13.4).

13.4.4.3 Metric 12: Proportion of Individuals with Deformities, Eroded Fins, Lesions, and Tumors (DELT) (All Stream Sizes)

The proportion of individuals with DELT anomalies correspond to the percent of diseased fish in Karr's (1981) original index. Studies of fish populations indicate that these anomalies are either absent or occur at very low rates naturally, but reach higher percentages at impacted sites (Mills et al., 1966; Berra and Au, 1981; Baumann et al., 1987). An increase in the frequency of occurrence of these anomalies is an indication of physiological stress due to environmental degradation, chemical pollutants, overcrowding, improper diet, excessive siltation, and other perturbations. Parasitic infections do not appear to be correlated with stream quality (Whittier et al., 1987; Steedman, 1988) and are therefore not included in this metric.

Common causes for DELT anomalies are a result of bacterial, fungal, viral, and parasitic infections; neoplastic diseases; and chemicals (Allison et al., 1977; Post, 1983; Ohio EPA, 1987).

In Ohio, the highest incidence of deformities, eroded fins, lesions, and tumors occurred in fish communities downstream from dischargers of industrial and municipal wastewater, and areas subjected to the intermittent stresses from combined sewers and urban runoff (see Sanders et al., Chapter 8). Leonard and Orth (1986) found this metric to correspond with increased degradation in streams in West Virginia. Karr et al. (1986) observed this metric to be most sensitive in low-quality streams.

The incidence of DELT anomalies was very low in the Lake Agassiz Plain ecoregion. However, since the potential for industrial development in this ecoregion exists, this metric may be a more important indicator of degraded conditions in future years. No relationship with DA was observed at stream or river sites for the proportion of individuals with DELT anomalies (Figure 13.4).

13.4.5 Alternative Metrics

Several metrics are considered as "alternative metrics" because they were originally tested for inclusion in the Lake Agassiz Plain IBI, but were not chosen for the final version. Alternative metrics can be used at the discretion of the researcher in situations where the final IBI score is close to the integrity classification cuttoffs or where a particular metric does not seem to be performing well.

Some alternative metrics were included in the Lake Agassiz Plain IBI for certain stream size classifications and have been expanded to include other stream size classifications (Table 13.1). Such is the case with the proportion of headwater species metric, which was used in headwater streams but may have applicability in moderate streams as well.

Two alternative metrics — the proportion of subterminal mouth minnows and the number of sucker species metrics — were not included in the final IBI. Because subterminal mouth minnows and suckers are primarily benthic insectivores, they depend on the availability of interstitial habitat to provide food and cover. A high proportion of subterminal mouth minnows and sucker species would be associated with high biotic integrity and a lack of siltation. There are eight subterminal mouth minnow species and nine sucker species that occur in the Lake Agassiz Plain ecoregion (Table 13.3). No drainage area relationship was observed for either metric (Figure 13.4).

13.4.6 Scoring Modifications

Samples with only a few species or extremely low numbers in the catch can present a scoring problem in some of the proportional metrics. Adjustments must be made to reduce the possibility of giving high scores to degraded sites. Aquatic habitats impacted by anthropogenic disturbances may exhibit a disruption in the food base and the sample will consist of very few species or numbers of individuals. At such low population sizes, the normal structure of the community is unpredictable (Ohio EPA, 1987). Based on Ohio EPA experiences, the proportion of omnivores, insectivorous fishes, and percent individuals affected by anomalies do not always match expected trends at these sample sizes (Rankin and Yoder, Chapter 23). Although scores are expected to deviate strongly from those of high-quality areas, this is not always observed. Rather, at these times, the opposite deviation of metric score is achieved due to low numbers of individuals or absence of certain taxa.

Similar to Ohio EPA experience, Bailey et al. (1993) found that in the Minnesota River Basin, many of the proportional metrics was unpredictable when the number of species at a site were extremely low. At these sites, some of the proportional metrics did not reflect the quality of the resource because percent composition was unpredictably influenced by the presence of a few species. Low score modifications were necessary for the proportional metrics when the "number of species" metric scored a "1," indicating severe impairment.

However, unlike Ohio, the "percent DELT anomalies" metric was not modified in the Minnesota River Basin because very few sites had species that exhibited these anomalies. This is probably because most sites were not directly impacted by industrial pollutants. Rather, habitat destruction through channelization and sedimentation were the primary anthropogenic influences on stream quality. Similar to the Minnesota River Basin, few DELT anomalies were found in fish from the Lake Agassiz Plain ecoregion.

Scoring very degraded sites without modifying scoring criteria for the proportional metrics can overestimate the total index score for these sites. The following scoring modifications, based on information from Bailey et al. (1993), Ohio EPA (1987), and Rankin and Yoder (Chapter 23), were adopted for evaluating sites in the Red River Basin.

13.4.6.1 Headwater Streams

Proportion of omnivores, insectivores, simple lithophils, and pioneer species metrics should be scored a "1" if less than 25 individuals are collected at a site or the number of species metric is scored a "1."

13.4.6.2 Moderate-Sized Streams and Large Rivers

The proportion of piscivore, omnivore, insectivore, and simple lithophil metrics should be scored a "1" if less than 50 individuals are collected at a site or when the number of species metric is scored a "1."

No scoring adjustments are necessary for proportion of tolerant species or the percent DELT anomalies. Further evaluation is needed to determine if scoring modifications will be needed for the other proportional metrics. In all cases, the biologist's best professional judgment should be used to decide when low score modifications are appropriate.

13.5 DISCUSSION

13.5.1 LAKE AGASSIZ PLAIN

13.5.1.1 Species Composition

A total of 111 sites were sampled in the Lake Agassiz Plain during 1993 and 1994: 63 species were collected, representing 16 families. This represents about 73% of the 86 species reported (Koel and Peterka, 1994) to occur in the Red River Basin fauna. Numerically, cyprinids dominated the catch (74%), followed by catostomids (9%). The most abundant species were fathead minnow (*Pimephales promelas*), common shiner (*Luxilus cornutus*), and creek chub, which comprised 20, 17, and 11% of the combined catch, respectively. Cyprinids also dominated the catch in terms of biomass (40%); however, common carp comprised 64% of the cyprinid biomass. If one excludes carp from the catch, catostomids were the dominant family by weight (35%), followed by cyprinids (14%). The most dominant species by weight was common carp (25%), followed by white sucker (*Catostomus commersoni*) (9%) and channel catfish (8%). Three species — the rainbow trout (*Oncorhynchus mykiss*), brook trout (*Salvelinus fontinalis*), and white bass (*Morone chrysops*) — were stocked for sport fisheries. The carp was the only non-game fish introduction collected during the study.

13.5.1.2 Trends in IBI Scoring

The IBI scores throughout the Lake Agassiz Plain were highly variable. There was virtually no correlation between IBI scores and DA ($r^2 = 0.0838$) (Figure 13.5). IBI scores ranged from 16 to 48, or from "very poor" to "good" using Karr's integrity classification modified for the Lake Agassiz Plain ecoregion (Table 13.2). The average overall score was 32 (Figure 13.6), while 40% of all sites were rated fair (Figure 13.7, IBI score between 31 and 40). With improved land and water management practices, one expects some sites to move into the excellent category (>50).

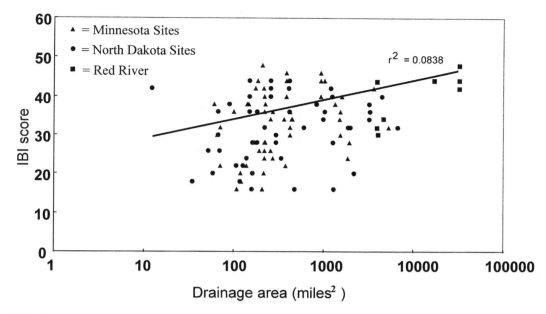

FIGURE 13.5 Plot of drainage area vs. IBI score for all sites sampled during the study period, 1993–1994. Triangle markers represent sites sampled in Minnesota, circle markers for North Dakota, and squre markers for sites sampled on the Red River.

Average metric scores can provide insight into fish community condition in the ecoregion. For example, the average metric score for the percent DELT anomalies metric was very high (4.56), indicating that pollution sources within the ecoregion are not negatively affecting fish health. Increases in the prevalence of DELT anomalies have been found in degraded stream habitats, often associated with industrial and municipal discharges (Sanders et al., Chapter 8). Since we attempted to avoid these areas, significant numbers of DELT anomalies were not expected. The proportion of omnivore and insectivore biomass also scored high relative to the other metrics, averaging 3.30 and 2.98, respectively. This indicates that streams within the ecoregion generally have a stable food base, allowing the more specialized feeders to thrive. The proportion of simple lithophilic spawners scored the lowest of all the metrics. Siltation is the most likely limiting factor affecting those species that rely on clean gravel and cobble substrates to successfully reproduce.

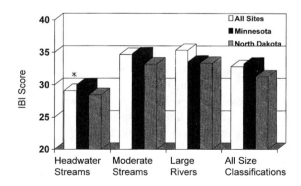

FIGURE 13.6 Mean IBI scores for headwater, moderate, and large rivers as well as all stream sizes. Scores for Minnesota and North Dakota exclude the sites located on the Red River, while the scores for all sites include them. Asterisks indicate a significant difference from other stream sizes ($p < 0.05$).

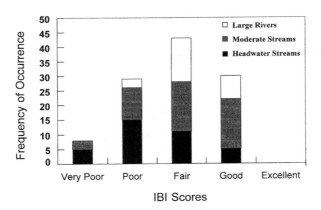

FIGURE 13.7 Integrity ranking from all sites sampled during the 1993–1994 study period, based on the modified integrity classification developed for the Lake Agassiz Plain ecoregion of Minnesota and North Dakota.

13.5.2 Headwater Streams, Moderate-Sized Streams, and Large Rivers

13.5.2.1 Species Composition

Community composition varied among headwater streams, moderate-sized streams, and large rivers. In headwater streams (<200 sq. mi. DA), 33 species, representing nine families were collected. The most abundant species in the catch at headwater streams were fathead minnows (41%), blacknose dace (12%), and brook stickleback (*Culaea inconstans*) (11%). Moderate-sized streams (200 to <1500 sq. mi. DA) had 47 species, representing 14 families. The most abundant species in moderate sized streams by number in the catch were common shiner (*Luxilus cornutus*) (23%), creek chub (13%), and fathead minnow (13%). There were 55 species representing 14 families, collected in large rivers (>1500 sq. mi. DA). The dominant species in large rivers numerically in the total catch were fathead minnow (16%), common shiner (14%), and spotfin shiner (*Cyprinella spiloptera*) (11%).

13.5.2.2 Trends in IBI Scoring

Average IBI score for headwater streams was 29.1, which was significantly lower (ANOVA, $p < 0.05$) than either moderate streams (mean = 34.1) or large rivers (mean = 35.3) (Figure 13.6). Some 55% of all headwater streams were rated "poor" or "very poor." This was in contrast to moderate-sized streams and large rivers, which had a poor or very poor score at only 28 and 12% of the sites, respectively (Figure 13.7).

There are a number of plausible explanations for the low biotic integrity of headwater streams in the Lake Agassiz Plain ecoregion. Stream channelization, particularly prevalent in headwater stream reaches, has undoubtedly led to some of the most irretrievable impairments to resource quality. Its negative effects are most pronounced on small headwater streams where many have been straightened and rerouted to serve as field drainage systems for row crops. Stream channelization reduces habitat quality by reducing pool depth, reducing substrate heterogeneity, altering riffle-run-pool sequences, increasing turbidity and the retention time for sediment in the stream channel, and reducing the retention time for water remaining in the stream channel (Rankin et al., 1992).

Due to the degree of channelization in this ecoregion, it was impossible to totally avoid the negative influence of channelization on habitat quality. Many headwater sites, although not channelized, were often surrounded upstream and downstream by large segments of channelized stream.

These unaltered stream segments, while certainly not pristine, represent the best conditions found in many of these streams and provide an important habitat refuge for biota.

In addition to channel modification, headwater sites often undergo extreme fluctuations in flow, creating temporal changes in habitat availability for fish. As expected, a relatively large proportion (39%) of the total number of species collected in headwater environments were pioneer species. Low-flow conditions at headwater sites may benefit pioneer species by forcing fish to move out of an area for a period of time and then recolonize the area later when conditions improve. Conversely, high-flow conditions also create unstable environments due primarily to increased velocity and turbidity. These conditions are exacerbated by the channelization that is so prevalent in the Lake Agassiz Plain ecoregion.

13.5.3 MINNESOTA AND NORTH DAKOTA

13.5.3.1 Species Composition

In general, streams in the eastern portion of the basin have a higher species richness than those in the west. Excluding the border waters of the Red River, a total of 52 species were collected in Minnesota streams, while 46 species were collected in North Dakota. Some of the rivers in the basin, like the Otter Tail River of Minnesota, are known to have a high species diversity (Koel and Peterka, 1994). In this study, 23 species were collected from a single site on the Otter Tail River, including 4 species (green sunfish, largemouth bass (*Micropterus salmoides*), greater redhorse (*Moxostoma valenciennesi*), and logperch (*Percina caprodes*)) not collected in any other tributary river during the study. The Otter Tail is like many of the other tributaries to the Red River in Minnesota, flowing through a number of ecoregions. Originating in the Northern Lakes and Forests ecoregion, the Otter Tail flows southward into the Central Hardwood Forests ecoregion before turning west into the Lake Agassiz Plain. The rolling hills and abundance of lakes and wetlands in the upper two ecoregions contrast sharply with the relatively featureless topography of the Lake Agassiz Plain ecoregion. This diversity of habitat, while common to tributaries in the eastern portion of the basin, is not present on the western side of the basin.

A number of species were unique to one state, either Minnesota or North Dakota. Species collected only in Minnesota included mooneye (*Hiodon tergisus*), goldeye (*Hiodon alosoides*), bigmouth buffalo (*Ictiobus cyprinellus*), silver lamprey (*Ichthyomyzon unicuspis*), green sunfish, pumpkinseed (*Lepomis gibbosus*), burbot (*Lota lota*), largemouth bass (*Micropterus salmoides*), greater redhorse, blacknose shiner (*Notropis heterolepis*), logperch, finescale dace (*Phoxinus neogaeus*), sauger (*Stizostedion canadense*), and central mudminnow (*Umbra limi*). Eight of the aforementioned species were collected from either the Otter Tail or Wild Rice Rivers of Minnesota. The central mudminnow was found extensively throughout the Minnesota portion of the basin (33 sites), but was not collected from North Dakota during the study. Eight species were collected only in North Dakota: largescale stoneroller (*Campostoma oligolepis*), banded killifish (*Fundulus diaphanus*), brown bullhead (*Ameiurus nebulosus*), brassy minnow (*Hybognathus hankinsoni*), bluegill (*Lepomis macrochirus*), orangespotted sunfish (*Lepomis humilis*), golden shiner (*Notemigonus crysoleucas*), and rainbow trout (*Oncorhynchus mykiss*). Six of the eight species collected exclusively from North Dakota were from either the Turtle or Sheyenne Rivers.

13.5.3.2 Trends in IBI Scoring

Overall, IBI scores were slightly higher for streams in Minnesota compared to those in North Dakota (Figure 13.6). Although significant differences in IBI scoring did not occur between the two states, there were some differences in individual metrics. For example, the species richness metric scored higher in Minnesota streams for all stream size classifications. Somewhat more perplexing is that most of the biomass metrics and the catch-per-unit effort metric scored higher in North Dakota. Only insectivore biomass in headwater streams was higher in Minnesota. The higher catch rates in North Dakota suggest that these streams may be more productive.

Pioneer species occurred more frequently in North Dakota headwater streams. With the exception of the green sunfish, every pioneer species was more common in North Dakota. The likely explanation for the variability might lie in the temporary nature of headwater streams in North Dakota related to climatic and topographical factors. Headwater streams in North Dakota typically originate in relatively dry prairie environments. Periods of drought may cause rapid and prolonged decreases in stream flow, making some sections of stream uninhabitable by fish. In contrast, Minnesota headwater streams are linked with wetland habitat, which has the effect of modifying these hydrologic extremes. In addition, the Minnesota portion of the Red River Basin receives more annual precipitation than North Dakota.

Sensitive species were more common in Minnesota. The average IBI score for this metric in Minnesota streams was 2.51. This is almost a full point higher than the average IBI score of 1.64 for streams in North Dakota. There were six sensitive species collected in Minnesota that were not found in North Dakota. These species included the goldeye, mooneye, greater redhorse, blacknose shiner, log perch, and finescale dace. Sensitive species such as the northern hogsucker (*Hypentelium nigricans*), rainbow darter (*Etheostoma caeruleum*), pugnose shiner (*Notropis anogenus*), mimic shiner (*Notropis volucellus*), and blackchin shiner (*Notropis heterodon*) are known to historically occur in the ecoregion, but were not collected in either state during the study period.

13.5.4 THE RED RIVER

13.5.4.1 Species Composition

The Red River was sampled at eight locations, ranging from sites close to its source to sites near the Canadian border (Figure 13.8). In total, 36 species, representing 12 families, were collected from the Red River. The most abundant species based on proportion of the catch included spotfin shiner (38%), emerald shiner (33%), and silver chub (*Macrhybopsis storeriana*) (4%), while shorthead redhorse (*Moxostoma macrolepidotum*) (20%), silver redhorse (*Moxostoma anisurum*) (16%), and carp (15%) dominated the catch based on biomass. Three species — the white bass (*Morone chrysops*), silver chub, and river shiner (*Notropis blennius*) — were unique to the Red River. The results of this survey compare favorably with the 1983 to 1984 Red River survey by Renard et al. (1986). Renard et al. (1986) sampled at 40 stations and collected the same number of species (36) as in this study, suggesting that the 8 sites sampled in this study were sufficient to characterize the fish community of the Red River.

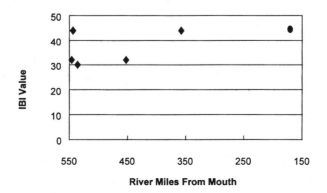

FIGURE 13.8 Longitudinal community trends in IBI scores for the Red River. The circle marker (River Mile 173) is an average.

13.5.4.2 Trends in IBI Scoring

Sites surveyed in the Red River ranged from "fair" (score of 32) to "good" (score of 48), based on IBI scoring criteria (Figure 13.8). IBI scores were generally higher in the Red River when compared to other large river sites in the ecoregion. The mean IBI score for the Red River was 39.5, while other large river sites averaged 33.4. Metrics that scored particularly well in the Red River included the number of large river species, proportion of omnivore species, and proportion of tolerant individuals. In fact, the tolerant species metric attained a maximum score of 5 at every site sampled. In contrast, the number of simple lithophilic individuals and proportion of round-bodied suckers metrics scored poorly in the Red River (mean scores of both were 1.5). This suggests that while many fish species are doing well in the Red River, siltation may be limiting those fish species that require clean gravel substrates to successfully reproduce.

The Red River sites were widely distributed and provide some insight into the biotic integrity of this river along a longitudinal gradient (Figure 13.8). Despite the input from potential pollution sources, IBI scores did not decrease with increasing distance downstream. Rather, the highest scoring sites were located nearest the Canadian border. At these most downstream sites, the evenness and the CPUE metrics scored well, showing an abundant, well-balanced fish population that was not dominated by opportunistic or tolerant species.

13.5.5 Variability

Two types of variability were examined to determine if any bias in sampling occurred. Although variability throughout the Lake Agassiz Plain was quite evident, variability in sampling technique between agencies needed to be eliminated as a source. Temporal variability was measured at eight sites. These sites were sampled during different years (1993 and 1994) and sampled the same year during different time periods. Sampling procedures remained the same between time periods; however, many of the sites were repeated with different crews. Differences between years were not significantly different (ANOVA, $p > 0.05$). Differences for sites repeated during the same year were also not significantly different (ANOVA, $p > 0.05$). A coefficient of variation (CV) compared the relative amounts of variation each site exhibited. The mean CV value for all 8 sites and all 20 samples was 9.06%.

Five sites, each with three consecutive reaches, were sampled by a single crew on the same day to examine the adequacy of the reach length. The mean CV for these 5 sites was 7.3%. Less than 2% difference existed between temporal variability with multiple samplers and the reach variability. It is concluded, therefore, that reach lengths were sufficient to characterize the fish community at each site and that no bias existed among samplers, sampling time, or sampling equipment for this project.

ACKNOWLEDGMENTS

We appreciate the efforts of P. Johnson (formerly U.S. EPA Region VIII), M. Ell (North Dakota Department of Health), and P. Bailey (Minnesota Pollution Control Agency), who initially planned and implemented this project. U.S. EPA Headquarters, Office of Science and Technology supplied financial assistance. Field assistance was provided by J. Enblom, K. Schmidt, and M. Feist (Minnesota Department of Natural Resources), L. Schluetter, A. Thompson, B.J. Kratz, and P. Seyer (North Dakota Game and Fish Department), and P. Wax (North Dakota Department of Health). Thanks to L. Hotka, S. Bissonette, and N. Proulx (MPCA) for their support, and to A. Johnson and D. Bain for solving computer-related problems. Our appreciation to Dr. John Peterka (North Dakota State University) and to Dr. James Underhill and Dr. Jay Hatch (Bell Museum) for taxonomic identification and vouchering fish specimens. Willis Mattison and local SWCD staff kept us informed about river conditions. Maps were made available by Carrie Bartz, MPCA, and personnel from USGS. Although this study may have been funded wholly or in part by the U.S. Environmental Protection Agency, the expressed opinions do not necessarily reflect the agency's position.

REFERENCES

Allison, L.N., J.G. Hnath, and W.G. Yoder. 1977. Manual of Common Diseases, Parasites, and Anomalies of Michigan Fishes. Michigan Department of Natural Resources, Lansing. Fisheries Management Report No. 8.

Angermeier, P.L. and J.R. Karr. 1986. Applying an Index of Biotic Integrity based on stream fish communities: considerations in sampling and interpretation, *North American Journal of Fisheries Management,* 6, 418–429.

Bailey, P.A., J.W. Enblom, S.R. Hanson, P.A. Renard, and K. Schmidt. 1993. *A Fish Community Analysis of the Minnesota River Basin.* Minnesota Pollution Control Agency, St. Paul, MN.

Balon, E.K. 1975. Reproductive guilds of fishes: a proposal and definition, *Journal of the Fisheries Research Board of Canada,* 32, 821–864.

Bauman, P.C., W.D. Smith, and W.K. Parland. 1987. Tumor frequencies and contaminant concentrations in brown bullhead from an industrialized river and a recreational lake, *Transactions of the American Fisheries Society,* 116, 79–86.

Becker, G.C. 1983. *Fishes of Wisconsin.* University of Wisconsin Press: Madison, WI.

Berkman, H.E. and C.F. Rabeni. 1987. Effect of siltation on stream fish communities, *Environmental Biology of Fishes,* 18, 285–294.

Berra, T.M. and R. Au. 1981. Incidence of teratological fishes from Cedar Fork Creek, Ohio, *Ohio Journal of Science,* 81, 225.

Burr, B.M. and M.L. Warren, Jr. 1986. A Distributional Atlas of Kentucky Fishes. Kentucky Nature Preserves Commission Scientific and Technical Series No. 4. Frankfort.

Davis, W.S. and A. Lubin. 1989. Statistical validation of Ohio EPA's invertebrate community index, in W.S. Davis and T.P. Simon (Eds.), *Proceedings of the 1989 Midwest Pollution Control Biologists Meeting.* Chicago, Illinois. U.S. Environmental Protection Agency, Region V, Chicago, IL. EPA 905/9-89-007.

Enblom, J.W. 1982. Fish and Wildlife Resources of the Roseau River. Minnesota Department of Natural Resources Special Report No. 130.

Fausch, K.D., J.R. Karr, and P.R. Yant. 1984. Regional application of an index of biotic integrity based on stream-fish communities, *Transactions of the American Fisheries Society,* 113, 39–55.

Forbes, S.A and R.E. Richardson. 1920. The Fishes of Illinois. (2nd ed.) *State Natural History Survey of Illinois,* 3, 1–357.

Gerking, S.D. 1955. Key to the fishes of Indiana, *Investigations of Indiana Lakes and Streams,* 4, 49–86.

Goldstein, R.M., T.P. Simon, P.A. Bailey, M. Ell, E. Pearson, K. Schmidt, and J.W. Enblom. 1994. Concepts for an Index of Biotic Integrity for Streams of the Red River of the North Basin, *Proceedings of the North Dakota Water Quality Symposium.* March 30–31, 1994, Fargo, ND. 169–180.

Goldstein, R.M. 1995. Aquatic Communities and Contaminants in Fish from Streams of the Red River of the North Basin, Minnesota and North Dakota. U.S. Geological Survey Water-Resources Investigations Report 95-4047.

Hanson, S.R., P.A. Renard, N.A. Kirsch, and J.W. Enblom. 1984. Biological survey of the Otter Tail River. Minnesota Department of Natural Resources Special Publication No. 137.

Hughes, R.M. 1995. Defining acceptable biological status by comparing with reference conditions, in W.S. Davis and T.P. Simon (Eds.), *Biological Assessment and Criteria: Tools for Water Resource Planning and Decision Making.* Lewis, Boca Raton, FL. 31–48.

Hughes, R.M. and J.M. Omernik. 1981. A proposed approach to determine regional patterns in aquatic systems, in *Acquisition and Utilization of Aquatic Habitat Inventory Information, Proceedings of a Symposium.* American Fisheries Society, Western Division, Portland, OR. 92–102.

Hughes, R.M., J.H. Gakstatter, M.A. Shirazi, and J.M. Omernik. 1982. An approach for determining biological integrity in flowing waters, in T.B. Braun (Ed.), *Inplace Resource Inventories: Principles and Practices: A National Workshop.* Society American Foresters, Bethesda, MD. 877–888.

Hughes, R.M., D.P. Larsen, and J.M. Omernik. 1986. Regional reference sites: a method for assessing stream pollution, *Environmental Management,* 10, 629–635.

International Red River Pollution Board, 1995. Thirty-Fifth Progress Report to the International Joint Commission.
Jung, C.O. and J. Libovarsky. 1965. Effect of size selectivity on population estimates based on successive removals with electrofishing gear, *Zoologicke Listy,* 14, 171–178.
Karr, J.R. 1981. Assessment of biotic integrity using fish communities, *Fisheries,* 6, 21–27.
Karr, J.M. and D.R. Dudley. 1981. Ecological perspectives on water quality goals, *Environmental Management,* 5, 44–68.
Karr, J.R., K.D. Fausch, P.L. Angermeier, P.R. Yant, and I.J. Schlosser. 1986. Assessing Biological Integrity in Running Waters: A Method and Its Rationale. Illinois Natural History Survey Special Publication 5.
Koel, T.M. and J.J. Peterka. 1994. Distribution and dispersal of fishes in the Red River of the North Basin: a progress report, in *Proceedings of the North Dakota Water Quality Symposium.* March 30–31, 1994, Fargo, ND. 159–168.
Kuehne, R.A. and R.W. Barbour. 1983. *The American Darters.* University of Kentucky Press, Lexington.
Leonard, P.M. and D.J. Orth. 1986. Application and testing of an Index of Biotic Integrity in small, cool water streams, *Transactions of the American Fisheries Society,* 115, 401–414.
Lyons, J. 1992. Using the Index of Biotic Integrity (IBI) to Measure Environmental Quality in Warmwater Streams of Wisconsin. U.S. Department of Agriculture, Forest Service, General .
Mills, H.B., W.C. Starrett, and F.C. Bellrose. 1966. Man's Effect on the Fish and Wildlife of the Illinois River. Illinois Natural History Survey Biological Notes 57.
Miller, D.L., P.M. Leonard, R.M. Hughes, J.R. Karr, P.B. Moyle, L.H. Schrader, B.A. Thompson, R.A. Daniels, K.D. Fausch, G.A. Fitzhugh, J.A. Gammon, D.B. Halliwell, P.L. Angermeier, and D.J. Orth. 1988. Regional applications of an index of biotic integrity for use in water resource management, *Fisheries,* 13, 12–20.
Minnesota Pollution Control Agency. 1994. *Minnesota's Nonpoint Source Management Program 1994.* (Variously paged.)
Neel, J.K. 1985. *A Northern Prairie Stream.* University of North Dakota Press, Grand Forks.
Ohio Environmental Protection Agency (OEPA). 1987. *Biological Criteria for the Protection of Aquatic Life. Vol. II. Users Manual for Biological Field Assessment of Ohio Surface Waters.* Ohio Environmental Protection Agency, Columbus.
Ohio Environmental Protection Agency (OEPA). 1989. *Biological Criteria for the Protection of Aquatic Life. Vol. III. Standardized Biological Field Sampling and Laboratory Methods for Assessing Fish and Macroinvertebrate Communities.* Ohio Environmental Protection Agency, Columbus.
Omernik, J.M. 1987. Ecoregions of the conterminous United States, *Anals of the Association of American Geographers,* 77, 118–125.
Omernik, J.M. and A.L. Gallant. 1988. Ecoregions of the Upper Midwest States. EPA/600/3-88/037. U.S. EPA, Environmental Research Laboratory, Corvallis, OR.
Page, L.M. 1983. *Handbook of Darters.* TFH Publications, Neptune, NJ.
Peterka, J.J. 1978. Fishes and fisheries of the Sheyenne River, North Dakota, *Annals of the Proceedings of the North Dakota Academy Science,* 32, 29–44.
Peterka, J.J. 1991. Survey of Fishes in Six Streams in Northeastern North Dakota, 1991. Mimeograph Report.
Pflieger, W.L. 1975. *The Fishes of Missouri.* Missouri Department of Conservation, Columbia.
Phillips, G.L. and J.C. Underhill. 1971. Distribution and Variation of the Catostomidae of Minnesota. Bell Museum of Natural History, University of Minnesota, Occasional Papers Number 10.
Pielou, E.C. 1975. *Ecological Diversity.* John Wiley & Sons, New York.
Plafkin, J.L., M.T. Barbour, K.D. Porter, S.K. Gross, and R.M. Hughes. 1989. Rapid Bioassessment Protocols for Use in Streams and Rivers: Benthic Macroinvertebrates and Fish. EPA 444/4-89-001. U.S. Environmental Protection Agency, Monitoring and Data Support Division, Washington, DC.
Post, G. 1983. *Textbook of Fish Health.* TFH Publications, Neptune, NJ.
Rankin, E.T., C.O. Yoder, and D. Mishne. 1992. *Ohio Water Resource Inventory. Vol. 1. Summary, Status, and Trends.* Ohio Environmental Protection Agency, Division of Water Quality Planning and Assessment, Columbus.
Renard, P.A., S.R. Hanson, and J.W. Enblom. 1983. Biological Survey of the Red Lake River. Minnesota Department of Natural Resources Special Publication No. 134.
Renard, P.A., S.R. Hanson, and J.W. Enblom. 1986. Biological Survey of the Red River of the North. Minnesota Department of Natural Resources Special Publication No. 142.

Simon, T.P. 1991. Development of Index of Biotic Integrity Expectations for the Ecoregions of Indiana. I. Central Corn Belt Plain. EPA 905/9-91/025. U.S. Environmental Protection Agency, Region V, Chicago.

Simon, T.P. 1992. Development of Biological Criteria for Large Rivers with an Emphasis on an Assessment of the White River Drainage, Indiana. EPA 905-R-92-006. U.S. Environmental Protection Agency, Region 5, Chicago.

Simon, T.P. and E.B. Emery. 1995. Modification and assessment of an Index of Biotic Integrity to quantify water resource quality in great rivers, *Regulated Rivers: Research and Management,* 11, 283–298.

Simon, T.P. and J. Lyons. 1995. Application of the Index of Biotic Integrity to evaluate water resource integrity in freshwater ecosystems, in W.S. Davis and T.P. Simon (Eds.), *Biological Assessment and Criteria: Tools for Water Resource Planning and Decision Making.* Lewis, Boca Raton, FL. 245–262.

Smith, P.W. 1971. Illinois Streams: A Classification Based on Their Fishes and an Analysis of Factors Responsible for the Disappearance of Native Species. Illinois Natural History Survey Biological Notes 76.

Smith, P.W. 1979. *The Fishes of Illinois.* University of Illinois Press: Champaign.

Steedman, R.J. 1988. Modification and assessment of an Index of Biotic Integrity to quantify stream quality in southern Ontario, *Canadian Journal of Fisheries and Aquatic Sciences,* 45, 492–501.

Stoner, J.D., D.L. Lorenz, G.J. Wiche, and R.M. Goldstein. 1993. Red River of the North Basin, Minnesota, North Dakota, and South Dakota, *Water Resources Bulletin,* 29, 575–615.

Tornes, L.H. and M.E. Bringham. 1994. Nutrients, Suspended Sediment, and Pesticides in Waters of the Red River of the North Basin, Minnesota, North Dakota, and South Dakota, 1970–90. U.S. Geological Survey Water-Resources Investigations Report 93-4231.

Trautman, M.B. 1981. *The Fishes of Ohio.* The Ohio State University Press, Columbus.

Underhill, J.C. 1989. The distribution of Minnesota fishes and late Pleistocene glaciation, *Journal of the Minnesota Academy of Science,* 55, 32–37.

U.S. Environmental Protection Agency (U.S. EPA). 1988. Standard Operating Procedures for Conducting Rapid Assessment of Ambient Surface Water Quality Using Fish. U.S. EPA, Region V, Central Regional Laboratory, Chicago.

U.S. Geological Survey. 1974. Drainage Ditch Inventory–1973: Minnesota District Report for Minnesota Department of Natural Resources, (variously paged).

Waters, T.F. 1977. *The Streams and Rivers of Minnesota.* University of Minnesota Press, Minneapolis, Minnesota.

Whittier, T.R., D.P. Larsen, R.M. Hughes, C.M. Rohm, A.L. Gallant, and J.M. Omernik. 1987. The Ohio Stream Regionalization Project: A Compendium of Results. U.S. EPA, Environmental Research Laboratory, Corvallis, OR.

14 Applications of Indices of Biotic Integrity to California Streams and Watersheds

Peter B. Moyle and Michael P. Marchetti

CONTENTS

14.1 Introduction 367
14.2 IBI Problems 368
    14.2.1 Low Species Richness 368
    14.2.2 High Endemism 370
    14.2.3 Introduced Species 370
    14.2.4 Dams and Diversions..371
    14.2.5 Watershed Evaluation 371
14.3 Use of the IBI in California 372
    14.3.1 Sierra Nevada Watersheds 372
    14.3.2 Putah Creek 372
    14.3.3 Dye Creek 375
14.4 Conclusion 376
Acknowledgments 378
References 379

14.1 INTRODUCTION

Indices of Biotic Integrity (IBIs) are measures of the health of streams and other aquatic systems that have been developed as an alternative to physical and chemical measures of water quality (Karr, 1981; Karr et al., 1986; Karr, 1993). Biotic integrity is defined as "the ability to support and maintain a balanced, integrated, adaptive community of organisms having a species composition, diversity, and functional organization comparable to that of the natural habitat of the region" (Karr and Dudley, 1981). Despite this broad definition, in practice IBIs have mostly used measures of fish abundance and diversity to determine biotic integrity; the assumption behind IBIs is that the responses of an integrated community of fishes adequately reflect both major environmental insults, such as a large pollution event, and more subtle long-term effects, such as chronic nonpoint source pollution and changes in land use.

    IBIs are most widely used in eastern North America, where fish communities are complex and largely composed of native species (Miller et al., 1988; Steedman, 1988). In western North America, IBIs have been difficult to develop for a combination of reasons: (1) fish species richness is low, making the development of adequate numbers of metrics for use in IBIs difficult; (2) endemism is high, so it is difficult to transfer IBIs from one region to another; (3) introduced fishes are often abundant, even in relatively "pristine" waters; (4) dams and diversions are a major cause of environmental change; and (5) measures of biotic integrity are often needed that cover entire watersheds, yet there is typically inadequate site-specific information to make a measure based on mean IBI scores.

The purpose of this chapter is to discuss these problems and their solutions, and to demonstrate the application of IBIs to three very different situations in California: (1) evaluation of watersheds in the entire Sierra Nevada range for setting conservation priorities; (2) long-term evaluation of a regulated stream (Putah Creek); and (3) evaluation of a fairly pristine stream (Dye Creek) that has been extensively invaded by exotic species (Figure 14.11).

14.2 IBI PROBLEMS

Three very different case studies were used as situations in California to demonstrate the application of the IBI: (1) evaluation of watersheds in the entire Sierra Nevada range for setting conservation priorities; (2) long-term evaluation of a regulated stream (Putah Creek); and (3) evaluation of a fairly pristine stream (Dye Creek) that has been extensively invaded by exotic species (Figure 14.1). California streams are characterized by low species richness, high endemism, increased numbers of introduced species, and increased anthropogenic disturbance by the construction of dams and diversions. These stream effects have created an emerging need in California for the ability to assess entire watersheds rather than reach specific areas.

14.2.1 Low Species Richness

A general assumption of most IBIs is that there is a positive relationship between species richness and IBI score. However, west of the Rocky Mountains, it is unusual for undisturbed streams to contain more than five or six native fishes. At high elevations, it is typical for streams to contain only one or no native fish species, usually a trout (*Oncorhynchus* spp.). In coastal streams, many of the species are anadromous and might be present only seasonally or have their abundances difficult to assess. The low diversity and shifting nature of the fish faunas make it difficult to develop metrics based on species diversity, trophic specialization, and reproductive strategy that are typical of IBIs for eastern streams (Miller et al., 1988). Three solutions to this problem have been suggested: use standard IBI metrics anyway, reduce the number of metrics, or develop metrics based on aquatic organisms other than fish.

The first solution was used by Lyons et al. (1995), who developed an IBI for streams of west-central Mexico with 2 to 10 species per sampling site. Their 10-metric IBI was correlated with environmental quality at their sites. Their success could have been partly related to high taxonomic diversity among the three basins sampled: a pool of 23 native fish species in 10 families. In California, poor success was obtained in developing IBIs using standard metrics (Miller et al., 1988, but see Putah Creek example below) despite considerable trophic specialization among the native fishes (Moyle and Herbold, 1987). The causes of this appeared to be related to: (1) low taxonomic diversity (pools of 5 to 15 species in 4 to 8 families) with a strong domination by cyprinid and salmonid species; (2) the near universality of spring-time spawning of the native fishes in response to high flows, resulting in similar responses to many environmental perturbations; (3) the abundance and diversity of nonnative species at many sites; and (4) the strong segregation of native species in response to gradients of elevation and temperature, reducing the taxonomic pool at each site even further.

One response to these problems is to reduce the number of metrics used to calculate the IBI, although there is presumably a relationship between the number of metrics and the sensitivity of the IBI as a measure of environmental change. This approach was used (five metrics) by Lyons et al. (1996) for Wisconsin coldwater streams because species diversity decreased as habitat and water quality improved. Hughes and Gammon (1987) reduced the 12 metrics of Karr (1981) to seven for

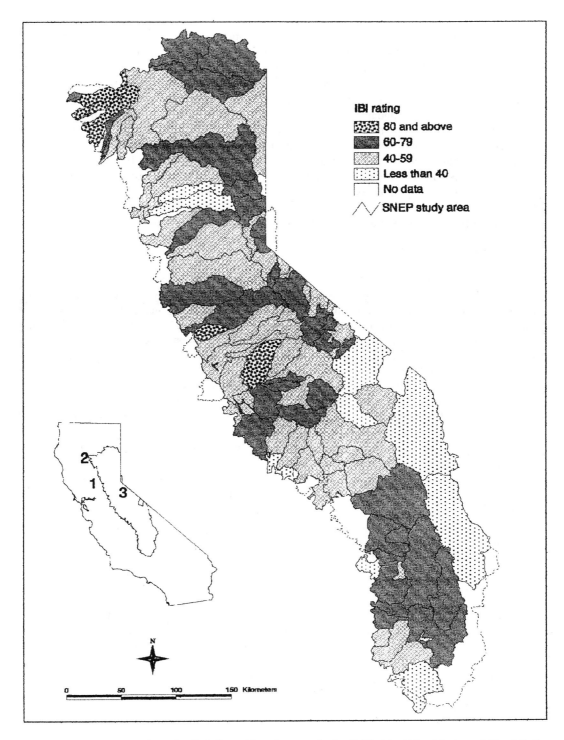

FIGURE 14.1 W-IBI ratings for 100 Sierra Nevada watersheds, California, from Moyle and Randall (in press). The insert map shows the general location of the three study areas: (1) Putah Creek, (2) Dye Creek, and (3) the Sierra Nevadas.

the moderately rich (19 native species) Willamette River drainage of Oregon, and found that the new IBI was a good predictor of water and habitat quality. For many California streams, just two metrics based on number of native fish species and the abundance of native fish is a good descriptor of biotic integrity of the fish community (Miller et al., 1988), but they may say little about habitat or water quality because native fish can be displaced from good habitat by nonnative fishes. In order to increase the number of metrics, metrics based on taxa other than fish may be needed. In California, diversity and abundance of amphibians with aquatic larvae may make especially appropriate measures because two to four species are often present in or near streams and they show a strong sensitivity to environmental change. The use of amphibians has not yet been attempted systematically, but the possibility of their use is demonstrated in examples below. It may also be possible to use benthic invertebrates for additional metrics. For California, stoneflies (Plecoptera), mollusks, and crayfish might be especially useful for this purpose because they are easy to collect along with fish and are sensitive to environmental change. The use of benthic invertebrates for an IBI has been proposed by Kerans and Karr (1994), but as an independent "B-IBI."

14.2.2 HIGH ENDEMISM

Coupled with the low species richness found in the drainage basins of western North America is high endemism. Adjacent drainages may share only a few species or may contain different species combinations because of zoogeographic events (Moyle, 1976a). This makes the development of widely applicable regional IBIs difficult and increases the difficulty of cross-drainage comparisons. One solution is to essentially have a pool of possible metrics for regional use, choose the appropriate metrics for a particular watershed, and then use standardized scores for comparisons (Miller et al., 1988). Metric "pools," for example, could be developed for each of the ecoregions discussed by Hughes et al. (1994). Another solution would be to use some measure of endemism as a metric. Such measures may have to be below the level of species. Brown et al. (1992) point out that California roach (*Lavinia symmetricus*) populations in the San Joaquin River drainage of central California show morphological divergence among tributary drainages, and indicate that a conservation strategy for this watershed should take this diversity into account. Li et al. (1995) point out the value of using genetic information in setting up any system of aquatic reserves in western North America, especially to protect the diversity of anadromous fishes that have frequently hybridized with introduced hatchery strains. Perhaps, a metric could be developed that gives a high score for the continued presence in abundance of all endemic forms (including genetically or morphologically distinct forms that are not formally described as a named taxon) and a low score for their absence. Such a metric, of course, would presuppose a high level of knowledge of the genetics and morphology of species throughout the region to which it was being applied.

14.2.3 INTRODUCED FISHES

Introduced fishes present a particular problem in the West because they frequently dominate local faunas. They are particularly likely to be dominant in highly disturbed warmwater streams at low elevations and in lightly disturbed montane streams, although their presence is pervasive (Moyle, 1976). Because introduced fishes tend to predominate in highly disturbed and polluted streams, it makes sense to use their diversity and abundance as a measure of loss of biotic integrity (as defined by Karr and Dudley, 1981, using native fishes in natural habitats) (Hughes and Gammon, 1987; Miller et al., 1988). Measures of introduced fish abundance are particularly likely to be useful for reflecting the degree to which dams and diversions have altered flow regimes. In California, natural flow regimes strongly favor native fishes adapted to the extreme seasonal patterns (Baltz and Moyle, 1993). In high mountain streams, introduced salmonids represent a strong disturbance to biotic integrity, especially in waters that were originally fishless, where introduced trout may eliminate large invertebrates and amphibians (Moyle et al., 1996).

The problem with using introduced fish as an entirely negative factor is that their abundance might provide a great deal of information about water and habitat quality, particularly if the native species have been displaced or eliminated through competition and predation, rather than habitat change. This represents a conflict between using an IBI to measure diversity and abundance of native organisms versus using an IBI to measure water and habitat quality. Thus, a montane stream in which native cutthroat trout (*Oncorhynchus clarki*) have been replaced by introduced rainbow trout (*O. mykiss*) will presumably respond to most environmental insults in a similar fashion to the way it would have if cutthroat trout were still present. Other species, especially brown trout (*Salmo trutta*), may respond quite differently (e.g., Strange et al., 1992). One solution for trout streams is to have one metric that gives a high score for the abundance of native trout and other metrics that score total trout abundance and abundance of large trout, regardless of species (Miller et al., 1988; Lyons et al., 1996). As the Putah Creek example below shows, a similar approach may have to be taken as well to warmwater streams and lowland rivers, especially where some native species are extinct and are therefore not available to be used as measures of habitat recovery. A California river or stream that supports some native fishes, a diversity of nonnative fishes, and a good fishery for introduced centrarchid bass, sunfish, and catfish would probably obtain fairly high IBI scores if metrics developed for eastern streams were used to evaluate it. The historic biotic integrity of such a system, of course, would be low because of the nearly complete disintegration of the original fish communities.

14.2.4 Dams and Diversions

Most of the larger streams in California and the West have flows regulated by dams and diversions to some degree. While in some cases a dam may cause a stream to dry up completely, the effect of most dams is to alter the amount and pattern of flow. In California, typically, peak winter and spring flows are captured and summer flows are augmented; the augmentation is often provided in order to improve or sustain fisheries for salmonids. The more uniform flow regimes that typically result do not favor native fishes, which need high spring flows for spawning and can survive through summer low-flow conditions (Moyle, 1976; Baltz and Moyle, 1993; Strange et al., 1992). The native fishes are also well adapted for persisting through winter flood events, which may greatly reduce abundances of nonnative warmwater species. Occasionally, as in Putah Creek, releases below dams may favor native fishes and be important in their conservation, because they allow native fishes to become abundant in areas in which they may have originally been present only in small numbers, seasonally, or during wet years. Thus, there is the possibility that a site below a dam, requiring flows that depend on dam operation, can receive as high an IBI score as a site in an unregulated stream, which is less dependent on human decisions for persistence of its fish community. Likewise, a stream below a dam that has become intermittent because of dam operations can receive an IBI score comparable to that of naturally intermittent streams and not reflect the major change that has taken place (Miller et al., 1988). There are no easy solutions to these problems of applying an IBI to regulated streams, except to provide caveats about the precarious nature of fish communities in such streams and to generally recognize that the more a stream has a flow regime resembling its original regime, the more likely it will have its original fish assemblage (Stanford et al., 1996).

14.2.5 Watershed Evaluation

IBIs are typically site-specific measures of biotic integrity. However, a growing need in the West is to evaluate entire watersheds in order to make decisions about priorities for conservation and restoration (Li et al., 1995; Frissell et al., 1996; Frissell and Bayles, 1996; Stanford et al., 1996). One way to produce a watershed-level IBI is to take the mean of all site-specific IBIs, which Steedman (1988) found had a high correlation with factors such as forest cover. Another approach, especially useful when adequate site-specific data are in short supply, is to develop an IBI using watershed-wide metrics, as given below for Sierra Nevada watersheds. These metrics are developed using expert opinion to evaluate broad distribution and abundance information. Because an IBI of this type is different in scoring and application than standard IBIs, Moyle and Randall (in press) follow the convention of Kerans and Karr (1994) in putting an additional letter in front of IBI, in this case W-IBI, to indicate the differences.

14.3 USE OF THE IBI IN CALIFORNIA

In earlier work (Miller et al., 1988), there was considerable skepticism about the ability to apply the IBI concept to California, for many of the reasons discussed above. The three examples below represent our first attempt to use IBIs since that time. The three situations are very different from one another, ranging from 100 watersheds over a large area (Sierra Nevada), to a regulated stream with a large number of native and introduced fish species (Putah Creek), to a small unregulated stream that has been extensively invaded by nonnative fish and frogs (Dye Creek). Future work will determine if the idiosyncratic IBIs developed for these situations have broader applicability.

14.3.1 SIERRA NEVADA WATERSHEDS

The Sierra Nevada is a range of mountains that makes up much of the eastern third of California and is the source of much of the state's water developed for human use. A congressionally mandated evaluation of the range and its biota found that aquatic habitats and ecosystems were among the most altered (Sierra Nevada Ecosystem Project, 1996). This finding was based in part on an analysis of 100 large watersheds by Moyle and Randall (in press), using a Watershed Index of Biotic Integrity (W-IBI). The W-IBI consists of six metrics, using both fish and amphibians (Table 14.1). It attempts to deal with both the problems of trout introductions into fishless waters, the absence of anadromous fish from many areas because of dams, the major differences between the watersheds on the west and east sides of the range (e.g., there were no anadromous fish on the east side), and the disappearance of native frogs due to a variety of factors. Of the 100 watersheds evaluated, 7 were found to be in excellent condition, 36 in good condition, 48 in fair condition, and 9 in poor condition (Moyle and Randall, in press). These ratings correlated well with various measures of disturbance to the watersheds and with subjective evaluations by agencies and environmental groups. They were used to develop recommendations for the systematic protection of aquatic biodiversity in the Sierra Nevada (Moyle, 1996), along the general lines recommended by Moyle and Yoshiyama (1994).

14.3.2 PUTAH CREEK

Putah Creek drains a large watershed in Lake, Napa, Solano, and Yolo Counties, on the east side of the Sacramento Valley, in central California. The creek feeds Berryessa Reservoir, backed up by Monticello Dam. Below Monticello Dam, the water flows in its original channel for about 13 km before being diverted into Solano County at Putah Diversion Dam. Below the diversion dam, only minimal releases have been made, mainly to satisfy needs for groundwater recharge and water for riparian landowners for 3 to 4 km below the dam (Marchetti and Moyle, 1996; Moyle et al., 1998). During drought years, releases are reduced even further, and much of the 35 km of the creek has gone dry, except for pools maintained by wastewater effluent and other sources of water. Despite this release schedule, a remarkably diverse assemblage of native fishes has maintained itself below the dam, and a fishery for nonnative fishes has developed in the lower and middle reaches of the creek. During a period of extended drought, loss of both these assemblages seemed imminent, and they were saved mainly through emergency releases of water from the dam (Moyle et al., 1998).

TABLE 14.1
Metrics and Scoring System for an Index of Biotic Integrity for Sierra Nevada Watersheds

I. Native ranid frogs
   1 Absent or rare
   3 Present
   5 Abundant and widely distributed
II. Native fishes
   1 Absent or rare *or* introduced where not native
   3 Present in much of native range
   5 Abundant, in most of native range within watershed
III. Native fish assemblages (excluding trout-only assemblage)
   1 Largely disrupted
   3 Present but scattered or containing exotic species
   5 Largely intact
IV. Anadromous fishes (if historically present)
   1 Absent or rare
   3 Present mainly below dams or uncommon
   5 Found in original range
V. Trout
   1 Range greatly expanded, mixture of nonnative and native species *or* range greatly reduced
   3 Range expanded but includes native species *or* range about the same but native populations reduced, exotics present.
   5 Mostly native species in original range
VI. Stream fish abundance
   1 Substantially lower than presumed historic levels *or* widespread and abundant in originally fishless areas
   3 Somewhat lower overall than historic levels *or* present in fishless areas
   5 About the same as or higher than historic levels

IBI score = [Total points possible/number of metrics] × 20

80–100	Aquatic communities in very good to excellent condition
60–79	Aquatic communities in good condition
40–59	Aquatic communities in fair condition
<40	Aquatic communities in poor condition

Note: See Moyle and Randall, in press, for detailed explanations of the metrics.

In order to develop a more permanent strategy for the conservation of the native fishes and maintenance of the fishery, sampling was performed 35 km below Putah Diversion Dam for 3 years. An IBI was developed along the lines of standard IBIs, but recognizing both the positive and negative aspects of introduced species (Table 14.2). A preliminary analysis of the data indicates a strong upstream-downstream trend in biotic integrity, with assemblages made up mostly of native species having the highest scores (Table 14.3). The IBI scores demonstrate (1) the rapid loss of native fishes below the point where releases from the dam provide significant flows to the creek; (2) the presence of a diverse assemblage of nonnative fishes in the lower reaches of the creek, and (3) downward shifts in biotic integrity during drought years. likewise, an analysis of a long-term (15 years) sampling program at one site on the creek indicates that IBI scores tend to decline during extended periods of drought, which favor nonnative fishes that can live in stagnant, eutrophic pools (Table 14.4). Low flows during drought periods apparently also reduce the ability of native fishes to spawn and decrease survival of their young.

TABLE 14.2
Metrics for an Index of Biotic Integrity for Putah Creek, Yolo Co, CA

I. Percentage native fish species
   1 <20%
   3 20–80%
   5 >80%
II. Number of native species present
   1 0–1
   3 2–4
   5 >4
III. Number of age classes, native cyprinids and suckers
   1 0–1
   3 2
   5 3+
IV. Total number of fish species present
   1 <5
   3 5–7
   5 >7
V. Total fish abundance
   1 Low numbers present
   3 Common, small numbers captured without difficulty
   5 Abundant, easy to capture in large numbers
VI. Percentage top carnivores
   1 <1%
   3 1–5%
   5 >5%
VII. Percent tolerant species[a]
   1 >20%
   3 5–20%
   5 <5%
VIII. Percent introduced lentic or "pond type" species[b]
   1 >40%
   3 10–40%
   5 <10%

IBI Score = [Total points/number of metrics] × 20

Note: Scoring classes are as presented in Table 14.1.

[a] Tolerant species include the following: common carp, goldfish, fathead minnow, green sunfish, black bullhead, red shiner, and golden shiner.
[b] Introduced lentic and "pond type" species were chosen due to their tendency to dominate backwaters, impoundments, and stagnant pools in altered warmwater streams. They include the following species: white crappie, black crappie, bluegill, largemouth bass, and inland silverside.

This analysis indicates that at least in some situations, a more or less conventional IBI can be used in California to demonstrate changes in fish communities as the result of human disturbances. The relatively high diversity of both native and introduced fishes in Putah Creek, however, is unusual, and this or similar IBIs will probably only be usable in lowland streams of the Central Valley.

TABLE 14.3
Scores for Individual Metrics and the IBI for Eight Sampling Sites on Lower Putah Creek, for a Drought Year (1994), an Average Water Year (1995), and a Wet Year (1996)

Site	Year	IBI Metrics								IBI
		I	II	III	IV	V	VI	VII	VIII	
0.0	1994	5	5	5	5	5	1	5	5	90
	1995	5	5	5	5	5	5	5	5	100
	1996	5	5	5	5	5	3	5	5	95
3.5	1995	5	5	3	5	5	5	5	5	95
	1996	5	5	3	5	5	5	5	5	95
6.4	1994	1	3	3	3	1	5	5	1	55
	1995	1	1	3	3	3	5	5	1	55
16.0	1994	1	1	1	3	1	5	3	1	40
	1995	1	1	1	5	3	5	3	1	50
18.4	1994	1	1	1	3	1	5	3	1	40
	1995	3	3	1	5	1	5	5	1	60
	1996	3	3	3	5	5	5	3	1	70
23.2	1994	1	3	1	5	5	1	5	1	55
	1995	1	1	1	5	1	3	3	1	40
	1996	1	3	1	5	1	5	5	1	55
26.4	1994	1	1	1	3	5	3	5	1	50
	1995	3	3	1	5	1	1	3	1	45
32.8	1994	1	1	1	3	5	1	1	1	35
	1995	1	1	1	5	5	1	3	1	45
	1996	3	3	1	5	1	3	3	1	50

Note: All samples were taken in August. Sites are listed by approximate river kilometers downstream from Putah Diversion Dam. All sites were not sampled in 1994 and 1996. Explanations of metrics and the calculation of the IBI score are given in Table 14.2.

14.3.3 DYE CREEK

Dye Creek, Tehama County, is a small Sierra Nevada tributary to the Sacramento River in northcentral California that is almost entirely within a preserve of The Nature Conservancy (TNC). The preserve is managed for seasonal livestock grazing and for hunting, as well as for natural values. The creek flows through rugged lava canyons that have limited accessibility, even for livestock, so the creek has remained lightly disturbed. Flows are low (1 to 5 cfs in summer), and some sections of its two major forks become intermittent by late summer. To assist TNC in developing a management plan for the watershed, a survey of the fish and amphibians was conducted in August and September 1996 (P. Crain and P.B. Moyle, unpublished data). The principal native fish in both forks was California roach, although small numbers of Sacramento sucker (*Catostomus occidentalis*) and rainbow trout were also present, and speckled dace (*Rhinichthys osculus*) and Sacramento squawfish (*Ptychocheilus*

TABLE 14.4
Annual Variation of IBI Scores over a 15-yr Period at One Sampling Site on Putah Creek, Located on the UC Davis Campus about 23 River Kilometers Downstream from the Putah Diversion Dam

Year	\multicolumn{8}{c}{IBI Metrics}	IBI Score
	I	II	III	IV	V	VI	VII	VIII	
1980	3	3	3	5	3	3	3	1	60
1981*	3	3	1	5	3	1	3	1	50
1984#	3	3	5	5	5	5	5	3	85
1985	3	5	3	5	5	5	3	1	75
1986#	3	3	3	5	5	5	5	1	75
1987	3	3	2	5	3	3	5	1	63
1988*	1	5	1	5	3	3	3	1	55
1989*	1	3	1	5	3	5	3	1	55
1990*	1	3	3	5	5	3	5	1	65
1991*	1	3	1	5	5	3	5	1	60
1992*	1	3	3	5	5	3	5	1	65
1993*	1	3	1	5	5	5	5	1	65
1994*	1	3	1	5	5	1	5	1	55
1995	1	3	1	5	5	3	5	1	60
1996#	1	3	3	5	5	5	5	1	70

Note: Explanations of metrics are given in Table 14.2. Note decreases in scores during drought years (*) and increases in scores during or following wet years (#). All samples taken in October.

grandis) were common in the mainstem. Remarkably, the south fork of the creek was dominated by nonnative green sunfish (*Lepomis cyanellus*) and bullfrogs (*Rana catesbeiana*), while the north fork contained almost entirely native fish and amphibians, especially the foothill yellow-legged frog (*Rana boylei*). This dichotomy was particularly evident in the middle reach of the south fork, which becomes a series of rockbound pools in summer, with no connecting flow, leaving habitats in which there were few refuges for California roach from predation by green sunfish. Green sunfish, as a consequence, were the only fish in many of these pools and even bullfrog populations seem to be suppressed here. Low abundances of bullfrogs at some sites in the south fork were associated with a die-off (Crain and Moyle, unpublished observations) of unknown cause. The mainstem below the confluence of the forks was largely dominated by native fishes, but the principal amphibians present were bullfrogs. The number of both native and introduced fish species increased in a downstream direction, although, except for the green sunfish, the nonnative fishes seemed to originate from farm ponds, rather than being a permanent part of the creek fauna. An IBI developed for the watershed, using both amphibian and fish data, reflects these trends (Tables 14.5 and 14.6). In this case, low IBI scores mainly reflect successful invasions by nonnative species and only partially reflect environmental degradation (ponds, ranch roads, and heavy grazing along the lowermost reaches). The mean IBI score for the entire Dye Creek watershed was 71 (good condition), which was only slightly lower than the W-IBI score (80, very good condition) assigned to it (Moyle and Randall, 1996) before the more detailed surveys were done that revealed the surprising extent of the invasions.

14.4 CONCLUSION

It is possible to develop IBIs for California streams for different purposes and situations. Whether the IBIs presented here are transferrable to other situations remains to be seen. The W-IBI seems

TABLE 14.5
Metrics for Scoring the Index of Biotic Integrity for Dye Creek, California

I. Percentage native fish species
   1 0%
   3 1–99%
   5 100%
II. Percentage native fish individuals
   1 0–10%
   3 11–89%
   5 90–100%
III. Overall fish abundance (1–5 rating system)[a]
   1 Rare, only one or two individuals present
   2 Small numbers present
   3 Common, fish easy to find, small numbers captured without difficulty
   4 Very common
   5 Abundant, visible in large numbers, easy to capture by the 100s.
IV. Ranid frogs[b]
   1 Bullfrogs only species present
   3 Bullfrogs and yellow legged frogs both present
   5 Foothill yellow legged frogs only species present
V. Overall amphibian abundance
   1–5 As for III
VI. Number of native fish and amphibian species present[c]
   1 0–1
   3 2–3
   5 4–5

IBI score = [Total points possible/number of metrics] × 20

80–100	Aquatic communities in very good to excellent condition
60–79	Aquatic communities in good condition
40–59	Aquatic communities in fair condition
<40	Aquatic communities in poor condition

[a] A 1–5 scoring system is used in the IBI because this system was also used in the field.
[b] Bullfrogs are introduced, while foothill yellow-legged frogs (*Rana boylei*) are native.
[c] A combined fish and amphibian metric was used because amphibian species numbers tend to decrease in a downstream direction, while native fish species numbers tend to increase.

to work as a tool for making a "first cut" at determining which watersheds should have the highest priority for management for conservation of aquatic biodiversity. The IBI developed for Putah Creek works like IBIs for eastern streams, perhaps because so much of its fauna is composed of species from the eastern U.S. Whether reaches of stream containing almost entirely nonnative fishes should receive fair to good IBI ratings is still a matter for debate. By the standard definition, based on the resemblance of the present fauna to the original fauna, virtually all biotic integrity in such reaches has been lost. On the other hand, an abundance of nonnative fishes in an assemblage containing multiple trophic groups does say something positive about the quality of the water and aquatic habitats. The application of an IBI to low-diversity situations like Dye Creek appears to be less useful because the IBI provides few insights beyond those provided by examining the distri-

TABLE 14.6
Values for Metrics and IBI Scores for Selected Sites on Dye Creek, Tehama County, California

Site	I % Native Fish Species	II % Native Fish Numbers	III Fish Ratings	IV Frog Species	V Amphibian Rating	VI No. Species	IBI Score
North Fork							
Upper	100	100	4	YLF	5	3	83
Middle	100	100	5	YLF	5	4	100
Lower	100	100	5	YLF	5	4	100
South Fork							
Upper	100	100	3	BF	3	3	66
Middle 1	50	<1	2	BF	3	1	36
Middle 2	75	78	3	BF	1	3	47
Lower	75	98	3	BF	3	4	67
Main Stem							
Upper	80	99	5	BF	1	4	73
Middle	80	99	4	BF	3	4	76
Lower (above ranch, near pond)	60	61	3	BF	3	3	56
Lower (below ranch house)	95	90	4	BF	3	5	77

Note: The metrics (I–VI) and method of calculating the IBI score are explained in Table 14.5. Upper = upper limit of fish distribution; only amphibians are found higher; YLF = foothill yellow-legged frog; BF = bullfrog.

bution and abundance of a few key species. For situations like Dye Creek, one needs to understand whether the invading vertebrates have caused wholesale changes to the aquatic invertebrate communities to know if a fairly high IBI rating based on vertebrates is in fact justified.

Despite continued misgivings about the use of IBIs in California, we believe they are a tool worth developing further. In particular, recommendations include greater use of amphibian metrics in IBIs and the development of some metrics with aquatic invertebrates to increase the number of metrics used in calculating each IBI. Serious thought needs to be given to developing two tiers of IBIs — one tier based on native species, the other tier based on community diversity, whether or not species are native. One of the prime reasons for developing IBIs is the increased interest of citizen's groups in monitoring local streams, and IBIs like those presented here have considerable potential for use by trained amateurs.

ACKNOWLEDGMENTS

The Sierra Nevada portion of this study was supported by the Sierra Nevada Ecosystem Project as authorized by Congress (HR 5503) through a cost-reimbursement agreement No. PSW-93-001-CRA between the U.S. Forest Service, Pacific Southwest Research Station, and the Regents of the University of California, Wildland Resources Center. The work on Putah Creek was largely supported by special grants from the University of California, Davis. We thank Dr. A. S. England, Office of Planning and Budget, for support and encouragement. The Dye Creek work was supported by The Nature Conservancy and much of the data collection was carried out under the direction of Patrick Crain.

REFERENCES

Baltz, D. M. and P. B. Moyle. 1993. Invasion resistance to introduced species by a native assemblage of stream fishes, *Ecological Applications,* 3, 246–255.

Brown, L. R., P. B. Moyle, W. A. Bennett, and B. D. Quelvog. 1992. Implications of morphological variation among populations of California roach *Lavinia symmetricus* (Cyprinidae) for conservation policy, *Biological Conservation,* 62, 1–10.

Frissell, C. A. and D. Bayles. 1996. Ecosystem management and the conservation of aquatic biodiversity and ecological integrity, *Water Resources Bulletin,* 32, 229–240.

Frissell, C. A., J. Doskocil, J. T. Gangemi, and J. A. Stanford. 1996. Identifying Priority Areas for Protection and Restoration of Aquatic Biodiversity: A Case Study in the Swan River Basin, Montana, U.S.A. Open File Report 136-95, Flathead Lake Biological Station, University of Montana.

Hughes, R. M. and J. R. Gammon. 1987. Longitudinal changes in fish assemblages and water quality in the Willamette River, Oregon, *Transactions of the American Fisheries Society,* 116, 196–209.

Hughes, R. M., S. A. Heiskary, W. J. Matthews, and C. O. Yoder. 1994. Use of ecoregions in biological monitoring, in S. L. Loeb and A. Spacie, Eds., *Biological Monitoring of Aquatic Systems.* Lewis, Boca Raton, FL. 125–149.

Karr, J. R. 1981. Assessment of biotic integrity using fish communities, *Fisheries (Bethesda),* 6, 21–27.

Karr, J. R. 1993. Measuring biological integrity: lessons from streams, in S. Woodley, J. Kay, and G. Francis (Eds), *Ecological Integrity and the Management of Ecosystems.* St. Lucie, Boca Raton, FL. 83–104.

Karr, J. R., K. D. Fausch, P. L. Angermeier, P. R. Yant, and I. J. Schlosser 1986. Assessing Biological Integrity in Running Waters: A Method and Its Rationale. Illinois Natural Historical Survey Special Publication 5, Champaign.

Karr, J. R. and D. R. Dudley. 1981. Ecological perspective on water quality goals, *Environmental Management,* 11, 249–256.

Kerans, B. L. and J. R. Karr. 1994. A benthic index of biotic integrity (B-IBI) for rivers of the Tennessee Valley, *Ecological Applications,* 4, 768–785.

Li, H. W., K. Currens, D. Bottom, S. Clarke, J. Dambacher, C. Frissell, P. Harris, R. M. Hughes, D. McCullough, A. McGie, K. Moore, R. Nawa, and S. Thiele. 1995. Safe havens: refuges and evolutionary significant units, in J.L. Nielsen (Ed.), *Evolution and the Aquatic Ecosystem.* American Fisheries Society Symposium 17, Bethesda, MD. 371–380.

Lyons, J., S. Navarro-Perez, P. A. Cochran, E. Santana, and M. Guzman-Arroyo. 1995. Index of Biotic Integrity based on fish assemblages for the conservation of streams and rivers in west-central Mexico, *Conservation Biology,* 9, 569–584.

Lyons, J., L. Wang, and T. D. Simonson. 1996. Development and validation of an Index of Biotic Integrity for coldwater streams in Wisconsin, *Transactions of the American Fisheries Society,* 16, 241–256.

Marchetti, M. P. and P. B. Moyle. 1995. Conflicting values complicate stream protection, *California Agriculture,* 49(6), 73–78.

Miller, D. L., P. M. Leonard, R. M. Hughes, J. R. Karr, P. B. Moyle, L. H. Schrader, B. A. Thompson, R. A. Daniels, K. D. Fausch, G. A. Fitzhugh, J. R. Gammon, D. B. Halliwell. P. L. Angermeier, and D. J. Orth. 1988. Regional applications of an index of biotic integrity for use in water resource management, *Fisheries (Bethesda),* 13, 12–20.

Moyle, P. B. 1976. *Inland Fishes of California.* University of California Press, Berkeley.

Moyle, P. B. 1996. Potential aquatic diversity management areas, in *Sierra Nevada Ecosystem Project: Final Report to Congress. Vol. II. Assessments, Commissioned Reports, and Background Information.* University of California, Centers for Water and Wildland Resources, Davis. 1493–1503.

Moyle, P. B. and B. Herbold. 1987. Life-history patterns and community structure in stream fishes of western North America: comparisons with eastern North America and Europe, in W. J. Matthews and D. C. Heins (Eds.), *Community and Evolutionary Ecology of North American Stream Fishes.* University of Oklahoma Press, Norman. 25–32.

Moyle, P. B., M. P. Marchetti, J. Baldrige, and T. L. Taylor. 1998. Fish health and biodiversity: justifying flows for a California stream, *Fisheries (Bethesda),* 23, 7: 6–15.

Moyle, P. B. and P. J. Randall. 1996. Biotic integrity of watersheds, in *Sierra Nevada Ecosystem Project: Final Report to Congress. Vol. II. Assessments, Commissioned Reports, and Background Information.* University of California, Centers for Water and Wildland Resources, Davis. 975–985.

Moyle, P. B. and P. J. Randall. (in press). Evaluating the biotic integrity of watersheds in the Sierra Nevada, California, *Conservation Biology.*

Moyle, P. B. and R. M. Yoshiyama. 1994. Protection of aquatic biodiversity in California: a five-tiered approach, *Fisheries (Bethesda),* 19, 6–18.

Moyle, P. B., R. M. Yoshiyama, and R. A. Knapp. 1996. Status of fish and fisheries, in *Sierra Nevada Ecosystem Project: Final Report to Congress. Vol. II. Assessments, Commissioned Reports, and Background Information.* University of California, Centers for Water and Wildland Resources, Davis. 953–973.

Sierra Nevada Ecosystem Project Team. 1996. *Sierra Nevada Ecosystem Project, Final Report to Congress, 1996. Vol. 1. Assessment Summaries and Management Strategies.* University of California Davis: Wildland Resources Center Report 36.

Steedman, R. J. 1988. Modification and assessment of an index of biotic integrity to quantify stream quality in southern Ontario, *Canadian Journal of Fisheries and Aquatic Sciences,* 45, 492–501.

Stanford, J.A., J. V. Ward, W. J. Liss, C. A. Frissell, R. N. Williams, J. A. Lichatowich, and C. C. Coutant. 1996. A general protocol for restoration of regulated rivers, *Regulated Rivers,* 12, 391–413.

Strange, E. M., P. B. Moyle, and T. C. Foin. 1992. Interactions between stochastic and deterministic processes in stream fish community assembly, *Environmental Biology of Fishes,* 36, 1–15.

Section IV

Application to Freshwater Resource Types Other Than Wadeable Warmwater Streams

15 Development and Application of an Index of Biotic Integrity for Coldwater Streams of the Upper Midwestern United States

Neal D. Mundahl and Thomas P. Simon

CONTENTS

- 15.1 Introduction 384
- 15.2 Methods for IBI Development, Validation, and Testing 385
  - 15.2.1 Data Collection 385
  - 15.2.2 Metric Selection 385
  - 15.2.3 Index Development and Scoring of Metrics 387
  - 15.2.4 Range of Metric Sensitivity and Metric-to-IBI Score Relationships 387
  - 15.2.5 IBI Validation and Testing 387
- 15.3 Results 388
  - 15.3.1 Metric Selection 388
  - 15.3.2 Scoring of Metrics 389
  - 15.3.3 Description of Metrics 393
    - 15.3.3.1 Metric 1: Number of Species 393
    - 15.3.3.2 Metric 2: Number of Coldwater Species 393
    - 15.3.3.3 Metric 3: Number of Tolerant Species 396
    - 15.3.3.4 Metric 4: Number of Minnow Species 397
    - 15.3.3.5 Metric 5: Number of Benthic Species 397
    - 15.3.3.6 Metric 6: Percent Coldwater Individuals 398
    - 15.3.3.7 Metric 7: Percent Intolerant Individuals 398
    - 15.3.3.8 Metric 8: Percent of Salmonids as Brook Trout 398
    - 15.3.3.9 Metric 9: Percent White Suckers 398
    - 15.3.3.10 Metric 10: Percent Top Carnivores 399
    - 15.3.3.11 Metric 11: Number of Coldwater Individuals per 150 m and Metric 12: Number of Warmwater Individuals per 150 m 399
  - 15.3.4 IBI Validation 400
    - 15.3.4.1 IBI Rating vs. Fish Habitat Rating 400
    - 15.3.4.2 Temporal Variability 400
  - 15.3.5 IBI Testing 401
    - 15.3.5.1 Ecoregion Comparisons 401
    - 15.3.5.2 "Natural" vs. "Managed" Coldwater Streams 402
- 15.4 Discussion 403
  - 15.4.1 Biological Integrity vs. Coldwater Stream Management 403

15.4.2 Comparison to Other Coldwater IBIs 404
15.4.3 Applications of the Coldwater IBI..407
15.4.4 Future Considerations and Refinements 408
Acknowledgments 409
References 409

15.1 INTRODUCTION

The Index of Biotic Integrity (IBI) was developed originally to assess the environmental health of warmwater streams (maximum daily mean temperatures >24°C) within the upper midwestern United States (Karr, 1981; Karr et al., 1986). By integrating many characteristics (e.g., species composition and richness, trophic structure, fish abundance and health) of the existing fish assemblage and comparing these to expectations based on high-quality (i.e., subjected to minimal human influence) regional reference assemblages, the IBI can effectively detect and often lead to the identification of a wide range of stressors negatively impacting stream ecosytems (Karr et al., 1986; Karr, in press). Since its inception, the IBI has been modified and adapted successfully for use in warmwater streams and rivers in many other regions of the U.S., as well as several other countries (Fausch et al., 1984; Miller et al., 1988; Oberdorff and Hughes, 1992; Lyons et al., 1995; Simon and Lyons, 1995; Hughes and Oberdorff, Chapter 5).

Despite the recreational and commercial importance of coldwater streams and rivers (maximum daily mean temperatures <22°C) in many regions of the world, the modification and application of the Index of Biotic Integrity to coldwater systems has not been as widespread as its use in warmwater streams (Simon and Lyons, 1995). Fish assemblages in coldwater systems are much different than those in warmwater streams, making many of the metrics used frequently in warmwater versions inappropriate for assessment of coldwater streams (Steedman, 1988; Lyons, 1992; Lyons et al., 1996). In addition, the reduced taxa richness characteristic of coldwater fish assemblages has made it difficult to devise very many potential metrics that successfully detect impairment within coldwater streams (Simon and Lyons, 1995; Lyons et al., 1996). Consequently, some investigators have developed versions of the IBI that are being used to assess both warmwater and coldwater fish assemblages within the same region (Hughes and Gammon, 1987; Langdon, 1988; Steedman, 1988; Oberdorff and Hughes, 1992; Halliwell et al., Chapter 12). Others have devised coldwater versions that include metrics that assess taxa other than fish, such as amphibians or aquatic invertebrates (Moyle et al., 1986; Fisher, 1989; Moyle and Mardetti, Chapter 14). Some have simply developed a coldwater IBI based on only a few metrics (Lyons et al., 1996). These approaches can create several problems. Since coldwater fish assemblages respond differently to impairment than do warmwater assemblages (Lyons, 1992; Lyons et al., 1996), the combination of warmwater/coldwater IBIs might not be well-suited to detect impairment in coldwater streams. Including taxa other than fish in the IBI may lead to improved sensitivity and better ability to detect disturbance (Moyle et al., 1986; Fisher, 1989), but collection and identification of additional taxa increases the time, effort, and cost of assessments, and expertise with these additional taxa may not be routinely available within many natural resource agencies. Finally, coldwater versions of the IBI with only a few metrics may not be as sensitive at detecting various levels of impairment as would versions containing more metrics. Some metrics may exhibit gradual change over the whole range of stream impairment, whereas others may respond over only a short range of impairment (Angermeier and Karr, 1986; Karr et al., 1986; Karr, in press). An IBI with only a few metrics runs a much greater risk of being insensitive to certain levels of disturbance when compared to a more metric-rich version (Angermeier and Karr, 1986).

This chapter describes the development, validation, and testing of a metric-rich coldwater IBI based solely on fish assemblages and designed for use within the upper midwestern U.S., specifically the states of Minnesota, Wisconsin, Michigan, and Indiana. Although this region is highly diverse in terms of geology, climate, vegetation, land use, and human impacts (Omernik and Gallant, 1988), the features common to coldwater fish assemblages in many different regions (Moyle and Herbold, 1987; Lyons, 1989; Poff and Allan, 1995; Lyons et al., 1996) suggest that a coldwater IBI may be applicable over a much wider area (e.g., several ecoregions) than is typical of warmwater versions.

15.2 METHODS FOR IBI DEVELOPMENT, VALIDATION, AND TESTING

15.2.1 DATA COLLECTION

Appropriate data for coldwater IBI development and testing were obtained from the collection records of various natural resource agencies, individuals, and other organizations in Minnesota, Wisconsin, Michigan, and Indiana. To be considered appropriate, data had to include the identity and numbers of all fish collected, the length of stream section sampled, and one or more measurements of stream width. All fish were collected by single-pass electrofishing (backpack, tote-barge, or boomshocker), a method reported to adequately assess fish abundance, species richness, and assemblage structure in small streams in the upper midwestern U.S. (Simonson and Lyons, 1995). Most collections covered approximately 150 m of stream length. Although longer stream reaches probably should be sampled in larger streams to provide high-quality data on the fish assemblages present (Simonson and Lyons, 1995), sample reaches of 150 m or 500 feet (152.4 m) were the most frequently encountered within the dataset available. Collection records for stream sections that differed in length by more than 5% from 150 m were adjusted accordingly to standardize values to a 150-m stream reach.

From the more than 300 sets of data gathered, groups of reference and impaired sites were selected for use in IBI development. Sites included in these two groups were chosen primarily based on coldwater stream classifications, as determined by state natural resource agencies. Reference sites were those located on stream sections receiving the highest coldwater stream classification possible (e.g., Class 1A in Minnesota, Class I in Wisconsin). Impaired sites were located on streams that were classified as marginal for supporting coldwater fisheries, or were known to be impacted by some specific human activity (i.e., agriculture, mining, construction, timber harvest).

15.2.2 METRIC SELECTION

Candidate metrics for the coldwater IBI were selected by examining the wide range of metrics used previously in various versions of the IBI (reviewed by Simon and Lyons, 1995; Lyons et al., 1996). Several investigators (e.g., Steedman, 1988; Lyons, 1992; Lyons et al., 1996) have shown that many of the metrics used in warmwater versions of the IBI are inappropriate for use in coldwater streams. Consequently, only metrics that were specifically related to characteristics of coldwater stream fish assemblages were considered (Table 15.1). Although other investigators (e.g., Moyle et al., 1986; Fisher, 1989) have attempted to increase the number of metrics used by incorporating invertebrate and amphibian data, only fish metrics were chosen for this study. When developing the pool of potential metrics, included were one or more metrics that corresponded with each of Karr's (1981) original index metrics so that as many fish assemblage characteristics (species richness and composition, indicator species, trophic function, reproductive function, abundance and condition) as possible were considered. Finally, based on the results of previous investigators (Leonard and Orth, 1986; Moyle et al., 1986; Hughes and Gammon, 1987; Miller et al., 1988; Steedman, 1988; Langdon, 1989; Lyons, 1992; Lyons et al., 1996) as well as prior knowledge of coldwater fish assemblages, the expected response of each metric to stream impairment (Table 15.1) was predicted. The classification of species into taxonomic, thermal, tolerance, and trophic categories (Appendix 15A) was based on Ohio EPA (1987), Karr et al. (1986), Lyons (1992), and Lyons et al. (1996).

TABLE 15.1
Potential Coldwater IBI Metrics and Their Predicted Response to Stream Impairment

Metric	Predicted Response
Species richness and composition	
Number of fish species	Increase
Number of coldwater species	Decrease
Number of benthic species	Decrease
Number of water column species	Variable
Number of minnow species	Increase
Percent of individuals that are coldwater species	Decrease
Indicator species	
Number of intolerant species	Decrease
Number of tolerant species	Increase
Percent of individuals that are intolerant	Decrease
Percent of individuals that are tolerant	Increase
Percent of individuals that are salmonids	Decrease
Percent of individuals that are brook trout	Decrease
Percent of individual salmonids that are brook trout	Decrease
Percent of individuals that are salmonids and sculpin	Decrease
Percent of individuals that are sculpin	Decrease
Percent of individuals that are creek chub and blacknose dace	Increase
Percent of individuals that are white sucker	Increase
Trophic function	
Percent of individuals that are generalist feeders	Increase
Percent of individuals that are invertivores	Decrease
Percent of individuals that are top carnivores	Decrease
Reproductive function	
Percent of individuals that are simple lithophilic spawners	Decrease
Abundance and condition	
Number of individuals per 150 m	Variable
Number of coldwater individuals per 150 m	Decrease
Number of warmwater individuals per 150 m	Increase
Percent of individuals with DELT[a] anomalies	Increase

[a] Deformities, disease, eroded fins, lesions, tumors.

The sensitivity of each metric for detecting impairment was determined by comparison of reference and impaired sites. Values for each metric were determined for all reference and impaired sites, and mean values for each metric were compared between reference and impaired sites with a t-test to determine whether metric values differed significantly between these site groups. Prior to comparisons, data were transformed when necessary to meet the assumptions of normality. Metrics in which mean values did not differ between reference and impaired sites were eliminated from further consideration.

Remaining metrics were examined to detect redundancies because one metric highly correlated with another contributes little new information to the overall assessment (Angermeier and Karr, 1986; Karr et al., 1986; Karr, 1991; Barbour et al., 1996). A correlation matrix was constructed to compare the relationships among the metrics for reference sites only. If the correlation coefficient (r) between two metrics was 0.80 or greater, and if a scatterplot of the two metrics displayed a linear relationship, then the metric that was the poorer discriminator between reference and impaired sites was eliminated.

15.2.3 Index Development and Scoring of Metrics

The coldwater IBI was developed by simply summing the scores for all metrics that displayed sensitivity to impairment and that were not strongly correlated with other metrics. Scoring for metrics was accomplished via a two-step process. First, values from each metric from reference sites only were plotted against the log of stream width to assess whether scoring needed to be stream-size dependent. For those metrics displaying such a relationship, scores were assigned to ranges of metric values using the trisection method developed by Lyons (1992), where trisections are constructed to account for stream-size variations in maximum possible metric value. Second, for metrics not displaying a stream-size relationship, the scoring procedures described by Lyons et al. (1996) were followed. These procedures differ slightly from those that have been used most frequently to score bioassessment metrics (Gerritsen, 1995). Metric values from the reference sites were arranged in order from high to low values. Either the upper 5% (for metrics where high values are best) or the lower 5% (for metrics where low values are best) of values were eliminated, and the remaining 95% of the value distribution was divided into equal thirds. A score of 10 was assigned to values in the third of the distribution representing the best reference condition; a score of 5 was assigned to values in the middle third of the distribution representing some reduced condition; and a score of 0 was assigned to values in the third of the distribution representing the greatest deviation from expected values. The total IBI score for a coldwater stream site was determined by summation of scores for all metrics. Ratings for total IBI scores were determined by subdividing the range of possible scores into five categories: excellent, good, fair, poor, and very poor.

All 93 reference sites and 23 impaired sites used in coldwater IBI development were then scored for each final metric, and these metric scores were summed to produce total site IBI scores. These scores were then compared between reference and impaired sites with a nonparametric median test to determine whether the coldwater IBI was able to distinguish between these site categories.

15.2.4 Range of Metric Sensitivity and Metric-to-IBI Score Relationships

Total IBI scores and individual metric values were compared in two different ways to provide information on the overall structure and function of the coldwater IBI. First, the ranges of stream degradation to which the individual metrics were most sensitive were examined graphically by plotting the total IBI scores for the 116 reference and impaired sites used in IBI development versus the individual metric values for each site. The range of total IBI scores across metric values that changed most rapidly was determined to be the range of primary sensitivity for that metric (Angermeier and Karr, 1986). Second, the overall strength of the relationships between individual metrics and total IBI scores for the same 116 sites was assessed numerically with the nonparametric Spearman's rank correlation coefficient (r_s).

15.2.5 IBI Validation and Testing

The ability of the coldwater IBI to detect stream impairment was examined by comparing total IBI scores from a separate data set of 35 stream sites in southeastern Minnesota with fish habitat rating scores (Simonson et al., 1994) determined for the sites at the time of fish collections. The dominant impacts on streams in this region result from erosion of agricultural lands and grazing by cattle (Omernik and Gallant, 1988), both of which degrade fish habitat. The fish habitat rating system for streams less than 10 m wide rates seven riparian zone and instream features: bank land use, bank stability, pool area, width-to-depth ratio, riffle-to-riffle or bend-to-bend ratio, silt substrate, and cover for fish (Simonson et al., 1994). Values determined for each rating item are summed to produce a total site score, which is then used to determine a qualitative site rating of excellent, good, fair, or poor. Individual sites were surveyed from one to three times during the period 1994 to 1996, for a total of 64 IBI fish habitat rating sets. Mean IBI scores were compared among habitat rating categories with a single-factor analysis of variance and Tukey's multiple range test. Although it also would be preferable to examine index response to additional types of physical and chemical perturbations, these types of data were not available for the majority of the stream sites.

We examined temporal variability in coldwater IBI scores by comparing scores from 14 sites in southeastern Minnesota sampled during the same time period in each of three successive years, 1994–1996. IBI scores were compared between years, and mean site scores were compared with site range and standard deviation (Spearman's rank correlation) to assess the relationship between biotic integrity and year-to-year variability (Karr et al., 1987; Lyons et al., 1996).

The coldwater IBI was used to compare the environmental health of coldwater streams within six ecoregions in Minnesota, Wisconsin, Michigan, and Indiana: Northern Lakes and Forests, North Central Hardwood Forests, Driftless Area, Western Cornbelt Plains, Central Corn Belt Plains, and Southern Michigan-Northern Indiana Till Plains (Omernik and Gallant, 1988). The number of stream site datasets used in this analysis varied from 15 to 88 per ecoregion. Although the datasets selected do not represent a truly random sample from all coldwater streams within each ecoregion, they probably are representative since no attempt was made to select either only the best or worst sites. Fish assemblage data from each site were used to calculate a single IBI score, and mean IBI scores were compared among ecoregions with a single-factor analysis of variance and Tukey's multiple range test. IBI scores within each ecoregion also were assigned to their respective rating categories, and the distribution of sites within these categories were compared across ecoregions with a Chi-square contingency table.

The coldwater IBI also was used to compare stream sites with different types of salmonid assemblages to determine whether the type of assemblage present was related in any way to biotic integrity. The 93 stream sites used as reference sites for coldwater IBI development were separated into three categories based on the type of salmonid assemblage present: native salmonids only, mixed native and nonnative salmonids, and nonnative salmonids only. The first category represents the "natural" salmonid assemblage expected within this region in the absence of human intervention (it is understood that this assemblage may exist today at some locations only because of human intervention). The remaining two categories represent "managed" assemblages, present today because of past introductions of nonnatives (although current assemblages may be self-sustaining and require little or no management). Mean IBI scores, metric values, and other characteristics of the salmonid assemblages were calculated and compared among the three site categories with nonparametric Kruskal-Wallis tests.

15.3 RESULTS

15.3.1 Metric Selection

Twenty-five potential metrics were tested for inclusion in the coldwater IBI (Table 15.1). The percent anomalies metric was eliminated prior to testing because most datasets did not include this information. Rather than replacing missing information with some intermediate value for this metric as has been done in other studies, it was deemed better to exclude this metric for the present and revisit it when sufficient data are available to make a more objective assessment of its importance in coldwater stream biotic integrity. The remaining 24 metrics were compared between reference and impaired sites, with 20 displaying significant differences in mean values between these two categories (Table 15.2). These comparisons are shown graphically in Figures 15.1 to 15.3. The four metrics (number of intolerant taxa, percent creek chub and blacknose dace, percent sculpin, and

TABLE 15.2
Mean (±SD) Values for Potential Coldwater IBI Metrics at Reference (n = 93) and Impaired (n = 23) Sites, and Results of *t*-Tests Between Reference and Impaired Sites

Metric	Reference Sites	Impaired Sites	t	p
Species richness and composition				
Number of fish species	5.3 (2.3)	11.4 (3.2)	86.2	<0.001
Number of coldwater species	2.4 (1.0)	1.8 (0.9)	7.4	0.007
Number of benthic species	1.1 (1.1)	2.2 (1.3)	18.2	<0.001
Number of water column species	3.6 (1.6)	7.0 (2.7)	60.7	<0.001
Number of minnow species	2.0 (1.4)	4.4 (1.5)	53.9	<0.001
Percent coldwater individuals	62.5 (34.4)	6.9 (7.8)	59.2	<0.001
Indicator species				
Number of intolerant species	1.5 (0.8)	1.9 (1.5)	3.2	0.090
Number of tolerant species	1.7 (1.3)	4.0 (1.7)	49.3	<0.001
Percent intolerant individuals	36.4 (34.0)	8.4 (8.2)	15.3	<0.001
Percent tolerant individuals	27.6 (30.3)	59.9 (26.6)	21.3	<0.001
Percent salmonids	51.1 (36.0)	4.1 (6.8)	40.0	<0.001
Percent brook trout	26.8 (32.1)	1.2 (2.6)	14.5	<0.001
Percent salmonids that are brook trout	54.6 (42.2)	31.0 (40.4)	5.4	0.021
Percent salmonids and sculpin	59.9 (35.7)	5.2 (7.0)	53.2	<0.001
Percent sculpin	8.5 (17.7)	1.2 (3.6)	3.8	0.058
Percent creek chub and blacknose dace	21.3 (27.2)	20.0 (23.5)	0.04	0.838
Percent% white sucker	3.3 (7.8)	20.4 (18.6)	46.6	<0.001
Trophic function				
Percent generalist feeders	29.1 (30.8)	70.6 (20.5)	36.7	<0.001
Percent invertivores	17.5 (22.6)	21.5 (17.7)	0.8	0.437
Percent top carnivores	54.8 (35.3)	5.8 (8.2)	43.3	<0.001
Reproductive function				
Percent simple lithophilic spawners	28.9 (29.5)	44.0 (28.0)	5.0	0.029
Abundance and condition				
Number of individuals	154.7 (120.1)	236.0 (243.2)	5.3	0.022
Number of coldwater individuals	92.3 (100.4)	17.7 (36.5)	12.2	0.001
Number of warmwater individuals	62.4 (83.4)	218.3 (230.3)	28.3	<0.001

percent invertivores) that did not differ between reference and impaired sites were eliminated from further consideration. In addition, the number of individuals per 150-m metric was eliminated because it displayed too large an overlap between reference and impaired sites (Figure 15.3).

Seven of the remaining 19 metrics, exhibited strong correlation with other metrics that were stronger discriminators between impaired and reference sites (Table 15.3). Five of these were correlated either positively or negatively with the percent coldwater individuals metric, whereas the remaining two were correlated positively with the number of species metric or the percent intolerant individuals metric (Table 15.3). After eliminating these seven metrics to reduce redundancy within the index, 12 metrics remained for the coldwater IBI.

15.3.2 Scoring of Metrics

Five of the 12 metrics (number of species, number of coldwater species, number of minnow species, number of benthic species, and number of tolerant species) selected for the coldwater IBI displayed a decrease in maximum species richness when stream widths were less than 5 m. Consequently, the tri-section method of Lyons (1992) was used to develop a scoring system for these metrics,

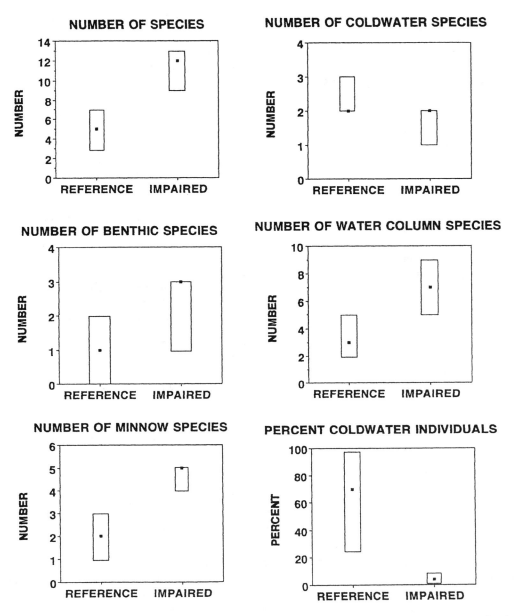

FIGURE 15.1 Sensitivity of potential coldwater IBI species richness and composition metrics at discriminating between reference (n = 93) and impaired (n = 23) stream sites. Dots are medians, and boxes represent interquartile ranges (25th to 75th percentiles).

and scores of 10 (expected condition for high-quality coldwater streams), 5 (moderate deviation from expected condition), or 0 (great deviation from expected condition) were assigned to the various sections of the species richness-stream width plots (Figure 15.4). Four of these metrics were assigned reverse scoring, with higher species richness resulting in a lower score.

Scoring for the remaining seven metrics, which did not display a relationship with stream width, is shown in Table 15.4. Two of these metrics (percent white suckers, number of warmwater individuals) were assigned reverse scoring.

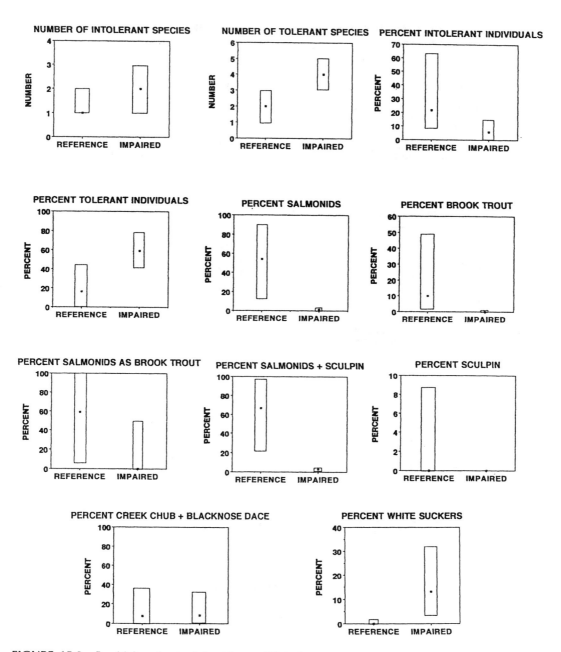

FIGURE 15.2 Sensitivity of potential coldwater IBI indicator species metrics at discriminating between reference (n = 93) and impaired (n = 23) stream sites. Dots are medians, and boxes represent interquartile ranges (25th to 75th percentiles).

The sum of the scores for all 12 metrics is the total coldwater IBI score. Possible IBI scores range from a maximum of 120 (excellent biotic integrity) to a minimum of 0 (very poor biotic integrity). Integrity ratings for total IBI scores and a general description of the fish community expected for each rating category are shown in Table 15.5.

The coldwater IBI based on the 12 metrics clearly distinguished between the impaired and reference sites used in metric selection and IBI development. Impaired sites had significantly

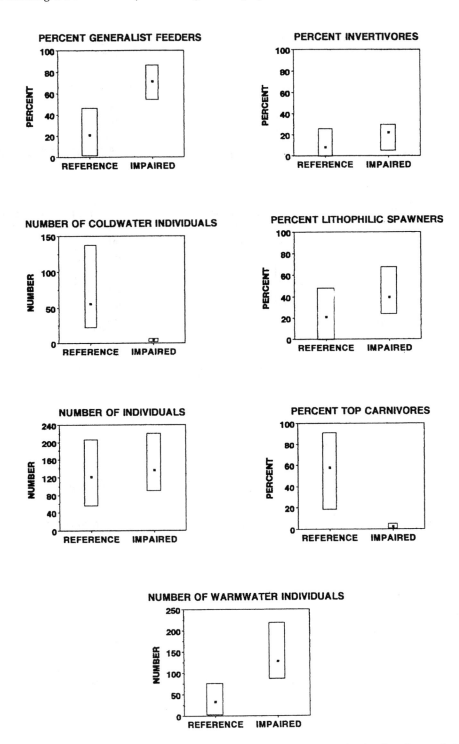

FIGURE 15.3 Sensitivity of potential coldwater IBI trophic function, reproductive function, and abundance and condition metrics at discriminating between reference (n = 93) and impaired (n = 23) stream sites. Dots are medians, and boxes represent interquartile ranges (25th to 75th percentiles).

Development and Application of an IBI for Coldwater Streams

TABLE 15.3
Metrics Eliminated Because of Strong Correlation (r) with Other Metrics That Were Stronger Discriminators Between Reference and Impaired Sites

Metric Eliminated	Metric Retained	r
Percent tolerant individuals	Percent coldwater individuals	−0.844
Percent salmonids	Percent coldwater individuals	+0.851
Percent generalists	Percent coldwater individuals	−0.878
Percent lithophilic spawners	Percent coldwater individuals	−0.815
Percent salmonids and sculpin	Percent coldwater individuals	+0.980
Number of water column species	Number of fish species	+0.846
Percent brook trout	Percent intolerant	+0.846

(nonparametric median test $x^2 = 26.25$, $p < 0.001$) lower IBI scores (mean = 19.3, median = 15) than did reference sites (mean = 67.9, median = 70). IBI scores for reference and impaired sites were correlated significantly with values for 11 of the 12 metrics (Table 15.6).

15.3.3 DESCRIPTION OF METRICS

15.3.3.1 Metric 1: Number of Species

In warmwater versions of the IBI, a decline in the total number of fish species present is used as an indicator of impairment (Karr et al., 1986; Simon and Lyons, 1995). However, high-quality coldwater streams generally have few fish species (Moyle and Herbold, 1987; Lyons, 1992; Lyons et al., 1996), and may exhibit an increase in species richness in response to increased human disturbance (Fisher, 1989; Lyons, 1992; Lyons et al., 1996). Lyons et al. (1996) hypothesized that the depauperate fish fauna found in coldwater streams results from the environmental harshness of these streams and the lack of native fishes with life histories adapted for success in these systems. Poor land-use practices and increased erosion in coldwater stream watersheds tend to produce elevated summer water temperatures through a combination of reduced groundwater inputs, decreased shading as riparian vegetation is removed, and wider and shallower stream channels (Lyons, 1992). Increased temperatures reduce the harshness of these streams, opening them to colonization by greater numbers of coolwater- or warmwater-adapted fish species if these species have access to these streams, as they often do in the upper midwestern U.S. (Lyons et al., 1996). Coldwater streams in regions lacking a warmwater fish fauna, or those with steeper gradients (>10 m/km), might not display an increase in species richness with impairment (Lyons et al., 1996). Because upper midwestern coldwater streams generally are low-gradient and usually are in close proximity to warmer aquatic habitats, impairment usually leads to increased fish species richness. Consequently, reverse scoring was used for this metric. This metric was one of only two in the index that exhibited sensitivity across the entire range of impairment (Figure 15.5). A very broad range of sensitivity also is exhibited by the number of species metric in the warmwater IBI (Karr et al., 1986).

15.3.3.2 Metric 2: Number of Coldwater Species

Despite the relatively low fish species richness in coldwater streams, a clear relationship is evident between stream impairment and the number of fish species well-adapted to life in cold water. As impairment increases in coldwater streams, total fish species richness tends to increase (see above), but the number of species of fish with strong affinities to cold water declines. Lyons et al. (1996)

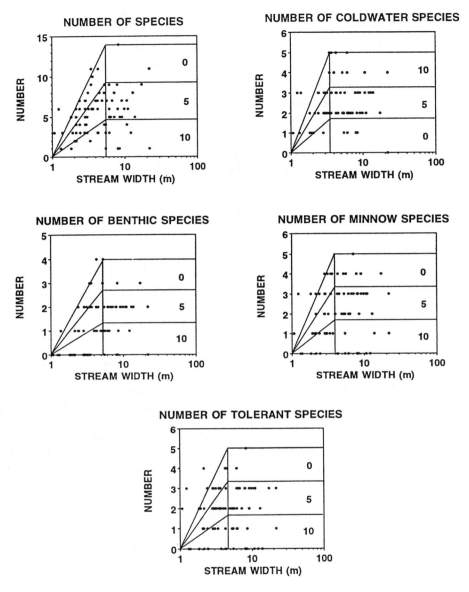

FIGURE 15.4 Maximum species richness plots for total number of species, number of coldwater species, number of benthic species, number of minnow species, and number of tolerant species vs. stream width for 93 reference coldwater stream sites. Numbers (0, 5, 10) on the right side of each plot are scoring criteria for values within each region of the plot.

reported this same phenomenon. Although the numbers of coldwater species differed significantly but not greatly between reference (mean = 2.4 species per site) and impaired (mean = 1.8 species per site) coldwater sites (probably the result of low numbers of species adapted to these habitats; Lyons, 1992; Lyons et al., 1996), this metric was not strongly correlated with any other metric used in the coldwater IBI, and therefore provided new information to the index. This metric was most sensitive to intermediate levels of impairment (Figure 15.5). Lyons et al. (1996) observed a similar relationship between stream impairment and the number of native stenothermal coldwater and coolwater species, but this metric was not included in their coldwater IBI because differences

TABLE 15.4
Scoring Criteria for the 12 Metrics Included in the Coldwater IBI

Metrics	Scoring		
	0	5	10
Number of species[a]	>9	5–9	<5
Number of coldwater species[a]	0–1	2–3	>3
Number of minnow species[a]	>3	2–3	<2
Number of benthic species[a]	>2	2	<2
Number of tolerant species[a]	>3	2–3	<2
Percent salmonids as brook trout	<12	12–92	>92
Percent intolerant individuals	<10	10–43	>43
Percent coldwater individuals	<42	42–88	>88
Percent white suckers	>1.5	>0–1.5	0
Percent top carnivores	<30	30–72	>72
Number of coldwater individuals	<32	32–75	>75
Number of warmwater individuals	>60	16–60	<16

[a] Scoring only for streams >5 m wide; scoring varies for streams <5 m wide.

between impaired and reference sites were "much less dramatic" than the differences found for other metrics. The number of native and exotic stenothermal coldwater and coolwater species did not display a consistent difference between high-quality and low-quality coldwater stream sites in Wisconsin (Lyons et al., 1996).

TABLE 15.5
Integrity Rating Categories for Coldwater IBI Scores and General Fish Community Characteristics for Each Category

Total IBI Score	Integrity Rating	Fish Community Characteristics
105–120	Excellent	Comparable to the best situations with little human disturbance; 3 or 4 coldwater species present; dominated (>75%) by brook trout; exotic salmonids absent or limited to few individuals; sculpin present; lampreys often present; white suckers absent; warmwater species absent or very uncommon.
70–100	Good	Some impairment present; coldwater intolerant species (sculpin, brook trout) reduced in abundance; white suckers present in low numbers; often dominated by brown trout or other exotic salmonids; higher species richness resulting from presence of more tolerant warmwater minnows and darters.
35–65	Fair	Moderate impairment; coldwater intolerant species rare or absent; brown trout and more tolerant coldwater species (e.g., brook stickleback) may be common; relatively high species richness; warmwater species relatively common.
10–30	Poor	High impairment; more tolerant warmwater species usually dominant; white suckers often abundant; salmonids very rare or absent; relatively high species richness.
0–5	Very poor	Severe impairment; coldwater fish absent; only warmwater species present.
No score	No score	Too few fish (<25 individuals) to calculate IBI score.

Modified from Karr et al. (1986) and Lyons et al. (1996).

TABLE 15.6
Correlation Coefficients (Spearman r_s) Between the Coldwater IBI and Individual Metrics for Data from Reference and Impaired Sites (n = 116)

Metrics	r_s	p
Number of species	−0.818	<0.001
Number of coldwater species	0.153	0.102
Number of minnow species	−0.834	<0.001
Number of benthic species	−0.610	<0.001
Number of tolerant species	−0.820	<0.001
Percent salmonids as brook trout	0.281	0.002
Percent intolerant individuals	0.631	<0.001
Percent coldwater individuals	0.884	<0.001
Percent white suckers	−0.697	<0.001
Percent top carnivores	0.797	<0.001
Number of coldwater individuals	0.568	<0.001
Number of warmwater individuals	−0.793	<0.001

15.3.3.3 Metric 3: Number of Tolerant Species

Karr (1981) originally proposed using the proportion of individuals as green sunfish as a metric to assess the increasing dominance of a fish assemblage by tolerant organisms as stream impairment increased. Most subsequent versions of the IBI included this metric or some variation of it (Karr et al., 1986; Simon and Lyons, 1995), including the coldwater IBI developed for use in Wisconsin (Lyons et al., 1996). In developing the present coldwater IBI, the potential percent tolerant indi-

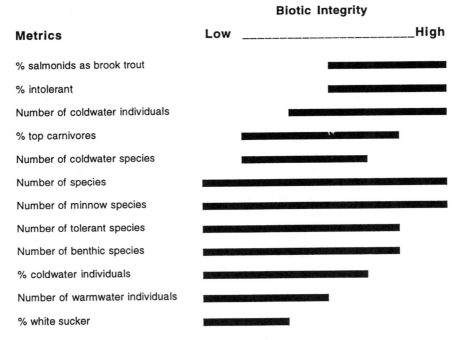

FIGURE 15.5 Ranges of primary sensitivity for the 12 metrics of the coldwater IBI (after Angermeier and Karr, 1986; Karr et al., 1986).

viduals metric varied significantly between impaired and reference sites (Table 15.2), but was strongly inversely correlated with the percent coldwater individuals metric (Table 15.3), a much stronger discriminator between impaired and reference sites. The number of tolerant species also was examined as a potential metric, and performed better than the percent tolerant individuals metric in discriminating between impaired and reference sites (Table 15.2). In addition, the number of tolerant species metric was not correlated strongly with any other metric, adding unique information to the index. Lyons et al. (1996) did not examine the number of tolerant species as a potential metric for their Wisconsin coldwater IBI. The number of tolerant species metric was found to be sensitive to all but the slightest levels of impairment (Figure 15.5). In contrast, the original percent green sunfish metric used in the warmwater IBI is most sensitive across a relatively narrow range of moderate to severe stream impairment (Karr et al., 1986).

15.3.3.4 Metric 4: Number of Minnow Species

The number of minnow species often has been used as a replacement metric for Karr's (1981) original number of catostomid species metric when there were few sucker species within the area or ecosystem type of interest (Simon and Lyons, 1995). Such is the case in coldwater streams, where suckers often are represented by only a single species (Moyle and Herbold, 1987; Lyons et al., 1996). Lyons et al. (1996) reported significant but small differences in the number of sucker species between high-quality and marginal coldwater streams in Wisconsin, but did not include this metric in their coldwater IBI. Cyprinids also are often poorly represented in high-quality coldwater streams (Moyle and Herbold, 1987); but as in the case of the total number of species metric (see above), increased coldwater stream impairment apparently results in increased minnow species richness. Consequently, a reversed scoring system was used for this metric (Figure 15.4). Lyons et al. (1996) reported that environmental degradation of coldwater streams results in increased species richness within some families of fishes, but they did not examine the number of minnow species as a potential metric for their coldwater IBI. Unlike the warmwater IBI number of sucker species metric, which is most sensitive to slight to moderate levels of stream impairment (Karr et al., 1986), the coldwater IBI number of minnow species metric is sensitive to all levels of impairment (Figure 15.5).

15.3.3.5 Metric 5: Number of Benthic Species

The number of darter species was an original warmwater IBI metric (Karr 1981), providing information on a group of fishes particularly sensitive to degradation of benthic habitats (Karr et al., 1986). Subsequent versions of the IBI have either used this metric without modification, or have added or substituted other fish taxa that also have a benthic orientation (Karr et al., 1986; Lyons, 1992; Simon and Lyons, 1995). Coldwater streams may contain many different kinds of benthic fishes (e.g., sculpins, darters, suckers, and benthic dace), but seldom are many of these present together in high-quality systems (Moyle and Herbold, 1987). Lyons et al. (1996) observed that both darter and sucker species richness increased with coldwater stream degradation in Wisconsin. A similar pattern was observed when all benthic species were combined in the development of the present version of the coldwater IBI; impaired coldwater sites on average contained twice as many benthic species as did higher-quality reference sites (Table 15.2). High-quality coldwater sites most often contained a single species of sculpin, whereas degraded coldwater sites often contained additional species of darters and suckers. A reverse scoring system was employed for this metric to support this trend (Figure 15.4). This metric was most sensitive in detecting moderate to severe coldwater stream impairment (Figure 15.5), whereas the number of darter species metric used in the warmwater IBI usually is most sensitive to slight to moderate levels of degradation (Karr et al., 1986). The final version of the Wisconsin coldwater IBI does not contain a benthic fishes metric (Lyons et al., 1996).

15.3.3.6 Metric 6: Percent Coldwater Individuals

As low- and medium-gradient coldwater streams in the upper midwestern U.S. are degraded, the native coldwater fish assemblage is joined and gradually replaced by coolwater or warmwater fishes, if these fishes have access to the coldwater streams (Lyons, 1992; Lyons et al., 1996). In general, the proportion of the assemblage comprised of coldwater-adapted species declines with increasing impairment (Lyons et al., 1996). This metric is one of the strongest discriminators between impaired and reference coldwater streams (Tables 15.2 and 15.3), and is especially sensitive to moderate to severe levels of stream degradation (Figure 15.5). Lyons et al. (1996) included this metric in the Wisconsin coldwater IBI, with a scoring system (scoring breaks at 42% and 86%) virtually identical to that developed independently for the present coldwater IBI (scoring breaks at 42% and 88%; Table 15.4).

15.3.3.7 Metric 7: Percent Intolerant Individuals

The number of intolerant species metric has been an important component of most versions of the IBI developed previously (Simon and Lyons, 1995). This metric was one of five used in the coldwater IBI for Wisconsin (Lyons et al., 1996), but it was unable to distinguish between impaired and reference sites during development of the present version (Table 15.2). However, the *percent intolerant individuals metric* was able to separate reference from degraded sites (Table 15.2). Athough the species assigned intolerant status for development of the coldwater IBI far exceeded the 5 to 10% of all species recommended by Karr et al. (1986), the percent intolerant metric was sensitive primarily to slight to moderate impairment of coldwater streams (Figure 15.5), the same range of primary sensitivity displayed by the number of intolerant species metric in the warmwater IBI (Karr et al., 1986).

15.3.3.8 Metric 8: Percent of Salmonids as Brook Trout

The brook trout is the only trout native to coldwater streams of the upper midwestern U.S. (Hendricks, 1980; Hunter, 1991). Because of its intolerance to stream degradation, Steedman (1988) used the presence of brook trout as an indicator of high-quality coldwater habitat. As coldwater streams are degraded, brook trout may be joined and eventually replaced by exotic and nonnative salmonids such as brown trout and rainbow trout, which are more tolerant of slightly elevated water temperatures and other environmental changes that may accompany degradation (Hunter, 1991). This change in the salmonid community is used as a metric in the Wisconsin coldwater IBI (Lyons et al., 1996), as well as in the present version.

Lyons et al. (1996) reported that the proportion of salmonids that were brook trout was the most sensitive metric in the Wisconsin coldwater IBI. In the present coldwater IBI, this same metric was not as effective at discriminating between impaired and reference sites as many other potential metrics, including the percent brook trout (of all fish) metric (Table 15.2). However, this latter metric was correlated with another metric and the former was not (Table 15.3). Consequently, the percent salmonids as brook trout metric was retained, and it had a scoring system (Table 15.4) and a range of sensitivity (Figure 15.5) virtually identical to that reported by Lyons et al. (1996).

15.3.3.9 Metric 9: Percent White Suckers

The proportion of the fish assemblage comprised of white suckers has been used as a substitute for the percent green sunfish metric in several versions of the IBI (Schrader, 1986; Langdon, 1988; Miller et al., 1988). White suckers are considered tolerant of many different types of stream degradation (Trautman, 1981), and tend to occur in higher densities and compose a larger proportion of the fish assemblage at degraded stream sites (Langdon, 1988; Lyons, 1992). The percent white suckers metric was a very strong discriminator of stream impairment (Table 15.2), as this species' presence even in very low numbers appeared to be indicative of some level of disturbance. The scoring system breaks for this metric were set extremely low (0% and 1.5%; Table 15.3), in comparison to those (10% and 25%) used for this metric in other versions of the IBI (Fausch and Schrader, 1987; Langdon, 1988). A similar, large difference in scoring system breaks is evident for the percent tolerant individuals metric between warmwater (20% and 50%) and coldwater (5% and 23%) versions of the IBI developed for use in Wisconsin (Lyons, 1992; Lyons et al., 1996). As with the percent green sunfish metric in the warmwater IBI (Karr et al., 1986), the percent white suckers metric was most sensitive to moderate to severe levels of coldwater stream impairment (Figure 15.5). Lyons et al. (1996) did not examine the percent white suckers metric during development of the Wisconsin coldwater IBI, although it was considered (but not selected) during development of that state's warmwater IBI (Lyons, 1992).

15.3.3.10 Metric 10: Percent Top Carnivores

The percent top carnivores or piscivores metric has appeared in most versions of the IBI (Simon and Lyons, 1995). Top carnivores are particularly susceptible to stream impairment because of their position within the trophic structure of stream communities, and their populations tend to decline and disappear as stream quality declines (Karr, 1981). In coldwater streams, salmonids are the primary top carnivores, but pike, bass, walleye, and burbot also may be present in some systems (Lyons et al., 1996). This metric generally is considered to be most sensitive to slight to moderate levels of impairment (Karr et al., 1986; Lyons et al., 1996). However, in the present coldwater IBI, this metric's primary range of sensitivity shifts away from slight impairment and more toward severe impairment (Figure 15.5). Scoring system breaks for this metric also differ considerably between the present IBI (30% and 72%; Table 15.4) and that for Wisconsin's coldwater streams (14% and 46%; Lyons et al., 1996).

15.3.3.11 Metric 11: Number of Coldwater Individuals per 150 m and Metric 12: Number of Warmwater Individuals per 150 m

Abundance or catch-per-effort metrics are common in most versions of the IBI (Simon and Lyons, 1995) because degraded stream sites generally are expected to yield fewer fish than are higher-quality sites (Karr et al., 1986). However, this trend is not always obvious in coldwater streams. Coldwater streams tend to lose coldwater-adapted species, but gain coolwater-or warmwater-adapted forms as stream impairment worsens. Thus, although the abundances of coldwater species follow expected changes relating to stream impairment, abundances of other species do not, thus confounding simple application of this metric to coldwater streams. Consequently, this metric was subdivided into two metrics: one to assess changes in abundance of coldwater forms, and the other to follow changes in non-coldwater species.

Both coldwater and warmwater fish abundance metrics were able to discriminate between impaired and reference sites (Table 15.2), and both responded to impairment as expected (Tables 15.1 and 15.2). The warmwater fish abundance metric was reverse scored because it increased with stream impairment (Table 15.4). In warmwater versions of the IBI, the fish abundance metric generally is most sensitive to moderate to severe levels of stream degradation (Karr et al., 1986). In the present coldwater IBI, the number of warmwater fishes metric is most responsive to these same levels of impairment (Figure 15.5), but the number of coldwater fishes metric is most sensitive to slight to moderate impairment (Figure 15.5). Lyons et al. (1996) examined a total catch-per-effort metric during coldwater IBI development, but it did not display a consistent difference between high- and low-quality sites, similar to the situation observed with the total number of individuals metric during development of the present coldwater IBI (significant difference between reference and impaired sites, but high variability; Figure 15.3, Table 15.2).

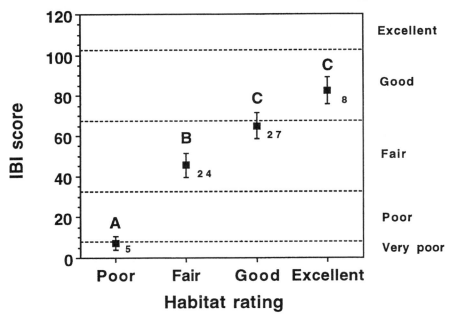

FIGURE 15.6 Mean (±SE) coldwater IBI scores vs. fish habitat rating categories for 63 stream sites sampled during 1994–1996 for the Whitewater Watershed Project, southeastern Minnesota. Dashed lines separate IBI rating categories (excellent to very poor). Mean values not sharing a common letter are significantly different from each other (ANOVA and Tukey's multiple range test). Numbers represent sample sizes.

15.3.4 IBI Validation

15.3.4.1 IBI Rating vs. Fish Habitat Rating

The present version of the coldwater IBI was highly sensitive to changes in the quality of available fish habitat within the southeastern Minnesota streams examined. Coldwater IBI scores for the 64 stream samples displayed a strong relationship to the fish habitat ratings for these streams (Figure 15.6). IBI scores for these samples ranged from 0 (very poor) to 100 (good), and habitat ratings ranged from poor to excellent. Mean IBI scores differed significantly (ANOVA $F = 8.91, p < 0.001$, df = 3, 60) among the habitat rating categories, with those sites having the best habitat rating also having the highest IBI scores. Mean IBI scores were significantly different among good, fair, and poor habitat rating categories, but were not distinctive between good and excellent habitat rating categories (Tukey's multiple range test; Figure 15.6).

15.3.4.2 Temporal Variability

The coldwater IBI developed for use in the midwestern U.S. appears to be a very consistent assessor of stream quality when multiple samples of the fish assemblage at a given site are examined over several years. No change in overall stream quality was evident at any of the stream sites during this period. Variation in IBI scores was low within the 14 Minnesota sites sampled each year for 3 successive years (Table 15.7). Two of these sites showed no change in IBI score during the three years examined, and six others varied by only five points among the years. No site exhibited more than a 20-point change in IBI score during this time period. With the exception of the two sites where IBI scores changed by 20 points, IBI ratings for the sites did not change during the three

TABLE 15.7
Variation in Coldwater IBI Scores at 14 Stream Sites in Southeastern Minnesota During a 3-year Period (site means, standard deviations (SD), 3-year ranges, and IBI ratings also are shown)

Stream/Site	IBI Score			Mean	SD	3-Year Range	IBI Rating
	1994	1995	1996				
Garvin Brook							
Site 1	100	95	100	98.3	2.9	5	Good
Site 2	100	100	85	95.0	8.7	15	Good
Trout Run							
Site 1	95	90	95	93.3	2.9	5	Good
Site 2	95	95	100	96.7	2.9	5	Good
East Indian Creek							
Site 1	75	80	60	71.7	10.4	20	Good/fair
Site 2	70	70	70	70.0	0.0	0	Good
Beaver Creek	90	85	90	88.3	2.9	5	Good
Tributary #9	100	85	95	93.3	7.6	15	Good
Whitewater River-North Br.							
Site 1	25	25	25	25.0	0.0	0	Poor
Site 2	25	30	20	25.0	5.0	10	Poor
Whitewater River-Middle Br.							
Site 1	70	50	65	61.7	10.4	20	Good/fair
Site 2	40	45	45	43.3	2.9	5	Fair
Whitewater River-South Br.							
Site 1	10	15	10	11.7	2.9	5	Poor
Site 2	25	15	30	23.3	7.6	15	Poor

years (Table 15.7). The general lack of yearly variation among IBI scores was further evidenced by the strong correlations of site scores between years for all possible year-to-year comparisons (1994 vs. 1995: Spearman $r_s = 0.951$, $p < 0.001$; 1995 vs. 1996: $r_s = 0.894$, $p < 0.001$; 1994 vs. 1996: $r_s = 0.915$, $p < 0.001$).

Mean IBI scores for the 14 stream sites rated these sites as poor (four sites), fair (two sites), and good (eight sites) (Table 15.7). These mean scores were not correlated significantly either with standard deviations (Spearman $r_s = 0.059$, $p = 0.842$) or with ranges ($r_s = 0.032$, $p = 0.913$) of these scores.

15.3.5 IBI TESTING

15.3.5.1 Ecoregion Comparisons

Over 300 stream sites scattered throughout six ecoregions were evaluated with the coldwater IBI. Of these sites, 4.2% were rated by the IBI as excellent, 18.6% as good, 29.9% as fair, 32.0% as poor, and 15.3% as very poor or no score. Mean IBI score was highest (58.1, fair rating) for sites within the Northern Lakes and Forests ecoregion, stretching across the northern portions of Minnesota, Wisconsin, and Michigan, and lowest (19.2, poor rating) for sites within the Central Corn Belt Plains (CCBP) ecoregion, in northern Indiana just south of Lake Michigan. Mean IBI score varied significantly (ANOVA $F = 12.27$, $p =< 0.001$, df = 5, 299) among the six ecoregions, with sites from the CCBP and Southern Michigan/Northern Indiana Till Plains (SMNITP) ecoregions having much lower mean scores (19.2 and 33.0, respectively) than those (47.7 to 58.1) from the other four ecoregions (Figure 15.7). This difference was most apparent in the proportion of sites

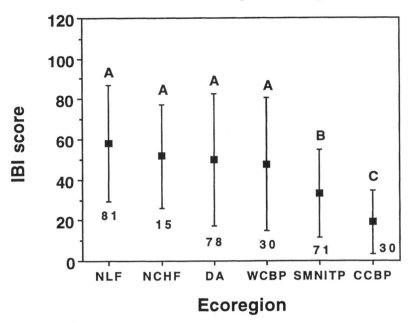

FIGURE 15.7 Mean (±SD) coldwater IBI scores for six ecoregions within the upper midwestern U.S. Mean values not sharing a common letter are significantly different from each other (ANOVA and Tukey's multiple range test). Numbers represent sample sizes. NLF = Northern Lakes and Forests, NCHF = North Central Hardwood Forests, DA = Driftless Area, WCBP = Western Corn Belt Plains, SMNITP = Southern Michigan/Northern Indian Till Plains, CCBP = Central Corn Belt Plains.

rated as poor or very poor/no score, with 70.0% of CCBP and SMNITP sites but only 34.6% of sites from the other four ecoregions receiving these ratings (Figure 15.8). The proportions of sites receiving the various ratings differed significantly ($X^2 = 72.2$, $p < 0.001$) among the six ecoregions.

15.3.5.2 "Natural" vs. "Managed" Coldwater Streams

Fish assemblages at stream sites where there were no nonnative salmonids ("natural" sites) exhibited many differences when compared to assemblages at sites where nonnative salmonids were present ("managed" sites). Sites with nonnative salmonids tended to have lower proportions of intolerant individuals, greater proportions of white suckers, more coldwater individuals, and fewer warmwater individuals than did sites without nonnative salmonids (Table 15.8). Sites with both native and nonnative salmonids tended to have more coldwater species and greater proportions of coldwater individuals and top carnivores than sites with either only native or only nonnative salmonids (Table 15.8). Consequently, overall IBI scores were higher at sites with native salmonids only and sites with mixed native and nonnative salmonids than at locations with only nonnative salmonids (Table 15.8); however, IBI scores did not differ between native-only and mixed salmonid sites ($t = 1.47$, $p = 0.145$). Sites with nonnative salmonids tended to have more total salmonids than those where only native salmonids were present, and salmonids comprised a greater proportion of the fish assemblage at locations with mixed native and nonnative salmonids than at sites with only native or nonnative salmonids (Table 15.8). The average number of brook trout per site did not differ between sites where brook trout were the only salmonids present and those sites where brook trout were present along with one or more species of nonnative salmonid (Table 15.8).

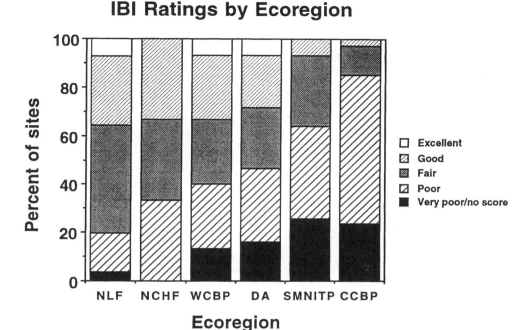

FIGURE 15.8 Distribution of sites among coldwater IBI rating categories for six ecoregions within the upper midwestern U.S. Ecoregion abbreviations and sample sizes are as in Figure 15.5.

15.4 DISCUSSION

15.4.1 BIOLOGICAL INTEGRITY VERSUS COLDWATER STREAM MANAGEMENT

Coldwater streams are highly regarded by the general population because of the recreational value of trout and salmon. In order to meet the requirements of a demanding sport fishery, the coldwater stream manager is often confronted with the need to increase trout production. Karr (1995) indicated that management agencies have been lulled into a false sense of security, believing that the Clean Water Act, wild and scenic river designations, and local laws and regulations are protective of water resources. Although coldwater fish are among the most sensitive to contaminants and water quality agencies afford trout streams added water quality protection, the management and water quality agencies are often at odds with watershed resource management decision-making (Simon et al., Chapter 3).

Karr and Dudley (1981) defined biological integrity as "the capacity of supporting and maintaining a balanced, integrated, adaptive community of organisms having a composition, diversity, and function comparable to that of the natural habitats of the region." Management decisions causing the decline of biological integrity should be carefully reviewed in light of preserving the structure and function of a balanced, integrated, and adaptive coldwater community. Thus, the responsible coldwater stream manager will take into consideration the impacts to other trophic levels of the coldwater community when stocking or enhancing top level carnivore abundance. Simon and Stewart (in press) showed that stocking of nonnative trout species on public lands in the Lake Michigan drainage affected and lowered biological integrity in these areas when compared with unmanaged private lands. Coldwater stream managers should review the need to supplement trout populations with exotic brown trout or nonnative Pacific salmonids and instead utilize native brook trout and lake trout. If these native species are incapable of supporting a sustainable fishery, perhaps

TABLE 15.8
Mean (± SD) IBI Scores, Metric Values, and Other Salmonid Characteristics for Three Categories of Reference Sites Based on Salmonid Assemblages

Site Characteristics	Site Categories			Kruskal-Wallis Statistic	p
	Native Only	Mixed	Exotic Only		
IBI score	66.1 (24.7)	74.9 (26.7)	50.3 (20.5)	10.82	0.004
IBI metrics					
Number of species	4.8 (2.4)	5.4 (3.0)	5.9 (2.6)	1.29	0.525
Number of coldwater species	1.9 (0.8)	2.8 (1.0)	2.1 (1.0)	15.30	<0.001
Number of minnow species	2.2 (1.4)	1.6 (1.5)	2.4 (1.1)	4.53	0.104
Number of benthic species	1.1 (1.0)	1.0 (1.1)	1.4 (0.9)	2.05	0.358
Number of tolerant species	1.9 (1.3)	1.5 (1.3)	2.1 (0.9)	4.84	0.089
Percent salmonids as brook trout	100.0 (0.0)	38.8 (30.3)	0.0 (0.0)	80.35	<0.001
Percent intolerant individuals	48.6 (37.8)	38.0 (30.3)	5.2 (7.6)	24.67	<0.001
Percent coldwater individuals	50.5 (37.6)	75.2 (29.6)	48.9 (27.2)	12.17	0.002
Percent white suckers	2.5 (6.4)	3.5 (8.5)	4.3 (8.3)	1.11	0.574
Percent top carnivores	41.8 (36.9)	66.1 (31.5)	44.1 (29.8)	10.18	0.006
Number of coldwater individuals	48.3 (50.5)	123.2 (118.5)	91.1 (88.7)	12.46	0.002
Number of warmwater individuals	94.3 (114.4)	38.7 (55.5)	67.1 (51.4)	7.66	0.022
Salmonid characteristics					
Number of salmonids	34.9 (37.8)	97.8 (109.9)	79.5 (93.3)	10.76	0.005
Percent salmonids	41.8 (36.9)	61.4 (33.9)	41.6 (31.3)	7.18	0.028
Percent brook trout	41.8 (36.9)	25.1 (27.6)	0.0 (0.0)	41.04	<0.001
Number of brook trout	34.9 (37.8)	32.3 (43.9)	0.0 (0.0)	39.10	<0.001

Note: Native salmonids only (n = 32), mixed native and exotic salmonids (n = 46), and exotic salmonids only (n = 15). Results of Kruskal-Wallis comparisons among site categories also are shown.

the stream is not suitable as a coldwater trout fishery but might be an important area for other obligate coldwater species.

15.4.2 Comparison with Other Coldwater IBIs

The coldwater IBI developed here stands in contrast to other versions developed previously to assess the biotic integrity of coldwater streams. Such a metric-rich index, relying only on fish-assemblage characteristics and designed solely for use in coldwater streams, is a new approach in coldwater bioassessment. The broad geographic range (six ecoregions within four states) across which the index can be applied also makes it unique among most versions developed to date.

The reduced taxa richness typical of fish assemblages in coldwater streams has made it difficult to develop an IBI specifically tailored to just coldwater systems (Simon and Lyons, 1995; Lyons et al., 1996). Investigators confronted with this problem have approached it in several different ways: developing combination warmwater/coolwater/coldwater IBIs, including taxa other than fish in assessments, and devising extremely simple IBIs based on only a few metrics (Simon and Lyons, 1995). While each of these approaches can result in a functional bioassessment tool for coldwater systems, they can also have several limitations that could reduce their usefulness.

A common approach in IBI development has been to create single versions of the IBI designed to assess warmwater and cool- or coldwater systems within the same geographic area. Hughes and Gammon (1987) were the first to take this approach, using their modified IBI to assess a continuum of coolwater to warmwater fish assemblages in the mainstem Willamette River in Oregon. Hughes et al. (in review) also have developed a combination warm/coolwater IBI for use in wadeable streams in the Willamette Valley eceoregion. Langdon (1988) and Steedman (1988) independently devised different versions of the IBI to be used in coldwater, coolwater, and warmwater streams in Vermont and southern Ontario, respectively. Finally, Oberdorff and Hughes (1992) developed a combined warmwater/coolwater/coldwater IBI to assess fish assemblages in the Seine River basin in France.

All of these combined versions of the IBI were reported to be successful at detecting impairment within their geographic areas of concern. However, if any of these versions are used to assess fish assemblages at historically coldwater sites, their abilities for detecting impairment may be compromised by several of the metrics used in calculating the IBIs. Coldwater fish assemblages display a different response to impairment than do their warmwater counterparts, with impairment typically leading to increased fish species richness at coldwater sites (Fisher, 1989; Lyons, 1992; Lyons et al., 1996; present study). All the combined versions of the IBI have used various species richness metrics (e.g., number of native species, number of cyprinid species, number of benthic species, number of water column species) based on the warmwater pattern of reduced species richness in response to impairment (Karr, 1981). Consequently, if these combined metrics are used to assess the biotic integrity of coldwater sites, they may be either rewarding impairment (higher scores for more species) or penalizing higher quality (lower scores for fewer species) at these coldwater sites. Since all these combined IBIs include two or more of these species richness metrics, high-quality coldwater sites could be underrated, and impaired sites could be overrated by this approach. As a result, the ability of these combined IBIs to detect impairment at coldwater locations could be limited. Coolwater fish assemblages also are not assessed well with either warmwater- or coldwater-based IBIs (Lyons, 1992; Lyons et al., 1996), suggesting that separate versions might be needed for warmwater, coolwater, and coldwater systems within the same geographic region (Lyons et al., 1996).

Another IBI development approach aimed at compensating for reduced fish species richness in coldwater streams has been to include metrics that use information from non-fish taxa. Moyle et al. (1986) included a metric for the number of amphibian species in an IBI developed for coastal streams in northern California; and Fisher (1989) included the number of amphibian species, amphibian biomass, and invertebrate density as metrics in various versions of the IBI developed for use in headwater streams in northern and central Idaho. Inclusion of metrics such as these may serve to make the coldwater IBI more sensitive to varying levels of disturbance (Moyle et al., 1986; Fisher, 1989), with multiple taxa providing for a more comprehensive assessment of biotic integrity (Karr, 1991). However, any improvement in sensitivity of the index must be compared to the additional effort, time, and costs required to collect, identify, and process these other taxa. Also, the investigator must determine if it would be appropriate to use a different environmental indicator other than fish. Multiple organism indices are often unable to adequately reflect the structure and function of reference conditions for any of the individual groups. In the case of amphibians, data for a number-of-species metric may be very easy to collect with little additional effort, as long as the number of species is low, identification is easy, and the amphibians can be captured by normal fish survey procedures. Measurement of amphibian biomass would require only a slight increase in effort, but collecting and counting invertebrates makes the process of data collection more intensive (Fisher, 1989). This increase in effort would serve to drive up the cost of evaluating each stream site (Karr, 1991), and likely lead to a reduction in the number of sites that could be surveyed, given time and budgetary constraints. Consequently, a coldwater IBI that focuses only on fish, or other taxa (e.g., amphibians) that are easily collected during fish sampling, would likely be the most cost-effective approach to biological assessment of coldwater streams.

In coldwater versions of the IBI, the number of metrics usually has been reduced when compared with warmwater versions, owing to the more simplified nature of coldwater fish assemblages (Simon and Lyons, 1995). Lyons et al. (1996) developed the IBI for Wisconsin's coldwater streams with the fewest metrics to date, using five metrics to characterize biotic integrity and the response of fish assemblages to disturbance. Although it met all the criteria for a valid IBI (Karr et al., 1986; Fausch et al., 1990; Simon and Lyons, 1995; Lyons et al., 1996), Lyons et al. (1996) speculated that the reduced number of metrics might limit the ability of the index to distinguish slight differences in biotic integrity. Others (Angermeier and Karr, 1986; Miller et al., 1988) have expressed similar reservations toward reduced-metric versions of the IBI.

The IBI is based on the premise that, while all metrics are effective in evaluating stream quality, the range of degradation to which a metric is sensitive varies considerably among metrics (Angermeier and Karr, 1986; Karr et al., 1986; Karr, 1991). Although some metrics may be sensitive over the whole range of stream impairment, others may respond over a much more limited range of impairment (Angermeier and Karr, 1986; Karr et al., 1986; Karr, in press). This allows some metrics to serve as important indicators of attainment of biotic integrity, whereas others are important in discriminating among impaired systems (Barbour et al., 1995). In a metric-rich IBI, all levels of stream impairment should be monitored simultaneously and independently by several individual metrics (Angermeier and Karr, 1986). Such is the case with the original warmwater IBI (Karr et al., 1986), where each level of stream degradation is covered by the range of primary sensitivity of at least four different metrics. The new version of the coldwater IBI developed here displays a similar coverage of all levels of impairment by several individual metrics (Figure 15.5); two metrics are sensitive across the entire range of degradation, and anywhere from three to eight additional metrics also provide coverage at any given level of disturbance. In contrast, the Wisconsin coldwater IBI has two metrics that best distinguish between fair- and good-quality sites and two others that best distinguish between poor- and fair-quality sites; the fifth metric apparently displayed no consistent pattern of distinguishing between quality groupings of sites (Lyons et al., 1996). Although the entire range of stream degradation could be covered by these five metrics, the limited redundancy of coverage within portions of the impairment range may compromise the overall effectiveness of this IBI, since overlap in the sensitivity ranges of several metrics helps to reinforce the overall assessment of biotic integrity (Karr, 1991). Natural conditions and the types of human impacts change from region to region, resulting in changes in the relative sensitivities of many metrics (Angermeier and Karr, 1986; Karr et al., 1986; Steedman, 1988). Although Lyons et al. (1996) eliminated potential coldwater metrics if they displayed different reponses between different ecoregions, there is still an increased probability that an IBI based on only a few metrics will not provide reliable assessments over a broad geographic area (Karr, 1991).

To contrast the present version of the coldwater IBI to that developed by Lyons et al. (1996), fish asemblages collected in 1996 at 32 coldwater stream sites in southeastern Minnesota were assessed with both indices. Sites were selected to represent as wide a range of biotic integrity as possible, from relatively undisturbed streams surrounded by state forest lands to streams draining areas of intensive row-crop agriculture and livestock grazing. Both IBIs demonstrated that the sites spanned a wide range of impairment, and a strong linear relationship was evident between the IBIs (Figure 15.9). Initial observations suggest that both IBIs generally rated sites similarly, but closer examination reveals important differences. Twenty (62.5%) of the sites were placed in the same rating category by both indices, but the remaining sites (37.5%) were rated differently by each index. Although the two indexes appeared to be in good agreement for most of the highest-quality sites, disagreements between indices became more pronounced as site quality declined (Figure 15.9). This suggests that the two indices perform equally well at distinguishing the better-quality sites, but that differences in each index cause them to respond somewhat differently as impairment increases. Since the present coldwater IBI has seven metrics that are sensitive to low levels of biotic integrity (Figure 15.5), it might be better equipped than a reduced-metric version for detecting impairment and distinguishing between sites with more severe levels of impairment. Such an ability to discriminate between impaired sites could be very important to coldwater fishery managers, providing them with information needed to establish management priorities (Barbour et al., 1995). Further testing and comparison of these two indices are needed to more clearly understand these differences and the mechanisms responsible for causing them.

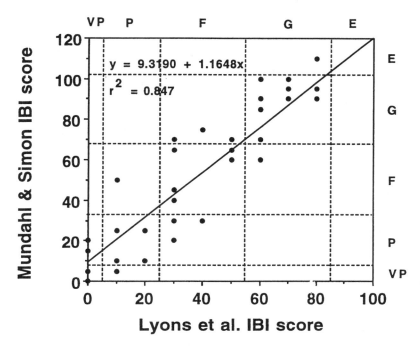

FIGURE 15.9 Comparison of coldwater IBIs calculated for 32 stream sites from southeastern Minnesota using two different methods. Dashed lines separate rating categories (E = excellent, G = good, F = fair, P = poor, VP = very poor) for each IBI method. The best-fit linear regression line and its equation and correlation coefficient describing the relationship between the two IBIs also are shown.

15.4.3 APPLICATIONS OF THE COLDWATER IBI

Because this coldwater IBI was developed from a dataset that included coldwater streams from Minnesota, Wisconsin, and Michigan, it should be appropriate for use in these states, as well as adjacent areas such as northeastern Iowa, northern Indiana, and Ontario. Most of the coldwater streams within these areas generally lie within six ecoregions: Northern Lakes and Forests, North Central Hardwood Forests, Driftless Area, Western Corn Belt Plains, Central Corn Belt Plains, and Southern Michigan/Northern Indiana Till Plains (Omernik and Gallant, 1988). In many instances, warmwater IBIs have been developed for specific ecoregions to account for specific differences in fish assemblages among different geographic regions (Omernik, 1995; Simon and Lyons, 1995). However, coldwater fish assemblages appear to exhibit much greater similarities over much broader geographic areas than are typical of warmwater fish assemblages (Moyle and Herbold, 1987; Lyons, 1989; Poff and Allan, 1995; Lyons et al., 1996). Consequently, Lyons et al. (1996) devised a coldwater IBI appropriate for all of Wisconsin, and hypothesized that such a coldwater IBI developed from a midwestern database might be appropriate for low-gradient (<1%) coldwater streams in regions with a fairly diverse warmwater fish fauna, including Wisconsin, Michigan, parts of Minnesota, Iowa, Ontario, and the northeastern U.S. The present coldwater IBI appears to work well over many of these areas, and as additional databases (e.g., Iowa, Ontario, northeastern U.S.) become available, the range of usefulness of this IBI will be further tested.

Lyons et al. (1996) summarized many of the considerations necessary for successful application of a coldwater IBI, including the need for a representative sample collected from an adequate length of stream, the size of fish to be collected, the minimum number of fish needed, the best season for sampling, the need to exclude stocked salmonids, and others. The present study strongly supports these considerations to ensure the most effective use of the IBI and the accuracy of its assessments of biotic integrity. In particular, strong emphasis is placed on the need to apply the coldwater IBI only to coldwater streams (either current or historic). Inappropriate application of the coldwater IBI to warmwater or coolwater streams will result in a false assessment of their true biotic integrity (Lyons et al., 1996), unless these systems historically were coldwater and the intent is to demonstrate their current status as coldwater resources. Furthermore, it is emphasized that a minimum number of fish must be collected from a stream reach before applying the index, to avoid potential problems that might arise when metrics (especially those that are proportion based) are scored with very low numbers of fish (Lyons et al., 1996). Lyons et al. (1996) recommend a minimum requirement of 25 fish per site, a number found at all high-quality reference sites in the present study. If fewer than 25 fish were collected from a site, it is recommended that an IBI rating not be applied until a resampling is conducted (Table 15.5).

15.4.4 FUTURE CONSIDERATIONS AND REFINEMENTS

This coldwater IBI should be viewed as an effort toward development of a metric-rich tool for assessing coldwater streams over a much broader geographic area than has been attempted previously. Many modifications of the IBI have occurred on an ecoregion-by-ecoregion basis to account for regional differences in stream fish assemblages (e.g., Hughes and Larsen, 1988; Hughes et al., 1994). The resulting ecoregion-specific indices are useful for assessing biotic integrity within the regions for which they were designed, but the multiple indices can make comparison and interpretation among them difficult (Yoder and Rankin, 1995). This may be especially troublesome when several different forms of the IBI are used in different ecoregions all lying within the same political boundaries (e.g., state or province). Since coldwater fish assemblages display so many similarities over broad geographic areas (Moyle and Herbold, 1987), development of a coldwater IBI applicable over a wide area is a logical step in biotic assessment in coldwater streams. Lyons et al. (1996) suspected that the coldwater IBI developed for Wisconsin may be applicable beyond Wisconsin's borders, but had no specific data with which to test this. During development of the present coldwater IBI, index calibration occured over a broad spatial scale, using data gathered from three states. Calibration on such a scale has been suggested as the "ideal" approach, but is seldom realized because of lack of cooperation among natural resource agencies across artificial political boundaries (Yoder and Rankin, 1995).

Because this coldwater IBI is designed for use over such a broad geographic area, future efforts should be directed toward development of expectation criteria on a regional or subregional basis, probably using ecoregions as the primary units. Following the lead of the Ohio EPA, tiered expectation criteria could be developed to document exceptional as well as acceptable coldwater stream resources on an ecoregion-by-ecoregion basis (analogous to the use designations for warmwater streams in Ohio; Yoder and Rankin, 1995). Criteria for exceptional status likely would be similar across all ecoregions (e.g., IBI score >100), whereas criteria for acceptable status likely would differ among ecoregions, reflecting differences that might exist in historical, irreversible modifications to physical stream habitat and watershed conditions. Development of such criteria would require a more thorough examination of fish-assemblage data from coldwater streams throughout each ecoregion, especially from reference-quality stream sites.

Although numerous potential metrics were examined during development of the coldwater IBI, there may be other characteristics of coldwater fish assemblages that could work just as well or better than those selected for final inclusion in the present IBI. Fausch et al. (1990) recommended several potential assemblage characteristics that might be useful for assessing coldwater streams, including population abundance and reproductive success and measures of stress or health for individual fish. Most of these were not examined during development of the coldwater IBI because information on them was generally lacking in the datasets used. Some measure of stress or fish health could be especially helpful in evaluating conditions at more severely degraded sites (Karr et al., 1986). The current dataset did not allow a proper examination of the usefulness of DELT anomalies as a metric for coldwater streams, and Lyons et al. (1996) did not examine this metric when they developed a coldwater IBI for Wisconsin. General observations, however, suggest that these anomalies might be too rare in fishes in degraded coldwater streams to serve as useful indicators of fish health or stress (N. Mundahl, personal observation). Steedman (1988; 1991) used the proportion of fish with a certain number of blackspot cysts as an indicator of fish health in Ontario streams that included coldwater sections. However, the health of some fishes might not be affected adversely by these parasites (e.g., Baker and Bulow, 1985; Hockett and Mundahl, 1989), and some investigators have recommended against using the prevalence of blackspot as an IBI metric (Ohio EPA, 1989). Additional metrics could also be examined for inclusion in a coldwater IBI, although the effectiveness of those already in the current version is quite good. In the current version, all metrics are able to distinguish between impaired and reference sites, and 11 of 12 metrics displayed values significantly correlated to total site IBI score. In working warmwater versions of the IBI, 10 or fewer metrics may exhibit significant correlation with total IBI score (Angermeier and Karr, 1986).

ACKNOWLEDGMENTS

We thank each of the numerous individuals that either provided field assistance, data, or support for this project. Special thanks to Jeffrey Quinn for initial coldwater IBI work in southeastern Minnesota, and to James Karr for his insightful comments that led to the improvement of this index. Funding for N.D.M. was provided in part by the Winona State University Professional Improvement Fund, the Minnesota Pollution Control Agency, and the Minnesota Department of Natural Resources Natural Heritage and Non-game Research Program. Although this project may have been funded wholly or in part by the U.S. Environmental Protection Agency, the opinions expressed by the authors do not necessarily reflect those of the agency.

REFERENCES

Angermeier, P.L. and J.R. Karr. 1986. Applying an index of biotic integrity based on stream fish communities: considerations in sampling and interpretation, *North American Journal of Fisheries Management,* 6, 418–429.

Baker, S.C. and F.J. Bulow. 1985. Effects of blackspot disease on the condition of stonerollers *Campostoma anomalum, American Midland Naturalist,* 114, 198–199.

Barbour, M.T., J. Gerritsen, G.E. Griffith, R. Frydenborg, E. McCarron, J.S. White, and M.L. Bastian. 1996. A framework for biological criteria for Florida streams using benthic macroinvertebrates, *Journal of the North American Benthological Society,* 15, 185–211.

Barbour, M.T., J.B. Stribling, and J.R. Karr. 1995. Multimetric approach for establishing biocriteria and measuring biological condition, in W.S. Davis and T.P. Simon (Eds.), *Biological Assessment and Criteria: Tools for Water Resources Planning and Decision Making.* Lewis, Boca Raton, FL. 63–77.

Fausch, K.D., J.R. Karr, and P.R. Yant. 1984. Regional application of an index of biotic integrity based on stream fish communities, *Transactions of the American Fisheries Society,* 113, 39–55.

Fausch, K.D., J. Lyons, J.R. Karr, and P.L. Angermeier. 1990. Fish communities as indicators of environmental degradation, in S. M. Adams (Ed.), *Biological Indicators of Stress in Fish.* American Fisheries Society, Symposium 8, Bethesda, MD. 123–144.

Fausch, K.D. and L.H. Schrader. 1987. Use of the Index of Biotic Integrity to Evaluate the Effects of Habitat, Flow, and Water Quality on Fish Communities in Three Colorado Front Range Streams. Final Report to the Kodak-Colorado Division and the Cities of Fort Collins, Loveland, Greeley, Longmont, and Windsor. Department of Fishery and Wildlife Biology, Colorado State University, Fort Collins, CO.

Fisher, T.R. 1989. Application and Testing of Indices of Biotic Integrity in Northern and Central Idaho Headwater Streams. M.S. Thesis, University of Idaho, Moscow.

Gerritsen, J. 1995. Additive indices for resource management, *Journal of the North American Benthological Society,* 14, 451–457.

Hendricks, M.L. 1980. *Salvelinus fontinalis* (Mitchill): brook trout, in D.S. Lee, C.R. Gilbert, C.H. Hocutt, R.E. Jenkins, D.E. McAllister, and J.R. Stauffer, Jr. (Eds.), *Atlas of North American Freshwater Fishes.* North Carolina State Museum of Natural History, Raleigh, NC. 114.

Hockett, C.J. and N.D. Mundahl. 1989. Effects of black spot disease on thermal tolerances and condition factors of three cyprinid fishes, *Journal of Freshwater Ecology,* 5, 67–72.

Hughes, R.M. and J.R. Gammon. 1987. Longitudinal changes in fish assemblages and water quality in the Willamette River, Oregon, *Transactions of the American Fisheries Society,* 116, 196–209.

Hughes, R.M., S.A. Heiskary, W.L. Matthews, and C.O. Yoder. 1994. Use of ecoregions in biological monitoring, in S.L. Loeb and A. Spacie (Eds.), *Biological Monitoring of Aquatic Systems.* Lewis, Boca Raton, FL. 125–151.

Hughes, R.M., P.R. Kaufmann, A.T. Herlihy, T.M. Kincaid, L. Reynolds, and D.P. Larsen. In review. Development and application of an index of fish assemblage integrity for wadeable streams in the Willamette Valley ecoregion, Oregon, USA, *Canadian Journal of Fisheries and Aquatic Sciences.*

Hughes, R.M. and D.P. Larsen. 1988. Ecoregions: an approach to surface water protection, *Journal of the Water Pollution Control Federation,* 60, 486–493.

Hunter, C.J. 1991. *Better Trout Habitat: A Guide to Stream Restoration and Management.* Island Press, Washington, DC.

Karr, J.R. 1981. Assessment of biotic integrity using fish communities, *Fisheries,* 6(6), 21–27.

Karr, J.R. 1991. Biological integrity: a long-neglected aspect of water resource management, *Ecological Applications,* 1, 66–84.

Karr, J.R. 1995. Protecting aquatic ecosystems: clean water is not enough, in W.S. Davis and T.P. Simon (Eds.), *Biological Assessment and Criteria: Tools for Water Resource Planning and Decision Making.* Lewis, Boca Raton, FL. 7–13.

Karr, J.R. (In press.) Rivers as sentinels: using the biology of rivers to guide landscape management, in R.J. Naiman and R.E. Bilby (Eds.), *The Ecology and Management of Streams and Rivers in the Pacific Northwest Coastal Ecoregion.* Springer-Verlag, New York.

Karr, J.R. and D.R. Dudley. 1981. Ecological perspective on water quality goals, *Environmental Management,* 5, 55–68.

Karr, J.R., K.D. Fausch, P.L. Angermeier, P.R. Yant, and I.J. Schlosser. 1986. Assessing Biological Integrity in Running Waters: A Method and Its Rationale. Illinois Natural History Survey Special Publication 5, Champaign, IL.

Karr, J.R., P.R. Yant, K.D. Fausch, and I.J. Schlosser. 1987. Spatial and temporal variability of the index of biotic integrity in three midwestern streams, *Transactions of the American Fisheries Society,* 116, 1–11.

Langdon, R. 1988. The development of fish population-based biocriteria in Vermont, in T.P. Simon, L.L. Holst, and L.J. Shepard (Eds.), *Proceedings of the First National Workshop on Biocriteria.* EPA 905-9-89-003. U.S. Environmental Protection Agency, Region 5, Environmental Sciences Division, Chicago. 12–25.

Leonard, P.M. and D.J. Orth. 1986. Application and testing of an index of biotic integrity in small, coolwater streams, *Transactions of the American Fisheries Society,* 115, 401–414.

Lyons, J. 1989. Correspondence between the distribution of fish assemblages in Wisconsin streams and Omernik's ecoregions, *American Midland Naturalist,* 122, 163–182.

Lyons, J. 1992. Using the Index of Biotic Integrity (IBI) to Measure Environmental Quality in Warmwater Streams of Wisconsin. U.S. Forest Service General Technical Report NC-149.

Lyons, J., S. Navarro-Perez, P.A. Cochran, E. Santana C., and M. Guzman-Arroyo. 1995. Development of an index of biotic integrity based on fish assemblages for the conservation of streams and rivers in west-central Mexico, *Conservation Biology,* 9, 569–584.

Lyons, J., L. Wang, and T.D. Simonson. 1996. Development and validation of an index of biotic integrity for coldwater streams in Wisconsin, *North American Journal of Fisheries Management,* 16, 241–256.

Miller, D.L., P.M. Leonard, R.M. Hughes, J.R. Karr, P.B. Moyle, L.H. Schrader, B.A. Thompson, R.A. Daniel, K.D. Fausch, G.A. Fitzhugh, J.R. Gammon, D.B. Halliwell, P.L. Angermeier, and D.J. Orth. 1988. Regional applications of an index of biotic integrity for use in water resource management, *Fisheries,* 13(5), 12–20.

Moyle, P.B., L.R. Brown, and B. Herbold. 1986. Final Report on Development and Preliminary Tests of Indices of Biotic Integrity for California. Report to U.S. Environmental Protection Agency, Corvallis Environmental Research Laboratory, Corvallis, OR.

Moyle, P.B. and B. Herbold. 1987. Life-history patterns and community structure in stream fishes of western North America: comparisons with eastern North America and Europe, in W.J. Matthews and D.C. Heins (Eds.), *Community and Evolutionary Ecology of North American Stream Fishes*. University of Oklahoma Press, Norman, OK. 25–32.

Oberdorff, T. and R.M. Hughes. 1992. Modification of an index of biotic integrity based on fish assemblages to characterize rivers of the Seine-Normandie basin, France, *Hydrobiologia*, 228, 117–130.

Ohio EPA. 1987. *Biological Criteria for the Protection of Aquatic Life. Vol. II. Users Manual for Biological Field Assessment of Ohio Surface Waters.* Ohio EPA, Division of Water Quality Monitoring and Assessment, Surface Water Section, Columbus, OH.

Omernik, J.M. 1995. Ecoregions: a spatial framework for environmental management, in W.S. Davis and T.P. Simon (Eds.), *Biological Assessment and Criteria: Tools for Water Resources Planning and Decision Making*. Lewis, Boca Raton, FL. 49–62

Omernik, J.M. and A.L. Gallant. 1988. Ecoregions of the Upper Midwest States. U.S. Environmental Protection Agency, EPA/600/3-88/037, Corvallis, OR.

Poff, N.L. and J.D. Allan. 1995. Functional organization of stream fish assemblages in relation to hydrological variability, *Ecology*, 76, 606–627.

Schrader, L.H. 1986. Testing of the Index of Biotic Integrity in the South Platte River Basin of Northeastern Colorado. M.S. thesis, Colorado State University, Fort Collins.

Simon, T.P. and J. Lyons. 1995. Application of the index of biotic integrity to evaluate water resources integrity in freshwater ecosystems, in W.S. Davis and T.P. Simon (Eds.), *Biological Assessment and Criteria: Tools for Water Resources Planning and Decision Making*. Lewis, Boca Raton, FL. 245–262.

Simon, T.P. and P.M. Stewart. (In press.) Structure and function of fish communities in the southern Lake Michigan basin with emphasis on restoration of native fish communities, *Natural Areas Journal.*

Simonson, T.D. and J. Lyons. 1995. Comparison of catch per effort and removal procedures for sampling stream fish assemblages, *North American Journal of Fisheries Management*, 15, 419–427.

Simonson, T.D., J. Lyons, and P.D. Kanehl. 1994. Guidelines for Evaluating Fish Habitat in Wisconsin Streams. U.S. Forest Service General Technical Report NC-164.

Steedman, R.J. 1988. Modification and assessment of an index of biotic integrity to quantify stream quality in southern Ontario, *Canadian Journal of Fisheries and Aquatic Sciences*, 45, 492–501.

Steedman, R.J. 1991. Occurrence and environmental correlates of black spot disease in stream fishes near Toronto, Ontario, *Transactions of the American Fisheries Society*, 120, 494–499.

Trautman, M.B. 1981. *The Fishes of Ohio,* 2nd ed. The Ohio State University Press, Columbus, OH.

Yoder, C.O. and E.T. Rankin. 1995. Biological criteria program development and implementation in Ohio, in W.S. Davis and T.P. Simon (Eds.), *Biological Assessment and Criteria: Tools for Water Resources Planning and Decision Making*. Lewis, Boca Raton, FL. 109–144.

Appendix 15A
Classification of Species into Tolerance, Feeding, Habitat, and Temperature Preference Categories for Calculation of Coldwater IBI Metrics

Family and Common Name	Scientific Name	Tolerance[a]	Feeding[b]	Habitat[c]	Temperature Preference[d]
Petromyzontidae					
northern brook lamprey	*Ichthyomyzon fossor*	I	—	—	C
silver lamprey	*Ichthyomyzon unicuspis*	I	—	—	—
American brook lamprey	*Lampetra appendix*	I	—	—	C
sea lamprey	*Petromyzon marinus*	I	—	—	C
Lepisosteidae					
spotted gar	*Lepisosteus oculatus*	—	TC	—	—
longnose gar	*Lepisosteus osseus*	—	TC	—	—
Amiidae					
bowfin	*Amia calva*	—	TC	—	—
Clupeidae					
skipjack	*Alosa chrysochloris*	—	TC	—	—
alewife	*Alosa pseudoharengus*	—	—	—	—
gizzard shad	*Dorosoma cepedianum*	—	—	—	—
Hiodontidae					
mooneye	*Hiodon tergisus*	—	—	—	—
Salmonidae					
coho salmon	*Oncorhynchus kisutch*	—	TC	—	C
rainbow trout	*Oncorhynchus mykiss*	—	TC	—	C
chinook salmon	*Oncorhynchus tshawytscha*	—	TC	—	C
brown trout	*Salmo trutta*	—	TC	—	C
brook trout	*Salvelinus fontinalis*	I	TC	—	C
Umbridae					
central mudminnow	*Umbra limi*	T	—	—	—
Esocidae					
grass pickerel	*Esox americanus*	—	TC	—	—
northern pike	*Esox lucius*	—	TC	—	—
muskellunge	*Esox masquinongy*	I	TC	—	C
Cyprinidae					
central stoneroller	*Campostoma anomalum*	—	—	—	—
goldfish	*Carassius auratus*	T	—	B	—
redside dace	*Clinostomus elongatus*	I	—	—	C
lake chub	*Couesius plumbeus*	—	—	—	C
spotfin shiner	*Cyprinella spiloptera*	—	—	—	—
steelcolor shiner	*Cyprinella whipplei*	—	—	—	—

Family and Common Name	Scientific Name	Tolerance[a]	Feeding[b]	Habitat[c]	Temperature Preference[d]
common carp	*Cyprinus carpio*	T	—	B	—
brassy minnow	*Hybognathus hankinsoni*	—	—	—	C
bigeye chub	*Hybopsis amblops*	I	—	—	—
striped shiner	*Luxilus chrysocephalus*	—	—	—	—
common shiner	*Luxilus cornutus*	—	—	—	—
silver chub	*Macrhybopsis storeriana*	—	—	—	—
pearl dace	*Margariscus margarita*	—	—	—	C
hornyhead chub	*Nocomis biguttatus*	—	—	—	—
river chub	*Nocomis micropogon*	I	—	—	—
golden shiner	*Notemigonus crysoleucas*	T	—	—	—
emerald shiner	*Notropis atherinoides*	—	—	—	—
bigmouth shiner	*Notropis dorsalis*	—	—	—	—
blacknose shiner	*Notropis heterolepis*	I	—	—	—
spottail shiner	*Notropis hudsonius*	I	—	—	—
sand shiner	*Notropis stramineus*	—	—	—	—
mimic shiner	*Notropis volucellus*	—	—	—	—
suckermouth minnow	*Phenocobius maribilis*	—	—	—	—
northern redbelly dace	*Phoxinus eos*	—	—	—	C
southern redbelly dace	*Phoxinus erythrogaster*	—	—	—	—
finescale dace	*Phoxinus neogaeus*	—	—	—	C
bluntnose minnow	*Pimephales notatus*	T	—	—	—
fathead minnow	*Pimephales promelas*	T	—	—	—
blacknose dace	*Rhinichthys atratulus*	T	—	—	—
longnose dace	*Rhinichthys cataractae*	—	—	B	—
creek chub	*Semotilus atromaculatus*	T	—	—	—
Catostomidae					
longnose sucker	*Catostomus catostomus*	—	—	B	C
white sucker	*Catostomus commersoni*	T	—	B	—
river carpsucker	*Carpiodes carpio*	—	—	B	—
quillback	*Carpiodes cyprinus*	—	—	B	—
lake chubsucker	*Erimyzon sucetta*	—	—	—	—
northern hog sucker	*Hypentelium nigricans*	I	—	B	—
smallmouth buffalo	*Ictiobus bubalus*	—	—	B	—
bigmouth buffalo	*Ictiobus cyprinellus*	—	—	B	—
spotted sucker	*Minytrema melanops*	—	—	B	—
silver redhorse	*Moxostoma anisurum*	—	—	B	—
river redhorse	*Moxostoma carinutum*	—	—	B	—
golden redhorse	*Moxostoma erythrurum*	—	—	B	—
shorthead redhorse	*Moxostoma macrolepidotum*	—	—	B	—
Ictaluridae					
black bullhead	*Ameiurus melas*	—	—	B	—
yellow bullhead	*Ameiurus natalis*	T	—	B	—
brown bullhead	*Ameiurus nebulosus*	—	—	B	—
channel catfish	*Ictalurus punctatus*	—	TC	B	—
stonecat	*Noturus flavus*	—	—	B	—
tadpole madtom	*Noturus gyrinus*	—	—	B	—
flathead catfish	*Pylodictus olivaris*	—	TC	B	—
Aphredoderidae					
pirate perch	*Aphredoderus sayanus*	—	—	—	—
Percopsidae					
trout—perch	*Percopsis omiscomaycus*	—	—	—	—

Family and Common Name	Scientific Name	Tolerance[a]	Feeding[b]	Habitat[c]	Temperature Preference[d]
Gadidae					
burbot	*Lota lota*	—	TC	—	C
Atherinidae					
brook silverside	*Labidesthes sicculus*	—	—	—	—
Gasterosteidae					
brook stickleback	*Culaea inconstans*	—	—	—	C
Moronidae					
white bass	*Morone chrysops*	—	TC	—	—
striped bass	*Morone saxatilis*	—	TC	—	—
Centrarchidae					
rock bass	*Ambloplites rupestris*	I	TC	—	—
green sunfish	*Lepomis cyanellus*	T	—	—	—
pumpkinseed	*Lepomis gibbosus*	—	—	—	—
warmouth	*Lepomis gulosus*	—	TC	—	—
orangespotted sunfish	*Lepomis humilis*	—	—	—	—
bluegill	*Lepomis macrochirus*	—	—	—	—
longear sunfish	*Lepomis megalotis*	I	—	—	—
redear sunfish	*Lepomis microlophus*	—	—	—	—
smallmouth bass	*Micropterus dolomieu*	I	TC	—	—
spotted bass	*Micropterus punctulatus*	—	TC	—	—
largemouth bass	*Micropterus salmoides*	—	TC	—	—
white crappie	*Pomoxis annularis*	T	—	—	—
black crappie	*Pomoxis nigromaculatus*	—	—	—	—
Percidae					
mud darter	*Etheostoma asprigene*	—	—	B	—
greenside darter	*Etheostoma blennioides*	—	—	B	—
rainbow darter	*Etheostoma caeruleum*	I	—	B	—
Iowa darter	*Etheostoma exile*	I	—	B	—
fantail darter	*Etheostoma flabellare*	—	—	B	—
johnny darter	*Etheostoma nigrum*	—	—	B	—
yellow perch	*Perca flavescens*	—	—	—	—
logperch	*Percina caprodes*	—	—	B	—
blackside darter	*Percina maculata*	—	—	B	—
sauger	*Stizostedion canadense*	—	TC	—	—
walleye	*Stizostedion vitreum*	—	TC	—	—
Sciaenidae					
freshwater drum	*Aplodinotus grunniens*	—	—	—	—
Cottidae					
mottled sculpin	*Cottus bairdi*	I	—	B	C
slimy sculpin	*Cottus cognatus*	I	—	B	C

[a] Tolerance categories: tolerant (T), intolerant (I), other (—).
[b] Feeding categories: top carnivore (TC), other (—).
[c] Habitat categories: benthic (B), other (—).
[d] Temperature preference categories: coldwater (C), warmwater (—).

16 Biological Monitoring and an Index of Biotic Integrity for Lake Erie's Nearshore Waters

Roger F. Thoma

CONTENTS

16.1 Introduction 418
16.2 Methods 419
    16.2.1 Sampling Sites 419
    16.2.2 Field Methods 419
        16.2.2.1 Gillnet Methods 419
        16.2.2.2 Trawl Methods 421
        16.2.2.3 Hoop Net Methods 421
        16.2.2.4 Beach Seine Methods 421
        16.2.2.5 Electrofishing 421
            16.2.2.5.1 General electrofishing methodologies 421
            16.2.2.5.2 Day sampling methods 422
            16.2.2.5.3 Night sampling methods 422
16.3 Sampling Results 423
    16.3.1 Gillnet 423
    16.3.2 Trawl 423
    16.3.3 Hoopnet 423
    16.3.4 Beach Seine 423
    16.3.5 Day Electrofishing 424
    16.3.6 Night Electrofishing 424
16.4 The Lake Erie Metrics 424
    16.4.1 Metric Selection Rationale 424
    16.4.2 Metrics 424
        16.4.2.1 Number of Native Species 424
        16.4.2.2 Number of Benthic Species 424
        16.4.2.3 Number of Sunfish Species 425
        16.4.2.4 Number of Cyprinid Species 425
        16.4.2.5 Number of Phytophilic Species 430
        16.4.2.6 Number of Intolerant Species 430
        16.4.2.7 Percent Tolerant Species 431
        16.4.2.8 Percent Omnivorous Species 431
        16.4.2.9 Percent Lake Individuals 431
        16.4.2.10 Percent Phytophilic Individuals 434
        16.4.2.11 Percent Top Carnivores 437
        16.4.2.12 Number of Individuals 437

		16.4.2.13	Percent Nonindigenous Species..437
		16.4.2.14	Percent Diseased Individuals..439
16.5	Scoring Considerations and Integrity Classifications 439
	16.5.1	Setting the 95th Percentile Line..439
	16.5.2	Integrity Classifications 444
16.6	Applications..447
	16.6.1	Lacustuary Assessments 447
		16.6.1.0	Ottawa River 447
		16.6.1.1	Maumee River..447
		16.6.1.2	Duck Creek 448
		16.6.1.3	Otter Creek...448
		16.6.1.4	Swan Creek 449
		16.6.1.5	Turtle Creek 449
		16.6.1.6	Toussaint River 450
		16.6.1.7	Portage River 450
		16.6.1.8	Sandusky River 451
		16.6.1.9	Muddy and Little Muddy Creeks 452
		16.6.1.10	Huron River 452
		16.6.1.11	Old Woman Creek 453
		16.6.1.12	Vermillion River...454
		16.6.1.13	Black River 454
		16.6.1.14	Rocky River 455
		16.6.1.15	Cuyahoga River..455
		16.6.1.16	Chagrin River...456
		16.6.1.17	Grand River..456
		16.6.1.18	Ashtabula River..456
		16.6.1.19	Conneaut Creek..457
	16.6.2	Lake Erie 458
	16.6.3	General...459
Acknowledgments 461
References 461

16.1 INTRODUCTION

Since Karr proposed the Index of Biotic Integrity (IBI) in 1981, numerous researchers have applied his assessment philosophy to the measurement of a myriad of ecosystem types. Minns et al. (1994) developed an IBI for the Great Lakes to assess nearshore environmental integrity. In the Minns IBI, the number of native species, number of native individuals, and the biomass of native species and the number of nonindigenous species, percent number of nonindigenous individuals, and percent biomass of nonindigenous species were used as metrics. This resulted in a reduction in the species number, trophic, and community health metrics by one each. Minns et al. also chose to use a 0–100 IBI scoring range and metric scoring continua instead of the traditional 5, 3, 1 methodology proposed by Karr. The IBI presented herein reduces the emphasis on the comparison of native and nonindigenous faunal components and attempts to remain faithful to Karr's original IBI concept.

In 1993, the Ohio EPA began a project designed to develop numerical biological criteria for shoreline waters of Lake Erie, including areas of tributary streams affected by lake levels, herein referred to as lacustuaries. The term "lacustuary" is a new word combination using the words lacustrine and estuary. A lacustuary is defined as a transition zone in a river that flows into a freshwater lake and is the portion of river affected by the water level of the lake. Lacustuaries begin where lotic conditions end in the river and end where the lake proper begins. They have hydrologic conditions similar to estuaries in that they are affected by tides (primarily wind driven, occasionally barometric) and are lentic habitats. Lacustuaries differ from estuaries in that their chemical properties are less saline, with salinity gradients going from higher upstream to lower at the lake interface (Brant and Herdendorf, 1972). It is felt that the term "lacustuary" is needed to avoid confusion of terms and concepts that ensue when estuary is used for freshwater systems. Although there are some similarities, estuaries and lacustuaries differ in numerous important functions and should not be confused with each other. To call a freshwater system an estuary dilutes the meaning of the word, making it less useful and could lead a non-scientist to believe that lacustuaries are the same as estuaries.

This IBI project was conducted in the following steps: (1) sampling of the general habitat types found in the Lake Erie ecosystem using the various sampling methodologies deemed appropriate for the specific area; (2) evaluation of sampler type efficiency and selection of the method to be used in each habitat type; (3) continued sampling using the selected methodology; (4) evaluation of potential metrics to be used in the development of a Lake Erie shoreline and lacustuary IBI; (5) selection and calibration of IBI metrics; (6) continued sampling; (7) calculation of Lake Erie shoreline and lacustuary IBI scores; and (8) evaluation of environmental conditions in Lake Erie and associated lacustuary areas. This study was built on data collected since 1982.

The development of numerical biological criteria for Lake Erie's shoreline and lacustuaries will greatly enhance the Ohio EPA's ability to understand and regulate pollution levels and their impacts in these little-studied areas. Communication with the public concerning these activities will be enhanced.

16.2 METHODS

16.2.1 SAMPLING SITES

Ninety sites (324 individual collections) were sampled in Lake Erie from 1993 through 1996. Site selection reflected the habitat types found in the lake's nearshore areas and provided a thorough coverage (approximately one site for every 5 miles or 8 km) of the area investigated. Sites were located along harbor breakwaters, sand/gravel beaches, the shores of the Lake Erie Islands, bedrock cliffs, and modified shore lines with numerous types of structures designed primarily to prevent shoreline erosion. Wetland/bay-like habitats were sampled in Sandusky Bay, East Harbor State Park, and Presque Isle PA (11 sites). Lacustuaries were sampled at 125 sites (593 individual collections) from 1982 through 1996. Sites were located at the mouth, head, and midsection of each lacustuary.

The Ohio EPA uses a system that utilizes river miles as a measure of site location on a stream reach or lakeshore. This chapter retains the Ohio EPA river mile system to report site location and uses metric measurements for all other distances.

16.2.2 FIELD METHODS

All fish collected by all methods (Figure 16.1) were identified to species, enumerated, examined for external anomalies, and either returned to the lake or preserved as voucher specimens and stored at the Ohio State University Museum of Biodiversity. Weights were taken on a representative subsample if more than 15 individuals of a species were captured. All fish were weighed if 15 or less individuals of a species were captured.

16.2.2.1 Gillnet Methods

Gill nets were used to sample river mouths that were dredged to accommodate interlake and international freighter traffic. Dredged areas were 7.6 m deep, with vertical walls constructed of

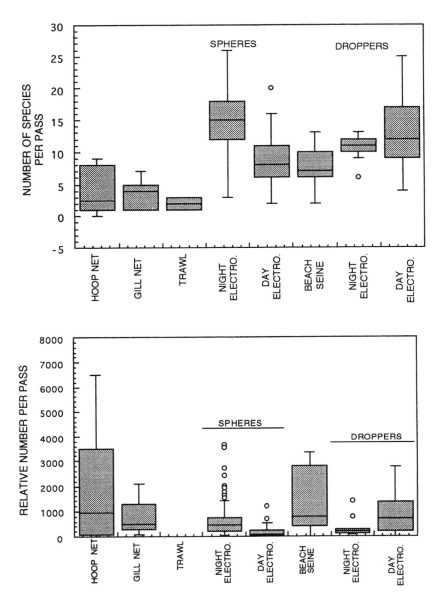

FIGURE 16.1 A comparison of the number of species and the relative number of individuals captured per sampling pass for the eight different sampler types tested in the Lake Erie area. Spheres and droppers refer to the shape of the anodes used in electrofishing efforts. Spheres were ball-shaped anodes and droppers were 1-m long cable anodes.

steel sheet piling, concrete, or railroad ties. Steel sheet piling was the most commonly used material. Structural habitat in these areas was extremely sparse due to the nature of the shoreline modifications. Gill nets were hung vertically, perpendicular to and against the seawall, with the smallest mesh size nearest the wall. Four mesh sizes were used, with each panel being 1.2 m wide and 7.6 m deep. Mesh sizes were 2.5, 5.0, 7.5, and 10 cm. Each net was 5 m wide and 7.6 m deep (or long), with a float at the surface end and cement anchors at the bottom. The near-shore end of the net was tied to the seawall to prevent it from being dragged away by strong currents in the ship channel. Nets were set in the evening and retrieved in the morning for a sampling time of approximately 12 hours. Sampling effort was recorded as number of fish per hour.

16.2.2.2 Trawl Methods

Samples were taken during daylight and night hours using standard bottom trawl methodologies and a 3-m trawl. The net was deployed on the windward side of the boat and towed with the wind at a distance of 9 to 12 m behind the boat at a speed of 0.5 knots and a depth of 7 m. Both time and distance trawled were recorded for each sample. An attempt was made to sample 500 m of bottom. Distance was estimated by the use of a G.P.S. unit, and depth was monitored with a depth finder. Sampling effort was recorded as number of fish per kilometer (fish/km).

16.2.2.3 Hoop Net Methods

The use of hoop nets was confined to shallow wetland embayment areas. This type of habitat was sampled in the Huron River and Old Woman Creek lacustuaries. Nets were set in 0.5 to 1.5 m of water perpendicular to shore in the evening and retrieved in the morning. A 15-m lead was set against the shore with the trap end distal to the shore. Two 8-m wings were set to the sides, each at 90° angles to the lead line. Nets were set in the evening and retrieved in the morning for a sampling time of approximately 12 hours. Sampling effort was recorded as number per hours sampled.

16.2.2.4 Beach Seine Methods

A 15-m bag seine was used to sample sandy shore areas. Leads were 2 m deep with 9-mm mesh and a bag with 4-mm mesh. One hundred meters of shoreline were sampled at each station using two 50-m hauls. The direction of the haul (whether east or west) was determined by the prevailing shoreline currents and was always conducted with the current. The shore side of the seine was kept on the beach and the lake side was approximately 7.5 m from shore, depending on water depth. All samples were taken during daylight hours. Sampling effort was recorded as numbers per kilometer (numbers/km).

16.2.2.5 Electrofishing

Electrofishing consistently caught more species and individuals in less time and with less effort than other sampling methods used in this study. It was also the only method that could be used under all habitat conditions, thus yielding a database that was easily comparable (in terms of catch per effort) under the variable conditions encountered. Previous Ohio EPA work indicated that night electrofishing would likely capture more species and individuals than day electrofishing. Both day and night collections were made for comparison purposes and are discussed below.

16.2.2.5.1 General electrofishing methodologies

A 5.8-m modified V-hull john boat was used for electrofishing. Electrical current was provided by a 7000-watt generator and Smith-Root pulsator. Controls were set on DC current, 60 pulses per second, 240–340 volts, and run at 5 to 6 amps. In low conductivity conditions, the voltage was increased to 360–500 volts, and the frequency was increased to 120 pulses per second in order to maintain 5 to 6 amps. Anodes were two separately charged 1-m circumference electrospheres. Two articulated booms supported by distal floats were positioned about 2.1 m in front of the boat on articulated booms (3 m total length), one to the port and one to the starboard side at angles of approximately 20° from the center line. This resulted in the two electrospheres being 4.3 m apart when deployed. The articulation of the booms allowed movement horizontally and vertically. Horizontal movement allowed the booms to flex if an obstacle (especially a submerged one) was encountered, and vertical movement allowed the electrosphere to ride with the wave action. Sampling was not conducted under wave conditions of 0.6 m or more. Swells or small waves less than 0.6 m cause nonarticulated booms to rise up and pull the anodes from the water, thus interrupting the current flow and allowing fish to escape. The placement of a flotation device at the distal end of the boom kept the electrosphere under the water's surface at the proper depth of a few centimeters as the mechanism rode up and down.

Each sampling site was 500 m long and within 1 m of the shore. A set sampling time was not used, and time varied between 2000 and 5000 seconds. The greater the number of fish to be captured in the zone and the greater the complexity of the shoreline, the longer it took to complete the sample. A crew of three individuals was used in all electrofishing efforts. During sampling, one individual was positioned on the bow of the boat with a dip net and served as the principal collector of fish captured in the electrical field; a second person was at mid-ship and served as an assistant to collect any fish missed by the principal netter; while the third person operated the outboard motor, pulsator controls (and spot lights at night), and collected any fish that surfaced at the back of the boat. All fish were placed in livewells supplied with fresh water from a pump. Common carp were placed in their own livewell to avoid excess oxygen consumption and the death of small fish that otherwise would frequently be trapped in common carp mouths and crushed.

The anode and cathode array deployments used in this study were different from those used in previous Ohio EPA sampling efforts. Anodes were two separately charged electrospheres, 1 m in circumference, and constructed with 12 strips of stainless steel 3 to 5 cm in width and approximately 2 mm thick. By using strips, it was possible to decrease the anode surface area when sampling in high conductivity waters by collapsing the strips on each other until a workable configuration had been achieved. The anodes were suspended 5 cm below the surface, 2.1 m in front of the boat on articulated booms, one to each side. Three sets of cathodes, each set to be used at different depths, were used. All had electrified portions 1.6 m in length. The cathode sets were designed to be deployed with the electrified surface at a maximum of 1.8, 3, and 7.3 m in depth. Cathodes, 1.8 m long, were used under all conditions where bottom depths were 2.5 m or less, and 3-m cathodes were used in depths greater than 3 m. Cathodes were deployed in one of two ways: (1) in most areas, eight cathodes were deployed from the sides of the boat at mid-ship (1.8- and 3-m cathodes), four on each side, or (2) in ship channel areas, four, 7.3 m depth cathodes were deployed from the front of the boat.

Sampling on lakeshores was conducted when winds were generally from the south, southeast or southwest, and wave action was 0.6 m or less. North winds resulting in waves greater than 0.6 m prevented effective sampling. If winds were from the southwest, zones were sampled from the west to the east. If winds were from the southeast, zones were sampled from the east to the west. This allowed the boat and stunned fish to move with the shoreline currents. After periods of sustained onshore winds and heavy wave action, sampling was avoided for at least a week.

16.2.2.5.2 Day sampling methods

All habitat types sampled in 1993 were sampled with day electrofishing. For the most part, the sampling techniques employed were the same as those used in Ohio EPA stream sampling efforts (Ohio EPA, 1988), except for the following differences: electrosphere anodes, cathode array and depth, three-person sampling crew (as described above), and site selection based on habitat type and the location of discharge pipes from industrial and municipal sources.

16.2.2.5.3 Night sampling methods

All night collections were made at least 30 minutes after sunset and before 5:00 AM. Submerged and above-surface lights were used during night sampling. Six submerged lights were mounted on the front of the boat as the primary illuminators of attracted fish. Three amber foglamps with 12-volt, 100-watt halogen bulbs were mounted on a post. Two posts were used, one on each side of the bow of the boat. The lamps were mounted one above the other and submerged just below the surface when deployed and positioned at 95° and 10° to the side and 10° toward the long axis of the boat. With all six lamps deployed on posts at the sides of the bow, an arc of light over 180° was achieved. If turbidity prevented the use of submerged lights, six 12-volt tractor flood lights mounted on the bow 1 m above the water were used. Four 12-volt tractor flood lights mounted at the stern, 1 m above the surface, two on each side of the boat, illuminated the sides of the boat aft of the bow. One light was directed forward and the other perpendicular to the side. A hand-held spotlight was used to search for stunned fish outside these illuminated areas (especially behind the boat) and to scan the shoreline. When sampling was completed, four 12-volt tractor floodlights, mounted on a transverse beam at the stern of the boat and directed at the sample processing area, were used to process the sample. All lights were powered by 12-volt, deep-cycle marine batteries in 1993, 1994, and 1995, while sampling lights were powered by the electrofishing generator in 1996.

16.3 SAMPLING RESULTS

16.3.1 GILLNET

Gillnet sampling efficiency, measured in terms of the number of species and individuals collected in ship channels, did not differ from that obtained by day or night electrofishing, but the level of effort required to collect gillnet samples was considerably higher. In addition, gillnets were susceptible to currents from shipping traffic and strong seiche action that would displace the net, making it ineffective. Furthermore, mortality in gillnets was often 100%, whereas by electrofishing, it was near 0%. Since electrofishing was a more reliable collection method, required less time, and was less damaging to fish, it was decided that electrofishing would be the method employed in ship channel areas.

16.3.2 TRAWL

This method was least effective for collecting fish in the nearshore area. Not more than three species or four adult fish were captured in a single sample. Most trawl sampling was hampered by snagging on boulders or debris. An effort was made in 1995 to trawl on sandy bottoms in the western basin (both day and night samples) in a final effort to evaluate this sampling method. Again, snagging problems hampered success.

16.3.3 HOOPNET

In the Huron River, both numbers of species and individuals captured in hoopnets were much lower than numbers obtained with electrofishing gear. In the Old Woman Creek lacustuary near Huron, Ohio, electrofishing and hoopnets were able to capture a similar number of species, but the hoopnets captured many fewer individuals and the level of effort required to collect a hoopnet sample was much greater than that required with electrofishing. Because of lower sampling efficiency and greater time requirements for hoopnets, it was decided to use electrofishing methods.

16.3.4 BEACH SEINE

This method could only be used in areas where no rubble, debris, or rocks were found. Mostly young-of-year fishes were captured and no adults of larger species were taken. The number of species captured by this method (when young-of-year fish were included) was similar to that collected with daylight electrofishing, whereas the number of individuals was much higher (due to the large number of young-of-year fish). Beach seining was not used because catches of adult fish were found to be lower than those obtained with electrofishing, and because of its limitation to sandy shores only.

16.3.5 DAY ELECTROFISHING

In lakeshore and breakwater habitats, electrofishing during the day captured fewer species and individuals than electrofishing at night. Also, fewer large individuals were captured. In rivers, day and night sampling were equally effective at collecting numbers of individuals and species. Thus, river areas were sampled during the day and lake areas were sampled at night to spread the workload out and allow for more collections during the week.

16.3.6 NIGHT ELECTROFISHING

At lake sites, night electrofishing captured more species and additional fish in all habitats than any other sampling method tested. It should also be noted that night electrofishing in particular was much more effective at avoiding young-of-year fish. Ohio EPA standard methods exclude young-of-year fish from numerical calculations because of their potential effect on index values (Angermeier and Schlosser, 1987) and the fact that their numbers in Lake Erie are highly variable from year to year (Trautman, 1981). Abundance of young-of-year fish can be more dependent on favorable weather conditions during the spawning season than on water quality factors. Further, numbers of young-of-year fish do not always translate (through recruitment) to adult fish. More importantly, night electrofishing could be employed effectively in all lakeshore habitat types sampled during this study.

16.4 THE LAKE ERIE METRICS

A large number of metrics (including Karr's 12 original metrics) were examined to determine the 12 metrics (Table 16.1) best suited for use in a Lake Erie IBI and lacustuary IBI. It was first determined that not all of Karr's 12 metrics were sensitive to changes in Lake Erie waters. Karr's metrics retained in their original form were: number of species, percent top carnivores, percent omnivores, and number of individuals. Examination of metrics for lake and lacustuary sites indicated that the relative abundances and percent composition of fish in the two types of habitat should be evaluated separately. The metrics with the best potential to assess environmental changes in Lake Erie and its lacustuaries are reported below.

16.4.1 METRIC SELECTION RATIONALE

Fidelity to Karr's original 12 metrics was accorded prime consideration in metric selection, retention of metrics with some modification was considered secondarily, and metric replacement was used only as a last recourse. When metrics were selected, an effort was made to use groupings that maximized the range of values possible. Metrics with low breadth can result in a yes–no, present–absent evaluation instead of the strongly–moderately–little deviation assessment intended by Karr. Metrics proposed by Karr that were not responsive to the range of environmental conditions encountered in this study were replaced. Relevance to the historic Lake Erie ecosystem was a consideration in selecting new metrics.

16.4.2 METRICS

16.4.2.1 Number of Native Species (Figures 16.2 and 16.3)

This metric was not modified and follows that of Karr (1981).

16.4.2.2 Number of Benthic Species (Figures 16.4, 16.5, and Table 16.2)

This metric is used in both lacustuary and lake areas. It is a modification of Karr's metric: number of darter species. It possess sufficient range and environmental responsiveness to be useful in a

TABLE 16.1
Metrics Used in the Ohio EPA's Two IBIs Developed to Evaluate Lake Erie Near-Shore Ecosystems and Those Developed by Minns et al. for the Great Lakes

Lake Erie Metrics	Lacustuary Metrics	Minns et al.[a]
Species number metrics		
# Species	# Species	# Natives
# Sunfish species	# Sunfish species	# Centrarchid species
# Phytophilic species	#Cyprinid species	# Cyprinid species
# Benthic species	# Benthic species	
Behavior/trophic guild metrics		
% Lake assoc. individuals	% Phytophilic individuals	% Specialist biomass
% Top carnivores	% Top carnivores	% Piscivore biomass
# Intolerant species	# Intolerant species	# Intolerant species
% Omnivore individuals	% Omnivore individuals	% Generalists biomass
% Nonindigenous ind.	% Nonindigenous ind.	% Nonindigenous ind.
% Tolerant individuals	% Tolerant individuals	
Community health metrics		
% DELT[b]	% DELT[b]	
Relative numbers[c]	Relative numbers[c]	# Native individuals
		# Nonindigenous species
		Biomass of natives species
		% Nonindigenous biomass

Note: Metrics are arranged correspondingly in rows.

[a] From Minns et al., 1994.
[b] Externally observable deformities, eroded fins, lesions, and tumors.
[c] Includes nonindigenous species and excludes gizzard shad.

Lake Erie IBI. It is thought to primarily respond to environmental disturbance from excess sedimentation and secondarily to toxicity and low oxygen levels. It comprises darters, sculpins, and madtoms. Expansion of the darter metric to a benthic species metric was necessary because of the naturally low number of darter species found in lentic environments in the Laurentian basin. Only sculpins and madtoms were added to this metric, as they most closely approximate the environmental sensitivities of the darters. Other benthic species of generally greater environmental tolerance, such as bullheads and suckers, were excluded to maintain Karr's original sensitivity level.

16.4.2.3 Number of Sunfish Species (Figures 16.2 and 16.3)

This metric is a modification of Karr's metric; sunfish species, and has been expanded to include members of the genera *Pomoxis* and *Micropterus*. This expansion increases the range of responsiveness to environmental conditions and habitats over which this metric can be used. Also, in Lake Erie, not all sites have equal potential to harbor all species of sunfish found in the basin.

16.4.2.4 Number of Cyprinid Species (Figure 16.5)

This metric replaces Karr's number of sucker species metric in lacustuaries. In lacustuaries, sucker species can be naturally low in numbers and abundance. Cyprinid species were historically a

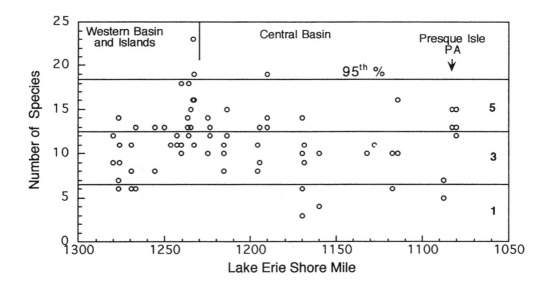

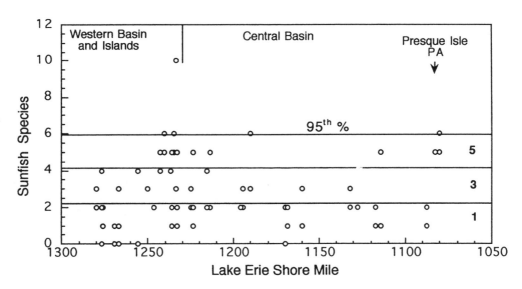

FIGURE 16.2 Lake Erie shoreline IBI metric scores plotted by shore mile (measured from east to west). Upper: number of species; Lower: number of sunfish species. Each graph displays the 95th percentile line and the 5, 3, and 1 scoring ranges. An absence of species scores zero in these Lake Erie IBI metrics.

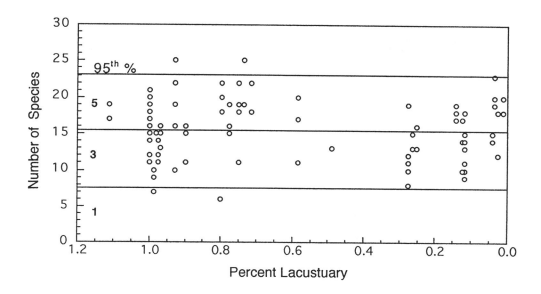
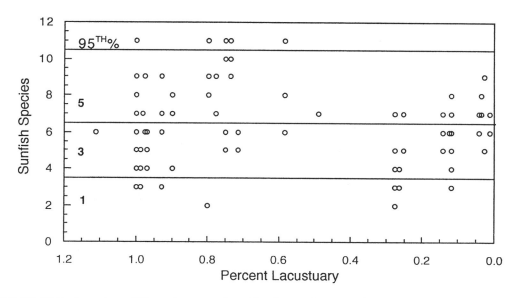

FIGURE 16.3 Lacustuary IBI metric scores plotted by distance from mouth as a measure of the percentage of the total lacustuary length. Upper: number of species; Lower: number of sunfish species. Each graph displays the 95th percentile line and the 5, 3, and 1 scoring ranges. An absence of species scores zero in these Lake Erie IBI metrics.

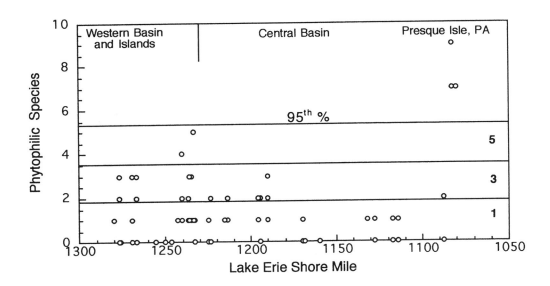

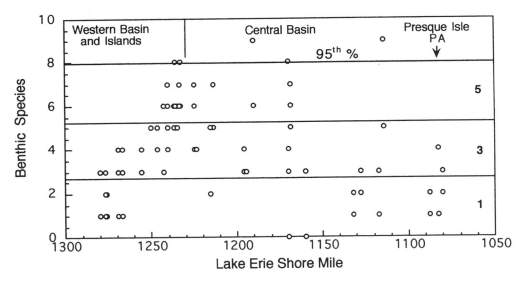

FIGURE 16.4 Lake Erie shoreline IBI metric scores plotted by shore mile (measured from east to west). Upper: number of phytophilic species; Lower: number of benthic species. Each graph displays the 95th percentile line and the 5, 3, and 1 scoring ranges. An absence of species scores zero in these Lake Erie IBI metrics.

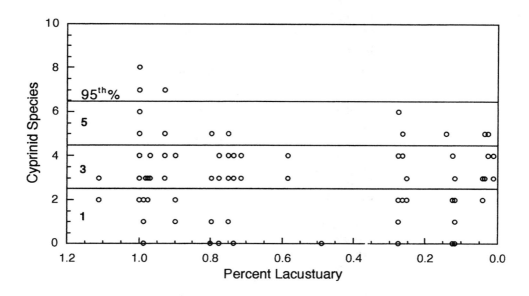
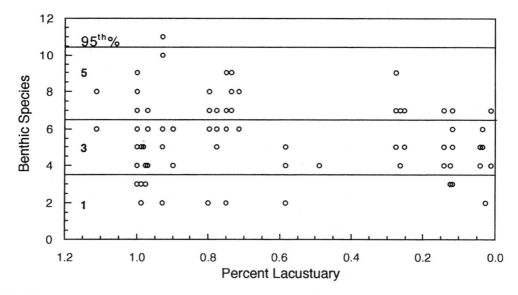

FIGURE 16.5 Lacustuary IBI metric scores plotted by distance from mouth as a measure of the percentage of the total lacustuary length. Upper: number of cyprinid species; Lower: number of benthic species. Each graph displays the 95th percentile line and the 5, 3, and 1 scoring ranges. An absence of species scores zero in these Lake Erie IBI metrics.

TABLE 16.2
Common and Scientific Names of Lake Erie Fish Species that Comprise the Benthic Species Metric

Common Name	Scientific Name
Stonecat madtom	*Noturus flavus*
Brindled madtom	*Noturus miurus*
Tadpole madtom	*Noturus gyrinus*
Blackside darter	*Percina maculata*
River darter	*Percina shumardi*
Channel darter	*Percina copelandi*
Logperch	*Percina caprodes*
Eastern sand darter	*Ammocrypta pellucida*
Johnny darter	*Etheostoma nigrum*
Greenside darter	*Etheostoma blennioides*
Iowa darter	*Etheostoma exile*
Rainbow darter	*Etheostoma caeruleum*
Orangethroat darter	*Etheostoma spectabile*
Fantail darter	*Etheostoma flabellare*
Least darter	*Etheostoma microperca*
Spoonhead sculpin	*Cottus ricei*
Mottled sculpin	*Cottus bairdi*

prominent community component that could be found in all lacustuary habitats, and several highly sensitive species (now apparently extirpated in Ohio) were primarily associated with Lake Erie nearshore areas. This new metric can accommodate future changes in the ecosystem if environmental conditions improve to the point that locally extirpated species become reestablished.

16.4.2.5 Number of Phytophilic Species (Figure 16.4 and Table 16.3)

This metric replaces Karr's number of sucker species metric in the lake proper. Sucker species are normally low in number and abundance in the nearshore lake areas, partly because of their tendency to occupy the deeper, colder offshore lake waters during the summer months. Historically, suckers were a dominant component of the Lake Erie fauna, but are presently much less abundant. They are not effectively sampled with this methodology. If Lake Erie should improve sufficiently to allow the resurgence of the sucker community and it is found that they can be sampled effectively under such conditions, consideration should be given to reinstating this metric.

Although the number of individuals of phytophilic species is usually low in the lake, the number of species is sufficiently high to illustrate responses to variations in environmental conditions. Variations in this metric are associated with increases in submerged aquatic vascular plants (especially *Potamogeton* and *Vallisneria*) that are found in high-quality, clear, low-polluted waters. It is important to capture the critical ecological element associated with aquatic vascular plant communities, an ecological parameter of substantial historical prominence.

16.4.2.6 Number of Intolerant Species (Figures 16.6, 16.7, and Table 16.4)

This is a slight modification of Karr's metric. Modifications follow existing Ohio EPA rationale and species listings (Ohio EPA, 1988). It is used in both lake and lacustuary areas.

TABLE 16.3
Common and Scientific Names of Lake Erie Fish Species that Comprise the Phytophilic Species Metric

Common Name	Scientific Name
Spotted gar	*Lepisosteus oculatus*
Longnose gar	*Lepisosteus osseus*
Bowfin	*Amia calva*
Central mudminnow	*Umbra limi*
Grass pickerel	*Esox americanus vermiculatus*
Chain pickerel	*Esox niger*
Northern pike	*Esox lucius*
Muskellunge	*Esox masquinongy*
N. pike x Muskellunge	Hybrid
Spotted sucker	*Minytrema melanops*
Lake chubsucker	*Erimyzon sucetta*
Golden shiner	*Notemigonus crysoleucas*
Pugnose minnow	*Opsopoeodus emiliae*
Blackchin shiner	*Notropis heterodon*
Blacknose shiner	*Notropis heterolepis*
Pugnose shiner	*Notropis anogenus*
Tadpole madtom	*Noturus gyrinus*
Western banded killifish	*Fundulus diaphanus menona*
Eastern banded killifish	*Fundulus diaphanus diaphanus*
Blackstripe topminnow	*Fundulus notatus*
Western mosquitofish	*Gambusia affinis*
Pirate perch	*Aphredoderus sayanus*
Black crappie	*Poxomis nigromaculatus*
Largemouth bass	*Micropterus salmoides*
Pumpkinseed sunfish	*Lepomis gibbosus*
Yellow perch	*Perca flavescens*
Least darter	*Etheostoma microperca*
Brook stickleback	*Culaea inconstans*
Three-spine stickleback	*Gasterosteus aculeatus*
Tubenose goby	*Proterorhinus marmoratus*

16.4.2.7 Percent Tolerant Individuals (Figures 16.8, 16.9, and Table 16.5)

This metric replaces Karr's percent green sunfish metric and follows existing Ohio EPA rationale. It is used in both lake and lacustuary areas.

16.4.2.8 Percent Omnivorous Species (Figures 16.6 and 16.7)

The rationale for this metric, as described by Karr (1981), is followed in this study.

16.4.2.9 Percent Lake Individuals (Figure 16.10 and Table 16.6)

This metric replaces Karr's percent insectivorous cyprinids metric in the lake. There are few primarily insectivorous cyprinids in Lake Erie's nearshore area and, consequently, such a metric always scores low regardless of the environmental quality. Karr originally designed this as a highly sensitive metric. The replacement metric (percent lake species) keeps this perspective by choosing

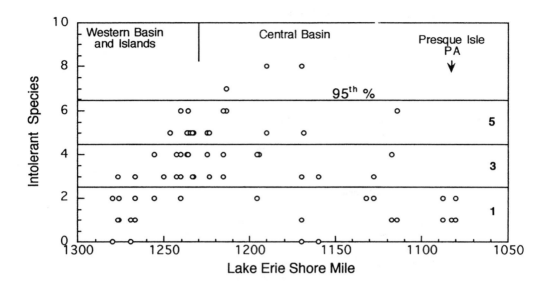
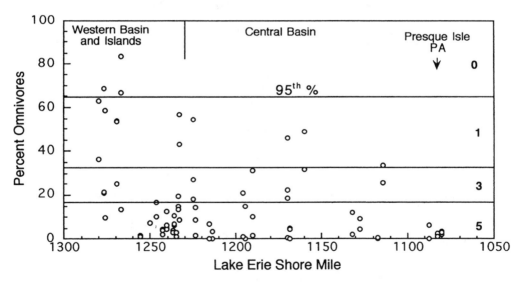

FIGURE 16.6 Lake Erie shoreline IBI metric scores plotted by shore mile (measured from east to west). Upper: number of intolerant species; Lower: percent omnivores. Each graph displays the 95th percentile line and the 5, 3, and 1 scoring ranges. An absence of species or individuals scores zero in the lake-associated, top carnivore, and intolerant metrics while omnivores (a negative metric) scores zero at values greater than the 95th percentile line.

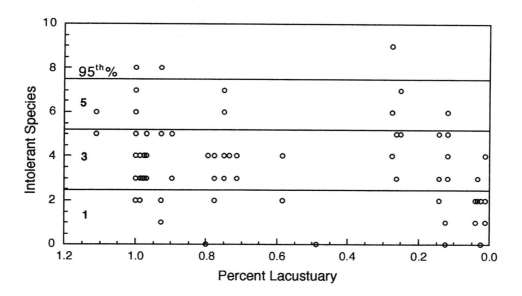

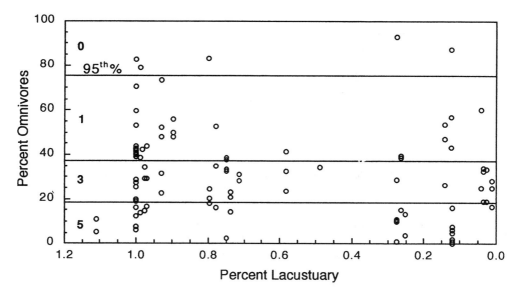

FIGURE 16.7 Lacustuary IBI metric scores plotted by distance from mouth as a measure of the percentage of the total lacustuary length. Upper: number of intolerant species; Lower: percent omnivores. Each graph displays the 95th percentile line and the 5, 3, and 1 scoring ranges. An absence of species or individuals scores zero in the intolerant metric, while omnivores (a negative metric) scores zero at values greater than the 95th percentile line.

TABLE 16.4
Common and Scientific Names of Lake Erie Fish Species that Comprise the Intolerant Species Metric

Common Name	Scientific Name
North brook lamprey	*Ichthyomyzon fossor*
Amer brook lamprey	*Lampetra appendix*
Paddlefish	*Polyodon spathula*
Mooneye	*Hiodon tergisus*
Black redhorse	*Moxostoma duquesnei*
Greater redhorse	*Moxostoma valenciennesi*
River redhorse	*Moxostoma carinatum*
Harelip sucker	*Lagochila lacera*
Hornyhead chub	*Nocomis biguttatus*
River chub	*Nocomis micropogon*
Bigeye chub	*Notropis amblops*
Streamline chub	*Erimystax dissimilis*
Longnose dace	*Rhinichthys cataractae*
Redside dace	*Clinostomus elongatus*
Pugnose minnow	*Opsopoeodus emiliae*
Silver shiner	*Notropis photogenis*
Rosyface shiner	*Notropis rubellus*
Blackchin shiner	*Notropis heterodon*
Bigeye shiner	*Notropis boops*
Mimic shiner	*Notropis volucellus*
Blacknose shiner	*Notropis heterolepis*
Pugnose shiner	*Notropis anogenus*
Popeye shiner	*Notropis ariommus*
Stonecat madtom	*Noturus flavus*
Northern madtom	*Noturus stigmosus*
Brindled madtom	*Noturus miurus*
Western banded killifish	*Fundulus diaphanus menona*
Channel darter	*Percina copelandi*
Gilt darter	*Percina evides*
Eastern sand darter	*Ammocrypta pellucida*
Rosyface shiner x silver shiner	Hybrid
Pallid shiner	*Notropis amnis*

a species guild that has proven sensitive to environmental disturbances in Lake Erie. Because sufficient numbers of lake-associated species still exist and much room for improvement is possible, this metric is ideal for measuring the long-term trends of Lake Erie fish communities.

16.4.2.10 Percent Phytophilic Individuals (Figure 16.11)

This metric replaces Karr's percent insectivorous cyprinids metric in lacustuary areas. As with percent lake species, this metric is designed to be highly sensitive to slight environmental change. Historically, lacustuaries exhibited high numbers of phytophilic species and very high numbers of individuals. Although numerous phytophilic species have disappeared from Lake Erie's lacustuaries, many species still subsist at very low numbers in almost all areas. As even the most polluted sites generally have the same phytophilic species (subsisting at very low numbers), it was decided to use the number of individuals, as sites of higher environmental quality exhibited much higher abundances than degraded sites. This allows discrimination between the very bad sites and fair

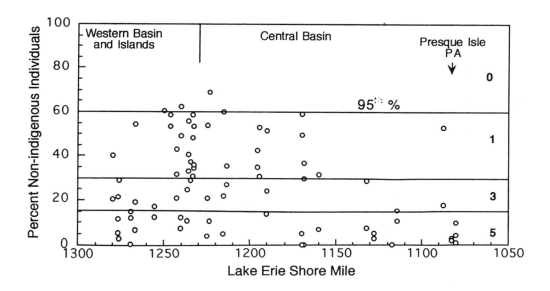

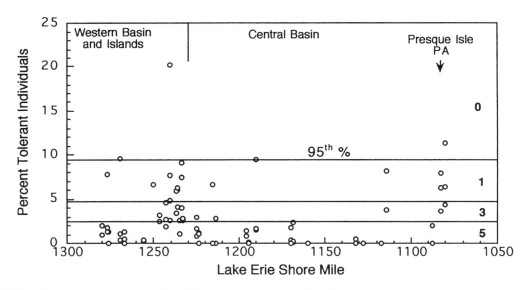

FIGURE 16.8 Lake Erie shoreline IBI metric scores plotted by shore mile (measured from east to west). Upper: percent non-indigenous individuals; Lower: percent tolerant individuals. Graphs display the 95th percentile line and the 5, 3, and 1 scoring ranges. The nonindigenous and tolerant metrics (both negative metrics) score zero at values greater than the 95th percentile line.

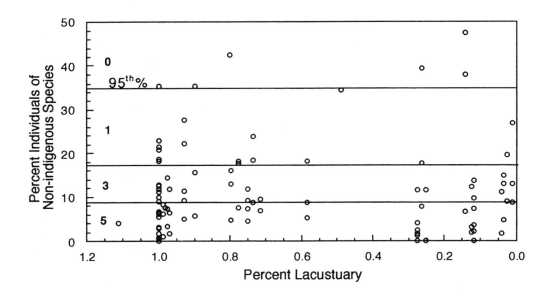
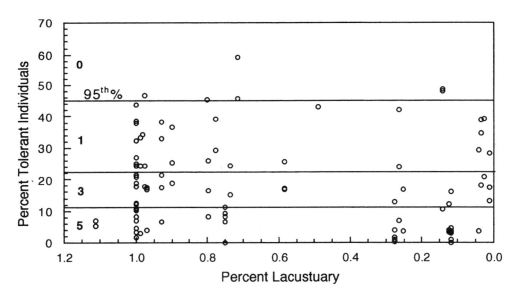

FIGURE 16.9 Lacustuary IBI metric scores plotted by distance from mouth as a measure of the percentage of the total lacustuary length. Upper: percent nonindigenous individuals; Lower: percent tolerant individuals. Graphs display the 95th percentile line and the 5, 3, and 1 scoring ranges. Nonindigenous and tolerant metrics (both negative metrics) score zero at values greater than the 95th percentile line.

TABLE 16.5
Common and Scientific Names of Lake Erie Fish Species that Comprise the Tolerant Species Metric

Common Name	Scientific Name
Central mudminnow	*Umbra limi*
White sucker	*Catostomus commersoni*
Common carp	*Cyprinus carpio*
Goldfish	*Carassius auratus*
Golden shiner	*Notemigonus crysoleucas*
Blacknose dace	*Rhinichthys atratulus*
Creek chub	*Semotilus atromaculatus*
Fathead minnow	*Pimephales promelas*
Bluntnose minnow	*Pimephales notatus*
Com. carp x goldfish	Hybrid
Yellow bullhead	*Ameiurus natalis*
Brown bullhead	*Ameiurus nebulosus*
Yellow bullhead x brown bullhead	Hybrid
Eastern banded killifish	*Fundulus diaphanus diaphanus*
Green sunfish	*Lepomis cyanellus*

sites. If lacustuary habitats should improve in the future, this metric may to be converted to a number of species metric.

16.4.2.11 Percent Top Carnivores (Figures 16.10 and 16.11)

This metric follows the rationale established by Karr (1981).

16.4.2.12 Number of Individuals (Figures 16.12 and 16.13)

This metric follows the rationale established by Karr (1981).

16.4.2.13 Percent Nonindigenous Species (Figures 16.8, 16.9, and Table 16.7)

This metric replaces Karr's percent hybrids metric. In Lake Erie, almost all hybridization occurs between common carp and goldfish and, to a lesser extent, among the sunfish species. Common carp x goldfish hybrids are found throughout the area studied and are primarily associated with the presence of common carp and, to a lesser extent, environmental conditions observed in Lake Erie. This is not to say that such hybrids are not indications of environmental stress; in fact, if Lake Erie's integrity were high, there would be few common carp and probably no goldfish in the system and, in turn, no common carp x goldfish hybrids. The problem with common carp x goldfish hybrids is that they move off shore during the summer and into nearshore areas in the spring. As a consequence, it is uncertain whether the hybrids were produced in the area where they were captured.

Sunfish hybrids have been found to increase in areas where sunfish numbers increase. Sunfish numbers were found to be highest at sites with healthy aquatic macrophyte populations. Healthy aquatic macrophyte populations, in turn, were associated with clear water sites that were not nutrient enriched. Cleaner sites without vegetation that tend toward the oligotrophic end of the scale, had lower sunfish abundances. The most common sunfish species associated with hybridization was the green sunfish. In shallow, highly turbid, nutrient-rich environments, areas that were previously heavily vegetated clear-water sites, sunfish hybridization was common although sunfish were not.

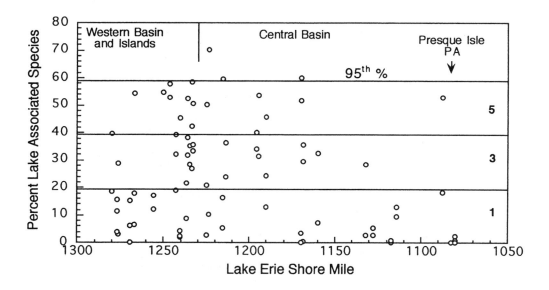
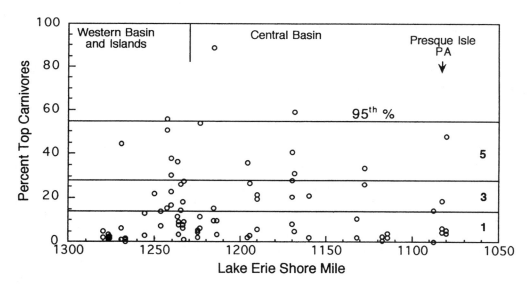

FIGURE 16.10 Lake Erie shoreline IBI metric scores plotted by shore mile (measured from east to west). Upper: percent lake-associated species; Lower: percent top carnivores. Each graph displays the 95th percentile line and the 5, 3, and 1 scoring ranges. An absence of species or individuals scores zero in the lake-associated, top carnivore, and intolerant metrics, while omnivores (a negative metric) scores zero at values greater than the 95th percentile line.

TABLE 16.6
Common and Scientific Names of Lake Erie Fish Species that Comprise the Lake Associated Species Metric

Common Name	Scientific Name
Lake sturgeon	*Acipenser fulvescens*
Spotted gar	*Lepisosteus oculatus*
Rainbow trout	*Oncorhynchus mykiss*
Lake trout	*Salvelinus namaycush*
Coho salmon	*Oncorhynchus kisutch*
Cisco: lake herring	*Coregonus artedii albus*
Lake whitefish	*Coregonus clupeaformis*
Rainbow smelt	*Osmerus mordax*
Longnose sucker	*Catostomus catostomus*
Silver chub	*Macrhybopsis storeriana*
Longnose dace	*Rhinichthys cataractae*
Blackchin shiner	*Notropis heterodon*
Pugnose shiner	*Notropis anogenus*
Burbot	*Lota lota*
White perch	*Morone americana*
Sauger	*Stizostedion canadense*
Walleye	*Stizostedion vitreum*
Channel darter	*Percina copelandi*
Spoonhead sculpin	*Cottus ricei*
Deepwater sculpin	*Myoxocephalus thompsoni*
Mottled sculpin	*Cottus bairdi*

These habitats were recently invaded by orangespotted sunfish and, as previously noted by Trautman (1981), orangespotted sunfish frequently hybridize with other sunfish species when invading new areas. Nonindigenous species have been found, in this study, to increase in areas of higher disturbance, especially that associated with extensive urban development.

16.4.2.14 Percent Diseased Individuals (DELT) (Figures 16.12 and 16.13)

This metric is a slight modification of Karr's metric and follows existing Ohio EPA rationale. It is a measure of the percent individuals that have externally observable deformities, eroded fins, lesions, or tumors. It is used in both lake and lacustuary areas.

16.5 SCORING CONSIDERATIONS AND INTEGRITY CLASSIFICATIONS

16.5.1 Setting the 95th Percentile Line

Because the fish community of Lake Erie has experienced pervasive negative impacts (Hartman, 1972; Regier and Hartman, 1973; Trautman, 1981; Van Meter and Trautman, 1970; White et al., 1975), the selection of reference sites and 95% lines is problematic. If one sets expectations at levels thought to be equivalent to the historic potential of Lake Erie, all sites would score so low that it would not be possible to distinguish among highly, moderately, and slightly polluted areas. Alternatively, if a straightforward 95th percentile line is employed, it becomes possible that sites will score in the exceptional range. This prospect is unacceptable in light of the present condition

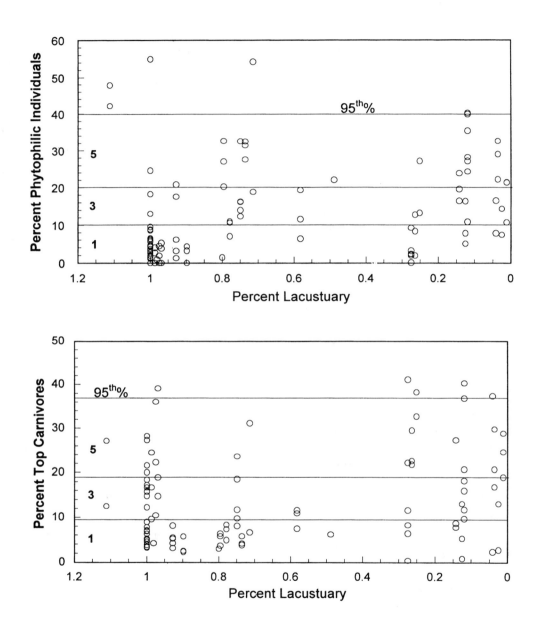

FIGURE 16.11 Lacustuary IBI metric scores plotted by distance from mouth as a measure of the percentage of the total lacustuary length. Upper: percent phytophilic individuals; Lower: percent top carnivores. Each graph displays the 95th percentile line and the 5, 3, and 1 scoring ranges. An absence of species or individuals scores zero in these metrics.

Biological Monitoring and an Index of Biotic Integrity for Lake Erie's Nearshore Waters 441

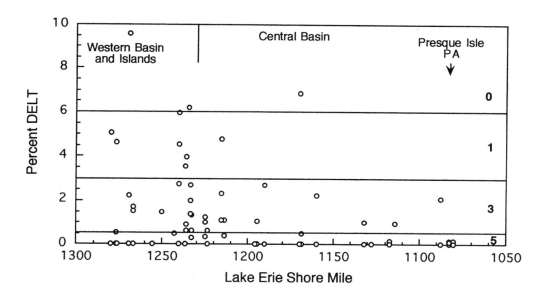

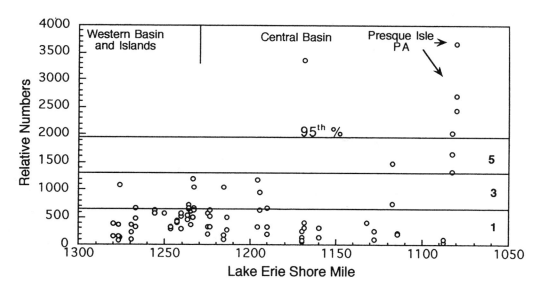

FIGURE 16.12 Lake Erie shoreline IBI metric scores plotted by shore mile (measured from east to west). Upper: percent DELT anomalies; Lower: relative number of individuals. Graph of numbers of individuals displays the 95th percentile line and the 5, 3, and 1 scoring ranges. An absence of individuals scores zero in the relative numbers metric. The scoring rationale for DELT anomalies follows existing Ohio EPA procedures for large rivers with the addition of a zero scoring category (DELT > 6%).

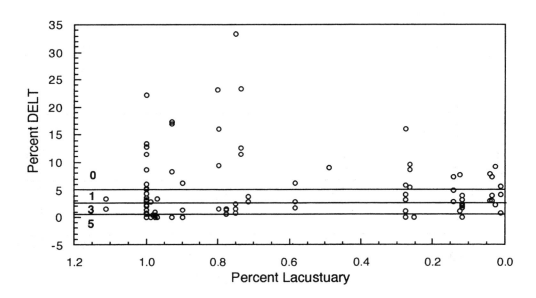

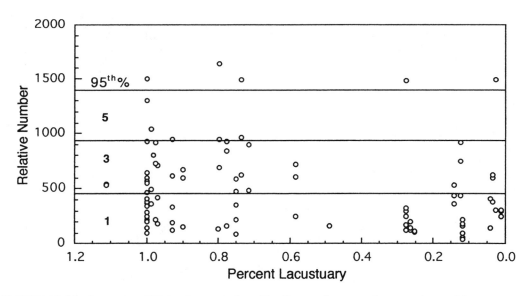

FIGURE 16.13 Lacustuary IBI metric scores plotted by distance from mouth as a measure of the percentage of the total lacustuary length. Upper: percent DELT anomalies; Lower: relative number of individuals. Graph of numbers of individuals displays the 95th percentile line and the 5, 3, and 1 scoring ranges. An absence of individuals scores zero in the relative numbers metric. The scoring rationale for DELT anomalies follows Ohio EPA procedures for large rivers, with the addition of a zero scoring category (DELT > 6%).

TABLE 16.7
Common and Scientific Names of Fish Species Recorded in Ohio and Considered in This Study to Be Nonindigenous to Lake Erie Waters

Common Name	Scientific Name
Ohio lamprey	*Ichthyomyzon bdellium*
Mountain brook lamprey	*Ichthyomyzon greeleyi*
Sea lamprey	*Petromyzon marinus*
Least brook lamprey	*Lampetra aepyptera*
Shovelnose sturgeon	*Scaphiryncus platorynchus*
Alligator gar	*Lepisosteus spatula*
Shortnose gar	*Lepisosteus platostomus*
Goldeye	*Hiodon alosoides*
Skipjack herring	*Alosa chrysochloris*
Alewife	*Alosa pseudoharengus*
Gizzard shad	*Dorosoma cepedianum*
Threadfin shad	*Dorosoma petenense*
Brown trout	*Salmo trutta*
Rainbow trout	*Oncorhynchus mykiss*
Coho salmon	*Oncorhynchus kisutch*
Chinook salmon	*Oncorhynchus tshawytscha*
Rainbow smelt	*Osmerus mordax*
Chain pickerel	*Esox niger*
Blue sucker	*Cycleptus elongatus*
Bigmouth buffalo	*Ictiobus cyprinellus*
Smallmouth buffalo	*Ictiobus bubalus*
Black buffalo	*Ictiobus niger*
River carpsucker	*Carpiodes carpio carpio*
Highfin carpsucker	*Carpiodes velifer*
Common carp	*Cyprinus carpio*
Goldfish	*Carassius auratus*
Speckled chub	*Macrhybopsis aestivalis*
Tonguetied minnow	*Exoglossum laurae*
Rosyside dace	*Clinostomus funduloides*
Rosefin shiner	*Lythrurus ardens*
River shiner	*Notropis blennius*
Steelcolor shiner	*Cyprinella whipplei*
Ghost shiner	*Notropis buchanani*
Mississippi silvery minnow	*Hybognathus nuchalis*
Bullhead minnow	*Pimephales vigilax*
Popeye shiner	*Notropis ariommus*
Grass carp	*Ctenopharyngodon idella*
Red shiner	*Cyprinella lutrensis*
Channel shiner	*Notropis wickliffi*
Blue catfish	*Ictalurus furcatus*
White catfish	*Ictalurus catus*
Mountain madtom	*Noturus eleutherus*
Scioto madtom	*Noturus trautmani*
American eel	*Anguilla rostrata*
Eastern banded killifish	*Fundulus diaphanus diaphanus*
Western mosquitofish	*Gambusia affinis*
Striped bass	*Morone saxatalis*

TABLE 16.7 *(continued)*
Common and Scientific Names of Fish Species Recorded in Ohio and Considered in This Study to Be Nonindigenous to Lake Erie Waters

Common Name	Scientific Name
White perch	*Morone americana*
Spotted bass	*Micropterus punctulatus*
Orangespotted sunfish	*Lepomis humilis*
Redear sunfish	*Lepomis microlophus*
Dusky darter	*Percina sciera sciera*
Longhead darter	*Percina macrocephala*
Slenderhead darter	*Percina phoxocephala*
Crystal darter	*Ammocrypta asprella*
Banded darter	*Etheostoma zonale*
Variegate darter	*Etheostoma variatum*
Spotted darter	*Etheostoma maculatum*
Bluebreast darter	*Etheostoma camurum*
Tippecanoe darter	*Etheostoma tippecanoe*
Whitetail shiner	*Notropis galactura*
Three-spine stickleback	*Gasterosteus aculeatus*
Round goby	*Neogobius melanostomus*

Note: All species not previously recorded in Ohio by Trautman (1981) are considered nonindigenous.

of Lake Erie. The intent of the IBI is to measure integrity, and Lake Erie presently exhibits very little integrity. A score of exceptional would be construed as an indication that Lake Erie is approaching full recovery — which it is not.

The approach employed in this IBI effort has been to use a modification of the 95th percentile methodology. When drawing the 95th percentile line, the line was always drawn between the 95% value and the next value point. This acknowledges the fact that if greater integrity existed, the 95% value would be more stringent while keeping scoring criteria at a level that allows discrimination of the present conditions. Using this methodology, none of the sites sampled in this study have scored 50 or higher.

Karr (in press) proposes the use of ecological dose-response curves to devise scoring criteria for IBI metrics. Such an approach may prove to be the best methodology to score Lake Erie's fish communities because of the extensive disturbances experienced and the lack of reference conditions. Future work on the Lake Erie and lacustuary IBI will examine ecological dose-response curves.

16.5.2 Integrity Classifications

Integrity ranges of exceptional (>50), good (>42), fair (>31), poor (>17), and very poor (≤17) have been set for Lake Erie and its lacustuaries (Figures 16.14 and 16.15, Table 16.8). The predicament of setting specific integrity ranges for Lake Erie and its lacustuaries is difficult because all sites sampled have been affected to some degree by dramatic ecological changes. One approach employed by the Ohio EPA has been to use the IBI value that occurs at the 25th percentile of the reference sites selected as representative of a habitat type as the level at which the "good" classification begins. It is incumbent in the 25th percentile approach that the reference site database be sites that very nearly approach biological integrity. In the Huron-Erie Lake Plain ecoregion (Omernik, 1987), where most sites have been impacted and do not display ecological integrity, the Ohio EPA elected

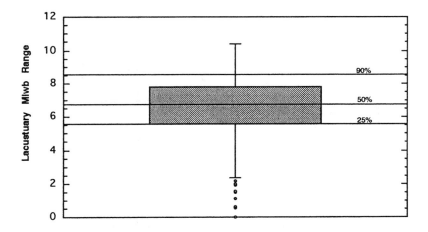

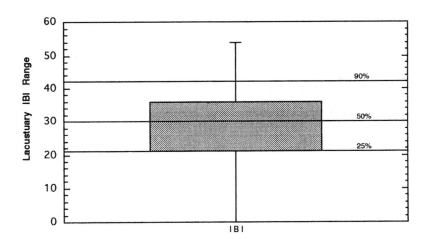

FIGURE 16.14 Ranges of IBI and MIwb scores for Lake Erie lacustuaries giving median, 75th percentile, and 25 percentile values. Outliers are points greater than 1.5× the distance of the interquartile distance. Horizontal lines give the 25th, 50th, and 90th percentiles.

to use the 90th percentile of all sites sampled to derive attainment criteria. Because the Lake Erie system displays pervasive negative environmental effects, an approach like the HELP ecoregion strategy is desirable. This work differs from the previous HELP effort by using only the least impacted sites to set the 90th percentile instead of all sites. Use of a 25th percentile in Lake Erie waters would result in criteria that accepts environmental degradation, while the 90th percentile of least impacted sets a goal that the data have demonstrated can be attained in a reasonable time frame with some environmental amelioration (even in light of pervasive impacts). Once the good attainment point was set, exceptional, fair, poor, and very poor integrity ranges were set based on this author's understanding of species composition at differing IBI levels.

The potential for this scoring system to change is great, as Lake Erie is currently in a state of dynamic flux. New nonindigenous species are invading at increasing rates (Mills et al., 1993),

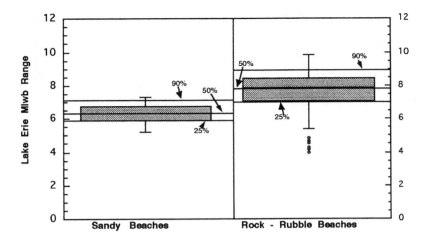

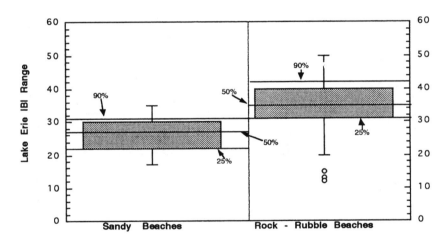

FIGURE 16.15 Ranges of IBI and MIwb scores for Lake Erie giving median, 75th percentile, and 25th percentile values. Outliers are points greater than 1.5× the distance of the interquartile distance. Horizontal lines give the 25th, 50th, and 90th percentiles.

TABLE 16.8
IBI Score Percentiles for Lake Erie Nearshore Areas

Habitat Type	50th Percentile		90th Percentile	
	IBI	MIwb[a]	IBI	MIwb
Lacustuary	39	8.0	42	8.6
Lakeshore (rubble)	36	7.7	42	8.9
Lakeshore (sand)	27	6.3	31	7.2

[a] Modified Index of well-being (follows Ohio EPA, 1988).

phosphorus levels are decreasing (Bertram, 1993; Makarewicz and Bertram, 1991), and the two are interacting in unpredictable ways to create considerable uncertainty. Continued monitoring will be required to track changing community conditions, and attainment criteria will need to be reviewed in light of future changes.

16.6 APPLICATIONS

16.6.1 LACUSTUARY ASSESSMENTS

16.6.1.0 Ottawa River

Index of Biotic Integrity scores (Figure 16.16) obtained from four years of biological monitoring in the Ottawa River have consistently remained in the poor to very poor range. Numerous combined sewer overflows, urban runoff, leaking landfills, and contaminated sediments combine to suppress communities to extremely low levels. Over the 10-year period of monitoring, none of the impact sources have been addressed and, consequently, no changes are detectable in fish communities. Restoration potential for this lacustuary is good as depths are still shallow enough to allow reestablishment of aquatic macrophyte communities, a factor critical to fish community integrity.

16.6.1.1 Maumee River

Eighteen sites throughout the 15-mile (24-km) long Maumee River lacustuary were sampled by the Ohio EPA in 1986 and 1993 (Figure 16.17). Potential sources of impact on the lacustuary include agriculture-related nutrient enrichment, sedimentation, and turbidity; municipal and industrial point source discharges; leaking landfills; habitat alterations; combined sewer overflows (CSOs), and polluted stormwater runoff from the surrounding metropolitan Toledo area. Most of the shoreline has been modified to some extent; sheet piling or riprap are commonly used for bank stabilization. A navigation channel is maintained in the lower 7.2 miles (11.6 km) of the Maumee River, but shallow-water areas are present along the shorelines. IBI scores for fish communities were consistently poor to very poor over the seven-year period. No change has occurred in the previously listed environmental insults. One site achieved a rank of fair in 1986. This site was located outside the main river channel in a sheltered backwater with aquatic macrophyte colonies (primarily *Myriophyllum* with some *Potamogeton*).

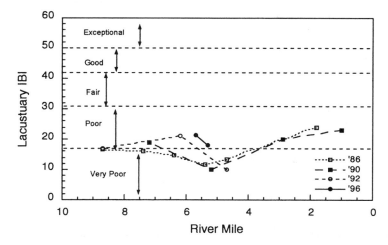

FIGURE 16.16 Ottawa River IBI scores for 1986, 1990, 1992, and 1996 with exceptional, good, fair, poor, and very poor classifications delimited by dashed lines.

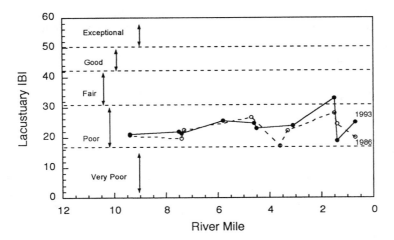

FIGURE 16.17 Maumee River IBI scores for 1986 and 1993 with exceptional, good, fair, poor, and very poor classifications delimited by dashed lines.

16.6.1.2 Duck Creek

Duck Creek is a small tributary to Maumee Bay with a heavily industrialized watershed. Sampling was conducted in the lacusturine portion of the stream in 1986. Principal identified sources of impact include lime slurry from the Toledo water treatment plant and runoff from adjacent railroad yards. Both sites had poor fish communities (Figure 16.18).

16.6.1.3 Otter Creek

Otter Creek is a small tributary to Maumee Bay with a heavily industrialized watershed. Sampling was conducted in the lacustuary at two sites in 1986 and at one site in 1993. Poor to very poor conditions were found in both years (Figure 16.19). Only the most tolerant of fish were collected. Communities were dominated by tolerant, omnivorous, and nonindigenous species.

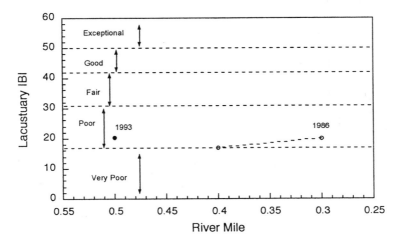

FIGURE 16.18 Duck Creek IBI scores for 1986 and 1993 with exceptional, good, fair, poor, and very poor classifications delimited by dashed lines.

Biological Monitoring and an Index of Biotic Integrity for Lake Erie's Nearshore Waters

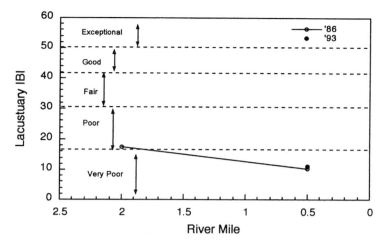

FIGURE 16.19 Otter Creek IBI scores for 1986 and 1993 with exceptional, good, fair, poor, and very poor classifications delimited by dashed lines.

16.6.1.4 Swan Creek

Swan Creek is an urbanized stream that flows through Toledo, Ohio, and is a tributary to the Maumee River. The primary sources of degradation in the lacusturine portion of the creek are urban runoff and CSOs. Fish communities were very poor in 1986 and poor in 1993 (Figure 16.20). Improvements were the result of decreasing CSO problems. The lacustuary continues to be plagued by urban-associated impacts.

16.6.1.5 Turtle Creek

The Turtle Creek lacustuary is 4.2 miles (6.8 km) long. The Ohio EPA staff sampled three locations in 1995. The upper site at river mile (RM) 3.0 was an extensive shallow mudflat. The bottom was comprised of a thick layer of silt and sediment that was easily disturbed and resulted in a continuously turbid water column. The fish community had low diversity, high numbers of tolerant species,

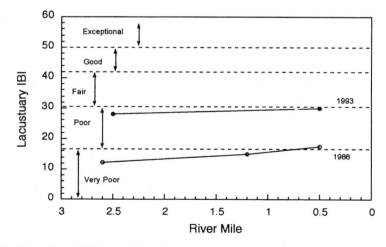

FIGURE 16.20 Swan Creek IBI scores for 1986 and 1993 with exceptional, good, fair, poor, and very poor classifications delimited by dashed lines.

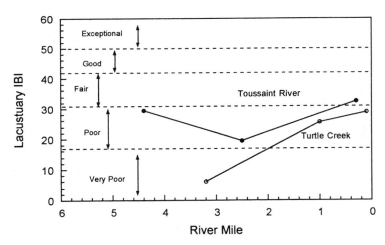

FIGURE 16.21 Toussaint River and Turtle Creek IBI scores for 1995 with exceptional, good, fair, poor, and very poor classifications delimited by dashed lines.

and was rated very poor to poor, indicating that nutrient enrichment was severely impacting the fish community (Figure 16.21). Better community conditions existed near the mouth of the creek (RMs, 0.3 = poor and 0.2 = fair) despite continued sediment impact. The higher IBI value near the mouth of Turtle Creek seems to be due to the mixing of river water with less turbid and less nutrient-rich Lake Erie water. Marinas are located at the mouth of the river in side channel areas, and there is considerable boat traffic in the lower 0.5 miles. The primary impact to Turtle Creek is from sediment and nutrient enrichment derived from nonpoint agricultural activities.

16.6.1.6 Toussaint River

The lacusturine portion of the Toussaint River encompasses the lower 7.9 miles (12.7 km) of the waterway. Ohio EPA staff sampled at RM 3.4, 1.4, and 0.3 in 1995 (Figure 16.21). All sampling locations were impacted by silt and sediment. The upper three-fourths of the lacustuary was extensive shallow mudflats (historic wetlands). The habitat presently comprises a thick layer of silt and sediment (as opposed to aquatic vegetation) that was easily disturbed and resulted in habitually turbid waters. The fish community had low diversity and high numbers of tolerant and nonindigenous species, indicating that nutrient enrichment was severely impacting the community (as they were in Turtle Creek). Habitat conditions in Toussaint River were similar to those found in Turtle Creek, except that marinas were less abundant, extended farther upstream, and located on the banks of the lacustuary. Impacts to the system were derived from nonpoint agricultural pollution, marinas, and boat traffic in declining order of importance.

16.6.1.7 Portage River

The lacustuary of the Portage River extends upstream to approximately RM 16.5 (26.5 km), but varies depending on Lake Erie water levels. The fish communities in this lacustuary are the best to be found among the western tributaries. Scores, for the most part, are marginally good (Figure 16.22). One site, located in the vicinity of a tributary heavily impacted by agricultural activities, drops into the poor category. Sites with higher IBI scores had higher abundances of top carnivores and species numbers. Abundances of spotted suckers, a species reported by Trautman to be sensitive to turbidity, were higher in this lacustuary than in any other area sampled in Ohio by the Ohio EPA.

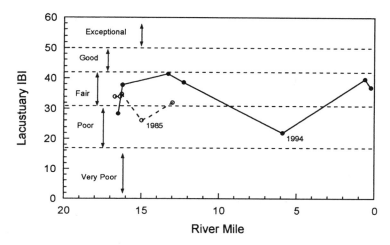

FIGURE 16.22 Portage River IBI scores for 1985 and 1994 with exceptional, good, fair, poor, and very poor classifications delimited by dashed lines.

16.6.1.8 Sandusky River

The Sandusky River was sampled at eight sites in 1988. Upstream portions of the lacustuary were intermittent in 1988 due to drought conditions and water removal by the city of Fremont. Some organic enrichment was contributed to the system near RM 14 by the Fremont wastewater treatment plant and a small tributary carrying waste from a food-processing facility. Agricultural pollutants in the form of nutrients, sediment, and silts were also affecting the lacustuary. Poor to fair IBIs existed throughout the lacustuary as a result of the numerous impacts (Figure 16.23). Habitat potential was good, as very little shoreline had been permanently modified and the channel was not dredged. A small section of the river between a lowhead dam on the lotic portion and the start of the lacustuary had been channelized (effectively eliminating most spawning habitat for upstream migrant fish). Much of the lacustuary that was historically wetland was open mudflat, providing very little usable habitat for aquatic organisms. If environmental impacts were reduced to the point

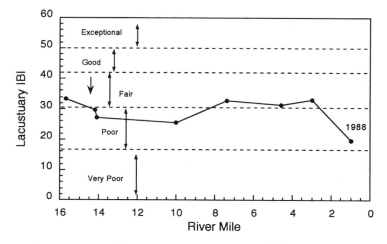

FIGURE 16.23 Sandusky River IBI scores for 1988 with exceptional, good, fair, poor, and very poor classifications delimited by dashed lines.

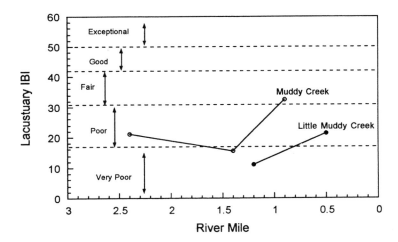

FIGURE 16.24 Muddy Creek and Little Muddy Creek IBI scores for 1988 with exceptional, good, fair, poor, and very poor classifications delimited by dashed lines.

that aquatic plant communities could reestablish, the fish community would return to its historically exceptional condition in which northern pike and muskellunge were abundant.

16.6.1.9 Muddy and Little Muddy Creeks

Muddy and Little Muddy Creeks are two of a number of Sandusky Bay tributaries that are negatively affected by pervasive mudflat habitat resulting from nutrient and sediment loads originating from intensively farmed watersheds in the HELP ecoregion. Sampling of these two streams was conducted by Ohio EPA staff in 1988. A thick layer of silt and sediment that was easily disturbed was present and kept the water column continually turbid. Lacustuary IBI scores were in the very poor and poor range (Figure 16.24). One site in Muddy Creek was located along a riprapped dike and attained a fair ranking. Fish communities displayed low diversity, with high abundances of tolerant species, omnivores, and nonindigenous species. This result was typical of western Lake Erie lacustuarine areas overwhelmed by sedimentation and nutrient enrichment.

16.6.1.10 Huron River

Fourteen sites in the Huron River lacustuary were sampled by the Ohio EPA from 1982 to 1993. All sites were distributed along the 10-mile (16-km) length of the lacustuary, downstream from the city of Milan, Ohio. Fish community conditions consistently ranged from fair to poor over the length of the river and the sampling interval (Figure 16.25). Little environmental perturbation of the lacustuary is derived from the lotic portions of the Huron River. Considerable disturbance originates from boat traffic traveling the lacustuary at high speeds, creating wave action, stirring of the bottom, and resuspension of bottom sediments. These conditions have remained consistent throughout the study period. Resuspension of sediments has led to the destruction of Huron River wetland communities, which has in turn diminished the quality of the lacustuarine habitat and its ability to process nutrients and sediment. The fish community reflects these disturbances in its structure. Common carp are abundant in the lower reaches, phytophilic species are reduced, historically abundant populations of northern pike and muskellunge are uncommon or absent, respectively, and nonindigenous species (i.e., common carp, smallmouth and bigmouth buffalo, ghost shiners, gizzard shad, and white perch) are abundant.

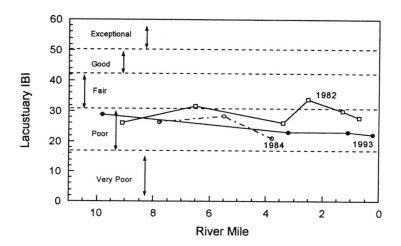

FIGURE 16.25 Huron River IBI scores for 1982, 1984, and 1993 with exceptional, good, fair, poor, and very poor classifications delimited by dashed lines.

16.6.1.11 Old Woman Creek

Old Woman Creek was one of a number of lacustuaries where the severity of impact of sedimentation (from a highway construction project) cannot be overstated. The Old Woman Creek lacustuary was wide and shallow. The bottom was composed of a thick layer of silt and sediment that is easily disturbed and keeps the water column turbid. Aquatic vegetation was largely limited to emergent varieties (*Nelumbo lutia*) as a consequence of water column turbidity. Prior to the construction project, the Old Woman Creek lacustuary was dominated by submerged aquatic vegetation, mostly of the genus *Potamogeton*. Predictably, present IBI scores were in the poor range at the two sampling locations (Figure 16.26). Fish communities were characterized by pollution-tolerant taxa, nonindigenous species, and low diversity. Historical fish communities contained significant populations of northern pike, largemouth bass, other sunfish, and native minnows.

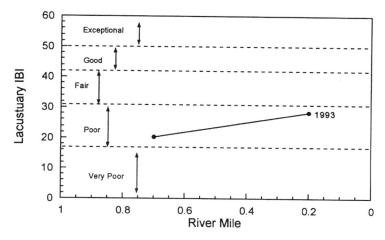

FIGURE 16.26 Old Woman Creek IBI scores for 1993 with exceptional, good, fair, poor, and very poor classifications delimited by dashed lines.

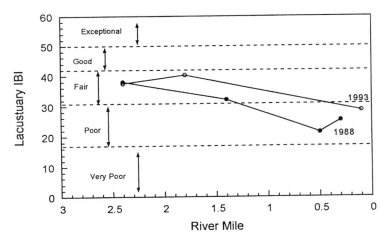

FIGURE 16.27 Vermilion River IBI scores for 1988 and 1993 with exceptional, good, fair, poor, and very poor classifications delimited by dashed lines.

16.6.1.12 Vermillion River

Vermilion River lacustuarine fish communities from unmodified habitats (sites from above RM 1.0) are among the best in the eastern lacustuarine group, scoring marginally good (Figure 16.27). Downstream, the lacustuary is lined by boat docks and marinas throughout its length. Communities decline to the poor range in response to the pervasive habitat modifications. Upper reaches with undisturbed habitat harbored such species as bigeye chub and black redhorse, two species found by the Ohio EPA to be sensitive to environmental disturbances. Downstream disturbed sites were characterized by common carp and white perch, species highly tolerant of environmental disturbance.

16.6.1.13 Black River

The Black River lacustuary was sampled by the Ohio EPA in 1992 and 1994. Scores for the IBI were consistently poor in 1982, and poor to fair in 1992 (Figure 16.28). East bank 1992 samples

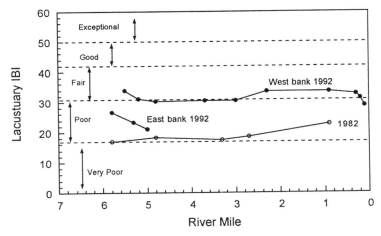

FIGURE 16.28 Black River IBI scores for 1982 and 1992 with exceptional, good, fair, poor, and very poor classifications delimited by dashed lines.

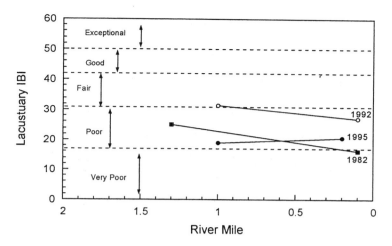

FIGURE 16.29 Rocky River IBI scores for 1982, 1992, and 1995 with exceptional, good, fair, poor, and very poor classifications delimited by dashed lines.

were located adjacent to a large, active slag pile. Community improvements over the 10-year period were due to upgrades at the upstream Elyria wastewater treatment plant that reduced loading to the Black River and its lacustuary. Removal of contaminated sediments after 1992 probably will lead to further fish community improvements. Presently, the lacustuary is limited by nutrient enrichment, primarily from upstream nonpoint pollution — both urban and rural. Very little submerged aquatic vegetation exists in the lacustuary, although habitat structure is suitable. With further reductions of pollutant loads and a resurgence of plant life, fish communities in the Black River lacustuary should recover and attain exceptional conditions.

16.6.1.14 Rocky River

The Rocky River lacustuary was sampled in 1982, 1992, and 1995. The river is in the Erie Ontario Lake Plain and is a predominately urbanized watershed. In 1982, IBI scores were poor (Figure 16.29). By 1992, after the closure of numerous wastewater treatment plants in the basin, IBI scores increased to marginally fair. Between 1992 and 1995, additional marinas were built in the lacustuary and IBI scores declined in response. Overall, impacts to the lacustuary originated from urban runoff, wastewater treatment discharges, and habitat loss.

16.6.1.15 Cuyahoga River

This lacustuary has continued to show improvements in its fish communities over the past 12 years as pollutant loadings have continued to be reduced. In 1984, fish communities were in the very poor range and limited by chemical and organic pollution loadings (Figure 16.30). They have now improved to the poor range and are presently limited by a lack of habitat. Most of the Cuyahoga River lacustuary has been dredged to 7.6 m to accommodate lake freighter traffic, and most of the shores are lined by seawalls constructed of steel sheet piling or concrete. One site (RM 4.0) sampled in 1996 had a small stretch of shoreline between two railroad trestles that lacked seawalls and had instead a sloping shoreline with logs, woody debris, and broken concrete slabs that afforded improved habitat conditions in which smallmouth bass, rock bass, and pumpkinseed sunfish were found — and the IBI attained near fair classification. With improved habitat conditions (i.e., greater structural diversity and substrate texture), the Cuyahoga River lacustuary could possibly improve to the good range.

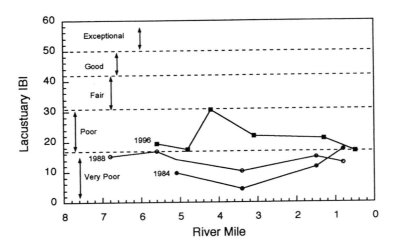

FIGURE 16.30 Cuyahoga River IBI scores for 1984, 1988, and 1996 with exceptional, good, fair, poor, and very poor classifications delimited by dashed lines.

16.6.1.16 Chagrin River

The Chagrin River lacustuary was sampled in 1988, 1993, and 1994. The stream is subject to increasing urbanization throughout the watershed. Most of the lacustuary has been converted to marinas and boat docks, especially in the side channels. IBI scores ranged from poor to fair (Figure 16.31). Three of the highest IBI scores came from sites with undisturbed shoreline habitat, while the fourth (near the mouth) was part riprap and part natural shoreline with no boat docks. All other sites were modified in some way to accommodate boat docking. The Chagrin River is not impacted immediately upstream from the lacustuary, and the only source of impact identifiable in the lacustuary is associated with boating activities.

16.6.1.17 Grand River

A total of 14 sites in the 4.6-mile (7.4-km) Grand River lacustuary have been sampled by the Ohio EPA since 1987. Potential sources of impact on the biota of the waterway included habitat alterations, and leaking chemical waste lagoons and landfills adjacent to the Grand River from RM 4.8 to 3.1 and the Painesville wastewater treatment plant discharge at RM 3.0. Boat docks were located along one or both shorelines from approximately RM 2.2 to 1.0. The lower mile was a ship channel with deep sheet piling along the banks.

Fish community IBIs ranged from good to poor (Figure 16.32). Good scores occurred in 1994 in the upper reaches of the lacustuary. Poor scores were recorded for sites in the ship channel area and near the Painesville wastewater treatment plant discharge. Variation in IBI scores between years appears to be a consequence of variable stream flows and lake levels, with higher flows providing greater dilution and lower lake levels increasing flushing rates, thus presumably reducing concentrations of contaminants and other pollutants.

16.6.1.18 Ashtabula River

Seven sites have been sampled by the Ohio EPA since 1989 in the 2.5-mile (4-km) length of the Ashtabula River lacustuary. Downstream RM 2.3, much of the waterway was lined with sheet piling and boat docks. A ship channel extended from the river mouth to RM 0.7. Fields Brook joins the Ashtabula River at RM 1.6. Sediment contamination has been documented downstream Fields Brook. In 1989, fish community sampling was conducted to evaluate the degree of impact associated

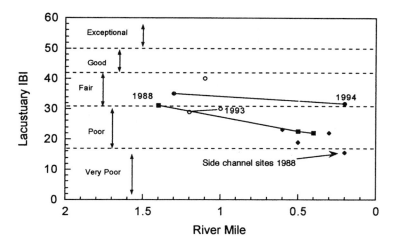

FIGURE 16.31 Chagrin River IBI scores for 1988, 1993, and 1994 with exceptional, good, fair, poor, and very poor classifications delimited by dashed lines. Nonconnected open circle was from the opposite shore in 1993.

with chemical degradation originating from Fields Brook and habitat alteration of the lacustuary. It was concluded that both shoreline development and chemical pollutants were impacting fish communities in the lower Ashtabula River. In general, IBIs were good to fair in upper reaches, fair to poor near Fields Brook, and fair to poor in the ship channel area (Figure 16.33).

16.6.1.19 Conneaut Creek

The Conneaut Creek lacustuary extends for 2.2 miles (3.5 km) upstream from the mouth. A total of five sites were sampled by the Ohio EPA since 1989. Very little environmental deterioration was seen in the lotic portions of the system; extensive areas of the basin are wooded. The lower 0.5 mile of the stream was a ship channel with deep sheet piling-lined banks; while upstream from RM 0.5, the channel was shallower and at least partially vegetated along the banks; most of this

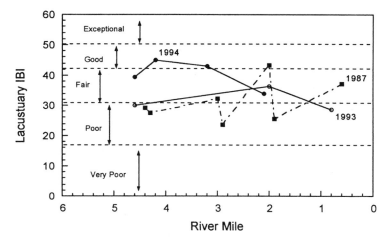

FIGURE 16.32 Grand River IBI scores for 1987, 1993, and 1994 with exceptional, good, fair, poor, and very poor classifications delimited by dashed lines.

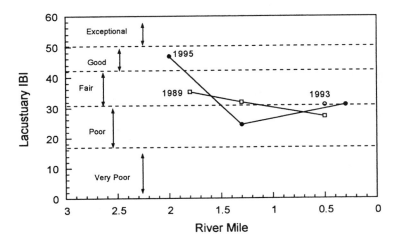

FIGURE 16.33 Ashtabula River IBI scores for 1989, 1993, and 1995 with exceptional, good, fair, poor, and very poor classifications delimited by dashed lines.

reach was relatively narrow with moderate accumulations of silt and sediment. An area of thick silt and sediment with a large expanse of emergent and submergent vegetation was present at RM 1.0. Upstream of the ship channel, in the area of vegetation, IBI scores were in the marginally good range, while ship channel sites (RM 1.3 and 0.6) had IBI scores in the poor range (Figure 16.34). The dichotomy of near good and poor community conditions found in this lacustuary illustrate the strong effect that habitat alterations can have on biological conditions, even in areas where no impacts from water column chemistry exist.

16.6.2 LAKE ERIE

In Lake Erie, three factors affect fish community structure: lake-wide trophic changes as a result of nutrient enrichment, habitat loss primarily in the form of wetland destruction or diking and shoreline modifications, and localized environmental impacts from industrial and municipal discharges. Of principal significance is the predominant effect of lake-wide trophic changes and

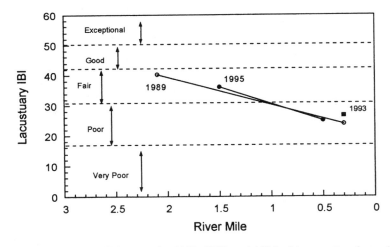

FIGURE 16.34 Conneaut Creek IBI scores for 1989, 1993, and 1995 with exceptional, good, fair, poor, and very poor classifications delimited by dashed lines.

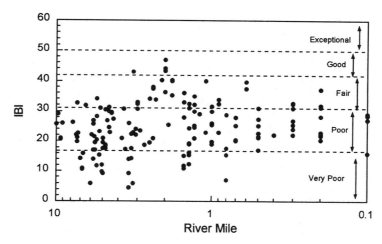

FIGURE 16.35 IBI scores for all lacustuary sites by river mile as measured from the river mouth (0) to the head waters. Exceptional, good, fair, poor, and very poor categories delimited by dashed lines.

associated species losses. These changes have resulted in most sites scoring as fair, with few good and no exceptional values attained (Figure 16.36). Four of the nine sites that clearly fall into the good range are from the shorelines of the Lake Erie Islands. Island sites score better, in part, due to their distance from lacustuaries and associated impacts. Habitat was also an important factor for island sites. The principal habitat type encountered around the islands was boulder rubble-strewn shorelines with high levels of substrate texture. It was observed in this study that the greater the habitat texture, the greater the relative abundance and number of species. Breakwater sites, at the mouths of lacustuaries, had habitat textures similar to island sites, but failed to reach the levels attained at island sites. This was due to lacustuaries experiencing environmental stress from higher loads of pollutants. Beaches were the area of lowest substrate texture and tended to score lower than other habitat types (in the absence of other environmental stresses). Examples of localized pollution impacts were found in the Maumee Bay and Cuyahoga River areas where, despite the fact that habitats were highly textured breakwaters, IBI values remained in the poor range. The only lake site in this study that fell in the very poor classification was just east of the Maumee Bay area. This site was a riprapped beach in an area where extensive settling of organic debris and urban waste was occurring. The dominant species at this site was goldfish, a highly tolerant fish.

16.6.3 GENERAL

None of the lake or lacustuary sites in this study attained an integrity level of exceptional and only a few attained the good level (Figures 16.35 and 16.36). This was reflective of the widespread and pervasive nature of environmental impacts in the region. Many species were missing (Trautman, 1981; Hartman, 1972) and trophic dynamics were radically changed (Regier and Hartman, 1973; Stoermer et al., 1987). Table 16.9 gives the 20 most commonly encountered species from all sites, lake only sites, and lacustuary only sites; 5 of the 20 most abundant species were nonindigenous species; 93 species were recorded; and the average relative abundance of individuals (number per kilometer) was 687.

At the good–fair integrity interface, similarities between Lake Erie and its lacustuaries begin to diverge. In the lake proper, environmental impacts are more widely dispersed and less intense; whereas, in the lacustuaries, they can be very intense and are always more concentrated and localized. In the lake only, 73 species were recorded and the average relative number of individuals (number per kilometer) was 934. Integrity levels of fair dominated the lake results (59%), poor to

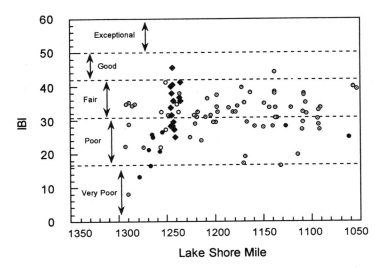

FIGURE 16.36 IBI scores for all Lake Erie sites including Sandusky Bay. Open circles = Sandusky Bay; solid diamonds = Bass islands area; open circles with dot = miscellaneous shore types (mostly rocky); and black dots = sandy beaches. Lakeshore miles are measured from east to west. Exceptional, good, fair, poor, and very poor categories delimited by dashed lines.

TABLE 16.9
The 20 Most Commonly Encountered Species for the Overall Study

All Sites	Lacustuaries	Lake Erie
Gizzard shad	**Gizzard shad**	**Gizzard shad**
White perch	Emerald shiner	**White perch**
Emerald shiner	Pumpkinseed	Emerald shiner
Pumpkinseed	**White perch**	Yellow perch
Common carp	**Common carp**	Rock bass
Bluegill	Brown bullhead	Spottail shiner
Freshwater drum	Bluegill	White bass
Spottail shiner	Freshwater drum	Logperch
Brown bullhead	Bluntnose minnow	**Common carp**
Rock bass	Green sunfish	Smallmouth bass
Yellow perch	**Goldfish**	Pumpkinseed
White bass	Largemouth bass	**Alewife**
Smallmouth bass	Spottail shiner	Bluegill
Largemouth bass	Smallmouth bass	Freshwater drum
Logperch	White crappie	Largemouth bass
Bluntnose minnow	White bass	Spotfin shiner
Green sunfish	Rock bass	Mimic shiner
Alewife	Black crappie	Brook silverside
Goldfish	**Orangespotted sunfish**	Brown bullhead
Spotfin shiner	Spotfin shiner	**Round goby**

Note: Lacustuary and lake areas are in order of decreasing abundance. Common names in boldface type are nonindigenous species.

very poor (24%) comprised the next largest classification, and good (17%) the least. In the lacustuaries, 87 species were recorded, and the average relative number of individuals (number per kilometer) was 552. Poor to very poor IBI scores dominated the results (71%), while fair comprised 23%, and good equaled only 6%.

ACKNOWLEDGMENTS

I would like to thank Ed Rankin, Chris Yoder, Marc Smith, Jeff DeShon, and Chuck McKnight for technical and spiritual support. Their reviews of the project and philosophical discussions were invaluable. Kelley Capuzzie, Steve White, Mat Raffinburg, Julie Sulivan, Gene Kim, Cory Timko, and Chris Winslow worked for me as college interns, helping with field collections and preparation. Julie Letterhos and Ava Hottman were instrumental in providing financial and administrative support. Monies for this project were provided by the Ohio EPA and came from state and federal sources. David Jude and Thomas Edsall provided critical reviews of the manuscript.

I would like to dedicate this chapter to two Ohio State University professors: Dr. Ronald L. Stuckey, who taught me about wetlands, coontail, and Lake Erie's nearshore plant communities; and the late Raymond F. Jezerinac, my mentor, who "learnt" me about fish, aquatic ecology, and the art of field work.

REFERENCES

Angermeier, P.L. and I.J. Schlosser. 1981. Assessing biotic integrity of the fish community in a small Illinois stream, *North American Journal of Fisheries Management,* 7, 331–338.
Bertram, P.E. 1993. Total phosphorous and dissolved oxygen trends in the Central Basin of Lake Erie, 1970–1991, *Journal of Great Lakes Research,* 19, 224–236.
Brant, R.A. and C.E. Herdendorf. 1972. Delineation of Great Lakes estuaries, *Proceedings 15th Conference, International Association for Great Lakes Research,* 15, 710–718.
Hartman, W.L. 1972. Lake Erie: effects of exploitation, environmental changes and new species on the fishery resource, *Journal of the Fisheries Research Board of Canada,* 29, 899–912.
Karr, J.R. 1981. Assessment of biotic integrity using fish communities, *Fisheries,* 6(6), 21–27.
Karr, J.R. (in press.) *Rivers as Sentinels: Using the Biology of Rivers to Guide Landscape Management.*
Makarewicz, J.C. and P. Bertram. 1991. Evidence for the restoration of the Lake Erie ecosystem, *BioScience,* 41(4).
Mills, E.L., J.H. Leach, J.T. Carlton, and C.L. Secor. 1993. Exoic species in the Great Lakes: a history of biotic crises and anthropogenic introductions, *Journal of Great Lakes Research,* 19, 1–54.
Minns, C.K., V.W. Cairns, R.G. Randall, and J.E. Moore. 1994. An Index of Biotic Integrity (IBI) for fish assemblages in the littoral zone of Great Lakes' areas of concern, *Canadian Jorunal of Fisheries and Aquatic Sciences,* 51, 1804–1822.
Ohio EPA (Environmental Protection Agency). 1988. *Biological Criteria for the Protection of Aquatic Life. Vol. I–III.* Ecological Assessment Section, Division of Water Quality Monitoring and Assessment, Ohio EPA, Columbus.
Omernik, J.M. 1987. Ecoregions of the conterminous United States, *Annals of the Association of American Geographers,* 77, 118–125.
Regier, H.A. and W.L. Hartman. 1973. Lake Erie's fish community: 150 years of cultural stress, *Science,* 180, 1248–1255.
Stoermer, E.F., J.P. Kociolek, C.L. Schelske, and D.J. Conley. 1987. Qualitative analysis of siliceous microfossils in the sediments of Lake Erie's central basin, *Diatom Research,* 2, 113–134.
Trautman, M.B. 1981. *The Fishes of Ohio.* Ohio State University Press. Columbus, Ohio.
Van Meter, H.D. and M.B. Trautman. 1970. An annotated list of the fishes of Lake Erie and its tributary waters exclusive of the Detroit River, *Ohio Journal of Science,* 70, 65–78.
White, A.M., M.B. Trautman, E.J. Foell, M.P. Kelty, and R. Gaby. 1975. *Water Quality Baseline Assessment for the Cleveland Area-Lake Erie. Vol. 2. The Fishes of the Cleveland Metropolitan Area Including the Lake Erie Shoreline.* U.S. Environmental Protection Agency, Region 5, Chicago.

17 Considerations for Characterizing Midwestern Large-River Habitats

Robin J. Reash

CONTENTS

17.1 Introduction 463
    17.1.1 Biological Assessment of Large Rivers 464
17.2 The Large River Paradigm 464
17.3 Unique Characteristics of Large Rivers 465
    17.3.1 Physical and Chemical Features 465
    17.3.2 Unique Biological Characteristics 466
17.4 Considerations for Fish Metric Development in Large Rivers 467
    17.4.1 Temporal Variability 467
    17.4.2 Spatial Variability and Habitat Effects 469
    17.4.3 Sampling Gear 469
17.5 Assessment of Point-Source Effects in Large Rivers 471
17.6 Conclusion 471
References 471

17.1 INTRODUCTION

Relative to streams and lakes, there has been a lag in the description of spatial and temporal axes of physical, chemical, and biological characteristics of large rivers. Hynes (1989) stated that less than 1% of the literature dealing with the ecology of running waters is concerned with large rivers. Johnson et al. (1995) argued that, historically, large river investigations have been hampered by sampling difficulties and the lack of an operational theoretical model. The historical information gap between large rivers and other aquatic habitats (streams, small rivers, and lakes) is closing, however. A considerable amount of biological and physicochemical information on large rivers has become available since 1989. The multifaceted differences between large rivers and other aquatic habitats must be understood, however, before biological expectations for large rivers are developed by water quality agencies. This chapter argues that the physical and biological characteristics of large rivers are unique to the extent that agencies must evaluate the underlying assumptions of these systems (rather than applying a stream paradigm) before defining narrative or numeric biological expectations. This chapter will focus on midwestern large rivers, operationally defined as river systems having a drainage area of greater than 20,000 sq. km. Prior to this discussion, the historical and fundamental factors of large rivers (within the larger realm) will be reviewed.

The barriers to a comprehensive understanding of large rivers are worth noting. First, many large rivers have a sparse historical record concerning faunal diversity and changes. Often, the scientific baseline is only recent; in many cases, the serious study of large rivers began only after pervasive anthropogenic modifications had occurred (Hynes, 1989). Although the historical record for some large rivers may be meager, the fundamental hydrological and physicochemical alterations that have plagued large rivers, for various societal benefits, have been well documented (Ward and Stanford, 1989; Regier et al., 1989). Dodge (1989) provides a global perspective of long-term faunal changes in large rivers. In many cases, large river faunal changes have been accelerated by hydrological modifications, essentially masking underlying effects of land use changes and water quality impairment.

Despite a clear understanding of what has happened to large rivers from their original condition, the concept of biological expectation (and ecological restoration) is complicated by the reality of hydrological modification. If a large river is converted from a lotic habitat to a lentic habitat by damming, what is the realistic target for development of biological criteria? Should the benchmark for adverse water quality impacts in large rivers be the same for those used in smaller systems? Gore and Shield (1995) conclude that only a portion of the ecological functions and values of large rivers can be rehabilitated because the restoration of large rivers to a pristine state is incompatible with present human population levels. Thus, it is my opinion that any numeric biological criterion developed for large rivers must be calibrated against the current and recent range of biological attributes.

The paucity of long-term studies on a global scale does not mean that some large river systems have been well-characterized both chemically and biologically. In the midwestern United States, there are examples of excellent long-term studies on some large rivers. Some of these include the Wabash River (Gammon, 1994), the Mississippi River (Fremling et al., 1989; Jahn and Anderson, 1986), and the Ohio River (Van Hassel et al., 1988; Pearson and Pearson, 1989). Most of these studies contain a strong fisheries emphasis. Long-term studies serve as important references for understanding historical changes in fauna and water quality, and for providing direct estimates of spatial and temporal variability.

17.1.1 Biological Assessment of Large Rivers

Increasingly, water resource agencies are expanding biological monitoring programs to include large rivers. Historically, aquatic life use attainment in large rivers was assessed using a chemical-specific approach. In some instances, large river mile segments (hundreds of miles) were designated as "not attaining" the aquatic life use because a single pollutant analysis at a single location exceeded an established water quality criterion. This pattern is changing such that biological assessments are being used as direct measures of aquatic resource health in large rivers, pursuant to Karr's (1995) guidelines. As an example, Schregardus (1994) demonstrated that chemical-based "impairment" of the Ohio River aquatic life use was an erroneous conclusion when actual fish community data indicated a balanced, diverse community. Although the movement toward a biological-based regulatory approach for large rivers is timely, there are certain fundamental hypotheses that must be tested before assessments of water quality effects can begin on a wholesale basis. It is my opinion the most important hypothesis is whether, and how much, a small-river paradigm can be applied to a large-river paradigm.

17.2 THE LARGE-RIVER PARADIGM

Although assessment and monitoring of large rivers has already been initiated by some water resource agencies, there is a conspicuous absence of a unifying theoretical model for large rivers. This is not viewed as a serious pragmatic problem, rather the lack of an empirical-based model for a given river should prompt a serious review of the subject. There are numerous examples of large rivers being described on the basis of biotic communities (community structure, species composition) and physical habitat. A model that describes basic functional attributes, however, is important in making certain predictions about how populations and communities respond to perturbations. For example, does an investigator assessing water quality impacts really understand the sources of energy (e.g., allocthonous versus autochthonous) in that particular stretch of river? Sedell et al. (1989) state that the problem of linking large river biotic communities with ecosystem metabolic interactions is due primarily to an inability to cope with complexities associated with spatial and temporal heterogeneity along the length of the river. Because of this, the authors state that ecologists simply cannot extrapolate concepts of riverine ecosystem function (applicable to upper river stretches) to downstream areas. Despite this caution, there are some biological responses observed in small streams or rivers that can be predicted to occur in large rivers (e.g., effects of organic enrichment). In these instances, a large river simply changes the scale of spatial effects.

The few empirical-based models that have been described for large rivers typically refute the application of lotic paradigms, such as the River Continuum Concept (RCC; Vannote et al., 1980). Thorp and Delong (1994) proposed the riverine productivity model (RPM) based on benthic macroinvertebrate trophic guild data for the middle Ohio River. Unlike the downstream transport of fine particulate organic matter embedded in the RCC, the RPM hypothesizes that carbon derived from authochthonous production and localized riparian zone production is the primary energy source for macroinvertebrates in large rivers. This paradigm has important implications for large rivers that are regulated by navigation dams; each navigation pool may be considered a quasi-independent system regarding energy production and processing. Based on the availability of certain structural components (islands, snags, streambank structures), the biological expectations of adjacent navigation pools may be significantly different.

Wehr and Thorp (1997) showed that, contrary to some models that predict a reduced phytoplankton community in large rivers, the Ohio River supports a diverse and substantial phytoplankton community throughout its length. The effects of navigation dams and tributaries on phytoplankton communities in the Ohio River were found to be insignificant.

While water resource agencies might, by necessity, be forced to postpone the academic debates of large river paradigms due to the higher priority of data collection and analysis, it should be noted that the Index of Biotic Integrity (IBI) developed for rivers and streams has strong theoretical assumptions (Fausch et al., 1990). These underlying assumptions drive the practical application of IBI as a tool to assess the health of fish communities and, hence, local water quality. The nine underlying assumptions of IBI are described by the authors, and a general caution is given: further IBI testing should focus on the validity of these assumptions in different settings ... "modification of the IBI for different habitats or different taxa would profit from development of a similar model of how the respective communities change with degradation" (Fausch et al., 1990; p. 135–136).

In summary, a set of theoretical underlying assumptions for the IBI is needed for large rivers, and these should be described before metric analysis is conducted. Gerritsen (1995) reiterates that a metric is defined as an ecological attribute of an assemblage *which is responsive to perturbation or disturbance*. It may be that several of the nine assumptions for rivers and streams (Fausch et al., 1990) are retained, or require only slight modifications for large rivers. Nonetheless, proposed IBI metrics for large rivers must be tested so that actual data provide the justification for an expected shift due solely to water quality degradation. Shields et al. (1995) found that IBI scores did not reflect local physical habitat conditions in the lower Mississippi River unless severe habitat degradation was present. The following section describes the characteristics of large rivers that, taken as a whole, separate these systems from small rivers and streams.

17.3 UNIQUE CHARACTERISTICS OF LARGE RIVERS

17.3.1 Physical and Chemical Features

A comprehensive review of physical and chemical features of large rivers is not within the scope of this chapter. The reader is referred to excellent reviews provided by Stalnaker et al. (1989),

Sedell et al. (1989), and Ryder and Pesendorfer (1989) for detailed information. Some of the more important factors, which provide support for the hypothesis that biological expectations of large rivers are different relative to small rivers and streams, are listed below:

1. Large rivers have distinct habitat zones and microhabitats, the result being that large river species often show a greater tendency toward habitat zonation and the use of specialized habitat types (Stalnaker et al., 1989).
2. In large rivers, the great volume of water has a buffering effect that reduces variation in temperatures and flows (Johnson et al., 1995); the zonation of velocity and temperature on vertical and horizontal axes, however, prohibits the use of a simplistic one-dimensional model to predict temperature and water quality in large rivers.
3. Large rivers, it is thought, receive the majority of their fine particulate organic carbon load from upstream processing of dead leaves and woody debris (Sedell et al., 1989); this general model has been shown to be invalid in some instances (e.g., Thorp and Delong, 1994).
4. River navigation (e.g., barge vessels) causes physical changes to water level and channel morphometry due to turbulence effects (Nielsen et al., 1986).

One of the most conspicuous differences in physical habitat between streams and large rivers is the amount of surface area subject to shading. In streams, liberal shading provides distinct microhabitats stratified by temperature gradients. In large rivers, fish utilize depth as a substitute for riparian cover (Stalnaker et al., 1989).

17.3.2 Unique Biological Characteristics

Biological production, diversity, and density in large rivers are profoundly affected by habitat zonation, much more so than in small rivers and streams. Some general biological characteristics of large rivers are:

1. Backwaters are biologically rich microhabitats; these habitats may represent less than 10% of the total surface area in large rivers yet support up to 90% of the fish biomass (Stalnaker et al., 1989). Larval fish densities may be higher in backwaters compared to main channel zones due to nursery habitats and nutrient enrichment (Holland, 1986; Sheaffer and Nickum, 1986). The success of nest-building fish species such as sunfish is highly dependent on the availability of backwaters for spawning and nursery purposes (Scott and Nielsen, 1989). Tributary mouths may contain biologically rich densities relative to the receiving river (Brown and Coon, 1994).
2. Macroinvertebrate density and diversity are enhanced in specialized, patchy habitats such as mid-river islands (Thorp, 1992). The macroinvertebrate fauna in large rivers shows significant spatial differences among different habitats within the same pool (Jahn and Anderson, 1986).
3. Distinct habitat types are found in large rivers, and these can be delineated on the basis of physical and chemical features. The Mississippi River was found to contain 13 aquatic habitats, with groupings of fish species somewhat unique for each habitat type (Baker et al., 1991).
4. Along an extended longitudinal gradient, fish composition and abundance are affected by various factors. In the upper and middle Ohio River, nearby tributary abundance, habitat preference, pollution tolerance, zoogeographic range, and optimal temperature were significantly correlated to catch data for 58 species (Reash and Van Hassel, 1988). Of these five factors identified, four showed stronger correlation statistics at sampling locations further downstream.

5. The importance of tributaries in recruitment of mainstem fish populations is species specific (Curry and Spacie, 1979).

17.4 CONSIDERATIONS FOR FISH METRIC DEVELOPMENT IN LARGE RIVERS

At a minimum, the development of multimetric indices for characterizing fish community integrity must incorporate the effects of temporal and spatial variability, and habitat influences. The actual methodologies that water resource agencies will choose to address these real-world influences will strongly depend on the purpose and objectives of assessments in large rivers. Osenberg and Schmitt (1996) emphasize that the ultimate goal of field assessments is to isolate and discriminate effects of human activities from other sources of variation. The authors summarize how assessments for detection of water quality impacts *should* be designed:

> ... fundamental goals of the assessment study are to estimate the state of the system that would have existed had the activity not occurred, estimate the state of the system that exists with the activity, and estimate the uncertainty associated with the difference between these estimates. (Osenberg and Schmitt, 1996; p. 6–7)

Water resource agencies should understand the potential magnitude of variability (temporal, spatial) and habitat effects. The regulatory scientist is concerned about the probability of incorrectly concluding that no adverse effect occurs when there truly is an effect (Type II error). In contrast, regulated industry is concerned with the probability of the agency incorrectly concluding that an adverse effect occurs when in fact this effect is actually within the bands of natural variation (Type I error). In monitoring studies, high statistical power is desired (i.e., reduce the probability of a Type II error), but the sample size must be sufficient to discriminate spatial and temporal variability from real water quality effects. The purpose of the next section is to cite examples that serve to describe the magnitude of these potential influences.

17.4.1 TEMPORAL VARIABILITY

In large rivers, temporal (within-year and among-year) variability of fish community characteristics is often significant. This is clearly not an unexpected result; however, it is interesting that temporal variability data are often not reported by authors who have studied large rivers. Simon and Emery (1995) present multimetric data for 15 navigation pools (total of 60 electrofishing sites) on the Ohio River within a 4-year period; however, no measures of temporal variability are included (it is unclear if sites were sampled on multiple occasions or just once). Likewise, Sylvester and Broughton (1983) document differential capture of several species among habitat types in the Upper Mississippi River; however, the effects of temporal variation were not studied. In contrast, Pierce et al. (1985) incorporated seasonal effects into a study of river stage effects on electrofishing in the Upper Mississippi River. Finally, Hughes and Gammon (1987) electrofished 26 sites on the Willamette River, each site being sampled twice. The authors present no estimates of the temporal variability of resulting IBI scores and metrics.

The magnitude of temporal variability in a large river can be illustrated by summarizing results of an annual fish monitoring study on the mainstem Ohio River. The Ohio River Ecological Research Program (ESE, 1997) is a long-term study of aquatic life near once-through cooled power plants on the Ohio River. Fish are sampled using several gear types and, because sampling is conducted three times from June to October, seasonal variability of population and community parameters can be documented. Within each year, catch data are log-transformed and analysis of variance is conducted to test for both temporal (seasonal) and spatial (upstream vs. downstream) effects. Table 17.1 indicates the effects of seasonal variability on community parameters at selected plant sites

TABLE 17.1
Temporal (Seasonal) Effects on Fish Community Parameters at Three Power Plant Study Sites on the Upper and Middle Ohio River, 1991–1996

Plant Site (RM)[a]	Sample Year	Significant ($p < 0.05$) Seasonal Effects Following ANOVA Analysis
Sammis (54)	1996	Species richness
	1995	CPE, biomass, MIWb, IBI
	1994	Biomass, IBI
	1993	CPE, sp. richness, MIWb, IBI
	1992	CPE, biomass
Cardinal (77)	1996	Biomass
	1995	CPE, biomass, MIWB
	1994	CPE
	1993	CPE, biomass, sp. richness, MIWb, IBI
	1992	CPE
Kyger Creek (260)[b]	1995	Biomass
	1993	CPE
	1992	CPE, biomass, IBI, MIWb
	1991	CPE, biomass

Note: Fishes were sampled during three seasonal surveys (June–October) each year. Abbreviations for community parameters are as follows: CPE = catch-per-effort (# fish/10 min.); biomass = total weight of all species; sp. richness = total no. of species collected; MIWb = Modified Index of Well-Being; IBI = Index of Biotic Integrity. MIWb and IBI followed Ohio EPA procedures (Ohio EPA, 1989).

[a] River miles form Pittsburgh, PA.
[b] No sampling was conducted at Kyger Creek Plant in 1994.

for years 1991–1995. These results correspond to nighttime near-shore electrofishing samples only. The results in Table 17.1 indicate that: (1) seasonal effects vary in magnitude from year to year per site; and (2) catch-per-effort and biomass show significant seasonal differences almost every year. For many community parameters that vary seasonally, the abundance of certain taxa (e.g., young-of-year [YOY] gizzard shad, round-bodied suckers) is largely responsible for temporal shifts.

Although many fish assessment studies do not address temporal variability by design, any resulting IBI metric should be regarded as preliminary until within-year and among-year variability data are incorporated. Regulatory agencies, in the process of developing IBI metrics for large rivers, can lessen the effects of within-year variability by choosing a standardized sampling season. The Ohio EPA (Sanders, 1995) sampled near-shore zones on the mainstem Ohio River during mid-September to mid-October only, effectively removing the influence of seasonal effects on IBI metrics. In the Ohio River, fall sampling typically results in the highest species richness, highest diversity-based indices (e.g., modified index of well-being), and highest IBI scores (ESE, 1996; ESE, 1995; Reash, 1990; Sanders, 1995). The Ohio EPA's standardized sampling minimizes seasonal effects; moreover, among-year variability of IBI scores at individual sites is relatively small at many zones (Randall Sanders, personal communication).

Selecting a standardized sampling season has been a successful strategy for minimizing potential seasonal effects on IBI metric development on the Ohio River. Water-resource agencies will not restrict fish assessment investigations for one season, however. Episodic spill events, anomalous river conditions (e.g., drought-induced low flows) and point-source investigations will require sampling at various times during the year. Because we know that seasonal effects are significant on the Ohio River, agencies need to use caution in applying numeric biological criteria (that were calibrated for one season) during times outside of the calibration season. In these cases, a more reasonable approach for detecting water quality impairment is to use a simple upstream vs. downstream approach (the control-impact design; Osenberg and Schmitt, 1996) with habitat and spatial effects removed or taken into account.

17.4.2 SPATIAL VARIABILITY AND HABITAT EFFECTS

In large rivers, spatial variability is closely associated with habitat effects. Spatial variability also occurs along longitudinal (i.e., upstream to downstream) and cross-sectional (bank to bank) axes. Spatial effects can be categorized by habitat types based on physical characteristics (main channel, main channel border, tailwaters, side channels, and backwaters/sloughs; Jahn and Anderson, 1986), or variability within a given habitat type. Because the distribution of fish in large rivers is highly discontinuous among zonation of habitat types, the development of multimetric indices should be stratified by habitat type. Sylvester and Broughton (1983) sampled fish in the Upper Mississippi River within three habitat types and found significant ($p < 0.001$) effects of habitat type on the number of species and individuals captured. Not surprisingly, total species richness of fish is highly correlated to diversity of habitat types (Jahn and Anderson, 1986). Simon and Emery (1995) found that several IBI metrics had significantly different distributions between riverine and lacustrine zones on the Ohio River. In response to these results, ORSANCO (1996) is developing multiple IBIs for the mainstem Ohio River that are segregated by habitat type, longitudinal position within a navigation pool, and longitudinal position from upper to lower Ohio River.

17.4.3 SAMPLING GEAR

The increased physical dimensions of large rivers (width, depth, discharge) reduce fish sampling effectiveness; the combined effects of habitat zonation and small proportion of surface area sampled account for this problem. Because the efficiency of capturing all the species present in a river segment is unknown (although assumed to be less than 100%), an investigator must select the most effective sampling gear to accomplish the objectives of the study. Multigear sampling will increase the probability of capturing all the species present in their true relative abundance. Simon and Lyons (1995) advocate the usage of multiple sampling techniques for the development of multimetric fish community indices in large rivers. There are two practical problems in using multigear sampling on large rivers: (1) many investigators, especially water resource agencies, do not implement multigear sampling; and (2) if multigear sampling is considered, the relative effectiveness of sampling gears needs to be determined.

Few published data document the relative effectiveness of multiple sampling gears used in large rivers. Annual fisheries sampling near coal-fired power plants on the Ohio River encompass multiple gears; these results can be used to document the effectiveness of the primary sampling method (shoreline electrofishing) relative to other methods. Table 17.2 indicates the number of species collected and number of individuals collected at these power plant sites on the upper and middle Ohio River during 1991 and 1992. These results clearly indicate that night electrofishing is the most effective sampling gear regarding species richness and total abundance of fishes. The effeectiveness of night electrofishing on the Ohio River was independently verified by Sanders (1990). Moreover, for species that are collected exclusively by one sampling technique, more of these species are captured by electrofishing than by any other gear. At most locations, gillnetting was the second most effective gear for capturing the actual fish species present. At many sites, beach seines captured more individuals. This factor, along with the prevailing discouragement of using gillnets for monitoring studies, seems to indicate that beach seining could be used as an adjunct sampling gear in large rivers.

TABLE 17.2
Relative Catches of Fish Using Four Sampling Gears at Three Power Plant Sites on the upper and middle Ohio River, 1991 and 1992 (EA, 1993, 1994)

Plant Site (RM)	Year	Total Species (all methods)	Electrofishing[a]		Beach Seines[b]		Gillnetting[c]		Trawling[d]	
			No. sp.	No. individ.	No. sp.	No. individ.	No. sp.	No. individ.	No. sp.	No. individ.
Sammis (54)	1991	51	44	14,429	23	1,156	29	2,521	21	1,183
	1992	42	38	3,066	11	892	15	325	13	122
Cardinal (76)	1991	48	40	6,171	29	2,989	20	875	12	553
	1992	40	37	2,513	11	234	18	212	5	136
Kyger Creek (260)	1991	41	39	5,881	14	938	21	1,187	6	254
	1992	38	31	997	16	432	18	297	7	69

[a] At each plant site, six 500-m zones were sampled during three seasons (N = 18).
[b] Two seine hauls at five sites during three seasons (N = 30 per plant site).
[c] Six stations with 48-hr sets during three seasons (N = 18 per plant site).
[d] Six stations sampled for 5 minutes (semi-balloon otter trawl) during three seasons (N = 18).

17.5 ASSESSMENT OF POINT-SOURCE EFFECTS IN LARGE RIVERS

Large rivers present some operational problems for assessing point-source discharges. The enlarged physical dimensions, combined with expected temporal and spatial variability of biological parameters, make the documentation of true adverse effects more difficult relative to streams and small rivers. This is not to say that detecting true adverse effects is so elusive that monitoring programs have impractical sensitivity. Rather, different strategies are required for large rivers.

A preliminary screening, followed by a definitive test, design is recommended for usage on large rivers. A calibrated multimetric index such as the IBI could be used to determine if the biological expectations for a given site might not be attained relative to an established reference condition. Once this preliminary screen is completed, definitive testing is conducted. The author suggests that multiple biological monitoring/water quality sites be clustered near the suspected stressor sources. Benthic, nonmobile organisms (e.g., Asiatic clam, native unionid mussel), or caged fish, are then placed within a gradient of effluent/receiving water dilutions. Standardized artificial substrate macroinvertebrate sampling could also be conducted. Appropriate ambient sites are also selected. Following a standard exposure period, indicative end-points of contaminant effects (growth rate, elevated tissue residues, etc.) are assessed using a balanced statistical design. This screening-definitive procedure would test the linkage between population and community effects, a process that may be difficult to elucidate in large rivers due to the inflated scale that must be reckoned.

17.6 CONCLUSION

Large rivers are the next type of water resources that will be assessed for the development of numeric biological criteria. The marked zonation of large river habitats requires stratified biological criteria based on habitat types. Standardized sampling protocols have already been established for night electrofishing (e.g., Sanders, 1990). Significant temporal and/or spatial variability of fish community parameters rule in large rivers, although among-year variability at selected sites may be small if sampling is conducted during a standardized period. Multimetric indices such as the IBI need to be tested for underlying assumptions, similar to streams and small rivers. Individual metrics need validation that they truly are responsive to water quality perturbations. Last, an evaluation of a point-source discharge should encompass a screening assessment, followed by a definitive assessment to confirm or refute actual adverse effects.

ACKNOWLEDGMENTS

The results presented for the Ohio River Ecological Research Program are provided courtesy of the sponsors: American Electric Power, Ohio Edison Company, Ohio Valley Electric Corporation, and Buckeye Power Company. EA Engineering, Science and Technology and Environmental Science and Technology, Inc. conducted the studies during 1991–1996. I thank Susan Foster and Gayle Pakrosnis for typing and editorial assistance. Comments from an anonymous reviewer improved the draft version of this manuscript.

REFERENCES

Baker, J. A., K. J. Killgore, and R. L. Kasul, R. L. 1991. Aquatic habitats and fish communities in the lower Mississippi River, *Reviews in Aquatic Science,* 3, 313.

Brown, D. J. and T. G. Coon. 1994. Abundance and assemblage structure of fish larvae in the lower Missouri River and its tributaries, *Transactions* of the *American Fisheries Society,* 123, 718.

Curry, K. D. and A. Spacie. 1979. The Importance of Tributary Streams to the Reproduction of Catostomids and Sauger, Water Resource Research Center Technical Report No. 32, Purdue University, West Lafayette, IN.

Dodge, D. P. 1989. *Proceedings of the International Large River Symposium (LARS),* Special Publication Canadian Journal of Fisheries and Aquatic Sciences 106.

EA (EA Engineering, Science, and Technology). 1993. 1991 Ohio River Ecological Research Program. EA Engineering, Science, and Technology, Deerfield, IL.

EA (EA Engineering, Science, and Technology). 1994. 1992 Ohio River Ecological Research Program. EA Engineering, Science, and Technology, Deerfield, IL.

ESE (Environmental Science & Engineering, Inc.). 1995. 1994 Ohio River Ecological Research Program. Environmental Science and Engineering, Inc., St. Louis, MO.

ESE (Environmental Science & Technology, Inc.). 1996. 1995 Ohio River Ecological Research Program. Environmental Science and Technology, Inc., St. Louis, MO.

ESE (Environmental Science & Technology, Inc.). 1997. 1996 Ohio River Ecological Research Program. Environmental Science & Technology, Inc., St. Louis, MO.

Fausch, K. D., J. Lyons, J. R. Karr, and P. L. Angermeier. 1990. Fish communities as indicators of environmental degradation, in S. M. Adams (Ed.), *Biological Indicators of Stress in Fish.* American Fisheries Society Symposium 8, Bethesda, MD.

Fremling, C. R., J. L. Rasmussen, R. E. Sparks, S. P. Cobb, C. F. Bryan, and T. O. Claflin. 1989. Mississippi River fisheries: a case history, in D.P. Dodge (Ed.), *Proceedings of the International Large River Symposium, Special Publication Canadian Journal of Fisheries and Aquatic Sciences* 106, p. 309.

Gammon, J. R. 1994. *The Wabash River Ecosystem.* DePauw University, Department of Biological Sciences, Greencastle, IN.

Gerritsen, J. 1995. Additive biological indices for resource management, *Journal of the North American Benthological Society,* 14, 451.

Gore, J. A. and F. D. Shield, Jr. 1995. Can large rivers be restored?, *BioScience,* 45, 142.

Holland, L. E. 1986. Distribution of early life history stages of fishes in selected pools of the Upper Mississippi River, *Hydrobiologia,* 136, 121.

Hughes, R. M. and J. R. Gammon. 1987. Longitudinal changes in fish assemblages and water quality in the Willamette River, Oregon, *Transactions of the American Fisheries Society,* 116, 196.

Hynes, H. B. N. 1989. Keynote address, in D. P. Dodge (Ed.), *Proceedings of the International Large River Symposium. Special Publication of the Canadian Journal of Fisheries and Aquatic Sciences* 106. p. 5.

Jahn, L. A. and R. V. Anderson. 1986. The Ecology of Pools 19 and 20, Upper Mississippi River: A Community Profile. U.S. Fish Wildlife Service Biological Report 85(7.6).

Johnson, B. L., W. B. Richardson, and T. J. Naimo. 1995. Past, present, and future concepts in large river ecology, *BioScience,* 45, 134.

Karr, J. R. 1995. Protecting aquatic ecosystems: clean water is not enough, in W. S. Davis and T. P. Simon (Eds.), *Biological Assessment and Criteria: Tools for Water Resource Planning and Decision Making.* Lewis, Boca Raton, FL. 7–13.

Nielsen, L. A., R. J. Sheehan, and D. J. Orth. 1986. Impacts of navigation on riverine fish production in the United States, *Pol. Arch. Hydrobiol.,* 33, 277.

Ohio Environmental Protection Agency (Ohio EPA). 1989. *Biological Criteria for the Protection of Aquatic Life. Vol. III. Standardized Biological Field Sampling and Laboratory Methods for Assessing Fish and Macroinvertebrate Communities.* Ohio EPA, Surface Water Division, Ecological Assessment Section, Columbus, OH.

Osenberg, C. W. and R. J. Schmitt. 1996. Detecting ecological impacts, in R. J. Schmitt, and C. W. Osenberg (Eds.), *Detecting Ecological Impacts: Concepts and Applications in Coastal Habitats.* Academic, New York.

ORSANCO (Ohio River Valley Water Sanitation Commission). 1996. Support Material for Second Annual Meeting of Ohio River Biological Experts. Ohio River Valley Water Sanitation Commission, Cincinnati, OH.

Pearson, W. D. and B. J. Pearson. 1989. Fishes of the Ohio River, *Ohio Journal of Science,* 89, 181.

Pierce, R. B., D. W. Coble, and S. D. Corley. 1985. Influence of river stage on shoreline electrofishing catches in the Upper Mississippi River, *Transactions of the American Fisheries Society,* 114, 857.

Reash, R. J. and J. H. Van Hassel. 1988. Distribution of upper and middle Ohio River fishes, 1973–1985. II. Influence of zoogeographic and physicochemical tolerance factors, *Journal of Freshwater Ecology,* 4, 459.

Reash, R. J. 1990. Results of Ohio River biological monitoring during the 1988 drought, in W. S. Davis (Ed.), *Proceedings of the 1990 Midwest Pollution Control Biologists Meeting.* EPA 905/9-90-005. U.S. Environmental Protection Agency, Region 5, Chicago, IL.

Regier, H. A., R. L. Welcomme, R. J. Steedman, and H. F. Henderson. 1989. Rehabilitation of degraded river ecosystems, in D. P. Dodge (Ed.), *Proceedings of the International Large River Symposium.* Special Publication *Canadian Journal of Fisheries and Aquatic Sciences,* 106, p. 86.

Ryder, R. A. and J. Pesendorfer. 1989. Large rivers are more than flowing lakes: a comparative review, in D. P. Dodge (Ed.), *Proceedings of the International Large River Symposium. Special Publication Canadian Journal of Fisheries and Aquatic Sciences,* 106, p. 65.

Sanders, R. E. 1990. A 1989 Night Electrofishing Survey of the Ohio River Mainstem (RM 280.8 to 442.5). Ohio EPA Ecological Assessment Section, Columbus, OH.

Sanders, R. E. 1995. Ohio's Near-Shore Fishes of the Ohio River: 1991–2000 (1994 Results). Ohio EPA Monitoring & Assessment Section, Columbus, OH.

Schregardus, D. R. 1992. Re-examining independent applicability, in *Water Quality Standards in the 21st Century,* EPA 823-R-92-009, U.S. Environmental Protection Agency, Office of Science and Technology, Washington, DC. p. 149.

Scott, M. T. and L. A. Nielsen. 1989. Young fish distribution in backwaters and main-channel borders of the Kanawha River, West Virginia, *Journal of Fish Biology,* 35, 21.

Sedell, J. R., J. E. Richey, and F. J. Swanson. 1989. The river continuum concept: a basis for the expected ecosystem behavior of very large rivers?, in D. P. Dodge (Ed.), *Proceedings of the International Large River Symposium, Special Publication Canadian Journal of Fisheries and Aquatic Sciences,* 106. p. 49.

Sheaffer, W. A. and J. G. Nickum. 1986. Backwater areas as nursery habitats for fishes in Pool 13 of the Upper Mississippi River, *Hydrobiologia,* 136, 131.

Shields, F. D., Jr., S. S. Knight, and C. M. Cooper. 1995. Use of the index of biotic integrity to assess physical habitat degradation in warmwater streams, *Hydrobiologia,* 312, 191.

Simon, T. P. and J. Lyons. 1995. Application of the index of biotic integrity to evaluate water resource integrity in freshwater ecosystems, in W. S. Davis and T. P. Simon (Eds.), *Biological Assessment and Criteria: Tools for Water Resource Planning and Decision Making.* Lewis, Boca Raton, FL. 245–262.

Simon, T. P. and E. B. Emery. 1995. Modification and assessment of an index of biotic integrity to quantify water resource quality in great rivers, *Regulated Rivers: Research and Management,* 11, 283.

Stalnaker, C. B., R. T. Milhous, and K. D. Bovee. 1989. Hydrology and hydraulics applied to fishery management in large rivers, in D. P. Dodge (Ed.), *Proceedings of the International Large River Symposium. Special Publication of the Canadian Journal of Fisheries and Aquatic Sciences,* 106, p. 13.

Sylvester, J. R. and J. D. Broughton. 1983. Distribution and relative abundance of fish in Pool 7 of the Upper Mississippi River, *North American Journal of Fisheries Management,* 3, 67.

Thorp, J. H. 1992. Linkage between islands and benthos in the Ohio River, with implications for riverine management, *Canadian Journal of Fisheries and Aquatic Sciences,* 49, 1873.

Thorp, J. H. and M. D. Delong. 1994. The riverine productivity model: an heuristic view of carbon sources and organic processing in large river ecosystems, *Oikos,* 70, 305.

Van Hassel, J. H., R. J. Reash, H. W. Brown, J. L. Thomas, and R. C. Mathews. 1988. Distribution of upper and middle Ohio River fishes, 1973–1985. I. Associations with water quality and ecological variables, *Journal of Freshwater Ecology,* 4, 441.

Vannote, R. L., G. W. Minshall, K. W. Cummins, J. R. Sedell, and C. E. Cushing. 1980. The River Continuum Concept, *Canadian Journal of Fisheries and Aquatic Sciences,* 37, 130.

Ward, J. V. and J. A. Stanford. 1989. Riverine ecosystems: the influence of man on catchment dynamics and fish ecology, in D. P. Dodge (Ed.), *Proceeding of the International Large River Symposium. Special Publication Canadian Journal of Fisheries and Aquatic Sciences,* 106, p. 56.

Wehr, J. D. and J. H. Thorp. 1997. Effects of navigation dams, tributaries and littoral zones on phytoplankton communities in the Ohio River, *Canadian Journal of Fisheries and Aquatic Sciences,* 54, 378.

18 Applying an Index of Biotic Integrity Based on Great-River Fish Communities: Considerations in Sampling and Interpretation

Thomas P. Simon and Randall E. Sanders

CONTENTS

18.1 Introduction 476
    18.1.1 Great River Reference Condition Development 476
    18.1.2 Stability and Consistency Needs 477
    18.1.3 Data Sources 478
18.2 Method Development Considerations 478
    18.2.1 Gear Limitations and Considerations 478
        18.2.1.1 Need for Multiple Gears 478
        18.2.1.2 Day vs. Night Sampling 479
        18.2.1.3 Complete Community Sampling 481
    18.2.2 Method Considerations 481
        18.2.2.1 Species Richness, Density, and Biomass Trends 481
        18.2.2.2 IBI and MIwb Trends 484
        18.2.2.3 Other Influences 485
            18.2.2.3.1 Downstream distance from dam 485
            18.2.2.3.2 Water temperature 486
18.3 Inherent Natural Variation 488
    18.3.1 Select Metric Comparison 488
        18.3.1.1 Exclusion of Gizzard Shad and Emerald Shiner 494
        18.3.2.2 Relative Number of Round-Bodied Suckers 494
        18.3.3.3 Percent DELT 495
    18.3.2 Stability/Precision/Accuracy 496
        18.3.2.1 Statistical Comparisons of Eight IBI/MIwb 496
        18.3.2.2 Minimum Detection Difference 497
    18.3.3 Habitat 497
        18.3.3.1 Major Substrate Differences 497
        18.3.3.2 QHEI Patterns 498
18.4 Conclusion 501
References 504

18.1 INTRODUCTION

Due in part to their enormous size, complexity of sampling requirements, and hypothesized intricate biological relationships, Large to Great Rivers have been one of the last surface water types to have survey and assessment techniques developed (Plafkin et al., 1989; Gibson et al., 1994; Karr and Dionne, 1991; Vannote et al., 1980). Although the size of a watershed necessary to define a "Great River" has varied, most riverine biologists agree that the United States' largest interjurisdictional rivers such as the Ohio and Mississippi Rivers, with watershed greater than 10,000 sq. mi., qualify as Great Rivers. Simon and Lyons (1995) reported Large and Great Rivers as streams with drainage areas between 1000 and 2300 sq. mi. and greater than 2300 sq. mi., respectively; while the Ohio EPA (1989) used the term "large rivers" to describe streams with watersheds 1000 to 6000 sq. mi.

Karr et al. (1985) have shown that Large and Great Rivers of the U.S. are not receiving adequate protection to conserve biological integrity. Numerous indicators reveal that perturbations caused by land use and human-induced chemical impacts from effluents are causing changes in water resource integrity (Karr et al., 1985b; Carlson and Muth, 1989; Ebel et al., 1989; Moyle and Williams, 1990). The reliance of water quality agencies on chemical-specific criteria as assessment indicators for large water bodies has been a surrogate for direct measurement of biotic communities; however, numerous shortcomings have been noted (Thurston et al., 1979; Gosz, 1980).

Movement toward a more comprehensive approach to assessment of water quality was improved by the development of the Index of Biotic Integrity (IBI) (Karr, 1981; Karr et al., 1986). The IBI was originally developed for small headwater and moderate-sized streams of the midwestern U.S. Simon and Emery (1995) modified the IBI to reflect measures of Great River fish communities appropriate for the Ohio River. Simon (1992) developed a large river version to address water resource issues in the White River drainage of Indiana.

The IBI is a composite measure of multiple attributes of fish communities that provides a basis for evaluating water resources. The IBI has been composed of 12 attributes (termed metrics) of fish communities. Attributes of fish communities for specific site assessment are compared to that of a series of sites that have been exposed to minimal human influence (Angermeier and Karr, 1986; Hughes, 1995). The IBI has been tested only on a limited basis for surface waters larger than wadeable streams. The purpose of the current study is to examine several properties of the Large and Great River IBI to provide guidelines for some of the more pressing questions regarding its application and interpretation. This chapter addresses the following questions: (1) What sampling decisions are relevant to minimizing contributions to inherent natural variation in the final IBI score? (2) How is the IBI affected by sampling effort, and what is a minimum sampling-reach length required for an adequate assessment? and (3) What are the effects of excluding select abundant species from IBI computations, and how important are these effects in interpreting the IBI? The last two questions concern problems that affect data collection and interpretation and are not specific to ecological assessment studies and IBI.

18.1.1 GREAT RIVER REFERENCE CONDITION DEVELOPMENT

Hughes (1995) and Hughes et al. (1986) indicated that there are specific issues in the development of reference conditions for Large and Great Rivers. Hughes (1995) identified the most obvious as the effect of size on the definition of reference condition. Horton (1945) demonstrated that there is an inverse relationship between stream order and stream abundance. The unique nature of Large and Great Rivers and the lack of comparative size classes hinder development of a reference condition based on a reference site approach (Hughes et al., 1986). Another attribute of Large and Great Rivers is that they do not necessarily resemble the ecoregion they drain; rather, they often resemble upstream conditions.

Two approaches have emerged for defining reference condition. The first approach uses minimally impaired reference sites after examining existing data, evaluating stressors, and reconnaissance of the river reach (Hughes et al., 1986; Yoder and Rankin, 1995). Sites are selected that are upstream of major sources of disturbance (Hughes and Gammon, 1987) or are located upstream of point sources, urban areas, and navigation dams. This approach requires that a few high-quality sites be sampled to calibrate expectations. The maximum species richness lines are quantified by these sites and, in turn, these determine the reference condition for designated uses. A second approach recognizes that few reference sites exist in Large and Great Rivers; thus, it is the cumulative dataset that reflects subtle attributes of the reference condition (Simon, 1991; 1992). All attempts are made to select sites based on the above criteria; however, more data is generally necessary to calibrate the reference condition for Large and Great Rivers. The benefit of this approach is that seldom does an increase in the number of sites in the dataset require recalibration of the reference condition.

Reference conditions for Great Rivers were developed from attributes of fish community structure and function based on single collections (i.e., never were samples combined) that estimate the relative abundances of all species. In order to calculate an Index of Biotic Integrity several steps are required. First, species are placed into various guild categories, such as "intolerants," "omnivores," "insectivores," "carnivores," and "simple lithophils." Criteria are developed for each metric such that values are converted to scores of 5 (deviates only slightly from least impacted condition), 3 (deviates moderately from least impacted condition), and 1 (deviates strongly from least impacted condition). Criteria are calibrated within a region and are based on stream size (Fausch et al., 1984). After all metric criteria are calibrated and validated, the individual metric scores are assigned and summed to compute a total IBI score (maximum score, 60 points). Based on the total score, the assessed fish community is assigned to various classifications of excellent, good, fair, poor, very poor, and no fish (Karr, 1981; Karr et al., 1986).

18.1.2 STABILITY AND CONSISTENCY NEEDS

The largest developmental critisicm of Large and Great River assessment techniques is a result of issues concerning stability and consistency of sampling (Gutreuter, 1996). The inability of previous researchers to develop consistent sampling techniques that arrived at repeat assessment answers was considered a method inconsistency. However, it is known that differences attributed to purpose and objectives of studies brought in much greater variation than the actual collection techniques (Simon, unpublished data). Gammon (1976; 1983) developed consistent assessment techniques for the Wabash River (Indiana); sampling techniques did not include smaller non-game species, but instead concentrated on larger members of the community. Simon (1992) showed that detection of subtle differences at Large River sites required collection of a "representative" sample of the entire fish community.

Assessment differences in Large and Great Rivers are often due to seasonal considerations. Numerous studies have shown that index period is an important consideration in sampling and should be conducted during stable flows after the community has been allowed to stabilize (Karr et al., 1986; Ohio EPA, 1989). Thus, for Large and Great Rivers, the best sampling periods would be between late summer and fall. Numerous 316 demonstration studies include seasonal sampling during peak flow or during times immediately following high-flow events. Data compiled during peak flows cause considerable confusion in interpretation. In addition, during the early spring, many species of large river fish migrate to tributary spawning grounds, which may cause bias in interpretation because of clumping and increased relative abundance. Thus, spring and early summer sampling periods should be avoided.

An equally important sampling consideration is how to calibrate effort. Many biologists sample for a prescribed time, while others sample for a prescribed distance with minimum time requirements. The experience with the present study has been that different individuals will cover a specific distance with different levels of timed effort. Also, significantly different distances can be sampled when only using a prescribed time. This causes significant differences in interpretation when one investigator spent either excessive or insufficient time in a river reach. Results suggest a select distance with minimum time requirements to evaluate a reach. Ohio EPA boat methods suggest that a minimum of 2000 to 3000 s should be sampled in a 0.5-km zone (Ohio EPA, 1989; Yoder and Smith, Chapter 2).

18.1.3 Data Sources

The most complete dataset available for Great Rivers is the Ohio River. The Ohio River has been investigated for over half a century by the Ohio River Valley Water Sanitation Commission (ORSANCO). ORSANCO has used biological information collected from lock chamber rotenone studies to assess the status of the river at 18 navigation dams. The Ohio River Ecological Research program, sponsored by American Electric Power, has collected annual data at a series of stations near electric generating facilities since the early 1970s (ESE, 1996). Finally, the work of the Ohio EPA (Sanders, 1990; 1991; 1992; 1993; 1994; 1995) has assessed several rounds of night fish collection at numerous reaches along the Ohio shore. The Ohio EPA dataset includes 141 sites distributed throughout the upper Ohio River above RM 487.2.

18.2 METHOD DEVELOPMENT CONSIDERATIONS

18.2.1 Gear Limitations and Considerations

Simon and Lyons (1995) identified long-known problems requiring further study for Large and Great Rivers. The most vital sample collection considerations included resolution of appropriate sampling techniques, defining temporal and spatial scales, and scope of sampling effort. The importance of standard collection methods cannot be overstated when discussing issues of consistency, variability, and gear selection (Ohio EPA, 1989).

Angermeier and Karr (1986) found that the length of distance sampled in a stream reach was an important consideration in wadeable streams in West Virginia, Illinois, and Ohio. They found that IBI scores from long reaches were less variable and more consistent than those from short reaches. The key habitat features of defining a stream reach included the presence of several pool-riffle sequences. Yoder and Rankin (1995) indicated that temporal "index period" must be defined in order to compare the collection to the calibrated reference condition. In addition, the importance of selecting the appropriate gear type has repeatedly been raised in order to ensure that complete sample collections are made for Large and Great Rivers, and whether the need to use multiple sampling gears is required to represent the fish community.

18.2.1.1 Need for Multiple Gears

The limitations of using a single gear type has been discussed by Nielsen and Johnson (1983). Utilizing multiple gears in order to get a representative snapshot of species richness for estimating biotic integrity results in problems of aggregating data and amount of effort (Simon and Lyons, 1995). During the early 1970s, numerous 316 demonstration studies of electric generating facilities located on Large and Great Rivers were required to collect monthly or seasonal information. Usually, multiple gears were used to provide near-shore estimates of adult and juvenile abundance, mid-channel estimates of species richness, and backwater estimates of select commercial and game species. The use of electroshockers, seines, gillnets or trawls, and hoopnets provided a rich dataset of information for evaluating the need for multiple gear types.

Multiple gear types were selected to determine whether any impact had occurred in the vicinity of the electric generating facilities. The fish community was measured by evaluating differences in structure between pre- and post-operation changes. The objective of these impact studies was to collect as many species over the course of the study to demonstrate that biodiversity had not been affected. The purpose of most water resource studies is not to collect every species, but rather

TABLE 18.1
Log-Transformed Analysis of Variance (ANOVA) Results Comparing Assemblage Characteristics of Electrofishing (E), Seining (S), Gillnetting (G), Trawling (T), and Hoopnetting (H) Results from the Ohio River Ecological Research Program During 1993

Parameter	F-Value Between Gears	p	Order of Means
CPE	1.997	0.140	E> S> T> G> H
Mean biomass	5.30	0.20	E> S> G> H> T
Species richness	253.65	0.001	E> S> T> G> H
Diversity	14.31	0.08	E = S> T> G> H

From EA, 1993; 1994; ESE, 1995.

to document a "representative" sample. Estimation of biotic integrity does not require intensive reach sampling, accruing resource expeditures to collect every last species that may occur at a site. Since rare species add very little to the total IBI score, the failure to collect a few rare species at a site will not detract from the assessment of biotic integrity (Yoder and Rankin, 1995).

Multiple gear type data was evaluated from adult fish collections using electrofishing, hoopnets, gillnets, trawls, and seines from data collected by the Ohio River Ecological Research Program between 1991 and 1993 (EA, 1993; 1994; ESE, 1995). Collection methods used by the Ecological Research Program were consistent with methods used by the Ohio EPA and ORSANCO. Each collection was analyzed by date, and samples were segregated into specific gear types and tested using Analysis of Variance (ANOVA, $p < 0.05$) to evaluate differences between specific community catch information (Table 18.1). The single gear type that provided the best estimate of numerical abundance, species richness, and proportion of frequently encountered common species was electrofishing. Relative abundance was usually lowest using gillnets, with the exception of a single site where trawl abundance was considerably lower. Species richness using gillnets was typically less than half that encountered while electrofishing, while trawls were intermediate between gillnets and electrofishing techniques. Catches were dominated by young-of-year fish. Species richness and total abundance using seines were less than in electrofishing methods; however, seines provided better estimates of species richness than trawls, gillnets or hoopnets.

A difference in the types of species captured using different collection gear was observed. Significant differences in species richness (ANOVA) between electrofishing and seine techniques were observed compared to all other collection techniques. Seining and electrofishing methods provided a representative collection of minnows, suckers, darters, and sunfish; while gillnets and hoopnets tended to capture sunfish, catfish, and suckers. Based on our analysis, multiple sampling gears were not warranted to accurately depict a Great River fish community; however, if two gear types need to be chosen, the two methods would be electrofishing and beach seining.

18.2.1.2 Day vs. Night Sampling

Sanders (1989) compared day and night sampling based on fish community collection from 18 stations in the Greenup, Meldahl, and Markland Dam pools. Sampling was conducted in a downstream direction, within 40 m of shore, during the fall 1988. Fishing in each zone was accomplished by a serpentine movement of the boat along the submerged "shelf" where depths gradually increased from shore to the vertical navigable waters (>2.8 m). Estimated depths of collection ranged from 0.7 to 1.3 m ($\bar{x} = 9.4$ m). For each station, a representative collection of fish was obtained from all available habitats, enumerated, weighted, and inspected for external anomalies (Ohio EPA, 1989;

TABLE 18.2
Comparison of Relative Catch Information Between Day and Night Electrofishing Surveys on the Ohio River Based on 10 Electrofishing Stations Repeat Sampled at the Same Locations Between July and September 1988

Attribute	Night	Day	p
Richness	48	34	0.026
Species richness	21.7	14	0.326
Relative abundance	5183	5154	0.255
Relative abundance (excluding gizzard shad)			0.047
Relative abundance (excluding gizzard shad and emerald shiner)			0.003
Relative weight (kg)	230	330	0.011
MIwb	9.4	7.1	0.003
IBI	43.0	32.2	0.003

Sanders et al., Chapter 9). Site distances varied between 0.5 to 1.18 km to evaluate species diversity correlates with changing distance.

Comparisons between repeated sampling at 10 Ohio River mainstem sites were conducted during daytime and nightime collections at select pools (Table 18.2). Studies conducted along the Ohio shoreline showed that many fish move from deeper offshore waters into the shallow, littoral areas at night. Sanders (1989) found night catches contained significantly more species, higher numbers and weights of fish (excluding gizzard shad and emerald shiners), and are compositionally more evenly distributed than day catches.

Total number of species from night electrofishing surveys was significantly different (ANOVA, $p > 0.05$) from day collections. Day and night sampling had a Jaccard's percent similarity of 0.653. Day and night sampling shared 32 fish species and a single hybrid. Day sampling found 9 species (i.e., paddlefish, ghost shiner, highfin carpsucker, blue sucker, brook silverside, white crappie, green sunfish, orangespotted sunfish, and dusky darter) not collected during night sampling, while night sampling had 17 unique species and 2 new hybrid combinations. Species collected during night electrofishing included redhorses, minnows, and darters, which had not been collected during day sampling. The two methods cumulatively accounted for 36.5% of the 159 species known to have existed in the Ohio River (Pearson and Pearson, 1989). Night surveys collected numerous rare species that are frequently overlooked during routine species assessments. Night electrofishing between 1989 and 1995 found 12 species afforded some special protection classification in Ohio.

Fish numbers and weight (excluding gizzard shad and emerald shiners) were statistically significant (ANOVA, $p > 0.05$), being higher for night than day catches. The total number of fish collected ranged from between 130.0 to 2264.7 fish/km ($\bar{x} = 367.5$ fish/km). The catch was dominated by gizzard shad (44.0%) and emerald shiner (16.5%); however, a single large gizzard shad sample downstream from a power station caused a highly skewed dominance. Cumulative relative weight ranged from 21.0 to 97.2 kg/km ($\bar{x} = 49.6$). Dominant biomass catch was dominated by channel catfish (16.6%), common carp (15.7%), smallmouth buffalo (15.4%), and gizzard shad (13.3%).

Sanders (1989) found night catch composition and weight to be more evenly distributed than in day catches. Evenness of both relative number and weight of catch was statistically higher (t-test, $p > 0.05$) in night samples. Relative number of individuals was similar between all years with the exception of 1993 (2221.0 individuals/km) when near-perfect sampling conditions existed (Table 18.2). Low water conditions concentrated fish and exposed less shelf habitat. Cumulative relative weight of collections was similar between 1990 and 1995, with the only exception being 1991 (108.5 kg/km; Table 18.2).

18.2.1.3 Complete Community Sampling

Resource management agencies often design sampling programs to answer questions regarding select indicator species, such as black basses or game species. These studies are designed to be frugal resource expenditures that maximize information return for the population of interest. Incidental occurrence of other non-target species may be recorded; however, identifications and enumerations are often incomplete. It can be argued that collection of information in this manner is not prudent since additional resource questions cannot be asked. This information has little "value-added" use since information content is limited. To understand the structure and function of Great River fish communities, including the population patterns of top carnivores, complete community data should be collected in order to understand limiting factors. Sample collection should be "representative," so that species are captured in a similar proportion as they are encountered in the reach (Hocutt and Stauffer, 1980). For example, netters should expend as much effort to collect small minnows and darters as they would to collect a trophy smallmouth bass.

18.2.2 METHOD CONSIDERATIONS

The rationale and sampling suggestions in the present study are based on 141 collections from seven years of effort on the Ohio River between 1989 and 1995. The primary sampling objectives are consistency and repeat site classification. These objectives influenced collection decisions and have proven that a consistent approach for Great River sampling is obtainable. In this section, pivotal decisions for sampling in Great Rivers based on our experiences in the Ohio River are discussed. How sample design decisions are made will determine precision and accuracy for development of an Index of Biotic Integrity for Great Rivers. Measurements of shoreline length, instream habitat, distances downstream from dams, and distance from flooded tributary mouths were recorded for analysis of variance of stations (Figure 18.1).

18.2.2.1 Species Richness, Density, and Biomass Trends

Species richness did not continue to increase with greater length of reach sampled by night electrofishing (Figure 18.1A). A scatterplot of increasing sampling effort were used to interprete whether adequate sampling distance and effort was spent in a stream reach. Catch results should show a sigmoid species area curve with an increase in sample distance (Angermeier and Karr, 1986). Yoder and Smith (Chapter 2) described the rationale for determining sampling distances for Large Rivers in Ohio. A sampling distance was arrived at when diminishing species increase was observed with increasing 0.25-km distance increments beyond 0.5 km. For Great Rivers, results show that species richness in the Ohio River was not statistically different (simple linear regression, $r^2 = 0.038$) between 0.4 and 1.2 km distances, nor was species richness significantly different ($r^2 = 0.002$) with increasing distance downstream from the navigation dams (Figure 18.1B). Our Ohio River results show that increasing species richness had already asymptoted at 0.4 km.

Species richness correlates with increasing sampling distance were evaluated using annual (Figure 18.2) and repeat site visits (Figure 18.3). These two measures were used to estimate how representative our collection efforts were on the Ohio River fish community. This seven-year effort in the Ohio River found that cumulative species richness for any specific year collected between 55 and 78% of the total number of Ohio River fish species (Figure 18.2). Two years of sampling effort (39 collection events) included 72% of the cumulative total number of species, and after three years (60 collection events), 82% of the cumulative number of species were collected. To collect a probability of 96% of the cumulative total number of Ohio River species required a total sample of 101 sites or five years of collection effort.

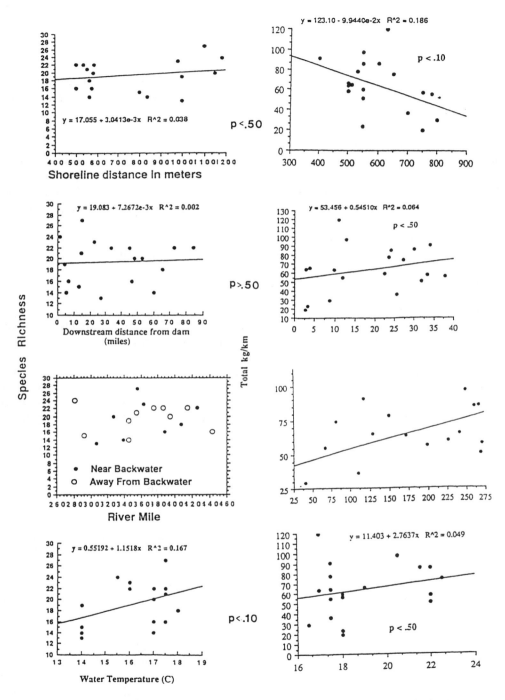

FIGURE 18.1 Species richness, total weight (kg/km), and relative total number (number/km) curves based on shoreline distance in meters, downstream distance from dam (miles), River Mile, and water temperature (°C).

Applying an IBI Based on Great-River Fish Communities

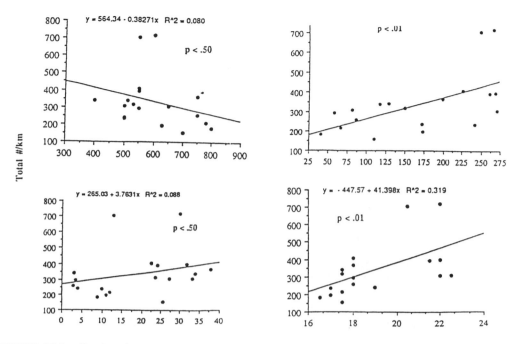

FIGURE 18.1 *Continued.*

Biological diversity at a site was evaluated by repeat sampling at nine individual sites (Figure 18.3). Results revealed that the number of new fish species added with increasing sampling effort reached an asymptote between the third and fourth event (Figure 18.3). Local species richness did not deviate strongly between navigation pools in the upper Ohio River, so that within any single pool, between 39 to 63 species were collected. Species richness did not show a relationship with drainage area in the upper Ohio River. Species richness within navigation pools was greatest in Meldahl (62 species), Greenup (56), Belleville (52), and Markland and Gallipolis (50). Differences in biological diversity are a result of anthropogenic and natural differences so that the number of species per collection event averaged 2.2 species per attempt, suggesting that adequate amount of collection effort had resulted in the majority of species being captured.

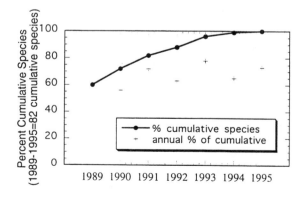

FIGURE 18.2 Percent cumulative species accumulating with increasing number of samples between 1989 and 1995. + = annual percentage of the colleceted species based on the total number of species within the Ohio River during the same year.

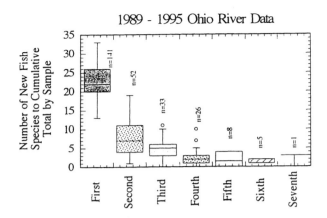

FIGURE 18.3 Number of new species added with increasing effort between 1989 and 1995.

Relative density (total number of individuals/km) was evaluated using catch results correlated with River Mile (a surrogate for drainage area), distance sampled, and downstream distance from navigation dam (Figures 18.1D–F). Density was significantly correlated with increasing drainage area ($r^2 = 0.330$, $p < 0.01$), distance downstream from a dam ($r^2 = 0.088$, $p < 0.05$); and was negatively correlated with increasing distance sampled ($r^2 = 0.080$, $p < 0.05$). Increased catches of gizzard shad contributed to statistically significant relationships.

Relative weight (kg/km) showed similar patterns as relative density with drainage area, distance sampled, and downstream distance from a dam, (Figures 18.1G–I). Relative weight was significantly correlated with increasing drainage area ($r^2 = 0.136$) and downstream distance from a dam ($r^2 = 0.064$, $p < 0.05$); and was negatively correlated with increasing distance sampled ($r^2 = 0.186$, $p < 0.10$). Significant differences in relative weight/km were due to increased weights contributed by channel catfish (ANOVA, $p < 0.02$).

18.2.2.2 IBI and MIwb Trends

Trends in cumulative IBI scores were compared with a variety of sampling and habitat factors (Figure 18.4). The IBI did not show any statistical difference with drainage area ($r^2 = 0.286$), distance of sampling zone ($r^2 = 0.003$), or seconds fished ($r^2 = 0.024$). A statistically significant difference was found between the IBI and proximity downstream of a dam ($r^2 = 0.622$, $p < 0.001$), increasing habitat heterogeneity and complexity (as measured by the QHEI, $r^2 = 0.444$, $p < 0.002$), decreasing water temperature ($r^2 = 0.315$, $p < 0.01$), and increasing water clarity ($r^2 = 0.194$, $p < 0.10$). The IBI did not show any pattern with increasing drainage area, distance sampled, or time sampled, suggesting that the amount of effort expended was adequate to characterize the biotic integrity of the Ohio River. The proximity of a sampling zone to a tributary mouth usually resulted in a lower IBI score. Tributaries are important backwater habitats for the Ohio River, and fish communities may migrate between the mainstem and tributary habitat. These areas generally have lower QHEI scores. A direct correlation was found between the IBI and increasing habitat complexity (Figure 18.4).

The Modified Index of Well-Being (MIwb) did not show statistically significant differences between sampling methods and habitat differences (Figure 18.5). The MIwb is an alternate biocriterion for Large and Great Rivers that determines patterns between biomass, species richness, evenness, and Shannon diversity (Gammon, 1976; 1983; Yoder and Smith, Chapter 2). The MIwb did not show any statistical difference with drainage area ($r^2 = 0.041$), seconds sampled ($r^2 = 0.042$), downstream distance from a dam ($r^2 = 0.004$), relative number/km ($r^2 = 0.016$), habitat complexity ($r^2 = 0.069$), or water temperature ($r^2 = 0.005$). The MIwb was statistically significant with increasing

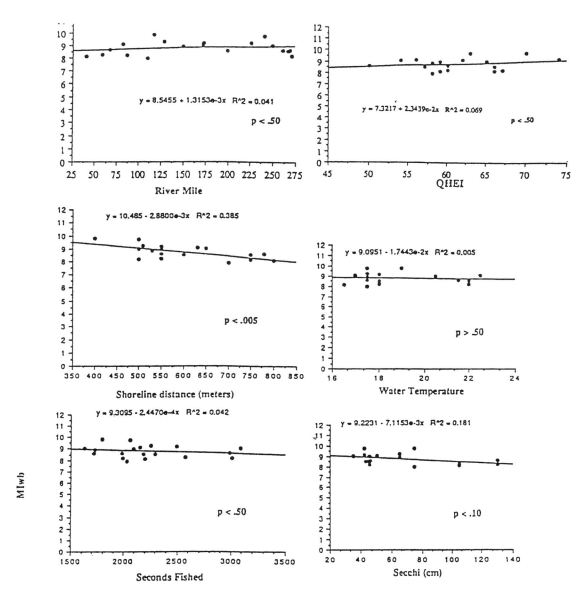

FIGURE 18.4 Relationship between the modified Index of Well-Being and (A) River Mile, (B) shoreline distance (m), (C) seconds fished, (D) downstream distance from dam (miles), (E) total number per km, (F) QHEI, (G) water temperature, (H) secchi depth (cm), (I) River Mile distance from a tributary, (J) total kg/km.

distance sampled ($r^2 = 0.385$), water clarity ($r^2 = 0.181$), and increasing relative weight ($r^2 = 0.381$). The relationship with relative weight was anticipated since this characteristic is autocorrelated with the MIwb.

18.2.2.3 Other Influences

18.2.2.3.1 Downstream distance from dam

The position of stations within pools related to species richness, density, biomass, the Modified Index of Well-Being, and the Index of Biotic Integrity were evaluated. No statistical relationship

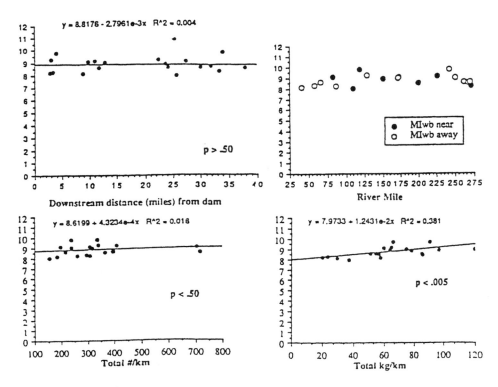

FIGURE 18.4 *Continued.*

was observed between downstream distance and species richness, density, biomass, and the Modified Index of Well-Being.

The distance downstream from a dam had a positive relationship with the relative number of gizzard shad ($p < 0.05$) and freshwater drum ($p > 0.01$) densities (Figure 18.6). Distance downstream had a negative relationship with emerald shiner ($p > 0.05$), mimic shiner ($p > 0.10$), and sauger ($p > 0.05$) densities (Figure 18.4). Gizzard shad and freshwater drum abundances are related to the lacustrine zone habitat, while minnows and sauger are more abundant in the riverine and transition zones.

Downstream distance from a dam did not show any statistical relationship for channel catfish ($p < 0.50$), carp ($p > 0.50$), or sauger ($p < 0.50$) mean relative weights (kg/km). Increased mean relative weight was observed with increasing downstream distance from a dam for smallmouth buffalo ($p < 0.20$), while a declining statistically significant relationship was observed with distance downstream of the dam for gizzard shad ($p < 0.10$) and freshwater drum ($p < 0.10$; Figure 18.7). The lacustrine zones of navigation pools showed lower IBI classifications as a result of impoundment, reduction in habitat diversity, and lack of substrate heterogenity.

18.2.2.3.2 Water temperature

Decreasing water temperature affects conductivity and ultimately electrofishing gear proficiency (Jacobs and Swink, 1982; Hardin and Connor, 1992). In order to determine if seasonal changes in species richness, density, biomass, MIwb, or IBI are a result of water temperature, catch based on thermal profiles during collection were evaluated. The Ohio EPA dataset was primarily collected during late fall index periods between mid-September and early November, 1989 to 1995. As water temperature increased, so did the total relative density of all species, the relative density of gizzard shad and freshwater drum, and the biomass of common carp. IBI values decreased.

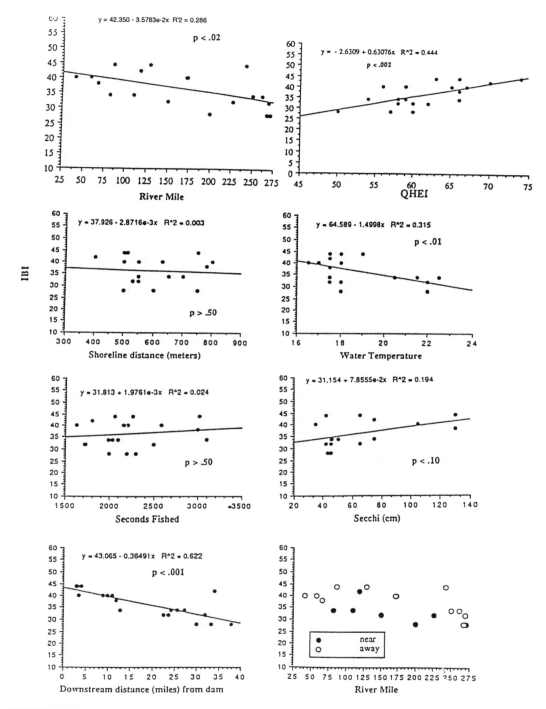

FIGURE 18.5 Relationship between the IBI and (A) River Mile, (B) shoreline distance (m), (C) seconds fished, (D) downstream distance from dam (miles), (E) QHEI, (F) water temperature, (G) secchi depth (cm), (H) River Mile distance from a tributary.

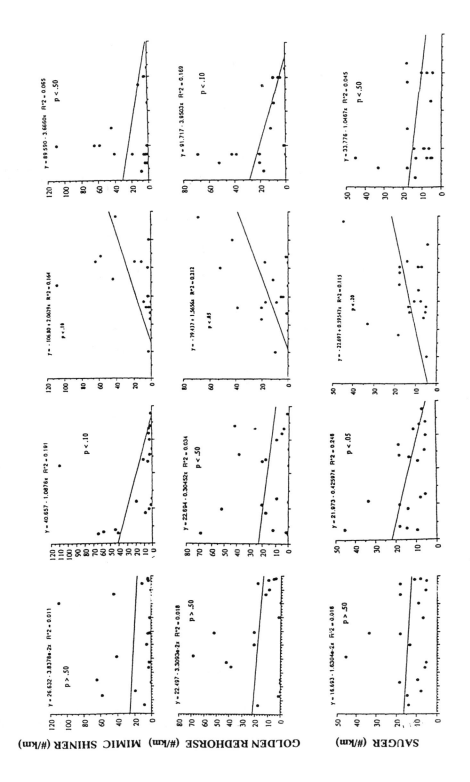

FIGURE 18.6 Species relationships for mimic shiner, golden redhorse, sauger, gizzard shad, emerald shiner, and freshwater drum number per km with (A) River Mile, (B) downstream distance from dam (mi.), (C) QHEI, and (D) water temperature.

Applying an IBI Based on Great-River Fish Communities

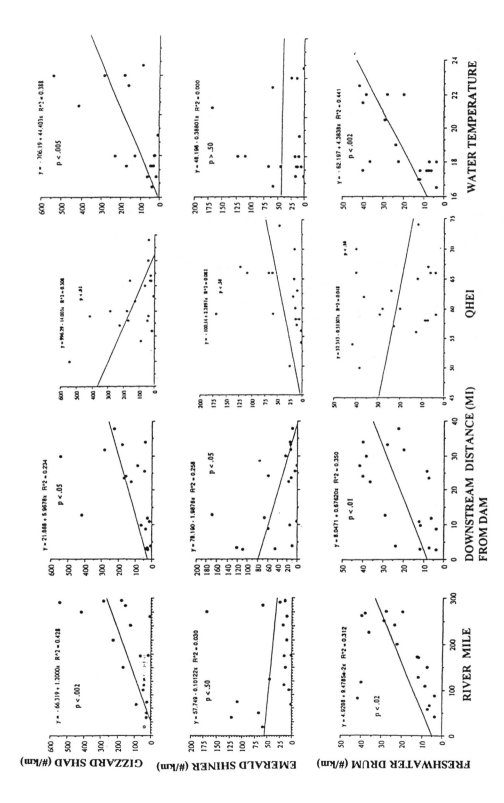

FIGURE 18.6 *Continued.*

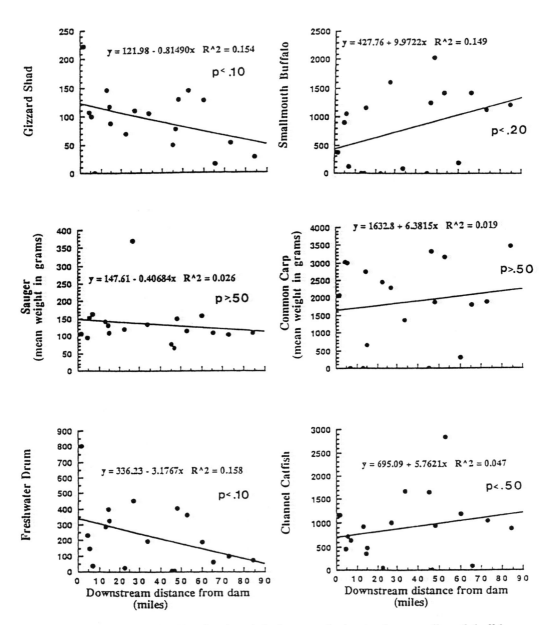

FIGURE 18.7 Species relationships for gizzard shad, sauger, freshwater drum, smallmouth buffalo, carp, and channel catfish mean weight (g) with downstream distance from dam (miles).

Water temperature was not statistically correlated for species richness, biomass, or the MIwb (Figure 18.6). Increasing water temperature was statistically significant for increasing density ($p > 0.01$) of cumulative number of species (Figure 18.1). Select species showed differential effects by water temperature. Both gizzard shad ($p < 0.005$) and freshwater drum ($p < 0.002$) showed an increase in abundance with increases in water temperature, while golden redhorse abundance declined ($p > 0.10$) with increasing water temperature. No relationship was seen between water temperature and abundance of emerald shiner, mimic shiner, or sauger.

Water temperature effects evaluated for four species and a few species, showed some biomass changes with temperature (Figure 18.8). Golden redhorse biomass declined significantly ($p < 0.10$)

Applying an IBI Based on Great-River Fish Communities

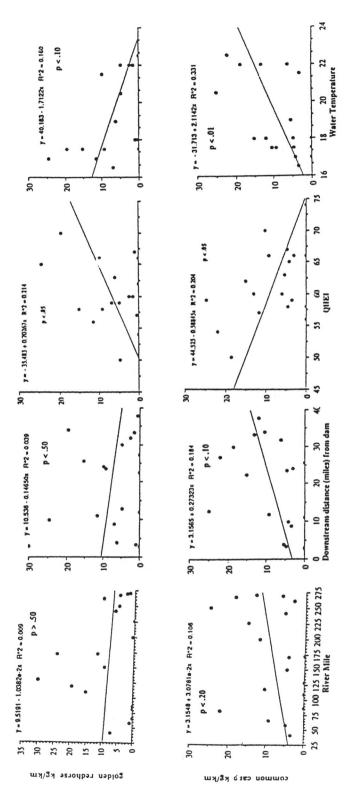

FIGURE 18.8 Relationships between weight (kg/km) for channel catfish, gizzard shad, golden redhorse, and common carp and (A) river mile, (B) downstream distance from dam (miles), (C) QHEI, and (D) water temperature.

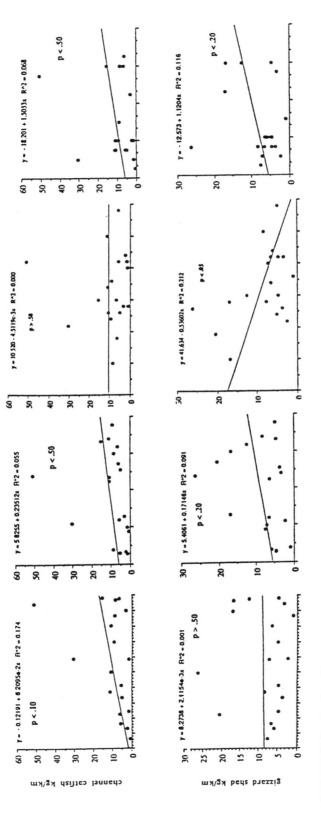

FIGURE 18.8 *Continued.*

Applying an IBI Based on Great-River Fish Communities

with increasing water temperature, while carp biomass increased significantly ($p > 0.01$) with increasing water temperature. Channel catfish and gizzard shad ($p < 0.20$) biomass did not show a statistical relationship with water temperature, although the biomass trend was generally positive with increasing temperature.

18.3 INHERENT NATURAL VARIATION

18.3.1 SELECT METRIC COMPARISON

Simon and Emery (1995) modified the Index of Biotic Integrity for the upper 300 miles of the Ohio River along the Western Allegheny Plateau ecoregion. The modified index uses 12 metrics that included 6 metrics from the original IBI (Karr et al., 1986): (1) total number of species, (2) number of sensitive species, (3) proportion of omnivores, (4) proportion of carnivores, (5) abundance, and (6) deformities, eroded fins, lesions, and tumors (DELT). Simon and Emery (1995) found that several metrics needed to be modified in order to accurately reflect community structure and function. A number of metrics were changed, including: number of centrarchid species (which included black basses), proportion of large river taxa, proportion of round-bodied suckers, proportion of tolerant species, proportion of insectivores, and proportion of simple lithophils. Sanders (1992) evaluated the performance of specific metrics to determine if characteristics showed expected patterns. Several metric changes needed to be incorporated into the IBI in order to depict subtle patterns in community structure.

18.3.1.1 Exclusion of Gizzard Shad and Emerald Shiner

Large schools of gizzard shad *Dorosoma cepedianum* and emerald shiner *Notropis atherinoides* are often sporadically collected in large numbers. These species do not exhibit random distributions so that usually when encountered, they can bias the site assessment. When these species are dominant in a collection (>40% of total number of individuals collected), subtle patterns exhibited in the IBI and MIwb are obscured (Figure 18.9). The data showed that the inclusion of gizzard shad and emerald shiner in the analysis could change the total metric score by as many as 8 IBI points. The inclusion of these schooling species in the IBI has the effect of raising point totals and

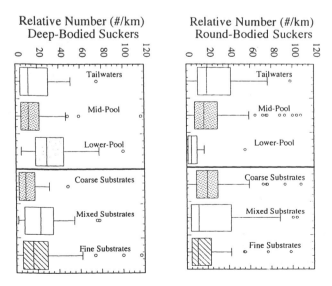

FIGURE 18.9 Relative number of round- and deep-bodied suckers with (A) pool position: tailwaters, mid-pool, and lower pool; and (B) substrate type: coarse substrates, mixed substrates, and fine substrates.

eliminating metric patterns of the less-dominant species. In order to avoid homogenizing site assessment, the proportional and abundance metrics should not include gizzard shad and emerald shiner. Other species that exhibit these same biases should also be tested and eliminated if they dominate the relative catch, including threadfin shad *Dorosoma petenense*, skipjack herring *Alosa chrysochloris*, and spotfin shiner *Cyprinella spiloptera*. Further evaluation of scoring modifications should be considered (see Rankin and Yoder, Chapter 24).

18.3.2.2 Relative Number of Round-Bodied Suckers

Modifications of the IBI have generally included the substitution of the number of sucker species with the number or proportion of round-bodied suckers (Simon and Lyons, 1995; Simon and Emery, 1995; Emery et al., Chapter 8). Differential sensitivities of the round-bodied and deep-bodied suckers have been considered an important distinction when evaluating biological integrity. The round-bodied suckers are considered more sensitive to environmental disturbance as a result of their feeding habits, spawning requirements, and thermal sensitivity (Emery et al., Chapter 8).

The use of relative number of round-bodied suckers is advocated for the upper Ohio River as a substitute for the number of sucker species (Figure 18.10). The round-bodied suckers reach their highest relative numbers in the riverine and transition zones of the pools. In the impounded lacustrine zone of the navigation pool, round-bodied suckers reach their lowest relative number (usually less than 20 round-bodied suckers/km), while deep-bodied suckers reflect their greatest relative abundance (100 deep-bodied suckers/km). The decline in round-bodied suckers demonstrates loss of biological integrity as a result of the navigation dams.

Patterns in round-bodied sucker distribution reach their highest relative abundance/km in riverine tailwaters of dams and decline significantly with degree of impoundment (Figure 18.10). No statistical difference in round-bodied sucker relative number/km was observed between coarse, mixed, or fine substrates. Although round-bodied sucker density declines over fine substrates, the number expected in the upper Ohio River is not statistically different from mixed substrates. Increases in deep-bodied sucker relative number/km are highest in lower pool impounded zones, although they also occur in the tailwaters of the riverine zones and mid-pool transitional zones. Deep-bodied suckers were found over all three substrate types; however, their lowest relative numbers occurred over coarse substrates.

18.3.3.3 Percent DELT

The number of individuals exhibiting deformities, eroded fins, lesions, and tumors (DELT) is increasing in Large and Great Rivers (Sanders et al., Chapter 9). The increase in DELT reflects the lowest extremes in biological integrity (Karr et al., 1986). Two approaches to using the DELT metric have been proposed to maximize information content. Simon and Emery (1995) used the percent of DELT anomalies, but proposed limiting the size of the specimens considered for this metric. They suggested only specimens that attained a minimum size of 200 mm total length (TL) should be considered in the analysis since juvenile and immature specimens would reflect the condition of the environment only after sufficient time for exposure had occurred. Although species such as minnows, darters, and other smaller species would not be considered for this analysis, it is often the long-lived benthic species that reflect the observed DELT anomalies. The shorter-lived, non-game species are generally not exposed to the stressor for sufficent time to cause gross external anomalies. Limiting the analysis to species that attain greater sizes will ensure that exposure time has been maximized in order to fully determine if DELT is a concern. Eliminating species in the proportional metrics that do not attain larger sizes has the effect of increasing the proportion of individuals that exhibit DELT.

The Ohio EPA (1989) based the percentage of DELT anomalies on the entire community catch for Large Rivers (Figure 18.11). The percent DELT anomalies observed in the Ohio River were below the established Large River criteria, with the exception of the Belleville, Gallipolis, and

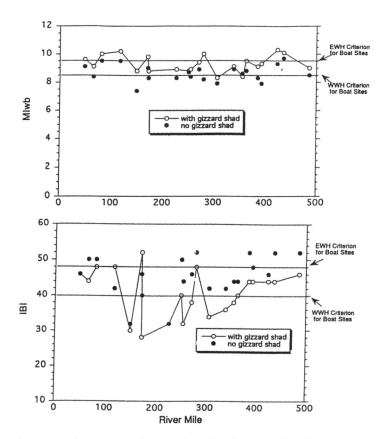

FIGURE 18.10 Influence of gizzard shad on the Modified Index of Well-Being (A) and Index of Biotic Integrity (B) for assessing aquatic life designated uses.

Greenup Pools. The New Cumberland Pool had the highest percent DELT anomalies observed (9.2% DELT). Problems associated with low catches, causing a bias in scoring procedures (Rankin and Yoder, Chapter 24), will require low-end scoring adjustments for the percent DELT metric. In order to avoid low-end scoring procedures, the relative number of DELT anomalies (number/km) could be used to show patterns or high frequency of DELT anomalies.

The alternate metric does not penalize a site if a minimum of two or three DELT individuals are collected, thus invoking low-end scoring for this metric (Rankin and Yoder, Chapter 24). For example, at sites where the number of non-game species may be very high, the dilution of the larger species susceptible to developing DELT is diminished. By using relative number of DELT anomalies, the number of DELTs that trigger a response will be known and will not be influenced by the swamping of the non-game species. Using relative number of DELTs, an increased relative number of DELT anomalies in the New Cumberland and Racine Pools becomes more apparent.

18.3.2 Stability/Precision/Accuracy

18.3.2.1 Statistical Comparisons of Eight IBI/MIwb

Eight sites longitudinally stratified along the Ohio River were repeat sampled five to seven times between 1989 and 1995 (Figure 18.12). MIwb and IBI scores from these eight sites were evaluated to determine patterns in stability, precision, and accuracy. MIwb scores seldom deviated between the collection events outside of the WWH aquatic life classification, with the exception of Gallipolis

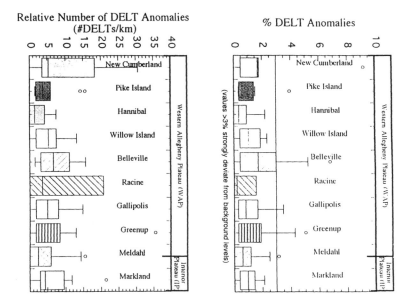

FIGURE 18.11 Distribution of DELT anomalies by pool in the upper Ohio River: (A) Percent DELT anomalies and (B) Relative number of DELT anomalies (number/km).

(RM 280.0) and Portsmouth (RM 356.0) sites. Similarly, a stable IBI score was typically observed, with the exception of Marietta upstream of the Muskingum River (RM 171.5), downstream Gallipolis dam (RM 280.0), Portsmouth (RM 356.0), White Oak Creek (RM 425.0), downstream Meldahl dam (RM 437.5), and North Bend (RM 487.2). Several sites ranged between not meeting the WWH to meeting the EWH aquatic life designated use. Differences in scoring are attributed to the inclusion of gizzard shad and emerald shiners in the community proportional metric calculations. Low-end scoring adjustments would reduce stability. It is imperative that inherent natural variation does not vary substantially when exposed to various hydrologic cycles. Ranges between drought (1991) and 100-year flood (1994) stages were observed in the Ohio River during the study period; however, the relatively small range of IBI difference between most sites show that collection procedures are adequate to assess the Ohio River.

Precision of site classification for both the IBI and MIwb biocriteria is very high. Site classifications are within an accurate classification range for all sites. The accuracy of site scoring was not significantly different between years for MIwb, while the differences in annual IBI scores at the site downstream of the Muskingum River (RM 172.7) were significant. Differences in gizzard shad catch influenced site classification (RM 172.7).

18.3.2.2 Minimum Detection Difference

Sampling between 1989 and 1995 showed that differences in IBI scores ranged from 4 (RM 437.5) to 16 (RM 172.7) points (Table 18.3). The central tendency of IBI scores had a standard deviation range between 2.00 to 5.89. The largest variance in IBI score was at RM 172.7, which had a value of 34.667.

18.3.3 Habitat

18.3.3.1 Major Substrate Differences

Differences in navigation pool zone patterns and three substrate classifications were evaluated with IBI, MIwb, QHEI, species richness, relative abundance (number/km), and relative weight

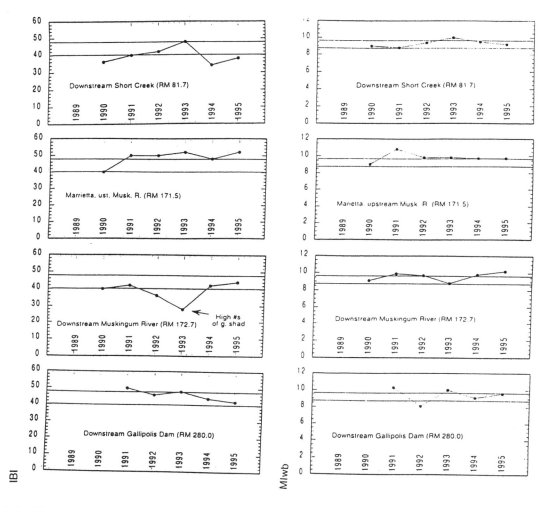

FIGURE 18.12 Distribution of IBI and MIwb scores for eight long-term monitoring sites on the Ohio River. (A) ORM 81.7, (B) ORM 171.5, (C) ORM 172.7, (D) 280.0, (E) ORM 356.0, (F) 425.0, (G) 437.5, and (H) ORM 487.2.

(weight/km) for fish collections between 1989 and 1995 (Figure 18.13). Navigation pools were divided into three zones: riverine or tailwaters, transitional or mid-pool, and lacustrine or lower pool (Stanford et al., 1988). Substrates were divided into three types, including coarse, mixed, and fine substrates. Coarse substrates included boulder, cobble, and gravel substrates; fine substrates included muck, sand, and silt; and mixed substrates incorporated ranges of both groups of materials.

Macrohabitat patterns based on 141 sample events in the Ohio River show that riverine or tailwater habitats generally performed best in IBI, MIwb, and total number of fish species (Figure 18.14). The tailwater habitat had the smallest range of the three macrohabitat zones for IBI. Although the lower pool or lacustrine zone scored the lowest on habitat complexity, no statistical difference was observed with the IBI or MIwb. The lacustrine or lower pool zone showed the highest results for relative number of fish/km and relative weight/km.

Microhabitat variables showed that mixed and coarse substrates scored best for all variables evaluated (Figure 18.13). Coarse substrates acheived the highest IBI and relative number of fish/km.

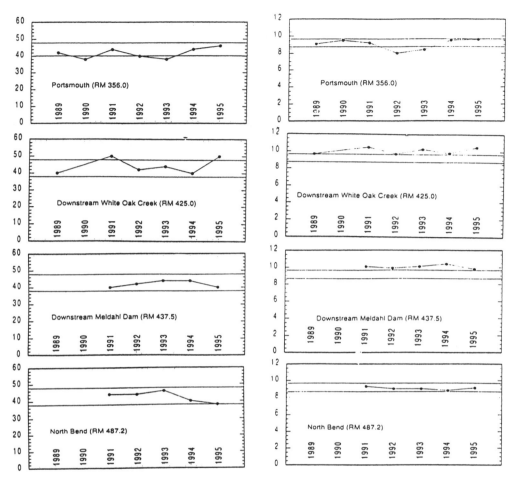

FIGURE 18.12 *Continued.*

TABLE 18.3
Descriptive Statistics for Ohio River Stations (Designated by River Mile) Repeat Sampled Between 1989 and 1995

	Ohio River Mile							
Statistic	487.2	437.5	425	356	280	172.7	171.5	81.7
Minimum IBI	38	40	40	38	42	28	40	34
Maximum IBI	46	44	50	46	50	44	52	48
Range	8	4	10	8	8	16	12	14
Mean	42.4	42	44.3	41.71	46	38.67	48.67	39.67
Median	44	42	43	42	46	41	50	39
Standard deviation	3.29	2	4.63	3.147	3.162	5.888	4.502	4.967
Variance	10.8	4	21.47	9.90	10.0	34.667	20.267	24.667
Standard error	1.47	0.894	1.89	1.19	1.414	2.404	1.838	2.028
Skewness	−0.35	0	0.415	−0.028	0	−1.078	−1.375	0.636
Kurtosis	−1.42	−1.75	−1.543	−145	−1.3	−0.193	0.445	−0.605

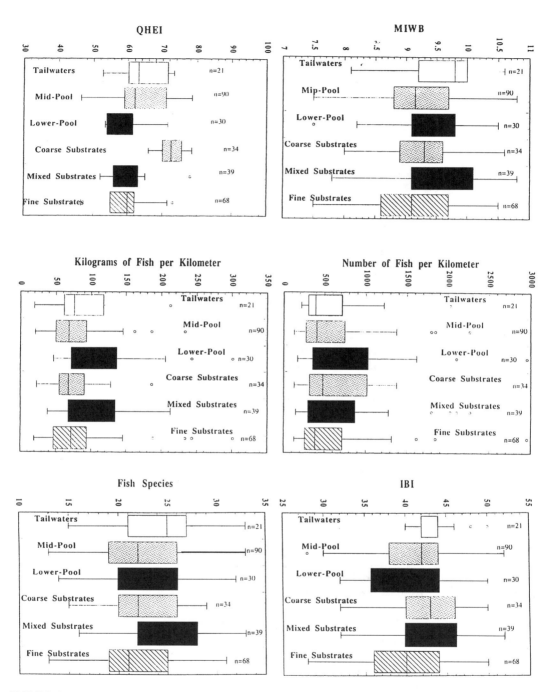

FIGURE 18.13 Relationships of IBI, MIwb, QHEI, Fish species, Relative number of fish (number/km), and relative weight (kg/km) with (A) pool position: tailwaters, mid-pool, and lower pool, and (B) substrate type: coarse substrates, mixed substrates, and fine substrates.

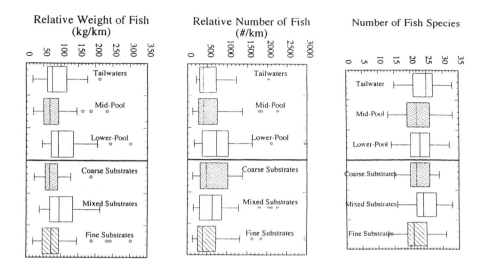

FIGURE 18.14 Number of fish species, relative number (number/km), and relative weight (kg/km) with (A) pool position: tailwaters, mid-pool, and lower pool, and (B) substrate type: coarse substrates, mixed substrates, and fine substrates.

Mixed substrates scored the highest on MIwb, number of fish species, and relative weight of fish/km. Fine substrates did not show statistically different results from coarse substrates for most variables, with the exception of mean IBI and QHEI. The IBI was not significantly different between coarse and mixed substrates. The QHEI was statistically different between coarse and mixed and fine substrates, while mixed and fine substrates were not significantly different.

18.3.3.2 QHEI Patterns

Patterns in fish community structure and function with increases in habitat complexity were evaluated, as measured by the Qualitative Habitat Evaluation Index (QHEI) (Rankin, 1995). Species richness, density, biomass, MIwb, and IBI were compared with changes in habitat quality (Figure 18.3). Habitat structure did not significantly affect community species richness, community density or biomass (Figure 18.13), or MIwb (Figure 18.5). Increases in habitat structure improved significantly ($p > 0.05$) the IBI (Figure 18.4) since habitat heterogeneity provides greater opportunities for additional species to inhabit a reach, increasing species composition, trophic guild, and abundance metrics.

Although habitat structure did not significantly improve species richness, community density, or biomass, specific patterns were observed in select large river species (Figure 18.6). Increased habitat structure did not affect emerald shiner or freshwater drum density; however, habitat structure negatively affected gizzard shad density. Increased habitat structure showed a statistically significant improvement in mimic shiner ($p > 0.10$) and golden redhorse ($p > 0.05$) densities. Community biomass and channel catfish biomass were not statistically affected by changes in habitat structure; however, statistically significant increases in carp and gizzard shad biomass ($p > 0.05$) corresponded with lowered habitat structure. Increased habitat structure showed a statistically significant increase in golden redhorse biomass ($p > 0.05$).

Relationships between downstream distance from a dam and habitat complexity declined in quality (Figure 18.15). Distance downstream from dams and QHEI values showed a statistically significant decline in habitat index scores ($p < 0.05$) and total warmwater habitat (WWH) attributes ($p < 0.20$). Increasing distance downstream from the dam showed a statistically significant increase

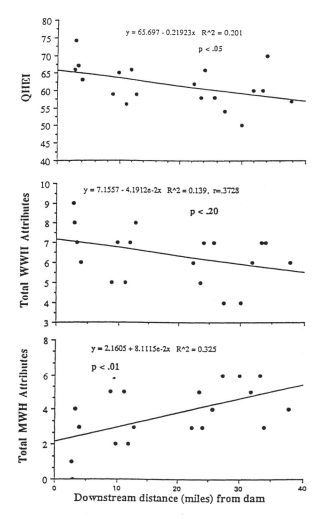

FIGURE 18.15 Effects of various habitat measures QHEI (A), total warmwater habitat attributes (B), and total modified warmwater habitat attributes (C) with downstream distance from dam (miles).

in total modified warmwater ($p < 0.01$) habitat attributes (MWH). The increase in the ratio of MWH to WWH attributes is used to classify designated uses for streams in Ohio. Rankin (1995) showed that increases in MWH attributes such as silt covered substrates; embeddedness; and muck, silt, and sand substrates would cause a lower biological integrity score. Declining QHEI values downstream of the dam, however, did not cause a reduction in MIwb values.

Distance downstream from a dam influences the potential IBI or MIwb score for sites along a river continuum, with the majority of the 141 sites along Ohio meeting the WWH or exceptional warmwater habitat (EWH) ecoregion criteria for Large Rivers (Figure 18.16). Mean MIwb scores for each of the navigation pools showed they were meeting WWH designated uses for aquatic life. IBI scores were more variable, and mean IBI values for several pools did not attain the WWH ecoregion biocriteria. The mean IBI scores for the Racine and Gallipolis Pools did not attain aquatic life designated uses, while the New Cumberland, Pike Island, Gallipolis, and Markland Pools did not have a single site that attained the EWH designated uses.

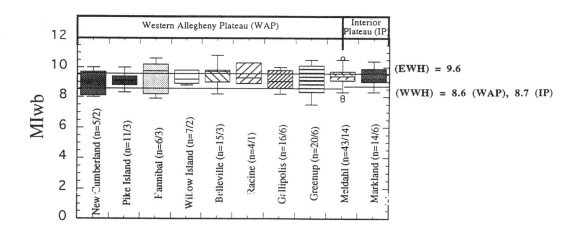
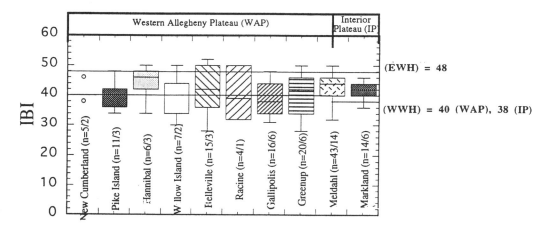

FIGURE 18.16 Distribution of Biocriteria scores by pool in the upper Ohio River: (A) MIwb and (B) IBI.

18.4 CONCLUSIONS

Sampling in Large and Great Rivers can be accomplished using the same sampling rationale as applied to streams and moderate-sized rivers. Gear bias and multiple gear issues need to be evaluated in order to select the single, best gear that will effectively sample the entire fish community in its true relative abundance. The efforts in the present study were developed with the expectation of collecting minnows, darters, madtoms, and other non-game species that may not be adequately represented in other surveys. This study on the Ohio River has demonstrated that a representative catch of all species at a site can be collected using DC night electrofishing for a distance of 0.5 km. Although a minimum of four collections is needed to achieve an asymptote of maximum species richness for biological diversity determinations in the Ohio River, this study sampled a sufficient portion of the community to assess biological integrity with a single collection.

Comparison of day and night collection efforts showed that statistically significant differences in night collection efforts were acheived. Index period is an important consideration in sampling, this study has shown that periods of low to stable flows should be targeted to enhance species catch. In the Ohio River drainage, this is between late summer and fall. The results show that habitat differences did not statistically influence species richness, density, or biomass trends in the Ohio River; however, IBI assessment classification was influenced by differences in habitat. Subtle differences in metric performance were enhanced with the removal of select schooling species that tended to clump together in distribution. Collection of gizzard shad and emerald shiner caused the swamping of select proportional metrics, such as the proportion of large river species; trophic guild metrics such as proportion of omnivores, insectivores, and carnivores; reduced patterns in abundance; and reproductive guild proportional metrics. Thus, the deletion of these species abundances before calculation of the proportion metrics is recommended, along with treatment of DELT anomalies using two specific approaches for Great River IBI development. Either long-lived fish need to be targeted using a minimum length interval to express the proportion of DELT in the community, or a relative number of individuals that possess DELT anomalies need to be indicated so that the presence of a minimum number of individuals possessing DELT is enough to cause low-end scoring adjustments. The precision and accuracy of the IBI showed that metric performance of select metrics did not differ statistically for eight sites sampled between 5 and 7 years. Minimum detection differences were well within the robustness of stream sampling approaches. Habitat structure differences did not significantly affect community species richness, community density, biomass, or the Modified Index of Well-Being. However, increases in habitat heterogeneity significantly ($p > 0.05$) improved the IBI. Specific Large River species responded differentially when habitat structure changed. Various components of the community changed with increasing habitat complexity; however, these changes caused a reduction response in other species.

REFERENCES

Angermeier, P.L. and J.R. Karr. 1986. Applying an Index of Biotic Integrity based on stream-fish communities: considerations in sampling and interpretation, *North American Journal of Fisheries Management,* 6, 418–429.

Carlson, C.A. and R.T. Muth. 1989. The Colorado River: lifeline of the American Southwest, *Canadian Special Publication Fisheries and Aquatic Sciences,* 106, 220–239.

EA Engineering, Science, and Technology (EA). 1993. *1991 Ohio River Ecological Research Program.* EA Engineering, Science, and Technology, Deerfield, IL.

EA Engineering, Science, and Technology (EA). 1994. *1992 Ohio River Ecological Research Program.* EA Engineering, Science, and Technology, Deerfield, IL.

Ebel, W.J., C.D. Becker, J.W. Mullan, and H.L. Raymond. 1989. The Columbia River: toward a holistic understanding, *Canadian Special Publication Fisheries and Aquatic Sciences,* 106, 205–219.

Environmental Science and Engineering, Inc. (ESE). 1995. *1993 Ohio River Ecological Research Program.* Environmental Science and Engineering, Inc., St. Louis, MO.

Gammon, J.R. 1976. The Fish Populations of the Middle 340 km of the Wabash River. Purdue University Water Resources Research Center Technical Report 86, LaFayette, IN.

Gammon, J.R. 1983. Changes in the fish community of the Wabash River following power plant start-up: projected and observed, in W.E. Bishop, R.D. Cardwell, and B.B. Heidolph (Eds.), *Aquatic Toxicology and Hazard Assessment: Sixth Symposium.* ASTM STP 802, Philadelphia. 350–366.

Goldstein, R.M., T.P. Simon, P.A. Bailey, M. Ell, E. Pearson, K. Schmidt, and J.W. Emblom. 1994. Concepts for an index of biotic integrity for the streams of the Red River of the North Basin, in *Proceedings of the North Dakota Water Quality Symposium,* Fargo, North Dakota. p. 169–180.

Guetreuter, S. 1996. Trends in fishes of the Upper Mississippi River system measured by the long-term resource monitoring program: 1990–1994. National Biological Service News Notes #2436.

Gibson, G.R. (Ed.). 1994. *Biological Criteria: Technical Guidance for Streams and Small Rivers.* EPA-822-B-94-001. U.S. Environmental Protection Agency, Office of Science and Technology, Washington, D.C.

Gosz, J.R. 1980. The influence of reduced stream flows on water quality, in W.O. Spofford, Jr., A.L. Parker, and A.V. Kneese, (Eds.), *Energy Development in the Southwest, Vol. 2. Resources for the Future.* Washington, DC. 3–48.

Hardin, S. and L.L. Connor. 1992. Variability of electrofishing crew efficiency, and sampling requirements for estimating reliable catch rates, *North American Journal of Fisheries Management,* 12, 612–617.

Hocutt, C.H. and J.R. Stauffer, Jr. 1980. *Biological Monitoring of Fish.* Lexington Books, Lexington, MA.

Horton, R.E. 1945. Erosional developments of streams and their drainage basins: hydrophysical approach to quantitative morphology, *Bulletin of the Geological Society of America,* 56, 275–370.

Hughes, R.M. 1995. Defining acceptable biological status by comparing with reference condition, in W.S. Davis and T.P. Simon (Eds.), *Biological Assessment and Criteria: Tools for Water Resource Planning and Decision Making.* Lewis, Boca Raton, FL. 31–47.

Hughes, R.M. and J.R. Gammon. 1987. Longitudinal changes in fish assemblages and water quality in the Willamette River, Oregon, *Transactions of the American Fisheries Society,* 116, 196–209.

Hughes, R.M., D.P. Larsen, and J.M. Omernik. 1986. Regional reference sites: a method for assessing stream potential, *Environmental Management,* 10, 629–635.

Jacobs, K.E. and W.D. Swink. 1982. Estimation of fish population size and sampling efficiency of electrofishing and rotenone in two Kentucky tailwaters, *North American Journal of Fisheries Management,* 2, 239–248.

Karr, J.R. 1981. Assessment of biotic integrity using fish communities, *Fisheries (Bethesda),* 6(6), 21–27.

Karr, J.R. and M. Dionne. 1991. Designing surveys to assess biological integrity in lakes and reservoirs, in *Biological Criteria: Research and Regulation, Proceedings of a Symposium.* 12–13 December 1990, Arlington, VA. EPA-440-5-91-005. U.S. Environmental Protection Agency, Office of Water, Washington, DC. 62–72.

Karr, J.R., K.D. Fausch, P.L. Angermeier, P.R. Yant, and I.J. Schlosser. 1986. Assessing Biological Integrity of Running Waters: A Method and Its Rationale. Illinois Natural History Survey Special Publication 5.

Karr, J.R., L.A. Toth, and D.R. Dudley. 1985b. Fish communities of midwestern rivers: a history of degradation, *BioScience,* 35, 90–95.

Moyle, P.B. and J.E. Williams. 1990. Biodiversity loss in the temperate zone: decline of the native fish fauna of California, *Conservation Biology,* 4, 275–284.

Nielsen, L.A. and D.L. Johnson (Eds.). 1983. *Fisheries Techniques.* American Fisheries Society, Bethesda, MD.

Ohio Environmental Protection Agency. 1989. *Biological Criteria for the Protection of Aquatic Life. Vol. III. Standardized Field Sampling and Laboratory Methods for Assessing Fish and Macroinvertebrate Communities.* Ohio EPA, Division of Water Quality Monitoring and Assessment, Columbus, OH.

Plakfin, J.L., M.T. Barbour, K.D. Porter, S.K. Gross, and R.M. Hughes. 1989. Rapid Bioassessment Protocols for Use in Streams and Rivers. Benthic Macroinvertebrates and Fish. EPA 440-4-89-001. Office of Water Regulations and Standards, U.S. Environmental Protection Agency, Washington, DC.

Rankin, E.T. 1995. Habitat indices in water resource quality assessments, in W.S. Davis and T.P. Simon (Eds.), *Biological Assessment and Criteria: Tools for Water Resource Planning and Decision Making.* Lewis, Boca Raton, FL. 181–208.

Reash, R.J. 1994. Biocriteria: a regulated industry perspective, in W.S. Davis and T.P. Simon (Eds.), *Biological Assessment and Criteria: Tools for Water Resource Planning and Decision Making.* Lewis, Boca Raton, FL. 153–166.

Sanders, R.E. 1990. A 1989 Night Electrofishing Survey of the Ohio River Mainstem (RM 280.8 to 442.5). Ohio EPA, Ecological Assessment Section, Columbus, OH.

Sanders, R.E. 1991. A 1990 Night Electrofishing Survey of the Upper Ohio River Mainstem (RM 40.5 to 270.8) and Recommendations for a Long-Term Monitoring Network. Ohio Environmental Protection Agency, Ecological Assessment Section, Columbus, OH.

Sanders, R.E. 1992. Ohio's Near-Shore Fishes of the Ohio River: 1991 to 2000 (Year One: 1991 Results). Ohio Environmental Protection Agency, Ecological Assessment Section, Columbus, OH.

Sanders, R.E. 1993. Ohio's Near-Shore Fishes of the Ohio River: 1991 to 2000 (Year Two: 1992 Results). Ohio Environmental Protection Agency, Ecological Assessment Section, Columbus, OH.

Sanders, R.E. 1994. Ohio's Near-Shore Fishes of the Ohio River: 1991 to 2000 (Year Three: 1993 Results). Ohio Environmental Protection Agency, Ecological Assessment Section, Columbus, OH.

Sanders, R.E. 1995. Ohio's Near-Shore Fishes of the Ohio River: 1991 to 2000 (Year Four: 1994 Results). Ohio Environmental Protection Agency, Ecological Assessment Section, Columbus, OH.

Simon, T.P. 1991. Development of Index of Biotic Integrity Expectations for the Ecoregions of Indiana. I. Central Corn Belt Plain. EPA 905-9-91-025. U.S. Environmental Protection Agency, Chicago, Illinois.

Simon, T.P. 1992. Biological Criteria Development for Large Rivers with an Emphasis on an Assessment of the White River Drainage, Indiana. EPA 905-R-92-006. U.S. Environmental Protection Agency, Chicago.

Simon, T.P. and E.B. Emery. 1995. Modification and assessment of an index of biotic integrity to quantify water resource quality in Great Rivers, *Regulated Rivers Research and Management,* 11, 283–298.

Simon, T.P. and J. Lyons. 1995. Application of the index of biotic integrity to evaluate water resource integrity in freshwater ecosystems, in W.S. Davis and T.P. Simon (Eds.), *Biological Assessment and Criteria: Tools for Water Resource Planning and Decision Making.* Lewis, Boca Raton, FL. 245–262.

Stanford, J.A., F.A. Hauer, and J.V. Ward. 1988. Serial discontinuity in a large river system, *Verh. Int. Verein. Theoret. Angew. Limnol.,* 23, 114–118.

Thurston, R.V., R.C. Russo, C.M. Fetterolf, Jr., T.A. Edsall, and Y.M. Baber, Jr. (Eds.), 1979. *A Review of the EPA Red Book: Quality Criteria for Water.* Water Quality Section, American Fisheries Society, Bethesda, MD.

Vannote, R.L., G.W. Minshall, K.W. Cummins, J.R. Sedell, and C.E. Cushing. 1980. The river continuum concept, *Canadian Journal of Fisheries and Aquatic Sciences,* 37, 130–137.

Yoder, C.O. and E.T. Rankin. 1995. Biological response signatures and the area of degradation value: new tools for interpreting multimetric data, in W.S. Davis and T.P. Simon (Eds.), *Biological Assessment and Criteria: Tools for Water Resource Planning and Decision Making.* Lewis, Boca Raton, FL. 263–286.

19 Tailwater Fish Index (TFI) Development for Tennessee River Tributary Tailwaters

Edwin M. Scott, Jr.

CONTENTS

19.1 Introduction 507
19.2 Methods 508
    19.2.1 Study Area 508
    19.2.2 Collection Procedures 509
    19.2.3 Data Analysis 510
    19.2.4 Rationale for Tailwater Fish Index 510
    19.2.5 Metrics, Scoring Criteria, and Species Designations 511
        19.2.5.1 Species Richness and Tolerance Composition Metrics 511
        19.2.5.2 Trophic Composition Metrics 516
        19.2.5.3 Abundance and Condition Metrics 516
19.3 Results and Discussion 517
    19.3.1 Metric Behavior 517
        19.3.1.1 Species Richness and Tolerance Composition Metrics 517
        19.3.1.2 Trophic Composition Metrics 518
        19.3.1.3 Abundance and Condition Metrics 519
    19.3.2 Comparison of TFI and IBI Analysis 519
        19.3.2.1 Douglas Tailwater 520
        19.3.2.2 Cherokee Tailwater 521
19.4 Conclusion 521
Acknowledgments 521
References 521

19.1 INTRODUCTION

Fish are useful biological indicators of stream health because they are relatively long-lived aquatic animals, integrating physical and chemical conditions of stream reaches in which they reside (Karr, 1981; Karr et al., 1986). Assessment of stream fish community condition was derived from an Index of Biotic Integrity (IBI) developed for small, warmwater streams in the early 1980s. The IBI compares 12 characteristics of a stream fish community to expectations of a reference fish community in a relatively undisturbed stream of similar size and physiogeographic region. Regional fisheries biologists develop expectations from their knowledge and experience of regional fish communities, allowing IBI assessments tailored to their respective regions. The Tennessee Valley Authority (TVA) first applied IBI concepts to existing fish survey data from 1984 (Saylor and Scott, 1987), and soon after incorporated IBI surveys as the preferred method for stream fish assessments in water quality monitoring studies across the Tennessee Valley by 1986 (Saylor et al., 1987; 1988; Saylor and Ahlstedt, 1990). Furthermore, IBI sampling methods were applied to tailwater fish community assessments associated with the TVA's Reservoir Releases Improvements (RRI) Program (Tomljanovich and Saylor, 1989; 1992; Yeager and Tomljanovich, 1990; Scott et al., 1996), which provides minimum flows and reaeration of tailwater flows.

Initially, stream IBI metrics and scoring criteria were applied to TVA tailwater fish surveys in order to assess the RRI program. But after 4 years of providing minimum flows, two dams failed to produce appreciable improvements in IBI scores downstream, calling into question the validity of comparing tailwater fish communities to reference conditions of free-flowing streams (Scott et al., 1996). Community improvements were occurring, but under IBI metrics and scoring criteria, fish community ratings remained in "very poor" to borderline "fair" status (Tomljanovich and Saylor, 1989; 1992; Yeager and Tomljanovich, 1990).

Dams have an enormous effect on streams by interrupting daily and seasonal streamflows, altering seasonal temperature patterns, reducing movement of sediment, and disrupting the natural processing and flow of organic materials and nutrients through the aquatic ecosystem (Yeager, 1993). Coarse and fine particulate organic matter (CPOM and FPOM), the energy source of streams on which the aquatic food web relies (Vannote et al., 1980), is trapped in reservoirs. Furthermore, as organic materials settle out in reservoirs and decompose, oxygen is removed from the water, degrading the quality of the water that is eventually released to tailwaters.

The IBI scoring criteria developed for unregulated streams cannot fairly represent attainable conditions in tailwaters. Alternative methods and scoring criteria to assess tailwater fish communities are needed. Since reference conditions for "pristine" or least impacted tailwaters do not exist (or could), tailwater fish community assessments in this study were based on "perceived potential" of tailwaters (suggested by Jim Ruane, Reservoir Environmental Management, Chattanooga, TN, and George Gibson, U.S. EPA, Washington DC, personal communication) and data from Douglas Tailwater collected between 1987 and 1994. The resulting Tailwater Fish Index (TFI) is a set of 12 metrics modified from the original IBI and rated by scoring criteria deemed acceptable for tailwaters with warmwater fisheries.

19.2 METHODS

19.2.1 STUDY AREA

Tailwater fish surveys were made below Douglas and Cherokee dams, two impoundments of the French Broad River and Holston River, respectively. These two rivers converge in East Tennessee to form the Tennessee River at Knoxville (Figure 19.1). Both tailwaters are characterized by long stretches of riverine conditions. The French Broad River below Douglas Dam flows approximately 28 miles before reaching the backwaters of Fort Loudoun Reservoir, while the Holston River flows about 48 miles from Cherokee Dam to Fort Loudoun Reservoir. Three sites in each tailwater were chosen to measure stages of biological recovery at various distances from the dams following establishment of minimum flows and reaeration of reservoir releases. Minimum flows at Douglas Dam and Cherokee Dam of 585 and 325 cfs, respectively, are provided by periodic turbine pulsing, and the target level of dissolved oxygen (DO) for reservoir release is 4 mg/L at both dams. Tailwater site selections were made based on the presence of adequate amounts of shallow riffles, runs, and deep pools to assure maximum physical habitat variability and associated fish species diversity. Sampling Douglas Tailwater began in 1987, prior to initiation of minimum flows, and continued in even-numbered years thereafter. Cherokee Tailwater samples were collected in odd-numbered years beginning in 1989, after minimum flows were established in 1988. An additional sample was taken in Cherokee Tailwater in 1996.

Reaeration devices and techniques for TVA storage reservoir releases were developed at Douglas Dam between 1987 and 1993. Surface water pumps were initially installed in the reservoir at the

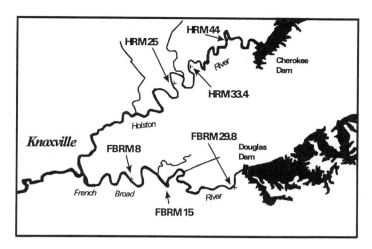

FIGURE 19.1 Sampling sites on Cherokee Tailwater (Holston River) and Douglas Tailwater (French Broad River) in east Tennessee, 1987 through 1996. (From Scott, E.M., Jr. and B.L. Yeager. 1997. Biological Responses to Improved Reservoir Releases, *Waterpower '97, Proceedings of the International Conference on Hydropower,* Aug. 5–8, Atlanta, 60–68. ASCE, with permission.)

face of the dam to force oxygenated surface water downward toward the hydroturbine intakes. Various modifications were tested until 1993, when an optimal system was placed in operation. A liquid oxygen injection system, composed of a 20,000-gallon storage tank, aluminum vaporizers, and 16 diffuser frames containing 12 miles of porous hoses submerged in the reservoir forebay, was developed from various prototypes tested between 1988 and 1993 and made operational in September 1993 (Mobley and Brock, 1995). Turbine venting was added in late summer 1995 to decrease the amount of high-cost liquid oxygen needed to meet target discharge DO concentration. Reaeration technology, developed at Douglas Dam, was transferred to Cherokee Dam, where it became operational during the summer of 1995. Thus, Douglas Tailwater has received varying amounts of reaeration since 1987, improving each year; but Cherokee Tailwater did not receive reaeration improvements until 1995. Historically, the oxygen content of discharges from both dams fell below 4 mg/L for more than 100 days during the summer, and was occasionally less than 1 mg/L. By 1995, reaeration efforts restored DO content of discharges to 4 mg/L for nearly the entire summer (Scott et al., 1996).

19.2.2 COLLECTION PROCEDURES

Backpack electrofishing units, seines, dipnets, and boat shockers were used to capture fish from deep pools and shallow shoals of riverine tailwater sites. Because many fish species are habitat specialists, efforts were made to sample all habitats present (e.g., riffles, runs, pools, undercut banks, vegetation, etc.). Samples were collected during daylight hours in spring or summer, while discharges from upstream dams were either zero or at minimum levels to ensure low flows for crew safety and effectiveness.

Shorelines and channel sections of deep pool areas were sampled with a shocker boat traveling in a downstream direction. Fish momentarily stunned by the electric shock were netted for identification, inspected for anomalies, and counted. Shocking runs continued until two successive runs failed to collect any new species for a given habitat (i.e., shoreline or mid-channel). Each 10-minute boat shocking run was counted as two units of sampling effort.

In shallow water throughout the river channel and along the shorelines, a crew of five persons collected fish using a backpack shocker, dip nets, and seines. A 20-foot seine with 4.8-mm mesh was routinely positioned perpendicular to the river's flow at a distance of 20 feet downstream from the person operating the backpack shocker (Saylor and Ahlstedt, 1990). As direct current (DC) was applied to the area equal to the width of the seine, the operator moved downstream to the stationary seine. Fish stunned by the electric field drifted with the river's flow into the waiting seine or were dipped by a crew member accompanying the person carrying the backpack shocker. Allowing for sag in the seine, each run sampled an area of approximately 300 square feet, and constituted a unit of sampling effort. Care again was taken to sample all habitats present with respect to depth, current velocity, and substrate. In areas of little or no current (e.g., sluggish pools and backwaters), seine hauls were made. Additional backpack shocking along shorelines sampled brush, boulder, undercut banks, and tree root habitats, and each 5-minute duration was counted as one unit of effort. Sampling was repeated in each habitat until three successive runs produced no additional new species.

All fish collected were identified and counted, with the exception of young-of-year (age 0) fishes, which are not counted as part of the fish community for biomonitoring purposes. Each fish was examined for anomalies (i.e., parasites, diseases, deformities, lesions, etc.). Incidence of five or more "black spot" grubs (*Neascus* spp.) on a fish was recorded as an anomaly. Some fish were preserved in formaldehyde and kept for voucher specimens or for later laboratory verification.

19.2.3 Data Analysis

While field survey methods in TVA tailwaters were identical to IBI methods used extensively by the TVA in large, unregulated streams (Saylor et al., 1987; 1988; Saylor and Ahlstedt, 1990; Tomljanovich and Saylor, 1989; 1992), slightly different metrics and modified scoring criteria were employed during analysis of tailwater data (Scott et al., 1996). Modifications of typical IBI analysis for warm tailwater applications include a less stringent set of scoring criteria (Table 19.1), expansion of one metric, and species (tolerant/intolerant) designations and expectations somewhat different from those used for IBI samples in unregulated streams (Table 19.2). The resultant index value is termed the Tailwater Fish Index (TFI).

Warm TFI criteria are based primarily on Douglas Tailwater fish surveys conducted between 1987 and 1994, as minimum flows and reaeration techniques at Douglas Dam were developed and made operational. During these years, Douglas Tailwater fish community metrics responded positively to improvements in water quality (Figures 19.2A–J), providing clues to setting meaningful expectations for tailwater situations (i.e., perception of tailwater potential for high-quality fish communities). Comparison of year-to-year and site-to-site species occurrences and trophic compositions indicated improvements over time and with increasing distance from the dam. In other words, better fish communities were found further downstream from the dam, and improvements spread upstream in later years, after reaeration devices were installed. Meanwhile, in Cherokee Tailwater, fish community responses were somewhat less consistent, following only minimum flow improvements.

19.2.4 Rationale for Tailwater Fish Index

IBI methods to measure biological integrity of unregulated warmwater streams were found to be inadequate for Douglas and Cherokee tailwater fish community assessments. Fish assemblages found in regulated tailwater stretches could not be effectively measured against IBI metrics and scoring criteria. Recognizing that tailwaters are unnatural river stretches lacking biotic integrity, their fish communities can still be used to measure relative ecological health of tailwaters. Perceptions of what desirable tailwater fish communities could be in light of human modifications to the natural ecosystem were the basis for modifications to IBI metrics and scoring criteria to develop the TFI (Hughes and Oberdorff, in press).

TABLE 19.1
Modifications of Index of Biotic Integrity (IBI) Metrics and Scoring Criteria for Warm Tailwater Fish Index (TFI) for Douglas and Cherokee Tailwaters, Tennessee

Metric	IBI			TFI — Warmwater		
	1	3	5	1	3	5
Species Richness	95% Max. Species Richness[a]			80% Max. Species Richness[b]		
1. Total species	<26	26–51	>51	<19	19–37	>37
2. Darter species	<5	5–8	>8	<4	4–6	>6
3. Sunfish species	<3	3–4	>4	<3	3–4	>4
4. Sucker species	<5	5–8	>8	<3	3–5	>5
Tolerance/Intolerance						
5. Intolerant species[c]	<2	2–3	>3	<4	4–8	>8
6. Percent tolerant individuals	>20	10–20	<10	>40	20–40	<20
Trophic Composition						
7. Percent omnivores (+ generalists)[e]	>30	15–30	<15	>50	25–50	<25
8. Percent specialist insectivores	<25	25–50	>50	<15	15–30	>30
9. Percent piscivores	<2	2–5	>5	No change		
Abundance and Condition						
10. Catch per unit effort	<9	9–16	>16	<6	6–11	>11.0
11. Percent hybrids	>1	Trace–1	0	>1	No change	<0.5
12. Percent diseased, injured, anomalous	>5	2–5	<2	No change		

[a] Native species.
[b] Total species.
[c] Number of intolerant species increased from 5 (IBI) to 15 species in TFI.
[d] Sculpins included as tolerant in TFI.
[e] Generalists (sculpins) added to omnivores in TFI.

(From Scott, E.M., Jr. and B.L. Yeager. 1997. Biological Responses to Improved Reservoir Releases, *Waterpower '97, Proceedings of the International Conference on Hydropower*, Aug. 5–8, Atlanta, 60–68. ASCE, with permission.)

19.2.5 Metrics, Scoring Criteria, and Species Designations

19.2.5.1 Species Richness and Tolerance Composition Metrics

Visual inspection of the total numbers of species collected in Douglas Tailwater samples between 1987 and 1994 (Figure 19.2A) was the basis for determining maximum species richness expectations in warm tailwaters. As tailwater conditions improved, greater biodiversity (i.e., species richness) was observed. The distribution of total species over time for the three Douglas Tailwater sampling sites indicated greatest diversity downstream (FBRM 8), along with generally increased diversity at all three sites over time. Based on the best conditions observed at FBRM 8, TFI scoring criteria for the total species metric were reduced from the IBI 95% maximum species richness to 80% of maximum expectations (Table 19.1). In addition, while IBI criteria are based only on native species, TFI criteria include introduced species (usually the result of reservoir fisheries management activities). Very rare or seldom collected species, such as spotfin chub (*Cyprinella monacha*) and blotchside logperch (*Percina burtoni*), which were included in previous IBI studies, were omitted from expectations in the present TFI analysis. Also, species that are apparently isolated from sources of recruitment into Douglas Tailwater by reservoirs or tributaries with regions of culturally diminished fish communities (e.g., lower Little Pigeon River) were not included in TFI expectations.

TABLE 19.2
Fish Species Expected to Occur in Douglas and Cherokee Tailwaters and Tolerance, According to IBI and TFI Analysis, Trophic Status, and Taxonomic Groupings

Common Name/Scientific Name	Expected		Tolerance		Trophic Status	Taxonomic Grouping
	IBI	TFI	IBI	TFI		
Lampreys — Petromyzontidae						
Ohio lamprey, *Ichthyomyzon bdellium*	X				PA	
Chestnut lamprey, *Ichthyomyzon castaneus*	X	X			PA	
American Brook lamprey, *Lampetra appendix*	X				HB	
Gars — Lepisosteidae						
Spotted gar, *Lepisosteus oculatus*	X	X			PI	
Longnose gar, *Lepisosteus osseus*	X	X	TOL	TOL	PI	
Mooneyes — Hiodontidae						
Mooneye, *Hiodon tergisus*	X	X		INT	IN	
Herrings — Clupeidae						
Skipjack herring, *Alosa chrysochloris*	X	X			PI	
Gizzard shad, *Dorosoma cepedianum*	X	X	TOL	TOL	OM	
Threadfin shad, *Dorosoma petenense*	X	X			PK	
Minnows — Cyprinidae						
Central stoneroller, *Campostoma anomalum*	X	X			HB	
Whitetail shiner, *Cyprinella galactura*	X	X		INT	IN	
Spotfin chub, *Cyprinella monacha*	X				SP	
Spotfin shiner, *Cyprinella spiloptera*	X	X	TOL	TOL	IN	
Steelcolor shiner, *Cyprinella whipplei*	X				IN	
Common carp, *Cyprinus carpio*	I	X	TOL	TOL	OM	
Blotched chub, *Erimystax insignis*	X	IS		INT	SP	
Bigeye chub, *Hybopsis amblops*	X	X			SP	
Striped shiner, *Luxilus chrysocephalus*	X	X			OM	
Warpaint shiner, *Luxilus coccogenis*	X	IS			SP	
Speckled chub, *Macrhybopsis aestivalis*	X	X			SP	
Silver chub, *Macrhybopsis storeriana*	X				SP	
River chub, *Nocomis micropogon*	X	X	TOL		OM	
Golden shiner, *Notemigonus crysoleucas*	X		TOL	TOL	OM	
Emerald shiner, *Notropis atherinoides*	X				SP	
Tennessee shiner, *Notropis leuciodus*	X	IS			SP	
Silver shiner, *Notropis photogenis*	X	IS			SP	
Rosyface shiner, *Notropis rubellus*	X	X			SP	
Telescope shiner, *Notropis telescopus*	X	X	INT	INT	SP	
Mimic shiner, *Notropis volucellus*	X	X			SP	
Stargazing minnow, *Phenacobius uranops*	X	IS			SP	
Bluntnose minnow, *Pimephales notatus*	X	X			OM	
Bullhead minnow, *Pimephales vigilax*	X	X			SP	
Blacknose dace, *Rhinichthys atratulus*					GE	
Suckers — Catostomidae						
River carpsucker, *Carpiodes carpio*	X	X			OM	Suckers
Quillback, *Carpiodes cyprinus*	X	X			OM	Suckers
Highfin carpsucker, *Carpiodes velifer*	X	R			OM	Suckers
Blue sucker, *Cycleptus elongatus*	X	R			IN	Suckers
Northern hog sucker, *Hypentelium nigricans*	X	X	INT	INT	IN	Suckers
Smallmouth buffalo, *Ictiobus bubalus*	X	X			OM	Suckers
Black buffalo, *Ictiobus niger*	X	X			OM	Suckers
Spotted sucker, *Minytrema melanops*	X	R			IN	Suckers

TABLE 19.2 *(continued)*
Fish Species Expected to Occur in Douglas and Cherokee Tailwaters and Tolerance, According to IBI and TFI Analysis, Trophic Status, and Taxonomic Groupings

Common Name/Scientific Name	Expected		Tolerance		Trophic Status	Taxonomic Grouping
	IBI	TFI	IBI	TFI		
Silver redhorse, *Moxostoma anisurum*	X	X			IN	Suckers
River redhorse, *Moxostoma carinatum*	X	X			IN	Suckers
Black redhorse, *Moxostoma duquesnei*	X	X		INT	IN	Suckers
Golden redhorse, *Moxostoma erythrurum*	X	X			IN	Suckers
Shorthead redhorse, *Moxostoma macrolepidotum*	X	X		INT	IN	Suckers
Freshwater Catfishes — Ictaluridae						
Black bullhead, *Ameiurus melas*	X	X	TOL	TOL	OM	
Yellow bullhead, *Ameiurus natalis*	X	X	TOL	TOL	OM	
Blue catfish, *Ictalurus furcatus*	X	X			PI	
Channel catfish, *Ictalurus punctatus*	X	X			OM	
Mountain madtom, *Noturus eleutherus*	X	X	INT	INT	SP	
Flathead catfish, *Pylodictis olivaris*	X	X		INT	PI	
Topminnows — Fundulidae						
Northern studfish, *Fundulus catenatus*	X	IS			SP	
Blackstripe topminnow, *Fundulus notatus*	X	X			SP	
Livebearers — Poeciliidae						
Western mosquitofish, *Gambusia affinis*	I	X	TOL	TOL	IN	
Silversides — Atherinidae						
Brook silverside, *Labidesthes sicculus*	X	X			SP	
Sculpins — Cottidae						
Banded sculpin, *Cottus carolinae*	X	X		TOL	IN/GE[a]	
Temperate basses — Moronidae						
White bass, *Morone chrysops*	I	X			PI	
Yellow bass, *Morone mississippiensis*	I	X			PI	
Striped bass, *Morone saxatilis*	I				PI	
Sunfishes — Centrarchidae						
Rock bass, *Ambloplites rupestris*	X	X		INT	PI	Sunfish
Redbreast sunfish, *Lepomis auritus*	I	X			IN	Sunfish
Green sunfish, *Lepomis cyanellus*	X	X	TOL	TOL	PI	Sunfish
Warmouth, *Lepomis gulosus*	X	X			IN	Sunfish
Bluegill, *Lepomis macrochirus*	X	X			IN	Sunfish
Longear sunfish, *Lepomis megalotis*	X	X			IN	Sunfish
Redear sunfish, *Lepomis microlophus*	I	X			IN	Sunfish
Smallmouth bass, *Micropterus dolomieu*	X	X		INT	PI	
Spotted bass, *Micropterus punctulatus*	X	X			PI	
Largemouth bass, *Micropterus salmoides*	X	X			PI	
White crappie, *Pomoxis annularis*	X	X			PI	Sunfish
Black crappie, *Pomoxis nigromaculatus*	X	X			PI	Sunfish
Perches — Percidae						
Greenside darter, *Etheostoma blennioides*	X	X		INT	SP	Darters
Bluebreast darter, *Etheostoma camurum*	X	X	INT	INT	SP	Darters
Speckled darter, *Etheostoma stigmaeum*	X				SP	Darters
Stripetail darter, *Etheostoma kennicotti*	X	X			SP	Darters
Redline darter, *Etheostoma rufilineatum*	X	X			SP	Darters
Snubnose darter, *Etheostoma simoterum*	X	X			SP	Darters
Wounded darter, *Etheostoma vulneratum*	X	X			SP	Darters
Banded darter, *Etheostoma zonale*	X	X			SP	Darters

TABLE 19.2 *(continued)*
Fish Species Expected to Occur in Douglas and Cherokee Tailwaters and Tolerance, According to IBI and TFI Analysis, Trophic Status, and Taxonomic Groupings

Common Name/Scientific Name	Expected		Tolerance		Trophic Status	Taxonomic Grouping
	IBI	TFI	IBI	TFI		
Yellow perch, *Perca flavescens*	I	X			IN	Darters
Tangerine darter, *Percina aurantiaca*	X	IS			SP	Darters
Blotchside logperch, *Percina burtoni*	X	R			SP	Darters
Logperch, *Percina caprodes*	X	X			SP	Darters
Gilt darter, *Percina evides*	X	X	INT	INT	SP	Darters
Dusky darter, *Percina sciera*	X	X		INT	SP	Darters
River darter, *Percina shumardi*	X	X			SP	Darters
Snail darter, *Percina tanasi*	I	X			SP	Darters
Sauger, *Stizostedion canadense*	X	X			PI	
Walleye, *Stizostedion vitreum*	X	X			PI	
Drums — Sciaenidae						
Freshwater drum, *Aplodinotus grunniens*	X	X			IN	

[a] Banded sculpins classified as insectivores in IBI analysis; generalists; generalists in TFI analysis.

Note: Abbreviations for expected designations are as follows: I, introduced; X, expected; R, rare; IS, isolated from recruitment sources. For tolerance: INT, intolerant; TOL, tolerant. For trophic status: GE, generalist; HB, herbivore; IN, insectivore; OM, omnivore; PA, parasitic; PI, piscivore; PK, planktivore; SP, specialist insectivore. (Nomenclature follows Etnier and Starnes, 1993).

Pool-dwelling species (e.g., many *Notropis* species) are absent from tailwaters because their habitat is greatly reduced during hydro-peaking operations at the dam. As a result, maximum species richness expectations were 81 and 70 species, for previous IBI and present TFI analyses, respectively. TFI metric scoring criteria for darter, sunfish (less *Micropterus* species), and sucker species were also based on 80% of maximum expectations, and were consistent with field observations (Figures 19.2B–D). As with total species, diversity of these three major families was expected to increase with improved tailwater environments.

Designations of species as intolerant in TFI analysis were expanded from five species in the original IBI type of analysis used in TVA tailwaters to 15 species (Table 19.2) to allow better resolution in rating tailwater conditions. Inclusion of additional species as intolerant was based on their presence or absence over time and location. For example, species that were absent in early years of the study, but appeared in later years, were considered for intolerant designations. Similarly, species that only occurred at the lower Douglas Tailwater site in early years and expanded their ranges into middle and upper sites in later years, were considered potential intolerant designates. In other words, species that increased in abundance from absent or nearly so over time, concurrent with restoration of minimum flows and 4 mg/L DO content, were determined to be somewhat intolerant of tailwater conditions, although other factors, as yet undetermined, could also have influenced occurrences of these 15 species. As in scoring criteria for species richness metrics, 80% of maximum number of expected intolerant species was used, and generally agreed with actual occurrences (Figure 19.2E). Diversity of intolerant species was expected to increase as tailwater conditions improved.

Designations of tolerant species in TFI analysis were identical to those of previous IBI studies in these tailwaters, with one exception: banded sculpin (*Cottus carolinae*) were considered tolerant to tailwater conditions (e.g., rapid flow fluctuations, altered food webs) in TFI analysis. Sculpins are common inhabitants of both Douglas and Cherokee tailwaters, particularly at the upper and middle sampling sites, where they comprise up to 23% of total fish sampled (Scott et al., 1996). Their benthic behavior makes them less vulnerable to fluctuations in flow, and their large mouths

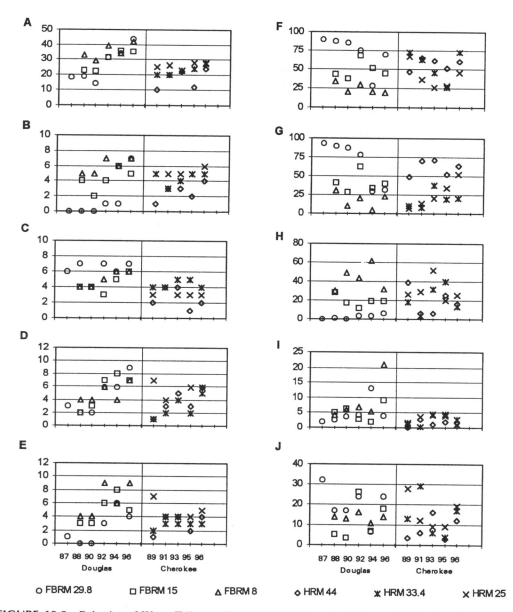

FIGURE 19.2 Behavior of Warm Tailwater Fish Index (TFI) metrics 1 through 10 in Douglas Tailwater (FBRM 8-29.8) and Cherokee Tailwater (HRM 25-44), 1987 through 1996: (A) total number of species; (B) number of darter species; (C) number of sunfish species; (D) number of sucker species; (E) number of intolerant species; (F) percentage tolerant individuals; (G) percentage omnivorous and generalist feeders; (H) percentage specialist insectivores; (I) percentage piscivores; and (J) catch per unit effort.

allow them to prey on a wide variety of benthic organisms, as well as other fish. In Tennessee, adult banded sculpins consume stonefly nymphs and other insects, small benthic riffle fish (e.g., darters) (Starnes, 1977), and even salamanders and crayfish (Etnier and Starnes, 1993). In this study, sculpins were occasionally collected with darter tails (e.g., logperch [*Percina caprodes*]) extending from their mouths. Smaller sculpins were also found in the stomachs of larger sculpins. Because sculpin adaptability to tailwater conditions was believed to be a competitive advantage over other benthic fish species, sculpins were designated as tolerant in TFI analysis.

With the inclusion of sculpins as tolerant, scoring criteria for the percentage of tolerant individuals were increased twofold over IBI criteria (Table 19.1). Higher percentages were considered indicative of poorer tailwater conditions. Occurrences of tolerant individuals in the 20 to 40% range were rated fair (3), while lesser percentages were rated good (5). Poor ratings (1) were given when over 40% of the sample was represented by tolerant individuals. Sculpins, gizzard shad (*Dorosoma cepedianum*), and common carp (*Cyprinus carpio*) were the predominant tolerant species encountered in these two tailwaters. A comparison of observed values for this metric in Douglas Tailwater samples indicated reasonable expectations (Figure 19.2F).

19.2.5.2 Trophic Composition Metrics

Two of the original IBI trophic composition metrics were modified for tailwater fish assessments. Individuals of species consuming a wide variety of prey organisms, termed generalist feeders, were combined with omnivores in assessment of this metric. Species designated as generalist feeders in tailwaters (e.g., banded sculpin and blacknose dace [*Rhinichthys atratulus*]) were combined with omnivores (e.g., gizzard shad, common carp, carpsuckers [*Carpiodes* spp.], buffaloes [*Ictiobus* spp.], bullheads [*Ameiurus* spp.] and channel catfish [*Ictalurus punctatus*]). TFI scoring criteria for this metric were increased 67% over IBI criteria to reflect inclusion of additional species and the altered ecosystems found in tailwaters, based on Douglas Tailwater observations (Figure 19.2G). Higher percentages of omnivores and generalist feeders indicated more disturbed communities.

Coupled with increased TFI scoring criteria for the percentage omnivore/generalist metric were decreased scoring criteria for the percentage specialist insectivore metric. Perturbations in the natural flow of organic materials to aquatic food webs caused by dams limit benthic macroinvertebrate communities in tailwaters (Yeager, 1993). In turn, those fish species designated as specialist insectivores (e.g., certain cyprinids [*Notropis* spp.], madtoms [*Noturus* spp.] and darters [*Etheostoma* and *Percina* spp.]) would also be impacted relative to fish community abundance. TFI scoring criteria for this metric were decreased 67% from previous IBI criteria. Greater percentages of specialist insectivores indicated better tailwater fish communities (Figure 19.2H).

No changes were made in scoring criteria or species designations for the percentage of piscivorous individuals metric. Frequency distribution of piscivore percentages in Douglas and Cherokee tailwaters showed that existing IBI scoring criteria for this metric were adequate for expectations of tailwater fish communities. Greater relative abundance of piscivores indicated healthier communities (Figure 19.2I).

19.2.5.3 Abundance and Condition Metrics

Overall abundance of fish, expressed as catch-per-unit effort (i.e., one seine haul, one backpack shocking run, or 5 minutes of boat shocking), was generally less in Douglas Tailwater than expectations for free-flowing streams (Figure 19.2J). Exceptions were relative to high occurrences of tolerant individuals (Figure 19.2F). TFI scoring criteria for the abundance metric were decreased from IBI criteria to satisfy perception of desirable tailwater expectations without high numbers of tolerant fishes.

Scoring criteria for the remaining two metrics assessing percentages of hybrids and percentages of diseased, injured, or anomalous individuals were unchanged from previous IBI studies in these tailwaters (Tomljanovich and Saylor, 1989; 1992). However, previous TVA studies modified original IBI criteria (Karr et al., 1986) for the percent hybrid metric, allowing up to 0.5% of individuals in total sample to be hybrids and the metric still receive a good rating (Tomljanovich and Saylor, 1989; 1992). Higher percentages of either hybrids or anomalous individuals were indicative of more damaged fish communities.

19.3 RESULTS AND DISCUSSION

19.3.1 METRIC BEHAVIOR

19.3.1.1 Species Richness and Tolerance Composition Metrics

Total species diversity in Douglas Tailwater showed a major positive trend during the study period (Figure 19.2A), which was consistent among all three sampling sites. Total diversity at FBRM 8 increased from 33 to 42 species between 1988 and 1996; while at FBRM 15, total species increased from 23 to 35. Fewer than 20 species were collected from FBRM 29.8 in 1987 to 1990; but after 1990, over 30 species were found, peaking at 44 species in 1996. On three (19%) of 16 occasions during the study period total species diversity met TFI scoring criteria for the best community rating, 13% were rated the worst, and otherwise (69%) was rated in the intermediate category. In Cherokee Tailwater, which did not have aeration improvements until 1995, species richness showed only modest improvements. Highest diversity was found at HRM 25, the most downstream site, where total species ranged from 23 to 28 between 1989 and 1996. At HRM 33.4, total species increased from 20 during the earliest two samples to 28 in the most recent sample. Total species at HRM 44 was most variable, ranging from 10 in 1989 to 24 in 1996, but falling from 22 in 1993 to 12 in 1995. Cherokee Tailwater never attained TFI criteria meeting the species richness expectations of a healthy community during the study period. On 13 (87%) of 15 occasions, total species diversity was rated in the intermediate category for community health.

Darter species diversity in Douglas Tailwater also increased during the study period (Figure 19.2B). At FBRM 8, five species in early samples increased to six and seven in the latter years. At the middle station, mixed results were seen, with darter species ranging from two in 1990 to six in 1994. The largest increase in darter diversity was observed at FBRM 29.8, where zero species were found during the initial three samples, one species appeared in two post-reaeration samples, and seven species occurred in 1996. Even the federally listed, threatened snail darter (*Percina tanasi*) was present at this site in 1996, as it was at the other Douglas Tailwater sites that year. Snail darters have been found at FBRM 8 during all samples 1988 to 1996, most likely a result of transplants from the Little Tennessee River to the Holston River (mile 14.4) in 1978 and 1979. On three occasions in Douglas Tailwater, observations of darter species were rated in the high quality range of expectations. Meanwhile, little increase in darter species diversity was observed at Cherokee Tailwater. Generally, more species (five) were collected at HRM 25, and a sixth species was found in 1996. Three to five species occurred at the middle station. A gradual improvement was seen at HRM 44, where only one species occurred in 1989, increasing to four in 1996. Number of darter species never met scoring criteria for high-quality fish communities, and usually fell in the intermediate category (67%). On five occasions (33%), darter diversity was rated in the lowest category.

Sunfish species (less *Micropterus* species) showed general improvement in Douglas Tailwater during the study period (Figure 19.2C). In contrast to most other metrics, numbers of sunfish species in this tailwater were usually highest at the upstream site, FBRM 29.8, where six or seven species were often present. At FBRM 8 and FBRM 15, sunfish species increased over time from four to six species. Good communities were indicated by sunfish diversity of at least five species, and this criterion was met in 10 of 16 occasions (63%). Little consistency was found in occurrences of sunfish species in Cherokee Tailwater, where the total number of sunfish species was generally about two species lower than in Douglas Tailwater. The middle Cherokee Tailwater site tended to have the most species, which met TFI's best criteria on two occasions. Otherwise, number of sunfish species was usually rated in the intermediate category (67%), except for three low ratings at HRM 44.

Sucker species showed a significant increase in Douglas Tailwater after 1990 (Figure 19.2D). Between two and four species were found at the three Douglas Tailwater sites between 1987 and

1990, but increased to six or seven species in 1992. The greatest improvement was observed at FBRM 29.8, where seven to nine species were found in 1996. All ratings for number of sucker species were rated in the poor or fair category prior to 1992; however, subsequent ratings increased to the highest category and remained there between 1992 and 1996, with the exception of one deviation during 1994. Sucker diversity in Cherokee Tailwater showed a general increase in number of species over time at two sites, although the improvement was less than in Douglas Tailwater. Diversity at HRM 44 increased from one species in 1989 to six in 1996, and from one to five at HRM 33.4. More species were usually found at HRM 25, ranging from four to seven ($\bar{x} = 5.4$), with six in 1996. On four (27%) occasions, good sucker communities were observed, four (27%) times for poor communities and seven (46%) times for fair sucker communities.

The revised list of intolerant tailwater species showed significant increases in Douglas Tailwater after 1990 (Figure 19.2E). Similar to total species diversity and darter diversity, greatest improvement was seen at FBRM 8, where four species in 1988 and 1990 increased to nine during 1992 and 1996. Intolerant species at FBRM 15 increased from three prior to 1992 to five or more in later years. At FBRM 29.8, never was more than one intolerant species found before 1992, while six have been collected since then. Under TFI criteria, two occasions (both at FBRM 8) met highest expectations. Five of seven observations for 1987 to 1990 fell in the poor category. Intolerant species showed little change over time in Cherokee Tailwater. Only three or four intolerant species were found at most sites. About half the observations were rated as fair tailwater fish communities, while the remainder fell in the poor category.

The percentages of tolerant individuals in Douglas Tailwater showed improving fish communities during the study period (Figure 19.2F). The more favorable conditions were seen at the lower-most sampling site, FBRM 8, where percentages declined from 34 to 19% between 1988 and 1996. The percentages of tolerant individuals at the upper site declined from 90 to 69% during the same period, with a decline to 28% in 1994. Mixed results were seen at the middle site, where low tolerant percentages were found in 1988 and 1990, and highest site percentages were found in 1992. Declines have been observed, however, since 1992. On only one occasion did observations meet expectations for a good fish community, that being FBRM 8 in 1996. Most often, poor ratings were found at the upper and middle stations, under TFI criteria. In Cherokee Tailwater, percentage tolerant individuals showed highly variable results. Most observations (73%) were in the poor ratings expectations; only four observations were rated as fair, and three of them were at HRM 25.

19.3.1.2 Trophic Composition Metrics

In Douglas Tailwater, the percentage of individuals classified as omnivores or generalist feeders declined during the study period. This was taken to be an indication of improvement in fish communities over time (Figure 19.2G). The most significant improvement occurred at FBRM 29.8, where omnivores and generalist feeders exceeded 75% of all fish collected between 1987 and 1992. In the last two sampling periods during 1994 and 1996, the percent omnivores and generalists declined to about 30%. At FBRM 8, the percentage of omnivores and generalists was the lowest of all sample sites, declining from just above 25% in 1988 to below 25% in later years. Composition of omnivores and generalist feeders observed at FBRM 15 ranged from 28 to 42%, except for a high value of 62% in 1992. On four occasions at FBRM 8, this metric was rated in the highest category for tailwater expectations. Five times before 1994, this metric was rated in the poorest category; once at FBRM 15 and four times at FBRM 29.8. Less improvement in omnivores and generalist feeders was observed in Cherokee Tailwater where the percentage increased at two sites overall. Carp and banded sculpin at HRM 44 caused inflated percentages above expectations for a good fish community. Low percentages were found at miles 33.4 and 25 during the first two years of sampling during 1989 and 1991, but increased in the latter two samples. While the percentage of omnivores and generalists remained in the desired range (i.e., less than 25%) at HRM 33.4 in 1995 and 1996, the percentage at HRM 25 increased into the intermediate category of expectations.

Percentages of specialist insectivores showed increases in abundance over time in Douglas Tailwater (Figure 19.2H), as predicted for improving fish communities. The best conditions occurred at FBRM 8, where over 30% of fish sampled were designated specialists. The decline in 1996 was partially due to a large increase in the piscivore indicator group. At FBRM 15, specialist insectivore abundance was around 20%, and gradually increased from nearly 0 in 1987 to 6% in 1996 at FBRM 29.8. TFI scoring criteria for good tailwater fish communities were met on four occasions, all at the lower-most sampling site. In Cherokee Tailwater, no discernible patterns in specialist insectivore percentages were evident, with wide ranges (30% or more) occurring at all three sites. This metric was influenced by large numbers of carp, spotfin shiners (*Cyprinella spiloptera*, a generalist insectivore), and banded sculpins, which reduced metric scores.

Percentages of piscivorous individuals showed signs of improvement in Douglas Tailwater (Figure 19.2I) with best overall percentages found at FBRM 8, ranging from 4.2 to 6.8% during the period 1988 to 1994, and increasing to 21% during 1996. Yellow bass (*Morone mississippiensis*) caused the increase in 1996, along with increased population numbers of three black bass species (*Micropterus*). Piscivore abundance at FBRM 15 and FBRM 29.8 showed a similar trend, but percentages were usually lower than at FBRM 8. Under TFI criteria, percentages of piscivores indicated good fish communities on 50% of the sampling occasions in Douglas Tailwater. Low piscivore percentages were usually observed in Cherokee Tailwater samples. No occurrences in excess of 5% were observed, and no trends were evident.

19.3.1.3 Abundance and Condition Metrics

Catch rates were generally high in Douglas Tailwater during the sample period (Figure 19.2J), indicating good fish community conditions in 75% of the samples. Unfortunately, large numbers of gizzard shad and carp, both undesirable species in assessing community health, were responsible for the highest catch rates, especially at FBRM 15 and FBRM 29.8. This metric showed better stability at FBRM 8, and still met TFI criteria for good expectations (i.e., more than 11 fish per sampling effort unit). Catch rates at Cherokee Tailwater were often high, usually a result of a large abundance of undesirable species. No trends were apparent other than lower abundance was observed at HRM 44 than at two downstream sites. In eight of 15 samples (53%), overall fish abundance met TFI criteria for healthy fish communities.

Percentages of individuals as hybrids never exceeded 1% of total fish samples in either tailwater. Hybridization is not a common problem in tailwaters, as it can be in smaller streams (Karr et al., 1986). On all but one occasion, this metric indicated good fish communities.

Percentages of anomalous individuals (i.e., injured, diseased, fin damage, skeletal deformities, etc.) indicated healthy fish communities in both tailwaters. Exceptions occurred at FBRM 15 in 1988 and HRM 44 in 1989 and 1995, when anomalies were excessive (6.6, 5.6, and 5.3%, respectively). A high incidence of leeches on sunfish and darters, along with fin erosion, lesions, and deformities on other species, were problems at FBRM 15 in 1988. At HRM 44, carp were found with fin erosion and swirled scales in 1989, and channel catfish had leeches in 1995.

19.3.2 COMPARISON OF TFI AND IBI ANALYSES

Tailwater fish communities are subjected to a series of habitat-altering events associated with the operation of hydroelectric dams (Yeager, 1993) that diminish biotic integrity. TFI metrics and scoring criteria compensate for habitat impairment and provide alternative expectations for assessing the ecological health of tailwaters. Attempting to classify tailwater communities according to IBI metrics and scoring criteria will invariably result in low and perplexing community ratings.

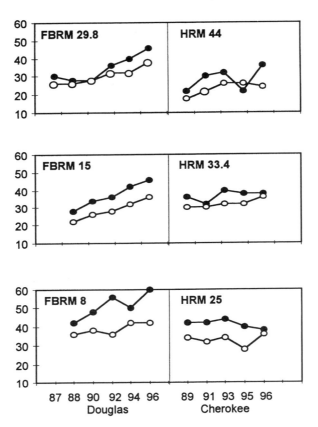

FIGURE 19.3 Comparison of Tailwater Fish Index (TFI, solid circles) and Index of Biotic Integrity (IBI, open circles) for Douglas and Cherokee tailwaters, Tennessee, 1987 through 1996.

19.3.2.1 Douglas Tailwater

Fish communities in Douglas Tailwater have shown relatively consistent improvement since 1987, based on both TFI and IBI analyses (Figure 19.3); however, TFI classifications result in more favorable assessments of community health than IBI classifications. For example, at FBRM 29.8, the TFI classifies fish communities as poor in 1987, 1988, and 1990, with TFI values ranging between 28 and 30. After 1990, TFI values increased to 36 (poor/fair), 40 (fair), and 46 (fair/good), in 1992, 1994, and 1996, respectively. Using the IBI, fish communities at FBRM 29.8 rated very poor/poor in 1987 and 1988 (IBI values of 26), and improved only to poor (IBI values of 28, 32, and 32) during 1990, 1992, and 1994, respectively. In 1996, an IBI value of 38 showed slight improvement to the poor/fair category.

Better improvement was indicated at FBRM 15 under TFI criteria. TFI values increased steadily over time, with values of 28 (poor), 34 (poor), 36 (poor/fair), 44 (fair), and 48 (good) between 1988 and 1996. Consistent improvement was also shown in IBI analysis; however, the final IBI value indicated only a poor/fair fish community.

The best fish communities in Douglas Tailwater were found at FBRM 8. TFI analysis showed improvement during all years between 1988 and 1996, except 1994, with values of 42 (fair), 48 (good), 56 (good/excellent), 50 (good), and 60 (excellent). The 1996 value was the highest possible score, indicating all 12 metrics satisfied expectations for a good tailwater fish community. For the same time period under IBI analysis, the fish community at FBRM 8 remained in poor/fair condition from 1988 to 1992 (IBI values of 36 and 38), rising to fair (IBI value of 42) in 1994 and 1996.

19.3.2.2 Cherokee Tailwater

In Cherokee Tailwater, where reaeration improvements were not installed until the summer of 1995, little consistency in fish community trends was evident using either TFI or IBI criteria. Minimum flows alone, beginning in 1988, had no dramatic influence on tailwater fish communities.

The expected downstream trend in fish community improvement was evident in Cherokee Tailwater, as index scores generally increased with distance from the dam. TFI scores at HRM 44 ranged from 22 (very poor) to 36 (poor) between 1989 and 1996. IBI values ranged from 18 (very poor) to 26 (very poor/poor). At HRM 33.4, TFI values were slightly better, ranging between 32 (poor) and 40 (fair), while IBI values ranged from 30 (poor) to 36 (poor/fair). TFI values at HRM 25 fell between 36 (poor/fair) and 44 (fair), although the lowest value occurred in 1996. IBI scores ranged from 28 to 36, reaching the highest score in 1996.

19.4 CONCLUSION

Because tailwaters are highly altered ecosystems, TFI analysis modified IBI metrics and scoring criteria to more accurately assess ecological health of tailwater fish communities based on perceived potentials. Much improved fish communities were found in Douglas Tailwater following the establishment of minimum flows and reaeration of Douglas Dam discharges to 4 mg/L DO between 1987 and 1996. Better fish community health occurred over time with increasing distance from the dam. Four species richness metrics, two tolerance composition metrics, three trophic composition metrics, and the overall abundance metric illustrated improved Douglas Tailwater fish communities under TFI criteria, and the most downstream site supported an excellent fish community in 1996. The lack of substantial improvement in Cherokee Tailwater fish communities is believed related to much less time that reaeration of Cherokee Dam releases has been provided.

ACKNOWLEDGMENTS

The author wishes to express appreciation to Charlie Saylor, Dave Tomljanovich, Bruce Yeager, and Steve Ahlstedt for their contributions of field collections, data analysis, and reporting tailwater fisheries surveys in the initial years of the RRI program; and to Jack Davis and Gary Brock, former RRI program managers, for supporting biological investigations in tailwaters.

REFERENCES

Etnier, D.A. and W.C. Starnes. 1993. *The Fishes of Tennessee*. The University of Tennessee Press. Knoxville.

Hughes, R.M. and Oberdorff. 1998. Applications of IBI concepts and metrics to waters outside the United States, in T.P. Simon (Ed.), *Assessing the Sustainability and Biological Integrity of Water Resource Quality Using Fish Assemblages*. In press.

Karr, J.R. 1981. Assessment of biotic integrity using fish communities, *Fisheries,* 6(6), 21–27.

Karr, J.R., K.D. Fausch, P.L. Angermier, P.R. Yant, and I.J. Schlosser. 1986. Assessing Biological Integrity in Running Waters. A Method and Its Rationale. Illinois Natural History Survey, Special Publication 5.

Mobley, M.H. and W.G. Brock. 1995. Aeration of reservoirs and releases: TVA porous hose line diffuser, *American Society of Civil Engineers,* submitted.

Saylor, C.F. and S.A. Ahlstedt. 1990. Application of the Index of Biotic Integrity (IBI) to Fixed Station Water Quality Monitoring Sites. Tennessee Valley Authority, Aquatic Biology Dept. Rpt. No. TVA/WR/AV-90/12, Chattanooga.

Saylor, C.F., G.D. Hickman, and A.M. Brown. 1987. Biological Assessment, Middle Fork Holston Watershed. Tennessee Valley Authority Division of Air and Water Resources and Division of Services and Field Operations, Norris, TN.

Saylor, C.F., G.D. Hickman, and M.H. Taylor. 1988. Application of the Index of Biotic Integrity (IBI) to Fixed Station Water Quality Monitoring Sites — 1987. Technical Report. Tennessee Valley Authority, Division of Water Resources, Chattanooga.

Saylor, C.F. and E.M. Scott. 1987. Application of the Index of Biotic Integrity to Existing TVA Data. Tennessee Valley Authority Division of Air and Water Resources, TVA/ONRED/AWR87/32, Chattanooga.

Scott, E.M., K.D. Gardner, D.S. Baxter, and B.L. Yeager. 1996. Biological and Water Quality Responses in TVA Tributary Tailwaters to Dissolved Oxygen and Minimum Flow Improvements — Implementation of the Reservoir Releases Improvements Program/Lake Improvement Plan. Tennessee Valley Authority, Resource Group, Water Management, Norris.

Tomljanovich, D.A. and C.F. Saylor. 1989. Results of Fisheries Investigations in Douglas Tailwater, August 1987–September 1988. Tennessee Valley Authority, Water Resources, Aquatic Biology Department, TVA/WR/AB-89/6.

Tomljanovich, D.A. and C.F. Saylor. 1992. Results of Fisheries Investigations in Douglas and Cherokee Tailwaters — FY 1990 and 1991. Tennessee Valley Authority, Water Resources, TVA/WR-92/2.

Vannote, R.L., G.W. Minshall, K.W. Cummins, J.R. Sedell, and C.E. Cushing. 1980. The river continuum concept, *Canadian Journal of Fisheries and Aquatic Science*, 37, 130–137.

Yeager, B.L. 1993. Impacts of Reservoirs on the Aquatic Environments of Regulated Rivers. Tennessee Valley Authority, River Basin Operations, Water Resources, TVA/WR-93/1, Norris.

Yeager, B.L. and D.A. Tomljanovich. 1990. Biological Investigations in Douglas and Cherokee Tailwaters, FY 1989. Tennessee Valley Authority, Water Resources, Aquatic Biology Department, TVA/WR/AB-90/7.

20 Reservoir Fish Assemblage Index Development: A Tool for Assessing Ecological Health in Tennessee Valley Authority Impoundments

Thomas A. McDonough and Gary D. Hickman

CONTENTS

20.1 Introduction 523
20.2 Background 524
20.3 Reservoir Classification 524
20.4 Reference Conditions and Scoring Criteria 525
20.5 Reservoir Fish Assemblage Index Metrics..527
      20.5.1 Taxa Richness and Composition 529
      20.5.2 Trophic Composition 533
      20.5.3 Reproductive Composition 534
      20.5.4 Abundance 534
      20.5.5 Fish Health 534
20.6 Sampling Methods 534
20.7 Reservoir Fish-Assemblage Index Variability 535
20.8 Reservoir Fish-Assemblage Index Results in TVA Reservoirs 537
20.9 Validity of the Reservoir Fish Assemblage Index 538
20.10 Conclusion 539
Acknowledgments 539
References 539

20.1 INTRODUCTION

Much of the Tennessee River and its major tributaries were impounded by the Tennessee Valley Authority (TVA) between 1933 and 1950 to provide flood control, navigation, and hydroelectric power generation, and to develop and conserve valley resources. Ecological impacts of impoundment were dramatic, but construction and operation of TVA reservoirs were primarily driven by economics. Ecological consequences were at best a secondary consideration. In recent years, increasing public awareness of the importance of water quality and the passage of the Clean Water Act have resulted in increased emphasis on TVA's charge to conserve valley resources. To obtain a holistic view of reservoir environmental quality, the TVA initiated its "vital signs" monitoring program in 1990 (Dycus and Meinert, 1991). This program included measurements of sediment toxicity, chlorophyll, and physical/chemical parameters, as well as assessments of fish and benthic macroinvertebrate communities. This chapter describes the Reservoir Fish Assemblage Index (RFAI) component of this program.

20.2 BACKGROUND

The Index of Biotic Integrity (IBI) was initially developed on small wadeable streams in the Midwest to provide a broadly based and ecologically sound tool to evaluate aquatic resource quality (Karr, 1981). Subsequently, this index has been widely adapted for use in other aquatic environments (Thompson and Fitzhugh, 1986; Miller et al., 1988; Hickman and McDonough, 1996). This tool often detects environmental quality problems when chemical and sediment toxicity tests indicate acceptable conditions (Karr et al., 1986; Ohio EPA, 1987). The TVA recognized a need to develop reservoir biomonitoring tools to assess progress toward meeting corporate goals of cleaning up the river system and attaining the fishable and swimmable standards outlined in the Clean Water Act. A fish community indexing system similar to the IBI was needed to assist in addressing environmental quality in reservoirs.

In 1988, the TVA contracted with James Karr to aid in the development of a Reservoir Index of Biotic Integrity. Dionne and Karr (1992) first analyzed long-term cove rotenone data from Tennessee Valley reservoirs. They determined that sampling protocols that restricted samples to shallow water at the back of coves and purposely omitted coves disturbed by local human influence provided biased data unsuitable for indexing reservoir fish populations. Jennings et al. (1995) examined data from the TVA's annual fall electrofishing sampling program. The conceptual framework of the stream IBI was maintained, but an alternate method was required to derive reference conditions. A preliminary index was developed with potential to provide a reliable assessment of fish assemblage status; however, index scores exhibited high annual variation. Examination of variability in supplemental studies revealed sampling error as the primary cause of this high variability. Analysis of 1989 to 1992 data suggested increased sampling effort and continued refinement of metrics would likely lead to a more reliable index.

Refinement of the index in 1993 and 1994 led to modifications that included: increased sampling effort, substitution of new metrics for those found to be unreliable, change of approach used to determine reference conditions, and addition of gillnet sampling to the analysis. Hickman and McDonough (1996) found RFAI variability to be reduced to an acceptable level following implementation of these refinements.

20.3 RESERVOIR CLASSIFICATION

Separation of reservoirs into biologically meaningful classes or groups to reduce variability in metric scores is a critical initial step in developing reference conditions. Care must be taken to compare only those reservoirs that would have similar communities under ideal conditions. McDonough and Barr (1977) used cluster analysis of historical cove rotenone data to classify Tennessee and Cumberland River reservoirs. Their classification scheme reflected geographic region (ecoregion) as well as physical and operational characteristics of reservoirs proceeding from the mouth of the river system to the headwaters (Figure 20.1). The 30 reservoirs included in the TVA's reservoir monitoring program differ somewhat from those examined by McDonough and Barr (1977); however, the classification scheme is similar. Reservoirs are currently classified into four groups, based on physical and operational characteristics and ecoregion (Table 20.1).

The first group includes run-of-river reservoirs on the Tennessee River, along with two reservoirs on the lower reaches of tributaries to the Tennessee River. These reservoirs have short retention times and small fluctuation zones during winter drawdown. These run-of-river reservoirs are located

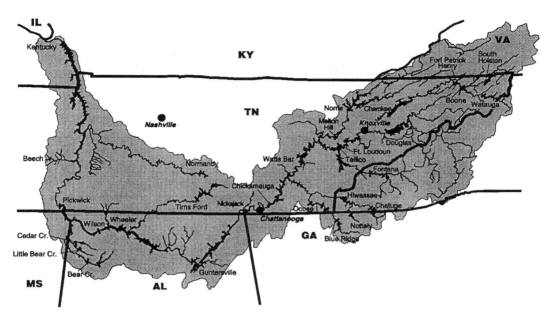

FIGURE 20.1 Map of the Tennessee Valley.

in three ecoregions: Central Appalachian Ridge and Valley, Interior Plateau, and Southwestern Appalachia. Since these reservoirs are interconnected, and each reservoir receives water from two to four upstream ecoregions, further classification of run-of-river reservoirs by ecoregion is not warranted.

Tributary reservoirs in the Tennessee River system have long retention times and substantial winter drawdowns. These tributary reservoirs are categorized into three groups (Blue Ridge, Central Appalachian Ridge and Valley, and Interior Plateau Ecoregions), based on physical/chemical features generally owing to both geological characteristics of different ecoregions and varying hydrological characteristics. Tributary reservoirs tend to differ among ecoregion in water chemistry, morphometry, and operation.

Within reservoirs, sites were classified into three zones: inflow, transition, and forebay (Figure 20.2), as described by Kimmel and Groeger (1984). Inflow zones are riverine. Water velocity decreases as cross-sectional area increases in the transition zone, causing suspended materials to settle, increasing water clarity with resultant increases in primary production. The forebay zone is the lacustrine area near the dam.

20.4 REFERENCE CONDITIONS AND SCORING CRITERIA

The IBI is a multimetric index based on comparisons of existing community characteristics to those present at relatively undisturbed reference sites. These reference sites provide a benchmark against which regional expectations for a healthy community are derived. Reservoirs require a different approach than streams, rivers, or natural lakes for determining biotic integrity. As artificial systems, they lack natural reference sites for determining characteristics that would be expected in waters unaffected by human impacts. Karr and Dionne (1991) and Jennings et al. (1995) discussed various conditions that make reservoirs unnatural systems, including their relative temporary nature on an evolutionary time scale, which precludes the expectation of adaptive communities. Due to these concerns, Jennings et al. (1995) suggested the term "biotic integrity" as inappropriate in reservoir applications and recommended the index be entitled the Reservoir Fish Assemblage Index (RFAI).

TABLE 20.1
Reservoir Classification Scheme Used by TVA's Vital Signs Monitoring Program along with Physical and Operational Characteristics and Ecoregion

Reservoir	Drainage Area (sq. km)	Reservoir Area (sq. km)	Elevation (m)	Mean Depth (m)	Residence Time (days)	Ecoregion
Run-of-River Reservoirs						
Chickamauga	53860	143	208	5.4	10	CARV
Fort Loudoun	24741	59	248	7.6	12	CARV
Guntersville	63342	275	181	4.6	13	IP
Kentucky	104145	649	110	5.4	23	IP
Melton Hill	8661	23	242	6.4	14	CARV
Nickajack	56658	42	193	7.1	4	SWA
Pickwick	85026	174	126	6.5	9	IP
Tellico	6806	64	248	8.0	31	CARV
Watts Bar	44819	158	226	7.9	19	IP
Wheeler	76658	272	170	4.8	11	IP
Wilson	79663	63	155	12.5	6	IP
Central Appalachian Ridge and Valley						
Boone	4767	17	422	13.4	38	CARV
Cherokee	8881	123	327	14.9	168	CARV
Douglas	11764	123	305	14.1	105	CARV
Fort Patrick Henry	4930	4	385	9.1	5	CARV
Norris	7544	138	311	18.2	244	CARV
South Holston	1821	31	527	26.4	337	CARV
Blue Ridge						
Blue Ridge	601	13	515	17.8	159	BR
Chatuge	490	28	588	10.2	261	BR
Fontana	4070	43	543	12.1	186	BR
Hiwassee	2508	25	465	21.1	103	BR
Nottely	554	17	542	12.3	213	BR
Ocoee #1	1542	8	253	13.6	30	BR/CARV
Watauga	1212	26	597	27.1	412	BR
Interior Plateau						
Beech	41	4	140	4.0	280	IP
Bear Creek	598	26	176	4.0	8	IP
Cedar Creek	464	170	177	6.4	127	IP
Little Bear Creek	158	6	189	8.9	193	IP
Normandy	505	13	267	10.5	162	IP
Tims Ford	1371	43	271	15.2	273	IP

Note: Abbreviations for ecoregions are: BR = Blue Ridge, CARV = Central Appalachian Ridge and Valley, IP = Interior Plateau, and SWA = Southwestern Appalachia.

The artificial nature of reservoirs, with no acceptable reference sites, requires that expected conditions be obtained through other sources. Alternative approaches include using historical or preimpoundment conditions, predictive models, best observed conditions, or professional judgment. Hickman and McDonough (1996) noted that preimpoundment conditions are inappropriate because of the significant habitat alterations resulting from impoundment. An incomplete under-

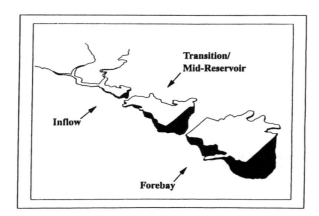

FIGURE 20.2 Hypothetical reservoir cross-section illustrating forebay, transition, and inflow zones.

standing of fish communities in reservoirs precludes effective predictive modeling of species richness, composition, and abundance. Jennings et al. (1995) established reference conditions based on best observed conditions for similar reservoirs. This method assumed the dataset included a wide range of conditions from good to poor for each community characteristic or metric. This assumption was not met in the reservoir system discussed here because of the low number of reservoirs within each group. Rather, the current approach relies on examination of observed conditions and professional judgment.

Scoring criteria for species richness metrics were determined from a composite list of species collected throughout the 1990 to 1994 period from each group of reservoirs and zones within reservoirs. Species richness expectations were adjusted using inferences of experienced biologists knowledgeable of the reservoir system, resident fish species, susceptibility of individual species to electrofishing and gillnetting, and effects of human-induced impacts on these species (Tables 20.2 to 20.6). Figure 20.3 compares the observed species richness range in each class of reservoir sites with maximum species richness expectations using this approach. This approach results in higher expectations than would be obtained using maximum observed species richness based on individual samples. This conservative approach to establishing scoring criteria is warranted owing to the lack of traditional reference sites. Once maximum species richness was determined for each reservoir type and longitudinal zone, the range of individual metric scores (zero to 95% of the expected maximum score) was trisected, as suggested by Fausch et al. (1984).

Scoring criteria for abundance and proportional metrics were determined by trisecting observed ranges after omitting outliers (Tables 20.2 to 20.6). The narrow range of conditions within some classes required adjustments by a team of biologists and environmental scientists based on knowledge of metric responses to human-induced conditions observed in other reservoir groups. Scoring criteria for the fish health metric are those described by Karr et al. (1986).

For each metric, these scoring criteria separated sites into three categories hypothesized to represent relative degrees of degradation. Sample results are compared to reference conditions and assigned a corresponding value: least degraded–5; intermediate–3; and most degraded–1 (Jennings et al., 1995). These categories are based on "expected" fish community characteristics in the absence of human-induced impacts other than impoundment. Individual metric scores for a site are summed to obtain the RFAI score.

20.5 RESERVOIR FISH ASSEMBLAGE INDEX METRICS

The RFAI uses 12 fish community metrics that can be broken down into five general categories. Metrics address the following assemblage characteristics:

TABLE 20.2
Scoring Criteria for Forebay and Transition Sections of Mainstream Reservoirs in the Tennessee River Valley

Metric	Reservoir Group	Gear	Forebay 1	Forebay 3	Forebay 5	Transition 1	Transition 3	Transition 5
1. Number of species	Lower mainstream	Combined	<14	14–27	>27	<16	16–30	>30
	Upper mainstream	Combined	<14	14–27	>27	<15	15–29	>29
	Other reservoirs	Combined	<13	13–24	>24	<13	13–26	>26
2. Number of Lepomid sunfish species	Lower mainstream	Combined	<2	2–3	>3	<2	2–3	>3
	Upper mainstream	Combined	<2	2–4	>4	<2	2–4	>4
	Other reservoirs	Combined	<2	2–4	>4	<2	2–4	>4
3. Number of sucker species	Lower mainstream	Combined	<4	4–6	>6	<4	4–7	>7
	Upper mainstream	Combined	<4	4–7	>7	<4	4–7	>7
	Other reservoirs	Combined	<4	4–6	>6	<4	4–6	>6
4. Number of intolerant species	Lower mainstream	Combined	<2	2–4	>4	<3	3–4	>4
	Upper mainstream	Combined	<2	2–4	>4	<2	2–4	>4
	Other reservoirs	Combined	<2	2–3	>3	<2	2–4	>4
5. Percent tolerant individuals	All	Electrofishing	>45%	20–45%	<20%	>50%	25–50%	<25%
	All	Gillnetting	>40%	20–40%	<20%	>40%	20–40%	<20%
6. Percent dominance	All	Electrofishing	>60%	40–60%	<40%	>60%	40–60%	<40%
	All	Gillnetting	>50%	30–50%	<30%	>50%	30–50%	<30%
7. Number of piscivore species	Lower mainstream	Combined	<4	4–7	>7	<4	4–7	>7
	Upper mainstream	Combined	<4	4–7	>7	<4	4–7	>7
	Other reservoirs	Combined	<4	4–7	>7	<4	4–7	>7
8. Percent omnivores	All	Electrofishing	>45%	20–45%	<20%	>50%	25–50%	<25%
	All	Gillnetting	>45%	30–45%	<30%	>45%	30–45%	<30%
9. Percent invertivores	All	Electrofishing	<35%	35–70%	>70%	<30%	30–60%	>60%
	All	Gillnetting	<5%	5–15%	>15%	<7%	7–15%	>15%
10. Number of lithophilic spawning species	Lower mainstream	Combined	<4	4–6	>6	<4	4–7	>7
	Upper mainstream	Combined	<3	3–6	>6	<4	4–7	>7
	Other reservoirs	Combined	<4	4–7	>7	<4	4–7	>7
11. Total number of individuals	All	Electrofishing	<50	50–100	>100	<50	50–100	>100
	All	Gillnetting	<15	15–35	>35	<15	15–35	>35
12. Percent anomalies	All	Combined	<2%	2–5%	<5%	>2%	2–5%	>5%

Note: Lower mainstream reservoirs include Kentucky, Pickwick, Wilson and Wheeler Reservoirs. Upper mainstream reservoirs include Guntersville, Nickajack, Chickamauga, Watts Bar, and Fort Loudoun Reservoirs. Other reservoirs include Melton Hill and Tellico Reservoirs.

TABLE 20.3
Scoring Criteria for Inflow Sections of Mainstream Reservoirs in the Tennessee River Valley

			Scoring Criteria		
Metric	Reservoir Group	Gear	1	3	5
1. Number of species	Lower mainstream	Electrofishing	<14	14–27	>27
	Upper mainstream	Electrofishing	<14	14–27	>27
	Melton Hill	Electrofishing	<13	13–24	>24
2. Number of Lepomid sunfish species	Lower mainstream	Electrofishing	<2	2–4	>4
	Upper mainstream	Electrofishing	<3	3–4	>4
	Melton Hill	Electrofishing	<3	3–4	>4
3. Number of sucker species	Lower mainstream	Electrofishing	<4	4–7	>7
	Upper mainstream	Electrofishing	<3	3–6	>6
	Melton Hill	Electrofishing	<3	3–6	>6
4. Number of intolerant species	Lower mainstream	Electrofishing	<3	3–6	>6
	Upper mainstream	Electrofishing	<2	2–4	>4
	Melton Hill	Electrofishing	<2	2–4	>4
5. Percent tolerant individuals	All	Electrofishing	>55%	30–55%	<30%
6. Percent dominance	All	Electrofishing	>60%	40–60%	<40%
7. Number of piscivore species	Lower mainstream	Electrofishing	<4	4–7	>7
	Upper mainstream	Electrofishing	<3	3–6	>6
	Melton Hill	Electrofishing	<4	4–7	>7
8. Percent omnivores	All	Electrofishing	>55%	30–55%	<30%
9. Percent invertivores	All	Electrofishing	<25%	25–50%	>50%
10. Number of lithophilic spawning species	Lower mainstream	Electrofishing	<4	4–7	>7
	Upper mainstream	Electrofishing	<4	4–7	>7
	Melton Hill	Electrofishing	<3	3–5	>5
11. Total number of individuals	All	Electrofishing	<50	50–100	>100
12. Percent anomalies	All	Electrofishing	<2%	2–5%	>5%

Note: Lower mainstream reservoirs include: Kentucky, Pickwick, Wilson and Wheeler Reservoirs. Upper mainstream reservoirs include: Guntersville, Nickajack, Chickamauga, Watts Bar and Fort Loudoun Reservoirs.

20.5.1 TAXA RICHNESS AND COMPOSITION

Total number of species: Higher numbers of species represent healthier aquatic ecosystems. Species introduced both inadvertently or deliberately were included only if they maintained a self-sustaining population.

Number of Lepomid sunfish species: Lepomid sunfish are basically invertivores, and a diverse sunfish community is indicative of a high-quality littoral zone.

Number of sucker species: This taxa richness metric provides a reflection of benthic habitat quality as this group generally inhabits deeper water and is particularly intolerant of substrate degradation.

Number of intolerant species: Intolerant species (such as black redhorse, longear sunfish, and skipjack herring) are sensitive to various forms of degradation. They are often rare and therefore inefficiently sampled, so presence/absence was used rather than relative abundance.

Percentage of tolerant individuals: This group becomes disproportionately abundant as aquatic conditions deteriorate. Examples of tolerant species include longnose gar, gizzard shad, common carp, and green sunfish.

Percent dominance: Numerical dominance of the fish assemblage by one species is considered indicative of degraded conditions.

TABLE 20.4
Scoring Criteria for Reservoirs in the Interior Plateau Ecoregion of the Tennessee River Valley

			Scoring Criteria						
			Forebay			Transition			
Metric	Reservoir Group	Gear	1	3	5	1	3	5	
1. Number of species	Normandy	Combined	<8	8–17	>17	<8	8–17	>17	
	Tims Ford	Combined	<10	10–20	>20	<11	11–20	>20	
	Other reservoirs	Combined	<10	10–19	>19				
2. Number of Lepomid sunfish species	Normandy	Combined	<2	2–3	>3	<2	2–3	>3	
	Tims Ford	Combined	<2	2–3	>3	<2	2–3	>3	
	Other reservoirs	Combined	<2	2–3	>3				
3. Number of sucker species	Normandy	Combined	<3	3–4	>4	<2	2–2	>2	
	Tims Ford	Combined	<4	4–6	>6	<4	4–6	>6	
	Other reservoirs	Combined	<3	3–5	>5				
4. Number of intolerant species	Normandy	Combined	<2	2–2	>2	<2	2–2	>2	
	Tims Ford	Combined	<2	2–2	>2	<2	2–2	>2	
	Other reservoirs	Combined	<2	2–2	>2				
5. Percent tolerant individuals	All	Electrofishing	>30%	15–30%	<15%	>30%	15–30%	<15%	
	All	Gillnetting	>35%	20–35%	<20%	>35%	20–35%	<20%	
6. Percent dominance	All	Electrofishing	>60%	40–60%	<40%	>60%	40–60%	<40%	
	All	Gillnetting	>50%	30–50%	<30%	>50%	30–50%	<30%	
7. Number of piscivore species	Normandy	Combined	<3	3–6	>6	<3	3–6	>6	
	Tims Ford	Combined	<4	4–6	>6	<4	4–6	>6	
	Other reservoirs	Combined	<3	3–6	>6				
8. Percent omnivores	All	Electrofishing	>25%	10–25%	<10%	>25%	10–25%	<10%	
	All	Gillnetting	>60%	40–60%	<40%	>60%	40–60%	<40%	
9. Percent invertivores	All	Electrofishing	<60%	60–80%	>80%	<50%	50–70%	>70%	
	All	Gillnetting	<3%	3–6%	>6%	<3%	3–6%	>6%	
10. Number of lithophilic spawning species	Normandy	Combined	<3	3–6	>6	<3	3–6	>6	
	Tims Ford	Combined	<4	4–6	>6	<4	4–6	>6	
	Other reservoirs	Combined	<3	3–6	>6				
11. Total number of individuals	All	Electrofishing	<40	40–80	>80	<40	40–80	>80	
	All	Gillnetting	<10	10–18	>18	<10	10–18	>18	
12. Percent anomalies	All	Combined	<2%	2–5%	>5%	<2%	2–5%	>5%	

Note: Other reservoirs include: Beech, Bear Creek, Little Bear Creek, and Cedar Creek Reservoirs.

TABLE 20.5
Scoring Criteria for Reservoirs in the Ridge and Valley Ecoregion of the Tennessee River Valley

		Scoring Criteria					
		Forebay			Transition		
Metric	Gear	1	3	5	1	3	5
1. Number of species	Combined	<10	10–19	>19	<11	11–20	>20
2. Number of Lepomid sunfish species	Combined	<2	2–3	>3	<2	2–3	>3
3. Number of sucker species	Combined	<3	3–5	>5	<3	3–6	>6
4. Number of intolerant species	Combined	<2	2–2	>2	<2	2–2	>2
5. Percent tolerant individuals	Electrofishing	>30%	15–30%	<15%	>30%	15–30%	<15%
	Gillnetting	>50%	30–50%	<30%	>50%	30–50%	<30%
6. Percent dominance	Electrofishing	>60%	40–60%	<40%	>60%	40–60%	<40%
	Gillnetting	>50%	30–50%	<30%	>50%	30–50%	<30%
7. Number of piscivore species	Combined	<3	3–6	>6	<4	4–6	>6
8. Percent omnivores	Electrofishing	>25%	10–25%	<10%	>25%	10–25%	<10%
	Gillnetting	>60%	40–60%	<40%	>60%	40–60%	<40%
9. Percent invertivores	Electrofishing	<60%	60–80%	>80%	<50%	50–70%	>70%
	Gillnetting	<3%	3–6%	>6%	<3%	3–6%	>6%
10. Number of lithophilic spawning species	Combined	<2	2–4	>4	<3	3–6	>6
11. Total number of individuals	Electrofishing	<40	40–80	>80	<40	40–80	>80
	Gillnetting	<15	15–30	>30	<15	15–30	>30
12. Percent anomalies	Combined	<2%	2–5%	>5%	<2%	2–5%	>5%

TABLE 20.6
Scoring Criteria for Reservoirs in the Blue Ridge Ecoregion of the Tennessee River Valley

		Scoring Criteria						
		Forebay				Transition		
Metric	Gear	1	3	5	1	3	5	
1. Number of species	Combined	<8	8–15	>15	<8	8–15	>15	
2. Number of Lepomid sunfish species	Combined	<2	2–3	>3	<2	2–3	>3	
3. Number of sucker species	Combined	<2	2–3	>3	<2	2–3	>3	
4. Number of intolerant species	Combined	<2	2–2	>2	<2	2–2	>2	
5. Percent tolerant individuals	Electrofishing	>30%	15–30%	<15%	>30%	15–30%	<15%	
	Gillnetting	>20%	10–20%	<10%	>20%	10–20%	<10%	
6. Percent dominance	Electrofishing	>60%	40–60%	<40%	>60%	40–60%	<40%	
	Gillnetting	>50%	30–50%	<30%	>50%	30–50%	<30%	
7. Number of piscivore species	Combined	<3	3–5	>5	<3	3–5	>5	
8. Percent omnivores	Electrofishing	>10%	5–10%	<5%	>10%	5–10%	<5%	
	Gillnetting	>30%	15–30%	<15%	>30%	15–30%	<15%	
9. Percent invertivores	Electrofishing	<75%	75–85%	>85%	<75%	75–85%	>85%	
	Gillnetting	<3%	3–6%	>6%	<3%	3–6%	>6%	
10. Number of lithophilic spawning species	Combined	<3	3–4	>4	<3	3–4	>4	
11. Total number of individuals	Electrofishing	<30	30–60	>60	<30	30–60	>60	
	Gillnetting	<10	10–18	>18	<10	10–18	>18	
12. Percent anomalies	Combined	<2%	2–5%	>5%	<2%	2–5%	>5%	

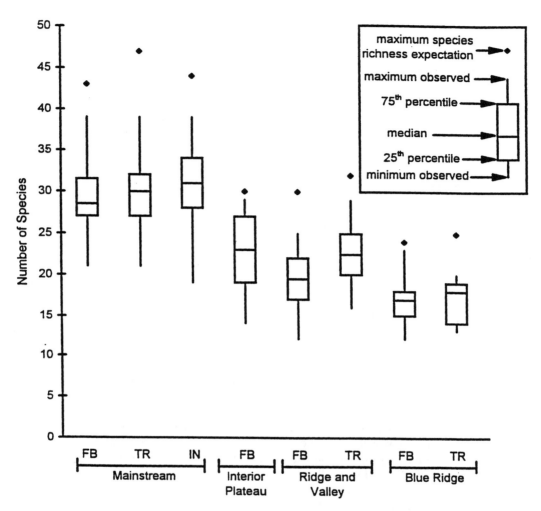

FIGURE 20.3 Box and whisker plot of the species richness metric, showing observed range and 25th and 75th percentiles for each class of reservoir sites compared with maximum species richness expectations (FB = forebay, TR = transition, and IN = inflow). (See text for explanation of method used to determine maximum species richness expectations.)

20.5.2 TROPHIC COMPOSITION

Number of piscivore species: This functional group is composed of specialist species (such as black bass, sauger, walleye, and white bass), which decline in diversity as habitat quality deteriorates. Number of resident species is used instead of relative abundance due to population manipulation by state fish agencies. Species whose populations are maintained by stocking are excluded.

Percentage of individuals as omnivores: Omnivores (such as common carp, bluntnose minnow, and gizzard shad) are less sensitive to environmental stresses due to the ability to vary their diets. As trophic links are disrupted due to degraded conditions, specialist species such as invertivores decline while opportunistic omnivorous species increase in relative abundance.

Percentage of individuals as invertivores: Due to the special dietary requirements of this group of species (such as spotted sucker, river redhorse, most shiners, bluegill, and logperch) and the limitations of their food source in degraded environments, proportion of invertivores increases with environmental quality.

20.5.3 REPRODUCTIVE COMPOSITION

Number of lithophilic spawning species: This metric includes species (such as logperch, walleye, redhorse species, and sauger) that lay eggs over rocky substrates and provide no subsequent parental care. This metric is included because of the sensitivity of this group of fish to sedimentation and siltation, a common symptom of degradation in reservoirs.

20.5.4 ABUNDANCE

Total number of individuals: Optimal environmental conditions generate high-quality fish assemblages that support large numbers of individuals.

20.5.5 FISH HEALTH

Percent of fish with diseases, parasites, anomalies, and/or natural hybrids: Health of the individual fish is a reflection of environmental quality. Incidence of natural hybridization is also indicative of degraded conditions.

20.6 SAMPLING METHODS

Soballe et al. (1992) noted that general patterns of reservoir structure and function can be determined through representative sampling of resident fish communities. However, one of the most difficult aspects of indexing environmental conditions based on resident fish communities is obtaining a representative sample. A representative sample includes the majority of resident species with individual species in approximate proportion to their true existence (Karr et al., 1986).

All methods of sampling fish communities are selective. Because reservoir habitats are diverse horizontally, longitudinally, and vertically, a cost-efficient representative sample of the entire community is not possible using a single gear. Meador et al. (1993) and Weaver et al. (1993) recommended a multigear approach to take advantage of differences in gear selectivity and efficiency to achieve a more accurate representation of the fish community structure. Simon and Lyons (1995), in discussing development of a big river IBI, recognized the need to include multiple sampling techniques to get a representative snapshot of the large river fish community. However, they noted that problems with aggregating results from different collection methods must be resolved before large river IBI development could move forward.

For this study, shoreline electrofishing was selected for the shallow littoral zone and experimental gillnetting for the limnetic bottom zone. The procedure for calculating a combined gear index is based on pooled electrofishing and experimental gillnet sampling results for specific metrics (taxa richness, reproductive composition, and fish health), with abundance and proportional metrics computed jointly (electrofishing and gillnetting results worth one half weight for each metric value).

Beginning in 1993, the standard sampling effort included 15 electrofishing transects (300 m in length) and 12 overnight experimental gillnet sets (five 6.1-m panels with mesh sizes of 2.5, 3.8, 5.1, 6.4, and 7.6 cm) in each appropriate reservoir zone; 12 gillnets are set at each site to assure 10 successful net sets. A Global Positioning System (GPS) was used to measure distance during electrofishing runs and identify standard sampling locations. Electrofishing and gillnet sampling were performed in all major habitats in approximate proportion to their occurrence within the zone. Hickman at al. (1994) and Hickman and McDonough (1996) provide more detailed descriptions of sampling procedures.

Reservoir Fish Assemblage Index Development

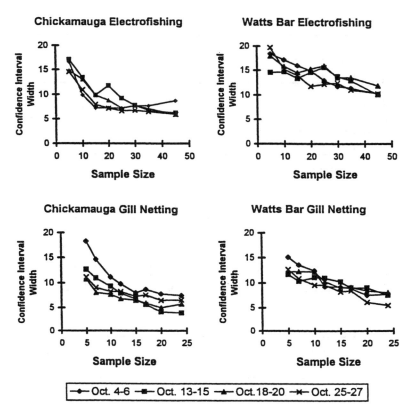

FIGURE 20.4 Confidence interval widths (±2 standard deviations) of RFAI scores from bootstrap samples of 100 randomly drawn sets of varying numbers of electrofishing runs and gillnet samples taken from Chickamauga and Watts Bar Reservoir forebays, 1993.

20.7 RESERVOIR FISH-ASSEMBLAGE INDEX VARIABILITY

Jennings et al. (1995) found RFAI variability to be high when samples were limited to 10 electrofishing runs and recommended sampling effort be increased. To further examine RFAI variability and determine the number and types of samples necessary to obtain a representative sample, TVA biologists intensively sampled the forebay zones of Chickamauga and Watts Bar Reservoirs. This included weekly collections of 45 standard electrofishing runs and 24 overnight experimental gillnet sets during October 1993.

Variability of RFAI scores based on bootstrap resampling analysis (Efron, 1981; Fore et al., 1995) of those special studies, decreased with increasing numbers of samples for both collection methods (Figure 20.4). As sample size increased, the rate of variability decrease lessened. Increased sampling effort beyond 15 to 20 electrofishing runs or 10 to 12 gillnet samples provided little increase in precision. Confidence interval widths (±2 standard deviations) included about 96% of samples; it was desirable to have these error bands no wider than the range of individual RFAI categories (10), as described by Hickman et al. (1994). This level of precision was not consistently achieved for either collection method with the sampling effort of 15 electrofishing transects and 10 overnight experimental gillnet sets. However, the desired level of precision was achieved with 15 electrofishing samples in all four sampling periods in Chickamauga Reservoir, but none of the four sample periods in Watts Bar Reservoir. Ten gillnet samples yielded this level of precision in three periods in Chickamauga Reservoir, but in only one sample period in Watts Bar Reservoir.

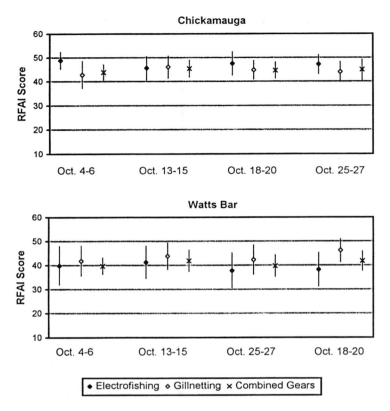

FIGURE 20.5 Mean and confidence intervals (±2 standard deviations) of RFAI scores from bootstrap samples of 100 randomly selected sets of 15 electrofishing runs, 10 gillnet samples, and combined electrofishing and gillnetting samples taken from Chickamauga and Watts Bar Reservoir forebays, 1993.

Even with doubling current effort, the desired level of precision was never reached using electrofishing in Watts Bar Reservoir.

RFAI variability was lowered significantly when sample results from both gear types were combined (Figure 20.5). The desired level of precision was achieved in both reservoirs in all four sample periods for the combined gear index. Between-sample variability was also lower for the combined gear index (Figure 20.5). Standard deviation (SD) of combined gear RFAI (pooling data from all four sample periods) was 42% lower than the electrofishing RFAI and 28% lower than the gillnet RFAI in Chickamauga Reservoir. In Watts Bar Reservoir, the SD of the combined gear index was 44% lower than the SD of the electrofishing index and 30% lower than the SD of the gillnet index. Within-sample period variability of index scores obtained using individual gear could be reduced by considerably increasing effort; however, for the purposes here, this would have been cost prohibitive. Combining sample results of the two gears provided a more cost-effective approach to reducing RFAI variability, both within and between sample periods.

Examination of species accumulation curves revealed that combining collections of the two gears resulted in a higher percentage of resident fish species being sampled (Figure 20.6). In both reservoirs, combining results of electrofishing and gillnet samples provided a more efficient means of sampling fish species than was obtained from a comparable level of effort using individual gear. In Chickamauga Reservoir, doubling the current level of effort for each gear would provide 25 to 28 species in electrofishing (30 samples) and 18 to 21 species in gillnetting (20 samples). Combining collection methods at the current level of effort would result in 30 to 32 species. In Watts Bar Reservoir, doubling current sampling effort provided 24 to 27 species in electrofishing and 20 to

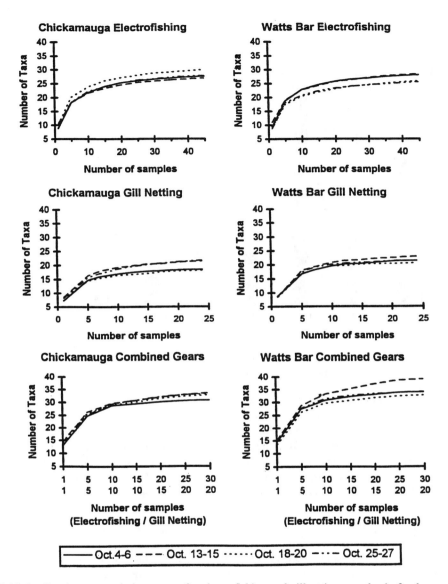

FIGURE 20.6 Species accumulation curves for electrofishing and gillnetting samples in forebay sections of Watts Bar and Chickamauga Reservoirs.

23 species in gillnetting. Combining collection methods at the current level of effort resulted in 32 to 37 species.

20.8 RESERVOIR FISH-ASSEMBLAGE INDEX RESULTS IN TVA RESERVOIRS

RFAI precision was examined both between and within years. Since no major changes to the river system or its watershed occurred during the study period, observed variability can be attributed to differences in meteorologic or hydrologic conditions and sampling error. In 1993 and 1994, approximately 20% of all sites were resampled within a few weeks of the initial survey for precision and accuracy variation determination purposes. Maximum observed variance in these samples was 10

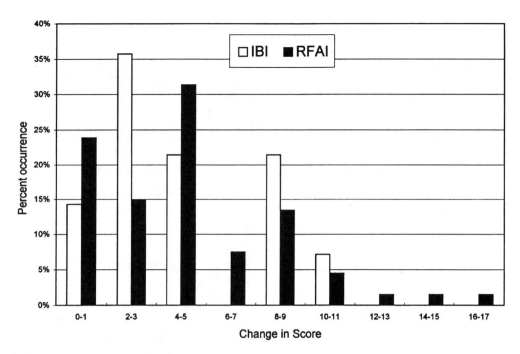

FIGURE 20.7 Comparison of annual variability between 1993 and 1994 RFAI scores with annual variability in IBI scores from large (greater than 1000-sq. mi. drainage area) rivers in the Tennessee Valley.

points. Adverse weather conditions during one of the samples affected results at this site. Over 95% of the repeated samples had differences of 6 points or less and most differed by 2 points or less.

The maximum difference between 1993 and 1994 RFAI scores at a station was 16 points. Almost 95% of the stations differed from the desired level of precision by 10 points or less. Kendell's indicated that rankings of sites were consistent between years ($p < 0.001$). Annual variability in RFAI scores was only slightly higher than annual variability in IBI scores from large (greater than 2500 sq. km drainage area) rivers in the Tennessee Valley (Figure 20.7).

20.9 VALIDITY OF THE RESERVOIR FISH-ASSEMBLAGE INDEX

There has been rapidly increasing interest in biomonitoring during the last two decades. Biomonitoring has often been found to be more cost effective than physical and chemical monitoring, and has been shown to detect environmental degradation when chemical and sediment toxicity tests indicate no problem (Karr et al., 1986; Ohio EPA, 1987). Herricks and Schaeffer (1985) defined six criteria that valid biomonitoring programs should satisfy. Karr et al. (1986) discussed the extent to which the IBI met these same criteria. This section evaluates the extent to which the RFAI meets each of these criteria.

> *Criterion 1: The measure must be biological.* The RFAI meets this criterion since it is based entirely on biological parameters.
>
> *Criterion 2: The measure must be amenable to application at other trophic levels, reflect effects at other levels of the biological-ecological hierarchy, or provide an experimentally verified connection to other organisms or trophic levels.* The RFAI meets this criterion since three of its composite metrics relate to trophic structure. Fish occupy niches at several trophic levels and have a long life cycle that allows them to integrate conditions over time. These characteristics make fish communities a desirable component of aquatic biomonitoring programs.

Criterion 3: The measure must be a good indicator, sensitive to the environmental conditions being monitored. The present studies showed the RFAI as partially meeting this criterion. In the Tennessee River Valley, reservoirs with low RFAI scores have known impacts. The true test of how well the RFAI meets this criterion, however, will require application in other river systems under a wider range of conditions.

Criterion 4: The response range of the measure must be suitable for the intended application. The IBI has been shown to be sensitive to slight changes in a wide range of environmental disturbances (Karr et al., 1986). The RFAI, like the IBI, is made up of individual metrics that were selected to reflect varying ranges of sensitivity. Combination of these diverse metrics allows detection across a wider range of anthropogenic resource degradation.

Criterion 5: The measure must be reproducible and precise within defined and acceptable limits. The RFAI detected significant differences among reservoirs and ranked reservoirs consistently on an annual basis (Hickman and McDonough, 1996). RFAI variability in bootstrap resampling during four sampling periods from two reservoirs was within acceptable levels (Hickman and McDonough, 1996).

Criterion 6: Variability of the measure must be low. If variability is high, the distributional characteristics of the data must be known. RFAI variability was found to be acceptable for the primary purposes of providing screening level assessments of reservoir ecological health, and targeting detailed assessment studies. Decisions to initiate expensive reservoir or watershed remediation activities should be based on repeated sampling over at least a three-year period. Sustained monitoring would determine year-to-year variability at a site, and provide a solid baseline against which to evaluate response of the fish community to remediation.

20.10 CONCLUSION

The most difficult aspects of developing the RFAI included determining the field protocols required to obtain a representative sample of the fish community, defining reference conditions, and establishing scoring criteria. The use of multiple gear types (electrofishing and gillnetting) provides a more complete representation of the fish assemblage than can be obtained using a single collection method. In the absence of reference sites, professional judgment and knowledge of resident fish assemblages are essential for determining reference conditions and scoring criteria, especially in groups of reservoirs with narrow ranges of observed conditions.

ACKNOWLEDGMENTS

The authors express appreciation to Al Brown, Gary Jenkins, Donny Lowery, George Peck, and Anita Masters, who collected the majority of the field data; Linda Torbett, who processed and managed the data; Steve Bloom, who provided statistical analysis and graphics; and Charlie Saylor, Ed Scott, Jr., and Donald Dycus, who provided considerable ecological inferences using their extensive knowledge of the reservoir system and resident fish distribution. Special thanks are also due to Dr. James Karr and key researchers at one time or another under his guidance, including Michele Dionne, Martin Jennings, and Leska Fore. All provided useful input and ideas and laid the foundation of the RFAI.

REFERENCES

Dionne, M. and J. R. Karr, 1992. Ecological monitoring of fish assemblages in Tennessee River reservoirs, in D. H. McKenzie, D. E. Hyatt, and V. J. McDonald (Eds.), *Ecological Indicators*. Vol. I. Elsevier Applied Science, London. 259–281.

Dycus, D. L. and D. L. Meinert. 1991. Reservoir Monitoring, 1990 Summary of Vital Signs and Use Impairment Monitoring on Tennessee Valley Reservoirs. Tennessee Valley Authority, Water Resources, Chattanooga.

Efron, B. 1981. Nonparametric estimates of standard error: the jackknife, the bootstrap and other methods, *Biometrika,* 68, 589–599.

Fausch, K. D., J. R. Karr, and P. R. Yant. 1984. Regional application of an index of biotic integrity based on stream fish communities, *Transactions of the American Fisheries Society,* 113, 39–55.

Fore, L. S., J. R. Karr, and L. L. Conquest. 1994. Statistical properties of an index of biological integrity used to evaluate water resources, *Canadian Journal of Fisheries and Aquatic Sciences,* 51, 1077–1087.

Herricks, E. E. and D. J. Schaeffer. 1985. Can we optimize biomonitoring?, *Environmental Management,* 9, 487–492.

Hickman, G. D., A. M. Brown, and G. Peck. 1994. Tennessee Valley Reservoir and Stream Quality 1993 Summary of Reservoir Fish Assemblage Results. Tennessee Valley Authority, Norris.

Hickman, G. H. and T. A. McDonough. 1996. Assessing the Reservoir Fish Assemblage Index — a potential measure of reservoir quality, in D. DeVries (Ed.), *Reservoir Symposium — Multidimensional Approaches to Reservoir Fisheries Management.* Reservoir Committee, Southern Division, American Fisheries Society, Bethesda, MD. 85–97.

Jennings, M. J., L. S. Fore, and J. R. Karr. 1995. Biological monitoring of fish assemblages in Tennessee Valley reservoirs, *Regulated Rivers: Research and Management,* 11, 263–274.

Karr, J. R. 1981. Assessment of biotic integrity using fish communities, *Fisheries (Bethesda),* 6(6), 21–27.

Karr, J. R., K. D. Fausch, P. L. Angermeier, P. R. Yant, and I. J. Schlosser. 1986. Assessing Biological Integrity in Running Waters: A Method and Its Rationale. Illinois National Historic Survey Special Publication 5.

Karr, J. R. and M. Dionne. 1991. Designing surveys to assess biological integrity in lakes and reservoirs, in *Biological Criteria. Research and Regulation.* Proceedings of a symposium. EPA-44015-91-005. Office of Water, U.S. Environmental Protection Agency, Washington, DC. 62–72.

Kimmel, B. L. and A. W. Groeger. 1984. Factors controlling primary production in lakes and reservoirs: a perspective, in *Lake and Reservoir Management.* EPA 440/5/84-001. U.S. Environmental Protection Agency, Washington, DC. 277–281.

Meador, M. R., T. F. Cuffney, and M. E. Gurtz. 1993. Methods for Sampling Fish Communities as Part of the National Water-Quality Assessment Program. U. S. Geological Survey OFR03-104. Raleigh, NC.

McDonough, T. A. and W. C. Barr. 1977. An analysis of fish associations in Tennessee and Cumberland drainage impoundments, *Proceedings of the Southeastern Association of Fish and Wildlife Agencies,* 31, 555–563.

Miller, D. L., P. M. Leonard, R. M. Hughes, J. R. Karr, P. B. Moyle, L. H. Schrader, B. A. Thompson, R. A. Daniel, K. D. Fausch, G. A. Fitzhugh, J. R. Gammon, D. B. Halliwell, P. L. Angermier, and D. J. Orth. 1988. Regional applications of an Index of Biotic Integrity for use in water resource management, *Fisheries,* 13(5), 12–20.

Ohio Environmental Protection Agency. 1987. *Biological Criteria for the Protection of Aquatic Life. Vol. II. Users Manual for Biological Field Assessment of Ohio Surface Waters.* Ohio Environmental Protection Agency, Division of Water Quality Monitoring and Assessment, Columbus, OH.

Simon, T. P. and J. Lyons. 1995. Application of the index of biotic integrity to evaluate water resource integrity in freshwater ecosystems, in W. S. Davis and T. P. Simon (Eds.), *Biological Assessment and Criteria: Tools for Water Resource Planning and Decision Making.* Lewis, Boca Raton, FL. 245–262.

Soballe, D. M., B. L. Kimmel, R. H. Kennedy, and R. F. Gaugush. 1992. Reservoirs, in C. T. Hackney, S. M. Adams, and W. H. Martin (Eds.), *Biodiversity of the Southeastern United States Aquatic Communities.* John Wiley & Sons, Inc., New York. 421–474.

Thompson, B. A. and G. R. Fitzhugh. 1986. A Use Attainability Study and an Evaluation of Fish and Macroinvertebrate Assemblages of the Lower Calcasieu River, Louisiana. Report LSU-CFI-29. Louisiana State University, Center for Wetland Resources, Coastal Fisheries Institute, Baton Rouge.

Weaver, M .J., J. J. Magnuson, and M. K. Clayton. 1993. Analyses for differentiating littoral fish assemblages with catch data from multiple sampling gears, *Transaction of the American Fisheries Society,* 122, 1111–1119.

21 Toward the Development of an Index of Biotic Integrity for Inland Lakes in Wisconsin

Martin J. Jennings, John Lyons, Edward E. Emmons,
Gene R. Hatzenbeler, Michael A. Bozek,
Timothy D. Simonson, T. Douglas Beard, Jr.,
and Don Fago

CONTENTS

21.1 Introduction...541
21.2 Methods and Materials..543
    21.2.1 Metrics..543
    21.2.2 Data..543
    21.2.3 Analyses...548
        21.2.3.1 Sampling Variation...548
        21.2.3.2 Relation to Lake Environmental Quality...549
        21.2.3.3 Effects of Lake Size and Ecoregion...549
21.3 Results..549
    21.3.1 Sampling Variation...549
    21.3.2 Relations with Lake Environmental Quality...552
    21.3.3 Effects of Lake Size and Ecoregion..553
21.4 Discussion..556
    21.4.1 Metrics Considered for Further Development..556
    21.4.2 Sampling Procedures...558
21.5 Conclusion...560
Acknowledgments..560
References..560

21.1 INTRODUCTION

Fish assemblages reflect the overall integrity of biological communities and are excellent indicators of the environmental quality of aquatic ecosystems (Fausch et al., 1990). Consequently, fish assemblage data are increasingly used to assess and monitor the condition of surface waters. Several types of environmental indices based on fish assemblages have been developed, with the most widely used and most effective based on the Index of Biotic Integrity or IBI (Fausch et al., 1990; Simon and Lyons, 1995). The IBI consists of several structural, compositional, and functional attributes of the assemblage, termed "metrics," that are chosen and applied based on empirical data from waters encompassing a wide range of environmental quality. A key feature, and strength, of the IBI concept is that specific metrics and metric scoring criteria are developed for the particular fish assemblages that occur in each different type of aquatic habitat (i.e., lentic vs. lotic, marine vs. freshwater) and in different geographic areas (i.e., ecoregions, drainage basins).

Several variants of the IBI have been formulated and applied to wadeable streams throughout North America, but little work has been done on IBIs for lakes (Simon and Lyons, 1995). The bias toward streams can be attributed to several factors, including the original development of the IBI in a stream setting (Karr et al., 1981), the more widespread and obvious nature of environmental degradation in streams vs. lakes (U.S. EPA, 1990), and the relative difficulty of characterizing fish assemblages in lakes (e.g., Weaver et al., 1993). To completely sample lakes requires methods effective in littoral or pelagic habitats. In addition, sampling should account for variation in fish distribution patterns on diel or seasonal time scales. Nonetheless, management of lake ecosystems would benefit from an effective IBI, which would provide another tool to assess resource condition and track changes associated with degradation or restoration.

Because of important ecological differences between streams and lakes, one expects that metrics in reliable IBIs will differ between these systems. Retention time and water volume affect the rate at which silt, nutrients, and contaminants are distributed through and removed from the water column. In comparison to stream fish that are confined to a narrow channel, lake resident fish may be better able to find refuges from highly localized impacts. Many of the impacts on streams lead to rapid and direct response of the biota, in contrast to lakes, in which the effects of impacts can be more gradual. The response time for reversing problems such as nonpoint sources is also greater in lakes than in streams. Finally, inshore-offshore movements of fishes affect catch rates in lakes, but are unrelated to resource quality, adding imprecision to estimates of community composition. To be useful, metrics must have a detectable, consistent, and predictable response to changes in resource condition. One would expect that the most promising lake metrics will be based on responses of species that are highly specialized, found in the littoral zone, and sensitive to water quality.

Only three published efforts to construct an IBI-type index for standing waters were found. Jennings et al. (1995) examined data from reservoirs in the Tennessee Valley and evaluated what they termed a "Reservoir Fish Assemblage Index" (RFAI). Structurally and conceptually, the RFAI was very similar to a typical IBI. However, the RFAI displayed substantial temporal and spatial fluctuations in scores that did not reflect real variations in biotic integrity or environmental quality. Jennings et al. (1995) concluded that increased fish sampling effort, changes in some metrics, and data from a greater number and variety of reservoirs were needed before a valid RFAI could be developed. Hughes et al. (1992) began assessment of potential metrics for an IBI for lakes in the northeastern United States. Many of their metrics appeared promising as indicators of biotic integrity and environmental quality, but metric selection and calibration apparently are not yet complete, and a functional IBI has not been developed (Larsen and Christie, 1993). Finally, Minns et al. (1994) presented a fully developed IBI for near-shore areas of the Great Lakes. This IBI accurately and precisely characterized relative environmental quality among three areas of Lake Huron and Lake Ontario, and appears to be a good indicator of ecosystem health. Many of the metrics used in this IBI probably will be broadly useful, but will need to be recalibrated for smaller lakes.

Efforts to begin development of an IBI for littoral-zone fish assemblages in Wisconsin lakes are described in this chapter. Wisconsin has a rich lacustrine fish fauna and a large number and wide variety of lakes. Existing lake IBIs need modification to adequately encompass this diversity of lake types. An IBI for Wisconsin lakes would likely be useful for lakes in adjacent states and perhaps parts of Ontario, which share many similarities in lake characteristics and fish assemblages. The primary goal is to identify metrics that warrant further assessment and calibration. Specifically, this chapter examines potential metrics for their correlations with independent measures of the environmental quality of Wisconsin lakes and for their temporal and spatial variability when environmental quality is stable. Also determined are the sampling techniques that yield the most accurate and precise estimates of values for particular metrics. The recommended metrics for further development have a strong monotonic association with environmental quality, vary relatively little in the absence of significant environmental change, and can be estimated with standardized sampling methods and reasonable sampling effort.

21.2 METHODS AND MATERIALS

21.2.1 METRICS

Based on of the littoral-zone fishes of Wisconsin lakes, 13 metrics were selected for analysis, 5 of which measured species richness: numbers of native species, centrarchid species, cyprinid species, intolerant species, and small benthic species; and 8 of which measured proportional abundance: percentages of total individuals caught for native benthic (large and small) fish, intolerant fish, tolerant fish, exotic fish, omnivores, top carnivores, simple lithophilic lake spawners, and vegetation-dwelling fish.

Names and metric classifications (Table 21.1) of all fishes occurring in this study lakes are from Lyons (1992). Native and exotic fishes are defined based on occurrence in Wisconsin prior to European settlement. It is recognized that many "native" species have been newly introduced into Wisconsin lakes since European settlement, but a lack of sufficient data to determine the origin of many species in many lakes (e.g., Magnuson et al., 1994) prevents making the designation more precise. Intolerant species are sensitive to a wide range of environmental degradation and usually persist only in relatively high-quality areas. Conversely, tolerant species are able to withstand high levels of degradation and may dominate in low-quality areas. Benthic species are closely associated with the lake bottom at all times; small species rarely exceed 150 mm total length. Omnivores have a diet that is typically 25% or more plant material and 25% or more animal material, whereas top carnivores feed largely on fish or crayfish as adults. Simple lithophilic spawners broadcast their eggs over clean gravel or cobble and do not build a nest or provide parental care. This metric is restricted to simple lithophils that spawn in the lake and excludes species such as redhorse (*Moxostoma* species) that may spend significant time in lakes but spawn in streams and rivers. Vegetation-dwelling species are those that are either almost always closely associated with aquatic vegetation as adults or that require aquatic vegetation for spawning.

21.2.2 DATA

Five different datasets on littoral-zone fish assemblages were employed to assess the 13 metrics. The lakes in these datasets were between 80 and 2800 ha in surface area and occurred over much of the state, with emphasis on the north, where most Wisconsin lakes are located. Excluded were dystrophic and acidic lakes and lakes subject to periodic winterkill because such lakes have distinctive fish assemblages (Tonn and Magnuson, 1982; Rahel, 1984) and may require unique metrics. However, impoundments were not excluded. Although it is recognized that these are artificial habitats and differ from natural lakes in some ways (Jennings et al., 1995), they were included for three reasons. First, they are perceived by the public and managed by the Wisconsin Department of Natural Resources (WDNR) essentially the same as natural lakes; that is, public and agency expectations for a healthy impoundment ecosystem are similar to those for natural lakes. Second, many Wisconsin lakes have had low-head dams built on their outlets to raise and regulate water levels, so the distinction between impoundment and natural lake is often unclear. Finally, preliminary surveys and analyses revealed few consistent differences in fish assemblages between most Wisconsin natural lakes and impoundments. The one exception was for mainstem impoundments on the state's larger rivers, such as the Mississippi and Wisconsin, which may have a relatively high proportion of large river specialist fish species and other distinctive fish assemblage attributes (Simon and Emery, 1995). Although one of the datasets contains information from two

TABLE 21.1
Names and Metric Classifications for Species Encountered in This Lake Study

Common Name	Scientific Name	Family	Origin	Tolerance	Habitat	Feeding	Spawning
Longnose gar	*Lepisosteus osseus*	Lepisosteidae	Native	Other	Other	Top carnivore	Other
Bowfin	*Amia calva*	Amiidae	Native	Other	Veg	Top carnivore	Other
Gizzard shad	*Dorosoma cepedianum*	Clupeidae	Native	Other	Other	Other	Other
Lake trout	*Salvelinus namaycush*	Salmonidae	Native	Intolerant	Other	Top carnivore	Lithophilic
Rainbow smelt	*Osmerus mordax*	Osmeridae	Exotic	Other	Other	Other	Other
Central mudminnow	*Umbra limi*	Umbridae	Native	Tolerant	Veg	Other	Other
Grass pickerel	*Esox americanus*	Esocidae	Native	Other	Veg	Top carnivore	Other
Northern pike	*Esox lucius*	Esocidae	Native	Other	Veg	Top carnivore	Other
Muskellunge	*Esox masquinongy*	Esocidae	Native	Intolerant	Veg	Top carnivore	Other
Largescale stoneroller	*Campostoma oligolepis*	Cyprinidae	Native	Other	SB	Other	Other
Spotfin shiner	*Cyprinella spiloptera*	Cyprinidae	Native	Other	Other	Other	Other
Common carp	*Cyprinus carpio*	Cyprinidae	Exotic	Tolerant	Other	Omnivore	Other
Brassy minnow	*Hybognathus hankinsoni*	Cyprinidae	Native	Other	Other	Other	Other
Common shiner	*Luxilus cornutus*	Cyprinidae	Native	Other	Other	Other	Lithophilic
Pearl dace	*Margariscus margarita*	Cyprinidae	Native	Other	Other	Other	Other
Hornyhead chub	*Nocomis biguttatus*	Cyprinidae	Native	Other	Other	Other	Other
Golden shiner	*Notemigonus crysoleucas*	Cyprinidae	Native	Tolerant	Other	Omnivore	Other
Pugnose shiner	*Notropis anogenus*	Cyprinidae	Native	Intolerant	Veg	Other	Other
Emerald shiner	*Notropis atherinoides*	Cyprinidae	Native	Other	Other	Other	Lithophilic
Blackchin shiner	*Notropis heterodon*	Cyprinidae	Native	Intolerant	Veg	Other	Other
Blacknose shiner	*Notropis heterolepis*	Cyprinidae	Native	Intolerant	Veg	Other	Other
Spottail shiner	*Notropis hudsonius*	Cyprinidae	Native	Intolerant	Other	Other	Other
Rosyface shiner	*Notropis rubellus*	Cyprinidae	Native	Intolerant	Other	Other	Lithophilic
Sand shiner	*Notropis stramineus*	Cyprinidae	Native	Other	Other	Other	Other
Mimic shiner	*Notropis volucellus*	Cyprinidae	Native	Other	Veg	Other	Other
Northern redbelly dace	*Phoxinus eos*	Cyprinidae	Native	Other	Veg	Other	Other
Bluntnose minnow	*Pimephales notatus*	Cyprinidae	Native	Tolerant	Other	Omnivore	Other
Fathead minnow	*Pimephales promelas*	Cyprinidae	Native	Tolerant	Other	Omnivore	Other

Blacknose dace	*Rhinichthys atratulus*	Cyprinidae	Native	Tolerant	SB	Other	Lithophilic
Longnose dace	*Rhinichthys cataractae*	Cyprinidae	Native	Other	SB	Other	Lithophilic
Creek chub	*Semotilus atromaculatus*	Cyprinidae	Native	Tolerant	Other	Other	Other
Quillback	*Carpiodes cyprinus*	Catostomidae	Native	Other	Benthic	Omnivore	Other
White sucker	*Catostomus commersoni*	Catostomidae	Native	Tolerant	Benthic	Omnivore	Other
Lake chubsucker	*Erimyzon sucetta*	Catostomidae	Native	Other	Veg	Other	Other
Northern hogsucker	*Hypentelium nigricans*	Catostomidae	Native	Intolerant	Benthic	Other	Other
Silver redhorse	*Moxostoma anisurum*	Catostomidae	Native	Other	Benthic	Other	Other
Golden redhorse	*Moxostoma erythrurum*	Catostomidae	Native	Other	Benthic	Other	Other
Shorthead redhorse	*Moxostoma macrolepidotum*	Catostomidae	Native	Other	Benthic	Other	Other
Greater redhorse	*Moxostoma valenciennesi*	Catostomidae	Native	Intolerant	Benthic	Other	Other
Black bullhead	*Ameiurus melas*	Ictaluridae	Native	Other	Other	Other	Other
Yellow bullhead	*Ameiurus natalis*	Ictaluridae	Native	Tolerant	Other	Other	Other
Brown bullhead	*Ameiurus nebulosus*	Ictaluridae	Native	Other	Other	Other	Other
Channel catfish	*Ictalurus punctatus*	Ictaluridae	Native	Other	Other	Top carnivore	Other
Tadpole madtom	*Noturus gyrinus*	Ictaluridae	Native	Other	Veg, SB	Other	Other
Troutperch	*Percopsis omiscomaycus*	Percopsidae	Native	Other	SB	Other	Other
Burbot	*Lota lota*	Gadidae	Native	Other	Other	Top carnivore	Lithophilic
Banded killifish	*Fundulus diaphanus*	Fundulidae	Native	Other	Other	Other	Other
Starhead topminnow	*Fundulus dispar*	Fundulidae	Native	Other	Other	Other	Other
Brook silverside	*Labidesthes sicculus*	Atherinidae	Native	Other	Other	Other	Other
Brook stickleback	*Culaea inconstans*	Gasterosteidae	Native	Other	Other	Other	Other
Ninespine stickleback	*Pungitius pungitius*	Gasterosteidae	Native	Other	Other	Other	Other
White bass	*Morone chrysops*	Moronidae	Native	Other	Other	Top carnivore	Other
Yellow bass	*Morone mississippiensis*	Moronidae	Native	Other	Other	Top carnivore	Other
Rock bass	*Ambloplites rupestris*	Centrarchidae	Native	Intolerant	Other	Top carnivore	Other
Green sunfish	*Lepomis cyanellus*	Centrarchidae	Native	Tolerant	Other	Other	Other
Pumpkinseed	*Lepomis gibbosus*	Centrarchidae	Native	Other	Other	Other	Other
Warmouth	*Lepomis gulosus*	Centrarchidae	Native	Other	Other	Top carnivore	Other
Bluegill	*Lepomis macrochirus*	Centrarchidae	Native	Other	Other	Other	Other
Smallmouth bass	*Micropterus dolomieu*	Centrarchidae	Native	Intolerant	Other	Top carnivore	Other
Largemouth bass	*Micropterus salmoides*	Centrarchidae	Native	Other	Other	Top carnivore	Other
White crappie	*Pomoxis annularis*	Centrarchidae	Native	Other	Other	Top carnivore	Other
Black crappie	*Pomoxis nigromaculatus*	Centrarchidae	Native	Other	Other	Top carnivore	Other
Rainbow darter	*Etheostoma caeruleum*	Percidae	Native	Intolerant	SB	Other	Other
Iowa darter	*Etheostoma exile*	Percidae	Native	Intolerant	Veg, SB	Other	Other

TABLE 21.1 (continued)
Names and Metric Classifications for Species Encountered in This Lake Study

Common Name	Scientific Name	Family	Origin	Tolerance	Habitat	Feeding	Spawning
Fantail darter	*Etheostoma flabellare*	Percidae	Native	Other	SB	Other	Other
Least darter	*Etheostoma microperca*	Percidae	Native	Intolerant	Veg, SB	Other	Other
Johnny darter	*Etheostoma nigrum*	Percidae	Native	Other	SB	Other	Other
Banded darter	*Etheostoma zonale*	Percidae	Native	Intolerant	SB	Other	Other
Yellow perch	*Perca flavescens*	Percidae	Native	Other	Other	Other	Other
Logperch	*Percina caprodes*	Percidae	Native	Other	SB	Other	Lithophilic
Blackside darter	*Percina maculata*	Percidae	Native	Other	SB	Other	Other
River darter	*Percina shumardi*	Percidae	Native	Other	SB	Other	Other
Walleye	*Stizostedion vitreum*	Percidae	Native	Other	Other	Top carnivore	Other
Freshwater drum	*Aplodinotus grunniens*	Sciaenidae	Native	Other	Other	Other	Other
Mottled sculpin	*Cottus bairdi*	Cottidae	Native	Intolerant	SB	Other	Other

Note: Most metric designations are taken from Lyons (1992). "Other" indicates that the species is not classified into one of the relevant metric categories; "Veg" indicates vegetation and "SB" indicates small benthic.

impoundments on the upper Wisconsin River, datasets that did not emphasize large mainstem systems were selected.

The first dataset encompassed four lakes — Allequash, Big Muskellunge, Sparkling, and Trout — all located in northcentral Wisconsin. This dataset was used to evaluate statistical properties of metrics in systems from undisturbed, least-impacted watersheds. Data from these lakes were collected as part of the U.S. National Science Foundation funded Long Term Ecological Research (LTER) project for North Temperate Lakes, LTER program, National Science Foundation, and were made available courtesy of John Magnuson, Center for Limnology, University of Wisconsin–Madison. All four lakes are circumneutral, oligotrophic, or mesotrophic, and have excellent water quality and relatively little riparian or watershed human disturbance (Magnuson et al., 1984). They range in surface area from 81 (Sparkling) to 1605 (Trout) ha. They represent least-impacted regional reference sites (Hughes et al., 1986) for development and testing of potential IBI metrics (Simon and Lyons, 1995). Each lake was sampled annually for fish in August from 1981 through 1995. Three sampling methods were used: a standard WDNR "boomshocker" (a boat-mounted electrofishing unit; Novotny and Priegel, 1974), small-mesh fyke nets, and a bag seine. The boomshocker produced pulsed DC output from a 5000-W generator, and two netters in the bow of the boat collected all fish observed. The boomshocker was operated for 30 min. at night along the shoreline in 1 to 2 m of water. Four stations were sampled on each lake. The fyke nets were 4 m long with an 8×1.25-m lead net, consisted of 6.4-mm stretch mesh throughout, had 51-mm throats, and were comprised of two 0.9×0.8-m steel frames followed by three 0.8-m diameter steel hoops toward the cod end of the net. The nets were set perpendicular to the shoreline such that the anterior frame was on the bottom in about 2 m of water. Six overnight sets were made on each lake during each year. The bag seine had dimensions of 12.2×1.2 m with 6.4-mm stretch mesh throughout. It was fished at night, with 18 hauls per lake per year. Each haul was made parallel to shore, with the offshore edge of the seine following the 1-m contour for 33 m.

The second dataset was from Sparkling Lake during 1982 and 1983, and is described more fully in Lyons (1987). This dataset was used to evaluate statistical properties of metrics in a lake with relatively undisturbed, or least-impacted conditions. Although this dataset was collected on the same lake and during the same time period as the LTER data, different techniques were used and different locations were sampled, and the two datasets do not overlap. Fish were collected during the day with a bag seine and small-mesh fyke nets. The bag seine was 15.2×1.8 m with 6.4-mm stretch mesh. Hauls were made perpendicular to shore, starting at the 1-m contour and pulling directly into the shoreline. The fyke nets were similar in construction to those used by LTER, although some of the frame, hoop, and lead net dimensions varied slightly among different nets. Fyke nets were fished parallel to shore between 1 and 4 m depth, and were set in the morning and retrieved approximately 8 hr later in the afternoon. Sparkling Lake was sampled by seine in early June, late June, mid-July, and early October 1982, and in early June, mid-July, mid-August, and early October 1983, and by fyke net in early June, late June, mid-July, mid-August, and early October 1983. During each sampling period, 21 to 30 seine hauls and 48 to 54 fyke net lifts were done.

The third dataset included 20 lakes and impoundments located throughout the state, sampled as part of a study to determine the effects of shoreline erosion control devices (particularly "seawalls") on littoral-zone habitat and biota. All of the lakes had significant human development in their riparian zone, but water quality and watershed development varied among the lakes. Each lake was sampled once between early May and early October at 18 locations. Six locations had seawalls along the shoreline for erosion control, six had rock riprap, and six had unaltered shorelines. Each location was sampled by wading with a combination of a standard Wisconsin Department of Natural Resources two-anode DC stream shocker (Lyons and Kanehl, 1993) and a $7 \text{ m} \times 1.8 \text{ m}$, 6.4-mm stretch-mesh, bag seine. Many of the locations were structurally complex and could not be sampled effectively with a seine alone. Each location was first enclosed with a 6.4-mm stretch mesh block net to form a 15×5-m rectangular area, the shoreline forming one side of the rectangle.

The area within the blocknet was then sampled for two passes parallel to the shore with the stream shocker, followed by a single pass with the seine.

Fyke net samples were taken in all shoreline study lakes except Winneconne, and used the same stations as the index samples (shock and seine). Samples were taken within a week of the index samples. The minifyke nets had two 0.91 × 0.91-m frames, four 0.61-m diameter fiberglass hoops spaced 0.61-m apart, and a 0.76 × 0.91-m lead. All netting was 4.8-mm bar mesh except the exclusion netting over the opening, which was 25-mm bar mesh. Two nets used in two lakes had 6.4-mm bar mesh.

Nets were set by mid-afternoon and collected during the next day. When numbers of fish in a net were in the thousands (composed primarily of young-of-the-year yellow perch, bluegills, smallmouth bass, and largemouth bass), a subsample of all fish was randomly taken, identified, counted, and average weight per specimen determined.

The fourth dataset was from 11 impoundments in northern Wisconsin that were sampled to help designate "Outstanding Resource Waters" (ORW) for stricter environmental protection. Some of the impoundments were selected because habitat and watershed were representative of the best possible environmental conditions for the region, whereas others had significant water or habitat quality problems. Each impoundment was sampled once between mid-June and late August 1994 using boomshockers, fyke nets, and bag seines. Boomshocking equipment and techniques were similar to those used for the LTER dataset. Ten randomly selected, 100-m stations were sampled on each impoundment. Fyke net characteristics and procedures were also similar to the LTER dataset, except that the fyke nets were slightly smaller, having frames and hoops with a height of 0.76-m and width of 1.2 m; 10 overnight fyke net sets were made on each impoundment. The bag seine used was 12.2 × 1.8-m with 4.8-mm stretch mesh. It was fished in the same manner as for the Lyons dataset, with 30 hauls made per lake.

The fifth dataset consisted of cumulative fish species lists (converted to presence–absence data) for 242 lakes and impoundments between 80 and 1600 ha from throughout the state. Data were collected between 1975 and 1990 using a variety of sampling techniques, especially seining and boomshocking, largely as part of a WDNR statewide fish distribution survey (FDS; Fago, 1988). Some lakes were sampled during more than 1 year, and species lists for individual lakes can be cumulative. The goal of the sampling was to determine distribution of fish species in the state. The exact sampling gears, techniques, and level of effort varied from lake to lake, but in all cases at least two different methods were used and the sampling effort was judged sufficient to determine the fish species present.

21.2.3 ANALYSES

21.2.3.1 Sampling Variation

Each dataset was used to investigate different attributes of the 13 potential metrics. The LTER data was used to quantify annual variation in metric values in the absence of significant changes in environmental quality. Any variation observed was presumably from a combination of sampling error and natural environmental variation (e.g., climatic fluctuations). For each metric, lake, and sampling method combination, the mean and coefficient of variation (CV) across all 15 years were calculated. Exotic species were extremely rare in this dataset (a few rainbow smelt, a pelagic species, were captured after 1986 in Sparkling Lake), so it was not possible to evaluate the percentage of individuals as exotic species metric. Metrics were deemed acceptable if they had relatively low CVs and mean values that were similar among lakes or that varied in a manner consistent with a well-known ecological gradient, such as increasing species richness with increasing lake size (Eadie et al., 1986; Rahel, 1986). Proportional abundance metrics were also judged on how well their values corresponded to expectations for a least-impacted site; metrics indicative of poor environmental conditions were expected to have consistently low values, and metrics indicative of good conditions were expected to have high values. These same criteria were used to compare the mean and CV among the three sampling methods to determine the best method for measuring a particular metric. For the species richness metrics, methods were also judged based on mean values, with the method that produced the highest mean value for a particular metric (i.e., greatest proportion of the species known to be present) considered the best.

The Sparkling Lake data were used to quantify within-year variation in metric values in the absence of significant changes in environmental quality. For each metric and method, the mean and CV were calculated across all sampling periods within each year; also plotted were metric values vs. sampling period to determine the presence of seasonal trends in metric values. To determine how recruitment of young-of-year (YOY) fish over the course of the year affected metric values and variability, each statistic and each plot were determined for all fish caught, and then done again for catches with YOY fish excluded. Stream studies suggest that inclusion of YOY fish in calculations may reduce the sensitivity and precision of IBI metric values in some instances (Angermeier and Karr, 1986). The same criteria were used as in the analysis of the LTER data to evaluate which metrics and sampling methods were best.

21.2.3.2 Relation to Lake Environmental Quality

The ORW and shoreline datasets were used to assess how well each metric reflected patterns of environmental quality among lakes. Lakes with anthropogenic impacts often exhibit increases in nutrient levels. A Trophic State Index (TSI; Carlson, 1977) of each lake was based on total P, or where data were available, the TSI was based on a mean from chlorophyll a, secchi, and total phosphorus (Table 21.4). Within each dataset, lakes were ranked for Trophic State Index (TSI; Carlson, 1977) and then tested for correlation between metric rankings and TSI. Those metrics with strong monotonic relationships with this measure of environmental quality, and that were also consistent with an understanding of how fish assemblages respond to environmental degradation (Fausch et al., 1990), were considered the best prospects for use in an IBI.

These datasets were also used to further evaluate the applicability of alternative sampling methods for each metric. Within each dataset, results from each sampling method were examined separately to determine which method yielded the best relations between a particular metric and environmental quality.

21.2.3.3 Effects of Lake Size and Ecoregion

The FDS dataset was used to determine relationships between values for the five species richness metrics and lake surface area (ha) and ecoregions (as delineated by Maxwell et al., 1995). Most lakes in Wisconsin fall into one of two major ecoregions, one in the southeastern portion of the state and the other in the northern third. Little data was available from the southwestern part of the state, which has very few lakes or impoundments. Analysis of covariance (ANCOVA) was used to test for statistical differences between the two ecoregions in the relationships between species richness metrics and lake area (SAS, 1990). Differences were considered significant at $p < 0.05$. Because the FDS data are cumulative species lists from multiple methods and time periods, they are not directly comparable with the method-specific metric values generated from the other datasets.

21.3 RESULTS

21.3.1 SAMPLING VARIATION

For all four lakes in the LTER dataset, the among-year variation was lower for the species richness metrics than for the proportional abundance metrics (Table 21.2). For the species richness metrics, CVs ranged from 12 to 39, and were similar among lakes for individual metrics. Mean values of species richness metrics varied substantially among lakes, with Trout Lake (the largest lake) usually

TABLE 21.2
Annual Mean and Coefficient of Variation (CV) from 1981 to 1995 for the Four Long Term Ecological Research Lakes and the 13 Metrics

Metric	Allequash			Big Muskellunge			Sparkling			Trout		
	Mean	CV	Method	Mean	CV	Method	Mean	CV	Method	Mean	CV	Method
Species richness metrics												
Native species	14.6	18	Seine	13.9	12	Seine	9.7	22	Seine	16.2	14	Seine
Centrarchid species	4.9	12	Fyke	4.7	10	Fyke	2.2	29	Fyke	4.8	15	Fyke
Cyprinid species	4.4	22	Seine	2.7	23	Seine	2.1	30	Seine	4.7	31	Seine
Intolerant species	3.9	27	Seine	3.5	28	Seine	2.4	31	Fyke	4.5	30	Seine
Small benthic species	2.8	39	Seine	2.9	32	Seine	2.5	36	Seine	4.3	24	Seine
Proportional abundance metrics (%)												
Native benthic fish	5	138	Seine	10	68	Seine	6	99	Seine	10	94	Seine
Exotic fish	0	—	—	0	—	—	2	224	Seine	0	—	—
Intolerant fish	5	44	Seine	14	47	Boom	4	40	Fyke	19	64	Boom
Tolerant fish	23	42	Seine	8	75	Seine	33	71	Seine	8	112	Boom
Omnivores	23	42	Seine	8	79	Seine	33	72	Seine	1	108	Fyke
Top carnivores	27	51	Boom	63	29	Boom	66	36	Fyke	32	52	Boom
Lithophilic spawners	17	84	Boom	57	32	Boom	23	85	Boom	25	53	Boom
Vegetation dwellers	2	68	Seine	25	87	Seine	41	75	Seine	35	80	Seine

Note: For each lake and metric, the values for method type with the lowest CV are shown. "Boom" indicates boomshocker.

having the highest values, and Sparkling Lake (the smallest) having the lowest values. For the proportional abundance metrics, CVs ranged from 29 to 138 (excluding the 224 for percentage exotic fish in Sparkling Lake, the only lake with exotic species), with most values between 40 and 80 percent. The CVs and mean values for many of the individual metrics varied substantially among lakes, and there was not a consistent relationship between lake size and CV or mean value. The CVs and mean values for the percentages of omnivores and tolerant fish were nearly identical within each lake, largely because both metrics were almost completely determined by the abundance of bluntnose minnows.

The sampling method that yielded the most precise estimates over time varied among metrics and in some cases among lakes for the LTER dataset. For most of the species richness metrics, seining provided both the lowest CV and the highest mean value in each lake (Table 21.2). The fyke net had a slightly lower CV than the seine for the number of intolerant species in Sparkling Lake (31 vs. 36%), but the seine had a higher mean value (3.2 vs. 2.4). The fyke net had the lowest CV for the number of centrarchid species in all four lakes, although the CVs for the boomshocker and seine were usually no more than 10% higher. The fyke net also had the highest mean value for number of centrarchid species in three lakes; the boom shocker had a slightly higher mean (5.2 vs. 4.9) in Allequash Lake. The method that yielded the lowest CV varied more among the proportional abundance metrics. Seines yielded the most precise estimates in all four lakes for the percentages of native benthic fish and vegetation dwellers, whereas the boomshocker had the most precise estimates for percentage of lithophilic lake spawners in all four lakes. For each of the remaining four metrics, the method with the lowest CV varied among the lakes. For percentage top carnivores, the boomshocker was best in three of four lakes, as was the seine for percentages of tolerant fish and of omnivores. For percentage intolerant fish, the boomshocker was best in two lakes, the fyke net in one, and the seine in the other.

Within-year variation in the Sparkling Lake data set (Lyons) was lower for the species richness metrics than for the proportional abundance metrics (Table 21.3). For the species richness metrics, CVs ranged from 13 to 55, with most less than 30. The CVs were similar between 1982 and 1983, and between the seine and fyke net. Mean values were higher for the seine except for the number of centrarchids species, which was higher for the fyke net. For the proportional abundance metrics, CVs ranged from 45 to 190. For all of the metrics, CVs differed substantially between 1982 and 1983, or between the seine and fyke net, or both. Most mean values were less than 10% for each of the years and methods, except for the percentages of tolerant fish and omnivores, which were each 61% for the seine in 1982, and 42% for the seine and 29% for the fyke net in 1983. Nearly all of the individuals caught that were classified into these two metrics were bluntnose minnows.

In Sparkling Lake, inclusion of YOY fish in calculations had different effects on species richness and proportional abundance metrics as estimated by seining. Relatively few YOY fish were caught with the fyke net, so YOY fish had little influence on metrics for this method. For seining, YOY fish were not caught in significant numbers until July, so they did not influence metric values in the early part of the summer. Over the entire early June through early October period, inclusion of YOY fish usually increased the mean, and decreased or had little effect on the CV for the species richness metrics. Conversely, for proportional abundance metrics, including YOY fish typically had little effect on the mean in 1982 but tended to lower it in 1983. In both years, CVs either changed little or increased when YOY fish were included.

Consistent seasonal trends in metric values were not evident in Sparkling Lake. In 1982, most metric values declined after June, with the exception of the percentages of omnivores and of tolerant fish, which increased. Declines in species richness metrics were generally less pronounced when YOY fish were included (Figure 21.1). However, no declines or increases were apparent during 1983 for either the seine or the fyke net data, and metric values fluctuated across the sampling period.

TABLE 21.3
Mean per Sampling Period and Coefficient of Variation (CV; %) Among Periods Within a Year (Early June through Early October) in Sparkling Lake for the 13 Metrics

	Seine				Fyke net	
	1982		1983		1983	
Metric	Mean	CV	Mean	CV	Mean	CV
Species richness metrics						
Native species	11.8	18	11.3	14	9.0	14
Centrarchid species	1.8	29	1.8	55	2.6	21
Cyprinid species	3.8	25	4.5	29	2.0	35
Intolerant species	5.3	18	5.3	18	3.4	16
Small benthic species	3.8	13	3.5	16	3.0	24
Proportional abundance metrics (%)						
Native benthic fish	6	70	3	94	2	134
Exotic fish	0	—	0	—	0	—
Intolerant fish	4	69	2	64	6	85
Tolerant fish	61	45	42	72	29	44
Omnivores	61	45	42	72	29	44
Top carnivores	1	80	0.4	59	5	100
Lithophilic spawners	3	72	4	116	1	190
Vegetation dwellers	7	82	16	70	2	140

Note: Data are for all fish captured, including young of the year.

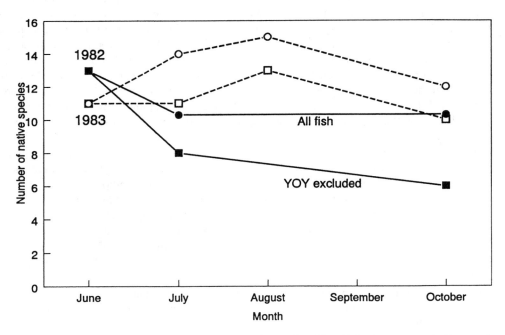

FIGURE 21.1 Changes in the number of native species in Sparkling Lake between early June and early October 1982 and 1983, as estimated by seining. Circles are data points where all fish are included, and squares are points where young-of-the-year fish have been excluded.

TABLE 21.4
Trophic State Index (TSI) and Lake Morphometric Variables for the Shoreline Dataset and Outstanding Resource Waters (ORW) Dataset

Lake	TSI	Area (ha)	Mean Depth (m)
Shoreline			
Big Silver	50.5	139	21
Camp	44.0	187	5
Chetek	70.6	312	13
Clark	34.6	351	7
Fox	78.5	1070	7
Kawaguesaga	47.3	271	18
Mead	78.4	130	5
Minocqua	44.7	550	23
Minong Flowage	57.6	633	9
Nagawicka	44.1	371	36
Nebagamon	44.7	370	20
Park	75.8	126	7
Ripley	50.1	169	18
Shawano	55.4	2454	9
Silver	47.3	188	NA
Wind	55.1	379	10
Winneconne	66	1824	NA
ORW			
Big Eau Pleine	64	2764	16
Brule River Flowage	47	120	20
Caldron Falls	52	412	15
Clam River Flowage	56	145	11
Gile Flowage	52	1369	NA
Minong Flowage	57	633	9
Rainbow Flowage	55	824	NA
Rice Lake	52	380	9
St. Croix Flowage	48	774	7
Willow Flowage	54	2552	10
Wissota	58	2550	NA

21.3.2 Relations with Lake Environmental Quality

For both fyke net and index samples from the shoreline dataset, number of intolerant species was negatively correlated with TSI, although the index method relation was stronger (Table 21.4). The proportion of individuals as intolerant species had a stronger correlation with TSI than the number of intolerant species for both methods; the index sample relation with TSI was stronger than the fyke sample's for this metric as well. Intolerant species, whether evaluated as a proportional metric or as a species richness metric, are associated with anthropogenic impacts in this dataset.

Index sampling more efficiently captured benthic fishes, several of which are intolerant. For index samples, number of small benthic species and proportion of individuals as benthic species were also correlated with TSI (Table 21.5); however, these relations were not observed for fyke nets.

Other species richness metrics showed comparatively weak and inconsistent relations with TSI. Number of native species and number of centrarchid species approached a significant negative

TABLE 21.5
Spearman Correlation Coefficients for Metrics and Trophic State Index for 17 (Index Sample) and 16 (Fyke) Lakes in the Shoreline Dataset

Metric	Index Sample (DC shock + seine)		Fyke Net	
	Rho	P	Rho	P
Species richness metrics				
Native species	0.01	0.96	−0.51	0.04
Centrarchid sp.	0.37	0.13	−0.47	0.06
Cyprinid sp.	−0.17	0.50	−0.30	0.25
Intolerant sp.	−0.67	0.003	−0.49	0.05
Sm. benthic sp.	−0.52	0.03	−0.38	0.14
Proportional metrics				
Native benthic	−0.39	0.13	−0.08	0.75
Exotic	—	—	—	—
Intolerant	−0.85	0.0001	−0.55	0.03
Tolerant	−0.43	0.08	−0.37	0.15
Omnivores	−0.49	0.05	−0.40	0.12
Top carnivores	0.20	0.44	−0.28	0.29
Lithophilic spawners	0.08	0.76	−0.37	0.15
Vegetation dwellers	−0.37	0.15	−0.38	0.14

correlation with TSI for fyke catches, but no correlation was observed with index samples. This may be related to the fyke nets more efficiently sampling centrarchids. No relation between cyprinids and TSI was detected with either method.

Other proportional metrics, including proportion of individuals as tolerant species, proportion of individuals as omnivores, and proportion of individuals as top carnivores, had relations to TSI that were inconsistent with predictions. It was predicted that proportion of individuals as top carnivores would decrease with increasing TSI, while both the tolerant and omnivore metrics would be positively correlated with TSI.

The ORW dataset yielded similar results, with method biases apparently affecting the strength of relations between metrics and TSI. No metric was significantly related to TSI for each of the three methods (Table 21.6). Metrics based on intolerant or habitat specialist groups (e.g., vegetation specialists, small benthic species) showed associations with TSI for at least one method type. Number of intolerant species was correlated with TSI for fyke samples, and proportion of individuals as intolerant species was correlated with TSI for seine and fyke samples. Proportion of individuals as vegetation dwellers was correlated with TSI for seine and fyke samples.

21.3.3 Effects of Lake Size and Ecoregion

Analyses of the FDS dataset indicated that values for species richness metrics were strongly influenced by ecoregion as well as by lake size (Figure 21.2). Total native species richness increased significantly with increasing lake size for both ecoregions, but the rate of increase was significantly greater for the southern ecoregion ($f = 16.68$, $p < 0.0001$). At a lake surface area of 80 ha, lakes in the south had about 20 species, whereas those in the north had 15. At a lake area of 1600 ha, lakes in the south had about 38 species, for a rate of increase of one species every 66 ha. In the north, a lake of 1600 ha usually had 28 species, for an increased rate of one species per 116 ha.

The number of centrarchid species also had a significant positive relation with lake size, and the rate of change differed between the northern and southern ecoregions ($f = 18.86$, $p < 0.0001$).

TABLE 21.6
Spearman Correlation Coefficients for Metrics and TSI for 11 Lakes in Outstanding Resource Waters Dataset

Metric	Boom		Seine		Fyke	
	Rho	P	Rho	P	Rho	P
Species richness metrics						
Native species	−0.02	0.93	0.04	0.91	0.01	0.98
Centrarchid species	−0.49	0.12	−0.49	0.12	−0.49	0.12
Cyprinid species	0.18	0.60	0.05	0.88	0.05	0.88
Intolerant species	−0.02	0.94	−0.21	0.52	−0.64	0.03
Sm. benthic species	−0.22	0.52	−0.24	0.48	−0.12	0.57
Proportional metrics						
Native benthic	0.79	0.004	0.49	0.12	0.51	0.10
Exotic	—	—	—	—	—	—
Intolerant	−0.09	0.79	−0.64	0.03	−0.64	0.03
Tolerant	−0.25	0.46	−0.12	0.72	−0.13	0.71
Omnivores	−0.17	0.61	0.02	0.96	−0.03	0.94
Top carnivores	0.38	0.24	0.52	0.10	0.52	0.10
Lithophilic spawners	0.12	0.72	0.30	0.36	0.18	0.60
Vegetation dwellers	−0.22	0.51	−0.74	0.009	−0.72	0.01

In both ecoregions, lakes of 80 ha usually had between five and six species. However, species number increased more rapidly with lake size in the south, so that at 1600 ha, lakes in the south had nine species, an increase of one species per 500 ha; whereas lakes in the north had between six and seven species, an increase of one species per 1500 ha.

For the number of cyprinid species, the relationship with lake size was more complex. Numbers of cyprinids were positively correlated with lake surface area in the northern ecoregion ($f = 16.64$, $p < 0.0001$), but not in the southern ecoregion ($f = 0.01$, $p = 0.9138$). For lakes of 80 ha, the northern ecoregion averaged about five species and the southern about six. At 1600 ha, lakes in the south ecoregion remained at about six cyprinids, whereas lakes in the north had increased to nine species, an addition of one species per 380 ha.

The number of intolerant species and the number of small benthic species were not significantly correlated with lake size for either ecoregion. In both ecoregions, all sizes of lakes averaged four to five intolerant species and four to six small benthic species.

21.4 DISCUSSION

21.4.1 METRICS CONSIDERED FOR FURTHER DEVELOPMENT

Based on statistical properties of metrics in the absence of degradation, 8 of the 13 metrics showed promise for use in an Index of Biotic Integrity: numbers of native, centrarchid, cyprinid, intolerant, and small benthic species, and percentages of exotics, top carnivores, and simple lithophilic lake spawners. These metrics had reasonably low CVs within and among years, and little among-lake variation in the sampling method that best estimated them. However, the associations with the one available measure of environmental quality (TSI) varied among the eight metrics. Results for number of native species were ambiguous, with fyke samples in the shoreline datasets showing some association, although the Lyons and LTER data suggested that seining would best estimate the number of native species. Shoreline index and all ORW samples showed no discernible trend in the relation between number of species and TSI. This metric will require additional evaluation,

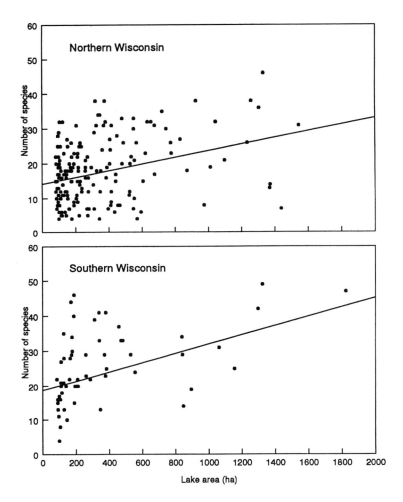

FIGURE 21.2 Plot of regression lines for native species richness vs. lake surface area for the northern and southern ecoregions.

although it will probably be less sensitive to changes in lake condition than metrics focusing on the more sensitive species. A fairly weak, yet consistent relation was observed for centrarchid species in the ORW dataset and in shoreline dataset fyke samples. The number of cyprinid species sampled with these methods did not appear to be related to TSI for any of the samples, nor was the proportion of lithophilic spawners; therefore, both of these should be dropped from further consideration. The number of intolerant species was negatively correlated with TSI for at least one method in both the shoreline and ORW datasets. The shoreline index samples, which are effective at sampling small benthic fishes, showed a negative correlation of small benthic species with TSI. The trend in the ORW dataset was for top carnivores to increase in relative abundance in more productive systems; this reflects high catches of largemouth bass in some highly productive systems. Although TSI reflects an incomplete view of potential impacts to lakes, the responses of the top carnivore, cyprinid, and lithophilic spawner metrics suggest that they will not provide an interpretable response and should not be considered for further development. The datasets contained few exotic species, so this metric was not adequately evaluated.

Among the other five metrics, percentages of native benthic, intolerant, tolerant, omnivore, and vegetation-dwelling fish, three had weaknesses that preclude further development. The percentage of native benthic fish was not associated with TSI in the shoreline dataset. Although a fairly strong statistical association was present for boomshocking in the ORW dataset, the relationship was the opposite of that predicted. The species driving the relation in the ORW boom sample were white suckers and logperch, both of which were captured in large numbers in some highly productive impoundments. In addition, the proportion of native benthic fish had a high CV within and among years in the Lyons dataset, indicating relatively large fluctuations in relative abundance even in the absence of major environmental changes. The percentages of tolerant fish and omnivores were very similar within lakes and redundant in the information they provided. They did not correlate well with environmental quality. These metrics had high values in many good-quality lakes, the opposite of what was expected. In these lakes, both metrics were almost completely determined by the abundance of a single species, bluntnose minnow. However, Whittier and Hughes (in press) assessed tolerance of fish species in northeastern U.S. lakes, and concluded that bluntnose minnows should be considered intolerant. Fine-tuning of metrics by adjusting tolerance classifications based on data from lakes may improve the performance of metrics based on tolerance classifications.

The percentage of vegetation-dwelling fish was correlated with TSI in ORW fyke nets, and a similar trend was observed in shoreline dataset fyke nets. The fish species that appeared to drive the relation was Iowa darter, which is also an intolerant species. However, this metric also had high within-year variation in the absence of changes in environmental quality, and should be further evaluated. The percentage of intolerant fish was also relatively low in some high-quality lakes in the LTER dataset, but this metric had a strong correlation with TSI in the shoreline dataset and a moderately strong correlation with TSI in the ORW dataset. The best method for estimating this metric varied among lakes in the Lyons dataset, suggesting that values for this metric might be particularly sensitive to differences in sampling technique. The strength of correlation with environmental conditions also varied, depending on methods for the shoreline dataset; however, this metric appears to have potential for further development.

Several metrics were not considered in the analyses because of a lack of data, but these might be worth considering for Wisconsin lakes. Minns et al. (1994) found good relationships between environmental quality and the relative biomass of exotics, piscivores, generalist feeders, and specialist feeders. They found biomass-based metrics to be generally less variable than abundance-based metrics. Most versions of the IBI have incorporated a metric dealing with the physical condition of individual fish, focusing on the presence of deformities, eroded fins, lesions, and tumors (Simon and Lyons, 1995; Simon and Sanders, Chapter 18). Such a metric is particularly useful in identifying extremely degraded ecosystems, where toxic substances in the water or sediment are a problem. Hughes et al. (1992) and Larsen and Christie (1993) proposed using the size or age structure of selected indicator species, particularly top carnivores. Similarly, Ryder and Edwards (1985) proposed using population parameters for lake trout and possibly walleye to assess ecosystem quality in the Great Lakes; however, the population status of such species, most of which are popular sportfish, is often confounded by fishery management activities, such as stocking and harvest regulation. Finally, assessment of the abundance of planktivorous fishes has been suggested as a potential metric (Jennings et al., 1995; Lyons et al., 1995), as these species may have a major influence on trophic function and, ultimately, water quality in lakes (Carpenter and Kitchell, 1993). However, in Wisconsin lakes, the most important planktivores occur in the pelagic zone all or much of the time, and a different set of sampling procedures is required to estimate their abundance (Rudstam et al., 1987).

21.4.2 SAMPLING PROCEDURES

We have taken a different approach to fish-assemblage sampling from what has been used for other versions of the IBI. In stream settings, a single method typically is used to estimate the structure and composition of the entire fish assemblage, and all metrics are calculated from the same dataset. Although a single method can effectively sample the entire assemblage in small streams (e.g.,

TABLE 21.7
Summary of Evaluations and Recommendations for 13 Metrics Evaluated for a Lake Index of Biotic Integrity

Metric	Evaluation	Recommendation	Method
Native species	Low CV, stable, but weak and inconsistent relation with env. qual.	Further evaluation	Fyke, seine
Centrarchid species	Low CV, stable. Consistent rel. to env. qual.	Use	Fyke
Cyprinid species	Low CV, stable, no relation to env. qual.	Drop	
Intolerant species	Low CV, stable, strong consistent relation to env. qual.	Use	Seine
Small benthic species	Low CV, stable, relation with env. qual. in shoreline dataset	Use	Seine
Proportion native benthic	High CV, not predictably related to env. qual.	Drop	
Proportion exotics	Insufficient data	Further evaluation	Boomshocker
Proportion intolerant	High CV, but strong relation with env. qual. in ORW and shoreline data	Use	Seine
Proportion tolerant	High CV, no relation with env. qual.	Drop	
Proportion omnivore	High CV, no relation with env. qual.	Drop	
Proportion top carnivore	Not predictably related to env. qual.	Drop	
Proportion lith. spawn.	No relation with env. qual.	Drop	
Proportion veg. dwellers	High CV, weak but consistent relation with env. qual.	Further evaluation	Seine, fyke

Simonson and Lyons, 1995), this is not the case in lakes (Lyons, 1987; Weaver et al., 1993; Magnuson et al., 1994). In lakes, each method catches a substantially different subset of the assemblage, and samples from different methods are not directly comparable. In some instances, lake IBIs have been developed based on a single method — the boomshocker (Minns et al., 1994; Jennings et al., 1995). Although this is the simplest approach, the sampling selectivity of the boomshocker limits the number and types of metrics that can be developed, and may reduce the overall sensitivity of the IBI to certain types of environmental degradation. Alternatively, in another lake version of the IBI, results from standardized levels of effort for multiple methods have been combined (e.g., Larsen and Christie, 1993). Although this approach provides more options for IBI development and may yield a more sensitive index, it requires that sampling efforts always be similar, which limits application of the IBI to datasets specifically collected for IBI analysis.

It is proposed here that multiple methods be used to generate the IBI, but that data from different methods not be combined (Table 21.7). Instead, the development of method-specific metric values is recommended, with different methods used to calculate different metrics. This will allow incorporation of the advantages of a multiple-method sampling regime without adding the complication of combining results from different methods. It is also recommended that each lake be sampled with fyke netting (which is effective for medium- to large-sized fish in water deeper than 1 m) and another method, such as seining, electroshocking by wading, or both (which is effective for small benthic and midwater fish in water less than 1 m deep). Sampling effort for each method should be set at a level where catches become asymptotic; that is, where adding an additional unit of effort does not substantially increase species number or change proportional abundances. Sample effort below this level introduces unacceptable variation that makes detection of impacts difficult or impossible (Jennings et al., 1995). Based on extrapolation of results from Lyons (1987), Magnuson et al. (1994), and our own preliminary analyses of the ORW dataset, 12 overnight fykes net sets, plus 20 seine hauls or wading electroshocking samples are recommended. This level of sampling effort will require a minimum of 2 to 3 days of work per lake for a two- or three-person crew.

The number of stations recommended should be considered a minimum because habitat varies within a lake, and this variation will affect both the species composition at the site level (Jennings et al., 1995) as well as sampling efficiency (Jennings et al., 1996). Randomly selecting a sufficient number of stations is probably the most practical approach to ensuring that the full range of habitat types is adequately represented. An alternative approach of defining habitat types for a stratified design requires a thorough habitat inventory, and may require new sample site selection to maintain appropriate levels of replication for habitat types when habitats are altered. From these results, it is proposed that the following methods and metrics be matched together for further IBI development. The seine should be used for the number of intolerant and small benthic species, and perhaps for the proportion of individuals as intolerant species. If sample sites contain cover types that make seining impractical, wading electrofishing should be substituted for, or used in addition to, seining. The fyke net should be used for the number of centrarchid species and proportion of vegetation specialists, although more data are needed to fully evaluate the behavior of these metrics over a range of conditions. Number of native species also needs additional evaluation with both seine and fyke nets. If the proportion of individuals as exotic species were calculated, it would probably be best estimated with a boomshocker, which is effective for sampling the most abundant and ubiquitous exotic encountered — carp.

If present, YOY fish should be included in species richness metric calculations. No consistent seasonal trends in metric values were observed in this study. Inclusion of YOY fish generally had little effect or actually improved the performance of species richness metrics for the seine, and had minimal effects on fyke net results from the LTER and Sparkling Lake (Lyons) datasets. Although it was not possible to calculate CVs for ORW and shoreline data, the patchy distribution of YOY fish that was observed might be problematic for calculating proportional metrics.

Expectations for metric values should incorporate either an ecoregional or limnologically based classification, and should also be scaled for lake size. An objective, limnologically based classification system provides a framework within which lakes can be stratified, thereby facilitating development of more finely tuned expectations for assemblage structure. One of the important issues in development of a reliable index is validation of predicted responses within each class of lakes.

The fact that many metrics did not show properties that would make them viable candidates for further development is not surprising, nor should this invalidate the concept of biological assessment for lakes. The relations between several metrics and trophic state were similar to those reported by Bachmann et al. (1996) for Florida lakes. Many species increase in abundance along a gradient from oligotrophic to hypereutrophic, and only a few sensitive species show a decline. Bachmann et al. (1996) also observed relatively constant values across a range of conditions for some trophic groups, such as top carnivores. In Wisconsin lakes, the largemouth bass is an abundant top carnivore and yet is fairly tolerant of a range of conditions. Species that are both abundant and tolerant of degradation are not expected to provide reliable response variables for an index designed to indicate degradation. Similarly, when such species are dominant in a group used in a metric, the metric will tend to be insensitive to changes in conditions. The most widespread sources of degradation in lakes are land use practices that increase nonpoint source inputs of nutrients and sediment, and alteration of littoral habitat. The metrics that show the most predictable responses to these impacts are likely to be most broadly applicable.

An IBI for lakes will be structured somewhat differently than the original IBI developed for streams. These differences are necessary because ecological processes are different between the two types of systems. Not only do important differences necessitate changing the structure of an IBI, but recognition of these differences also leads to slightly different applications for the index. Although all habitat types in streams can be effectively sampled with a single method, sampling the full range of habitats in lakes requires multiple methods. This factor, along with active management of several abundant species, make a lake index less sensitive in detecting perturbations.

Many of the anthropogenic impacts on lakes would not be expected to lead to immediate response in the fish community. A lake is capable of buffering some impacts, such as nonpoint source inputs, at least temporarily. Other impacts, such as modification of littoral zone habitat, tend to be very gradual and cumulative. Time lags in effects (temporal scale) and the difficultly in detecting the effects of small modification at a whole lake scale (spatial scale) should be considered when interpreting fish community data in relation to resource quality. Changes in many fish community metrics may not be immediate in a lake with real impacts. For rapid diagnosis of emerging water resource problems in lakes, assessment of habitat quality, land use practices, and some water quality parameters may be more appropriate than fish community-based indices. A lake IBI has practical application as part of an integrated approach to resource assessment, and as a tool in tracking response of the biological community for restoration efforts. Metrics based on intolerant species or habitat specialists show the greatest potential for use in such an index.

21.5 CONCLUSIONS

No reliable biomonitoring index has been developed for inland lakes. Initial efforts to develop an Index of Biotic Integrity (IBI) for Wisconsin lakes and reservoirs are described. We used several datasets to evaluate potential IBI metrics for spatial and temporal variation, appropriate sampling, and response to anthropogenic activity. In addition, the relation between species richness and lake area and ecoregion was evaluated with a statewide fish distribution database. Several metrics had relatively low variation and provided consistent scores in the absence of impacts to the lake. In addition, some of these metrics were related to independent measures of resource condition in a predictable manner. The metrics exhibiting greatest potential for use in a lake IBI include those based on intolerant or habitat specialist species. Choice of sampling methods will influence reliability of metrics. Seining and backpack or towed DC-electrofishing were most effective for sampling intolerant and benthic species; whereas fyke nets were more effective for sampling centrarchids. Several metrics had unacceptably high variation in the absence of changes in resource condition; these include proportion of tolerant, omnivore, and native benthic fish. The numbers of cyprinid and top carnivore species, and the proportion of lithophilic spawners were not predictably related to resource condition in our datasets. Species richness was related to lake area, although the rate of species addition is different between ecoregions. Therefore, metric scoring criteria must be adjusted for both lake size and location in the state.

ACKNOWLEDGMENTS

The authors thank John Magnuson for making the LTER data available and Barb Benson for providing them in the proper form for analyses. The manuscript was improved by comments from two anonymous reviewers. Funding for this study was provided by Federal Aid in Sportfish Restoration Grant F-95-P and the Wisconsin Department of Natural Resources.

REFERENCES

Angermeier, P. L., and J. R. Karr. 1986. Applying an index of biotic integrity based on fish communities: considerations in sampling and interpretation, *North American Journal of Fisheries Management,* 6, 418–429.

Bachmann, R.W., B.L. Jones, D.D. Fox, M. Hoyer, L.A. Bull, and D.E. Canfieds, Jr. 1996. Relations between trophic state indicators and fish in Florida (U.S.A.) lakes, *Canadian Journal of Fisheries and Aquatic Science,* 53, 842–855.

Carlson, R.W. 1977. A trophic state index for lakes, *Limnology and Oceanography,* 22, 361–369.
Carpenter, S. R. and J. F. Kitchell (Eds.). 1993. *The Trophic Cascade in Lakes.* Cambridge University Press, New York.
Eadie, J. McA., T. A. Hurley, R. D. Montgomerie, and K. L. Teather. 1986. Lakes and rivers as islands: species-area relationships in the fish faunas of Ontario, *Environmental Biology of Fishes,* 15, 81–89.
Fago, D. 1988. Retrieval and Analysis System Used in Wisconsin's Statewide Fish Distribution Survey, 2nd ed. Wisconsin Department of Natural Resources, Madison, Research Report 148.
Fausch, K. D., J. Lyons, J. R. Karr, and P. L. Angermeier. 1990. Fish communities as indicators of environmental degradation, *American Fisheries Society Symposium,* 8, 123–144.
Hughes, R. M., D. P. Larsen, and J. M. Omernik. 1986. Regional reference sites: a method for assessing stream potential, *Environmental Management,* 10, 629–635.
Hughes, R. M., T. R. Whittier, S. A. Thiele, J. E. Pollard, D. V. Peck, S. G. Paulsen, D. McMullen, J. Lazorchak, D. P. Larsen, W. L. Kinney, P. R. Kaufman, S. Hedtke, S. S. Dixit, G. B. Collins, and J. R. Baker. 1992. Lake and stream indicators for the United States Environmental Protection Agency's environmental monitoring and assessment program, in D. H. McKenzie, D. E. Hyatt, and V. J. McDonald (Eds.), *Ecological Indicators.* Vol. I. Elsevier Applied Science, New York. 305–335.
Jennings, M. J., L. S. Fore, and J. R. Karr. 1995. Biological monitoring of fish assemblages in Tennessee Valley reservoirs, *Regulated Rivers: Research and Management,* 11, 263–274.
Jennings, M.J., M. Bozek, G. Hatzenbeler, and D. Fago. 1996. Fish and habitat, in *Wisconsin Department of Natural Resources, Shoreline Protection Study: A Report to the Wisconsin State Legislature.* WDNR PUBL-RS-921-96. 1–44.
Karr, J. R. 1981. Assessment of biotic integrity using fish communities, *Fisheries,* 6, 21–27.
Larsen, D. P. and S. J. Christie (Eds.). 1993. EMAP — Surface Waters 1991 Pilot Report. EPA/620/R-93/003. U.S. Environmental Protection Agency, Washington, DC.
Lyons, J. 1987. Distribution, abundance, and mortality of small littoral-zone fishes in Sparkling Lake, Wisconsin, *Environmental Biology of Fishes,* 18, 93–107.
Lyons, J. 1992. Using the Index of Biotic Integrity (IBI) to Measure Environmental Quality in Warmwater Streams of Wisconsin. U.S. Department of Agriculture, Forest Service, North Central Forest Experiment Station, St. Paul, MN. General Technical Report NC-149.
Lyons, J. and P. Kanehl. 1993. A Comparison of Four Electroshocking Procedures for Assessing the Abundance of Smallmouth Bass in Wisconsin Streams. U.S. Department of Agriculture, Forest Service, North Central Forest Experiment Station, St. Paul, MN. General Technical Report NC-159.
Lyons, J., S. Navarro-Pérez, P. A. Cochran, E. Santana, and M. Guzmán-Arroyo. 1995. Index of biotic integrity based on fish assemblages for the conservation of streams and rivers in west-central Mexico, *Conservation Biology,* 9, 569–584.
Magnuson, J. J., B. J. Benson, and A. S. McLain. 1994. Insights on species richness and turnover from long-term ecological research: fishes in north temperate lakes, *American Zoologist,* 34, 437–451.
Magnuson, J. J., C. J. Bowser, and T. F. Kratz. 1984. Long-term ecological research (LTER) on north temperate lakes of the United States. *Internationale Vereinigung für Theoretische und Angewandte Limnologie Verhandlungen,* 22, 533–535.
Maxwell, J. R., C. J. Edwards, M. E. Jensen, S. J. Paustian, H. Parrott, and D. M. Hill. 1995. A Hierarchical Framework of Aquatic Ecological Units in North America (Nearctic Zone). U.S. Department of Agriculture, Forest Service, North Central Forest Experiment Station, St. Paul, MN. General Technical Report NC-176.
Minns, C. K., V. W. Cairns, R. G. Randall, and J. E. Moore. 1994. An index of biotic integrity (IBI) for fish assemblages in the littoral zone of Great Lakes' areas of concern, *Canadian Journal of Fisheries and Aquatic Sciences,* 51, 1804–1822.
Novotny, D. W. and G. R. Priegel. 1974. Electrofishing Boats: Improved Designs and Operational Guidelines to Increase the Effectiveness of Boom Shockers. Wisconsin Department of Natural Resources, Madison. Technical Bulletin 73.
Rahel, F. J. 1984. Factors structuring fish assemblages along a bog lake successional gradient, *Ecology,* 65, 1276–1289.
Rahel, F. J. 1986. Biogeographic influences on fish species composition of northern Wisconsin lakes with applications for lake acidification studies, *Canadian Journal of Fisheries and Aquatic Sciences,* 43, 124–134.

Rudstam, L., G. C. S. Clay, and J. J. Magnuson. 1987. Density and size estimates of cisco (*Coregonus artedii*) using analysis of echo peak PDF from a single-transducer sonar, *Canadian Journal of Fisheries and Aquatic Sciences,* 44, 811–821.

Ryder, R. A. and C. J. Edwards (Eds.). 1985. A Conceptual Approach for the Application of Biological Indicators of Ecosystem Quality in the Great Lakes Basin. Report to the Great Lakes Science Advisory Board from the Great Lakes Fishery Commission and the International Joint Commission, Windsor, Ontario, Canada.

SAS. 1990. *SAS/STAT User's Guide.* Version 6, 4th ed. SAS Institute, Inc., Cary, NC.

Simon, T. P. and E. B. Emery. 1995. Modification and assessment of an index of biotic integrity to quantify water resource quality in great rivers, *Regulated Rivers: Research and Management,* 11, 283–298.

Simon, T. P. and J. Lyons. 1995. Application of the index of biotic integrity to evaluate water resource integrity in freshwater ecosystems, in W. S. Davis and T. P. Simon (Eds.), *Biological Assessment and Criteria. Tools for Water Resource Planning and Decision Making.* Lewis, Boca Raton, FL. 245–262.

Simonson, T. D. and J. Lyons. 1995. Comparison of catch per effort and removal procedures for sampling stream fish assemblages, *North American Journal of Fisheries Management,* 15, 419–427.

Tonn, W. M. and J. J. Magnuson. 1982. Patterns in the species composition and richness of fish assemblages in northern Wisconsin lakes, *Ecology,* 63, 1149–1166.

U.S. EPA (United States Environmental Protection Agency). 1990. The Quality of Our Nation's Water: A Summary of the 1988 National Water Quality Inventory. EPA/440/4-90-005. U.S. Environmental Protection Agency, Washington, D.C.

Weaver, M. J., J. J. Magnuson, and M. K. Clayton. 1993. Analyses for differentiating littoral fish assemblages with catch data from multiple sampling gears, *Transactions of the American Fisheries Society,* 122, 1111–1119.

Whittier, T. R. and R. M. Hughes. (In press.) Evaluation of fish species tolerances to environmental stressors in Northeast USA lakes, *North American Journal of Fisheries Management.*

22 Development of IBI Metrics for Lakes in Southern New England

Thomas R. Whittier

CONTENTS

22.1 Introduction 563
22.2 Impediments to Developing Lake IBIs 564
    22.2.1 Scope of Assessment 564
    22.2.2 Sampling Issues 564
    22.2.3 Public Perceptions and Management of Lakes 565
    22.2.4 Ecological Factors 565
22.3 Methods 566
    22.3.1 Sample Design and Field Methods 566
    22.3.2 Quantitative Methods 566
22.4 Results 568
    22.4.1 Species Richness Metrics 569
    22.4.2 Trophic Composition Metrics 575
22.5 Discussion 577
Acknowledgments 580
References 580

22.1 INTRODUCTION

Among the principles implicit in the field of ecology are assumptions that: (1) living organisms react to changes in their ecosystems, (2) one should be able to detect and quantify changes in biological assemblages due to adverse stress caused by human activity, and (3) these changes should be predictable and generalizable to broad classes of ecosystems and assemblages. In the early 1980s, Karr and colleagues successfully applied these principles to fish assemblages in warmwater streams in the Midwest (Karr, 1981; 1991; Karr et al., 1986; Lyons, 1992; Ohio EPA, 1988). The resulting Index of Biotic Integrity (IBI) has been repeatedly modified for other assemblages, regions, and ecosystems (e.g., Leonard and Orth, 1986; Miller et al., 1988; Steedman, 1988), with varying degrees of success.

To date, there has been only limited effort in developing IBIs for lakes. Minns et al. (1994) developed an eight-metric IBI for littoral areas in the Great Lakes. Dionne and Karr (1992) and Jennings et al. (1995) presented preliminary results for a Reservoir Fish Assemblage Index for the Tennessee Valley, but concluded that additional work was needed. The only other published research for inland lakes is in Wisconsin (Jennings et al., Chapter 21). They evaluated metric variability related to sampling methods, and how the metrics performed as indicators of human-induced stress. They were satisfied with the performance of only four metrics.

In this chapter, some of the issues that may account for the apparent difficulty in developing inland lake IBIs are reviewed. Several candidate metrics for lakes in southern New England are presented, and some of the implications of the results and needs for further research are discussed.

22.2 IMPEDIMENTS TO DEVELOPING LAKE IBIS

The reasons that development of lake IBIs lags behind that for warmwater streams can be divided into four groups: (1) scope of assessment, (2) sampling issues, (3) perceived values of lakes and the management practices that support those values, and (4) ecological factors. The issues discussed below are generalizations meant to emphasize differences; exceptions to each point can be found. Clearly, these concerns are not the exclusive domain of lake assessments.

22.2.1 SCOPE OF ASSESSMENT

For streams, the conceptual and actual units for biological assessments are not entire streams, but rather separated points or reaches along the stream length. In reporting results, the points or reaches are plotted or tabled as individual samples, and the assessment of biotic condition usually integrates the individual samples and changes along the stream length. For lakes, the conceptual units are usually entire lakes, although sampling is usually at stations or portions of shoreline; and the data are generally combined to represent the entire lake. Only when lakes are sufficiently large are sampling, analysis and assessment done on subunits (e.g., coves, bays, arms).

Another facet of this issue is one of "apples and oranges." Biological assessments need to be developed for classes of reasonably comparable ecosystems (Plafkin et al., 1989). An IBI is based on expected characteristics of a particular assemblage type, in a particular size and type of water body, in a particular ecoregion or basin. For streams, assemblage expectations are usually defined for ecoregion (or basin), temperature type (cold or warm), and stream size. The classes of ecoregions and temperature types usually can encompass a very large number of streams in the same assessment framework.

For lakes, classing water bodies into reasonably comparable groups is more complex. Stream size describes habitat volume. Habitat volume in lakes is more multidimensional. Lake volume defines only the maximum possible habitat volume. The usable volume may be limited by oxygen depletion (either natural or amplified by anthropogenic eutrophication) and temperature extremes in summer or oxygen depletion in winter, and thus may vary greatly within and among years. Natural differences in stratification regimes affect expectations for the fish assemblages, as does basin shape — which determines the proportion of the habitat that is littoral vs. pelagic. Watershed position and relative connectedness are also important; a large headwater lake or a large isolated lake could be expected to have a fish assemblage that differs from a large well-connected lake closer to a mainstem river.

22.2.2 SAMPLING ISSUES

Biological assessments require data, collected in a consistent manner, that represent or index the resident assemblages (Plafkin et al., 1989). For wadeable streams, electrofishing provides appropriate data for IBI assessments (e.g., Ohio EPA, 1988). For lakes, no single gear is sufficient to sample in all habitat types, nor for all species. Numerous studies have examined the issue of gear selectivity, sufficiency of effort and sampling variability in northern lakes (e.g., Weaver et al., 1993). Whenever more than one gear or method is used, questions arise as to how and whether to combine data, and what constitutes a unit of sampling effort. Jennings et al. (Chapter 21) propose evaluating and using data from different methods for different metrics, as well as limiting the data collection to littoral assemblages.

Problems with using multiple gear are compounded further when certain effective methods cannot be applied in all lakes. Night beach seining is often the only effective method for collecting cyprinids, darters and sculpins, but many lakes lack beaches clear of obstacles and suitable for seining. Electrofishing is a good single method for sampling the littoral zone, but loses its effectiveness in low-conductivity lakes common in northern regions, and in lakes with very narrow littoral zones. Likewise, gillnets are the only reliable method for collecting salmonids and other pelagic species; but resource managers of low-productivity northern lakes often restrict or forbid the use of gillnets.

22.2.3 Public Perceptions and Management of Lakes

The public, and therefore management agencies, value different things about lakes than they value for streams. The public often values a park-like aesthetic for lakes, as a place for homes, vacation cabins, and parks for camping, picnics, and swimming. People like the water; they like to see it, be near it, on it, or in it, and often they want to use the water for drinking. Except for drinking water, lakes are more amenable to these things than are streams. Activities in, on, or near the water lead people to want water clear of obstacles (vegetation, snags, etc.) for boating, fishing, and swimming. And, of course, people want good fishing, regardless of whether the desired species is native or even suited for "their" lake.

Lakes are often managed to enhance the values listed above, usually to the detriment of biotic integrity. Snags and weeds are cleared for boating, swimming, docks, and front-yard aesthetics. Retaining walls or riprap are added for erosion control (Jennings et al., 1996; Christensen et al., 1996). Lakes have been subjected to stocking and other manipulations of fish assemblages for well over a century. In southern New England, much of the public is not aware that most of the currently widespread game fish are not native (e.g., largemouth and smallmouth basses, bluegill, northern pike, black crappie, and brown, rainbow, and lake trouts).

Management goals are closely linked with who does the managing. Streams tend to be managed by, or under the authority of state departments of environmental quality, who are concerned with issues defined by the Clean Water Act and subject to oversight by the U.S. EPA. This is in part due to historic problems from point-source pollution on streams and rivers. As the obvious water quality problems in streams are corrected, many of these agencies are shifting their regulatory emphasis to the biological condition of streams, for which the IBI is well suited.

Lakes, on the other hand, are managed by various legal entities. In some lake-rich regions where lakes receive more recreational use than streams, lakes in the public domain are often managed by state fish and game departments. Their mandates emphasize management for fishability, not biotic integrity. Unfortunately, in some instances, interagency politics has led to lakes and streams becoming the exclusive territory of different agencies wherein environmental quality agencies often only become involved with lakes over eutrophication and fish tissue contaminants issues. Finally, unlike streams and rivers, lakes often have what amounts to owners: private individuals and families, lake and homeowners associations, sports clubs, organization camps, resorts, and water districts. Technically, many of these lakes are public; but in reality, the owners of the land surrounding lakes limit access to and activities on "their" lakes, and the public agencies often choose not to press their legal mandate. It is reasonable to assume that most lake owners do not manage for biotic integrity.

22.2.4 Ecological Factors

In general, lakes tend to be new geographic features compared with rivers and streams, and fishes have had longer to speciate in, and adapt to flowing water. Thus, there tend to be fewer lake-dwelling species. For example, in southern New England, the regional species pool of lake dwellers is considerably smaller than the already depauperate species pool for lotic systems (Miller et al., 1988; Halliwell et al., Chapter 12). There are only three native sunfish: pumpkinseed is nearly ubiquitous, and the other two are rather uncommon. Depending on the location, there are one or two each of suckers, darters, and native top carnivores, and three to five possible lake-dwelling minnows (Whittier et al., 1997a). In addition, for more than a century, the fish fauna in New England has been liberally augmented with nonnatives. In the Northeast, an estimated 74% of all lakes have at least one non-native species, with non-native individuals outnumbering the natives in 31.5% of lakes (T. Whittier, unpublished data). Introductions are especially common in southern New England where nearly all lakes have nonnatives.

Fish species tolerances to stress (whether natural or anthropogenic) in lake ecosystems often differ from their tolerances in stream ecosystems. Whittier and Hughes (1998) evaluated 45 species tolerances to five anthropogenic stressors in Northeast lakes. They found that eight species usually classified as tolerant or moderately tolerant of disturbance in streams appear to be intolerant or moderately intolerant of degraded conditions in lakes. Five species usually classified as intermediately tolerant in streams were very tolerant in northeastern lakes.

22.3 METHODS

22.3.1 SAMPLE DESIGN AND FIELD METHODS

The data used here were from the U.S. Environmental Protection Agency's Environmental Monitoring and Assessment Program (EMAP) Northeast Lakes Pilot (Larsen et al., 1991; 1994; Stevens, 1994). EMAP sampled fish assemblages, as well as water chemistry, zooplankton, physical habitat, and riparian birds in 179 lakes and reservoirs during the summers of 1992 to 1994 in the Northeast (New England, New York, and New Jersey). The lakes were selected, using a systematic random design, from all lakes larger than 1 ha. For this study, the 50 lakes in southern New England (MA, CT, RI, and the southern third of NH) were used, along with five additional lakes selected and sampled in 1991 for methods evaluation (Larsen and Christie, 1993).

Fish assemblages were sampled with overnight sets of gillnets, trapnets, and minnow traps, and by night seining (Whittier et al., 1997b). The level of effort was determined by lake size. The sampling objective was to collect a representative sample of the fish assemblage at each lake. Sites were selected using a stratified random design. Littoral fish sampling was performed at random stations within each macrohabitat class. Pelagic sample sites were chosen in random directions from the deepest location. The collected fish were identified to species and counted. As part of the EMAP Quality Assurance procedures (Chaloud and Peck, 1994), specimens of all species were vouchered with the Museum of Comparative Zoology at Harvard University to confirm identifications — and for permanent archival. To develop the metrics evaluated in this study, data from all gear were combined. Water samples were collected at 1.5 m at the deepest part of the lake. Field methods are detailed by Baker et al. (1997). Field and laboratory data from the Northeast Lakes Pilot can be found on the EMAP's website (http://www.epa.gov/emap/html/dataI/surfwater/data/nelakes).

22.3.2 QUANTITATIVE METHODS

Two measures of anthropogenic stress to evaluate metric performance were chosen: total phosphorus (TP) as a measure of eutrophication stress, and the extent of human activity in the watershed as a measure of generalized human-induced stress. The latter is an indirect measure of stress, but is based on the premise that increased human activity in the watershed increases the frequency and strength of a multitude of anthropogenic perturbations to the lake ecosystem. Although acidification is also a stressor in some areas of the Northeast, only four of the sampled lakes in

TABLE 22.1
Principal Components Analysis of Watershed-Level Human Influence Measures: Three Land Use/Land Cover Variables (% forest, % urban, % agricultural), Human Population Density (#/sq. km), and Road Density (km/ha)

Principal Component	Eigenvalue	Proportion of Variance	Cumulative
PC-1	3.09	0.62	0.62
PC-2	0.88	0.17	0.79

	Eigenvectors	
	PCA-1	PCA-2
% Forests	−0.42	0.34
% Urban	0.49	0.31
% Agricultural	0.32	−0.81
Population density	0.52	0.21
Road density	0.46	0.30

Note: All variables except % forest were $\log_{10}(x + 1)$ transformed. The first Principal Component (PCA-1) scores estimate generalized human influence.

southern New England had pH < 6.0. Preliminary examination of the fish assemblage data did not reveal any distinct qualitative acidification effects; thus, pH stress was used only to aid interpretation of metric behavior.

Watershed-scale measures of human disturbance were developed from digitized coverages of human population and road density (Census Bureau, 1991; U.S. EPA, 1992), and land use/land cover (USGS, 1990), from which proportions of the watershed in urban, industrial/commercial, residential, forested, wetlands, and agricultural categories were calculated (C. Burch Johnson, unpublished data). The first axis scores of a principal components analysis (PCA) (Gauch 1982, PROC PRINCOMP; SAS 1985) of the land cover (% forest, % urban, % agricultural), human population density, and road density were used to estimate human disturbance in the watershed, as described in Whittier et al. (1997a). The first principal component accounted for 62% of the variability in these five variables, with % forests loading negatively, and all human activity variables loading positively (Table 22.1).

To determine species native ranges, species maps and descriptions in a variety of fisheries texts (Kendall, 1914; Hubbs and Cooper, 1936; Hubbs and Lagler, 1964; Scott and Crossman, 1973; Lee et al., 1980; Trautman, 1981; Becker, 1983; Smith, 1985; Schmidt, 1986; Underhill, 1986; Page and Burr, 1991), and state biological survey reports from New York (NYSDC 1927-39) and New Hampshire (NHFGC 1937-39) were examined. Trophic guild and habitat preferences were based on these sources, along with the summary tables in Halliwell et al. (Chapter 12), Karr et al. (1986), Ohio EPA (1989), Lyons (1992), and Minns et al. (1994). Most tolerance classifications were from Whittier and Hughes (1998); for the remaining uncommon species, information contained in all of the above sources was employed to make tolerance classifications.

Due to the relatively low native species richness in New England, it was anticipated that several commonly used metrics would not be effective. Thus, a large number of potential metrics were examined (e.g., Miller et al., 1988; Simon and Lyons, 1995). To evaluate candidate metrics, consideration was given to the following: How many species contributed to the metric? What was the statistical distribution of raw scores among lakes? Was there a lake size effect? How do raw metric scores relate to the two measures of stress? In particular, do the raw metric values distinguish between the most degraded and the least degraded lakes? What ecological characteristics do the metrics represent, and how does one expect these to change over a range of natural conditions and human-induced stress? Primarily, a graphical-based approach was used (Fore et al., 1996), examining scatterplots of raw metric data and residuals from regressions.

22.4 RESULTS

In the summers of 1991 to 1994, EMAP sampled 55 lakes and reservoirs in southern New England (SNE; Figure 22.1). All, except three, lakes were less than 300 ha (Figure 22.2). Eighteen of the lakes are in the Northeastern Highlands ecoregion (NHE), with the remainder in the Northeastern Coastal Zone (NCZ) ecoregion (Omernik, 1987). Ten lakes were eutrophic or hypereutrophic, with the remainder split between mesotrophic and oligotrophic (Figures 22.2A, B). Phosphorus values tended to be higher in the NCZ and in lakes less than 100 ha. There was no association between lake size and watershed disturbance (Figures 22.2C, D). Watersheds were less disturbed in the NHE. Water level control structures are common on northeastern lakes (personal observation); 10 of the SNE sample lakes have "Reservoir" in their names. Named SNE reservoirs tend to be the larger lakes and have relatively low stressor values.

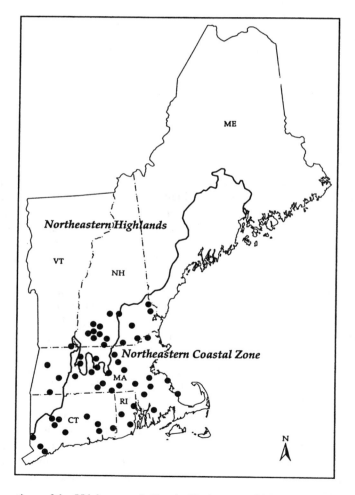

FIGURE 22.1 Locations of the 55 lakes sampled by the Environmental Monitoring and Assessment Program (EMAP) in southern New England during the summers of 1991 to 1994.

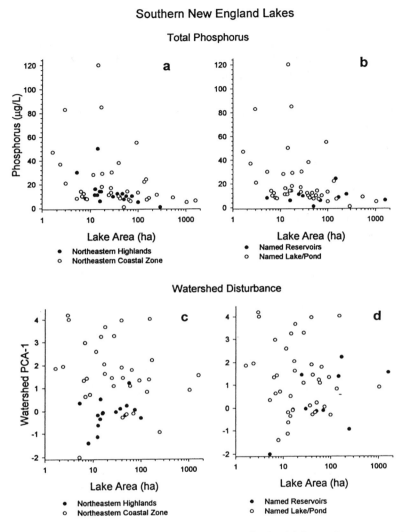

FIGURE 22.2 Relationship between sampled southern New England lake surface areas and two stressor measures: total phosphorus (μg/L), and watershed disturbance (1st-axis PCA scores for % forest, % urban, % agricultural land uses, road density, and human population density), plotted by ecoregions (A and C: solid circles = Northeast Highlands ecoregion, open circles = Northeast Coastal Zone ecoregion), and lake type (B and D: solid circles = named Reservoirs, open circles = named Ponds or Lakes).

Forty fish species were collected, 18 of which are native to the region, 20 are introduced, and two are native to some of the lakes (Table 22.2). Of the 40 species collected, 18 were found in three or fewer lakes. Assemblages were characterized by centrarchids, yellow perch, chain pickerel, golden shiner, and bullheads. Bluegill tended to dominate in the south, being replaced by yellow perch to the north.

22.4.1 SPECIES RICHNESS METRICS

Nearly all of the original (Karr, 1981) and variants (Miller et al., 1988; Simon and Lyon, 1995) of the species richness metrics were problematic in the SNE. The total number of species EMAP collected was similar to the total (44) used by Minns et al. (1994) in the Great Lakes. But slightly

TABLE 22.2
Fish Species Native Status, Tolerance Ranks, and Trophic Guild as Used in This Study, Collected at 55 Southern New England Lakes by EMAP During Summers 1991 to 1994

Species (Scientific Name)	# of Lakes as Natives	# of Lakes as Introduced or Stocked	Tolerance Rating	Trophic Guild
American eel				
Anguilla rostrata	8		9/M	TC
Alewife				
Alosa pseudoharengus	1	1	10/M	
Gizzard shad				
Dorosoma cepedianum		1	—/M	
Common carp				
Cyprinus carpio		3	5/T	GF
Common shiner				
Luxilus cornutus	1		12/M	IN
Golden shiner				
Notemigonus crysoleucas	37	2	5/T	GF
Bridle shiner				
Notropis bifrenatus	2		13/MI	IN
Spottail shiner				
N. hudsonius	1		14/MI	IN
Fathead minnow				
Pimephales promelas		1	15/I	GF
Creek chub				
Semotilus atromaculatus	2		12/M	IN
Fallfish				
S. corporalis	5		11/M	IN
White sucker				
Catostomus commersoni	23		8/MT	GF
Creek chubsucker				
Erimyzon oblongus	1		9/M	IN
White catfish				
Ameiurus catus		2	—/M	
Yellow bullhead				
A. natalis		20	6/MT	GF
Brown bullhead				
A. nebulosus	42		5/T	GF
Channel catfish				
Ictalurus punctatus		2	—/M	
Tadpole madtom				
Noturus gyrinus		1	—/MI	IN
Northern pike				
Esox lucius		3	10/M	TC
Chain pickerel				
E. niger	34		7/M	TC
Rainbow smelt				
Osmerus mordax		2	11/M	GF
Rainbow trout				
Oncorhynchus mykiss		7	—/MI	
Atlantic salmon				
Salmo salar		1	15/I	TC*

TABLE 22.2 *(continued)*
Fish Species Native Status, Tolerance Ranks, and Trophic Guild as Used in This Study, Collected at 55 Southern New England Lakes by EMAP During Summers 1991 to 1994

Species (Scientific Name)	# of Lakes as Natives	# of Lakes as Introduced or Stocked	Tolerance Rating	Trophic Guild
Brown trout				
S. trutta		3	—/M	
Brook trout				
Salvelinus fontinalis		6	13/MI	
Lake trout				
S. namaycush		2	14/MI	TC*
Banded killifish				
Fundulus diaphanus	11		10/M	IN
Mummichog				
F. heteroclitus		1	—/T	GF
White perch				
Morone americana	4	8	5/T	
Rock bass				
Ambloplites rupestris		2	12/M	TC
Banded sunfish				
Enneacanthus obesus	6		9/M	IN
Redbreast sunfish				
Lepomis auritus	5		10/M	IN
Pumpkinseed				
L. gibbosus	50		4/T	GF
Bluegill				
L. macrochirus		35	4/T	GF
Smallmouth bass				
Micropterus dolomieu		15	10/M	TC
Largemouth bass				
M. salmoides		50	5/T	
Black crappie				
Pomoxis nigromaculatus		21	4/T	
Swamp darter				
Etheostoma fusiforme	3		—/M	IN
Tesselated darter				
E. olmstedi	2		11/M	IN
Yellow perch				
Perca flavescens	44		6/MT	IN

Note: Tolerance rating scores from Whittier and Hughes (1998) range 4–16, I = Intolerant, MI = Moderately Intolerant, M = Moderate, MT = Moderately Tolerant, T = Tolerant. For Trophic Guild: TC = Top Carnivore, GF = Generalist Feeder, IN = Insectivore.

* All collected fish were stocked.

more than half of the SNE species are not native to New England, compared to 25% nonnative species collected by Minns et al. Many of the New England nonnatives were introduced 100+ years ago and are firmly established (NHFGC, 1937-39). Some authors suggest that these species are now "naturalized" and should be considered as part of the resident species pool (Halliwell et al., Chapter

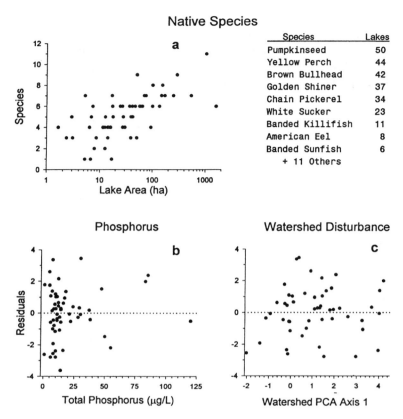

FIGURE 22.3 Native species richness. (A) Number of native species by lake size. (B) Residuals of the native species by lake size regression plotted against total phosphorus. (C) Residuals plotted against watershed disturbance PCA axis 1 (higher axis 1 value = increased human activity in the watershed).

12). Under that proposal, only the two non-North American species (common carp and brown trout) and populations maintained by stocking (nearly always salmonids in SNE) could be considered nonnative. There are currently no state-run warmwater species stocking programs in the SNE (D. Halliwell, personal communication). A strict interpretation of native status was employed for this assessment.

The number of native species in SNE lakes increased with lake size ($r^2 = 0.41$, $p = 0.0001$; Figure 22.3A), ranging from 2 to 9 species (11 in one lake). Two features of these data differed from those in midwestern streams. First, there were apparently no fishless lakes in the SNE, while some fairly large streams (watersheds up to 2000 sq. km) in Ohio were fishless (Whittier and Rankin, 1992). Second, the data formed a band rather than a wedge shape (i.e., variance about the regression line was fairly constant rather than increasing with lake size). To be a useful metric, native species richness residual scores, from the regression with lake size, should generally decrease with increasing stressor scores (i.e., fewer than "expected" natives in more stressed lakes, and more native species in less stressed lakes). However, there was no apparent pattern in these data (Figures 22.3B, C). Jennings et al. (Chapter 21) also found no relationship between native species richness and human impact, as measured by a Trophic State Index. Total species richness showed similar patterns and lack of relationship to stressor measures. Treating "naturalized" (non-stocked) North American species as natives produced a pattern similar to that for total species richness.

The number of introduced species was a useful metric in Great Lakes littoral areas (Minns et al., 1994). In the SNE, all (except three) lakes had between one and six nonnative species (Figure 22.4A), generally increasing with lake size ($r^2 = 0.32$; $p = 0.0001$). Residual scores plotted against

Development of IBI Metrics for Lakes in Southern New England

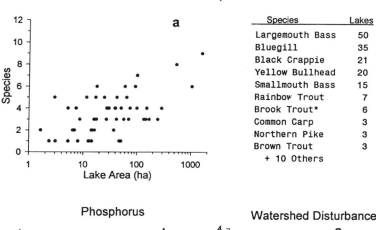

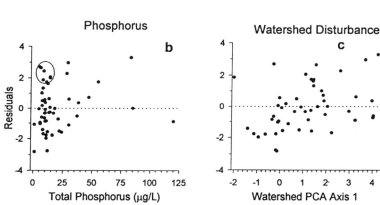

FIGURE 22.4 Nonnative species richness. (A) Number of nonnative species by lake size. (B) Residuals of the nonnative species by lake size regression plotted against total phosphorus. (C) Residuals plotted against watershed disturbance PCA axis 1 (higher axis 1 value = increased human activity in the watershed).

TP showed a lack of pattern with increased stress, except that several of the relatively low TP lakes with high residuals (more nonnative species than expected) were stocked with one to three salmonids (Figure 22.4B). Nonnative species residuals showed the expected pattern with watershed disturbance (Figure 22.4C); a number of lakes with high residual scores and intermediate watershed disturbance were stocked. Thus, the number of nonnative species adjusted for lake size should be a useful metric.

Most species richness and composition metrics used in other IBIs did not appear to be applicable in the SNE. There were only two rarely collected darters (and no sculpins) (Table 22.2). Likewise, there were only two suckers, with the creek chubsucker collected only once. Cyprinid species richness was somewhat higher; however, four of five native minnows were uncommonly collected. Golden shiner was widespread (72% of sampled SNE lakes) and tolerant of degraded conditions. Only 8 of the 55 sampled lakes had native cyprinids other than golden shiner; 6 of these had only one other minnow. Native minnow species richness may be a useful metric, but it would be essentially a binary metric. Of the three native sunfish in SNE (Table 22.2), pumpkinseed was ubiquitous (91% of sampled lakes). The other two sunfish were uncommon and were not collected in the same lakes, preferring very different habitats. Therefore, a native sunfish metric contained very little information.

Atlantic salmon and fathead minnow were the only SNE lake fish rated intolerant of anthropogenic stress in northeast lakes by Whittier and Hughes (1998). Lowering the threshold to tolerance scores of 11 (of 16) added 13 uncommon species (11 of these were collected in only one or two lakes each; Table 22.2). Only five lakes had more than one species from the expanded list of intolerants and most

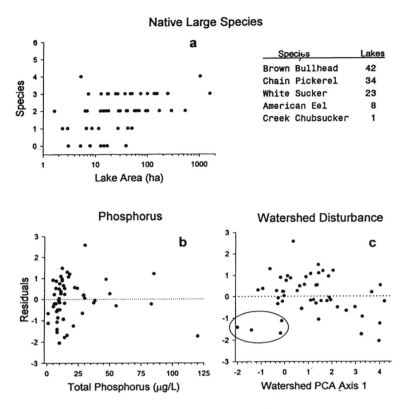

FIGURE 22.5 Native large species richness. (A) Number of native large species by lake size. (B) Residuals of the native large species by lake size regression plotted against total phosphorus. (C) Residuals plotted against watershed disturbance PCA axis 1 (higher axis 1 value = increased human activity in the watershed).

had none. There was a very weak relationship between lake size and intolerant species ($r^2 = 0.14$, $p = 0.0045$). Maximum intolerant species (uncorrected for lake size) tended to decrease with increased TP; most lakes with intolerant species greater than 0 had moderate to low watershed disturbance. Thus, this metric may be useful, although most lakes would get the minimum metric score.

One of the features of the original sucker species metric was longevity, providing a "multiyear integrative perspective" (Karr et al., 1986). To address the large-bodied, long-lived component of lake fish assemblages, five relatively large native species were selected to compose an analogous metric. There was a significant, but weak, relationship between large species richness and lake size ($r^2 = 0.19$, $p = 0.009$; Figure 22.5A). There was little association between increased TP and fewer large native species (Figure 22.5B). Both the regression residuals (Figure 22.5C) and the raw species counts tended to decrease with increased watershed disturbance. The four lakes circled in the lower left of Figure 22.5C are subject to manipulations (e.g., reclamation, draining) that are not detected by either stressor measures. This appears to be a useful metric. A similar analysis using 10 native small species (4 minnows, 2 sunfish, 2 darters, and killifish) showed no associations with any of the stressor measures.

Perhaps the most frequently replaced metric is % green sunfish (Simon and Lyons, 1995). In SNE, % bluegill would be the most likely candidate. Bluegill is not native and sometimes dominates the assemblages, being most abundant in CT and RI, becoming less dominant in MA, and being replaced by yellow perch as a dominant in NH. It occurred in only two NHE lakes. In the NCZ ecoregion where it was most widespread, % bluegill was not associated with higher stressor measures. A variant metric might be % individuals of the most abundant nonnative species (often bluegill). This did not improve the metric performance. Somewhat better was the proportion of

Development of IBI Metrics for Lakes in Southern New England

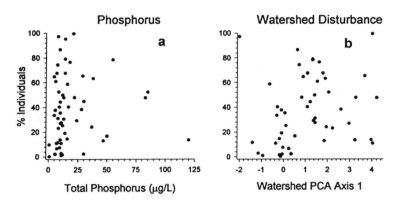

FIGURE 22.6 Percent of individuals of nonnative species. (A) % nonnative individuals by total phosphorus, (B) % nonnative individuals by watershed disturbance PCA axis 1 (higher axis 1 value = increased human activity in the watershed).

individuals of all nonnative species (or conversely proportion of native individuals). This metric showed no lake size effect, and no relationship with TP (Figure 22.6A). However, there was a tendency for the % nonnative individuals to increase with watershed stress. Most of the lakes in the upper left of Figure 22.6B had small watersheds and were probably more stressed than indicated by the watershed analysis. This metric is probably useful.

There were nine tolerant species collected (Table 22.2). Eleven of the 12 smallest lakes had greater than 90% tolerant individuals (Figure 22.7A). Pumpkinseed were fairly abundant in many of these ponds. Removing pumpkinseed from this metric had the greatest effect in smaller lakes. There was a clear association between % tolerants and TP. With pumpkinseed included, all eutrophic and hypereutrophic lakes (except one) had greater than 80% tolerant individuals (Figure 22.7B). The pattern was less dramatic with pumpkinseed removed (Figure 22.7D). For watershed disturbance, the pattern of increasing proportion of tolerant individuals with increasing stress was stronger with pumpkinseed removed (Figures 22.7C, E).

22.4.2 Trophic Composition Metrics

As with the species richness metrics, the trophic composition metrics were problematic in SNE lakes. For example, chain pickerel and American eel were the only native top carnivores (Table 22.2). Other natives such as yellow perch, white perch, and brook trout are piscivorous as large adults. However, white perch is native in only about a third of the SNE lakes and tolerant of stressed conditions. Brook trout is native to SNE streams, but is nearly always maintained by stocking in lakes. The most widespread top carnivore was the nonnative largemouth bass (in 50 of 55 sampled lakes), which was often found in quite degraded lakes, as was black crappie. Eight species were selected that are strongly piscivorous and not highly tolerant of stressed conditions (Table 22.2). This metric was also examined with largemouth bass included. Any species maintained by stocking was not included for that lake, which effectively eliminated Atlantic salmon and lake trout.

For the restricted species list (excluding largemouth bass), there was a slight tendency for percent top carnivores to increase with lake size. The highest scores occurred in the lakes with TP < 20 and low to moderate watershed disturbance (Figures 22.8A, B). The one outlier lake had 19% of individuals as rock bass; removing rock bass would still leave this lake with one of the highest percent top carnivores scores. Including largemouth bass tended to decrease the resolution of this metric (Figures 22.8C, D). That is, the peaks of the scatterplots moved to the right on the stressor

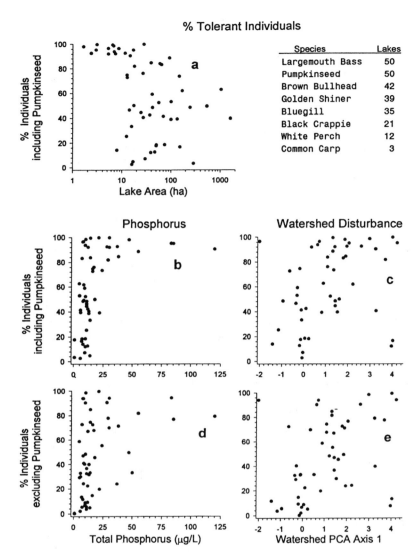

FIGURE 22.7 Percent of individuals of tolerant species (A) % tolerant individuals by lake size, including pumpkinseed. (B) % tolerant individuals by total phosphorus, including pumpkinseed. (C) % tolerant individuals by watershed disturbance including pumpkinseed. (D) % tolerant individuals by total phosphorus, excluding pumpkinseed. (E) % tolerant individuals by watershed disturbance excluding pumpkinseed.

plots, and the range of scores approximately doubled. There is concern that removing a highly tolerant species changes the metric from a trophic guild metric into a tolerance metric. I believe that the original intent of the metric is maintained with the reduced list.

A percent insectivorous individuals metric also required some adjustments. Only seven uncommonly collected species met the strict definitions of insectivory of Halliwell et al. (Chapter 12). In the present study, six additional native generalist feeder species were selected, species that tend toward the insectivorous end of the trophic spectrum and that are not highly tolerant of degraded conditions (Table 22.2). With this species list, percent insectivorous individuals scores related very well to TP and watershed disturbance (Figures 22.9B, C). The three outlier lakes in the upper right of the watershed disturbance plot all had yellow perch as the most abundant species. Also problematic for this metric was that 11 of the 12 smallest lakes had less than 10% insectivores (Figure

Development of IBI Metrics for Lakes in Southern New England

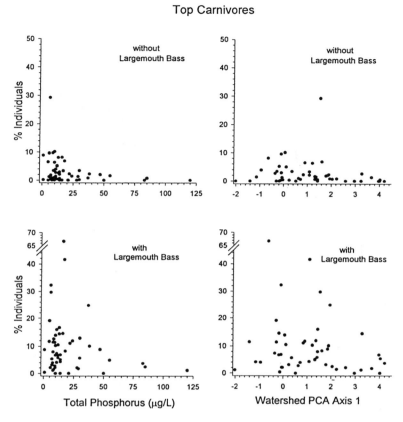

FIGURE 22.8 Percent of individuals of top carnivore species. (A) % Top carnivore individuals by total phosphorus, excluding largemouth bass. (B) % Top carnivore individuals by watershed disturbance excluding largemouth bass. (C) % Top carnivore individuals by total phosphorus, including largemouth bass. (D) % Top carnivore individuals by watershed disturbance, including largemouth bass.

22.9A). Eight of these small lakes had pumpkinseed as a dominant species. Adding pumpkinseed to the insectivore species list for small lakes removed the size effect, but did not improve the relationship with the stressor measures.

There are few true omnivore species (with a substantial portion of their diet including plant material) in the Northeast (Halliwell et al., Chapter 12), but a fairly large number of generalist feeder species (feeding at several trophic levels rather than primarily from one trophic level [Minns et al., 1994]). Ten species were chosen for a generalist feeder metric (Table 22.2). The maximum scores for % generalist feeder individuals decreased with lake sizes greater than about 30 ha (Figure 22.10A). Most of the small lakes had high scores, due generally to dominance by bluegill and pumpkinseed. The lowest scores (<30%) were all in relatively low-productivity lakes (TP < 20). The highest scores tended to be in lakes with moderate levels of stress (Figures 22.10B, C).

22.5 DISCUSSION

Seven of the candidate metrics appear to be successful and should provide a framework for future refinement of lake metrics and development of an IBI for inland lakes in southern New England. Two are species richness metrics: nonnative species richness adjusted for lake size as a negative metric, and large species richness (which might not need to be adjusted for lake size) as a positive

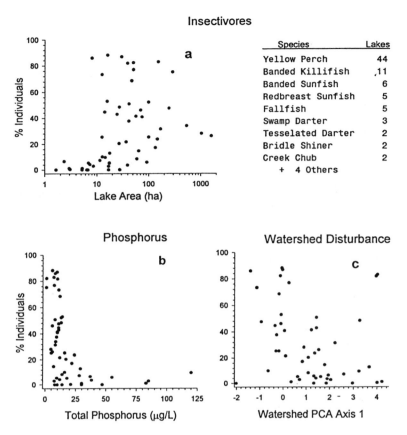

FIGURE 22.9 Percent of individuals of insectivore species. (A) % Insectivore individuals by lake size. (B) % Insectivore individuals by total phosphorus. (C) % Insectivore individuals by watershed disturbance PCA axis 1 (higher axis 1 value = increased human activity in the watershed).

metric. The percent nonnative individuals and percent tolerant individuals metrics could be considered in the richness and composition category (Karr et al., 1986) or in the indicator species category of Simon and Lyons (1995) as negative metrics. Minns et al. (1994) placed the former metric in the abundance and condition category. It might be possible to develop an intolerant species metric, or a small species metric. With judicious selection of species, all three trophic composition metrics appear to be useful.

The results presented here also illustrate some of the conceptual and ecological issues important for assessing biotic integrity of inland lakes, and differences between lotic and lentic fish assemblages. Lakes in New England are naturally species depauperate (Schmidt, 1986). In this way, they are somewhat analogous to coldwater streams, where increased total species richness is usually an indication of degradation (Lyons et al., 1996). In New England lakes, the source of additional species is introductions, rather than immigration from warmwater streams. The relationship between native species richness and introductions is complex for SNE lakes. Moyle and Light (1996) proposed that most successful invasions occur without loss of native species, and that nonnative predators should have greater effect on native assemblages than nonpredators. A substantial proportion of nonnative species in SNE are predators. For the Northeast as a whole, Whittier et al. (1997a) demonstrated an association between increased predator richness (usually from nonnatives) and lower minnow richness. However, preliminary analyses could not demonstrate an overall decrease in native species richness with increased numbers of nonnatives species (T. Whittier,

Development of IBI Metrics for Lakes in Southern New England

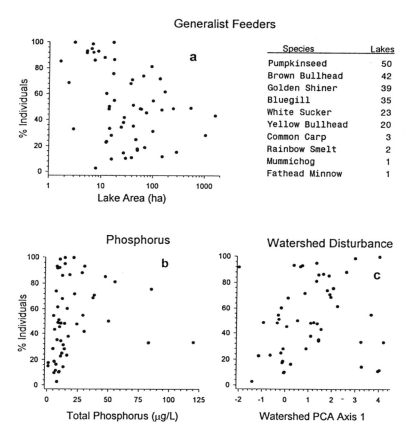

FIGURE 22.10 Percent of individuals of generalist feeder species. (A) % Generalist feeder individuals by lake size. (B) % Generalist feeder individuals by total phosphorus. (C) % Generalist feeder individuals by watershed disturbance PCA axis 1 (higher axis 1 value = increased human activity in the watershed).

unpublished data). Thus, in SNE, there appears to be adequate ecological "space" for the additional nonnative species in general, but some portions of the native assemblage may have been extirpated. It probably cannot be known whether SNE lakes had more minnow species prior to the introduction of additional littoral predators.

A third rule for species invasion (Moyle and Light, 1996) was that increased invasion success occurs when native assemblages are depleted or disrupted. The greater introduced species richness and higher proportion of nonnative individuals in more stressed lakes support this idea. However, it appears that these stress levels have not generally been high enough to eliminate native species, because native species richness showed no association with anthropogenic stress.

Fish species tolerance of, or sensitivity to degraded lake conditions appears to differ from their tolerances to impaired lotic systems. When using species tolerance classifications from stream IBIs, the proportion of tolerant individuals was lowest in the most stressed lakes (T. Whittier, unpublished data). This led Whittier and Hughes (1998) to evaluate individual species tolerances to five stressor measures in northeastern lakes, and assign new tolerance ratings. Applying these revised classes to the SNE assemblage data generally produced the expected associations with stressor measures, except in the smallest lakes and ponds (≤10 ha). These results, coupled with the larger species pool for lotic fish compared with lentic fish, pose some interesting questions relative to how lake ecosystems stress fish compared to stream ecosystems, and whether there are any truly intolerant warmwater lake species.

The trophic composition metrics presented a number of interesting challenges, in addition to the guild membership issues (e.g., lack of objective criteria, dietary plasticity, and changes with age) discussed by Minns et al. (1994). In SNE lakes, a number of species with obvious trophic guild membership react to degraded conditions in the opposite manner to what the metric should indicate. The clearest examples are the piscivores. For most of SNE, chain pickerel and American eel are the only native top carnivores. Largemouth bass is clearly a top carnivore, but is tolerant of degraded lake conditions. Black crappie and white perch are usually placed in the piscivore guild, but are two of the most tolerant species in SNE lakes (Whittier and Hughes, 1998), and among the dominant species in many stressed lakes. Finally, 16 of 40 species collected by EMAP in SNE lakes could be considered as piscivores (Halliwell et al., Chapter 12), producing very high percent piscivorous individuals scores (40–100% for more than half of the lakes). In selecting species for this metric, I did not try to choose the most sensitive piscivores, but limited the list to the most carnivorous species that were also not highly tolerant, regardless of the native status. This produced metric values in the 0 to 10% range and the expected association between the metric and stressor scores. Similar adjustments were needed for the other two trophic composition metrics.

Additional work is needed in a number of areas. First, biomass data may have the potential to improve all of the trophic guild metrics, and provide additional abundance and condition metrics (Minns et al., 1994). However, EMAP, which was not primarily a fisheries survey, did not collect those data. Second, some standard unit of sampling effort needs to be established. Multiple gear sampling makes that a complex problem. A sampling scheme combining electrofishing and gillnets might provide abundance data with lower variability. Third, additional work on lake classification schemes is needed. Even within a relatively homogenous area like southern New England, there are questions about whether lakes in the Northeast Coastal Zone ecoregion should be held to the same standard as those in the Northeast Highlands ecoregion. Finally, work is needed to develop metric scoring and an overall IBI, and to validate both the metrics and the index with additional data.

ACKNOWLEDGMENTS

I thank Bob Hughes, Dave Halliwell, and Bob Daniels for numerous insightful discussions of these issues. Martin Jennings, Melissa Drake, Dave Peck, and Bob Hughes provided useful reviews of an earlier draft. This research was supported by the U.S. Environmental Protection Agency through Contract Number 68-C5-0005 to Dynamac International, Inc.

REFERENCES

Baker, J. R., D. V. Peck, and D. W. Sutton, (Eds.). 1997. *Environmental Monitoring and Assessment Program — Surface Waters: Field Operations Manual for Lakes.* EPA/620/R-97/001. U.S. Environmental Protection Agency, Washington, DC.

Becker, G. C. 1983. *Fishes of Wisconsin.* University of Wisconsin Press. Madison, WI.

Census Bureau. 1991. Census of population and housing, 1990: Summary tape file 1 on CD-ROM (name of State). Washington, DC.

Chaloud, D. J. and D. V. Peck. (Eds.). 1994. *Environmental Monitoring and Assessment Program: Integrated Quality Assurance Project Plan for the Surface Waters Resource Group, 1994 Activities.* EPA 600/X-91/080 (Revision 2.00). U.S. Environmental Protection Agency, Las Vegas.

Christensen, D. L., B. R. Herwig, D. E. Schindler, and S. R. Carpenter, 1996. Impacts of lakeshore residential development on coarse woody debris in north temperate lakes, *Ecological Applications*, 6, 1143–1149.

Dionne, M. and J. R. Karr. 1992. Ecological monitoring of fish assemblages in Tennessee River reservoirs, in *Ecological Indicators.* D. H. McKenzie, D. E. Hyatt and V. J. McDonald (Eds.), Elsevier Applied Science. London. 259–281.

Fore, L. S., J. R. Karr, and R. W. Wisseman, 1996. Assessing invertebrate responses to human activities: evaluating alternative approaches, *Journal of the North American Benthological Society,* 15, 212–231.
Gauch, H. G., Jr. 1982. *Multivariate Analysis in Community Ecology.* Cambridge University Press, New York.
Halliwell, D. B. 1989. *A Classification of Streams in Massachusetts — To Be Used as a Fisheries Management Tool.* Ph.D. dissertation. University of Massachusetts. Amherst.
Hubbs, C. L. and G. P. Cooper. 1936. *Minnows of Michigan.* Bulletin of the Cranbrook Institute of Science. Number 8. Bloomfield Hills, MI.
Hubbs, C. L. and K. F. Lagler. 1964. *Fishes of the Great Lakes Region.* Bulletin of the Cranbrook Institute of Science. Number 26. Bloomfield Hills, MI.
Jennings, M. J., L. S. Fore, and J. R. Karr. 1995. Biological monitoring of fish assemblages in Tennessee Valley reservoirs, *Regulated Rivers: Research and Management,* 11, 263–274.
Jennings, M. J., M. Bozek, G. Hatzenbeler, and D. Fago. 1996. Fish and habitat, in *Shoreline Protection Study: A Report to the Wisconsin State Legislature.* PUBL-RS-921-96. Wisconsin Department of Natural Resources. Madison, WI. 1–48.
Karr, J. R. 1981. Assessment of biotic integrity using fish communities, *Fisheries,* 6(6), 21–27.
Karr, J. R. 1991. Biological integrity: a long-neglected aspect of water resource management, *Ecological Applications,* 1, 66–84.
Karr, J. R., K. D. Fausch, P. L. Angermeier, P. R. Yant, and I. J. Schlosser. 1986. Assessing Biological Integrity in Running Waters: A Method and Its Rationale. Illinois Natural History Survey, Special Publication No. 5. Champaign, IL.
Larsen, D. P. and S. J. Christie. (Eds.) 1993. EMAP-Surface Waters 1991 Pilot Report. EPA/620/R-93/003. U.S. Environmental Protection Agency. Corvallis, OR.
Larsen, D. P., D. L. Stevens, A. R. Selle, and S. G. Paulsen. 1991. Environmental Monitoring and Assessment Program, EMAP-Surface Waters: A northeast lakes pilot, *Lake and Reservoir Management,* 7, 1–11.
Larsen, D. P., K. W. Thornton, N. S. Urquhart, and S. G. Paulsen. 1994. The Role of sample surveys for monitoring the condition of the Nation's lakes, *Environmental Monitoring and Assessment,* 32, 101–134.
Lee, D. S., C. R. Gilbert, C. H. Hocutt, R. E. Jenkins, D. E. McAllister, and J. R. Stauffer, Jr. 1980. *Atlas of North American Freshwater Fishes.* North Carolina Biological Survey. Raleigh.
Leonard, P. M. and D. J. Orth. 1986. Application and testing of an index of biotic integrity in small, coolwater streams, *Transactions of the American Fisheries Society,* 115, 401–415.
Lyons, J. 1992. Using the Index of Biotic Integrity (IBI) to Measure Environmental Quality in Warmwater Streams of Wisconsin. General Technical Report NC-149. U.S. Department of Agriculture, Forest Service, North Central Experimental Station.
Lyons, J., L. Wang, and T. D. Simonson. 1996. Development and validation of an index of biotic integrity for coldwater streams in Wisconsin, *North American Journal of Fisheries Management,* 16, 241–256.
Miller, D. L. et al. 1988. Regional application of an index of biotic integrity for use in water resource management, *Fisheries,* 13(5), 12–20.
Minns, C. K., V. W. Cairns, R. G. Randall, and J. E. Moore. 1994. An index of biotic integrity (IBI) for fish assemblages in littoral zone of Great Lakes' areas of concern, *Canadian Journal of Fisheries and Aquatic Sciences,* 51, 1804–1822.
Moyle, P. B. and T. Light. 1996. Biological invasions of fresh water: empirical rules and assembly theory, *Biological Conservation,* 78, 149–161.
New Hampshire Fish & Game Commission (NHFGC). 1937–1938. *Biological Surveys,* 3 volumes. Concord, NH.
New York State Department of Conservation (NYSDC). 1927–1939. *Biological Surveys,* 12 volumes. Albany, NY.
Ohio Environmental Protection Agency (OEPA). 1988. *Biological Criteria for the Protection of Aquatic Life. Vol. I. The Role of Biological Data in Water Quality Assessment.* OEPA Division of Water Quality Monitoring and Assessment. Columbus.
Ohio Environment Protection Agency (OEPA). 1989. *Biological Criteria for the Protection of Aquatic Life. Vol. III. Standardized Biological Field Sampling and Laboratory Methods for Assessing Fish and Macroinvertebrate Communities.* OEPA Division of Water Quality Planning and Assessment. Columbus.
Omernik, J. M. 1987. Ecoregions of the conterminous United States, *Annals of the Association of American Geographers,* 77, 118–125.
Page, L. M. and B. M. Burr. 1991. *A Field Guide to Freshwater Fishes.* Houghton Mifflin. Boston.

Plafkin, J. L., M. T. Barbour, K. D. Porter, S. K. Gross, and R. M. Hughes. 1989. *Rapid bioassessment protocols for use in streams and rivers: benthic macroinvertebrates and fish*. EPA/444/4-89-001. U.S. Environmental Protection Agency, Washington, DC.

SAS Institute. 1985. *SAS User's Guide: Statistics, Version 6*. Cary, NC.

Schmidt, R. E. 1986. Zoogeography of the Northern Appalachians, In *The Zoogeography of North American Freshwater Fishes*. Hocutt, C. H. and E. O. Wiley (Eds.). John Wiley & Sons, New York. 137–159.

Scott, W. B. and E. J. Crossman. 1973. *Freshwater Fishes of Canada*. Bulletin of the Fisheries Research Board of Canada.

Simon, T. P. and J. Lyons. 1995. Application of the Index of Biotic Integrity to evaluate water resource integrity in freshwater ecosystems, in W.S. Davis and T.P. Simon (Eds.), *Biological Assessment and Criteria: Tools for Water Resource Planning and Decision Making*. Lewis. Boca Raton, FL. 245–262.

Smith, C. L. 1985. *The Inland Fishes of New York State*. Department of Environmental Conservation. Albany, NY.

Steedman, R. J. 1988. Modification and assessment of an index of biotic integrity to quantify stream quality in southern Ontario, *Canadian Journal of Fisheries and Aquatic Science*, 45, 492–501.

Stevens, D. L., Jr. 1994. Implementation of a national monitoring program, *Environmental Management*, 42, 1–29.

Trautman, M. B. 1981. *The Fishes of Ohio*. Ohio State University Press. Columbus.

Underhill, J. C. 1986. The Fish Fauna of the Laurentian Great Lakes, the St. Lawrence Lowlands, Newfoundland and Labrador, in C. H. Hocutt and E. O. Wiley (Eds.), *The Zoogeography of North American Freshwater Fishes*. John Wiley & Sons, New York. 105–136.

U.S. Environmental Protection Agency (U.S. EPA), Office of Information Resources Management. 1992. *1990 Block/Precinct Level Coverage User Documentation*. Washington, DC.

U.S. Geological Survey (USGS) 1990. *Land Use and Land Cover Digital Data from 1:250,000- and 1:100,000-scale Maps. Data Users Guide 4*, Reston, VA.

Weaver, M. J., J. J. Magnuson, and M. D. Clayton. 1993. Analyses for differentiating littoral fish assemblages with catch data from multiple sampling gears, *Transactions of the American Fisheries Society*, 122, 1111–1119.

Whittier, T. R. and R. M. Hughes. (1998). Evaluation of fish species tolerances to environmental stressors in Northeast USA lakes. *North American Journal of Fisheries Management* (in press).

Whittier, T. R. and E. T. Rankin. 1992. Regional patterns in three biological indicators of stream condition in Ohio, in D. H. McKenzie, D. E. Hyatt, and V. J. McDonald (Eds.), *Ecological Indicators*. Elsevier Applied Science. London. 975–995.

Whittier, T. R., D. B. Halliwell, and S. G. Paulsen. 1997a. Cyprinid distributions in Northeast USA lakes: evidence of regional-scale minnow biodiversity losses, *Canadian Journal of Fisheries and Aquatic Sciences*, 54, 1593–1607.

Whittier, T. R., P. Vaux, G. D. Merritt, and R. B. Yeardley, Jr. 1997b. Fish sampling, in *Environmental Monitoring and Assessment Program — Surface Waters: Field Operations Manual for Lakes*. Baker, J. R., D. V. Peck, and D. W. Sutton (Eds.), EPA/620/R-97/001. U.S. Environmental Protection Agency, Washington, DC.

Section V

Data Validation

23 Relations Between Fish Metrics and Measures of Anthropogenic Disturbance in Three IBI Regions in Virginia

Roy A. Smogor and Paul L. Angermeier

CONTENTS

23.1 Introduction 585
23.2 Methods 587
    23.2.1 Data Source 587
    23.2.2 Habitat Variables 587
    23.2.3 Fish Sampling 588
    23.2.4 Potential Fish Metrics 588
    23.2.5 Statistical Tests and Considerations 592
23.3 Results 594
    23.3.1 Relations Between Disturbance Measures and Metrics in Coastal Plain 594
    23.3.2 Relations Between Disturbance Measures and Metrics in Piedmont 594
    23.3.3 Relations Between Disturbance Measures and Metrics in Mountain 602
    23.3.4 Interrelations among Disturbance Measures 604
23.4 Discussion 604
23.5 Conclusion 607
Acknowledgments 608
References 608

23.1 INTRODUCTION

The utility of any bioassessment index depends on its practicality and ability to adequately reflect disturbance (e.g., Pratt and Bowers, 1992). For the Index of Biotic Integrity (IBI: Karr, 1981; Karr et al., 1986; Karr, 1991), practical issues include choosing the number and types of metrics and the spatial framework (e.g., physiographies, ecoregions, drainages, and ranges of stream size and anthropogenic disturbance) within which to use the index. Several investigations of the IBI for stream fish have been bound, for practical reasons, to intrastate spatial scales (e.g., Ohio EPA, 1988; Lyons, 1992).

    The ability of the IBI to reflect anthropogenic disturbance (hereafter referred to simply as "disturbance") depends directly on the validity of the assumptions used to score each metric (Table 23.1). Despite their importance, these assumptions have received little critical examination (Fausch et al., 1990), especially with regard to how relations between metrics and disturbance may vary with region, stream size, or type and range of disturbance. The natural variation in fish-assemblage

TABLE 23.1
Assumptions of Metrics Commonly Used in Versions of the IBI Adapted for Freshwater Stream Fish Assemblages

Taxonomic and richness metrics
    The number of native species decreases
    The number of native species in particular taxa decreases
    The number of intolerant species decreases
    The number of nonnative species or proportion of nonnative individuals increases
Trophic metrics
    The proportion of individuals that are trophic specialists (e.g., insectivorous cyprinids, top carnivores) decreases
    The proportion of individuals that are trophic generalists (e.g., omnivores) increases
Reproductive metrics
    The proportion of individuals that require silt-free, mineral spawning substrates decreases
Tolerance and fish-condition metrics
    The proportion of individuals that are tolerant increases
    The proportion of individuals that are intolerant decreases
    The incidence of externally evident disease, parasites, and morphological anomalies increases

Note: Assumptions describe how metrics are expected to change with increasing degradative anthropogenic disturbance of stream systems.

attributes across regions and stream sizes (e.g., Smogor and Angermeier, Chapter 10) strongly implies that the utility of individual metrics also varies spatially. For example, metrics most diagnostic of disturbance in small, upland streams may not be as useful in large, lowland streams.

Several studies have addressed how individual metrics are related to the IBI (Angermeier and Karr, 1986; Karr et al., 1987; Steedman, 1988) or how the IBI relates to measures of water or habitat quality (Steedman, 1988; Berkman et al., 1986; Leonard and Orth, 1986; Angermeier and Schlosser, 1987; Hughes and Gammon, 1987; Lyons, 1992; Hall et al., 1994; Rankin, 1995; Roth et al., 1996; Scott and Hall, 1997; Stauffer and Goldstein, 1997; Wang et al., 1997). However, very few of these studies provided explicit evidence of the validity of individual IBI-metric assumptions with respect to relevant regions, stream sizes, or other environmental gradients. Moreover, studies that used the same data to determine IBI metric scoring criteria as to calculate IBIs (e.g., Steedman, 1988; Leonard and Orth, 1986; Hughes and Gammon, 1987; Shields et al., 1995) merely showed that the IBI could reflect how its individual metrics were *assumed* to vary with disturbance. Neither relating individual metric values or scores to total IBI scores nor relating total IBI scores to habitat measures represents a valid investigation of how each metric reflects disturbance. A better approach to examine IBI metric assumptions would be to (1) define relevant and practicable bioassessment regions, (2) assess disturbance, in non-IBI terms, at sites in each region, and (3) determine how each IBI metric is related to disturbance in each region. Ideally, the data used for such an investigation should be independent of that used for any future IBI assessment. We know of only two studies that have approximately met these criteria: Ohio EPA (1988; for Ohio streams) and Lyons (1992; for Wisconsin streams). Although both studies accounted for points (1) and (2) above, neither explicitly examined point (3).

In Virginia, anthropogenic effects on stream habitat and fish assemblages are likely to differ among the state's major physiographic regions: Coastal Plain, Piedmont, Blue Ridge, and Ridge and Valley. Even disregarding effects due to disturbance, many watershed, riparian, and instream features vary naturally among these regions (Smock and Gilinsky, 1992). Not surprisingly, fish metrics of potential use for the IBI also vary naturally (i.e., at least-disturbed sites) across these physiographic regions (Smogor, 1996). Region-specific effects of disturbance on fish assemblages complicate attempts to assess these effects on a statewide scale. Therefore, a region-specific,

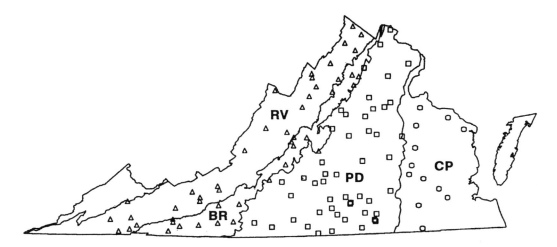

FIGURE 23.1 Map of Virginia showing locations of 108 sample sites in three physiographic regions: Coastal Plain (CP), Piedmont (PD), and Mountain (MT) = Blue Ridge (BR) + Ridge and Valley (RV). For CP, number of sites (N) = 12 (circles); for PD, N = 49 (squares); and for MT, N = 47 (triangles).

multivariate-based investigation of relations between fish metrics and disturbance measures provides a biologically realistic starting point for developing IBIs for streams in Virginia.

This study examines uni-, bi-, and multivariate relations between measures of disturbance and potential IBI fish metrics for wadeable warmwater streams in each of three physiographic regions in Virginia (Figure 23.1). Specifically, it investigates (1) if fish metrics relate to selected habitat variables in ways consistent with prevailing IBI assumptions, (2) how these relations differ among IBI regions, and (3) which fish metrics may be most useful for stream-fish IBIs in Virginia.

23.2 METHODS

23.2.1 Data Source

A subset of data from a 1987 to 1990 survey of Virginia stream fishes (Angermeier and Smogor, 1992) was used. The data subset included catch-per-effort of fish species and estimates of selected instream- and riparian-habitat measures at each of 108 sampling sites; sites were sampled from June to September 1988 to 1990. Sites occurred in 3rd- through 6th-order streams across three major physiographic regions that represent each of three potential regions (i.e., IBI regions) within which to use a distinct version of the IBI (Figure 23.1; Smogor 1996). Sites were distributed among physiographies as follows: for Coastal Plain (CP), N = 12; for Piedmont (PD), N = 49; and for Mountain (MT), N = 47. The Mountain region comprised the Blue Ridge and Ridge and Valley physiographic provinces (Figure 23.1). Sites ranged from about 50 to 250 m long and drained areas 7 to 454 sq. km; sampling effort (i.e., stream surface area sampled) was approximately proportional to stream size.

23.2.2 Habitat Variables

Variables were employed that presumably reflected anthropogenic disturbance manifested throughout watersheds, along riparian zones, and in stream channels. These variables, collectively, are referred to as "habitat variables" or "disturbance measures." Although these habitat variables did not encompass all possible anthropogenic effects on fish assemblages, they did include measures that reflected the most-documented and pervasive anthropogenic effects on stream systems in Virginia (Virginia Department of Environmental Quality, 1994).

For each site, three watershed-scale disturbance variables were estimated: (proportional area of) watershed as urban, (proportional area of) watershed as forest, and number of pollution point sources (Table 23.2). Also at each site, we visually estimated the following (Table 23.2; see Smogor [1996] for further details): (1) riparian width, (2) riparian forest: proportion of riparian area forested, (3) bankside woody cover: proportion of bank length as trees or shrubs, (4) maximum (wetted) depth, and (5) instream woody cover or instream cover: proportion of 1 m wide, bank-to-bank transects that contained woody debris or logs (for CP and PD) or woody debris, logs, or rock crevices/ledges (for MT). Site rankings of overall disturbance (Smogor and Angermeier, Section 10.2.3 in Chapter 10) were used to define the approximate upper third of all ranked sites as least-disturbed and the lowest third as most-disturbed, for comparisons herein.

23.2.3 FISH SAMPLING

At each site, the authors and two or three co-workers used an electric seine (Angermeier et al., 1991) to collect fish in a series of representative channel units (e.g., riffles, runs, pools). For most sites, each habitat unit was blocked bank-to-bank with upstream and downstream nets (0.64-cm mesh). Workers made two seine passes in an upstream direction in each blocked habitat unit. Dipnets (0.64-cm mesh) were employed to capture all stunned fish, and each fish was identified to species. For further details, see Smogor (1996) or Smogor and Angermeier (Section 10.2.2 in Chapter 10).

23.2.4 POTENTIAL FISH METRICS

We examined metrics that presumably reflect effects of typical anthropogenic disturbance on fish communities *and* are relatively easy to determine from field data (Karr et al., 1986; Miller et al., 1988; Fausch et al., 1990; Simon and Lyons, 1995). Some of the metrics examined have been used widely in IBI analyses; others were judged potentially useful, although they had not been considered previously. Fish metrics were grouped into three classes: taxonomic, trophic, and reproductive (Table 23.3). Due to statistical constraints (see below, Section 23.2.5), metrics were limited to those that showed adequate variation across sites within each physiography, were approximately normally distributed without extreme outliers (after transformation), and were not highly intercorrelated with many others (i.e., not having many pairwise Pearson correlations greater than 0.80). For example, excluded from all analyses were the number of native sucker species because too few sucker species occur in Virginia's CP and because PD and MT distributions of this metric were asymmetric with extreme outliers.

In total, 14 fish metrics were used: five taxonomic, four trophic, and five reproductive (including one tolerance metric). Taxonomic, trophic, reproductive, and tolerance classifications of species (Smogor and Angermeier, Table 10.4 in Chapter 10) were based on various regional texts (e.g., Pflieger, 1975; Jenkins and Burkhead, 1994) and on personal experience. Native vs. nonnative status (by major river drainage) was based on Jenkins and Burkhead (1994).

The following five metrics, which have been used widely and whose rationales have been discussed elsewhere (e.g., Karr, 1981; Karr et al., 1986; Ohio EPA, 1988; Lyons, 1992; Simon and Lyons, 1995), were used: number of native species, number of nonnative species, number of native sunfish species, number of native darter or sculpin species, proportion of individuals that are members of tolerant species. Numbers of native minnows and of native sunfishes comprised all native species in the families Cyprinidae and Centrarchidae, respectively. Number of native darter or sculpin species comprised all native species of *Percina*, *Etheostoma*, or *Cottus*. Twenty sites in MT had no native sunfish species; therefore, nonnative species were included in this metric for analyses of MT sites.

Number of native minnow species has been used as an IBI metric in several studies (e.g., Ohio EPA, 1988; Hughes and Gammon, 1987; Bramblett and Fausch, 1991; Hall et al., 1994; see Table 2 in Simon and Lyons, 1995). Native cyprinid species are a widely distributed and ecolog-

TABLE 23.2
Values of Habitat Variables at Sites Within Each Physiography

Variable	MD	$\bar{x}$	SE	CV	Categ.	N	Categ. or range	N	Categ.	N
Coastal Plain (N = 12; 23–147 sq. km)										
Discrete										
Riparian width	50.0	50.0	0.00	0.0	<50 m	0			≥50 m	12
Riparian forest	1.000	0.975	0.0179	6.375	<1.00	2			1.00	10
Number of point sources	0.00	0.33	0.188	195.40	0	9			>0	3
Watershed as urban	0.033	0.070	0.0272	135.428	0	4			>0	8
Continuous										
Bankside woody cover	0.73	0.75	0.043	19.68			0.50–0.95			
Instream woody cover	0.41	0.48	0.077	54.96			0.08–0.87			
Maximum depth	84	82	5.8	25			54–112 cm			
Watershed as forest	0.60	0.61	0.040	22.65			0.41–0.85			
Piedmont (N = 49; 7–454 sq. km)										
Discrete										
Riparian width	50.0	40.2	2.02	35.3	<50 m	18			≥50 m	31
Riparian forest	0.95	0.77	0.042	37.84	<1.00	26			1.00	23
Number of point sources	0.00	0.16	0.061	260.67	0	42			>0	7
Watershed as urban	0.030	0.051	0.0110	151.345	0.00	15	0.00<x<0.05	13	≥0.05	21
Continuous										
Bankside woody cover	0.800	0.778	0.0228	20.506			0.30–1.00			
Instream woody cover	0.250	0.293	0.0294	70.149			0.00–0.92			
Maximum depth	80	74	3.2	31			30–120 cm			
Watershed as forest	0.595	0.584	0.0212	25.347			0.24–0.90			

TABLE 23.2 (continued)
Values of Habitat Variables at Sites Within Each Physiography

Variable	MD	$\bar{X}$	SE	CV	Categ.	N	Categ. or range	N	Categ.	N
				Mountain (N = 47; 14–363 sq. km)			Category or Range of Variable			
Discrete										
Riparian width	27.5	27.1	2.46	62.2	<20 m	16	20≤ x <50	20	≥50 m	11
Riparian forest	0.40	0.46	0.053	79.87	<0.20	11	0.20≤ x <0.80	22	≥0.80	14
Number of point sources	0.00	0.13	0.058	310.64	0	42			>0	5
Watershed as urban	0.014	0.041	0.0098	164.475	0.00	14	0.00<x<0.05	21	≥0.05	12
Continuous										
Bankside woody cover	0.80	0.72	0.031	29.64			0.05–1.00			
Instream cover	0.080	0.103	0.0140	92.689			0.00–0.35			
Maximum depth	63	69	3.5	35			35–130 cm			
Watershed as forest	0.63	0.59	0.039	45.86			0.00–1.00			

Note: Medians (MD), means (X), standard errors (SE), and coefficients of variation (CV) are shown. Categories (Categ.) of discrete variables and ranges of continuous variables are also shown. Number of sites in each physiography (N) is followed by the range of stream sizes (shown as watershed area) sampled.

TABLE 23.3
Fish Metrics Used in Analyses of Fish-vs.-Disturbance Relations

Metric	Code
Taxonomic metrics	
Number of native species	NATSP
Number of nonnative species	NONNATSP
Number of native minnow species	MINSP
Number of native sunfish species	SUNSP
Number of native darter or sculpin species	DARSCLSP
Trophic metrics	
Proportion as generalist feeders	GENPRP
Proportion as invertivores	INVPRP
Proportion as benthic, specialist invertivores	BINVPRP
Proportion as specialist carnivores	CARNPRP
Reproductive metrics	
Proportion as mineral-substrate, simple spawners (i.e., simple lithophils)	SLITHPRP
Proportion as mineral-substrate spawners (i.e., lithophils)	LITHPRP
Proportion as various-substrate, manipulative spawners (i.e., generalist spawners)	VMANPRP
Number of late-maturing (>2 yr) species	AGE3SP
Proportion as members of tolerant species	TOLPRP

Note: Metrics are arranged in three classes: taxonomic, trophic, and reproductive. "Proportion" refers to proportion of individuals. Subsequent tables refer to fish metrics by their codes. For MT sites only, SUNSP comprised native plus nonnative sunfishes. SLITHPRP and LITHPRP excluded individuals of species classified as "tolerant."

ically diverse group in Virginia (Jenkins and Burkhead, 1994). The 48 native Virginian cyprinids represented by our data include five tolerant and two intolerant species (Smogor and Angermeier, Table 10.4 in Chapter 10). This metric was expected to decrease with increased disturbance in Virginia streams.

"Intolerant" species were classified as those whose ranges or abundances have decreased, presumably due to anthropogenic effects, and "tolerant" species as those known to be affected least detrimentally by typical anthropogenic disturbances to streams and watersheds; many of these species' historical ranges or local proportional abundances have increased with increases in anthropogenic disturbance (e.g., common carp, *Cyprinus carpio*; bluntnose minnow *Pimephales notatus*; and green sunfish, *Lepomis cyanellus*). Each species' tolerance rating was determined before ranking sites for disturbance and before choosing IBI regions, thus minimizing the bias that otherwise would result from classifying species after having examined their occurrence patterns relative to disturbance in the study streams.

Karr et al. (1986) recommended that less than 10% of the species in a region be classified as "intolerant." This limit ensures that an intolerance metric contributes exclusively to the highest IBI scores; that is, it only reflects sites at the highest end of the biotic integrity continuum. Our classifying 5.6% (8 of 143 species statewide) as intolerant seems reasonable; however, fewer than 5.6% occurred in some IBI regions because tolerance classes were determined before we determined IBI regions. Similarly, classification as "tolerant" was limited to a small percentage of the included species; this ensures that the tolerance metric reflects exclusively the lowest end of the biotic integrity continuum; that is, only those severely degraded sites dominated by tolerant species or individuals. This study classified 12% (17 of 143 statewide) of species as tolerant; however, fewer than 12% occurred in some IBI regions. Because few intolerant species occurred in PD and in CP, the metric (number of intolerant species) varied little among sites in these regions. For MT sites, most intolerant species occurred in the Clinch River drainage, and the metric did not equally represent all MT sites. For these reasons, this metric was excluded from our analyses.

For trophic metrics, each species was classified based on three factors: number of food types typically eaten, feeding behavior, and feeding group. Four food types were designated: (1) detritus, (2) algae or vascular plants, (3) invertebrates, and (4) fish (including fish blood) or crayfish. "Generalist feeders" were species in which adults eat three or four food types; "specialists" eat one or two food types. Also designated were two mutually exclusive feeding behaviors: benthic and nonbenthic. Benthic species feed, as adults, mostly along the stream bottom and have specialized morphology (e.g., inferior mouth) for doing so. Fish species were assigned to one of five feeding groups based on the primary food type(s) of subadults and adults: (1) detritivore/algivore/herbivore, (2) algivore/herbivore/invertivore, (3) invertivore, (4) invertivore/piscivore, or (5) piscivore or fish parasite. Group 4 comprised species in which subadults eat primarily invertebrates, but adults eat primarily fish or crayfish (e.g., American eel, *Anguilla rostrata*; crappies, *Pomoxis* spp.; yellow perch, *Perca flavescens*). "Carnivores" were species in groups 4 or 5 (Table 23.3).

The four trophic metrics represented a continuum from generalist feeders to specialist carnivores or specialist benthic invertivores; invertivores represented the middle of this continuum. Trophic specialists were expected to be most abundant at least-disturbed sites, and vice versa (Table 23.1). Using similar trophic categories for Virginia fishes, Angermeier (1995) found that specialized species were more likely to be extirpated than were generalized ones, suggesting that trophic specialists are sensitive to anthropogenic disturbance.

For reproductive metrics, one life-history trait and two classification factors were considered: species age at maturity, spawning substrate, and spawning behavior. "Late-maturing species" were defined as those in which females typically do not spawn before 3 years of age. Number of late-maturing species may be indicative of chronic disturbance because such species may be slower to recover from disturbance than those with shorter generation times (e.g., Niemi et al., 1990; Schlosser, 1990). Four spawning-substrate categories were designated: (1) uses no substrate or has pelagic eggs, (2) typically uses vegetation or organic debris, (3) known to use various substrates and not restricted to unsilted mineral substrates, (4) obligately uses unsilted mineral substrates, i.e., lithophilic. Also, we designated "manipulative" vs. "simple" (nonmanipulative) spawners (Table 23.3). "Manipulative spawners" build nests, depressions, or cavities *or* actively guard eggs or young (e.g., lampreys, catfishes, and sunfishes). "Simple spawners" exhibit relatively little nest preparation or parental care. Because manipulative spawners can alter spawning substrates or provide extended care to eggs or young, we presume that they would be less vulnerable to effects of habitat disturbance (e.g., excessive siltation) than simple spawners. Alternatively, simple lithophils represented the reproductive group likely to be most sensitive to disturbance effects, especially excessive siltation of streambeds (Muncy et al., 1979; Ohio EPA, 1988; Lyons, 1992; Rabeni and Smale, 1995), which is one of the most common and widespread disturbances in Virginia streams (Virginia Department of Environmental Quality, 1994; Jenkins and Burkhead, 1994). Excluded were any species classified as "tolerant" from the "lithophil" classification to minimize contrary information contributed by species originally classified as both.

The four trophic metrics and three of the reproductive metrics were chosen to represent a broad continuum of trophic or reproductive susceptibility to anthropogenic disturbance (Table 23.3). Consistent with prevailing IBI metric assumptions, one would expect that more-disturbed sites will have more trophic or reproductive "generalists" and fewer "specialists" than less-disturbed ones, and vice versa.

23.2.5 STATISTICAL TESTS AND CONSIDERATIONS

For sites in each of the three physiographic regions, this study examined all pairwise correlations (Pearson's r for PD and for MT; Spearman's rho [r_s] for CP) between fish metrics and the four continuously distributed habitat variables: bankside woody cover, instream (woody) cover, maximum depth, and watershed as forest (Table 23.2). The following habitat variables had highly skewed or disjunct distributions: riparian width, riparian forest, presence of point sources, and watershed as urban (Table 23.2); therefore, each of their distributions of values was divided into categories and treated as discretely distributed (hereafter referred to as the "discrete habitat variables").

For PD and for MT sites, multivariate analysis of covariance (MANCOVA; SAS, 1990), with watershed area as covariate, was used to examine how fish metrics or the four continuously distributed habitat variables varied among categories of each discrete habitat variable. Pillai's trace statistic (V; SAS, 1990) divided by one less than the number of statistical groups (s; e.g., one less than the number of categories, for each discrete habitat variable) was used as a measure of multivariate differences. The value V/s represents generalized variance and is directly analogous to R^2, the coefficient of determination in regression (Thorndike, 1978; Cramer and Nicewander, 1979; Serlin, 1982). For CP sites, small sample sizes precluded multivariate tests; therefore, Wilcoxon two-sample tests were employed to assess differences in metrics or in continuous habitat variables among categories of each discrete habitat variable.

Also, for PD and for MT sites, canonical correlation analysis (hereafter referred to as "COR"; Thompson, 1984; Gittins, 1985) was used to examine relations between fish metrics and the four continuous habitat variables. COR represents a more realistic, comprehensive, concise, and (potentially) easily interpretable investigation of relations than would multiple uni- or bivariate analyses.

Because fish metrics were correlated with stream size (Smogor and Angermeier, Chapter 10) and because an assessment of relations between metrics and habitat that were independent of stream-size effects was desired, standardized fish-metric residuals in CORs were used. Residuals were obtained from general linear regression models (SAS, 1990) of (natural log of) watershed area as a function of transformed fish metrics (see below, this section). For sites within each physiography, three preliminary CORs were performed that used residuals of a model that included only the fish metrics of each metric class (i.e., taxonomic, trophic, or reproductive). For two final CORs (one for PD sites, one for MT sites), residuals from a model that included a selected, multiclass set of metrics were used. Metrics for final CORs were those shown by preliminary CORs to be most related with habitat variables. Using fish metric residuals removed most of the (linear) statistical effects of stream size on fish metrics, and allowed for interpretation less confounded by these effects. Hereafter, COR and its results, the fish-metric residuals are referred to simply as "fish metrics."

Proper use of multivariate tests and the reliability of their results require that the variable-to-sample size ratio (p/n) be small and that variables have few high intercorrelations (Thorndike, 1978; Williams, 1983; Gittins, 1985; ter Braak, 1987). For COR specifically, the total number of variables (i.e., summed across both sets) must be less than the number of samples; as p/n approaches 1.00, the value of the first canonical correlation rapidly approaches 1.00, thereby rendering results meaningless (Gittins, 1985; ter Braak, 1987). Thorndike (1978) suggested that p/n be less than 0.10 (but preferably much smaller, 0.02) to allow for a conclusive COR; however, Gittins (1985) showed that valid interpretation of ecological data was possible with p/n as high as 0.50. To limit p/n via p, a separate COR was performed for each of the three classes of fish metrics (i.e., preliminary CORs as described in preceding paragraph); all p/n were less than 0.20. For each of two final CORs (one for PD, one for MT), $p/n < 0.27$. Therefore, p/n values seemed sufficiently low to allow valid and meaningful interpretations of results based on COR.

To complement the two final CORs, MANOVA and descriptive discriminant analysis (i.e., canonical analysis: Thorndike, 1978; Williams, 1983; Gittins, 1985; Huberty, 1986) were employed to examine how selected fish metrics (i.e., residuals) varied among least-disturbed, most-disturbed, and moderately disturbed sites in PD and in MT. Consistencies in results between COR and the other analyses indicated robustness of results.

Each variable was transformed so that its distribution best exhibited univariate normality, which is necessary but not sufficient to maximize the likelihood of multivariate normality (Tabachnick and Fidell, 1996). Data represented as proportions were arcsine transformed (arcsine[$x^{0.5}$]), and data represented as counts were square-root transformed ($[x + 0.5]^{0.5}$). Watershed area and maximum depth were natural-log transformed. Despite transformations, some sites remained extreme outliers for some fish metrics and, therefore, potentially biased COR results; COR is based on least-squares procedures that can be overly sensitive to outliers (Gittins, 1995). One site with no native minnow species and one site with an extremely high proportion as tolerants were deleted from CORs of PD. For CORs of MT sites, one site with an extremely high and another with an extremely low number of native darters or sculpins were deleted.

The conclusions based on multivariate results reflect a best-possible and realistic representation of the multiple relations between habitat variables and fish metrics. Nonetheless, general applicability of these multivariate results is unjustifiable because the findings were not validated. Because of potential data limitations in the multivariate analyses, the authors did not rely solely on multivariate results; data was also examined uni- or bivariately to facilitate multivariate-based interpretations. This combined approach is believed to have provided the most comprehensive analysis.

23.3 RESULTS

23.3.1 RELATIONS BETWEEN DISTURBANCE MEASURES AND METRICS IN COASTAL PLAIN

Fish metrics were related with disturbance measures, but most relations were contrary to IBI assumptions. For example, sites in more-forested watersheds had fewer native sunfish species ($r_s = -0.40$) and more tolerants ($r_s = 0.67$) than did less-forested sites (Table 23.4). Also, sites in more urbanized watersheds had more native sunfish species (median = 5.0 for high urban) than did less-disturbed sites (median sunfish = 2.0 for low urban; Wilcoxon two-sample test, $p = 0.09$). Except for instream woody cover, disturbance measures were not related with stream size ($r_s = 0.62$, $p = 0.0303$ for instream woody cover vs. watershed area).

The strongest bivariate correlations (i.e., absolute values of $r_s \geq 0.55$, $p \leq 0.06$) among fish metrics revealed two contrasting groups. One group comprised a number of native species, number of native minnow species, number of late-maturing species, and proportion as benthic invertivores. Again, inconsistent with IBI assumptions, metrics in this group tended to be least at least-disturbed sites; that is, those with more bankside and instream woody cover (Table 23.4). A second group, comprising proportions as invertivores and as generalist spawners, had greater values at less-disturbed sites. The many bivariate correlations among fish metrics, 19 of them with absolute values of $r_s > 0.50$ ($p < 0.10$), prevented comprehensive and concise interpretation of results.

23.3.2 RELATIONS BETWEEN DISTURBANCE MEASURES AND METRICS IN PIEDMONT

Fish metrics were related with the four continuously distributed disturbance measures. About one fifth of the variance in selected fish metrics could be accounted for and reasonably interpreted via these relations (sum of RED = 0.229 for PD sites, Table 23.5). Taxonomic and trophic metrics were most related with habitat variables, whereas reproductive metrics were least related: the first two canonical correlations explained from 0.19 (COM2 for proportion as specialist carnivores) to 0.40 (COM2 for number of native minnow species) of the variance in nonreproductive metrics (Table 23.5).

Relations revealed by the first canonical correlation (CAN1) were contrary to IBI assumptions, due largely to more-disturbed sites (i.e., those with less bankside and instream woody cover) having more native minnow species (FC1 = 1.13 and intraset structure coefficient = 0.92; Table 23.5; Figure 23.2). A graphical display showed little separation of most- and least-disturbed sites in the space defined by CAN1, except for four left-most sites (mean of CAN1 scores = −0.14 for least-disturbed sites and 0.13 for most-disturbed sites; Figure 23.2A).

TABLE 23.4
Spearman (r_s, for CP sites) and Pearson (r, for PD and MT sites) Correlations Between Taxonomic, Trophic, or Reproductive Metrics and Continuously Distributed Habitat Variables

	Coastal Plain (N = 12)[a]				Piedmont (N = 47)[b]				Mountain (N=45)[c]			
	Bankside Woody Cover	Instream Woody Cover	Maximum Depth	Watershed as Forest	Bankside Woody Cover	Instream Woody Cover	Maximum Depth	Watershed as Forest	Bankside Woody Cover	Instream Cover	Maximum Depth	Watershed as Forest
NATSP			0.39		−0.31	−0.32						
NONNATSP						−0.31		−0.32				
MINSP		−0.53			−0.54	−0.43			0.38			
SUNSP		0.42	0.40	−0.40								
DARSCLSP						−0.33						
GENPRP				−0.39		−0.43		−0.39				
INVPRP		0.42	−0.34							0.40		
BINVPRP	−0.37	−0.36		0.32								
CARNPRP			0.38		0.34	0.39	0.31		−0.36			−0.31
TOLPRP				0.67				−0.35				
SLITHPRP		—	—	—					0.46	0.37		
LITHPRP		—	—	—	−0.30							
VMANPRP	0.48	0.45			0.35				0.41			
AGE3SP	−0.40	−0.37										

[a] If $|r_s| = 0.67$, P = 0.02
if $0.53 \geq |r_s| > 0.39$, $0.08 \leq P < 0.20$
if $0.39 \geq |r_s| > 0.30$, $0.20 \leq P < 0.33$

[b] If $0.55 > |r| > 0.45$, $0.0001 \leq P < 0.001$
if $0.45 \geq |r| > 0.38$, $0.001 < P < 0.01$
if $0.38 \geq |r| > 0.30$, $0.01 < P < 0.05$

[c] If $0.46 \geq |r| > 0.37$, $0.001 < P < 0.01$
if $0.37 \geq |r| > 0.30$, $0.01 < P < 0.05$

Note: Number of sites = N. Only absolute values (e.g., |r|) of r_s or $r > 0.30$ are shown. Dashes (—) denote that a fish metric was not included in analyses for the particular physiography. See Table 23.3 for fish-metric codes. For Piedmont correlations, two sites with extremely outlying fish-metric values were excluded: one for MINSP and one for GENPRP and TOLPRP. Similarly, two Mountain sites with extreme DARSCLSP were excluded.

TABLE 23.5
Summary of Canonical Correlations (CC) of Selected Fish Metrics, for Piedmont or for Mountain Sites

Piedmont (N = 47)

	DF	W	F	P	r_c	RED	TOTAL
CAN1	28,131	0.26	2.11	0.0026	0.66	0.152	0.236
CAN2	18,105	0.47	1.81	0.0328	0.59	0.061	
CAN3	1,076	0.71	1.42	0.1884	0.48	0.016	

	FC1	FC2	FC3	FSH1	FSH2	FSH3	HAB1	HAB2	HAB3	COM1	COM2	COM3
NATSP	−0.13	−0.36	−0.45	0.67	−0.01	−0.04	0.45	−0.01	−0.02	0.20	0.20	0.20
MINSP	1.13	0.47	0.22	0.92	−0.28	0.14	0.61	−0.17	0.07	0.37	0.40	0.40
GENPRP	−0.19	−0.60	0.31	0.36	−0.78	0.22	0.24	−0.46	0.11	0.06	0.27	0.28
CARNPRP	−0.10	0.53	0.96	−0.50	0.48	0.36	−0.33	0.28	0.18	0.11	0.19	0.22
TOLPRP	−0.48	−0.81	0.53	−0.49	−0.54	0.18	−0.32	−0.32	0.09	0.10	0.21	0.21
VMANPRP	0.50	0.32	−0.79	−0.56	−0.12	−0.21	−0.37	−0.07	−0.10	0.14	0.14	0.15
AGE3SP	0.24	0.16	0.62	0.46	0.06	0.44	0.30	0.03	0.21	0.09	0.09	0.14
Bankside w. cover	−0.82	−0.20	−0.31				−0.89	0.02	−0.05			
Instream w. cover	−0.45	0.48	0.30				−0.53	0.71	0.26			
Maximum depth	0.29	0.27	0.77				−0.04	0.37	0.79			
Wtrshd. as forest	0.28	0.68	−0.60				0.14	0.82	−0.51			

Mountain (N = 45)

	DF	W	F	P	r_c	RED	TOTAL
CAN1	32,123	0.17	2.33	0.0005	0.74	0.111	0.216
CAN2	2,198	0.39	1.82	0.0263	0.65	0.071	
CAN3	1,270	0.68	1.26	0.2597	0.50	0.022	

MINSP	0.27	0.76	-0.45	-0.19	0.54	-0.01	-0.14	0.35	0.00	0.02	0.14	0.14
DARSCLSP	-0.32	-0.67	0.70	-0.11	-0.23	0.53	-0.09	-0.15	0.27	0.01	0.03	0.10
BINVPRP	0.59	0.23	-0.04	0.73	-0.11	-0.06	0.54	-0.07	0.00	0.30	0.30	0.30
CARNPRP	0.01	0.08	-0.54	0.42	-0.02	-0.14	0.31	-0.01	-0.07	0.10	0.10	0.10
TOLPRP	-0.74	-0.67	-0.85	0.18	-0.55	-0.49	0.14	-0.36	-0.25	0.02	0.15	0.21
LITHPRP	-0.49	0.09	-0.66	-0.53	0.54	0.10	-0.40	0.35	0.05	0.16	0.28	0.28
VMANPRP	0.52	0.07	0.24	0.69	-0.11	0.03	0.51	-0.07	0.02	0.26	0.27	0.27
AGE3SP	-0.39	0.08	0.49	-0.23	0.61	0.39	-0.17	0.40	0.20	0.03	0.19	0.23
Bankside w. cover	-0.33	0.92	-0.16				-0.37	0.92	-0.09			
Instream cover	0.86	0.29	-0.48				0.75	0.14	-0.64			
Maximum depth	0.33	0.14	0.48				0.33	0.01	0.61			
Wtrshd. as forest	0.47	0.28	0.56				0.26	0.38	0.68			

Note: "CAN1" to "CAN3" are the first through third CCs; "FSH1" to "FSH3" and "HAB1" to "HAB3" are the fish and habitat composites, respectively, of each CC. For each CC, redundancy (RED) or total redundancy (TOTAL) is the variance in fish metrics accounted for by the habitat composite or by all (= 4) of the habitat composites, respectively. Wilks' lambda (W), its F-test approximation (F), statistical probability (P), degrees of freedom (DF = degrees for numerator, degrees for denominator), and canonical correlation coefficient (r_c) are shown for only those successive CCs that accounted cumulatively for > 0.90 of TOTAL. For individual fish metrics or habitat variables of each CC, "FC1" to "FC3" are the standardized function coefficients; numbers under composite names are the standardized intraset and interset (for fish metrics only) structure coefficients for each CC. For fish metrics, "COM1" to "COM3" (= communality) are the accumulated amounts of each fish metric's variance accounted for by each successive habitat composite. Number of sites is "N." Two Piedmont sites with extremely outlying metric values were excluded: one site for GENPRP and TOLPRP, and one for MINSP. Similarly, two mountain sites with extreme DARSCLSP were excluded. See Table 23.3 for fish-metric codes. "Wtrshd." = Watershed and "w." = woody.

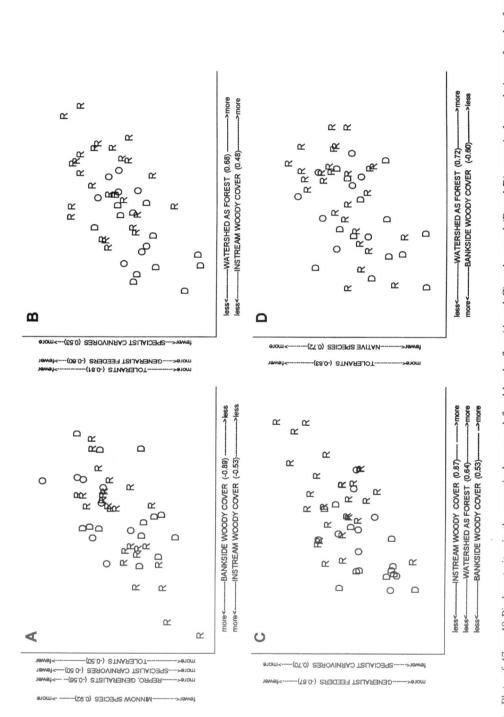

FIGURE 23.2 Plots of 47 or 48 Piedmont sites in the canonical spaces defined by the first (A and C) and second (B and D) canonical correlation analyses of fish metrics vs. four disturbance measures. Least- (R), most- (D), and moderately (O) disturbed sites are shown. Plots C and D (N = 48) are for an analysis that excluded number of native minnow species and number of late-maturing species, two metrics shown to be nonmonotonically related with disturbance. Intraset structure coefficients are shown after each metric or disturbance measure name. Vertical- and horizontal-axis scales are identical among plots.

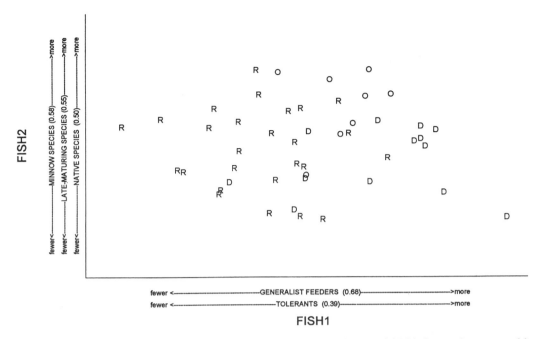

FIGURE 23.3 Plot of a descriptive discriminant (i.e., canonical) analysis of 47 Piedmont sites grouped by disturbance. Least- (R), most- (D), and moderately (O) disturbed sites are shown in the space defined by the first (FISH1) and second (FISH2) canonical composites of the fish metrics. Intraset structure coefficients are shown after each metric name. See Table 23.6 for summary of analysis. Vertical and horizontal axes span equal canonical distances.

Relations revealed by CAN2 were much more consistent with IBI assumptions than were those of CAN1; however, they accounted for much less fish metric variance (RED = 0.06 for CAN2 vs. RED = 0.15 for CAN1; Table 23.5). For CAN2, less-disturbed sites (i.e., those in more forested watersheds and with more instream woody cover) had fewer trophic generalists, fewer tolerants, and more trophic specialists (as carnivores) than did more-disturbed sites. Most-disturbed sites (mean of CAN2 scores = –0.83) and least-disturbed sites (mean = 0.47) were largely separate in the space of CAN2, and their location was consistent with how the habitat composite reflected disturbance (Figure 23.2B).

Discriminant-analysis and MANOVA results were similar to those of COR, except emphases of the first two canonical relations were reversed. Sites were separated along the first canonical composite of the discriminant analysis (i.e., FISH1 in Figure 23.3 and Table 23.6), similar to their separation depicted by CAN2 of the COR: most-disturbed sites had more generalist feeders and more tolerants than did least-disturbed sites and tended to group in canonical space accordingly (Figures 23.2 and 23.3; see Table 23.6 for group means along FISH1 and FISH2). Discriminant-analysis, MANOVA, and univariate results showed that number of native minnow species and number of late-maturing species were greatest at moderately disturbed sites and thus unimodally, rather than monotonically, related with disturbance (e.g., FISH2 in Figure 23.3; Table 23.6). A rerun COR with these two metrics removed accounted for nearly as much fish metric variance as did the original COR (sum of RED = 0.213; Table 23.6); and both CAN1 (RED = 0.122) and CAN2 (RED = 0.074) of the rerun COR were largely consistent with IBI assumptions. Specifically, fewer generalists and more specialist carnivores and more native species and fewer tolerants occurred at less-disturbed sites; that is, those with more bankside and instream cover and more forested watersheds (mean of CAN1 scores = 0.48 and mean of CAN2 scores = 0.23 for least-

TABLE 23.6
Summaries of (1) a Rerun Canonical Correlation Analysis (COR) and (2) a Descriptive Discriminant (i.e., canonical) Analysis of Selected Fish Metrics Among Disturbance Categories, for 48 or 47 Piedmont Sites, Respectively

					Canonical Correlation Analysis (N = 48)			
	DF	W	F	P	r_c	RED	TOTAL	
CAN1	20,130	0.36	2.37	0.0020	0.62	0.122	0.225	
CAN2	12,106	0.58	1.99	0.0317	0.49	0.074		
CAN3	682	0.77	1.90	0.0910	0.44	0.017		

	FC1	FC2	FC3	FSH1	FSH2	FSH3	HAB1	HAB2	HAB3	COM1	COM2	COM3
NATSP	-0.35	0.56	0.17	-0.45	0.72	0.28	-0.28	0.35	0.12	0.08	0.20	0.22
GENPRP	-0.48	-0.22	0.26	-0.87	-0.19	0.14	-0.54	-0.10	0.06	0.29	0.30	0.31
CARNPRP	0.47	-0.22	1.13	0.70	-0.27	0.35	0.44	-0.13	0.15	0.19	0.21	0.23
TOLPRP	-0.44	-0.85	0.48	-0.07	-0.83	-0.17	-0.04	-0.41	-0.07	0.00	0.17	0.17
VMANPRP	0.17	0.44	-1.35	0.34	-0.46	-0.44	0.21	-0.23	-0.19	0.05	0.10	0.13
Bankside w. cover	0.32	-0.56	-0.42				0.53	-0.60	-0.17			
Instream w. cover	0.63	-0.34	0.12				0.87	-0.20	0.21			
Maximum depth	0.02	0.18	0.95				0.29	-0.05	0.89			
Wtrshd. as forest	0.44	0.84	-0.25				0.64	0.72	-0.22			

					Discriminant Analysis (N = 47)		
					Least-	Moderately-	Most-
	DF	W	F	P	(N = 26)	(N = 8)	(N = 13)
FISH1	14,76	0.44	2.76	0.0024	-0.76	0.56	1.18
FISH2	639	0.79	1.70	0.1476	-0.10	1.04	-0.44

	FC1	FC2	FISH1	FISH2
NATSP	-0.49	-0.51	-0.06	0.50
MINSP	0.01	1.40	0.22	0.58
GENPRP	0.79	-0.51	0.68	0.03
CARNPRP	-0.06	-0.28	-0.21	-0.33
TOLPRP	0.57	-0.77	0.39	-0.40
VMANPRP	0.06	1.39	0.03	-0.27
AGE3SP	0.79	0.81	0.25	0.55

Note: For the COR, see Table 23.5 for explanations and definitions of terms; the analysis depicted here is identical to that in Table 23.5 except "MINSP" and "AGE3SP" were omitted, and no sites were excluded. For the discriminant analysis, "FISH1" and "FISH2" are canonical composites, of the fish metrics, that define the space in which 26 least-, 13 most-, and 8 moderately-disturbed sites are maximally separated (also see Figure 23.4). "FC1" and "FC2" are the standardized function coefficients as in COR, and numbers in columns labelled "FISH1" and "FISH2" are the intraset structure coefficients, which are Pearson coefficients for each fish metric vs. the first or second canonical-composite scores, respectively. Means of "FISH1" and "FISH2" for each group of sites are also shown under "Least-," "Moderately-," and "Most-." As in the original COR, two sites with extremely outlying fish metric values were excluded from the discriminant analysis: one site for GENPRP and TOLPRP, and one for MINSP. For the rerun COR only one site with extreme GENPRP and TOLPRP was excluded.

disturbed sites vs. CAN1 mean = –0.81 and CAN2 mean = –0.55 for most-disturbed sites; Figures 23.C, D; Table 23.6).

Examination of plots similar to those in Figure 23.2, but with PD sites depicted by drainage or by year of sample, revealed (except for CAN1 of the original COR) little separation or clustering of sites by drainage or by year, which suggested that effects of drainage or of year had little confounding influence on our interpretations. For CAN1 of the original COR (Figure 23.2A), the four left-most sites largely defined the overall correlation (i.e., fewer native minnows occurred at sites with more bankside cover). Bankside cover might be an unimportant "effect"; fewer native minnows at these sites could have resulted from a reservoir effect not accounted for in our disturbance measures. All four sites were in the Roanoke River drainage upstream of Kerr Reservoir, a major impoundment and barrier to stream fish dispersal (Jenkins and Burkhead, 1994). For small, nonvagile, predation-prone, obligately riverine species (e.g., many cyprinids), stream sites upstream of large reservoirs are less likely than unisolated sites to be recolonized or repopulated after local extirpations or declines in abundance (e.g., Winston et al., 1991).

Fish metrics differed little among discrete habitat categories. The most pronounced differences were for trophic metrics in the presence vs. absence of point sources (one-way MANCOVA, P/s = 0.24, P = 0.05). This result was largely due to sites with no point sources having the most tolerant individuals, contrary to IBI assumptions; however, this effect was indistinguishable from a stream-size effect (one-way MANCOVA, P/s = 0.28, P = 0.02; for point source × stream size). Sites with no point sources were smaller than those with point sources (Wilcoxon two-sample test, P = 0.1033), and proportion as tolerants tended to be greatest at smallest sites (r_s = –0.45, P = 0.0142 for 29 least-disturbed PD sites; Smogor and Angermeier, Chapter 10, Figure 10.2). Overall, after accounting for fish metric-vs.-stream size relations, metrics differed little among discrete habitat categories.

23.3.3 Relations Between Disturbance Measures and Metrics in Mountain

Fish metrics at MT sites were related moderately with the four continuously distributed disturbance measures. About one fifth of the variance in selected fish metrics could be attributed to and reasonably interpreted via these relations (sum of RED = 0.204, Table 23.5). Reproductive metrics were related more with habitat variables than they were at PD sites (COM2 = 0.19–0.28 for three reproductive metrics).

Relations revealed by CAN1 were not readily interpretable as consistent or not with IBI assumptions because the habitat composite did not depict a collective gradient of disturbance, but rather showed a mostly singular gradient of instream cover. Sites with much instream cover had more trophic specialists (as benthic invertivores and as carnivores), but also had fewer reproductive specialists (i.e., lithophils) and more reproductive generalists (Figure 23.4A; Table 23.5). Graphic display of sites in CAN1 space showed little separation of most-disturbed (CAN1 mean = –0.11) from least-disturbed sites (CAN2 mean = 0.23; Figure 23.4A).

Relations revealed by CAN2 were consistent with IBI assumptions. Less-disturbed sites (i.e., those with more bankside woody cover and [to a much lesser extent] more forested watersheds) had more native minnow and late-maturing species, more reproductive specialists, and fewer tolerants than did more-disturbed sites. In CAN2 space, least-disturbed sites (CAN2 mean = 0.57) were separate from most-disturbed sites (mean = –0.44), three of which comprised the lower left tail of the canonical correlation (Figure 23.4B). Unlike results for PD sites, MANOVA and discriminant analysis showed few differences in fish metrics among disturbance categories for MT sites (V/s = 0.17 and P = 0.5176 for omnibus MANOVA). However, some univariate results were consistent with CAN2 results: number of late-maturing species differed more among sites than any other metric and was greatest at least-disturbed and least at most-disturbed sites (R^2 = 0.14, F = 3.34, P = 0.0452; see Figure 23.4B).

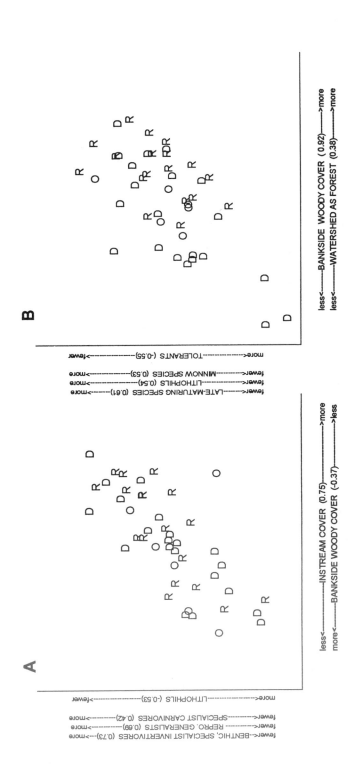

FIGURE 23.4 Plots of 45 Mountain sites in the canonical spaces defined by the first (A) and second (B) canonical correlations of a single canonical correlation analysis of fish metrics vs. four disturbance measures. Least- (R), most- (D), and moderately (O) disturbed sites are shown. Intraset structure coefficients are shown after each metric or disturbance measure name. Vertical- and horizontal-axis scales are identical among plots.

Plots similar to those in Figure 23.4, but with MT sites depicted by drainage or by year of sample, were examined. There was little separation or clustering of sites by drainage or by year, suggesting that effects of drainage or year did not confound our interpretations.

In MT, fish metrics were related slightly more with discrete habitat variables than in PD. However, similar to PD results, interpretations were confounded by stream-size effects. Most-pronounced differences were for reproductive metrics at sites in least- vs. most-urbanized watersheds and for taxonomic metrics at sites with the narrowest versus widest riparian zones (one-way MANCOVAs: V/s = 0.50, P = 0.0006 for urban effect; V/s = 0.49, P = 0.0051 for riparian effect). Two metrics dominated these results: proportion as lithophils was greatest at less-disturbed (i.e., less-urbanized) sites, and number of nonnative species was greatest at sites with the widest riparian zones, contrary to IBI assumptions. As mentioned above, these relations were dependent, in part, on relations between fish metrics and stream size as well (one-way MANCOVAs: V/s = 0.45, P = 0.0024 for urban × stream size; V/s = 0.48, P = 0.0065 for riparian × stream size).

23.3.4 RELATIONS AMONG DISTURBANCE MEASURES

Most of our interpretations of fish-vs.-habitat relations were not obfuscated by relations among disturbance measures. Compared with those of fish metrics, interrelations of habitat variables were fewer and easily interpretable. In each of CP, PD, and MT, the four continuously distributed habitat variables mostly were positively and weakly intercorrelated (all Pearson r > –0.20 and <0.31), suggesting that, when considered collectively, variables neither over- nor misrepresented disturbance. The presence of any strong negative correlations would have diminished the adequacy of these habitat variables as accordant measures of disturbance. Alternatively, the lack of strong positive intercorrelations suggested that each variable reflected distinct aspects of disturbance.

The continuously distributed habitat variables were weakly related with stream size (all Pearson r > –0.18 and <0.30) except that sites in larger watersheds in MT tended to be deeper (Pearson r = 0.43, P = 0.0031) and, in CP, had more instream woody cover (r_s = 0.62, P = 0.0303) than did other sites. Continuous habitat variables differed little among habitat categories as well (V/s < 0.21 for every one-way MANCOVA); therefore, these relations did not confound our interpretations. Moreover, the most-pronounced relations between continuous and discrete habitat variables reflected disturbance concordantly. Specifically, in PD and in MT, least-urbanized sites tended to have the most-forested watersheds and the most instream and bankside woody cover (V/s = 0.16, P = 0.0651 for variables among urban categories in PD; V/s = 0.20, P = 0.0285 for variables among urban categories in MT); although, these relations were not entirely independent of stream-size effects (one-way MANCOVAs, V/s = 0.37, P = 0.0513 for urban × stream size in PD; V/s = 0.17, P = 0.0751 for urban × stream size in MT). Overall, the few relatively weak relations among habitat variables did not confound our interpretations of fish-vs.-disturbance relations.

23.4 DISCUSSION

Despite the apparently small amounts of variance accounted for in these analyses (i.e., redundancies), they did reveal meaningful relations between fish metrics and disturbance measures. Large amounts of explained variation may be the exception for analyses of assemblage-level ecological relations at large geographic scales (e.g., Gittins, 1985). For COR specifically, Gittins (1985) showed that relations explaining as little as 0.10 of the shared variance between two sets of variables could be ecologically meaningful. To our knowledge, few ecological studies that have included COR have also provided measures of redundancy (i.e., amount of variance in one set of variables accounted for by the other set), and many have had p/n near or even greater than 1.00, thereby precluding meaningful or reliable interpretation (Gittins, 1985; ter Braak, 1987).

There is one COR-based study (on stream invertebrates) and one multivariate study of fish-assemblage structure that parallel our analyses; each of these studies evidence amounts of shared variance similar to those we found. For stream sites across a large area of northwestern North America, Corkum and Cibrowski (1988) reported redundancies of 0.13 to 0.27 for relations between taxonomic, morphobehavioral, and body-size measures of stream invertebrates and 10 environmental variables. For a statewide sample of stream sites in Arkansas, Matthews et al. (1992) found a Pearson correlation of 0.39 between a multivariate composite of fish-species abundances and one of water-quality measures, thus accounting for 0.15 of the variance shared by the composites. Based on principal components analysis, the results accounted for a maximum of 0.26 of the variance in any single water-quality variable, and a maximum of 0.10 of that in any single species' abundance. Although initially appearing to be low, amounts of explained variance in fish metrics and their canonical composites in this study are similar to those obtained in previous large-scale multivariate studies. Perhaps more importantly, the relations revealed by our analyses were readily interpretable and meaningful with respect to the questions asked of the data.

For CP sites, most relations between fish metrics and disturbance were contrary to typical IBI assumptions. Native minnow species, trophic specialists, and late-maturing species were fewest at least-disturbed sites; whereas, tolerants and reproductive generalists were greatest at least-disturbed sites. Typical IBI metrics and their assumed relations with anthropogenic disturbance are based largely on studies of non-lowland streams and may be inappropriate for CP streams.

Based on our experience in Virginia and North Carolina, the physical, chemical, and biotic features of CP flowages tend to differ widely among sites, while also remaining distinct from those of more-upland streams in PD and MT. Therefore, an IBI for CP streams probably would require metrics or metric scoring criteria different from those used in most previous versions of the IBI. For upland, warmwater streams, less-disturbed sites typically have more fish species than disturbed sites (Karr, 1981; Karr et al., 1985; Karr et al., 1986; Ohio EPA, 1988; Lyons, 1992); however, increased species richness may not necessarily reflect increased biotic integrity in CP streams. Prior to anthropogenic disturbance, many sites in Atlantic CP probably were dystrophic, low-flow blackwaters (Smock and Gilinsky, 1992) with low species richness compared to that in river mainstems or in more-upland streams. Typical disturbances in the CP — such as draining, forest-clearing, and fertilizing of watersheds — are likely to cause localized increases in instream productivity and fish-species richness, analogous to patterns observed for natural, less-speciose coldwater streams (e.g., Lyons, 1992).

For CP streams, there was little utility for metrics that included proportions as lithophils, a commonly used IBI-metric category (Miller et al., 1988; Simon and Lyons, 1995). Scott and Hall (1997) found that proportion as lithophils was a useful IBI metric for assessing disturbance in 1st- through 3rd-order Maryland CP streams; however, their CP streams had steeper gradients and a riffle-pool structure that resembled upland streams more than the typical low-gradient CP streams of the southern Atlantic slope. Relatively few CP streams in Virginia and North Carolina contain substantial amounts of mineral substrates coarser than sand; therefore, fishes requiring such substrates are extremely localized or rare. Similarly, taxonomic metrics that included numbers of sucker, darter, or sculpin species had limited utility because the richness of these taxa is naturally low in CP (Jenkins and Burkhead, 1994; personal observations). Additional basic research on how disturbance affects fish assemblages in CP streams is needed to allow one to choose definitive IBI metrics for the CP.

Despite their inter- and intraregion distinctiveness, CP streams remain relatively understudied; consequently, biologists understand less about the fish-vs.-disturbance relations in CP than in more-upland streams. Similar to the situation for CP streams, Bramblett and Fausch (1991) found that typically used IBI metrics (i.e., those developed for use in midwestern upland streams) inadequately reflected disturbance in western Great Plains streams. These setbacks should not be misconstrued as evidence of the inadequacies of the IBI. One major strength of an IBI-based approach is that the index can be modified to apply not only to each of a wide variety of fish assemblages (Simon and Lyons, 1995), but also to assemblages of other organisms (e.g., stream invertebrates; Ohio EPA, 1988; DeShon, 1995). As Simon and Lyons (1995) explain, although metrics and their scoring criteria differ among the many current versions of fish-based IBIs, all versions share the same multimetric process that has been proven useful in assessing how anthropogenic disturbance affects stream fishes and their habitats.

For PD and MT streams, fish metric-vs.-disturbance relations matched prevailing IBI metric assumptions more than for CP, probably because the fish assemblages of PD and MT streams more resembled those for which typical IBI metrics have been most fully validated. However, the present results may be attributable partly to fish-vs-disturbance relations having been more detectable in MT or PD than in CP; values of the four continuously distributed habitat variables ranged least in CP (Table 23.2). Alternatively, the habitat variables simply may have represented disturbance less adequately in CP than in PD or MT. The effects of anthropogenic disturbance (e.g., clearing of vegetation, nonpoint runoff, impoundment) have been studied little for CP, compared to more upland streams (Smock and Gilinsky, 1992); therefore, upland-based preconceptions of these effects may be misleading when applied to lowland streams.

Fish metrics reflected disturbance at PD sites differently from what they showed at MT sites. In PD, relations between disturbance measures and trophic metrics were most consistent with prevailing IBI assumptions; whereas, in MT, relations for reproductive and taxonomic metrics were most consistent. Only one metric was related with disturbance similarly in both PD and MT: tolerants were most abundant at most-disturbed sites in each region.

These differences in fish-vs.-disturbance relations across CP, PD, and MT strongly suggest that the utility of particular metrics will vary among IBI regions. Whereas previous researchers suggested the same, and then adapted versions of the IBI to particular regions (Miller et al., 1988; Simon and Lyons, 1995), very few have evidenced explicitly the abilities of individual metrics to reflect disturbance in particular regions. Although results presented here might suggest that only a few metrics would be needed for an adequate IBI in each region, limiting Virginia IBIs to only those metrics is not recommended. Collectively, IBI metrics should have a broad ecological basis to compensate for incomplete knowledge of how anthropogenic disturbance affects aquatic systems and for the inability to account completely for all types and ranges of past, present, and future disturbances.

Because of data limitations, the metrics found to be most related with disturbance measures likely represent an incomplete set of those that would be useful in each region. Highly disturbed sites were underrepresented in our samples, and disturbance measures provided a limited representation of the possible types of disturbances to stream fish assemblages. Karr (1991) discussed five major classes of environmental factors that can affect aquatic biota via changes due to anthropogenic disturbance of streams and watersheds: food (energy) source, water quality, habitat structure, flow regime, and biotic interactions; in this study, disturbance measures represented mostly habitat factors. Compared to our study, those using a more comprehensive and relevant set of disturbance measures, and including sites more representative of the entire range of disturbances, could provide better guidance for identifying especially effective sets of metrics.

In addition to enhanced disturbance measures and samples, an increased understanding of region-specific disturbance effects on fish assemblages would allow one to develop metrics with improved ability (i.e., effectiveness) to reflect those disturbances. For example, although species are classified as "tolerant" or "intolerant" for the entire state, tolerance classifications may improve metric effectiveness if determined by region. In fact, some species apparently "intolerant" of the typical disturbances in regions predominated by lower-gradient streams (e.g., northern hog sucker, rock bass, fantail darter, and blacknose dace in Illinois; see Bertrand et al., 1996) may justifiably be considered as moderately to highly "tolerant" in regions with typically higher-gradient streams (e.g., mountain regions of West Virginia and Virginia; Leonard and Orth, 1986; personal observation). Regional tolerance classifications may be further complicated by species sensitivity varying among particular types of disturbance in a region. For example, blacknose dace, although widespread and generally tolerant of typical habitat-related disturbances, appear to be sensitive to stream acidification (Baker and Christensen, 1991; Hall et al., 1994). Despite these complicating factors, if species or taxa are classified appropriately for tolerance by region, one could improve the effectiveness of taxonomic metrics for each region by omitting tolerant species or taxa from richness counts, thereby limiting potentially contrary information contained in some of the taxonomic metrics.

One can increase the effectiveness of trophic or reproductive metrics by explicitly classifying species according to functional (e.g., trophic, reproductive) roles that relate most to disturbance effects in each IBI region. Meeting this requirement limits the possibility that a metric will contribute superfluous, contrary, or overly redundant information to an IBI assessment. For example, for CP and for PD sites, two common, widespread darter species (tessellated darter and johnny darter) were classified as benthic, specialist invertivores; in contrast, they were classified (statewide) as tolerants and as reproductive generalists. Although "benthic, specialist invertivore" represents a specialized functional role, "invertivore" or "insectivore" may be still too general a category to accurately reflect disturbance across all relevant regions or stream types. These two darter species feed largely on midge larvae (Jenkins and Burkhead, 1994) and may benefit from moderate anthropogenic disturbance via increased midge abundance, a common indicator of degraded stream conditions (Ohio EPA, 1988; Berkman et al., 1986; Plafkin et al., 1989; Lenat and Crawford, 1994).

"Finer-tuned" functional categorizations of species, coupled with multivariate analyses, would likely provide further insights. For example, by accounting for contrary or excessively redundant interrelations among species' functional traits or functional IBI metrics, a canonical correlation analysis of fish-vs.-disturbance measures would help one develop suites of functional characteristics that could be used as IBI metrics. Potentially, these new metrics would more accurately and precisely reflect region-specific disturbance than currently available metrics, thereby improving the IBI's ability to reflect solely anthropogenic effects. Given adequate samples, a carefully chosen set of meaningful disturbance measures, and a set of metrics based on well-defined functional categorizations of species, such an approach could yield practical bioassessment tools such as Yoder and Rankin's (1995) "Biological Response Signatures," while also providing insights for conceptual models of fish and their relations with habitat (e.g., habitat-templet model; Townsend and Hildrew, 1994).

23.5 CONCLUSION

Relations between fish metrics and measures of anthropogenic disturbance differ across three physiographic regions in Virginia; the ability of region-specific IBIs to accurately and reliably reflect disturbance will depend on how well their metrics represent these relations. From an idealistic viewpoint, a current lack of information prohibits choosing a conclusively complete and well-validated set of metrics for each IBI region in Virginia. However, from a more practical viewpoint, conservation, management, and regulation of Virginia's stream resources would be enhanced by incorporating biotic assessment that uses somewhat imperfect, but nonetheless adequate IBIs. Further studies in Virginia should try to validate and expand on the metric-vs.-disturbance relations revealed in this study. New findings should be used to regularly reevaluate and refine taxonomic, trophic, reproductive, and tolerance classifications of species and to develop new metrics that improve the ability to detect region-specific disturbance effects. Until then, it is recommended that an IBI for each region in Virginia include at least two metrics from each of a taxonomic, trophic, and reproductive class of metrics, as well as a tolerance metric — at least for PD and MT IBIs.

For a CP IBI, no particular metrics are recommended because the results of this study provided little unequivocal information, given such a small sample of sites. However, less reliance should be placed on traditional IBI-metric expectations that are based on assumed monotonic relations between metrics and disturbance. Virginia's CP streams warrant special research attention because traditional IBI metrics and their assumptions appear least tenable there.

For PD sites, more reliance on trophic metrics than on others is warranted, especially because traditional expectations for some taxonomic metrics seem inappropriate for PD streams. Similar to the CP, high richness at some PD sites may not necessarily reflect high biotic integrity: number of native minnow species and number of native species tended to be highest at moderately disturbed PD sites. Current metrics that may be especially useful for a PD IBI are: proportion as generalist feeders, proportion as specialist carnivores, and proportion as tolerants. We also encourage development and consideration of alternative trophic categorizations (of species) that better reflect region-specific relations between trophic function and disturbance than do current metrics.

For MT sites, we recommend relying most on reproductive and some taxonomic metrics, especially proportion as lithophils, number of late-maturing species, and number of native minnow species. We also encourage the study of additional or alternative reproductive metrics; we expect that metrics developed to reflect how disturbance affects species' life-history traits (Balon, 1975; 1984; see Simon, Chapter 6) would be especially useful.

ACKNOWLEDGMENTS

The authors thank P. Braaten, L. Calliott, G. Galbreath, T. Hash, M. Kosala, P. Lookabaugh, D. Magoulick, B. Mueller, T. Richards, S. Smith, J. Tousignant, and C. Weiking for fieldwork assistance, and D. Orth, E. Smith, A. Dolloff, and two anonymous reviewers for helpful criticism. Fieldwork and some data analysis were funded by Federal Aid in Sport Fish Restoration grant F-86-R from the Virginia Department of Game and Inland Fisheries. Additional analysis was supported by funding from the U.S. Environmental Protection Agency.

REFERENCES

Angermeier, P. L. 1995. Ecological attributes of extinction-prone species: loss of freshwater fishes of Virginia, *Conservation Biology,* 9, 143–158.

Angermeier, P. L. and J. R. Karr. 1986. An index of biotic integrity based on stream fish communities: considerations in sampling and interpretation, *North American Journal of Fisheries Management,* 6, 418–429.

Angermeier, P. L. and I. J. Schlosser. 1987. Assessing biotic integrity of the fish community in a small, Illinois stream, *North American Journal of Fisheries Management,* 7, 331–338.

Angermeier, P. L. and R. A. Smogor. 1992. Development of a Geographic Information System for Stream and Fisheries Data in Virginia. Final report for Federal Aid Project F-86, Virginia Department of Game and Inland Fisheries.

Angermeier, P. L., R. A. Smogor, and S. D. Steele. 1991. An electric seine for collecting fish in streams, *North American Journal of Fisheries Management,* 11, 352–357.

Balon, E. K. 1975. Reproductive guilds of fishes: a proposal and definition, *Journal of the Fisheries Research Board of Canada,* 32, 821–864.

Balon, E. K. 1984. Patterns in the reproductive styles in fishes, in G. W. Potts and R. J. Wootton (Eds.), *Fish Reproduction: Strategies and Tactics.* Academic Press, New York. 35–54.

Baker, J. P. and S. W. Christensen. 1991. Effects of acidification on biological communities in aquatic ecosystems, in D. F. Charles (Ed.), *Acidic Deposition and Aquatic Ecosystems.* Springer-Verlag, New York. 83–106.

Berkman, H. E., C. F. Rabeni, and T. P. Boyle. 1986. Biomonitors of stream quality in agricultural areas: fish versus invertebrates, *Environmental Management,* 10, 413–419.

Bertrand, W. A., R. L. Hite, D. M. Day, W. Ettinger, W. Matsunaga, S. Kohler, J. Mick, and R. Schanzle. 1996. Biological Stream Characterization (BSC): Biological Assessment of Illinois Stream Quality through 1993. Illinois Environmental Protection Agency, Bureau of Water, IEPA/BOW/96-058.

Bramblett, R. G. and K. D. Fausch. 1991. Variable fish communities and the index of biotic integrity in a western Great Plains river, *Transactions of the American Fisheries Society,* 120, 752–769.

Corkum, L. D. and J. J. H. Cibrowski. 1988. Use of alternative classifications in studying broad-scale distributional patterns of lotic invertebrates, *Journal of the North American Benthological Society,* 7, 167–179.

Cramer, E. M. and W. A. Nicewander. 1979. Some symmetric, invariant measures of multivariate association, *Psychometrika,* 44, 43–54.

DeShon, J. E. 1995. Development and application of the Invertebrate Community Index (ICI), in W. S. Davis and T. P. Simon (Eds.), *Biological Assessment and Criteria: Tools for Water Resource Planning and Decision Making.* Lewis, Boca Raton, FL. 217–243.

Fausch, K. D., J. Lyons, J. R. Karr, and P. L. Angermeier. 1990. Fish communities as indicators of environmental degradation, *American Fisheries Society Symposium,* 8, 123–144.

Gauch, H. G. 1981. *Multivariate Analysis in Community Ecology.* Cambridge University Press, Cambridge, England.

Gittins, R. 1985. *Canonical Analysis. A Review with Applications in Ecology.* Springer-Verlag, New York.

Hall, L. W. Jr., S. A. Fischer, W. D. Killen, Jr., M. C. Scott, M. C. Ziegenfuss, and R. D. Anderson. 1994. Status assessment in acid-sensitive and non-acid-sensitive Maryland coastal plain streams using an integrated biological, chemical, physical, and land-use approach, *Journal of Aquatic Ecosystem Health,* 3, 145–167.

Huberty, C. J. 1986. Questions addressed by multivariate analysis of variance and discriminant analysis, *Georgia Educational Researcher,* 5, 47–60.

Hughes, R. M. and J. R. Gammon. 1987. Longitudinal changes in fish assemblages and water quality in the Willamette River, Oregon, *Transactions of the American Fisheries Society,* 116, 196–209.

Jenkins, R. E. and N. M. Burkhead. 1994. *The Freshwater Fishes of Virginia.* American Fisheries Society, Bethesda, MD.

Karr, J. R. 1981. Assessment of biotic integrity using fish communities, *Fisheries (Bethesda),* 6(6), 21–27.

Karr, J. R. 1991. Biological integrity: a long neglected aspect of water resource management, *Ecological Applications,* 1, 66–84.

Karr, J. R., L. A. Toth, and D. R. Dudley. 1985. Fish communities of midwestern rivers: a history of degradation, *BioScience,* 35, 90–95.

Karr, J. R., P. R. Yant, K. D. Fausch, and I. J. Schlosser. 1987. Spatial and temporal variability of the index of biotic integrity in three midwestern streams, *Transactions of the American Fisheries Society,* 116, 1–11.

Karr, J. R., K. D. Fausch, P. L. Angermeier, P. R. Yant, and I. J. Schlosser. 1986. Assessing Biological Integrity in Running Waters: A Method and Its Rationale. Illinois Natural History Survey Special Publication 5, Champaign.

Lenat, D. R. and J. K. Crawford. 1994. Effects of land use on water quality and aquatic biota of three North Carolina Piedmont streams, *Hydrobiologia,* 294, 185–199.

Leonard, P. M. and D. J. Orth. 1986. Application and testing of an index of biotic integrity in small, coolwater streams, *Transactions of the American Fisheries Society,* 115, 401–415.

Lyons, J. 1992. Using the Index of Biotic Integrity (IBI) to Measure Environmental Quality in Warmwater Streams of Wisconsin. General Technical Report, NC-149, U.S. Department of Agriculture, Forest Service, North Central Forest Experiment Station, St. Paul, MN.

Matthews, W. J., D. J. Hough, and H. W. Robison. 1992. Similarities in fish distribution and water quality patterns in streams of Arkansas: congruence of multivariate analyses, *Copeia,* 1992, 296–305.

Miller, D. L., P. M. Leonard, R. M. Hughes, J. R. Karr, P. B. Moyle, L. H. Schrader, B. A. Thompson,. R. A. Daniels, K. D. Fausch, G. A. Fitzhugh, J. R. Gammon, D. B. Haliwell, P. L. Angermeier, and D. J. Orth. 1988. Regional applications of an index of biotic integrity for use in water resource management, *Fisheries (Bethesda),* 13(5), 12–20.

Muncy, R. J., G. J. Atchison, R. V. Bulkley, B. Menzel, L.G. Perry, and R. C. Summerfelt. 1979. Effects of Suspended Solids and Sediment on Reproduction and Early Life of Warmwater Fishes: A Review. EPA-600/3-79-042. U.S. Environmental Protection Agency, Office of Research and Development, Environmental Research Laboratory, Corvallis OR.

Niemi, G.J., P. DeVore, N. Detenbeck, D. Taylor, A. Lima, and J. Pastor. 1990. Overview of case studies on recovery of aquatic systems from disturbance, *Environmental Management,* 14, 571–587.

Ohio Environmental Protection Agency (Ohio EPA). 1988. Biological Criteria for the Protection of Aquatic Life. Vol. II. Users Manual for Biological Field Assessment of Ohio Surface Waters. Ohio Environmental Protection Agency, Division of Water Quality Monitoring and Assessment, Surface Water Section, Columbus, Ohio.

Pflieger, W. L. 1975. *The Fishes of Missouri*. Missouri Department of Conservation, Jefferson City.

Plafkin, J. L., M. T. Barbour, K. D. Porter, S. K. Gross, and R. M. Hughes. 1989. Rapid Bioassessment Protocols for Use in Streams and Rivers: Benthic Macroinvertebrates and Fish. EPA/444/4-89/001. U.S. Environmental Protection Agency, Office of Water. Washington, DC.

Pratt, J. R. and N. J. Bowers. 1992. Variability of community metrics: detecting changes in structure and function, *Environmental Toxicology and Chemistry,* 11, 451–457.

Rabeni, C. F. and M. A. Smale. 1995. Effects of siltation on stream fishes and the potential mitigating role of the buffering riparian zone, *Hydrobiologia,* 303, 211–219.

Rankin, E. T. 1995. Habitat indices in water resource quality assessments, in W. S. Davis and T. P. Simon (Eds.), *Biological Assessment and Criteria: Tools for Water Resource Planning and Decision Making.* Lewis, Boca Raton, FL. 181–208.

Roth, N. E., J. David Allan, and D. L. Erickson. 1996. Landscape influences on stream biotic integrity assessed at multiple spatial scales, *Landscape Ecology,* 11, 141–156.

SAS. 1990. *SAS/STAT User's Guide*. Vols. 1 and 2. SAS Institute, Cary, NC.

Scott, M. C. and L. W. Hall, Jr. 1997. Fish assemblages as indicators of environmental degradation in Maryland coastal plain streams, *Transactions of the American Fisheries Society,* 126, 349–360.

Serlin, R. C. 1982. A multivariate measure of association based on the Pillai-Bartlett procedure, *Psychological Bulletin,* 91(2), 413–417.

Shields, F. D., Jr., S. S. Knight, and C. M. Cooper. 1995. Use of the index of biotic integrity to assess physical habitat degradation in warmwater streams, *Hydrobiologia,* 312, 191–208.

Simon, T. P. and J. Lyons. 1995. Application of the index of biotic integrity to evaluate water resource integrity in freshwater ecosystems, in W. S. Davis and T. P. Simon (Eds.), *Biological Assessment and Criteria: Tools for Water Resource Planning and Decision Making.* Lewis, Boca Raton, FL. 245–262.

Smock, L. A. and E. Gilinsky. 1992. Coastal Plain blackwater streams, in C. T. Hackney, S. M. Adams, and W. H. Martin (Eds.), *Biodiversity of the Southeastern United States: Aquatic Communities.* Wiley, New York. 271–313.

Smogor, R. A. 1996. Developing an Index of Biotic Integrity for Warmwater Wadeable Streams in Virginia. M.S. thesis. Department of Fisheries and Wildlife Sciences, Virginia Polytechnic Institute and State University, Blacksburg, VA.

Stauffer, J. C. and R. M. Goldstein. 1997. Comparison of three qualitative habitat indices and their applicability to prairie streams, *North American Journal of Fisheries Management,* 17, 348–361.

Steedman, R. J. 1988. Modification and assessment of an index of biotic integrity to quantify stream quality in southern Ontario, *Canadian Journal of Fisheries and Aquatic Sciences,* 45, 492–501.

ter Braak, C. J. F. 1987. Ordination, in R. H. Jongman, C. J. F. ter Braak, and O. F. R. van Tongeren (Eds.), *Data Analysis in Community and Landscape Ecology.* Pudoc, Wageningen, The Netherlands. 91–173.

Tabachnick, B. G. and L. S. Fidell. 1996. *Using Multivariate Statistics.* Harper and Row, New York.

Thompson, B. 1984. Canonical Correlation Analysis: Uses and Interpretation. Sage University Paper series on Quantitative Applications in the Social Sciences, 07-047. Sage Publications, Beverly Hills.

Thorndike, R. M. 1978. *Correlational Procedures for Research.* Gardner, New York.

Townsend, C. R. and A. G. Hildrew. 1994. Species traits in relation to a habitat templet for river systems, *Freshwater Biology,* 31, 265–275.

Virginia Department of Environmental Quality. 1994. Virginia Water Quality Assessment for 1994, 305(b) Report to EPA and Congress. Virginia Department of Environmental Quality Information Bulletin Number 597. Richmond.

Wang, L., J. Lyons, P. Kanehl, and R. Gatti. 1997. Influences of watershed land use on habitat quality and biotic integrity in Wisconsin streams, *Fisheries (Bethesda),* 22(6), 6–12.

Winston, M. R., C. M. Taylor, and J. Pigg. 1991. Upstream extirpation of four minnow species due to damming of a prairie stream, *Transactions of the American Fisheries Society,* 120, 98–105.

Williams, B. K. 1983. Some observations on the use of discriminant analysis in ecology, *Ecology,* 64, 1283–1291.

Yoder, C. O. and E. T. Rankin. 1995. Biological Response Signatures and the Area of Degradation Value: new tools for interpreting multimetric data, in W. S. Davis and T. P. Simon (Eds.), *Biological Assessment and Criteria: Tools for Water Resource Planning and Decision Making.* Lewis, Boca Raton, FL. 263–286.

24 Methods for Deriving Maximum Species Richness Lines and Other Threshold Relationships in Biological Field Data

Edward T. Rankin and Chris O. Yoder

CONTENTS

24.1 Introduction 611
24.2 Methods 612
    24.2.1 Data 612
    24.2.2 Calculations 612
24.3 Results and Discussion 613
    24.3.1 IBI Metrics: Total Species, Sucker Species 613
    24.3.2 Sensitive Species vs. the QHEI 614
    24.3.3 IBI vs. Dissolved Oxygen 615
    24.3.4 IBI vs. Total Recoverable Copper and Total Recoverable Cadmium 616
    24.3.5 General Considerations for Using 95th Percentile Regressions 618
24.4 Conclusion 620
References 621

24.1 INTRODUCTION

Scatterplots of environmental data often form a wedge shape where the upper border of the points is interpreted as representing an environmental threshold or limiting factor. The development of the IBI (Index of Biotic Integrity) for fishes (Karr, 1981; Karr et al., 1986) and other similar multimetric indices (Deshon, 1995) (e.g., Invertebrate Community Index, ICI), for example, has led to the use of what are termed "maximum species richness lines" (MSRLs) to derive scoring criteria for certain metrics of these indices. With the IBI, the MSRLs for total species represents the maximum number of species observed for streams of a given size (drainage area or stream order). Wedge-shaped relationships have also been described in models of fish standing stock vs. habitat variables (Terrell et al., 1996). Terrell at al. (1996) considered the wedge-shaped distributions to be "consistent with the hypothesis that habitat is capable of limiting populations and that numerous "other factors" often further limit populations below what habitat could support."

The interpretation of such wedge-shaped associations has important implications for resource agencies because they might have application (e.g., IBI calibration) to the protection and restoration of aquatic resources and are integral to many indices of ecological integrity used by such agencies. Much of the work of scientists charged with protecting or restoring aquatic resources is a quest to identify environmental factors that can affect or limit natural assemblages of organisms. The purpose of this chapter is to use an existing method of deriving such threshold relationships, summarized by Blackburn et al. (1992), to: (1) compare lines derived from this method to lines drawn by eye for IBI metrics, (2) describe and interpret threshold relationships for several different environmental datasets, and (3) examine the problems, limitations, and concerns when deriving and interpreting such threshold associations.

24.2 METHODS

24.2.1 DATA

The data used in these analyses are from the Ohio EPA intensive monitoring program for streams and rivers (Yoder, 1991; Yoder and Rankin, 1995). The biological data were collected between 1979 and 1995. Fish community data were collected with standardized towboat, longline, or boat-mounted pulsed-DC electrofishing gears used by the Ohio EPA in their stream monitoring program (Ohio EPA, 1989). MSRL plots were based on statewide "least impacted" reference sites from this database used to derive Ohio's numerical biological criteria (Yoder and Rankin, 1995; Ohio EPA, 1987). Also used were data on two IBI metrics from the Ohio EPA biological database (Ohio ECOS): species richness (which excludes exotic species) and native sucker species (Family Catostomidae) that have strong drainage area relationships. These metrics were used in Ohio EPA's deviation of the IBI for wadeable streams and had MSRLs drawn by eye (Yoder 1991, Ohio EPA 1987). Two other datasets from Ohio ECOS are associations between chemical variables (total recoverable copper, total recoverable cadmium, and dissolved oxygen) and the IBI. Chemistry data from individual grab samples were collected between 1982 and 1994 at or near fish assemblage sampling sites (mean values of the IBI), also sampled in the same year (15 June through 15 October). Water quality data (total recoverable copper, total recoverable cadmium, and dissolved oxygen) are used in the state monitoring program and were analyzed by the Ohio EPA water quality laboratory based on U.S. EPA-approved collection and analytical procedures (Ohio EPA, 1991). The toxicity of metals is affected by water hardness. Therefore, total recoverable copper and total recoverable cadmium data were standardized to a hardness of 300 using the hardness/toxicity relationship derived for Ohio's water quality standards for the chronic (outside mixing zone 30-day average) criteria as follows:

$$\text{Total recoverable copper } (\mu g/L): e(0.9231 * [\ln \text{hardness}] - 1.784)$$

$$\text{Total recoverable cadmium } (\mu g/L): e(0.7852 * [\ln \text{hardness}] - 3.283)$$

The fifth dataset, also from Ohio ECOS, is the number of sensitive fish species at sampling sites, plotted against a measure of habitat quality (the Qualitative Habitat Evaluation Index [QHEI]), a visual habitat index collected concurrently with fish samples (Ohio EPA, 1989; Rankin, 1989; Rankin, 1995). Sensitive fish species are those categorized as intolerant or moderately intolerant by the Ohio EPA (1987) and are listed in Appendix 24A.

24.2.2 CALCULATIONS

The 95th percentile regressions were calculated as described in Blackburn et al. (1992). This method involves dividing the x-axis into a number of equal-sized categories, and for each category taking the upper 10th percentile of y-values for each x-category. Ordinary least squares linear regressions were then calculated for this data. Blackburn et al. (1992) performed a simulation analysis for 2 to 100 x-axis categories, and plotted the upper bound slope vs. number of categories to decide how many classes to use. The effect of category number will vary with the density of data points along the upper wedge of points. For the datasets they used, Blackburn et al. (1992) found that the variability of the upper-bound slope estimates increased with fewer than five classes, but was relatively constant with more than five classes. They also found that the slope estimates were "reasonably accurate and similar" when less than 20 size classes were used. Blackburn et al. (1992) urged caution with datasets containing few data points because such datasets are more likely to include points away from the upper bound. In essence, there is a higher probability that for a given category, points that would define the upper bound would be "missing" in small datasets. In this chapter, regressions for 5, 10, and 15 categories were calculated. Where regressions were essentially the same, regressions for 10 categories are reported.

The number of data points used in these analyses ranged from 666 (sensitive species vs. QHEI) to 10,219 (total recoverable copper vs. IBI). All these datasets were much larger than the ambient data used by Blackburn et al. (1992) (range = 41–300 points), so they should not suffer from problems of small sample sizes. All of the MSRL of the metrics of Ohio EPA's IBI calibration were originally drawn by eye. Lines for these metrics were drawn so that 5% of the points were above the line and so that the slope: (1) followed the outer boundary of points, and (2) matched the known biological response of the metric with stream size (Ohio EPA, 1987). The range (of x-values) over which a line with a slope was drawn (as opposed to a horizontal scoring line) was estimated visually and also incorporated professional judgment; 95% regressions were run over this same range of data. Similar decisions were made for regressions with chemical variables. The implications of this for interpreting data are discussed later.

24.3 RESULTS AND DISCUSSION

24.3.1 IBI Metrics: Total Species, Sucker Species

Perhaps the best-known datasets that display wedge-shaped relationships are certain metrics of the IBI. The total number of species (excluding exotic species; Figure 24.1) captured in a sample vs. drainage area and the total number of sucker species vs. drainage area (Figure 24.2) from reference sites illustrate metrics with a sharp, distinct outer boundary. These distinct upper bounds make ecological sense because of the well-documented increase in species richness, both total and sucker species, with drainage size in warmwater streams (Rankin, 1989; Horwitz, 1978; Paller et al., 1994). Because of the distinct outer boundary and a large sample size, the statistical derivations of the 95% lines were insensitive to the number of categories (5 to 15), with all lines virtually identical (Figures 24.1 and 24.2).

For total species and sucker species, lines generated statistically were very similar to those drawn by eye (Figures 24.1 and 24.2). The 95% confidence interval around the slope of the statistically derived lines included the lines drawn by eye over the range of the metric used in the IBI scoring (Figure 24.3). Here, where sample sizes are ample and the outer bounds are distinct, statistically derived lines were essentially equivalent to lines derived by eye.

An additional 95% regression line was derived for total species from *all* data in the Ohio database, rather than just reference site data. This line was slightly outside the 95% confidence interval around the slope. At small drainage areas, the line was higher than the reference site lines; and at larger drainage areas, the line was lower than reference site lines. Drainage area is essentially a surrogate variable for flow; in some small streams, the added wastewater can result in the stream being functionally "larger"; and when impacts are negligible, more species may be present than predicted by drainage area alone. This larger database also contains more sites that have impaired aquatic life assemblages, some of which are caused by wastewater effluents. In larger streams, the larger number of degraded sites results in a lower 95th percentile. Therefore, the inclusion of a large number of nonreference sites could affect the 95% lines for IBI metrics. For this reason, the calibration of IBI metrics is done on least impacted reference sites (Karr et al., 1986; Yoder and Rankin, 1995).

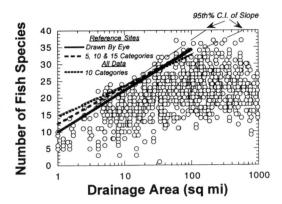

FIGURE 24.1 Total number of fish species (excluding exotic species) in a sample vs. drainage area (sq. mi.). Points represent least impacted reference sites from Ohio. Data shown as points collected from 1979 to 1989. Lines on graph are: drawn by eye (solid line); generated from 95th percentile regressions (based on 5, 10, and 15 x-axis categories) from reference sites (long, thick dashes); and generated from a 95th percentile regression (10 x-axis categories) of all data in Ohio EPA's database (short, thick dashes). A 95% confidence interval of the slope of the 10 x-axis category 95% regression from reference sites is indicated by a long, thin dashed line.

24.3.2 Sensitive Species vs. the QHEI

The maximum number of sensitive fish species in Ohio streams and rivers is strongly and clearly limited by physical habitat quality, as measured by the IBI (Figure 24.3). Sensitive species, as defined for Ohio streams and rivers by the Ohio EPA (1987), are those considered intolerant or moderately intolerant to a wide variety of environmental stresses for use in the IBI. A site with a QHEI near 80 can have as many as 17 sensitive fish species; whereas the maximum number of sensitive species expected with a QHEI of 40 is less than 7 (Figure 24.3). This type of statistical relationship can be useful for projecting likely limitations to fish communities from habitat degradation, and provides some insight into the ultimate restorability possible in high-quality streams. Clearly, a regression through the mean (thick solid line on Figure 24.3) might provide a very different interpretation of the effect of habitat on sensitive species.

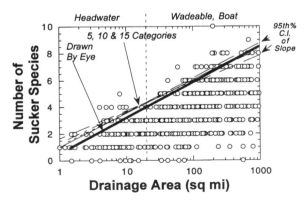

FIGURE 24.2 Total number of sucker species (Family Catostomidae) in a sample vs. drainage area (sq. mi.). Points represent least impacted reference sites from Ohio. Data shown as points collected from 1979 to 1989. Lines on graph are: drawn by eye (solid line); and generated from 95th percentile regressions (based on 5, 10, and 15 x-axis categories) from reference sites (long, thick dashes). A 95% confidence interval of the slope of the 10 x-axis category 95% regression from reference sites is indicated by a long, thin dashed line.

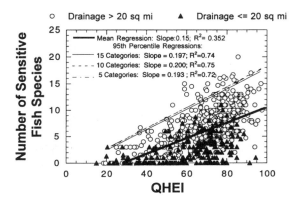

FIGURE 24.3 The number of sensitive fish species vs. QHEI. Points represent data from 1979 to 1994 where both fish data and habitat data were collected from the same site. Lines on graph are: generated from 95th percentile regressions based on 5, 10, and 15 x-axis categories (dashed lines); and a regression through the mean (solid line). Data are coded by drainage areas ≤ 20 sq. mi. (solid triangles) and drainage areas > 20 sq mi (open circles).

It is important to attempt to account for other variables known to affect aquatic life when interpreting 95% regression lines such as the QHEI/sensitive species association in Figure 24.3. Although the outer boundary of the points in Figure 24.3 clearly reflects a limitation of sensitive species by habitat quality, other parameters may be limiting sensitive species below this boundary. For example, it is reasonable to expect that number of sensitive species for a given QHEI may be affected by variables such as drainage area and stream gradient. On Figure 24.3, sites of less than 20 sq. mi. drainage area are denoted with a solid triangle. Obviously, there is a similarly distinct, but different outer boundary of points for streams with drainage areas less than 20 sq. mi. This indicates the threshold relationship defined by the 95th percentile plot from the entire dataset should not be applied to headwater streams, but rather a separate line should be derived.

24.3.3 IBI vs. Dissolved Oxygen

Dissolved oxygen (DO) is a well-known limiting parameter for fish in streams and rivers, and is one of the primary causes of aquatic life impairment in Ohio streams (Ohio EPA, 1997). DO criteria in Ohio's water quality standards criteria differ and are stratified by aquatic life use (see Figure 24.4). The most sensitive aquatic life use in Ohio is the Exceptional Warmwater Habitat stream classification (EWH, DO criteria 6.0 mg/L average; 5.0 mg/L minimum), which comprises about 10% of Ohio's perennial streams and rivers. Most streams (85%) fall into the Warmwater Habitat aquatic life use (WWH, DO criteria 5.0 mg/L average; 4.0 mg/L minimum), and a few physically modified streams, with lower aquatic potential, are classified as Modified Warmwater Habitat (MWH, DO criteria 4.0 mg/L average; 3.0 mg/L minimum) and Limited Resource Water (LRW, DO criteria 3.0 mg/L average; 2.0 mg/L minimum). A plot such as Figure 24.4 of the IBI vs. dissolved oxygen (from daytime grab samples) was used in the development of the Ohio dissolved oxygen criteria that are tiered by aquatic life use for Ohio's streams.

The range of x-axis values over which to run a regression can have effects on the slope and intercept of 95% lines. For DO, regressions were run for values between zero and 6.0 mg/L (Figure 24.4, dashed line) and for values between zero and 10.0 mg/L (Figure 24.4, solid line). These two ranges reflect values of dissolved oxygen that appear to have a slope greater than zero. The difference between these lines was relatively minor. The DO-axis is on a log scale, however, and the difference between a DO of 6 and 10 on such a scale is relatively modest; regressions for other data can be more affected by the choice of the x-axis range. An iterative process of calculating regressions over

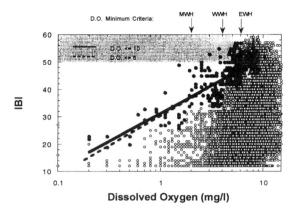

FIGURE 24.4 Index of Biotic Integrity vs. dissolved oxygen (mg/L). Points represent data from 1979 to 1994, where both dissolved oxygen data and fish data were collected from the same or nearby sites in the same year between June 15 and October 15. Lines on graph are: generated from 95th percentile regressions (based on 10 x-axis categories) using dissolved oxygen data ranging between 0 and 6 (dashed line); and generated from a 95th percentile regression (based on 10 x-axis categories) using dissolved oxygen data ranging between 0 and 10 (solid line). Minimum dissolved oxygen criteria for Ohio aquatic life uses are denoted on the top of the graph. The shaded box represents an area of sites with IBI values ≥ 50 and dissolved oxygen concentrations ≤ 5.

various ranges of the x-axis tempered with ecological judgment is probably the best solution to this concern.

24.3.4 IBI vs. Total Recoverable Copper and Total Recoverable Cadmium

Toxic metals have been a major cause of impairment in Ohio streams and rivers although the total miles attributable to metals has declined significantly from 24.7 to 6.2% of impaired miles over the last 10 years (Ohio EPA, 1997). The maximum expected IBI score for a given total recoverable copper or total recoverable cadmium concentration declines with increasing concentrations of these metals (Figure 24.5), although the outer boundary of points shows more variability than do the IBI metric vs. drainage area plots. Factors such as hardness (incorporated in Figure 24.5 through standardization of metals to a hardness of 300), total dissolved solids, and pH can affect the toxicity of metals (Sorenson, 1991). Thus, measurement of the dissolved form of metal is thought to be more reflective of the toxic proportion than the total recoverable metal (used here). This may explain some variation observed in the outer bound of the relationship between the IBI and total recoverable copper and total recoverable cadmium. Nevertheless, samples with high concentrations of total recoverable copper and total recoverable cadmium are more frequently associated with lower IBI values and less frequently associated with high IBI values (total recoverable copper: $X^2 = 402$, $P < 0.0001$; total recoverable cadmium: $X^2 = 435$, $P < 0.0001$; Table 24.1).

Although the upper boundary of the wedge of points in the plots made of total recoverable metals was variable relative to IBI metrics, this data is still useful in examining whether chemical criteria appear to offer adequate protection under ambient conditions. An alternate approach to a 95% regression line in situations with variable outer bounds is to divide the dependent variable (i.e., IBI) into meaningful ranges of biological condition and to examine the upper percentiles of the chemical variables (e.g., with a box and whisker plot) within these condition ranges. This information can be used to examine whether there are ranges of a chemical above which few sites achieve a specified range of biological condition. Such an exercise does not provide a definitive "safe" limit, but does indicate concentrations that have been repeatedly observed to be limiting to

TABLE 24.1
Two-Way Chi-Square Test of Independence for Total Recoverable Copper (μg/L) and Total Recoverable Cadmium and the IBI in Ohio Streams from 1982–1994

IBI Range	Total Recoverable Copper (μg/L) Standardized to 300 Hardness				
	<10	10–24.9	25–49.9	50–99.9	≥100
50–60	735 (677)	76 (113)	14 (24)	0 (7.3)	1 (5.4)
40–49	2006 (1881)	246 (314)	29 (66)	9 (20.2)	6 (15.1)
30–39	2276 (2194)	322 (366)	53 (77)	22 (23.6)	5 (17.6)
20–29	2348 (2421)	464 (404)	108 (85)	16 (26.0)	19 (19.4)
12–19	1006 (1199)	289 (200)	90 (42)	43 (12.9)	36 (9.6)

$X^2 = 402, P < 0.0001$

IBI Range	Total Recoverable Cadmium (μg/L) Standardized to 300 Hardness				
	<0.5	0.5–0.99	1.0–4.99	5.0–9.99	≥10
50–60	718 (690)	14 (22.4)	5 (20.8)	0 (2.0)	0 (1.4)
40–49	1929 (1858)	34 (60.4)	16 (55.9)	4 (5.4)	0 (3.8)
30–39	2225 (2163)	42 (70.3)	32 (65.1)	2 (6.2)	8 (4.4)
20–29	2422 (2436)	96 (79.2)	77 (73.3)	4 (7.0)	2 (5.0)
12–19	1013 (1160)	84 (37.7)	120 (34.9)	14 (3.4)	7 (2.4)

$X^2 = 435, P < 0.0001$

Note: Numbers reflect observed frequencies and expected values (in parentheses).

aquatic life. Figure 24.6 illustrates this analysis, as box and whisker plots of the upper 10% of total recoverable copper and total recoverable cadmium by ranges of the IBI.

The need to examine whether total recoverable copper or total recoverable cadmium can affect aquatic life under ambient conditions is especially important with the decision by the U.S. EPA to allow NPDES permit limits to be derived by translating dissolved metals-based criteria into total recoverable limits. This process assumes the dissolved form of these metals is the primary toxic form of the metal, and relies on the derivation of "translators," derived regionally or site specifically. Because of the potential for seasonal and spatial variation in the translators and uncertainty in the fate or long-term toxicity of total recoverable metals under ambient conditions, it is important to examine whether total recoverable metals are consistently associated with degraded aquatic life, especially where translators could result in higher loadings of total recoverable metals, particularly in the absence of other safeguards such as sediment criteria or biological criteria.

For both total recoverable copper (Figure 24.6, upper panel) and total recoverable cadmium (Figure 24.6, lower panel), sites with higher biological quality have fewer high total recoverable metals values than sites with low biological quality. One aspect of these plots is that the higher total recoverable copper concentrations observed at IBI values greater than 40 are near the existing Ohio EPA and Great Lakes Water Quality Initiative (GLWQI) total recoverable copper criteria (at 300 hardness). In contrast, higher total recoverable cadmium concentrations observed at IBI values greater than 40 are well below the existing Ohio EPA criteria, and even further from the proposed GLWQI criterion. We have observed very few sites that have biological values meeting WWH (IBIs > 40) and total recoverable cadmium values near the existing Ohio EPA criteria, and even fewer near the proposed GLWQI criteria. High total recoverable cadmium values are more frequently associated with degraded communities that fail to attain the Ohio biocriteria. At a minimum, this

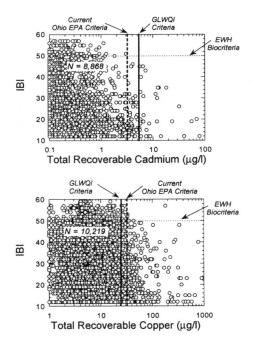

FIGURE 24.5 Scatter plots of IBI vs. total recoverable cadmium (top panel) and total recoverable copper (bottom panel) standardized to a hardness of 300 on the basis of Ohio EPA's outside mixing zone, 30-day average criteria equations. Points represent data from 1979 to 1994 where both water chemistry data (individual grab samples) and fish data (means for a site) were collected from the same or nearby sites in the same year between June 15 and October 15. Existing Ohio EPA outside mixing zone, 30-day average criteria at 300 hardness (dashed vertical line) and proposed GLWQI criteria at 300 hardness (solid vertical line) are denoted on each plot.

analysis should focus monitoring resources on streams where site-specific or regional translators would result in ambient total recoverable copper or total recoverable cadmium concentrations that will be near or above concentrations associated with lower than expected IBI values to ensure that discharge limits are protective. This is especially important because of the lack of sediment criteria for aquatic life. There is a great deal that is unknown about the toxicity of heavy metals and other toxic compounds in ambient sediments. It seems clear, for example, that the toxicity of metals in the sediments is related to the concentration of acid-volatile sulfide (AVS). The presence of AVS, however, may be affected by factors such as seasonality and the burrowing activities of certain aquatic invertebrates (Peterson et al., 1996). Given the uncertainties of the fate and toxicity of many toxic compounds, it seems reasonable to rely on indications of toxic effects revealed by the ambient aquatic biota as reality checks on the concentrations and forms of compounds discharged to streams.

24.3.5 GENERAL CONSIDERATIONS FOR USING 95TH PERCENTILE REGRESSIONS

A major consideration when calculating a 95th percentile regression is choosing the range of the independent variable. In this study, we used our biological interpretation of the relationship and a visual inspection of the data were used to make these decisions. In other cases, we ran several regressions varying the range of the independent variable to determine the effect of these choices. For IBI metrics, certain species groups or guilds may essentially disappear or be depauperate at certain drainage sizes (e.g., sunfish and suckers in small streams), or they may plateau at larger stream drainages. Although sequential 95% regressions over short ranges of the x-axis could be

Methods for Deriving Maximum Species Richness Lines and Other Threshold Relationships 619

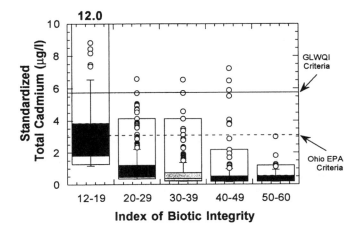

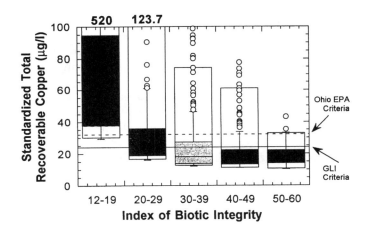

FIGURE 24.6 Box and whisker plots of the upper 10th percentile of total recoverable cadmium (top panel) and total recoverable copper (bottom panel) by ranges of the IBI. These metals are standardized to a hardness of 300 on the basis of Ohio EPA's outside mixing zone, 30-day average criteria equations. Points represent data from 1979 to 1994, where both water chemistry data (individual grab samples) and fish data (means for a site) were collected from the same or nearby sites in the same year between June 15 and October 15. Existing Ohio EPA outside mixing zone 30-day average criteria at 300 hardness (dashed horizontal line) and proposed GLWQI criteria at 300 hardness (solid horizontal line) are denoted on each plot.

calculated, sample sizes will typically be insufficient to provide a better estimate than can be had by incorporating professional judgment and visually selecting the range of the independent variable.

The 95th percentile line is considered an empirical estimate of an upper threshold effect of an independent parameter (e.g., copper, dissolved oxygen, stream size) on some dependent parameter (e.g., IBI, sensitive species). Several factors may affect this interpretation. The interpretation of a 95% percent line, for example, may vary with the distribution of the underlying data. The IBI values for given categories of an independent value can be significantly skewed from normal. The IBI at various ranges of dissolved oxygen used in the derivation of the 95% regression line of Figure 24.4 are significantly (Coefficient of Skewness, γ, $P < 0.01$) skewed, especially at lower concentrations of dissolved oxygen (Figure 24.7). Because of greater scatter or variability at the extremes of a given category in a skewed distribution, the 95% line would deviate further from a mean regression than a more normally distributed dataset. For the metals copper and cadmium,

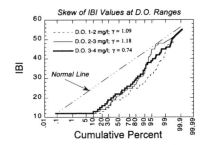

FIGURE 24.7 Normal probability plot of the IBI for three ranges of dissolved oxygen (mg/L) used in the derivation of 95% regression lines. Coefficients of skewness are denoted for each curve; all are signficant at $P < 0.01$. Straight lines illustrate a nonskewed, normal distribution.

95% regression lines do not ensure that concentrations below these lines are "safe"; rather, they provide information about when some criteria or other specified concentration may be unsafe, given the range of other impacts and natural background conditions that are part of the dataset. Although sites along the outer boundary may be strongly influenced by the level of the independent variable, sites below this boundary could be affected at some lower "level" of the independent variable. This was best illustrated for habitat vs. sensitive species (Figure 24.3), where the estimates of thresholds are improved by examining the data in light of other parameters that might also affect the response variable (e.g., stream size).

Another problem in using the 95% regression lines is that the accuracy of the lines decreases with decreasing sample size. Sample sizes for certain categories of an independent parameter may be low, usually at the extreme of an environmental parameter. For total copper (see Table 24.1), there are 8371 chemical samples with total copper values less than or equal to 10, but only 67 samples with total copper values greater than or equal to 100. Thus, the estimates at the extreme end of the regressions could be more variable. This also limits how many covarying factors can be used without reducing sample sizes too much that it affects the accuracy of regressions. Although the Ohio EPA has been monitoring water chemistry in streams for more than 20 years, many toxic parameters have too few "elevated" concentrations at sites with biological data to provide the resolution needed to permit calculation of 95% regression lines.

24.4 CONCLUSION

Although a line drawn by eye can inject a biologist's knowledge and experience into such a derivation, a line drawn with a statistical method could reduce potential bias in determining the slope or intercept of such a line. For the IBI metrics examined here, lines drawn by eye did not differ very much from statistically derived lines. It is likely, however, that there will be attempts to calibrate IBIs with much smaller databases than were used here. In these situations, statistically derived lines would be more likely to vary from lines drawn by eye. Where sample sizes are small, one would rely heavily on lines drawn by eye, largely because many species richness/stream size lines have been empirically derived and described from many locales, and this information can be useful for the determination of an accurate slope and intercept. One reasonable strategy, if data is sufficient, would be to derive a line or lines statistically and then adjust the slope of this line to improve accuracy based on what is known about the response of the metric. The IBI is actually very robust to the minor variations in the derivation of the 95% line when the eventual assessment biocriteria are derived objectively (e.g., using the 25th percentile of reference sites). For example, if 95% lines are drawn so that slightly more reference sites score values of 1 than 3, the 25th percentile of these sites will be slightly lower and the assessment of whether a site is impaired or not will not likely change significantly. Small sample sizes, however, will obviously result in more variation in a statistical or visual derviation of a 95% line. A visual derivation of such a line should not be a "solution" to a problem caused by an inadequate sample size.

REFERENCES

Blackburn, T. M., J. H. Lawton, and J. N. Perry. 1992. A method for estimating the slope of upper bounds of plots of body size and abundance in natural animal assemblages, *Oikos*, 65, 107–112.

DeShon, J. E. 1995. Development and application of the invertebrate community index (ICI), in W.S. Davis and T. P. Simon (Eds.), *Biological Assessment and Criteria: Tools for Water Resource Planning and Decision Making*. Lewis, Boca Raton, FL. 217–244.

Horwitz, R. J. 1978. Temporal variability patterns and the distributional patterns of stream fishes, *Ecological Monographs*, 48, 307–321.

Karr, J. R. 1981. Assessment of biotic integrity using fish communities, *Fisheries*, 6, 21–27.

Karr, J. R., K. D. Fausch, P. L. Angermier, P. R. Yant, and I. J. Schlosser. 1986. Assessing Biological Integrity in Running Waters: A Method and Its Rationale. Illinois Natural History Survey Special Publication No. 5, Champaign.

Ohio Environmental Protection Agency. 1989. *Biological Criteria for the Protection of Aquatic Life. Vol. III. Standardized Biological Field Sampling and Laboratory Methods for Assessing Fish and Macroinvertebrate Communities*. Division of Water Quality Monitoring and Assessment, Columbus, OH.

Ohio Environmental Protection Agency. 1987. *Biological Criteria for the Protection of Aquatic Life. Vol. II. Users Manual for Biological Field Assessment of Ohio Surface Waters*. Division of Water Quality Monitoring and Assessment, Surface Water Section, Columbus, OH.

Ohio Environmental Protection Agency. 1991. Ohio EPA Manual of Surveillance Methods and Quality Assurance Practices. Ohio EPA, Division of Environmental Services, Columbus, OH.

Paller, M. H. 1994. Relationships between fish assemblage structure and stream order in South Carolina coastal streams, *Transactions of the American Fishery Society*, 23, 150–161.

Peterson, G. S., G. T. Ankley, and E. N. Leonard. 1996. Effect of bioturbation on metal-sulfide oxidation in surficial freshwater sediments, *Environmental Toxicology and Chemistry*, 15, 2147–2155.

Rankin, E. T. 1989. The Qualitative Habitat Evaluation Index (QHEI), Rationale, Methods, and Application. Ohio Environmental Protection Agency, Division of Water Quality Planning and Assessment, Ecological Assessment Section, Columbus, OH.

Rankin, E. T. 1995. The qualitative habitat evaluation index (QHEI), in W.S. Davis and T. P. Simon (Eds.), *Biological Assessment and Criteria: Tools for Water Resource Planning and Decision Making*. Lewis, Boca Raton, FL. 181–208.

Sorenson, E. M. B. 1991. *Metal Poisoning in Fish*, CRC, Boca Raton, FL. Chap. 1.

Terrell, J. W., B. S. Cade, J. Carpenter, and J. M. Thompson. 1996. Modeling stream fish habitat limitations from wedge-shaped patterns of variation in standing stock, *Transactions of the American Fishery Society*, 125, 104–117.

Trautman, M. B. 1981. *Fishes of Ohio*. Ohio State University Press, Columbus.

Yoder, C. O. 1991. The integrated biosurvey as a tool for the evaluation of aquatic life use attainment and impairment in Ohio surface waters, *Biological Criteria: Research and Regulation, Proceedings of a Symposium*, December 12–13, 1990, Arlington, VA. EPA-440/5-91-005. U.S. Environmental Protection Agency, Office of Water, Washington, DC.

Yoder, C.O. and E.T. Rankin. 1995. Biological criteria program development and implementation in Ohio, in W.S. Davis and T. P. Simon (Eds.), *Biological Assessment and Criteria: Tools for Water Resource Planning and Decision Making*. Lewis, Boca Raton, FL. 109–144.

Appendix 24A
Ohio Fish Species Classified as "Sensitive"

These are a combination of Ohio EPA's intolerant (I) and moderately intolerant categories (M).

Ichthyomyzon fossor	North brook lamprey	I
Ichthyomyzon bdellium	Ohio lamprey	I
Ichthyomyzon greeleyi	Mountain brook lamprey	I
Lampetra appendix	American brook lamprey	I
Polyodon spathula	Paddlefish	I
Hiodon alosoides	Goldeye	I
Hiodon tergisus	Mooneye	I
Cycleptus elongatus	Blue sucker	I
Moxostoma anisurum	Silver redhorse	M
Moxostoma duquesnei	Black redhorse	I
Moxostoma erythrurum	Golden redhorse	M
Moxostoma macrolepidotum	Shorthead redhorse	M
Moxostoma valenciennesi	Greater redhorse	I
Moxostoma carinatum	River redhorse	I
Lagochila lacera	Harelip sucker	I
Hypentelium nigricans	Northern hog sucker	M
Nocomis biguttatus	Hornyhead chub	I
Nocomis micropogon	River chub	I
Notropis amblops	Bigeye chub	I
Erimystax dissimilis	Streamline chub	I
Erimystax x-punctata	Gravel chub	M
Macrhybopsis aestivalis	Speckled chub	I
Rhinichthys cataractae	Longnose dace	I
Exoglossum laurae	Tonguetied minnow	I
Clinostomus elongatus	Redside dace	I
Clinostomus funduloides	Rosyside dace	I
Opsopoeodus emiliae	Pugnose minnow	I
Notropis photogenis	Silver shiner	I
Notropis rubellus	Rosyface shiner	I
Lythrurus ardens	Rosefin shiner	M
Notropis heterodon	Blackchin shiner	I
Notropis boops	Bigeye shiner	I
Notropis stramineus	Sand shiner	M
Notropis volucellus	Mimic shiner	I
Notropis heterolepis	Blacknose shiner	I
Notropis anogenus	Pugnose shiner	I
Notropis ariommus	Popeye shiner	I
Notropis wickliffi	Channel shiner	I
Notropis amnis	Pallid shiner	I
Noturus flavus	Stonecat madtom	I
Noturus eleutherus	Mountain madtom	I

Noturus stigmosus	Northern madtom	I
Noturus trautmani	Scioto madtom	I
Noturus miurus	Brindled madtom	I
Fundulus diaphanus menona	Western banded killifish	I
Labidesthes sicculus	Brook silverside	M
Micropterus dolomieu	Smallmouth bass	M
Lepomis megalotis	Longear sunfish	M
Percina sciera sciera	Dusky darter	M
Percina macrocephata	Longhead darter	I
Percina phoxocephala	Slenderhead darter	I
Percina copelandi	Channel darter	I
Percina evides	Gilt darter	I
Percina caprodes	Logperch	M
Ammocrypta asprella	Crystal darter	I
Ammocrypta pellucida	Eastern sand darter	I
Etheostoma blennioides	Greenside darter	M
Etheostoma zonale	Banded darter	I
Etheostoma variatum	Variegate darter	I
Etheostoma maculatum	Spotted darter	I
Etheostoma camurum	Bluebreast darter	I
Etheostoma tippencanoe	Tippecanoe darter	I
Etheostoma caeruleum	Rainbow darter	M

25 Adjustments to the Index of Biotic Integrity: A Summary of Ohio Experiences and Some Suggested Modifications

Edward T. Rankin and Chris O. Yoder

CONTENTS

25.1 Introduction 625
25.2 Background 626
    25.2.1 Low-End Adjustments to the IBI 626
    25.2.2 Other Options for IBI Metric Scoring Adjustments 628
25.3 Results and Discussion 629
    25.3.1 Zero Scoring 629
    25.3.2 Predominant Species Swamping 633
    25.3.3 Affects of Young-of-Year 634
25.4 Conclusion 635
References 636

25.1 INTRODUCTION

The Index of Biotic Integrity (IBI; Karr, 1981; Fausch et al., 1984) has been used as a robust measure of the biological integrity of fish assemblages in the United States, Canada, and Europe (Karr et al., 1986; Miller et al., 1988; Steedman, 1988; Yoder, 1989; Karr, 1991; Oberdorff and Hughes, 1992; Barbour et al., 1995; Davis et al., 1996). The success of the IBI is partly due to the ability of users to adapt and calibrate the index to reflect regional conditions and expectations. Our application of modified IBIs to Ohio rivers and streams during the past 20 years has yielded a database of more than 15,000 samples. When properly derived, calibrated, and used, the IBI typically provides a reliable indication of biological integrity without the need for after-the-fact scoring adjustments. However, situations were encountered where the IBI required after-the-fact adjustments to more accurately reflect environmental quality. For example, when fish numbers are very low, the IBI becomes increasingly susceptible to yielding inflated scores that may not accurately reflect environmental quality. In other examples, the presence of a predominant species can overwhelm" certain metrics and diminish the contribution of the remainder of the sample to the composite IBI score. In yet other situations, predominance by young-of-year fish can result in distorted IBI scores (Angermeier and Karr, 1986). Although these situations have been relatively infrequent in Ohio (<3% of samples), the IBI may require after-the-fact adjustments in order to appropriately reflect the ambient environmental setting. This points up the need to recognize the role of trained biologists in using the IBI, a precaution emphasized by Karr et al. (1986).

While some of these influences can be addressed via quality control (e.g., not including young-of-year (YOY) fish less than 20 to 25 mm, elimination of selected predominant species; Ohio EPA 1989) or conditional statements (e.g, low-end scoring criteria; Ohio EPA, 1987), not every situation can be preempted in this manner. However, recognizing the need to make such adjustments may be paramount to an accurate assessment of a specific stream or river segment.

The purpose of this chapter is to: (1) describe the situations where IBI scores have been adjusted; and, (2) suggest additional modifications and criteria to further improve the responsiveness, sensitivity, and accuracy of the IBI.

25.2 BACKGROUND

The Ohio EPA has been collecting data on fish assemblages in Ohio rivers and streams since the late 1970s. In 1983 and 1984, the U.S. EPA funded a study of the chemical, physical, and biological attributes of the ecoregions in Ohio (Larsen et al., 1986; Whittier et al., 1987). The reference site data generated by this study, supplemented with other data collected during rotating basin surveys between 1980 and 1989, formed the basis of our derivation of modified IBIs applicable to headwater (<20 sq. mi. drainage area), wadeable, and boatable rivers and streams (Ohio EPA, 1987; 1989; Yoder, 1989; Yoder and Rankin, 1995a). The modified versions of the IBI generated from this work along with the Modified Index of Well-Being (MIwb; Gammon, 1976; Gammon et al., 1981; Ohio EPA, 1987) and the Invertebrate Community Index (ICI; Ohio EPA, 1987; DeShon, 1995) form the basis for numerical biological criteria (Yoder and Rankin, 1995a) that are a part of the Ohio Water Quality Standards (WQS; Ohio Administrative Code 3745-1). The success of these efforts have recently led to the development of IBIs for Lake Erie near-shore waters (Thoma, Chapter 16) and for the Ohio River mainstem (Simon and Emery, 1995).

The application of the IBI in Ohio rivers and streams has been partly focused on assessing the effects of point source discharges on aquatic life, which during the late 1970s and early to mid-1980s were sometimes quite severe. In rivers and streams with fish communities that were highly degraded due to especially severe toxic impacts, the proportional metrics of the IBI did not always respond in a predictable manner when comparatively few fish were collected in a sample. Without adjustment, the resultant IBI scores seemed to be inflated and not reflective of the severely degraded environmental conditions. As a result, low-end scoring guidelines were developed to correct for such aberrant IBI scores (Ohio EPA, 1987). This is by far the most common reason for needing to make after-the-fact scoring adjustments.

25.2.1 LOW-END ADJUSTMENTS TO THE IBI

Two ranges of numeric abundance where there was a need to adjust the proportional IBI metrics were identified: (1) when total relative numbers (CPUE) are less than 200 (per 0.3 km in headwater and wadeable streams; per 1.0 km in boatable rivers; Ohio EPA, 1987), adjustments are based on a mix of quantitative and qualitative criteria (the latter requiring the use of best professional judgment); and, (2) when CPUE is less than 50 (<25 in headwater streams of less than 8 sq. mi. drainage area), adjustments are made on a default basis (the proportional metrics automatically receive a metric score of 1). Out of 15,390 samples collected between 1978 and 1996, after-the-fact adjustments to IBI metric scores were made in 351 (2.3%) samples.

Table 25.1 summarizes the mean relative number of individuals (CPUE) collected at regional reference sites that reflect least impacted conditions and generally correspond to attainment of the biological criteria (Yoder and Rankin, 1995a). These CPUEs are much higher than that at which after-the-fact metric scoring adjustments are necessary (i.e., less than 50 to 200). Thus, it is only at sites with extreme departures from regional reference conditions that such IBI scoring adjustments are made. Table 25.2 lists the qualitative criteria presently used to adjust the proportional metric scores at CPUEs greater than 50, but less than 200 (Ohio EPA, 1987).

TABLE 25.1
Mean Relative Number of Fish Collected per 0.3 km (headwater and wadeable sites) or per 1.0 km (boatable sites) at Regional Reference Sites in Ohio

Headwater Streams	Wadeable Streams	Boatable Streams
1468.6 (85.3)	1175.1 (39.1)	545.1 (18.8)

25.2.2 OTHER OPTIONS FOR IBI METRIC SCORING ADJUSTMENTS

In utilizing the IBI in the mid-1980s, the scoring problems that arose with low CPUEs were immediately recognized. It was also recognized that changing the scoring of metrics to include zero scores (in addition to the traditional 5-3-1 metric scoring hierarchy) might resolve some of these problems without the need to make after-the-fact scoring adjustments. However, to remain consistent with the original versions of the IBI, it was decided to adhere to the 12–60 scoring scale (Ohio EPA, 1987). The problem is illustrated by an example from a small northwestern Ohio stream, Evans Ditch. This stream is typical of nearly all small, headwater streams in the Huron/Erie Lake Plain (HELP) ecoregion of Ohio, in that it has been extensively modified for agricultural drainage purposes. As such, the original fish communities have been severely altered, as has the contemporary reference condition for this ecoregion. A fish sample collected in 1986 yielded an IBI score of 32, which meets the biological criterion for the Warmwater Habitat use designation in the HELP ecoregion. Upon closer examination, the sample met several of the metric scoring adjustment criteria in Table 25.2. The predominance of blackstriped topminnow (*Fundulus notatus*), which is one of the species listed in Table 25.2, skewed three of the proportional metric scores and resulted in an inflated IBI score compared to the results obtained at adjacent sites (Figure 25.1). Using the criteria in Table 25.2, the proportion of tolerant fishes, pioneering fishes, and insectivores metric scores were adjusted, and resulted in an IBI score of 24. Additionally, when the site immediately downstream from the point source discharge was zero scored, it more appropriately reflected the severe character of the impact. Both adjustments resulted in an assessment that was much more consistent with the severely degraded water quality and modified habitat conditions and longitudinal trend of IBI scores in Evans Ditch (Figure 25.1).

Subsequent evaluation ascertained whether zero scoring of metrics could improve the behavior of the IBI at sites with low CPUEs, while at the same time improving the low-end resolution of the IBI. Theoretically, the latter would be accomplished by extending the minimum IBI score from 12 to 0. Comparisons were made between low-end adjusted IBIs, unadjusted IBIs, and zero-scored IBIs for selected datasets. The zero scoring methods tested included: (1) scoring the species richness metrics (e.g., total species, darter species, sunfish species, etc.) as zero when no species of that metric were collected; and (2) scoring the proportional metrics to zero when the CPUE is less than 50 (<25 at headwater sites of less than 8 sq. mi.). Including zero metric scores theoretically extends the lower range of the IBI to zero, which would improve the classification of severely degraded stream and river segments.

25.3 RESULTS AND DISCUSSION

25.3.1 ZERO SCORING

The overall quality of streams and rivers in Ohio, as reflected by the IBI, has improved substantially over the past 15 to 20 years (Ohio EPA, 1997). The reduction and elimination of most of the severe impacts from point sources of chemical pollution have been documented (Ohio EPA, 1997). In some cases, this has resulted in the full attainment of the biological criteria for both fish and

TABLE 25.2
Guidelines Used by the Ohio EPA for Making "Low-End" Scoring Adjustments to IBI Metrics When Samples Include Fewer than 200 Individual Fish CPUE (all individuals including tolerants)

IBI Metric	Narrative Guidelines for Scoring Modifications
% Omnivores	For wading and boat sites, a metric score of "1" is assigned if the number of individuals is <50. For numbers between 50 and 200, a metric score of "1" can be assigned when: 　Species considered as generalist feeders predominate (i.e., comprise >50% by numbers in aggregate); this includes creek chub, blacknose dace, and green sunfish. 　A single species comprises >50% of the sample; this metric can be rescored on a recalculated % omnivores minus the predominant species. For headwaters sites of less than 8 sq. mi. drainage area, the number of individuals criterion decreases to <25.
% Insectivores	For wadeing and boat sites, a metric score of "1" is assigned if the total number of individuals is <50 (<25 for headwater sites <8 sq. mi. drainage area). For numbers between 50 and 200, a metric score of "1" can be assigned when: 　The sample is dominated (>50% of the sample) either individually or in aggregate by striped shiner, common shiner, spotfin shiner, green sunfish, blackstripe topminnow, and/or young-of-the-year (YOY) of any insectivorous species that can function as omnivores under certain conditions (Angermeier, 1985).
% Top Carnivores	For wading and boat sites, a metric score of "1" is assigned if the number of individuals is <50 (<25 for headwater sites at less than 8 sq. mi. drainage area). For boat sites, a metric score of "1" is assigned for samples with 50 to 200 individuals if the sample is exclusively YOY and/or juvenile top carnivores. For wading sites, a metric score of "1" can be assigned when: 　The top carnivores are predominately grass pickerel and/or YOY or juvenile largemouth bass.
% Simple Lithophils	For wading and boat sites, a metric score of "1" is assigned if the number of individuals is <50 (<25 for headwater sites at less than 8 sq. mi. drainage area). This is seldom contrary to the score prior to the adjustment.
% DELT Anomalies	For wading and boat sites, a metric score of "1" is assigned if the number of individuals is <50 (<25 for headwater sites at less than 8 sq. mi. drainage area). For numbers between 50 and 200, a metric score of "1" may be assigned when: 　Circumstances suggest that the frequency of DELT anomalies is underestimated or not representative due to low numbers; this may happen when the sample is predominately YOY fish which have not yet had time to "accrue" anomalies.
% Pioneering Sp.	For headwaters sites, a metric score of "1" is assigned if the number of individuals is <50 (<25 for headwater sites at less than 8 sq. mi. drainage area). For numbers between 50 and 200, a metric score of "1" can be assigned when: 　The sample is dominated (>50% of the sample) by a single species; the metric score is based on the proportion of pioneering species less this single predominant species.
% Round-bodied Suckers	No adjustments are necessary for this metric.

Note: Number of individuals means the number of fish per 0.3 km for wadeing and headwater sites and per 1.0 km for boat sites.

macroinvertebrates. This has also reduced the need to make after-the-fact adjustments to IBI scores using the existing low-end scoring procedures through time, as illustrated by Figure 25.2. However, some seriously polluted segments still exist, and it is in these areas that the need to make manual adjustments to IBI scores remains and where the potential benefits of zero scoring are realized.

Three of the major objectives of biological monitoring include: (1) identification of impaired conditions; (2) quantification of the degree of impairment; and (3) delineation of the associated causes and sources of impairment. Fulfilling these objectives at sites with few individual fish may

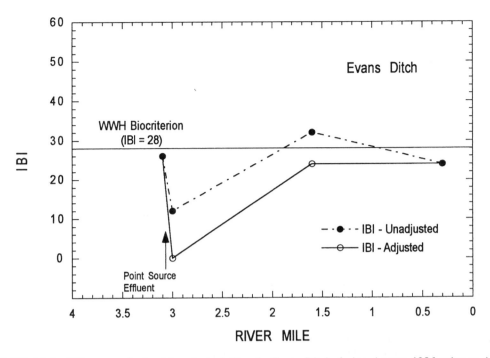

FIGURE 25.1 IBI scores obtained by electrofishing in Evans Ditch during August 1986 prior to (open circles) and following (closed circles) after-the-fact adjustments to the IBI scores using the criteria in Table 25.2.

be confounded by the unpredictable behavior of certain IBI metrics without making manual adjustments. Making such adjustments to IBI metric scores will not substantially improve the identification of impaired conditions (i.e., failure to attain the biological criteria) since unadjusted IBI scores rarely reach the values required to show full attainment of the biological criteria. However, improving the behavior of the IBI at highly degraded sites could improve the quantification of an impairment by extending the lower scoring range of the IBI and reducing the variance in IBI scores. This would also further enhance the characterization and diagnosis of associated causes and sources of impairment using the concept of biological response signatures (Yoder and Rankin, 1995b). It would also improve the ability to quantify incremental improvements in fish communities using tools such as the Area of Degradation Value (Yoder and Rankin, 1995b).

TABLE 25.3
Difference Between IBI Values Adjusted with Low-End Scoring Based on the Criteria in Table 25.2 and IBI Values Calculated with IBI_{z1}, IBI_{z2}, and IBI_{NO}

IBI_{z1}	IBI_{z2}	IBI_{NO}
1.46	−0.37	4.25

Note: IBI_{z1} = Species richness and proportional metrics scored to zero. IBI_{z2} = Species richness and proportional metrics scored to zero at sites with fewer than 50 individuals (<25 at headwater sites of less than 9 sq. mi.). IBI_{NO} = IBI values with no low-end scoring adjustments.

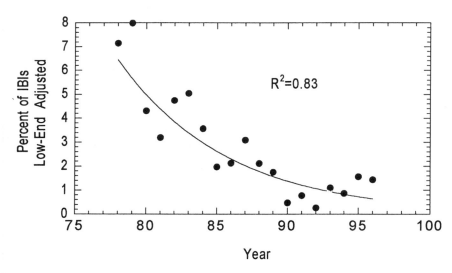

FIGURE 25.2 Proportion of Ohio EPA fish collections that required low-end, manual adjustments of proportional IBI metrics by year.

Data from the Ottawa River through Lima, Ohio, illustrates the improvement in the ability of the zero-scored IBI to more appropriately classify severely degraded sites (Figure 25.3). The upper panel of Figure 25.3 portrays the IBI as presently scored (solid line), and with zero scoring added (dashed line) from two sampling years, 1979 and 1996. Although data from both 1979 and 1996 indicate significant departures from the biological criterion, the impairment was much more severe in 1979. The magnitude and severity of the impairment observed in 1979 was further amplified by the zero-scoring approach in which minimum scores below 12 were the result. The differences between the currently used qualitative and zero-scoring IBI adjustments in the 1996 results were much less, and illustrate a reduction of the limiting conditions prevalent in 1979. This was further supported by the Modified Index of Well-Being (MIwb; lower panel, Figure 25.3) results and other chemical and pollutant-specific indicator information (Ohio EPA, 1992).

As illustrated in the Ottawa River example, the zero scoring of IBI metrics extends the lower scoring range of the IBI from 12 to 0 and improves the ability to illustrate increasingly severe degrees of impairment in the poor and very poor categories. As intended, this would have the greatest effect on IBIs less than 30 (Figure 25.4). However, zero scoring, as illustrated in Figure 25.3, does not necessarily address the scoring problems that occur at sites with CPUEs greater than 50 but less than 200. These still need to be addressed "manually" following the qualitative guidelines in Table 25.2. Figure 25.4 is a scatterplot of the current IBI (x-axis) vs. zero-scored IBIs (y-axis). Open circles are IBIs that did not require manual adjustments and points denoted as X are the 2.3% of points in our database that required manual adjustments for CPUEs greater than 50, but less than 200. Many of the zero-scored IBI samples still scored substantially higher than those IBI scores that were manually adjusted (X) following the guidelines in Table 25.2. Thus, zero scoring (i.e., for species richness metrics and proportional metrics at CPUE less than 50) alone did not compensate for the need to manually adjust the IBI using the qualitative criteria (Table 25.2).

Identification of stressors that are associated with impaired fish assemblages is an important benefit of biological monitoring. Habitat impacts, such as channelization, result in shifts in the species and trophic composition, but, except for total loss of habitat (e.g., dewatering, concrete channels) rarely result in the significant reduction of species or severe reductions in abundance (e.g., <200 CPUE). IBI scores for the most frequently encountered habitat impacts in Ohio are usually greater than 20 (Rankin, 1995). Most toxic chemical impacts have more severe effects (e.g., reductions in species richness, elimination of sensitive species, high incidence of deformities, eroded

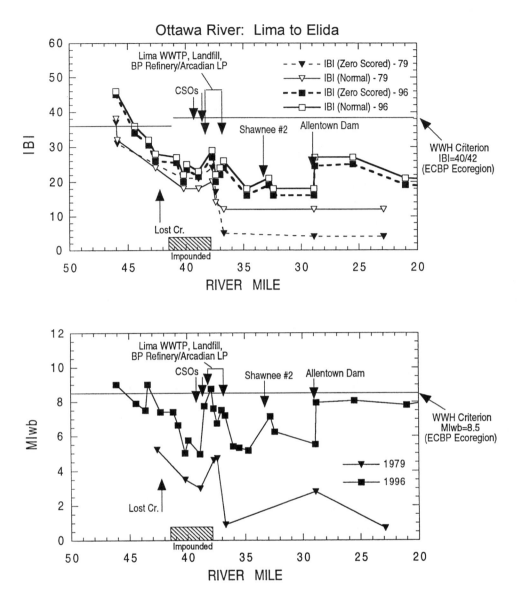

FIGURE 25.3 Upper Panel: A longitudinal plot of the IBI (zero scored, dash line/solid symbols; current scoring methods, solid line, open symbols) vs. river mile from the Ottawa River through Lima, Ohio. Triangles denote data from 1979, squares from 1996. Lower Panel: A longitudinal plot of the MIwb vs. river mile from the Ottawa River through Lima. Triangles denote data from 1979, squares from 1996.

fins, lesions, and tumors (DELT) etc.), than habitat impacts, but may have similar IBI scores in the 20s because the present 12–60 IBI lacks comparative resolution in the lower scoring ranges. Using the zero-scoring modifications explored in this study could add important resolution. While some IBI metrics (e.g., sensitive species, DELT anomalies) have been demonstrated to be individually diagnostic of toxic impacts (Anderson et al., 1991; Yoder and Rankin, 1995b), the diagnostic meaning of the composite IBI score could be improved with zero scoring. IBI scores based on the present low-end scoring methods (Ohio EPA, 1987) and zero scoring (both species richness and proportional metrics, as described above) at sites impacted primarily by habitat disturbances and sites where toxic impacts predominate (Yoder and Rankin, 1995b) were compared (Figure 25.5).

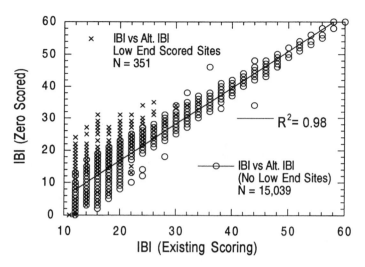

FIGURE 25.4 Scatterplot of the current Ohio EPA IBI vs. an alternate IBI with species richness metrics zero-scored (where species values for a metric were zero) and proportional metrics with relative numbers of fewer than 50 (25 in headwaters less than 8 sq. mi.) were scored zero. Samples with manual low-end adjustments denoted with an "x" and all others with an open circle.

Zero scoring had comparatively little effect on the distribution of IBI scores at reference sites or those impacted primarily by habitat modifications because the community response did not include those changes most affected by zero scoring (i.e., no severe reductions in species richness or abundance). However, zero scoring substantially increased the difference between the two IBI scoring methods at sites with toxic impacts, and further amplified and extended the gradient of biological quality indicated by IBI scores.

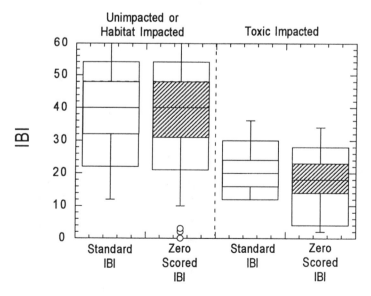

FIGURE 25.5 Box and whisker plots of selected fish community samples that were least impacted or affected by habitat impacts (left) or impacted by toxic chemical impacts (right). For each impact type, sites were scored with current Ohio EPA IBI and an alternate IBI with species richness metrics zero-scored (where species values for a metric were zero), and proportional metrics with relative numbers of fewer than 50 (25 in headwaters less than 8 sq. mi.) were zero-scored.

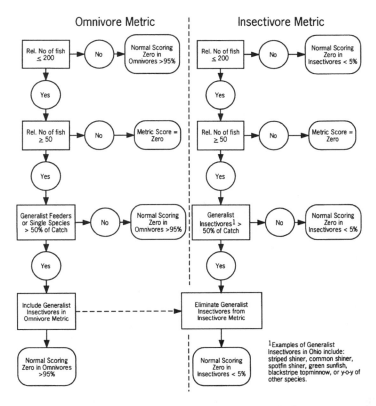

FIGURE 25.6 Flowchart illustrating possible conditional scoring criteria for two proportional IBI metrics: percent individuals as omnivores (left) and percent individuals as insectivores (right).

Figure 25.6 illustrates two examples of proposed modifications to proportional metrics: percent of individuals as omnivores and percent of individuals as insectivores. These modifications include conditional statements to further improve the resolution of the IBI, especially at sites with a CPUE greater than 50, but less than 200. In essence, these statements are used to capture the logic behind the qualitative adjustments that are currently applied to the IBI (Table 25.2). Such modifications could be used on any metric to improve the resolution of the IBI; however, the resolution attainable with the IBI is correlated to the robustness and the precision of the sampling methods and calibration framework. Failure to adequately control and minimize sources of error in field sampling and IBI calibration will handicap the capability to resolve some of the differences discussed here.

25.3.2 Predominant Species Swamping

In addition to the low-end adjustments at sites with few individuals, there are other infrequently encountered situations when the IBI might require other types of after-the-fact adjustment or modification. A single species, for example, may dominate a sample (usually >90% by numbers), and the phenomenon is unrelated to the environmental quality of the site. In Ohio, this occurs primarily in larger waterbodies such as Lake Erie (see Thoma, Chapter 16) and the Ohio River (Simon and Emery, 1995; Simon and Sanders, Chapter 18), but may occur in inland rivers and streams in the vicinity of large impoundments. The species most frequently involved are gizzard shad (*Dorosoma cepedianum*) and emerald shiner (*Notropis atherinoides*). Such occurrences can significantly alter the scoring of the affected proportional metrics and obscure the remainder of the sample, while yielding little ecological insight into conditions at a site. Where such occurrences are common, such as in Lake Erie near-shore and harbor areas, the predominant species (e.g., gizzard shad) may be excluded from IBI calculations or from certain metrics (Thoma, Chapter 16). In the Ohio River, conducting sampling at night greatly reduces this problem (Sanders, 1992). In inland waters, for those few samples where the problem has been evident, the IBI is calculated with and without the dominant species. The decision about which IBI to use is made based on which one is the most representative portrayal of biological quality. Examining the results of the MIwb, which is typically less affected by the predominance of a single dominant species when species richness is high, can be invaluable.

25.3.3 Affects of Young-of-Year

Another situation where the IBI may require manual after-the-fact adjustment is at sites dominated by young-of-year (YOY) or juvenile fish. Schlosser (1985) and Angermeier and Karr (1986) have demonstrated the effect that YOY fish can have on the IBI and the subsequent interpretation of site quality. The Ohio EPA excludes most YOY fish through the use of a sufficiently large dipnet mesh size so that most YOY fish are not captured. However, there are situations where YOY fish cannot always be excluded from 1+-year-old fish (i.e., late summer and fall samples). In following the recommendations of Angermeier and Karr (1985), the option of excluding YOY or small juveniles from the IBI calculations is available (Ohio EPA, 1987). However, this option has rarely been exercised in our database because few situations have been encountered where YOY and small juveniles have contributed to an inflated IBI score. A unique situation was recently experienced, where the fish community was essentially eliminated from a 17-mile stream segment by a coal mine release. Fish data collected in the first year following cessation of the acutely toxic conditions indicated that the IBI scores were based on samples almost completely comprised of YOY and juvenile fish that did not accurately portray the condition of the stream. As a result, achieving the ecological recovery end-points based on the IBI that were established for these streams could be misleading if this phenomenon was not adequately included. Considering this phenomenon, the ecological recovery end-points (Ohio EPA, 1994) required that: (1) IBI values from the recovery zones were to be calculated with and without YOY and that the difference between the scores for a site could not deviate more than at the upstream control site; and (2) the MIwb must concurrently demonstrate attainment of the recovery endpoints. Figure 25.7 illustrates the effect that the inclusion

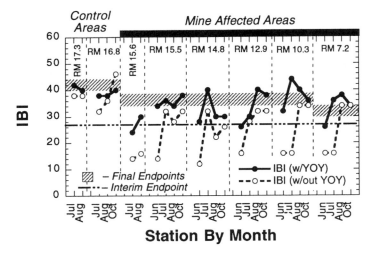

FIGURE 25.7 Calculation of IBI scores with (solid circles) and without (open circles) YOY fish at sites in Leading Creek during 1994, approximately 9 to 12 months after a mine spill eliminated the fish community downstream of River Mile 16.0. Shaded bar in mine-affected areas represent final ecological recovery endpoints for the IBI, the horizontal line represents an interim end-point for IBI scores outside of a "poor" narrative range.

TABLE 25.4
The Occurrence of After-the-Fact Manually Adjusted IBI Scores in the Ohio EPA Database During the Period 1978–1996 by Adjustment Category (N = 15,390)

Adjustment Category	Frequency of Occurrence	Percent of All Samples
Low-end	351	2.28
Species exclusion	5	0.03
YOY/Juvenile exclusion	11	0.07
ALL (Total)	367	2.39

of YOY fish can have on IBI scores in the recovery areas as described above. Eleven of 22 IBI scores calculated with and without YOY fish were significantly different from the upstream site (Figure 25.7, Table 25.4). In this case study, recovery of the original fish community to pre-release levels was much more dependent on reproduction and much less so on ingress from downstream areas or tributaries. This is likely a result of the overwhelming magnitude of the impact and the lack of similarly sized tributary streams that could have served as refugia or recolonization epicenters. Angermeier and Karr (1986) illustrated how the indiscriminant inclusion of YOY can result in inflated IBI values. The example here also illustrates a similar situation, albeit uncommon, where the predominance of YOY and small juvenile fish inappropriately inflates the IBI and results in the premature notion of ecological recovery.

25.4 CONCLUSION

The use of the IBI has provided the Ohio EPA with a robust and useful tool for water quality management in Ohio. An important caveat in the use of the IBI is *not* to apply the index "blindly" and without thoughtful consideration of the ecological principles underlying the metrics and the rich information each contains. In fact, such usage would qualify as "bad science" and be at odds with the original intent to have the IBI used in support of biological judgment (Karr et al., 1986), not the other way around. One popular criticism of IBI-type indices is the "loss" of information (Suter, 1993), which, in our experience, ignores the inherent process of constructing, calibrating, and using the IBI. Prior examination of both metric and overall IBI behavior identified the situations where the composite IBI score could be misleading.

Our experience shows that adjustments, such as zero scoring and conditional scoring criteria (Table 25.2) are needed on an infrequent basis (2.3%; Table 25.4) or, when CPUEs are adequate, have little significant effect on the IBI. The adjustments described and tested here result in an IBI-based approach that, in addition to detecting impairment, yields an improved ability to characterize and quantify the severity of impairment in a particular stream or river segment, and discriminate among various stressors associated with the impairment. Adopting the additional scoring adjustment and modification procedures described here will result in an improved capability to fulfill these objectives. The system of numerical IBI scores and the gradient of environmental quality to which it is correlated is essential to enhancing and improving communication with risk managers and decision-makers, many of whom are non-scientists. The qualities inherent to the IBI, when it is properly constructed, calibrated, and used, make it a reliable and easily interpreted indicator of environmental quality.

REFERENCES

Anderson, K., A. Boulanger, H. Gish, J. Kelly, J. Morril, and L. Davis. 1991. Using machine learning techniques to visualize and refine criteria for biological integrity, in *Biological Criteria: Research and Regulation, Proceedings of a Symposium.* EPA-440-5-91-005. U. S. Environmental Protection Agency, Office of Water, Washington DC. 123–128.

Angermeier, P. L. and J. R. Karr. 1986. Applying and index of biotic integrity based on stream-fish communities: considerations in sampling and interpretation, *North American Journal of Fisheries Management,* 6, 418–429.

Barbour, M. T., J. B. Stribling, and J. R. Karr. 1995. Multimetric approach for establishing biocriteria and measuring biological condition, in W.S. Davis and T. P. Simon (Eds.), *Biological Assessment and Criteria: Tools for Water Resource Planning and Decision Making.* Lewis, Boca Raton, FL. 63–80.

Davis, W. S., B. D. Snyder, J. B. Stribling, and C. Stoughton. 1996. Summary of State Biological Assessment Programs for Streams and Wadeable Rivers. EPA 230-R-96-007. U.S. Environmental Protection Agency, Office of Policy, Planning, and Evaluation, Washington, DC.

DeShon, J. E. 1995. Development and application of the invertebrate community index (ICI), in W. S. Davis and T. P. Simon (Eds.), *Biological Assessment and Criteria: Tools for Water Resource Planning and Decision Making.* Lewis, Boca Raton, FL. 217–244.

Fausch, K. D., J. R. Karr, and P. R. Yant. 1984. Regional application of an index of biotic integrity based on stream fish communities, *Transactions of the American Fishery Society,* 113, 39–55.

Gammon, J. R., A. Spacie, J. L. Hamelink, and R. L. Kaesler. 1981. Role of electrofishing in assessing environmental quality of the Wabash River, in J. M. Bates and C. I. Weber (Eds.)., *Ecological Assessments of Effluent Impacts on Communities of Indigenous Aquatic Organisms.* ASTM STP 730.

Gammon, J. R. 1976. The Fish Populations of the Middle 340 km of the Wabash River. Purdue University, Water Resources Research Center Technical Report 86.

Karr, J. R. 1981. Assessment of biotic integrity using fish communities, *Fisheries,* 6, 21–27.

Karr, J. R. 1991. Biological integrity: a long-neglected aspect of water resource management, *Ecological Applications,* 1, 66–84.

Karr, J. R., K. D. Fausch, P. L. Angermeier, P. R. Yant, and I. J. Schlosser. 1986. Assessing Biological Integrity in Running Waters: A Method and Its Rationale. Illinois Natural History Survey Special Publication 5, Champaign.

Larsen, D. P., J. M. Omernik, R. M. Hughes, C. M. Rohm, T. R. Whittier, A. J. Kinney, A. L. Gallant, and D. R. Dudley. 1986. The correspondence between spatial patterns in fish assemblages in Ohio streams and aquatic ecoregions, *Environmental Management,* 10, 815–828.

Miller, D. L., P. M. Leonard, R. M. Hughes, J. R. Karr, P. B. Moyle, L. H. Shrader, B. A. Thompson, R. A. Daniels, K. D. Fausch, G. A. Fitzhugh, J. R. Gammon, D. B. Halliwell, P. L. Angermeier, and D. J. Orth. 1988. Regional applications of an index of biotic integrity for use in water resource management, *Fisheries,* 13(5), 12–20.

Oberdorff, T. and R. M. Hughes. 1992. Modification of an index of biotic integrity based on fish assemblages to characterize rivers of the Seine-Normandie Basin, France, *Hydrobiologia,* 228, 116–132.

Ohio Environmental Protection Agency. 1987. Biological Criteria for the Protection of Aquatic Life. Vol. II. Users Manual for Biological Field Assessment of Ohio Surface Waters. OEPA, Division of Water Quality Monitoring and Assessment, Surface Water Section, Columbus, OH.

Ohio Environmental Protection Agency. 1989. Biological Criteria for the Protection of Aquatic Life. Vol. III. Standardized Biological Field Sampling and Laboratory Methods for Assessing Fish and Macroinvertebrate Communities. OEPA, Division of Water Quality Monitoring and Assessment, Columbus, OH.

Ohio Environmental Protection Agency. 1992. Biological and Water Quality Study of the Ottawa River, Hog Creek, Little Hog Creek, and Pike Run (Hardin, Allen, and Putnam Counties, Ohio). OEPA Technical Report EAS/1992-9-7. OEPA, Division of Water Quality Planning and Assessment, Columbus, OH.

Ohio Environmental Protection Agency. 1994. Ecological Endpoints for Streams Affected by the Meigs #31 Mine Discharges During July–September 1993. OEPA Technical Report. EAS/1994-1-1. OEPA, Division of Surface Water, Columbus, OH.

Ohio Environmental Protection Agency. 1997. in E. T. Rankin, C. O. Yoder, and D. Mishne (Eds.), *Ohio Water Resource Inventory. Vol. I. Summary, Status and Trends.* Division of Surface Water, Ecological Assessment Section. Columbus.

Rankin, E. T. 1989. The Qualitative Habitat Evaluation Index (QHEI), Rationale, Methods, and Application. Ohio EPA, Division of Water Quality Planning & Assessment, Ecological Assessment Section, Columbus.

Rankin, E. T. 1995. The qualitative habitat evaluation index (QHEI), in W. S. Davis and T. P. Simon (Eds.), *Biological Assessment and Criteria: Tools for Water Resource Planning and Decision Making*. Lewis, Boca Raton, FL. 181–208.

Sanders, R.S. 1992. Day versus night electrofishing catches from near-shore waters of the Ohio and Muskingum rivers, *Ohio Journal of Science,* 92(3), 51–59.

Schlosser, I. J. 1985. Flow regime, juvenile abundance, and the assemblage structure of stream fishes, *Ecology,* 66, 1484–1490.

Steedman, R. J. 1988. Modification and assessment of an index of biotic integrity to quantify stream quality in southern Ontario, *Canadian Journal of Fisheries and Aquatic Sciences,* 45, 492–501.

Simon, T. P. and E. B. Emery. 1995. Modification and assessment of an index of biotic integrity to quantify water resource quality in great rivers, *Regulated Rivers: Research and Management,* 11, 283–298.

Suter, G. W., II. 1993. A critique of ecosystem health concepts and indexes, *Environmental Toxicology and Chemistry,* 12, 1533–1539.

Whittier, T. R., D. P. Larsen, R. M. Hughes, C. M. Rohm, A. L. Gallant, and J. R. Omernik. 1987. The Ohio Stream Regionalization Project: A Compendium of Results. EPA-600/3-87/025. U.S. Environmental Protection Agency, Corvallis, OR.

Yoder, C. O. 1989. The development and use of biological criteria for Ohio surface waters, in G. H. Flock (Ed.), *Water Quality Standards for the 21st Century, Proceedings of a National Conference*. U.S. Environmental Protection Agency, Office of Water, Washington, DC. 139–146.

Yoder, C. O. 1995. Policy issues and management applications of biological criteria, in W. S. Davis and T. P. Simon (Eds.), *Biological Assessment and Criteria: Tools for Water Resource Planning and Decision Making*. Lewis, Boca Raton, FL. 327–344.

Yoder, C. O. 1991. The integrated biosurvey as a tool for the evaluation of aquatic life use attainment and impairment in Ohio surface waters, in *Biological Criteria: Research and Regulation, Proceedings of a Symposium,* December 12-13, 1990, Arlington, VA. U.S. EPA, Office of Water, Washington, DC. EPA-440/5-91-005: 110.

Yoder, C. O. and E. T. Rankin. 1995a. Biological criteria program development and implementation in Ohio, in W. S. Davis and T. P. Simon (Eds.), *Biological Assessment and Criteria: Tools for Water Resource Planning and Decision Making*. Lewis, Boca Raton, FL. 109–144.

Yoder, C. O. and E. T. Rankin. 1995b. Biological response signatures and the area of degradation value: new tools for interpreting multi-metric data, in W. S. Davis and T. P. Simon (Eds.), *Biological Assessment and Criteria: Tools for Water Resource Planning and Decision Making*. Lewis, Boca Raton, FL. 263–286.

26 Integrating Assessments of Fish and Macroinvertebrate Assemblages and Physical Habitat Condition in Pennsylvania

Blaine D. Snyder, James B. Stribling, Michael T. Barbour, and Carroll L. Missimer

CONTENTS

26.1 Introduction 639
    26.1.1 Background 640
    26.1.2 Objectives and Assessment Framework 640
26.2 Physical Setting 641
    26.2.1 The Codorus Creek Study Area 641
    26.2.2 Subecoregional Reference Locations 641
26.3 Survey and Assessment Methods 642
    26.3.1 Physical Habitat Assessment Methods 642
    26.3.2 Benthic Macroinvertebrate Survey and Assessment Methods 643
    26.3.3 Fish Survey and Assessment Methods 644
26.4 Condition of Physical Habitat Structure 646
26.5 Condition of the Benthic Macroinvertebrate Assemblage 646
26.6 Condition of the Fish Assemblage 647
26.7 Integrated Assessment 648
Acknowledgments 650
References 650

26.1 INTRODUCTION

Biological assessment approaches of the past often focused on a single indicator species, or a limited number of attributes such as species distribution or abundance trends. Karr et al. (1986) noted that the accurate assessment of biological condition requires a method that integrates biological responses through examination of patterns and processes from individual to ecosystem levels. Whereas classical approaches were valuable for measurement of narrow ranges of perturbation or selected anthropogenic effects, they often were not successful in screening for all types of degradation, including complex cumulative impacts (Karr, 1991). Approaches are now available, and have been widely tested and applied, that define an array of biological measures that individually provide information on specific biological attributes, and collectively provide an overall assessment of water resource condition (Karr et al., 1986; Ohio EPA, 1987; Plafkin et al., 1989; Lyons, 1992; Barbour et al., 1995; Simon and Lyons, 1995; Yoder and Rankin, 1995; Barbour et al., 1996; Fore et al., 1996).

This comprehensive, multiple measure or multimetric approach, first successfully applied as the fish Index of Biotic Integrity (IBI) (Karr, 1981), has become a valuable tool for detecting the effects of the broad range of anthropogenic impacts (Davis and Simon, 1995). The multimetric concept has also been developed and tested for other biological assemblages of the aquatic community, particularly benthic macroinvertebrates and periphyton (Plafkin et al., 1989; Barbour et al., 1992; Kerans et al., 1992; Kerans and Karr, 1994; DeShon, 1995; Rosen, 1995; Fore et al., 1996; Barbour et al., 1996). Just as each individual biological metric is expected to change in some predictable way with increased human influence (Karr et al., 1986; Fausch et al., 1990), each assemblage within the aquatic community is expected to have a response range to perturbation events or degraded conditions (Gibson et al., 1996). Biological assessments that include multiple attributes, and target multiple species or even multiple assemblages, are more likely to provide an increasingly comprehensive detection capability of a broader range of stressors than single-dimension measures such as diversity indices, species richness, or indicator species (Yoder and Rankin, 1995).

The primary goal of biological assessment is to determine the biological condition of a water resource, and an essential component of the assessment is the establishment of benchmarks for comparison and the determination of deviations from those benchmarks. To adequately assess the ecological health of a waterbody, a reference condition must be established as the foundation for developing regionalized biological expectations or benchmarks. This chapter presents the application of a regional reference approach used in a watershed-specific assessment in southern Pennsylvania. Surveys of the fish and macroinvertebrate assemblages and the technical framework of the IBI were used, in combination with physical habitat data, to provide an integrated assessment of the condition of the watershed and examine the effects of multiple stressors.

26.1.1 BACKGROUND

The case study described in this chapter uses a contemporary multimetric approach for biological assessments based on comparisons of Codorus Creek (York County, PA) biological condition with ecoregional reference conditions. Many of the early studies of Codorus Creek (most relating to point-source discharge effects of a pulp and paper manufacturing facility) relied on professional judgment, or individual ideas of what constituted a healthy/natural waterbody, which was based on years of field experience (Academy of Natural Sciences of Philadelphia, 1974; Denoncourt, 1992). The assessment concepts and methods used in this case study were developed by the U.S. EPA (Gibson et al., 1996) and have been endorsed by many state and federal regulatory agencies (Southerland and Stribling, 1995; Davis et al., 1996). This approach is advocated as a framework for biocriteria (U.S. EPA, 1993; Gibson et al., 1996; Davis and Simon, 1995) and is under development by the Pennsylvania Department of Environmental Protection (DEP) (R. Kime, PA DEP, personal communication).

26.1.2 OBJECTIVES AND ASSESSMENT FRAMEWORK

This case study was designed to: (1) characterize the fish and macroinvertebrate assemblages at selected locations in Codorus Creek compared with minimally impaired subecoregional reference locations; (2) relate differences in biological condition among sampling stations to: (a) point and nonpoint source impacts, and (b) spatial changes in the quality of stream physical habitat; and (3) integrate biological and physical habitat information for a comprehensive assessment of condition of the water resource.

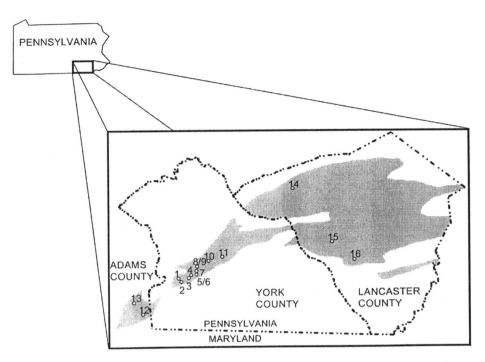

FIGURE 26.1 Location of the 11 Codorus Creek monitoring stations (stations 1–11) and subecoregional reference stations (stations 12–16). (Shaded area = Piedmont Limestone Lowland subecoregion).

26.2 PHYSICAL SETTING

26.2.1 THE CODORUS CREEK STUDY AREA

The Codorus Creek drains approximately 278 sq. mi. of land in southeastern Pennsylvania. The Codorus Creek study area (Figure 26.1; stations 1 through 11) is entirely contained within a unique subecoregion labeled the Piedmont Limestone Lowlands. Streams within the Piedmont Limestone Lowland subecoregion are characterized by glide-pool dominated, relatively low-gradient landscapes, and substrates typically of fine sediments and some exposed bedrock (Woods et al., 1996). Therefore, natural streams of this subecoregion tend to resemble streams of the Coastal Plain more so than the riffle/run (cobble/gravel substrate) dominated streams typical of most of the Piedmont.

This study of the Codorus Creek watershed was conducted during September and October 1995 and included a total of 16 sampling stations (Table 26.1): four located upstream from a point source paper mill outfall; seven downstream of the point source; and five representing reference locations within the same subecoregion as the Codorus Creek but outside of the watershed. Stations located upstream and downstream of the outfall were part of a historical monitoring program, with the exception of stations 6 and 9, which were added to the assessment program to enhance the examination of habitat quality.

26.2.2 SUBECOREGIONAL REFERENCE LOCATIONS

The quantitative expectations for attributes that constitute the biological integrity of a water resource system vary geographically (Karr, 1995), and ecoregions/subecoregions are recommended as a basic geographic unit for establishing reference conditions (NRC, 1992; Hughes et al., 1994). Ecoregions are characterized by relative homogeneity in a combination of geographic physical and chemical characteristics that are likely to be associated with spatial patterns in resource quality,

TABLE 26.1
Piedmont Limestone Lowland Biosurvey and Subecoregional Reference Locations

Station Number	County	Stream	Location
1	York	Oil Creek	Downstream from municipal wastewater treatment facility; 5.4 km upstream from paper mill discharge
2	York	West Branch Codorus Creek	Downstream from reservoir release and tailwaters; 5.3 km upstream from paper mill discharge
3	York	West Branch Codorus Creek	Municipality of Spring Grove; 1.9 km upstream from paper mill discharge
4	York	West Branch Codorus Creek	0.8 km upstream from paper mill discharge
5	York	West Branch Codorus Creek	Immediately downstream from paper mill treated effluent
6	York	West Branch Codorus Creek	0.2 km downstream from paper mill discharge
7	York	West Branch Codorus Creek	2.2 km downstream from paper mill discharge
8	York	West Branch Codorus Creek	3.5 km downstream from paper mill discharge
9	York	West Branch Codorus Creek	3.7 km downstream from paper mill discharge
10	York	West Branch Codorus Creek	9.9 km downstream from paper mill discharge
11	York	Codorus Creek	Downstream from South Branch Codorus Creek confluence; 16.3 km downstream from paper mill discharge
12	Adams	Conewago Creek	Subecoregional reference
13	Adams	Conewago Creek	Subecoregional reference
14	Lancaster	Little Chickies Creek	Subecoregional reference
15	Lancaster	Little Conestoga Creek	Subecoregional reference
16	Lancaster	Pequea Creek	Subecoregional reference

quantity, stressors, and their interrelationships (Omernik, 1987). Ecoregions are based on landscape-level features such as climate, physiography, geology, soils, and vegetation (Griffith et al., 1994). Omernik (1987) and Omernik and Gallant (1990) delineated ecoregions of the conterminous United States, and more recently, Woods et al. (1996) developed state-level maps for Pennsylvania that transfer ecoregions into higher resolution subecoregions.

All reference streams were located within the Piedmont Limestone Lowlands subecoregion, were of the same relative size, and were minimally disturbed areas of potentially natural landscapes within the subecoregion. The reference locations were delineated through use of maps, available data or photos, discussions with resource managers, and finally, field reconnaissance. More than 40 potential candidate reference locations were evaluated; however, only 5 satisfied our strict objectives to locate areas of minimal disturbance (e.g., least impairment of surrounding lands and riparian zones, or a riparian vegetative zone width of greater than 12 m). Reference stations included a total of five sites (stations 12 through 16) within four streams in two counties.

26.3 SURVEY AND ASSESSMENT METHODS

26.3.1 PHYSICAL HABITAT ASSESSMENT METHODS

An assessment of physical habitat characteristics was made concurrently with the macroinvertebrate and fish biosurveys. The habitat characterization followed a modification of the physical habitat assessment procedure (Barbour and Stribling, 1991; 1994) originally developed for U.S. EPA Rapid Bioassessment Protocols (Plafkin et al., 1989). This visual-based approach consists of scoring a continuum of conditions for several habitat parameters. The key to reducing subjectivity in this approach is to provide adequate training and address relevant parameters that represent the physical habitat structure of the region (Barbour, 1994; Hannaford et al., 1997). Variability among investigators can be relatively high (Barbour, 1994); however, this study involved an experienced team of two scientists for habitat assessments, employing the same team of investigators for all sampling locations.

Included in this approach is a point scale for each parameter, with 0 being the lowest (poor) and 20 being the highest (optimal) score. Physical habitat quality is scored by visually assessing 10 parameters along the stream reach at the time of the biosurvey (Barbour and Stribling, 1994).

The following habitat parameters were evaluated (i.e., scored) at each of the biosurvey stations:

1. *Bottom Substrate/Available Cover.* This characteristic refers to the availability of habitat for the support of aquatic organisms and cover for nesting, oviposition sites, or refugia. A variety of substrate materials and habitat types is desirable.
2. *Pool Substrate Characterization.* The relationship between the diversity in substrate material composition and the taxonomic diversity of the biological assemblage is well documented. For this parameter, pools with a diverse substrate are rated higher than those that are uniform.
3. *Pool Variability.* This parameter rates the mixture of pool sizes within a stream reach. This variability is essential for the habitat to support a healthy, diverse biological community.
4. *Channel Alteration.* This characteristic potentially allows the cursory estimation of stream system stability. Channelization involves reduction in sinuosity, and results in increased velocity and subsequent intensification of erosional effects.
5. *Sediment Deposition.* Deposition of sediment from large-scale watershed erosion can smother the instream habitat. The degree to which the stream bottom is filled with silt is an indication of the severity of this parameter.
6. *Channel Sinuosity.* This parameter assumes that a stream with bends provides more diverse habitat than a straight-run or uniform stream. Habitat can be produced by the amplified force of water at bends, resulting in well-developed run areas rather than pools and glides.
7. *Channel Flow Status.* This parameter is designed to evaluate the channel flow and substrate coverage. The degree to which the substrates are exposed is an indication of decreasing flow and decreasing available substrate area.
8. *Bank Vegetative Protection.* This parameter rates the quality and coverage of streambank vegetation. Disruption of native streambank vegetation results in reduced streambank stability, elevates the erosion potential, and reduces the area of bank vegetative cover and shading.
9. *Bank Stability.* This parameter rates the evidence of erosion or bank failure. Degraded bank condition is characterized by banks with frequent erosion scars and/or high erosion potential.
10. *Riparian Vegetative Zone Width.* This parameter rates the entire riparian buffer zone on both sides of the stream. Decreasing buffer zone width reduces stream shading, and reduces the effectiveness of these areas to filter particulate pollutants from stormwater runoff.

26.3.2 BENTHIC MACROINVERTEBRATE SURVEY AND ASSESSMENT METHODS

Benthic macroinvertebrate sampling consisted of a multihabitat composite from each station. The field technique was standardized by unit area to insure comparable effort at all sites, and followed a method developed for low-gradient streams (U.S. EPA, 1997). In accordance with U.S. EPA (1997) guidance, a 100-meter reach of stream (i.e., the same used for habitat assessment) was sampled at each station. A D-frame net (0.3 m width and 595-μ mesh) was jabbed for approximately 1 m in length, 20 times for each composite sample. The 20 jabs were distributed among the different, available habitat types (e.g., woody debris/snags, aquatic vegetation, vegetated banks, undercut banks, coarse/hard substrates, fine/soft substrates) in proportion to their occurrence within the reach.

The composited sample was returned to the laboratory for processing and identification, and macroinvertebrates were identified to the lowest feasible taxon (i.e., generally genus or species level). A macroinvertebrate subsampling procedure (i.e., a randomized 100-organism subsample) was conducted by laboratory personnel following the guidelines of the U.S. EPA Rapid Bioassessment Protocols (Plafkin et al., 1989) and Barbour and Gerritsen (1996).

Thirty-one metrics relevant to the structure and function of the macroinvertebrate assemblage were considered as candidate (potential) metrics. Except for the modified Hilsenhoff Biotic Index (HBI; Hilsenhoff, 1987) and the Shannon-Wiener Index, the candidate metrics were based on either relative abundance or number of taxa within a specific taxonomic or functional (trophic) grouping.

The process of metric selection was separated into two phases. First, an optimization phase was performed, whereby the metrics were evaluated for their ecological relevance and natural variability (Barbour et al., 1995). This was followed by a calibration phase to determine the discriminatory power and sensitivity of each candidate metric to perturbation. The 31 candidate metrics were evaluated for efficacy and validity for implementation into the bioassessment process. Core metrics were those remaining that provided useful information for discriminating between good- and poor-quality ecological conditions.

Eight of the 31 metrics examined showed strong potential for discrimination in a bioassessment: (1) total number of taxa; (2) number of intolerant Ephemeroptera, Plecoptera, and Trichoptera (EPT) taxa; (3) percent dominant taxon; (4) percent intolerant EPT; (5) Shannon-Wiener Index; (6) Hilsenhoff Biotic Index; (7) percent scrapers; and (8) percent collector-gatherers (Table 26.2).

Standardization of these measurements into a logical progression of scores is the typical means for comparing and interpreting metric values (Barbour et al., 1995). For each metric, the range of possible values is divided into scoring categories from selected percentiles of the population of reference sites. The establishment of scoring criteria made it possible to discriminate between impaired and unimpaired sites within the study area. For the macroinvertebrate bioassessment, metric values below the lower quartile of reference conditions were typically judged to be less than desirable for the metric. The distance from the lower quartile (25th percentile) can be termed a "scope for detection" for metrics that decrease with impairment (e.g., total taxa, EPT taxa, Shannon-Wiener Index). For those metrics that increase in value under impaired conditions (e.g., percent dominant taxon, HBI, percent collector-gatherers), the scope for detection would be from the 75th percentile to the maximum value for the metric.

Ratings of 5, 3, or 1 were assigned to each metric (Table 26.2) from each sampling station according to whether its value approximated (5), deviated somewhat from (3), or deviated strongly from (1) subecoregional reference values. Individual metric scores for each of the eight core metrics were summed for each station to yield the macroinvertebrate IBI for that station.

26.3.3 Fish Survey and Assessment Methods

Fish collection techniques were standardized to provide a representative sample, ensure data quality, and allow a comparison of fish assemblages along the length of the study area. The same 100-m reach of stream used for habitat assessment and macroinvertebrate sampling was sampled at each station, to allow an integrated assessment. A backpack electrofisher (pulsed DC from a variable-voltage pulsator, Coffelt Model Mark-10, CPS pulse pattern) was used to sample all available microhabitats within each reach. All fish were identified to species, and specimens less than 20 mm (total length) were not included in the bioassessment process.

Fish data were tabulated and summarized to examine individual fish assemblage attributes. Fish data were incorporated into an IBI (Karr et al., 1986) to characterize the biological condition of the specific stream segments. Twenty ecological measurements relevant to the structure and function of fish assemblages were considered as candidate metrics for the IBI. The metric selection process was similar to that described for the macroinvertebrate assessment, and only minor modifications of Karr's original metrics were required. The exceptions to this generalization were Karr's Metric

TABLE 26.2
Benthic Macroinvertebrate Metric Scoring Criteria Based on Subecoregional Reference Conditions

Category	Metric	Expected Response to Increasing Perturbation	Scope for Detecting Impairment	Scoring Criteria		
				5	3	1
Richness measures	Total number of taxa	Decrease	<25th percentile	≥17	9–16	≤8
	Number of intolerant EPT taxa	Decrease	<25th percentile	≥2	1	0
Composition measures	Shannon-Wiener Index	Decrease	<25th percentile	≥3.34	1.67–3.33	<1.67
	Percent dominant taxon	Increase	>75th percentile	≤26.0	26.1–43.0	>43.0
	Percent modified EPT	Decrease	<25th percentile	≥7.80	3.90–7.79	<3.90
Tolerance measures	Hilsenhoff Biotic Index	Increase	>75th percentile	≤5.02	5.03–6.26	>6.26
Trophic measures	Percent collector-gatherers	Increase	>75th percentile	≤29.0	29.1–55.0	>55.0
	Percent scrapers	Decrease	<25th percentile	≥22.0	11.0–21.9	<11.0

4 — the number and identity of sucker species; Metric 6 — the proportion of individuals as green sunfish; Metric 8 — proportion of individuals as insectivorous cyprinids; and Metric 11 — proportion of individuals as hybrids.

The sucker species metric was replaced with the number of minnow species (Ohio EPA, 1997), due to the relative low number of sucker species known to inhabit streams of the study area. Only two sucker species were collected from the Codorus Creek drainage and from the subecoregional reference streams; whereas, minnows often dominate the species richness and relative abundance of fish assemblages of the region. The green sunfish metric was replaced with the proportion of individuals as tolerant species (Ohio EPA, 1997). Several species meet the criteria for inclusion in the tolerant species metric (i.e., ability to survive in degraded waters and/or increase in abundance with increasing stream degradation). The inclusion of several species, rather than a single tolerant species, was intended to improve the sensitivity of the metric. The insectivorous cyprinid metric was modified to include all insectivores/invertivores (Karr, 1991). Because hybrids can be difficult to identify and might not be a consistent indicator of resource impairment, the hybrid metric was replaced with an alternative metric. The most promising alternative metrics are those that consider fish reproductive guilds, based on their sensitivity to physical habitat impairment. Fishes that exhibit simple spawning behavior (e.g., broadcast spawners) requiring clean gravel and/or cobble substrates for successful reproduction (i.e., lithophilous) appear to be an environmentally sensitive reproductive guild (Ohio EPA, 1987). Because of their sensitivity to environmental disturbances, the proportion of individuals as simple lithophilic spawners was used as an alternative IBI metric.

Twelve core metrics (Table 26.3) were chosen to represent elements and processes of the fish assemblages within the subecoregion: (1) total number of species; (2) number of darter species; (3) number of sunfish species; (4) number of minnow species; (5) number of intolerant species; (6) percent tolerant species; (7) percent omnivores; (8) percent invertivores; (9) percent piscivores (top carnivores); (10) catch per unit effort; (11) percent simple lithophils; and (12) DELT (Ohio EPA, 1987) anomalies.

For the fish bioassessment, the area below the 95th percentile threshold value was trisected for each metric to determine the scoring criteria for each. This method follows the Maximum Species Richness Line (MSRL) approach recommended by Fausch et al. (1984) and the Ohio EPA (1987) for fish IBI assessments. IBI metric ratings of 5, 3, and 1 (Table 26.3) were assigned to the trisected areas, and denoted whether the metric value approximated (5), deviated somewhat (3), or deviated strongly (1) from the subecoregional reference expectations. Scoring criteria for one metric —

TABLE 26.3
Fish Metric Scoring Criteria Based on Subecoregional Reference Conditions

Category	Metric	Expected Response to Increasing Perturbation	Scoring Criteria		
			5	3	1
Species richness and composition	Total number of fish species	Decrease	>12	6–12	<6
	Number and identity of darter species	Decrease	>1	1	0
	Number and identity of sunfish species	Decrease	>2	1–2	0
	Number of minnow species	Decrease	>7	4–7	<4
	Number of intolerant species	Decrease	>6	3–6	<3
	Proportion of individuals as tolerant species	Increase	<19	19–38	>38
Trophic composition	Proportion of individuals as omnivores	Increase	<10	10–20	>20
	Proportion of individuals as insectivores	Decrease	>66	33–66	<33
	Proportion of individuals as piscivores (top carnivores)	Decrease	>2	1–2	<1
Abundance and condition	Catch per effort (n/min)	Decrease	>13.4	6.7–13.4	<6.7
	Proportion of individuals as simple lithophils	Decrease	>48	24–48	<24
	Proportion of individuals with deformities, fin erosion, lesions, or tumors	Increase	<0.5	0.5–3	>3

percent DELT anomalies — followed those established by the Ohio EPA (1987). Individual metric scores for each of the 12 core metrics were summed at each station to yield the fish IBI for that particular station.

26.4 CONDITION OF PHYSICAL HABITAT STRUCTURE

An evaluation of physical habitat quality is critical to any assessment of biological condition (Plafkin et al., 1989). Observed physical habitat conditions are usually the result of complex interplay between hydrogeomorphological factors and anthropogenic landscape alterations (Gregory et al., 1991). The physical habitat assessment framework used here provided a basic evaluation of instream and near-stream physical habitat quality. The physical habitat index scores are summarized in Figure 26.2 as box plots for stations upstream of the paper mill discharge (four stations), downstream of the point source (seven stations), and subecoregional reference stations (five stations). Physical habitat assessment scores demonstrated considerable overlap between the upstream and downstream stations. Total scores also showed that ecoregional reference sites were distinct from (i.e., higher than) the Codorus Creek sites. The distinction was primarily the result of higher quality riparian habitats (e.g., bank vegetative protection, relatively undisturbed riparian zones). Station physical habitat scores ranged from 129 to 148 upstream of the outfall, 124 to 150 downstream of the outfall, and 151 to 173 for the reference sites. Results of an ordinal scale of ranking stations by habitat index score indicate that all Codorus Creek stations (stations 1–11) rate as fair, and all reference stations (stations 12–16) rate as good.

26.5 CONDITION OF THE BENTHIC MACROINVERTEBRATE ASSEMBLAGE

Taxa richness of the benthic macroinvertebrates ranged from 12 (at downstream station 5) to 24 taxa (at upstream station 4 and subecoregional reference stations 12 and 16). Composition and relative abundance data were used in computing the eight macroinvertebrate IBI metric values for

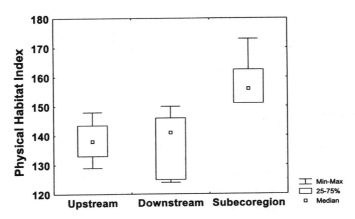

FIGURE 26.2 Physical habitat index scores for stations upstream of the paper mill outfall (4 stations), downstream from the outfall (7 stations), and subecoregional reference locations (5 stations).

each sampling station. The macroinvertebrate metric ratings tabulated for the assessment resulted in total scores ranging from 14 to 32 for stations upstream of the mill outfall, 12 to 28 for stations downstream of the outfall, and 36 to 40 for the subecoregional reference locations. Macroinvertebrate IBI scores demonstrated considerable overlap among the population of upstream and downstream sites. Scores of the subecoregional reference stations were distinct from (higher than) all Codorus Creek stations.

The index value criterion threshold is based on the IBI score distribution for the reference streams. Ohio EPA (1987) defined the attainable conditions for aquatic life based on a 25th percentile of the ecoregional reference site conditions of each measurable attribute related to drainage area. The 25th percentile of the subecoregional reference station scores was considered here to be the threshold for a rating of good biological condition, and ratings of fair and poor were derived by bisecting the range between the lowest possible score and the lower quartile of reference expectations. This approach resulted in the assignment of condition class categories of good (greater than or equal to 36), fair (23 to 35), and poor (less than or equal to 22), and allowed a ranking of stations. Application of these criteria to the IBI results indicates that biological condition based on the macroinvertebrate assemblage rated good at all reference stations. Stations 2 (upstream of the outfall), 5/6 (mill outfall and mixing zone), and 8/9 (a portion of the downstream study area) rated poor; whereas, the six remaining stations (stations 1, 3, 4, 7, 10, and 11) rated fair.

26.6 CONDITION OF THE FISH ASSEMBLAGE

The biosurvey resulted in the collection of 3487 fish, representing 35 species distributed among six families. Subecoregional reference stream collections yielded catches (at each station) of 13 to 19 fish species, of which 8 to 9 were intolerant fishes. Tolerant species and omnivores composed 57 and 31% of the catch, respectively, and insectivores/invertivores dominated the trophic guild abundances. Fish were largely free (less than 0.3%) of DELT anomalies (Ohio EPA, 1997). The subecoregional reference station results were used to establish expectations for all IBI metrics. The index value criterion threshold was based on the IBI value distribution in the reference streams. The IBI metric ratings (Section 26.3.3) resulted in total scores ranging from 26 to 38 for stations upstream of the paper mill outfall, 30 to 44 for stations downstream of the outfall, and 40 to 50 for the subecoregional reference locations.

The approach for using quartiles to establish IBI assessment criteria (Yoder and Rankin, 1995) was applied to interpret Codorus Creek bioassessment scores. This allowed assignment of condition class categories of good, fair, and poor based on fish IBI scores of greater than or equal to 40; 26

to 38; and less than or equal to 24, respectively. Application of these criteria to the IBI results indicates that biological condition based on the fish assemblage rated fair at upstream stations 1 through 4 and at downstream stations 5 through 10. The fish assemblage at downstream station 11 rated good (indicating that fish assemblage condition approximated subecoregional reference conditions), possibly reflecting the faunal contributions of a nearby tributary.

26.7 INTEGRATED ASSESSMENT

Interpretation of bioassessment results is based on the fact that all ecosystems have a biological potential (Southwood, 1977), and that biological communities cannot reflect that potential if stressors are present (Plafkin et al., 1989). In cases where physical and chemical habitat are nondegraded (i.e., rated good), biological potential is expected to be high. The biological potential of severely or moderately degraded physical habitat is substantially reduced from natural systems. However, in areas subjected to organic enrichment, an artificial elevation in biological condition of the macroinvertebrate assemblage (i.e., beyond its habitat potential) may be observed (Barbour and Stribling, 1991; 1994). Instances when areas of good physical habitat are shown to have an impaired biota are often indicative of some other type of stressor (e.g., toxicants) that is depressing the biological condition.

Compared with the five subecoregional reference locations, all sampling sites demonstrated some combination of physical habitat and biological impairment. The assessments of fish, macroinvertebrates, and physical habitat are presented simultaneously for each sampling station in Figure 26.3 to illustrate our current concept of an integrated assessment. The assessment scores are presented as a percentage of reference condition that allows direct comparison and presentation of all three components. The area enclosed by each plot (Figure 26.3) can be measured to provide a comparison of the integrated biological and physical habitat condition among the stations. Theoretically, a relatively equilateral triangle is expected; that is, biological assemblages reflecting the quality of the available habitat. This is the case with the reference stations 12 through 16. The stations shown to be least comparable to the reference sites were stations 2, 5, 8, and 9. Interpreted as an integrated assessment, the figure shows the influences of organic enrichment at station 1 resulting from the treatment and discharge of municipal waste upstream of the monitoring site. The plots also reflect the effects of the paper mill discharge and mixing zone at stations 5 and 6, and habitat degradation at stations 8 and 9. Station 2 proved to be an extreme outlier because physical habitat quality was high, yet the scores for both biotic indices were low, reflecting the "artificial" coldwater conditions created by the reservoir release. The reservoir tailwaters have created and sustained a valuable coldwater trout fishery; however, the bioassessment benchmarks or criteria used to rate the station were based on natural biological conditions of the subecoregion.

The relationships between physical habitat and the biotic indices are presented as linear regressions in Figure 26.4 ($r = 0.672$ for fish IBI; $r = 0.587$ for macroinvertebrate IBI). Station 2 was omitted from the regressions because of its aberrant condition influenced by a reservoir release. The effects of organic enrichment at station 1 and point source stressors at the paper mill outfall and mixing zone stations 5 and 6 are evident in that they are outside of the 95% confidence interval that is apparently predictive of habitat quality. Results indicate that biological condition is reduced at the "end of pipe" station 5; however, biological condition improved slightly immediately downstream from the paper mill outfall (i.e., as close as 0.2 RKm downstream at station 6). The improved downstream conditions are interrupted by a reach of degraded physical habitat at stations 8 and 9. The biological communities in this reach reflect the low habitat scores (i.e., the lowest scores for the entire survey). As habitat conditions improve downstream (e.g., stations 10 and 11), so does the condition of the fish and macroinvertebrate assemblages.

Results from this case study demonstrate the importance of including physical habitat quality and incorporating regional reference conditions in site-specific biological assessments. The combination of fish and macroinvertebrate assemblage and physical habitat information, when put into

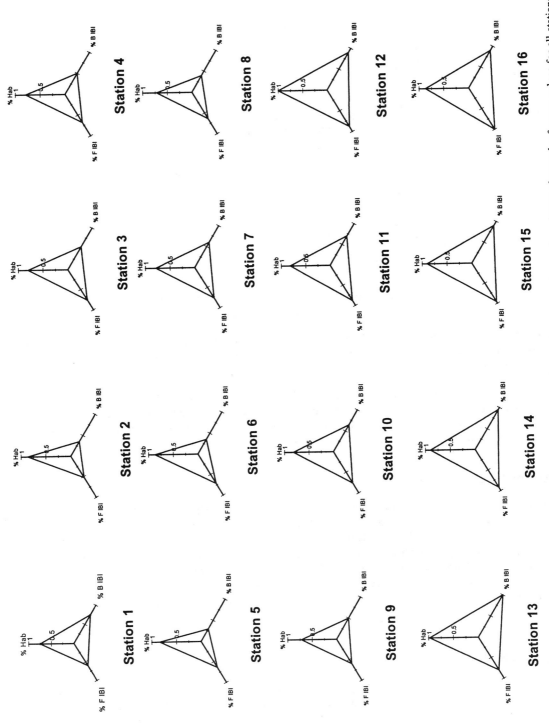

FIGURE 26.3 Fish IBI, benthic macroinvertebrate IBI, and physical habitat index results as percentages of maximum observed reference values for all stations.

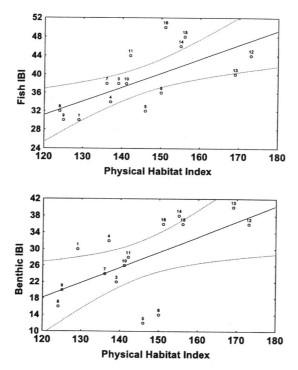

FIGURE 26.4 The relationship between Index of Biotic Integrity (IBI) scores and physical habitat index scores (fish top graph: r = 0.672; macroinvertebrate bottom graph: r = 0.587). Solid line indicates regression, and dashed lines are 95% confidence limits.

the context of subecoregional expectations, allowed the examination of the effects of multiple stressors within the watershed. In the case of the Codorus Creek watershed, the range and extent of point source discharges, reservoir releases, and impaired physical habitat stressors were reflected by using biological condition as the environmental indicator.

ACKNOWLEDGMENTS

Funding for this project was provided by the P.H. Glatfelter Company. The authors would like to sincerely thank ichthyologist, mentor, and friend Dr. Robert F. Denoncourt for discussions of pertinent historical data; Dr. C. Neal Carter, Robert Callahan, and Larry Metzger of the P.H. Glatfelter Company for their support; Pennsylvania Department of Environmental Protection biologists Rodney Kime, Robert Schott, and William Botts for discussions relevant to metrics and ecological knowledge of the area; and Tetra Tech scientists Jeroen Gerritsen, Erik Leppo, and Christiana Gerardi for their technical support for field sampling, biological sample processing, and data analysis aspects of this project.

REFERENCES

Academy of Natural Sciences of Philadelphia. 1974. Cursory survey of Codorus Creek for the P.H. Glatfelter Company. Academy of Natural Sciences Department of Limnology, Philadelphia, PA.

Barbour, M. T. 1994. An Approach to Assess the Condition of the Physical Habitat and the Benthic Macroinvertebrate Assemblage of Streams. Ph.D. dissertation. University of Maryland, Baltimore.

Barbour, M. T. and J. B. Stribling. 1991. Use of habitat assessments in evaluating the biological integrity of stream communities, in *Proceedings of the Symposium: Biological Criteria: Research and Regulation.* EPA-440/5-91-005. U.S. Environmental Protection Agency, Office of Water, Washington, DC. 25–38.

Barbour, M. T. and J. B. Stribling. 1994. A technique for assessing stream habitat structure, in *Proceedings of the Conference: Riparian Ecosystems of the Humid United States: Function, Values, and Management.* National Association of Conservation Districts, Washington, DC. 156–178.

Barbour, M.T. and J. Gerritsen. 1996. Subsampling of benthic samples: a defense of the fixed organism method, *Journal of the North American Benthological Society,* 15, 386–392.

Barbour, M. T., J. L. Plafkin, B. P. Bradley, C. G. Graves, and R. W. Wisseman. 1992. Evaluation of EPA's rapid bioassessment benthic metrics: metric redundancy and variability among reference stream sites, *Environmental Toxicology and Chemistry,* 11, 437–449.

Barbour, M. T., J. B. Stribling, and J. R. Karr. 1995. Multimetric approach for establishing biocriteria and measuring biological condition, in W. S. Davis and T. P. Simon (Eds.), *Biological Assessment and Criteria: Tools for Water Resource Planning and Decision Making.* Lewis, Boca Raton, FL. 63–77.

Barbour, M. T., J. Gerritsen, G. E. Griffith, R. Frydenborg, E. McCarron, J. S. White, and M. L. Bastian. 1996. A framework for biological criteria for Florida streams using benthic macroinvertebrates, *Journal of the North American Benthological Society,* 15(2), 185–211.

Davis, W. S., B. D. Snyder, J. B. Stribling, and C. Stoughton. 1996. Summary of State Biological Assessment Programs for Streams and Wadeable Rivers. EPA 230-R-96-007. U. S. Environmental Protection Agency, Office of Policy, Planning, and Evaluation, Washington, DC.

Davis, W. S. and T. P. Simon (Eds.). 1995. *Biological Assessment and Criteria: Tools for Water Resource Planning and Decision Making.* Lewis, Boca Raton, FL.

Denoncourt, R. F. 1992. Faunal Survey of the Codorus Creek Drainage in Relation to P. H. Glatfelter Company, September 1992. Report prepared for the P. H. Glatfelter Company, Spring Grove, PA.

DeShon, J. E. 1995. Development and application of the invertebrate community index (ICI), in W. S. Davis and T. P. Simon (Eds.), *Biological Assessment and Criteria: Tools for Water Resource Planning and Decision Making.* Lewis, Boca Raton, FL. 217–244.

Fausch, K. D., J. R. Karr, and P. R. Yant. 1984. Regional application of an index of biotic integrity based on stream fish communities, *Transactions of American Fisheries Society,* 113, 39–55.

Fausch, K. D., J. Lyons, J. R. Karr, and P. L. Angermeier. 1990. Fish communities as indicators of environmental degradation, *American Fisheries Society Symposium,* 8, 123–144.

Fore, L. S., J. R. Karr, and R. W. Wisseman. 1996. Assessing invertebrate responses to human activities: evaluating alternative approaches, *Journal of the North American Benthological Society,* 15(2), 212–231.

Gibson, G. R., M. T. Barbour, J. B. Stribling, J. Gerritsen, and J. R. Karr. 1996. Biological Criteria: Technical Guidance for Streams and Small Rivers. EPA 822-B-96-001. U.S. Environmental Protection Agency, Office of Water, Washington, DC.

Gregory, S. V., F. J. Swanson, W. A. McKee, and K. W. Cummins. 1991. An ecosystem perspective of riparian zones, *BioScience,* 41, 540–551.

Griffith, G. E., J. M. Omernik, C. M. Rohm, and S. M. Pierson. 1994. Florida Regionalization Project. EPA/Q-95/002. U.S. Environmental Protection Agency, Environmental Research Laboratory, Corvallis, OR.

Hannaford, M. J., V. H. Resh, and M. T. Barbour. 1997. Effect of training on the variability of visual-based stream habitat assessment, *Journal of the North American Benthological Society,* 16, xx–xx.

Hilsenhoff, W. L. 1987. An improved index of organic stream pollution, *The Great Lakes Entomologist,* 20, 31–39.

Hughes, R. M., S. A. Heiskary, W. J. Matthews, and C. O. Yoder. 1994. Use of ecoregions in biological monitoring, in S. L. Loeb and A. Spacie (Eds.), *Biological Monitoring of Aquatic Systems.* Lewis, Boca Raton, FL. 125–151.

Karr, J. R. 1981. Assessment of biotic integrity using fish communities, *Fisheries,* 6(6), 21–27.

Karr, J. R. 1991. Biological integrity: a long-neglected aspect of water resource management, *Ecological Applications,* 1, 66–84.

Karr, J. R. 1995. Protecting aquatic ecosystems: clean water is not enough, in W. S. Davis and T. P. Simon (Eds.), *Biological Assessment and Criteria: Tools for Water Resource Planning and Decision Making.* Lewis, Boca Raton, FL. 7–14.

Karr, J. R., K. D. Fausch, P. L. Angermeier, P. R. Yant, and I. J. Schlosser. 1986. Assessing Biological Integrity in Running Waters: A Method and Its Rationale. Illinois Natural History Survey Special Publication 5, Champaign.

Kerans, B. L. and J. R. Karr. 1994. A benthic index of biotic integrity (B-IBI) for rivers of the Tennessee Valley, *Ecological Applications*, 4, 768–785.

Kerans, B. L., J. R. Karr, and S. A. Ahlstedt. 1992. Aquatic invertebrate assemblages: spatial and temporal differences among sampling protocols, *Journal of the North American Benthological Society*, 11, 377–390.

Lyons, J. 1992. Using the Index of Biotic Integrity (IBI) to Measure Environmental Quality in Warmwater Streams of Wisconsin. North Central Forest Experiment Station, U.S. Department of Agriculture. General Technical Report NC-149.

NRC. 1992. *Restoration of Aquatic Ecosystems*. National Research Council, National Academy Press, Washington, DC.

Ohio Environmental Protection Agency. 1987. *Biological Criteria for the Protection of Aquatic Life. Vols. I–III*. Ohio Environmental Protection Agency, Division of Water Quality Monitoring and Assessment, Surface Water Section, Columbus.

Omernik, J. M. 1987. Ecoregions of the conterminous United States, *Annals of the Association of American Geographers*, 77, 118–125.

Omernik, J. M. and A. L. Gallant. 1990. Defining regions for evaluating environmental resources, in H. G. Lund and G. Preto (Coordinators). *Global Natural Resource Monitoring and Assessments: Preparing for the 21st Century*. American Society of Photogrammetry and Remote Sensing, Bethesda, MD. 936–947.

Plafkin, J. L., M. T. Barbour, K. D. Porter, S. K. Gross, and R. M. Hughes. 1989. Rapid Bioassessment Protocols for Use in Streams and Rivers: Benthic Macroinvertebrates and Fish. EPA/440/4-89-001. U.S. Environmental Protection Agency, Office of Water, Washington, DC.

Rosen, B. H. 1995. Use of periphyton in the development of biocriteria, in W. S. Davis and T. P. Simon (Eds.), *Biological Assessment and Criteria: Tools for Water Resource Planning and Decision Making*. Lewis, Boca Raton, FL. 209–216.

Simon, T. P. and J. Lyons. 1995. Application of the index of biotic integrity to evaluate water resource integrity in freshwater ecosystems, in W. S. Davis and T. P. Simon (Eds.), *Biological Assessment and Criteria: Tools for Water Resource Planning and Decision Making*. Lewis, Boca Raton, FL. 245–262.

Southerland, M. T. and J. B. Stribling. 1995. Status of biological criteria development and implementation, in W. S. Davis and T. P. Simon (Eds.), *Biological Assessment and Criteria: Tools for Water Resource Planning and Decision Making*. Lewis, Boca Raton, FL. 81–96.

Southwood, T. R. E. 1977. Habitat, the templet for ecological strategies?, *Journal of Animal Ecology*, 46, 337–365.

U.S. Environmental Protection Agency (U.S. EPA). 1993. An SAB Report: Evaluation of Draft Technical Guidance on Biological Criteria for Streams and Small Rivers. Prepared by the Biological Criteria Subcommittee of the Ecological Processes and Effects Committee. EPA-SAB-EPEC-94-003. U.S. Environmental Protection Agency, Science Advisory Board, Washington, DC.

U.S. EPA. 1997. Field and Laboratory Methods for Macroinvertebrate and Habitat Assessment of Low Gradient Nontidal Streams. Mid-Atlantic Coastal Streams Workgroup, Environmental Services Division, Region 3, Wheeling, WV.

Woods, A. J., J. M. Omernik, D. D. Brown, and C. W. Kiilsgaard. 1996. Level III and IV Ecoregions of Pennsylvania and the Blue Ridge Mountains, the Ridge and Valley, and the Central Appalachians of Virginia, West Virginia, and Maryland. EPA/600R-96/077. U.S. Environmental Protection Agency, National Health and Environmental Effects Laboratory, Corvallis, OR.

Yoder, C. O. and E. T. Rankin. 1995. Biological criteria program development and implementation in Ohio, in W. S. Davis and T. P. Simon (Eds.), *Biological Assessment and Criteria: Tools for Water Resource Planning and Decision Making*. Lewis, Boca Raton, FL. 109–144.

Index

A

Abundance metric
  description of, 88
  for fish assemblages, 18–19
  in northeastern United States, 325
  number of individuals per unit area or catch per effort, 325
  percent of hybrid individuals, 325
  percent of individuals displaying DELT, 325–326
  for tailwater fish index, 516, 519
Acantharchus pomotis, 185
Acipenser spp.
  A. brevirostrum, 141
  A. fulvescens, 141
  A. medirostris, 142
  A. oxyrhynchus, 142
  A. transmontanus, 142
Acrocheilus glutaceus, 127, 146
Agencies, collaboration among
  case study of, 59–61
  common ground for, 58–59
  description of, 58
  differences of opinion, 59
  guidance for development of, 61–62
Allopatric speciation, 70–72
Alosa spp.
  A. aestivalis, 130, 145
  A. alabamae, 130, 145
  A. chrysochloris, 145
  A. mediocris, 130, 145
  A. pseudoharengus, 130, 145
  A. sapidissima, 130
Ambloplites spp.
  A. ariommus, 185
  A. cavifrons, 185
  A. constellatus, 185
  A. rupestris, 185
Ameiurus spp.
  A. brunneus, 169
  A. catus, 169
  A. melas, 169
  A. natalis, 169
  A. nebulosus, 170
  A. platycephalus, 170
Amia calva, 132, 144
Ammocrypta spp.
  A. beani, 189
  A. clara, 189
  A. pellucida, 189
  A. vivax, 189
Anguilla rostrata, 145
Anthropogenic disturbances, index of biotic integrity and, studies in Virginia to determine relations between
  methods
    data source, 587
    fish sampling, 588
    habitat variables, 587–590
    statistical tests, 592–594
  metrics
    description of, 586
    disturbances measures and
      in coastal plain, 594, 606–607
      in mountains, 602–604, 606–607
      in Piedmont, 594, 596–602, 606–607
      region-specific findings, 605–606
      relations among, 604
    number of intolerant species, 591
    number of tolerant species, 591
    optimum methods to examine, 586
    reproductive, 592
    species richness, 588, 591–592
    trophic guild, 592
    types used, 588, 591–592
  overview, 585–586
  species classification, 607
Apeltes quadracus, 130, 181
Aphredoderus sayanus, 178
Aplodinotus gunniens, 202
Aquatic communities, *see* Fish communities
Aquatic systems
  fish communities for evaluating, *see* Fish communities
  man-made degradation of, 5
  perturbations of, 7–8
Ariadnophils, reproductive guild classification of, 110
Assemblages, *see* Fish assemblages

B

Benthic algae, 126
Benthic predators
  biological assessments of, in Pennsylvania
    assessment of condition of, 646–647
    metrics for assessing, 643–644
  crushers, 131
  description of, 130
  diggers, 131
  grazers, 130–131

hunters of mobile benthos, 131
lie-in-wait predators, 131
tearers, 131
Biocriteria, definition of, 59
Biological assessments
  case study of, in Pennsylvania
    background, 640
    benthic macroinvertebrates
      assessment of condition of, 646–647
      metrics for assessing, 643–644
    fish assemblage
      assessment of condition of, 647–648
      metrics for assessing, 644–646
    framework for, 640
    integrated assessments, 648–650
    objectives, 640
    physical habitat assessments, 646
    physical setting, 641–642
    survey and assessment methods
      benthic macroinvertebrate metrics, 643–644
      fish metrics, 644–646
      physical habitat, 642–643
  goal of, 640
Biological community
  flow levels of, effect on biological integrity, 8–9
  new evaluation methods, 36–37
Biological criteria, from fish assemblages
  decision criteria, 35–36
  definition of, 37
  description of, 35, 80
  evaluation methods, 36–37
  modified index of well-being, 39–40
  Ohio EPA modified index of biotic integrity, 37–39
  regionally referenced numerical, 36
Biological integrity
  biological monitoring programs to detect changes in, 6–7
  criteria for, 7
  critical flow values and, 8–9
  definition of, 4, 403
  factors that affect, 7–8
Biological monitoring
  changes in biological condition evidence by using, 6–7
  of Lake Erie, fish sampling methods for
    beach seine, 421, 423
    electrofishing, 421–424
    gillnet, 419–421, 423
    hoopnet, 421, 423
    sites of, 419
    trawl, 421, 423
Biological species concept, 70
Biotic integrity
  definition of, 367
  index of, see Index of biotic integrity
Brevoortia spp.
  B. sapidissima, 145
  B. tyrannus, 129, 145
Brood hiders, reproductive guild classification of, 109
BSC, see Biological species concept

C

California, index of biotic integrity evaluations in
  Dye Creek, 375–376
  Putah Creek, 372–375
  Sierra Nevada watershed evaluations, 372
Campostoma spp.
  C. anomalum, 127, 146
  C. oligolepis, 127, 146
Carassius auratus, 146
Carnivores
  description of, 132
  parasites
    blood suckers, 133
    description of, 133
  whole body (piscivore)
    ambush, 132
    chasing, 132
    protective resemblance, 132–133
    stalking, 132
Carpiodes spp.
  C. carpiodes, 128, 164
  C. cyprinus, 128, 164
  C. velifer, 128, 164
Catostomus spp.
  description of, 203–204
  feeding of, 203–204, 211
  members of
    C. catastomus, 164
    C. columbianus, 127, 164
    C. commersoni, 128, 165
    C. macroheilus, 165
    C. platyrhynchus, 165
  in Ohio River, studies to determine effect on index of biotic integrity metrics
    indirect effects on metrics
      catch per unit effort, 222
      percent DELT anomalies, 222
      percent insectivores, 218, 222
      percent omnivores, 222
      percent simple lithophils, 222
      total number of species, 218
    longitudinal trends, 207
    materials and methods
      sample methods, 204–206
      study area, 204
    prevalent areas, 207–211
    proportional metrics, 215, 218
    species composition, 206
    species richness, 215
    temporal trends, 212–214
Centrarchus macropterus, 185
Clinostomus
  C. elongatus, 132, 146
  C. funduloides, 147
Coldwater streams
  in upper midwestern North American, index of biotic integrity for
    applications of, 407–408
    assemblage characteristics, 408–409

Index

biological integrity vs. coldwater stream management, 403–404
comparison with other indices of biotic integrity, 404–406
description of, 384
development of, 385–387
metrics
number of benthic species, 397
number of coldwater individuals per 150 m, 399
number of coldwater species, 393–395
number of minnow species, 397
number of species, 393
number of tolerant species, 396–397
number of warmwater individuals per 150 m, 399
percent coldwater individuals, 398
percent of salmonids as brook trout, 398
percent tolerant individuals, 398
percent top carnivores, 399
percent white suckers, 398–399
scoring of, 387, 389–393
selection of, 385–387, 388–389
sensitivity range, 387
modifications of, 408–409
testing of
ecoregion comparisons, 401–402
methods, 387–388
"natural" vs. "managed" coldwater streams, 402
validation of, 387–388
IBI rating vs. fish habitat rating, 400
temporal variability, 400–401
warmwater streams and, comparisons between, 384
Connecticut specific index, 326
Coregonus spp.
C. artedi, 175
C. autumnalis, 175
C. clupeaformis, 175
C. hoyi, 175
C. huntsmani, 175
C. kiyi, 175
C. laurettae, 175
C. nasus, 175
C. reighardi, 175
C. sardinella, 175
C. zenithicus, 130, 176
Cottus spp.
C. aleuticus, 182
C. asper, 182
C. baileyi, 183
C. bairdi, 183
C. carolinae, 183
C. cognatus, 183
C. confusus, 183
C. girardi, 183
C. hypselurus, 184
C. rhotheus, 184
C. ricei, 184
Couesius plumbeus, 147
Crystallaria asprella, 189
Ctenopharyngodon idella, 127, 147
Culaea inconstans, 182

Cycleptus elongatus, 165
Cyprinella
C. analostana, 147
C. caerulea, 132, 147
C. callistia, 147
C. camura, 147
C. galactura, 131, 147
C. labrosa, 131, 148
C. lutrensis, 131, 148
C. spiloptera, 129, 148
C. trichroistia, 149
C. venusta, 149
C. whipplei, 132, 149
Cyprinus carpio, 128, 149

D

Dallia pectoralis, 131, 174
Dams
fish migration effects, 318, 508
in western North America, effect on index of biotic integrity metric use, 371
Darters, as index of biotic integrity metric, 82, 85, 323, 517
Deformities, erosion, lesions, and tumors metric
in bioassessments, 226–227
examples of, 226
fish community quality use, 244
in Lake Agassez Plain ecoregion, 356–357
pollution sites associated with, 225–226
stream quality indicator use, 243
study of, in fish assemblages in Ohio Rivers
area for, 227–228
conclusions, 243–244
description of, 226–227
methods
examination for DELT anomalies, 229–230
fish sampling, 229
severity classifications, 230
stream quality analyses, 230–231
results
anomalies predominantly seen, 235–236, 2312
effect of point source discharge, 238–239
index of biotic integrity and, 237–239
severity of anomalies, 239–243
species with anomalies, 236–237
DELT, see Deformities, erosion, lesions, and tumors metric
Detritivores
description of, 128
filter feeders
filters, 128
suction feeders, 128
particulate feeders
biters, 128
scoopers, 129
Dispersal, 66–67
Dorosoma spp.
D. cepedianum, 145
D. petenense, 130, 146

Drift feeders
  description of, 131
  surface, 131–132
  water column, 132
Dye Creek
  description of, 375–376
  index of biotic integrity evaluations, 375–376

E

Ecosystems
  biological integrity of
    biological monitoring programs to detect changes in, 6–7
    criteria for, 7
    critical flow values and, 8–9
    definition of, 4, 403
    factors that affect, 7–8
  health, criticisms of, 10–12
Elassoma zonatum, 186
Electrofishing
  equipment for, 25
  field conditions for, 34
  methods of, 26
    boat, 31–32
    general types, 421–422
    wading, 28, 30–31
  for reservoir fish assemblage index development, 534
  studies of, 24–25
Endemism, in western North America streams, 370
Ennecanthus spp.
  E. chaetodon, 186
  E. gloriosus, 186
  E. obesus, 186
Environmental indicators
  attributes of, 21
  definition of, 19
  fish assemblages as
    advantages of, 22
    biological criteria
      decision criteria, 35–36
      definition of, 37
      description of, 35
      evaluation methods, 36–37
      modified index of well-being, 39–40
      Ohio EPA modified index of biotic integrity, 37–39
      regionally referenced numerical, 36
    versus chemical/physical indicators, 22
    cost considerations, 34–35
    data management, 34
    description of, 19–21
    enumeration procedures, 32–34
    field conditions, 34
    fish handling, 32–34
    information processing, 34
    mobility concerns, 22–23
    sampling methods
      boat, 31–32
      considerations for, 23–24
      field, 24–26
      initial decisions, 24
      personnel to implement, 24
      wading, 26–31
    value of, 21–23
Environmental Protection Agency (EPA), 57–58
Epiphytic algae, 126
Erimystax spp.
  E. cahni, 149
  E. dissimilis, 149
  E. insignis, 149
  E. x-punctata, 149
Ermyzon spp.
  E. oblongus, 165
  E. sucetta, 165
Esox spp.
  E. americanus, 173
  E. americanus vermiculatus, 173
  E. lucius, 173
  E. masquinongy, 174
  E. niger, 174
Etheostoma spp.
  E. acuticeps, 190
  E. aquali, 190
  E. aspirgene, 190
  E. baileyi, 190
  E. barbouri, 190
  E. barrenense, 191
  E. bellum, 191
  E. blennioides, 191
  E. blennius, 191
  E. boschungi, 191
  E. caeruleum, 191
  E. camurum, 191
  E. chlorobranchium, 191
  E. chlorosomum, 191
  E. cinereum, 191
  E. colletei, 191
  E. collis, 192
  E. coosae, 192
  E. cragini, 192
  E. crossopterum, 192
  E. ditrema, 192
  E. duryi, 192
  E. etnieri, 192
  E. euzonum, 192
  E. exile, 192
  E. flabellare, 193
  E. flavum, 193
  E. fusiforme, 193
  E. gracile, 193
  E. histrio, 193
  E. jordani, 193
  E. juliae, 193
  E. kanawhae, 193
  E. kennicotti, 193
  E. longimanum, 194
  E. luteovinctum, 194
  E. lynceum, 194
  E. maculatum, 194
  E. microlepidum, 194
  E. microperca, 194
  E. moorei, 194

E. neopterum, 194
E. nigripinne, 194
E. nigrum, 194
E. obeyense, 194
E. olivaceum, 194
E. olmstedi, 195
E. osburni, 195
E. pallididorsum, 195
E. parvipinne, 195
E. podostemone, 195
E. proeliare, 195
E. punctulatum, 196
E. pyrrhogastor, 196
E. radiosum, 196
E. rufilineatum, 196
E. rupestre, 196
E. sagitta, 196
E. sanguifluum, 196
E. serrifer, 196
E. simoterum, 197
E. smithi, 197
E. spectabile, 197
E. squamiceps, 197
E. stigmaeum, 197
E. striatulum, 197
E. swaini, 197
E. swannanoa, 197
E. tippecanoe, 197
E. trisella, 198
E. tuscumbia, 198
E. variatum, 198
E. virgatum, 198
E. vitreum, 198
E. vulneratum, 198
E. wapiti, 198
E. whipplei, 198
E. zonale, 198
E. zonistium, 198
Exceptional warmwater habitat, 615
Exoglossum
   E. laurae, 131, 149
   E. maxillingua, 131, 149
Exposure indicators, definition of, 19

F

Federal agencies, collaboration among
   case study of, 59–61
   common ground for, 58–59
   description of, 58
   differences of opinion, 59
   guidance for development of, 61–62
Feeding strategies, for creating trophic guild structure of North American fishes
   carnivores
      description of, 132
      parasites
         blood suckers, 133
         description of, 133
      whole body (piscivore)
         ambush, 132
         chasing, 132
         protective resemblance, 132–133
         stalking, 132
   detritivores
      description of, 128
      filter feeders
         filters, 128
         suction feeders, 128
      particulate feeders
         biters, 128
         scoopers, 129
   generalists, 134
   herbivores
      classification of, 126–127
      description of, 126–127
      feeding anatomy, 127
      particulate feeders
         browsers, 127
         description of, 127
         grazers, 127
   information existing regarding, 125–126
   invertivores
      benthic predators
         crushers, 131
         description of, 130
         diggers, 131
         grazers, 130–131
         hunters of mobile benthos, 131
         lie-in-wait predators, 131
         tearers, 131
      drift feeders
         description of, 131
         surface, 131–132
         water column, 132
   multiple trophic states, 134–135
   opportunists, 134
   planktivores
      description of, 129
      nondiscriminate filter feeders
         gulping, 130
         mechanical sieve, 129
         mucus entrapment, 129
         pump filtration, 129–130
         ram filtration, 129
      particulate feeders
         description of, 130
         size-selective pickers, 130
Fish, *see also specific species*
   as environmental indicators, 4–5
   freshwater, *see* Freshwater fish
   sampling methods
      beach seine, 421, 423
      electrofishing, *see* Electrofishing
      gillnet, 419–421, 423
      hoopnet, 421, 423
      sites of, 419
      trawl, 421, 423
Fish assemblages
   abundance metric for, 18–19
   biological assessments of, case study in Pennsylvania
      assessment of condition of, 647–648

metrics for assessing, 644–646
coldwater *vs.* warmwater, 407
definition of, 274
as environmental indicator
  advantages of, 22
  bioassessment use
    habitat assessments, 44–47
    Ottawa River case example, 42–44
    overview, 41
    Scioto River case example, 41–42
    water quality management program evaluations, 41
  biological criteria
    decision criteria, 35–36
    definition of, 37
    description of, 35
    evaluation methods, 36–37
    modified index of well-being, 39–40
    Ohio EPA modified index of biotic integrity, 37–39
    regionally referenced numerical, 36
  vs. chemical/physical indicators, 22
  cost considerations, 34–35
  data management, 34
  description of, 19–21
  enumeration procedures, 32–34
  field conditions, 34
  fish handling, 32–34
  information processing, 34
  for long-term goal assessments, 48–49
  miscellaneous uses, 51
  mobility concerns, 22–23
  of non-game communities, 51
  sampling methods
    boat, 31–32
    considerations for, 23–24
    field, 24–26
    initial decisions, 24
    personnel to implement, 24
    wading, 26–31
  for statewide assessments, 47–48
  value of, 21–23
  for watershed assessments, 49–51
factors that affect, 274
historical studies of, 18
index of biotic integrity for studying, *see* Index of biotic integrity
reasons for using, 18
relative abundance data, 18–19
reservoir index, development of, 542
  metrics
    abundance, 534
    percent DELT, 534
    reproductive comparison, 534
    taxa richness and composition, 529
    trophic composition, 533–534
  overview, 524
  reference conditions, 525–527
  reservoir classification, 524–525
  results of, in Tennessee Valley Authority reservoirs, 537–538
  sampling methods, 534
  scoring criteria, 525–532
  validity of, 538–539
  variability, 535–537
in upper Snake River Basin, studies of
  description of, 274
  history of, 277, 279
  native fish species, 277, 279
  purpose of, 275
watershed assessments using, 49–51
Fish communities
  as biological integrity indicator, historical use, 4
  degradation of, water resource quality and, 49
  diversity of, 66
    alpha, 69
    beta, 69
    gamma, 69
    historical factors
      Ohio River Basin fishes as case study of, 72–74
      principles, 72
    regional factors of
      speciation, *see* Speciation
      species, 69–70
      stream capture, 74–75
    regional movements and, 69
  habitat affinity, 66
  local and regional factors
    dispersal, 68–69
    metapopulation
      model, 66–67
      mutation, 67
      in stream fishes, estimations of structure, 67–68
    migration, 66
    species' range, 68–69
  principles of, 65–66
Flow levels, effect on biological integrity of ecosystem, 8–9
Freshwater fish, index of biotic integrity for species classification in northeastern United States
  aquatic habitat classification, 319
  assumptions for, 316
  existing versions, constraints of, 326
  fish species tolerance classification, 319
  formulation considerations, 320
  northeast regional applications
    fish abundance and condition metrics, 325–326
    fish sampling methods, 320–321
    metrics used, 321–322
    overview, 320
    species richness and composition metrics, 322–325
    trophic guild structure metrics, 325
  regional modifications, 320
  resident species for developing, 316–317
  trophic categories
    benthic insectivores, 318
    generalist feeders, 317–318
    minor types of, 318
    northeast groups, 317
    top carnivores, 318
    water column insectivores, 318
  validation of, 326
Fundulus spp.
  F. catenatus, 179
  F. chrysotus, 179

Index

F. diaphanus, 179
F. dispar, 179
F. heteroclitus, 180
F. julisia, 180
F. lineolatus, 180
F. notatus, 180
F. notti, 180
F. olivaceus, 181
F. rathbuni, 181
F. stellifer, 181

G

Gambusia spp.
   G. affinis, 132, 181
   G. holbrooki, 181
Gasterosteus spp.
   G. aculeatus, 182
   G. wheatlandi, 182
Gill chamber brooders, reproductive classification of, 111
Great rivers
   definition of, 476
   index of biotic integrity, sampling considerations for
     data sources, 477
     gear limitations and considerations
       community sampling, 481
       day *vs.* night sampling, 479–481
       description of, 478
       multiple gears, 478–479
     habitat
       effect on metrics, 500
       qualitative habitat evaluation index, 500–501
       substrate differences, 496–500
     method considerations
       biomass, 484
       density, 484
       downstream distance from dam, 485–486
       species richness, 481–484
       water temperature, 486–493
     minimum detection difference, 496
     modified index of well-being, 484–485
     reference conditions for, 476–477
     select metrics
       description of, 493
       Emerald shiner exclusion, 493–494
       gizzard shad exclusion, 493–494
       percent DELT anomalies, 494–495
       relative number of round-bodied suckers, 494
     stability and consistency, 477–478
     statistical comparisons, 495–496
     trends, 484–485
Green plants, trophic classification of, 126
Green sunfish, as index of biotic integrity metric, 86–87, 325, 574–575

H

Habitat, *see also specific study,* habitat
   fish assemblages to study, 44–47
Hemitremia flammea, 150

Herbivores
   anatomy, 127
   classification of, 126–127
   description of, 126–127
   particulate feeders
     browsers, 127
     description of, 127
     grazers, 127
Hiodon
   H. alosoides, 132, 144
   H. tergisus, 132, 144
Hybognathus spp.
   H. hankinsoni, 129, 150
   H. hayi, 150
   H. nuchalis, 150
   H. placitus, 150
   H. regius, 150
Hypentelium spp.
   H. etowanum, 165
   H. nigricans, 165
   H. roanokense, 165
Hypomesus olidus, 174
Hypophthalmichthys spp.
   H. molitrix, 128, 151
   H. nobilis, 129, 151

I

IBI, *see* Index of Biotic Integrity
Ichthyomyzon
   I. bdellium, 139
   I. castaneus, 139
   I. fossor, 139
   I. gagei, 128, 139
   I. greeleyi, 139
   I. unicuspis, 139–140
ICI, *see* Invertebrate community index
Ictalurus spp.
   I. furcatus, 170
   I. punctatus, 170
Ictiobus spp.
   I. bubalus, 166
   I. cyprinellus, 166
   I. niger, 166
Index of Biological Integrity, *see* Index of biotic integrity
Index of biotic integrity
   adaptability of, 81
   adjustments to
     description of, 635
     fish assemblage stressors, 630–631
     low-end, 626
     metric
       modifications, 633
       scoring, 627
     predominant species swamping, 633–634
     for young-of-year fish, 634–635
   anthropogenic disturbances and, studies in Virginia to determine relations between
     methods
       data source, 587
       fish sampling, 588

habitat variables, 587–590
statistical tests, 592–594
metrics
   description of, 586
   disturbances measures and
      in coastal plain, 594, 606–607
      in mountains, 602–604, 606–607
      in Piedmont, 594, 596–602, 606–607
      region-specific findings, 605–606
      relations among, 604
   number of intolerant species, 591
   number of tolerant species, 591
   optimum methods to examine, 586
   reproductive, 592
   species richness, 588, 591–592
   trophic guild, 592
   types used, 588, 591–592
overview, 585–586
species classification, 607
for coldwater streams in upper midwestern North American
   applications of, 407–408
   assemblage characteristics, 408–409
   biological integrity vs. coldwater stream management, 403–404
   comparison with other indices of biotic integrity, 404–406
   description of, 384
   development of, 385–387
   metrics
      number of benthic species, 397
      number of coldwater individuals per 150 m, 399
      number of coldwater species, 393–395
      number of minnow species, 397
      number of species, 393
      number of tolerant species, 396–397
      number of warmwater individuals per 150 m, 399
      percent coldwater individuals, 398
      percent of salmonids as brook trout, 398
      percent tolerant individuals, 398
      percent top carnivores, 399
      percent white suckers, 398–399
      scoring of, 387, 389–393
      selection of, 385–387, 388–389
      sensitivity range, 387
   modifications of, 408–409
   testing of
      ecoregion comparisons, 401–402
      methods, 387–388
      "natural" vs. "managed" coldwater streams, 402
   validation of, 387–388
      IBI rating vs. fish habitat rating, 400
      temporal variability, 400–401
criticisms of, 10–12
definition of, 274, 302, 367, 476
description of, 3–4, 12, 80, 249–250, 625
development of, 80
for freshwater fish species classification in northeastern United States
   aquatic habitat classification, 319
   assumptions for, 316

existing versions, constraints of, 326
fish species tolerance classification, 319
formulation considerations, 320
northeast regional applications
   fish abundance and condition metrics, 325–326
   fish sampling methods, 320–321
   metrics used, 321–322
   overview, 320
   species richness and composition metrics, 322–325
   trophic guild structure metrics, 325
regional modifications, 320
resident species for developing, 316–317
trophic categories
   benthic insectivores, 318
   generalist feeders, 317–318
   minor types of, 318
   northeast groups, 317
   top carnivores, 318
   water column insectivores, 318
validation of, 326
for great rivers
   data sources, 477
   gear limitations and considerations
      community sampling, 481
      day vs. night sampling, 479–481
      description of, 478
      multiple gears, 478–479
   habitat
      effect on metrics, 500
      qualitative habitat evaluation index, 500–501
      substrate differences, 496–500
   method considerations
      biomass, 484
      density, 484
      downstream distance from dam, 485–486
      species richness, 481–484
      water temperature, 486–493
   minimum detection difference, 496
   modified index of well-being, 484–485
   reference conditions for, 476–477
   select metrics
      description of, 493
      Emerald shiner exclusion, 493–494
      gizzard shad exclusion, 493–494
      percent DELT anomalies, 494–495
      relative number of round-bodied suckers, 494
   stability and consistency, 477–478
   statistical comparisons, 495–496
   trends, 484–485
index of well-being and, relationship between, 40
for lakes, see also specific lake
   impediments to development of
      ecological factors, 565–566
      management agencies, 565
      public perceptions, 565
      sampling considerations, 564–565
      sampling gear, 564–565
      scope of assessment, 564
      water classification difficulties, 564
   in southern New England
      field methods, 566

Index

fish sampling methods, 566
metrics
    number of green sunfish species, 574–575
    number of intolerant species, 573–574
    number of introduced species, 572–573
    number of native species, 572
    number of sucker species, 574
    number of tolerant species, 575, 579–580
    percent insectivores, 576–577
    percent omnivores, 577
    percent top carnivores, 575–576
    species richness, 569–575, 578–579
    trophic composition, 575–577, 580
    types evaluated for use, 567–568
overview, 563–564, 577–580
quantitative methods, 566–567
results, 568–575
sample design, 566
in Wisconsin
  methods and materials for developing
    data sources, 543, 547–548
    metrics, 543–546
  metrics
    environmental quality of lake and, relationship between, 549, 553–554
    lake size and ecoregion effects, 549, 554–555
    method-specific, 558–559
    number of benthic species, 553–554
    number of cyprinid species, 555
    sampling variation among, 548–549
    species richness, 549–551
    trophic state and, 559
    types for further development, 555–558
  overview, 541–543
  sampling procedures, 557–560
metrics, *see also specific study,* metrics
  Catostomus spp. effect on, in Ohio River
    indirect effects on metrics
      catch per unit effort, 222
      percent DELT anomalies, 222
      percent insectivores, 218, 222
      percent omnivores, 222
      percent simple lithophils, 222
      total number of species, 218
    longitudinal trends, 207
    materials and methods
      sample methods, 204–206
      study area, 204
    prevalent areas, 207–211
    proportional metrics, 215, 218
    species composition, 206
    species richness, 215
    temporal trends, 212–214
  definition of, 80–81, 249, 302
  individual abundance, 88
  percent hybrid individuals, 88–89
  percent individuals with anomalies, 89
  species richness
    darter, 82, 85
    intolerant, 86
    percent green sunfish, 86–87
    sucker, 86
    sunfish, 85–86
  stream size and, studies in Virginia
    discussion
      general recommendations, 261, 263, 268
      stream size adjustments, 263
    functional metrics, 269
    methods
      data source, 251–252
      disturbances, 253–254
      fish sampling, 252
      habitat effects, 253–254
      physiography, drainage, and stream size, 254
    overview, 249–250
    potential metrics
      description of, 254
      reproduction, 256
      taxonomic, 254, 256
      tolerant species, 256
      trophics, 256
    results
      disturbance effects, 261
      least-disturbed sites, 260–261, 267
      statistical tests, 256, 260
      taxonomic metrics, 268–269
      watershed land use effects, 253
  trophic composition
    percent insectivorous individuals, 87
    percent omnivorous individuals, 87
    percent piscivorous individuals, 88
  in upper Snake River, *see* Upper Snake River
modifications
  definition of, 81
  description of, 5, 81, 250
  by Ohio EPA, 37–39
overview of, 303
principles of, 81, 249
purpose of, 124
scores, definition of, 249–250
uses of, 4
watershed
  description of, 377
  for Sierra Nevada watersheds, 372
in western North America
  California
    Dye Creek, 375–376
    Putah Creek, 372–375
    Sierra Nevada watershed evaluations, 372
  difficulties in using, 367–368
  problems associated with
    dams and diversions, 371
    description of, 368
    high endemism, 370
    introduced fishes, 370–371
    low species richness, 368–370
    watershed evaluation, 371–372
  watershed evaluation
    problems associated with, 371–372
    Sierra Nevada, 372
zero scoring, 627–633

Index of well-being
  development of, 36
  index of biotic integrity and, relationship between, 40
  modifications to, 39–40, 484–485
  principles of, 39
Insectivores
  assessment of, 87
  percent, as IBI metric, 218, 222, 318
Invertebrate community index, development of, 36
Invertivores
  benthic predators
    biological assessments of, in Pennsylvania
      assessment of condition of, 646–647
      metrics for assessing, 643–644
    crushers, 131
    description of, 130
    diggers, 131
    grazers, 130–131
    hunters of mobile benthos, 131
    lie-in-wait predators, 131
    tearers, 131
  drift feeders
    description of, 131
    surface, 131–132
    water column, 132
Iwb, see Index of Well-Being

L

Labidesthes sicculus, 132, 181
Lacustuary
  assessments, in Lake Erie region
    Duck Creek, 448
    Maumee River, 447–448
    Ottawa River, 447
    Otter Creek, 448–449
  definition of, 418
  estuaries and, comparison between, 419
Lagochila lacera, 131, 166
Lake Agassiz Plain ecoregion
  index of biotic integrity development
    materials and methods
      DELT examinations, 345
      fish sampling procedures, 344–345
      reference condition, 343–344
      reference site selection, 344
    metrics
      alternative, 357
      description of, 345
      modifications, 357–358
      number of individuals per meter, 356
      overview, 346
      proportion of individuals as simple lithophilic spawners, 356
      proportion of individuals with deformities, eroded fins, lesions, and tumors, 356–357
      species composition and richness evenness, 352
      in headwater streams, 360
      in Lake Agassiz Plain, 358
      in large streams, 360
      in Minnesota, 361
      in moderate-sized streams, 360
      in North Dakota, 361
      number of benthic insectivore species, 347
      number of minnow species, 352
      number of sensitive species, 353
      proportion of headwater species, 345, 347
      proportion of large river individuals, 352
      proportion of round-bodied suckers, 347, 352
      proportion of tolerant individuals, 353
      in Red River, 362
      total number of fish species, 345, 349–351
    trophic guilds
      proportion of insectivore biomass, 354
      proportion of omnivore biomass, 353–354
      proportion of pioneer species, 354–355
      proportion of piscivore biomass, 354
    scoring trends
      in headwater, moderate-sized, and large rivers, 360–361
      in Lake Agassiz Plain, 358–359
      in Minnesota, 361–362
      in North Dakota, 361–362
      in Red River, 363
    variability, 363
Red River Basin
  drainage features, 341
  historical data, 342–343
study area description of, 341–342
Lake Erie
  biological monitoring of, fish sampling
    beach seine methods, 421, 423
    electrofishing, 421–424
    gillnet method, 419–421, 423
    hoopnet methods, 421, 423
    sites of, 419
    trawl methods, 421, 423
  index of biotic integrity
    applications of, lacustuary assessments
      Duck Creek, 448
      Maumee River, 447–448
      Ottawa River, 447
      Otter Creek, 448–449
    classifications, 444–447
    metrics
      number of benthic species, 424–425, 428–430
      number of cyprinid species, 425, 430
      number of individuals, 437, 441–442
      number of native species, 424, 426–427
      number of phytophilic species, 430, 432–433
      number of sunfish species, 425–427
      percent diseased individuals, 439, 441–442
      percent lake individuals, 431, 434, 438
      percent nonindigenous species, 435–437, 439
      percent omnivorous individuals, 431–433
      percent phytophilic individuals, 434, 437, 440
      percent tolerant individuals, 431, 435–436
      percent top carnivores, 437–438, 440
      selection rationale, 424
    scoring considerations, 439, 444

Lakes, index of biotic integrity for, *see also specific lake*
  impediments to development of
    ecological factors, 565–566
    management agencies, 565
    public perceptions, 565
    sampling considerations, 564–565
    sampling gear, 564–565
    scope of assessment, 564
    water classification difficulties, 564
  in southern New England
    field methods, 566
    fish sampling methods, 566
    metrics
      number of green sunfish species, 574–575
      number of intolerant species, 573–574
      number of introduced species, 572–573
      number of native species, 572
      number of sucker species, 574
      number of tolerant species, 575, 579–580
      percent insectivores, 576–577
      percent omnivores, 577
      percent top carnivores, 575–576
      species richness, 569–575, 578–579
      trophic composition, 575–577, 580
      types evaluated for use, 567–568
    overview, 563–564, 577–580
    quantitative methods, 566–567
    results, 568–575
    sample design, 566
  in Wisconsin
    methods and materials for developing
      data sources, 543, 547–548
      metrics, 543–546
    metrics
      environmental quality of lake and, relationship between, 549, 553–554
      lake size and ecoregion effects, 549, 554–555
      method-specific, 558–559
      number of benthic species, 553–554
      number of cyprinid species, 555
      sampling variation among, 548–549
      species richness, 549–551
      trophic state and, 559
      types for further development, 555–558
    overview, 541–543
    sampling procedures, 557–560
Lampetra spp.
  L. aepyptera, 140
  L. appendix, 140
  L. ayresi, 129, 140
  L. japonica, 140
  L. richardsoni, 128, 140–141
  L. tridentata, 141
Large rivers
  barriers to understanding, 463–464
  biological assessment of, 464
  characteristics of
    biological, 466–467
    chemical, 465–466
    habitat types, 466–467
    physical, 465–466
  description of, 463–464
  fish metric development considerations
    description of, 467
    habitat effects, 469
    sampling gear, 469–470
    spatial variability, 469
    temporal variability, 467–469
  index of biotic integrity application to, 465
  paradigms, 464–465
  point-source discharge assessments, 471
  riverine productivity model, 465
Lepisosteus spp.
  L. oculatus, 143
  L. osseus, 143
  L. platostomus, 143
  L. spatula, 144
Lepomis spp.
  L. auritus, 186
  L. cyanellus, 186
  L. gibbosus, 186
  L. gulosus, 186
  L. humilis, 187
  L. macrochirus, 187
  L. marginatus, 187
  L. megalotis, 187
  L. microlophus, 131, 187
  L. punctatus, 188
  L. symmetricus, 188
Lithopelagophils
  characteristics of, 99
  reproductive guild classification of, 99
Lithophils
  characteristics of, 99
  percent of, as metric of Ohio River quality, 222
  reproductive guild classification of
    brood hiders, 109
    nest spawners, 110
    open substratum spawners, 99
    substratum choosers, 109
Lota lota, 132, 179
Luxilus spp.
  L. albeolus, 151
  L. cardinalis, 151
  L. cerasinus, 151
  L. chrysocephalus, 151
  L. coccogenis, 152
  L. cornutus, 132, 152
  L. pilsbryi, 152
  L. zonatus, 152
Lythurus spp.
  L. ardens, 152
  L. fumeus, 152
  L. lirus, 152
  L. snelsoni, 153
  L. umbratilis, 153

M

Machybopsis spp.
  M. aestivalis, 131, 153
  M. gelida, 153

M. meeki, 153
M. storeriana, 153
Macroinvertebrates
  benthic, biological assessments of
    assessment of condition of, 646–647
    metrics for assessing, 643–644
  reasons for not using in bioassessments, 23
Macrophytes, 126–127
Margariscus margarita, 153
Maximum species richness lines
  from biological field data, methods for deriving
    calculations, 612–613
    data sources, 612
    dissolved oxygen levels, 615–616
    index of biotic integrity metrics, 613
    95th percentile regressions, 618–620
    sensitive species *vs.* QHEI, 614–615
    total recoverable calcium levels, 616–618
    total recoverable copper levels, 616–618
  description of, 611
Menidia beryllina, 181
Metapopulation, of fish communities
  model, 66–67
  mutation, 67
  stream fishes, estimations of structure, 67–68
Metrics, of index of biotic integrity, *see also specific metric; specific study or investigation, metrics*
  applications outside United States and Canada
    individual abundance, 88
    percent hybrid individuals, 88–89
    percent individuals with anomalies, 89
    species composition
      darter species richness, 82, 85
      intolerant species richness, 86
      percent green sunfish individuals, 86–87
      sucker species richness, 86
      sunfish species richness, 85–86
    species richness, 82
    trophic composition
      percent insectivorous individuals, 87
      percent omnivorous individuals, 87
      percent piscivorous individuals, 88
  definition of, 80–81
  deformities, erosion, lesions, and tumors, *see* Deformities, erosion, lesions, and tumors metric
Microgadus tomcod, 179
Micropterus spp.
  M. coosase, 188
  M. dolomieui, 131, 188
  M. punctulatus, 188
  M. salmoides, 188
Midwestern North America
  fishes of, reproductive guild classification of
    bearers
      external, 111
      internal, 111
    guarders
      nest spawners, 110–111
      substratum choosers, 109–110
    nonguarders
      brood hiders, 109
      open substratum spawners, 98–99, 108
    overview, 100–107
  upper, index of biotic integrity for coldwater streams
    applications of, 407–408
    assemblage characteristics, 408–409
    biological integrity vs. coldwater stream management, 403–404
    comparison with other indices of biotic integrity, 404–406
    description of, 384
    development of, 385–387
    metrics
      number of benthic species, 397
      number of coldwater individuals per 150 m, 399
      number of coldwater species, 393–395
      number of minnow species, 397
      number of species, 393
      number of tolerant species, 396–397
      number of warmwater individuals per 150 m, 399
      percent coldwater individuals, 398
      percent of salmonids as brook trout, 398
      percent tolerant individuals, 398
      percent top carnivores, 399
      percent white suckers, 398–399
      scoring of, 387, 389–393
      selection of, 385–387, 388–389
      sensitivity range, 387
    modifications of, 408–409
    testing of
      ecoregion comparisons, 401–402
      methods, 387–388
      "natural" *vs.* "managed" coldwater streams, 402
    validation of, 387–388
      IBI rating *vs.* fish habitat rating, 400
      temporal variability, 400–401
Migration
  effect on local extinction threat, 66–67
  effect on species' range, 68–69
  gene flow and, 68–69
  pattern variations in metapopulations, 68
Minytrema melanops, 166
Monitoring programs, *see also* Biological monitoring
  agency collaboration
    common ground for, 58–59
    description of, 58
  challenges associated with, 58
  criteria for success, 58
  purpose of, 58
Morone spp.
  M. americana, 184
  M. chrysops, 185
  M. mississipiensis, 185
  M. saxatilis, 132, 185
Moxostoma
  M. anisurum, 166
  M. ariommum, 167
  M. atripinne, 167
  M. carinatum, 131, 167
  M. cervinum, 167
  M. duquesnei, 167

Index   665

M. erythrurum, 167
M. hamiltoni, 167
M. hubbsi, 167
M. macroiepidotum, 168
M. pappillosum, 168
M. poecilurum, 168
M. rhothoecum, 168
M. robustum, 168
M. valenciennesi, 131, 169
MSRL, *see* Maximum species richness lines
Mugil cephalus, 128, 202
Mylocheilus caurinus, 153
Myoxocephalus quadricornis, 184

N

Nest spawners, reproductive guild classification of, 110–111
New England lakes, index of biotic integrity for
    field methods, 566
    fish sampling methods, 566
    metrics
        number of green sunfish species, 574–575
        number of intolerant species, 573–574
        number of introduced species, 572–573
        number of native species, 572
        number of sucker species, 574
        number of tolerant species, 575, 579–580
        percent insectivores, 576–577
        percent omnivores, 577
        percent top carnivores, 575–576
        species richness, 569–575, 578–579
        trophic composition, 575–577, 580
        types evaluated for use, 567–568
    overview, 563–564, 577–580
    quantitative methods, 566–567
    results, 568–575
    sample design, 566
Nocomis spp.
    N. asper, 153
    N. biguttatus, 153
    N. effusus, 154
    N. leptocephalus, 154
    N. micropogen, 154
    N. playrhynchus, 154
    N. raneyi, 154
North America
    midwestern, *see* Midwestern North America
    northeastern, freshwater fish species classification studies for developing indices of biotic integrity
        aquatic ecoregions, 313
        fish species
            current distribution
                in New England, 314–315
                in New Jersey, 313
                in New York, 313–314
                in Vermont-West, 314
            historical environmental perturbations on, 311
            information species regarding, 335
            introduced, 316
            native, 315
            naturalized, 316
            rarely encountered types of, 337
            resident, 316–317
                distribution of, 311–313
            species richness and composition, glaciation effects on, 304
            types of, 305–310
        index of biotic integrity
            aquatic habitat classification, 319
            assumptions for, 316
            existing versions, constraints of, 326
            fish species tolerance classification, 319
            formulation considerations, 320
            northeast regional applications
                fish abundance and condition metrics, 325–326
                fish sampling methods, 320–321
                metrics used, 321–322
                overview, 320
                species richness and composition metrics, 322–325
                trophic guild structure metrics, 325
            regional modifications, 320
            resident species for developing, 316–317
            trophic categories
                benthic insectivores, 318
                generalist feeders, 317–318
                minor types of, 318
                northeast groups, 317
                top carnivores, 318
                water column insectivores, 318
            validation of, 326
        methodology
            description of, 303–304
            information sources, 304
    western, index of biotic integrity for
        California
            Dye Creek, 375–376
            Putah Creek, 372–375
            Sierra Nevada watershed evaluations, 372
        difficulties in using, 367–368
        problems associated with
            dams and diversions, 371
            description of, 368
            high endemism, 370
            introduced fishes, 370–371
            low species richness, 368–370
            watershed evaluation, 371–372
        watershed evaluation
            problems associated with, 371–372
            Sierra Nevada, 372
Northeastern United States, *see* North America, northeastern
Notemigonus crysoleucas, 155
Notropis spp.
    N. alborus, 132, 155
    N. altipinnis, 155
    N. amblops, 131, 155
    N. ammophilus, 155
    N. amnis, 155
    N. amoenus, 155

N. anogenus, 129, 155
N. ariommus, 155
N. asperifrons, 155
N. atherinoides, 130, 156
N. atrocaudalis, 156
N. bairdi, 156
N. bifrenatus, 130, 156
N. blennius, 156
N. boops, 130, 132, 156
N. buccatus, 156
N. buchanani, 157
N. chalybaeus, 157
N. chiliticus, 157
N. chrosomus, 157
N. dorsalis, 157
N. girardi, 157
N. greenei, 157
N. heterodon, 157
N. heterolepis, 157
N. hubbsi, 157
N. hudsonius, 158
N. hypsinotus, 158
N. leuciodus, 158
N. lineapunctatus, 158
N. maculatus, 158
N. nubilus, 127, 159
N. ortenburgeri, 159
N. ozarcanus, 159
N. perpallidus, 159
N. photogenis, 159
N. potteri, 159
N. procne, 159
N. rubellus, 160
N. rubricroceus, 160
N. rupestis, 160
N. sabinae, 160
N. scabriceps, 160
N. semperasper, 160
N. shumardi, 160
N. spectrunculus, 160
N. stilibius, 160
N. stramineus, 160
N. telescopus, 161
N. texanus, 129, 161
N. topeka, 161
N. volucellus, 161
N. wickliffi, 161
N. xaenocephalus, 161
Noturus spp.
 N. albater, 170
 N. baileyi, 170
 N. elegans, 170
 N. eleutherus, 170
 N. exilis, 170
 N. flavater, 170
 N. flavipinnis, 171
 N. flavus, 171
 N. gilberti, 171
 N. gyrinus, 171
 N. hildebrandi, 171
 N. insignis, 172
 N. lachneri, 172
 N. leptacanthus, 172
 N. miurus, 172
 N. munitus, 172
 N. nocturnus, 172
 N. phaeus, 172
 N. stanauli, 172
 N. stigmosus, 173
 N. taylori, 173
 N. trautmani, 173
Number of minnow species metric
 for coldwater streams in upper midwestern United States, 397
 for Lake Agassiz Plain ecoregion, 352
Number of species metric
 in coldwater streams, 393
 darter, 323
 sunfish, 323–324

O

Ohio River
 Basin, fish communities in, 72–74
 studies to determine effect on index of biotic integrity metrics of *Catostomus* spp. in
  indirect effects on metrics
   catch per unit effort, 222
   percent DELT anomalies, 222
   percent insectivores, 218, 222
   percent omnivores, 222
   percent simple lithophils, 222
   total number of species, 218
  longitudinal trends, 207
  materials and methods
   sample methods, 204–206
   study area, 204
  prevalent areas, 207–211
  proportional metrics, 215, 218
  species composition, 206
  species richness, 215
  temporal trends, 212–214
Omnivores
 classification of, 133–134
 definition of, 133
 feeding strategy of, 133–134
Oncorhynchus spp.
 O. clarki, 176
 O. gorbuscha, 176
 O. keta, 176
 O. kisutch, 176
 O. mykiss, 176
 O. nerka, 176
 O. tshawytscha, 176
Open substratum spawners, Balon's reproductive guild classification of
 description of, 98
 lithopelagophils, 99
 lithophils, 99
 pelagophils, 98–99
 phytolithophils, 99, 108

phytophils, 108
psammophils, 108
Opsopoeodus emiliae, 129, 161
Osmerus mordax, 174

P

Parapatric speciation, 72
Parasites
    blood suckers, 133
    feeding strategy of, 133
Pelagophils
    characteristics of, 98–99
    reproductive guild classification of, 98–99
Pennsylvania, case study of biological assessments in
    background, 640
    benthic macroinvertebrates
        assessment of condition of, 646–647
        metrics for assessing, 643–644
    fish assemblage
        assessment of condition of, 647–648
        metrics for assessing, 644–646
    framework for, 640
    integrated assessments, 648–650
    objectives, 640
    physical habitat assessments, 646
    physical setting, 641–642
    survey and assessment methods
        benthic macroinvertebrate metrics, 643–644
        fish metrics, 644–646
        physical habitat, 642–643
Perca flavescens, 199
Percent coldwater-adapted species, as metric, 295–296
Percent hybrid individuals
    description of, 88–89
    in northeastern United States, 325
Percent salmonids, as metric, 294–296, 398
Percent top carnivores
    for freshwater fish, 318
    in lakes, 437–438, 440, 575–576
Percina spp.
    P. antesella, 199
    P. auentiaca, 131
    P. aurantiaca, 199
    P. burtoni, 199
    P. caprodes, 131, 199
    P. copelandi, 199
    P. crassa, 199
    P. evides, 199
    P. gymnocephala, 199
    P. jenkinsi, 199
    P. macrocephala, 199
    P. maculata, 200
    P. nasuta, 200
    P. nigrofasciata, 200
    P. notogramma, 200
    P. oxyrhynchus, 200
    P. palmaris, 200
    P. pantherina, 200
    P. peltata, 200
    P. phoxocephala, 200
    P. rex, 200
    P. roanoka, 200
    P. sciera, 201
    P. shumaridi, 201
    P. squamata, 201
    P. tanasi, 131, 201
    P. uranidea, 201
    P. vigil, 201
Percopsis omiscomaycus, 178
Petromyzon marinus, 128, 141
Phenacobius spp.
    P. cataostomus, 161
    P. crassilabrum, 131, 161
    P. mirabilis, 131, 161
    P. teretulus, 131, 161
    P. uranops, 131, 161
Phoxinus spp.
    P. cumberlandensis, 127, 161
    P. eos, 162
    P. erythrogaster, 130, 162
    P. oreas, 129, 162
    P. tennessensis, 127, 162
Phytolithophils
    characteristics of, 99, 108
    reproductive guild classification of, 99, 108
Phytophils
    characteristics of, 108
    reproductive guild classification of
        nest spawners, 110–111
        open substratum spawners, 108
        substratum choosers, 109–110
Phytoplankton, 126
Pimephales spp.
    P. notatus, 129, 162
    P. promelas, 129, 162
    P. tenellus, 162
    P. vigilax, 162
Pioneer species metric, for index of biotic integrity for Lake Agassiz Plain ecoregion
    description of, 354–355
    in North Dakota headwater streams, 362
Planktivores
    description of, 129
    nondiscriminate filter feeders
        gulping, 130
        mechanical sieve, 129
        mucus entrapment, 129
        pump filtration, 129–130
        ram filtration, 129
    particulate feeders
        description of, 130
        size-selective pickers, 130
Platygobio gracilis, 132, 163
Point-source discharges
    DELT findings and, 238–239
    in large rivers, 471
Polyodon spathula, 129, 143
Polyphils, reproductive guild classification of, 110
Pomoxis spp.
    P. annularis, 189
    P. nigromaculatus, 189

Prosopium spp.
  P. coulteri, 177
  P. cylindraceum, 177
  P. williamsoni, 177
Psammophils
  characteristics of, 108
  reproductive guild classification of, 108
Ptychocheilus oregonensis, 163
Pungitius pungitius, 130, 182
Putah Creek
  description of, 372
  index of biotic integrity evaluations, 373–375
Pylodictis olivaris, 173

Q

Qualitative habitat evaluation index (QHEI), 45, 500–501

R

Relictual speciation, 72
Reproductive guild classification
  criticisms of, 98, 111–112
  of midwestern North American fishes
    bearers
      external, 111
      internal, 111
    guarders
      nest spawners, 110–111
      substratum choosers, 109–110
    nonguarders
      brood hiders, 109
      open substratum spawners, 98–99, 108
    overview, 100–107
  origins of, 97–98
  principles of, 97–98
Reservoir fish assemblage index, development of, 542
  metrics
    abundance, 534
    percent DELT, 534
    reproductive comparison, 534
    taxa richness and composition, 529
    trophic composition, 533–534
  overview, 524
  reference conditions, 525–527
  reservoir classification, 524–525
  results of, in Tennessee Valley Authority reservoirs, 537–538
  sampling methods, 534
  scoring criteria, 525–532
  validity of, 538–539
  variability, 535–537
Response indicators, definition of, 19
RFAI, *see* Reservoir fish assemblage index
Rhinichthys
  R. atratulus, 134, 163
  R. cataractae, 163
  R. falcatus, 163
  R. osculus, 163

Richardonius balteatus, 163
Riverine productivity model, 465
Rivers
  great, *see* Great rivers
  large, *see* Large rivers
  Ohio, *see* Ohio river
  Upper Snake River Basin, *see* Upper Snake River Basin

S

Salmo spp.
  S. salar, 177
  S. trutta, 177
Salvelinus spp.
  S. alpinus, 177
  S. fontinalis, 134, 177
  S. malma, 177
  S. namaycush, 178
Scaphirhynchus spp.
  S. albus, 142
  S. platorynchus, 142
Semotilus spp.
  S. atromaculatus, 163
  S. corporalis, 164
Sierra Nevada
  description of, 372
  watershed index of biotic integrity evaluations, 372
Spatial scales, 65
Speciation
  allopatric, 70–72
  parapatric, 72
  relictual, 72
  through peripheral isolation, 71
  vicariance, 70–71
Species, 70
Species richness
  darter
    description of, 82, 85
    in northeastern United States, 323
  intolerant
    description of, 86
    in northeastern United States, 324–325
  in lakes, 569–575
  maximum lines
    from biological field data, methods for deriving
      calculations, 612–613
      data sources, 612
      dissolved oxygen levels, 615–616
      index of biotic integrity metrics, 613
      95th percentile regressions, 618–620
      sensitive species *vs.* QHEI, 614–615
      total recoverable calcium levels, 616–618
      total recoverable copper levels, 616–618
    description of, 611
  percent green sunfish individuals
    description of, 86–87
    in northeastern United States, 325
  sucker
    description of, 86
    in northeastern United States, 324

sunfish
  description of, 85–86
  in northeastern United States, 323–324
Speleophils, reproductive guild classification of, 111
State agencies, *see* Agencies
Stenodus leucichthys, 178
Stizostedion spp.
  S. canadense, 201
  S. vitreum, 133, 202
Stream capture, as factor in diversity of fish community, 74–75
Streams
  coldwater, *see* Coldwater streams
  habitat structure of, factors that affect, 274
  size, and index of biotic integrity metrics, studies in Virginia
    discussion
      general recommendations, 261, 263, 268
      stream size adjustments, 263
    functional metrics, 269
    methods
      data source, 251–252
      disturbances, 253–254
      fish sampling, 252
      habitat effects, 253–254
      physiography, drainage, and stream size, 254
    overview, 249–250
    potential metrics
      description of, 254
      reproduction, 256
      taxonomic, 254, 256
      tolerant species, 256
      trophics, 256
    results
      disturbance effects, 261
      least-disturbed sites, 260–261, 267
    statistical tests, 256, 260
    taxonomic metrics, 268–269
    watershed land use effects, 253
  of upper Snake River Basin
    environmental variables, 285–286
    fish assemblages, 286
    fish metrics, 286
    methods, 284–286
    results
      canonical correspondence analysis, 289–292
      fish metrics, 289–291
      habitat degradation, 292
      native species, 292
      percent coldwater-adapted species, 295
      percent common carp, 294
      percent omnivores, 292–294
      percent salmonids, 294–295
      species collected, 286–288
    sampling sites, 284–285
Stressor indicators, definition of, 19
Substratum choosers, reproductive guild classification of, 109–110
Suckers, *see also Catostomus* spp.
  index of biotic integrity metrics, 85–86

Sunfish, as index of biotic integrity tolerant species metric, 85–86, 425–426
  green, 86–87, 325, 574–575
Surface water, human activities that alter, 274

T

Tailwater fish index, development of
  fish species, 512–514
  index of biotic integrity and, comparison between
    for Cherokee Tailwater, 521
    for Douglas Tailwater, 520–521
  methods
    collection procedures, 509–510
    data analysis, 510
    study area, 508–509
  metrics
    abundance, 516, 519
    species richness
      darter species, 517
      description of, 511–516
      sucker species, 517–518
      sunfish species, 517
      total species diversity, 517
    tolerant species, 514–518
    trophic composition
      description of, 516
      percent insectivores, 519
      percent omnivores, 518–519
      percent piscivorous, 519
  rationale for, 510
Thymallus arcticus, 178
Tinca tinca, 164
Total number of species metric, for freshwater fish of northeastern United States, 322–323
Total recoverable calcium levels, maximum species richness lines derived from, 616–618
Total recoverable copper levels, maximum species richness lines derived from, 616–618
Trophic guild metric
  food chain concept of, 125
  for freshwater fish species classification in northeastern United States
    benthic insectivores, 318
    generalist feeders, 317–318
    minor types of, 318
    northeast groups, 317
    top carnivores, 318
    water column insectivores, 318
  of green plants, 126
  for Lake Agassiz Plain ecoregion
    proportion of insectivore biomass, 354
    proportion of omnivore biomass, 353–354
    proportion of pioneer species, 354–355
    proportion of piscivore biomass, 354
  for lakes
    in southern New England, 575–577, 580
    in Wisconsin, 559
  for North American fishes, using fish strategies
    carnivores
      description of, 132

parasites
  blood suckers, 133
  description of, 133
  whole body (piscivore)
    ambush, 132
    chasing, 132
    protective resemblance, 132–133
    stalking, 132
detritivores
  description of, 128
  filter feeders
    filters, 128
    suction feeders, 128
  particulate feeders
    biters, 128
    scoopers, 129
generalists, 134
herbivores
  classification of, 126–127
  description of, 126–127
  feeding anatomy, 127
  particulate feeders
    browsers, 127
    description of, 127
    grazers, 127
information existing regarding, 125–126
invertivores
  benthic predators
    crushers, 131
    description of, 130
    diggers, 131
    grazers, 130–131
    hunters of mobile benthos, 131
    lie-in-wait predators, 131
    tearers, 131
  drift feeders
    description of, 131
    surface, 131–132
    water column, 132
multiple trophic states, 134–135
opportunists, 134
planktivores
  description of, 129
  nondiscriminate filter feeders
    gulping, 130
    mechanical sieve, 129
    mucus entrapment, 129
    pump filtration, 129–130
    ram filtration, 129
  particulate feeders
    description of, 130
    size-selective pickers, 130
of northeastern United States freshwater fishes, 325
for tailwater fish index
  description of, 516
  percent insectivores, 519
  percent omnivores, 518–519
  percent piscivorous, 519
Trophic state index, 549

U

Ultra-fine organic material feeders, 133–134
Umbra spp.
  U. limi, 174
  U. pygmaea, 174
United States, *see* North America
Upper Cumberland River, stream capture tests for, 74–75
Upper Snake River Basin
  description of, 273–274
  ecoregions, 277
  environmental description of, 275–277
  fish assemblage studies
    description of, 274
    history of, 277, 279
    native fish species, 277, 279
    purpose of, 275
  index of biotic integrity for, 295–296
  major streams, characterization of
    environmental variables, 285–286
    fish assemblages, 286
    fish metrics, 286
    methods, 284–286
    results
      canonical correspondence analysis, 289–292
      fish metrics, 289–291
      habitat degradation, 292
      native species, 292
      percent coldwater-adapted species, 295
      percent common carp, 294
      percent omnivores, 292–294
      percent salmonids, 294–295
      species collected, 286–288
    sampling sites, 284–285
  pollutant sources, 277
  reference streams, characterization of
    canonical correspondence analysis, 282–283
    description of, 279
    environmental variables, 281–283
    fish metrics, 281–282
    fish species collected, 280–281
    methods, 279
  surface-water quality in, 277
U.S. Fish and Wildlife Service, 59
USFWS, *see* U.S. Fish and Wildlife Service

V

Vicariance, 70–71
Viviparous fishes, reproductive classification of, 111

W

Water resources
  degradation of, 5
  statewide assessments of, 47–48
Watershed
  fish assemblages as assessment method for, 49–51
  index of biotic integrity
    description of, 377

Index

for Sierra Nevada watersheds, 372
Western North America, *see* North America, western
Wisconsin lakes, index of biotic integrity for
  methods and materials for developing
    data sources, 543, 547–548
    metrics, 543–546
  metrics
    environmental quality of lake and, relationship between, 549, 553–554
    lake size and ecoregion effects, 549, 554–555
    method-specific, 558–559
    number of benthic species, 553–554
    number of cyprinid species, 555
    sampling variation among, 548–549
    species richness, 549–551
    trophic state and, 559
    types for further development, 555–558
  overview, 541–543
  sampling procedures, 557–560

Y

Young-of-year fish, index of biotic integrity modifications for, 634–635

Z

Zoogeographic integrity coefficient (ZIC), 288–289

```
QH          136040
96.8    Simon, Thomas P., ed.
.B5       Assessing the
A77       sustainability and
1999      biological
          integrity of water
          resources using
          fish communities
```